Springer Theses

Recognizing Outstanding Ph.D. Research

Aims and Scope

The series "Springer Theses" brings together a selection of the very best Ph.D. theses from around the world and across the physical sciences. Nominated and endorsed by two recognized specialists, each published volume has been selected for its scientific excellence and the high impact of its contents for the pertinent field of research. For greater accessibility to non-specialists, the published versions include an extended introduction, as well as a foreword by the student's supervisor explaining the special relevance of the work for the field. As a whole, the series will provide a valuable resource both for newcomers to the research fields described, and for other scientists seeking detailed background information on special questions. Finally, it provides an accredited documentation of the valuable contributions made by today's younger generation of scientists.

Theses may be nominated for publication in this series by heads of department at internationally leading universities or institutes and should fulfill all of the following criteria

- They must be written in good English.
- The topic should fall within the confines of Chemistry, Physics, Earth Sciences, Engineering and related interdisciplinary fields such as Materials, Nanoscience, Chemical Engineering, Complex Systems and Biophysics.
- The work reported in the thesis must represent a significant scientific advance.
- If the thesis includes previously published material, permission to reproduce this must be gained from the respective copyright holder (a maximum 30% of the thesis should be a verbatim reproduction from the author's previous publications).
- They must have been examined and passed during the 12 months prior to nomination.
- Each thesis should include a foreword by the supervisor outlining the significance of its content.
- The theses should have a clearly defined structure including an introduction accessible to new PhD students and scientists not expert in the relevant field.

Indexed by zbMATH.

Yann Gouttenoire

Beyond the Standard Model Cocktail

A Modern and Comprehensive Review of the Major Open Puzzles in Theoretical Particle Physics and Cosmology with a Focus on Heavy Dark Matter

Doctoral Thesis accepted by
Hamburg University, Hamburg, Germany

Springer

Author
Dr. Yann Gouttenoire
Tel Aviv University
Tel Aviv, Israel

Supervisors
Prof. Geraldine Servant
Deutsches Elektronen-Synchrotron DESY
Theory Group
University of Hamburg
Notkestrasse, Germany

Prof. Filippo Sala
Sorbonne Université
Jussieu, Paris, France

ISSN 2190-5053 ISSN 2190-5061 (electronic)
Springer Theses
ISBN 978-3-031-11861-6 ISBN 978-3-031-11862-3 (eBook)
https://doi.org/10.1007/978-3-031-11862-3

This Springer imprint is published by the registered company Springer Nature Switzerland AG
The registered company address is: Gewerbestrasse 11, 6330 Cham, Switzerland

Supervisors' Foreword

This Ph.D. thesis manuscript provides a remarkable and complete survey of the important questions at the interface between theoretical particle physics and cosmology. It spans a very large range of physics in early universe cosmology. Out of the eight main chapters, four of them are reviews and the other four discuss new research results. It goes through all major open puzzles in theoretical particle physics beyond the Standard Model and presents the status of the art in the proposed solutions. The clear logic in the organization of the material, the fact that both standard and more advanced computations are carried out in detail, and the extensive bibliography, make this volume a highly valuable reference for both researchers and students entering this field.

The first topic reflects many aspects in Dark Matter physics, from the production mechanism with non-trivial effects (Sommerfeld, bound state formation, dilution by entropy injection) to precise phenomenology and indirect detection signatures. Over the past decades, Weakly-Interacting-Massive-Particles around the TeV scale have been among the most studied dark matter candidates. Motivated by the absence of experimental evidence for such particles, this thesis explores the possibility that dark matter is much heavier than what is conventionally assumed. The author shows that it is possible to have thermal dark matter with mass beyond the unitarity upper limit of about 100 TeV, and motivates data analysis in this mass range by a plethora of current telescopes, like ANTARES, and future ones, like CTA, LHAASO and KM3NeT.

The second topic is the analysis of supercooled confinement phase transitions during which the dark matter particle acquires its mass. It is massless outside the bubble and in the form of free quarks while it is composite and massive inside the bubble. This leads to very interesting dynamics that had never been addressed in this context before (string fragmentation, cosmological deep inelastic scatterings) and that have implications for the final relic abundance of dark matter as a function of the amount of supercooling. These effects also affect the calculation of the bubble wall velocity, which is in turn relevant for the determination of the associated gravitational-wave signal and for the production of other early-universe relics. Supercooled cosmological phase transitions have become a hot topic in the last few years. Yann Gouttenoire has contributed to bring new insights in this field.

The other line of investigation concerns gravitational waves from cosmic strings. While the broad literature on this topic over the last two decades mainly focuses on the predictions within the standard cosmological paradigm, this work shows that the shape of the spectrum can be drastically different from usual predictions due to a variety of physical effects, such as non-standard equations of state due to intermediate matter eras or intermediate inflationary eras as motivated by either heavy unstable particles or supercooled cosmological phase transitions respectively. It is shown how to use the potential measurement of a gravitational-wave spectrum produced by cosmic strings to extract information on very high energy physics and in particular derive bounds on the mass and lifetime of heavy unstable particles. These limits would extend well beyond what can presently be inferred from Big Bang Nucleosynthesis constraints.

Research in particle physics beyond the Standard Model has become highly interdisciplinary and much closer to cosmology. In this thesis, Dr Yann Gouttenoire presents original work that illustrates the non-trivial connections between many topics, and provides a very comprehensive review of this thriving field. We are, in fact, already using this text as a precious reference for both our research work, and for the new students that we are introducing to particle cosmology.

Notkestrasse, Germany
Jussieu, Paris
July 2022

Géraldine Servant
Filippo Sala

Abstract

This book provides a thorough survey of the important questions at the interface between theoretical particle physics and cosmology. After discussing the theoretical and experimental physics revolution that led to the rise of the Standard Model in the past century, this volume reviews all major open puzzles, like the hierarchy problem, the small value of the cosmological constant, the matter-antimatter asymmetry, or the dark matter problem, and presents the state of the art in the proposed solutions, with an extensive bibliography. This manuscript emphasises the fields of thermal dark matter, cosmological first-order phase transitions and gravitational-wave signatures. Comprehensive and encyclopedic, this book is a rich resource for both researchers and students entering the field.

Written after a 3-year doctoral program in DESY, in addition to the reviews composing two third of the material, one-third presents the original Ph.D. research work of the author. Weakly Interacting Massive Particles (WIMPs) around the TeV scale have since long been among the best-motivated and most studied Dark Matter (DM) candidates. However, the absence of experimental evidence for such particles either in colliders, at telescopes or in underground laboratories, has stimulated model-building and studies of new methods of detection beyond the WIMP paradigm. This thesis performs several steps in this direction.

We are in particular interested in the case where DM is much heavier than a TeV. A well-known obstacle for such a realization is the unitarity bound on the annihilation cross-section which constrains the mass of thermal DM to be smaller than 100 TeV. However, the unitarity bound can be evaded in presence of entropy injection which dilutes the DM abundance. In this thesis, we investigate two possible sources of entropy injection.

First, we study the entropy injection following reheating after an early matter era, when a heavy spectator field, which we choose to be the DM mediator, dominating the energy density of the universe, decays into radiation. We study in detail the corresponding constraints from indirect detection, Cosmic Microwave Background (CMB) and 21 cm, and we show that experimentalists have interests to extend the thermal DM constraints to DM masses beyond the 10/100 TeV range.

Second, we study the entropy injection following reheating after an early stage of vacuum domination generated by a supercooled confining first-order phase transition. Considering the well-motivated scenario where DM is a composite state of a new confining force, we found that a variety of new effects, e.g. string fragmentation and deep-inelastic-scattering in the early universe, play an important role for setting the final DM abundance. In both cases, we show that we can increase the DM mass up to the EeV scale, 4 orders of magnitude larger than the unitarity bound.

Such scenarios involve non-standard cosmologies (either matter era or inflationary era inside the radiation era), and we show that these can be probed using the would-be imprints on the Gravitational-Waves (GW) spectrum from Cosmic Strings (CS) if observed with future GW detectors. In this thesis, we study in detail the computation of the GW spectrum from CS in the presence of non-standard cosmology and the associated constraints on DM models responsible for such a change of cosmology.

Publications Related to This Thesis

[1] M. Cirelli, Y. Gouttenoire, K. Petraki and F. Sala, *Homeopathic Dark Matter, or how diluted heavy substances produce high energy cosmic rays*, *JCAP* **02** (2019) 014, [1811.03608].

[2] I. Baldes, Y. Gouttenoire and F. Sala, *String Fragmentation in Supercooled Confinement and Implications for Dark Matter*, *JHEP* **04** (2021) 278, [2007.08440].

[3] Y. Gouttenoire, G. Servant and P. Simakachorn, *Beyond the Standard Models with Cosmic Strings*, *JCAP* **07** (2020) 032, [1912.02569].

[4] Y. Gouttenoire, G. Servant and P. Simakachorn, *BSM with Cosmic Strings: Heavy, up to EeV mass, Unstable Particles*, *JCAP* **07** (2020) 016, [1912.03245].

Submitted after the completion of the first version of this thesis but directly linked to [2] and to the material discussed in this manuscript:

[5] I. Baldes, Y. Gouttenoire, F. Sala and G. Servant, *Supercool Composite Dark Matter beyond 100 TeV*, *JHEP* **07** (2022) 084, 07 [2110.13926]. https://doi.org/10.1007/JHEP07(2022)084

Acknowledgements

This is with a sincere feeling of gratitude, that I am acknowledging Géraldine Servant. She first played an important role in my academic life when I was completing my master of theoretical physics in Paris and that the number of Ph.D. subjects seemed infinite. Among the numerous researchers that I visited throughout that year, she was the one, whom enthusiasm for physics, as well as sympathy for the disoriented student I was, convinced me that I could confidently go into her supervision for a Ph.D., which I consider now as one of the best choices I have ever done in my life. I am indebted for the careful consideration she has manifested during the whole completion of the Ph.D., for instance by planning a numerous amount of conferences and summer schools per year or by always being a precious help with administration. I am indebted for the time she invested in crucial times, like during postdoc applications, and for dissecting every single word of this thesis manuscript with valuable comments and corrections added. On the scientific side, I have been impressed by her ambition and determination throughout the completion of projects and by the resilience that she has manifested in presence of difficulties. Her formidable experience and her remarkable mastery in model-building were rewarding numerous times. She has been an extraordinary project designer, with my three main independent projects all being connected to each other in an elegant and coherent manner. I am very grateful to her for having successfully transmitted to me her enthusiasm for fantastic fields of research like cosmological first-order phase transitions and gravitational waves of primordial origin.

I am also indebted to Géraldine for having introduced me to Filippo Sala, who supervised my master thesis and who co-supervised my Ph.D. thesis. The impact of Filippo on the young physicist I have become since I have met him is absolutely consequent. I have learnt so much from his critical thinking, constantly pushing me to be more skeptical about my claims and to think harder. Filippo is a careful thinker, a perfectionist, who gives a great attention to every single detail in his work. He has constantly been an inspiration, by showing me how to become more meticulous, more rigorous and more accurate. Besides that, I have always had a lot of fun thinking about physics with him, like when spending long days of confinement wondering about what happens when a quark approaches a wall boundary of confined phase. I also thank Filippo for having carefully read my thesis and for the numerous valuable comments added. Finally, I thank Filippo for having been helpful and diligent on the administrative side at multiple times.

I am deeply grateful to Géraldine, Filippo and Marco Cirelli for having invested so much time and energy—about 200 emails—into trying to build a cotutelle between Hamburg university and Sorbonne university. I would like to especially thank Marco for having moved heaven and earth in Sorbonne university and EDPIF doctoral school. I thank him for having addressed them a 11-page report to state that in 2020 it is impossible to have a Ph.D. affiliated with two universities because the two doctoral certificate templates do not look the same, and because in France the doctoral certificate needs to be printed on a special parchment paper. I thank Brando Bellazzini and Benoit Estienne for having accepted to play the role of scientific tutor and mentor at EDPIF doctoral school.

I would like to thank Peera Simakachorn for the absolutely admirable dedication into our projects, which ended up with very thorough and detailed studies, with two of our common papers being 96 and 151 pages. I thank Iason Baldes for many valuable physics discussions throughout our projects. I thank Kallia Petraki for teaching me Sommerfeld effects and bound state formation while hosting me in her office in LPTHE, and for many insightful physics discussions.

I thank Thomas Konstandin for having accepted to be a reviewer of the thesis, as well as Marco Cirelli, Dieter Horns and Günter Sigl for having accepted to be part of the jury. I also thank the anonymous third reviewer for having agreed to award the Summa Cum Laude. I thank Felix Giese and Marco Hufnagel for having translated the abstract in German.

To complete my Ph.D. in the extremely rich and dynamic environment of DESY in Hamburg was a golden opportunity. With about 100 members in the theory group, we are exposed to a tremendous amount of seminars, conferences and interesting discussions. I am grateful to every colleagues and friends who have made the atmosphere so friendly and my time in Hamburg so unique. The list would be long and I am sure that I don't need to cite their name for them to know what I think about them. I would simply thank my three partners in crime Adrien, Akshansh and Henrique for unforgettable moments, and Jorinde for her company during the difficult confinement phase. I also thank CrossFit WestGym and Erika for great times.

Je remercie mes amis Claude et Mathieu pour des moments incroyables durant notre master, merci en particulier à Claude pour les nuits passées à discuter de physique dans les escaliers du bâtiment G du campus de l'ENS Cachan. Je remercie mon ami Sylvain pour avoir été un excellent binôme de travail pendant nos 2 ans à Orsay et pendant notre année d'agrégation. Une interaction scientifique de qualité et une discipline exceptionnelle nous ont permis de tenir le rythme effréné des 4 mois de préparation des épreuves orales, et je pense que l'on peut en être vraiment fier. Une pensée pour mes amis agrégatifs pour avoir rendu la préparation du concours si formidable. Merci à Arnaud Le Diffon pour des corrections de leçons de qualité.

Je remercie l'équipe enseignante du Magistère de Physique Fondamentale d'Orsay pour m'avoir autant subjugué. Ces deux années étaient incroyablement passionnantes, bien au-delà de mes attentes. La liste de personnes qui m'ont marquées durant ces deux ans est longue. Je voudrais simplement remercier Pélini, Jérémy, Bacchus et Leila pour tous nos moments de rire et de complicité.

Je remercie mes professeurs Bruno Dernoncourt et Jean-Paul Roux pour m'avoir captivé pour la physique ainsi que mon professeur de mathématiques Pierre-Jean Hormière pour avoir cru en mon potentiel et m'avoir recommandé le Magistère d'Orsay. Je remercie ma bande de sup 1 crétins et l'internat Claude Fauriel pour avoir fait de nos deux ans de prépa une aventure humaine exceptionnelle, qui restera à jamais gravée dans nos mémoires. Merci Jonathan pour avoir transformé l'internat en école de spectacle. Je remercie mon professeur de physique-chimie de première et terminale au lycée Albert Triboulet, Mr. Thibault, pour ses qualités de pédagogie exceptionnelle.

Je remercie ma bande de riders pour des moments inoubliables dans les bois du Forez ou au cours des nombreux étés passés à Châtel. Je remercie Lois pour notre amitié sincère. Je remercie la bande des Rozierois, qui est comme une seconde famille. Je remercie Justine pour son soutien et son admiration sans faille alors que j'étais si focalisé sur les études. Je remercie mon frère, Houda, Édouard, Henrique, Sarah et Luis, mais aussi Eva, Élisa, Florian, Mathilde, Cyrielle, Clément, Simon et Antoine, pour leur compagnie durant le confinement à Rozier. Je remercie Bertand Phélut pour les cours de tennis. Je remercie Dounia pour sa grandeur d'âme.

Je dédie mon travail à ma famille. D'abord à ma maman, disparue en 2008, pour son amour inconditionnel et sa générosité inégalable. À mon papa pour son soutien inconditionnel, pour sa curiosité hors-du-commun, et pour m'avoir transmis son intêret pour les grandes questions. Ensuite à mon frère pour notre complicité sans égal et notre admiration inconditionnelle l'un pour l'autre. Je dédie également mon travail à Corinne, mes grand-parents, mes tantes, mes oncles, mes cousines et mes cousins pour leur gentillesse et leur soutien, et en particulier à mes grand-mères pour leur générosité infinie.

Contents

About the Author

Yann Gouttenoire During his Bachelor, Yann Gouttenoire had the opportunity to benefit from the excellent quality of courses at the Magistere of Fundamental Physics in Orsay, France. The next year in 2015, he received a 2-year scholarship at École Normale Supérieure Paris-Saclay. The first year, he got awarded the “Agrégation” while in the second year he earned his MS in Theoretical Physics from École Normale Supérieure rue d’Ulm. In 2017, he joined DESY in Hamburg, one of the world’s leading centers for high-energy particle-physics research, where he studied the topics presented in this book. He defended his PhD on the 29th of September 2020 after which he got awarded the highest distinction from the University of Hamburg (“Summa Cum Laude”). In 2021, he earned an international Postdoctoral Fellowship in Tel Aviv University founded by the Azrieli Foundation.

Chapter 1
Introduction

The current knowledge of particle physics has been theoretically achieved in the early 70s and summarized in a concise theoretical model called the Standard Model of Elementary Particles (SM). The discovery at the Large Hadron Collider (LHC) in 2012 [1, 2] of the last missing piece of the SM, the Higgs boson, quantum excitation of the field giving the particles their mass, has finally validated the entire SM. In spite of its elegant simplicity, the SM suffers from important theoretical issues. Why is gravity so weak with respect to the other forces while we naively expect quantum corrections to equilibrate their strengths? Why do neutrinos have non-vanishing masses while the SM predicts they should be massless? Why is the electron 300 000 times lighter than the top quark? Why does the strong force seem to be invariant under Charge-conjugation and Parity?

The production of elementary particles as heavy as the Higgs boson at colliders requires a huge energy, around the TeV scale. This corresponds to the plasma temperature in the very first moments of the universe, $\sim 10^{-13}$ s after the Big-Bang. Hence, the study of elementary particles at very small length scales is closely connected to the study of the early universe a fraction of a second after the Big-Bang. A good understanding of the phenomena taking place at the earliest epochs is necessary to explain the state of the universe today, as for instance the abundance of light elements.

In spite of its remarkable agreement with the data of Planck satellite, the Standard Model of Cosmology suffers from important theoretical issues. Only 5% of the energy content of our universe has been theoretically identified. We observe that 26% is made of Dark Matter (DM), a fluid responsible for making the galaxies spinning faster and whose nature is unknown. The remaining 69% constitutes the energy of the vacuum, responsible for the accelerated expansion of the universe, and for which the theoretical prediction from the SM of elementary particles is off by 120 orders of magnitude. Another important challenge is how to explain why our universe is only filled with matter and does not also contain anti-matter.

Y. Gouttenoire, *Beyond the Standard Model Cocktail*, Springer Theses,
https://doi.org/10.1007/978-3-031-11862-3_1

The discovery by LIGO in 2015 [3] of Gravitational Waves (GW) of astrophysical origin has opened a new avenue of investigation of our universe. With future GW experiments like SKA, LISA, CE or ET, there is hope for discovering GW of cosmological origin, produced during the early universe, at the time when the universe was still opaque to light.

The main goal of the research carried out during this thesis is to investigate the possibility that the DM of the universe is composed of particles much heavier than the TeV scale, to motivate such a scenario and to present distinct ways to probe it. As this is a very interdisciplinary topic, we will also discuss important topics in early universe dynamics such as cosmological phase transitions and gravitational-wave probes of new physics beyond the Standard Model. My main new contributions can be summarized as follows:

- Effect of entropy injection from the decay of the DM mediator on the DM abundance and implications for its mass prediction and searches at indirect detection experiments.
- Investigation of new dark matter production mechanisms during a confining supercooled first-order phase transition: string fragmentation and deep-inelastic scattering.
- Probing a non-standard matter era induced by DM mediators, or a second period of inflation induced by the supercooled first-order phase transition generating the DM mass, using gravitational waves from cosmic strings.

Out of the eight main chapters in this thesis, four of them (Chaps. 5, 7, 8 and 9) are reporting my new contributions that have led to four research articles. The other chapters (Chaps. 2, 3, 4 and 6) are detailed reviews which may a priori appear disconnected from my main work but they were included as they reflect work that I have been carrying, and on which future upcoming articles will be based. Most of the figures are mine, using existing results in the literature, although a large number of these figures have never appeared in the literature.

Plan of the thesis.
The plan of the thesis is described below.

Chapter 2: Standard Model of Elementary Particles.
Chapter 2 introduces the SM and tries to highlight the main historical details, either theoretical breakthroughs or experimental discoveries, leading to its completion. We introduce the different open problems and discuss the main proposed solutions.

Chapter 3: Standard Model of Cosmology.
Chapter 3 starts by introducing the Standard Model of cosmology and computing the abundance of baryons, neutrinos and the time of photon decoupling. We show the necessity for a period of inflation in order to explain the striking flatness and homogeneity of our universe. We motivate the use of Gravitational-Waves (GW) as a new search strategy for probing the early universe. Open problems are introduced together with the main proposed solutions.

Chapter 4: Thermal Dark Matter.
Chapter 4 is a review on thermal Dark Matter. After having discussed the two production mechanisms of thermal Dark Matter, freeze-in and freeze-out, we focus on the second on, which is motivated by the WIMP miracle. We compute the DM abundance from the eV range to the 100 TeV range and we present the constraints from Lyman-α, CMB, Big-Bang Nucleosynthesis (BBN), indirect detection, direct detection. We then give more attention to the case where DM is heavy, a limit where non-perturbative Sommerfeld effects are important and where unitarity is saturated, and which is the case of interest for Chap. 5 and follow-up studies of Chap. 7 (e.g. [4]). All the figures in the chapter are mine (Figs. 4.1, 4.2, 4.3, 4.4, 4.5 and 4.6).

Chapter 5: Homeopathic Dark Matter.
Based on studies performed between May 2017 and January 2019 with Marco Cirelli, Kalliopi Petraki and Filippo Sala.

Publication [5]:

Due to the **unitarity bound**, Dark Matter (DM) can not be heavier than ~100 TeV if its abundance is set by the standard freeze-out scenario, otherwise it overcloses the universe. Nevertheless, we show that we can extend the unitarity bound to the EeV range by introducing a long-lived heavy DM mediator which decays into radiation before BBN, hence **injecting entropy** and **diluting DM**. Large DM mass corresponds to large coupling constant, implying the presence of non-perturbative Sommerfeld effects which enhance the indirect-detection constraints. We consider the possibility of having heavy thermal DM in a benchmark toy model where a $U(1)_D$ is kinetically coupled to $U(1)_Y$. We also study the associated phenomenology. We find that entropy injection opens a large parameter space with DM masses up to **EeV scale**, far beyond the standard unitarity bound. This motivates future indirect-detection experiments aiming to explore cosmic rays beyond the TeV range, e.g. CTA, LHAASO, KM3NET, HERD, ISS- CREAM. However, the region far beyond 100 TeV (e.g. 10 PeV) is far beyond the reach of standard methods: colliders, telescopes and direct detection, which motivates new methods of detection, see Chap. 9. The Chap. 5 is a minimal adaptation of the publication [5].

Chapter 6: First-order Cosmological Phase Transition.
The Chap. 6 is devoted to 1st-Order cosmological Phase Transitions (1stOPT), and motivates the research work exposed in Chap. 7 and published in [6] as well as other ongoing works, e.g. [4]. We start by considering the physics involved during **bubble nucleation**: we present the standard techniques for computing an **effective potential** at finite-temperature, and we remind how to use it in order to compute the **bounce action**, numerically or analytically. Second, we consider the physics involved during **bubble propagation**: we present various methods, valid in different regimes, for computing the **bubble wall velocity**. A special attention is given to the computation of the friction pressure in the ballistic regime, at leading order and next-to-leading-order in gauge coupling constant. Finally, we discuss the physics involved during **bubble collision**. We remind how to compute the resulting GW emission, using

the most recent results in the literature. Particularly, we discuss how to compute the energy transfer to sound-waves with arbitrary speed of sound. we conclude the chapter by introducing the two classes of **nearly-conformal potential** present in the litterature leading to supercool first-order phase transitions. First, we introduce the **Coleman-Weinberg** potential, valid in the weakly-coupled regime. Second, we introduce the **light-dilaton** potential, which assumes a strongly-coupled regime, or its 5D-holographic dual, the Goldberger-Wise potential, which assumes the existence of a warped extra-dimension. For each of these models, we compute the nucleation temperature and GW parameters α and β , both numerically and analytically. All the figures in the chapter are mine (Figs. 6.1, 6.2, 6.3, 6.4, 6.6, 6.7, 6.8, 6.9, 6.10, 6.11, 6.12, 6.13 and 6.14).

Chapter 7: Supercooled Composite Dark Matter.
Based on studies performed between October 2018 and July 2020 with Iason Baldes, Filippo Sala and Geraldine Servant.

Publications [4, 6]**:**

First-order phase transitions driven by **nearly-conformal potential** are known to predict a potentially large hierarchy between the critical temperature, when the two minima coincide, and the nucleation temperature, when the phase transition (PT) completes. For intermediate temperatures, the universe is dominated by the vacuum energy of the PT, which leads to a short period of inflation, also known as **supercooling**. An interesting cosmological consequence is the dilution of any relic produced beforehand. The dilution of the DM abundance had previously only been studied in the case of a supercooled PT driven by a Coleman-Weinberg potential, generated by quantum corrections in the weakly-coupled regime. The aim of our study is to consider the case where the supercooled PT is driven by a nearly-conformal potential arising from strongly-coupled dynamics. A possible realization is the light-dilaton potential, motivated by holography. In contrast to the weakly-coupled scenario where DM already exists before the PT, in the strongly-coupled regime, the DM is **composite** and is only formed when the techni-quanta enter inside the expanding bubble of confined phase. In the limit of large supercooling, the distance between techni-quarks, set by the inverse temperature, is large with respect to the confining scale. Hence, it is energetically favorable for the techni-quarks to form a **flux tube** with the bubble wall instead of with their neighbors. This has three important consequences. First, due to the large kinetic energy of the incoming quark in the wall frame, the flux tube breaks into many hadrons, similarly to the **string fragmentation** which follows the decay of a Z-boson into a $q\bar{q}$ pair in the SM. Second, due to charge conservation, a quark is **ejected** from the bubble. Third, the rest frame of the gluon string between the incoming quark and the wall is boosted with respect to the plasma frame. Hence, the string fragments have large momenta and can undergo **deep-inelastic-scattering** (DIS) with the different mediums which they encounter, which enhances further the abundance of composite states. Such mediums can be the diluted bath, the particles from the preheated bath or the string fragments from the neighboring bubbles. We show that the second possibility is the most relevant. As a consequence of these new effects which we point out, the resulting DM abundance, assuming that DM is a composite state of the confining sector, is very different from the trivial case where it only receives a dilution factor. Due to the enhanced DM production, we find that in order to get the correct DM abundance, the required supercooling must be larger than in the weakly-coupled scenario. Chapter 7 is entirely devoted to this study and is almost entirely identical to the publication [6].

Chapter 8: Beyond the Standard Models with Gravitational Waves from Cosmic Strings.
Based on studies performed between October 2018 and April 2020 with Geraldine Servant and Peera Simakachorn.

Publication [7]:

Cosmic Strings (CS) are topological defects arising from the breaking of a $U(1)$ symmetry. The reason why they have been the subject of so many studies for the last 40 years is probably due to their universal behavior called **scaling regime**. The fraction of energy density stored in long cosmic strings, $\Omega_\infty \propto \rho_\infty/\rho_c$ with ρ_∞ the energy density of infinitely-long strings and ρ_c the critical density, is conserved through the evolution of the universe, from formation of the network until today. This results from a conspiracy between Hubble expansion and energy loss into loop formation. Hence, CS do not overclose the universe, as the other topological defects, domain walls and monopoles, would do, neither they redshift away: CS are always there.

An important consequence is that CS constitute a **long-standing** source of GW, which leads to a **scale-invariant** GW spectrum (a power law). Note that the GW are not emitted by infinitely-long strings which have a conserved topological charge, but they are instead emitted by **loops**. A second intriguing property is that in a radiation-dominated universe, the GW spectrum is **flat** in frequency. A property which we can impute to the presence of a fortuitous cancellation between the loop formation rate and the redshift factor. However, this is not the case anymore if the loop formation occurs in a different background, e.g. kination and matter domination where the slope turns to f^1 and f^{-1}, respectively. Also, if loop formation occurs during a period of inflation, the scaling regime is violated and loop formation freezes. In our study, we propose to use the would-be detection of GW spectrum from CS by future GW experiments to scan for the presence of a **non-standard cosmology** in the early universe.

In Chap. 8, we first review in great details the computation of the GW spectrum from CS including deviation from the scaling regime due to change of cosmology and deviation from the Nambu-Goto approximation due to quantum effects in the presence of small-scale structures. Then, we provide the energy scales of an early matter era, as well as the energy scale and duration of a second period of inflation, which we could detect with future GW interferometers. Chapter 8 is a minimal adaptation of the publication [7]. The sections dealing with kination and metastable

cosmic strings have been removed. The Figs. 8.1 and 8.3 are mine and have been added during the redaction of the manuscript.
Chapter 9: Probe Heavy Dark Matter with Gravitational Waves from Cosmic Strings.
Based on studies performed between October 2018 and April 2020 with Geraldine Servant and Peera Simakachorn.

Publication [8]:

In Chap. 9, we suppose that the early period of non-standard cosmology introduced in the previous chapter is induced by particle dynamics. First, we consider that the **early matter-domination era** is induced by a **heavy unstable relic**. We give model-independent constraints on the lifetime τ_X and the product mass times abundance $m_X Y_X$ of the relic, which extends the current BBN contraints up to $\tau_X \lesssim 10^{-17}$ s. Then, we focus on the model of **Homeopathic DM** presented in Chap. 5 and we show that the large DM mass region which was opened by the entropy dilution is within the reach of GW experiments if a string network of tension $G\mu \gtrsim 10^{-15}$ is detected. Finally, we apply our model-independent constraints on **second period of inflation** to the model of **Supercooled Composite DM** presented in Chap. 7. We show that the regions of the parameter space leading to the correct DM abundance are within the reach of future GW experiments. Being sensitive to large DM mass, this new method of detection offers an excellent complementary with standard techniques based on colliders, indirect-detection and direct-detection whose sensitivities generally fade away beyond the TeV scale. Chapter 9 is a minimal adaptation of the publication [8]. The sections dealing with scalar oscillating moduli, scalar particles produced gravitationally or though the Higgs portal have been removed. Section 9.2 together with its Fig. 9.7, were not included in the publication [8] but have instead been realized for the purpose of those notes.

We conclude the thesis manuscript in Chap. 10. Technical details are reported in a number of Appendices.

References

1. ATLAS collaboration, Aad G et al (2012) Observation of a new particle in the search for the standard model higgs boson with the ATLAS detector at the LHC. Phys Lett. B716:1–29. arXiv:1207.7214
2. CMS collaboration, Chatrchyan S et al (2012) Observation of a new boson at a mass of 125 GeV with the CMS experiment at the LHC. Phys Lett B716:30–61. arXiv:1207.7235
3. LIGO Scientific, Virgo collaboration, Abbott BP et al (2016) GW150914: first results from the search for binary black hole coalescence with advanced LIGO. Phys Rev D93:122003. arXiv:1602.03839
4. Baldes I, Gouttenoire Y, Sala F, Servant G, Supercool composite dark matter beyond 100 TeV. arXiv:2110.13926
5. Cirelli M, Gouttenoire Y, Petraki K, Sala F (2019) Homeopathic dark matter, or how diluted heavy substances produce high energy cosmic rays. JCAP 02:014. arXiv:1811.03608

6. Baldes I, Gouttenoire Y, Sala F (2021) String fragmentation in supercooled confinement and implications for dark matter. JHEP 04:278. arXiv:2007.08440
7. Gouttenoire Y, Servant G, Simakachorn P (2020) Beyond the Standard models with cosmic strings. JCAP 07:032. arXiv:1912.02569
8. Gouttenoire Y, Servant G, Simakachorn P (2020) BSM with cosmic strings: heavy, up to EeV mass. Unstable particles. JCAP 07:016. arXiv:1912.03245

Chapter 2
Standard Model of Elementary Particles

The Standard Model (SM) is the theory which describes at the quantum level three of the four fundamental forces—electromagnetism, weak interaction and strong interaction—and all the known elementary particles. It is currently the most rigorous theory of particle physics, with an unprecedented level of precision and accuracy in its predictions. It is the result of the collaborative work of the greatest physicists in the last century, whom at least seventy of them earned the Nobel prize, since the introduction of the photon by Einstein in 1905 [1] or the unification of Quantum Mechanics and Special Relativity by Dirac in 1928 [2, 3]. The SM has been theoretically achieved after three major breakthroughs in the end 60s/early 70s.

1. The unification of electromagnetism and weak interactions under the gauge group $SU(2) \times U(1)$ by Glashow in 1961 [4], followed by the incorporation of spontaneous symmetry breaking $SU(2)_L \times U(1)_Y \rightarrow U(1)_{\text{e.m.}}$ by Weinberg and Salam in 1967 [5, 6] in order to provide a mass for the vector bosons without violating gauge invariance. Weinberg and Salam used the Higgs mechanism developed by many authors in 1964 [7–11], which was itself inspired from the Nambu-Goldstone theorem [12, 13] formalized in Quantum Field Theory in 1962 by Goldstone, Weinberg and Salam [14]. The renormalizability of such a massive gauge theory is proved in 1972 by 't Hooft and his adviser Veltman who developed the technique of dimensional regularization [15–17].
2. Second, the theoretical justification by Gross, Wilczek and Politzer in 1973 [18, 19] that Quantum Chromodynamics, a Yang-Mills gauge theory with color triplet quarks and color octet gluons, proposed by Gell-Mann et al. the previous year [20, 21], manifests asymptotic freedom, in excellent agreement with proton-electron scattering, and hence is the correct gauge theory for describing strong interactions.
3. Third, the prediction of a third generation of quarks in order to explain CP violation by Kobayashi and Maskawa in 1973 [22].

Y. Gouttenoire, *Beyond the Standard Model Cocktail*, Springer Theses,
https://doi.org/10.1007/978-3-031-11862-3_2

The Standard Model of elementary particles is reviewed in many excellent books, e.g. [23–85]. For historical complements about the period giving rise to the SM, we call attention to [86–91].

2.1 Fields and Symmetries

2.1.1 *The Lorentz Representations*

When Dirac unifies Quantum Mechanics and Special Relativity, he adds the **spinor** ψ, with spin $j = 1/2$, to the list of fields compatible with Special Relativity

$$\textbf{Spin 0}: \text{ scalar } \phi(x), \qquad \textbf{Spin 1/2}: \text{ spinor } \psi_\sigma(x), \qquad \textbf{Spin 1}: \text{ vector } A_\mu(x), \tag{2.1}$$

with $\sigma = 1, 2$ and $\mu = 1..4$. The spin $j = 0,\ 1/2,\ 1, ..$ labels the different representations under the group $SU(2)$ of spatial rotation. Thanks to the spin-statistics connection,[1] j also tells how the corresponding states occupy the available phase space: even spins obey to the Bose-Einstein statistics whereas odd spins obey to the Fermi-Dirac statistics. Hence, we call them **bosons** and **fermions**, respectively.

Actually, there are two nonequivalent spinorial representations of the Lorentz group: the **right-handed Weyl spinors** ψ_R and the **left-handed Weyl spinors** ψ_L.[2] ψ_R and ψ_L are related by a **parity transformation** $\vec{x} \to -\vec{x}$, so they transform the same way under spatial rotation but in the opposite way under a Lorenz boost. It is often convenient to merge the two two-component Weyl spinors into a four-component **Dirac spinor**

$$\psi = \begin{pmatrix} \psi_L \\ \psi_R \end{pmatrix}, \tag{2.3}$$

and to introduce the **Dirac conjugate**

$$\bar{\chi} = \chi^\dagger \gamma^0, \qquad \text{with} \quad \chi^\dagger = {}^t\chi^*, \tag{2.4}$$

such that $\bar{\chi}\psi$ is a scalar and $\bar{\chi}\gamma^\mu\psi$ is a 4-vector, while $\bar{\chi}\gamma^5\psi$ is a pseudo-scalar (changes sign under parity) and $\bar{\chi}\gamma^5\gamma^\mu\psi$ is a pseudo-4-vector. We have introduced the Dirac matrices γ^μ which obey to the Clifford algebra

---

[1] Fierz and Pauli formulate the connection between spin and statistics in 1939/1940 [92, 93].

[2] The Lorentz algebra so(1, 3) can be decomposed as the tensor product of two commuting su(2) subalgebras

$$\text{so}(1,\ 3) = su(2) \oplus su(2). \tag{2.2}$$

Therefore, irreducible representations of the Lorentz group can easily be constructed from the irreducible representations of su(2), which are labeled by $j = 0,\ 1/2,\ 1,$ and so on. Hence, the scalar representation of the Lorentz group is (0, 0), the spinorial representations are $(0,\ \frac{1}{2})$ and $(\frac{1}{2},\ 0)$, the 4-vector representation is $(\frac{1}{2},\ \frac{1}{2})$, ...

$$\{\gamma^\mu, \gamma^\nu\} = 2\eta^{\mu\nu}, \qquad \text{and} \qquad \gamma^5 = i\gamma^0\gamma^1\gamma^2\gamma^3. \tag{2.5}$$

A convenient representation is the Weyl basis, already assumed when writing Eq. (2.3)

$$\gamma^0 = \begin{pmatrix} 0 & 1 \\ 1 & 0 \end{pmatrix}, \qquad \gamma^i = \begin{pmatrix} 0 & \sigma^i \\ -\sigma^i & 0 \end{pmatrix}, \qquad \gamma^5 = \begin{pmatrix} -1 & 0 \\ 0 & 1 \end{pmatrix}, \tag{2.6}$$

where σ^i are the Pauli matrices.

2.1.2 *The Gauge Interactions*

Simply speaking, the Standard Model is a set of fields among the list given in Eq. (2.1), interacting with each other. The list of quantum fields constituting the SM is shown in Fig. 2.1. A powerful aspect is that the field content as well as their interactions are entirely dictated by symmetries. The three forces are described by the gauge structure

$$SU(3)_c \times SU(2)_L \times U(1)_Y \quad \to \quad SU(3)_c \times U(1)_{\text{e.m.}} \tag{2.7}$$

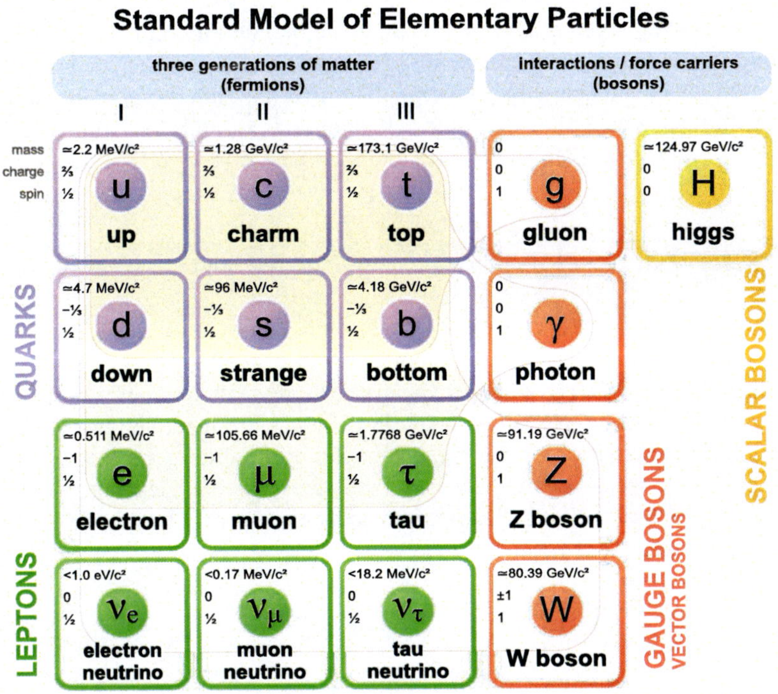

Fig. 2.1 Elementary particles of the standard model

where each generator of each Lie Group defines a force mediator, each of them being a vector boson of spin 1. More precisely, the 8 generators of $SU(3)_c$ define the 8 **gluons**

$$SU(3)_c: \quad G^a_\mu, \qquad a = 1...8, \tag{2.8}$$

mediators of the strong force responsible for the binding of the nucleus, with gauge coupling g_3. The three generators of $SU(2)_L$ as well as the unique generator of $U(1)_Y$, which we denote by

$$SU(2)_L: \quad W^a_\mu, \qquad a = 1..3, \quad \text{and} \quad U(1)_Y: \quad B_\mu, \tag{2.9}$$

mediate the electroweak interactions. The **vector bosons** $\mathbf{W^\pm}$, responsible for the beta decay are given by linear combinations of W^1_μ and W^2_μ. The **photon**, mediator of the electromagnetism force, and well as the **vector boson** $\mathbf{Z^0_\mu}$, responsible for neutrino elastic scattering, are given by linear combinations of W^3_μ and B_μ. We provide more details in Sect. 2.2.3.

2.1.3 *The Matter Content*

In addition to the gauge fields which mediate the interactions and are adjoint representations of the gauge groups, there are the matter fermions, which are fundamental representations of the gauge groups through which they interact. On the one hand, there are **quarks**, which transform as triplet under $SU(3)_c$ and come as three $SU(2)_L$ doublet Q^n_L and six $SU(2)_L$ singlets u^n_R and d^n_R where $n = 1, 2, 3$ is the generation index

$$Q^{1,2,3}_L = \begin{pmatrix} u_L \\ d_L \end{pmatrix}, \begin{pmatrix} c_L \\ s_L \end{pmatrix}, \begin{pmatrix} t_L \\ b_L \end{pmatrix}, \quad u^{1,2,3}_R = u_R,\ c_R,\ t_R \quad \text{and} \quad d^{1,2,3}_R = d_R,\ s_R,\ b_R, \tag{2.10}$$

Their **hypercharges Y**, eigenvalues under $U(1)_Y$, are $+1/6$, $+2/3$ and $-1/3$. The corresponding **electric charges Q**, after electroweak symmetry breaking, are given by the Gell-Mann–Nishijima formula $Q = \tau_3 + Y$, rederived in Eq. (2.75) of Sect. 2.2.3. The quantity τ_3 is the eigenvalue under the third generator of $SU(2)_L$, also known as the **weak isospin**. On the second hand, there are **leptons** which transform as singlets under $SU(3)_c$, and come as three $SU(2)_L$ doublet L^n_L and three $SU(2)_L$ singlets e^n_R

$$L^{1,2,3}_L = \begin{pmatrix} \nu_e \\ e_L \end{pmatrix}, \begin{pmatrix} \nu_\mu \\ \mu_L \end{pmatrix}, \begin{pmatrix} \nu_\tau \\ \tau_L \end{pmatrix} \quad \text{and} \quad e^{1,2,3}_R = e_R,\ \mu_R,\ \tau_R. \tag{2.11}$$

Their **hypercharges Y** are $-1/2$ and -1. A first curiosity is the absence of **right-handed neutrino** $\boldsymbol{\nu_R}$ in the SM. A second curiosity is the emergence of three **gener-**

Table 2.1 The fields of the Standard Model and their symmetry transformation

		$SU(3)_c$	$SU(2)_L$	$U(1)_Y$	Spin
Gauge fields	B_μ	**1**	**1**	0	1
	W^a_μ	**1**	**3**	0	1
	G^a_μ	**8**	**1**	0	1
Quarks	Q_L	**3**	**2**	$+\frac{1}{6}$	$\frac{1}{2}$
	u_R	**3**	**1**	$+\frac{2}{3}$	$\frac{1}{2}$
	d_R	**3**	**1**	$-\frac{1}{3}$	$\frac{1}{2}$
Leptons	L_L	**3**	**2**	$-\frac{1}{2}$	$\frac{1}{2}$
	e_R	**3**	**1**	-1	$\frac{1}{2}$
Higgs	H	**1**	**2**	$+\frac{1}{2}$	0

ations $n = 1, 2, 3$ of quarks (u, d), (c, s) and (t, b) as well as three generations of leptons (ν_e, e), (ν_μ, μ), (ν_τ, τ). As shown in Fig. 2.1, the first generation contains the lightest fermions while the third generation contains the heaviest ones.

2.1.4 The Higgs Field

Finally, the last particle in the SM is the **Higgs boson**. It comes as a complex scalar, doublet under $SU(2)_L$ with hypercharge $+1/2$ and spin 0 called the **Higgs multiplet**

$$H = \begin{pmatrix} \phi^+ \\ \phi^0 \end{pmatrix}, \tag{2.12}$$

where ϕ^+ and ϕ^0 are two complex scalar fields.

The arrow in Eq. (2.7) indicates that the vacuum of the universe in which we live, lies in a particular gauge configuration of $SU(2)_L \times U(1)_Y$, chosen arbitrarily when in the early universe, the **Higgs field** H acquires a non-zero **vacuum expectation value** $\langle H \rangle \neq 0$. We talk about **spontaneous** or **dynamical** breaking of the electroweak symmetry. This is the Higgs mechanism developed in 1964, which gives the masses to the gauge bosons $W^\pm$, Z^0, the quarks and the leptons,[3] while preserving the gauge symmetry at the quantum level. We provide more details in Sect. 2.2.3.

A summary of the field content of the SM is given in Table 2.1 and in Fig. 2.1.

[3] Except for the neutrinos which are considered massless and only left-handed in the Standard Model.

2.2 The Standard Model in a Nutshell

2.2.1 The Lagrangian

The golden rules:
The dynamical evolution of the fields, including both their free evolution and their interactions, is encoded in the Lagrangian density $\mathscr{L}$. It is a function of the quantum fields introduced in the previous section, whose construction in the SM follows two golden rules

1. $\mathscr{L}$ must be **invariant** under the gauge structure $SU(3)_c \times SU(2)_L \times U(1)_Y$.
2. $\mathscr{L}$ must be **renormalizable**, so it must only include operators up to dimension four.[4]

The Lagrangian: We obtain:

$$\mathscr{L}_{\rm SM} = -\frac{1}{4}B_{\mu\nu}B^{\mu\nu} - \frac{1}{4}W^a_{\mu\nu}W^{a\mu\nu} - \frac{1}{4}G^a_{\mu\nu}G^{a\mu\nu} \tag{2.13}$$
$$+ i\bar{L}^n_{\rm L}\not{D}L^n_{\rm L} + i\bar{e}^n_{\rm R}\not{D}e^n_{\rm R} + i\bar{Q}^n_{\rm L}\not{D}Q^n_{\rm L} + i\bar{u}^n_{\rm R}\not{D}u^n_{\rm R} + i\bar{d}^n_{\rm R}\not{D}d^n_{\rm R} \tag{2.14}$$
$$+ \mu^2|H|^2 + \lambda|H|^4 . \tag{2.15}$$
$$+ (D_\mu H)^\dagger(D^\mu H) - \left(Y^e_{mn}\bar{L}^m_{\rm L}He^n_{\rm R} + Y^d_{mn}\bar{Q}^m_{\rm L}Hd^n_{\rm R} + Y^u_{mn}\bar{Q}^m_{\rm L}\tilde{H}u^n_{\rm R} + h.c.\right) \tag{2.16}$$
$$-\theta_1\frac{g_1^2}{32\pi^2}B^{\mu\nu}\tilde{B}_{\mu\nu} - \theta_2\frac{g_2^2}{32\pi^2}W^{a\mu\nu}\tilde{W}^a_{\mu\nu} - \theta_3\frac{g_3^2}{32\pi^2}G^{b\mu\nu}\tilde{G}^b_{\mu\nu} \tag{2.17}$$

Gauge fields kinetic terms:
The first line in Eq. (2.13) contains the kinetic terms for the gauge fields X_μ with the strength tensors $F_{\mu\nu}$ being

$$F_{\mu\nu} = \partial_\mu X_\nu - \partial_\nu X_\mu + gf^{abc}X^b_\mu X^c_\nu, \tag{2.18}$$

where f^{abc} are the **structure constant** of the corresponding Lie group, defined from the algebra of the generators

$$\left[\tau^a, \tau^b\right] = i\, f^{abc}\,\tau^c. \tag{2.19}$$

Particularly, for $U(1)_Y$ and $SU(2)_L$, we have $f_1^{abc} = 0$ and $f_2^{abc} = \varepsilon^{abc}$, where ε^{abc} is the Levi-Civita tensor.

[4] Given an operator $\mathcal{O}$ with dimension $4 + \delta > 4$, since the lagrangian density $c\,\mathcal{O}$ must have dimension 4, the coupling constant c must have the appropriate scaling $c \propto \Lambda^{-\delta}$, where Λ is some energy scale. Hence the transition amplitude of any process occuring at energy E, at order n in perturbation, depends on the quantity $\left(\frac{E}{\Lambda}\right)^{\delta n}$, which is divergent for $E \gg \Lambda$, and even more divergent than the perturbation order n is large. Hence, the cancellation of the UV divergences would need an infinity of counterterms and the theory is not renormalizable.

Fermion kinetic terms:
The second line in Eq. (2.14) contains the kinetic terms for the quarks and leptons, assuming the Feynman notation $\not{D} = \gamma^\mu D_\mu$. Invariance under the gauge symmetries is satisfied thanks to the **gauge-covariant derivatives** which contain the interactions between the fermions and the gauge fields

$$D_\mu = \partial_\mu - i g_1 B_\mu Y - i g_2 W_\mu^a \tau_2^a - i g_3 G_\mu^a \tau_3^a, \tag{2.20}$$

where the X_μ^a are the gauge bosons, τ^a are the generators of the corresponding gauge symmetries and g are the corresponding gauge couplings.

Global symmetries:
In the absence of the Yukawa matrices Y^d, Y^u and Y^e, cf. Eq. (2.16), the Lagrangian of the SM is invariant under a large $U(3)^5$ global symmetry, where 3 is the number of flavors and 5 is the number of kinetic terms in Eq. (2.14), which can be decomposed as [94]

$$U(3)^5 = SU(3)_q^3 \times SU(3)_l^2 \times U(1)^5 \tag{2.21}$$

where

$$\begin{aligned} SU(3)_q^3 &= SU(3)_{Q_L} \times SU(3)_{u_R} \times SU(3)_{d_R}, \\ SU(3)_l^2 &= SU(3)_{L_L}^3 \times SU(3)_{e_R}, \\ U(1)^5 &= U(1)_{Q_L} \times U(1)_{u_R} \times U(1)_{d_R} \times U(1)_{L_L} \times U(1)_{e_R}. \end{aligned} \tag{2.22}$$

These global symmetries have not been imposed directly on the lagrangian, hence they are called **accidental symmetries**. The five $U(1)$ symmetries, which rotate the fermion fields in a flavor-blind way, can be reshuffled as $U(1)^5 = U(1)_B \times U(1)_L \times U(1)_Y \times U(1)_{d_R} \times U(1)_{e_R}$. In the presence of the Yukawa interactions, we can check that the global symmetry group gets broken to

$$U(3)^5 \longrightarrow U(1)_B \times U(1)_{L_e} \times U(1)_{L_\mu} \times U(1)_{L_\tau}, \tag{2.23}$$

where we dropped the $U(1)_Y$ since it is gauged. We have introduced the baryonic and leptonic symmetries defined by

$$l_{e,\mu,\tau} \xrightarrow{U(1)_{L_e, L_\mu, L_\tau}} e^{i\alpha_{e,\mu,\tau}} l_{e,\mu,\tau} \quad \text{and} \quad q_n \xrightarrow{U(1)_B} e^{i\frac{\alpha}{3}} q_n, \quad \text{with } n = 1, 2, 3. \tag{2.24}$$

From the **Noether theorem**, there exists corresponding baryonic and leptonic charges B, L_e, L_μ and L_τ, which must be conserved[5] in all processes. They are the **baryon number** and **lepton numbers**.

---

[5] The are however broken by non-perturbative processes, see Sect. 3.5.2.

2.2.2 *Quantum Chromodynamics*

The birth of QCD:
A lot of progress as been made since Yukawa predicted in 1935 the existence of the π meson as the mediator responsible for the force binding protons and neutrons together in the nucleus [95]. In 1961, Gell-Mann and Ne'eman interpreted the **hadrons** as **bound-states** of three 'fictitious leptons' [96, 97] which Gell-mann later called **quarks** u, d, s [98]. It was then possible to catalog all the possible hadrons as being different representations of the group $SU(3)_f$, where f denotes the number of **flavors**.[6] Gell-Mann received the Nobel prize in 1969. Seminal articles are collected in the book [99].

In 1965, the non-observation of hadrons based on the tensor product $3 \otimes 3$ but instead based on $3 \otimes \bar{3}$ led Han and Nambu to postulate that quarks appear under three **colors** and that hadrons are singlet under a second $SU(3)_c$ symmetry [100].

In 1968, Bjorken proposed that the structure functions for the electron-proton deep-inelastic scattering should scale at high-energy [101, 102]. Simply speaking, the energy dependence of the cross-section should mimick the one for electron-muon scattering. The scaling properties where observed at SLAC the same year [103]. This pushed Feynman to develop the **parton model** in which the proton is described by a collection of point-like constituents from which the electron scatters incoherently [104].

In 1973, using **renormalization group equations**, Gross, Wilczek and Politzer proved that the non-abelian $SU(3)_c$ Yang-Mills gauge theory [105] with color triplet quarks and color octet gluons, proposed by Fritzsch, Gell-Man and Leutwyler the previous year [20, 21], becomes weaker at high energy, a property called **asymptotic freedom** [18, 19]. Then, the theory of Quantum Chromodynamics (QCD) was universally recognized as the correct fundamental theory of hadrons and they shared the Nobel prize in 2004. The gluons were discovered in 1979 at the experiment PETRA taking place at DESY [106]. Their existence was inferred from the presence of a third jet in e^+e^- collision.

Running of the strong coupling:
Upon neglecting the quark masses, **renormalization group equations** at one loop read [107]

$$\beta_3(\mu) \equiv \mu \frac{d}{d\mu} g_3(\mu) = -\frac{g_3^3(\mu)}{16\pi^2}\left(\frac{11}{3}N_c - \frac{2}{3}n_f\right), \tag{2.25}$$

where μ is the renormalization scale, $N_c = 3$ is the number of colors and n_f is the number of quarks lighter than μ (up to 6 in the SM). The solution is

$$\alpha_3(\mu) \equiv \frac{g_3^2(\mu)}{4\pi} = \frac{12\pi}{(11N_c - 2n_f)\ln\left(\mu^2/\Lambda^2\right)}, \tag{2.26}$$

where Λ is an integration constant.

---

[6] In the SM, there are six flavors of quarks (d, u, s, c, b, t).

Asymptotic freedom:
At high energy $\mu \gg \Lambda$, the strong coupling constant α_s, solution of the renormalization group equation in Eq. (2.26), tends to zero. This is the seminal result of Gross, Wilczek and Politzer showing that QCD is **asymptotically free**. Hence, perturbation theory can be applied at energies $\mu \gg \Lambda$

Color confinement:
The solution of the renormalization group equation in Eq. (2.26) exhibits a second important feature of QCD. The dimensionless coupling g_3 is related to a free dimensionful parameter, the integration constant Λ, which is a free mass scale, fixed by measurements of α_s. This property is known as **dimensional transmutation**. At that scale, the coupling constant diverges and the theory becomes strongly-coupled. Below the scale Λ, we expect the quarks and gluons to confine, giving rise to a mass gap in the particle spectrum. The later is given by the vacuum expectation value of the correlator $\bar{q}q$, or **meson condensate**

$$\langle \bar{q}q \rangle \simeq \Lambda^3. \tag{2.27}$$

This is the flip side of asymptotic freedom, called **infrared slavery** or **color confinement**. However, even though it has been experimentally verified as a failure of free-quark searches, it has only been verified in lattice numerical simulations [108] and still remain to be theoretically proven. Proving the existence of mass gap in Yang-Mills theories is defined as one of the Millenium problem.

From injecting the measurement of the strong coupling at the Z pole [109]

$$\alpha_3(m_Z) = 0.1179 \pm 0.0010, \tag{2.28}$$

and from setting $n_f = 5$, we obtain $\Lambda \simeq 90$ MeV. Renormalization group equations at 3 loops [109] give $\Lambda \simeq 1.4$ GeV. Instead of relying on perturbation theory, the QCD scale Λ is better computed in lattice simulations [110]

$$\Lambda = 211(10) \text{ MeV}. \tag{2.29}$$

During the late 70s and early 80s, it is shown that the phenomenology of color confinement can be reproduced by modeling gluon field lines as Nambu-Goto strings [111–115]. The Lund group [116–118] gives a successful description of jet fragmentation in terms of string fragmentation, which is now at the core of the very widespread high-energy event generator code PYTHIA [119].

In 1999, Yang-Mills gauge theory in the large N super-conformal limit are shown to be dual to anti de Sitter space time [120]. A new method of investigation of strongly-coupled theories, as being dual of gravitational theories, is opened.

The theory of mesons:
Below the Landau pole $\mu \lesssim \Lambda_{\text{QCD}}$, the strong coupling constant diverges and perturbation theory can not be used to explain certain phenomena observed in Nature, as for instance nuclear physics. Instead, an Effective Field Theory (EFT) of nuclear interac-

tions can be constructed based on symmetries of the fundamental QCD Lagrangian. For introductory textbooks on effective field theory, the reader is referred to [121, 122]. Let us consider a simple theory with only up and down quarks, which in good approximation can be assumed to be massless,

$$\mathscr{L} = -\frac{1}{4}\left(F^a_{\mu\nu}\right)^2 + i\bar{u}\not{D}u + i\bar{d}\not{D}d. \tag{2.30}$$

Upon introducing left-handed and right-handed component $\psi^{R/L} = \frac{1}{2}(1 \pm \gamma_5)\psi$, we can see that the Lagrangian

$$\mathscr{L} = -\frac{1}{4}\left(F^a_{\mu\nu}\right)^2 + i\bar{u}_L\not{D}u_L + i\bar{d}_R\not{D}d_R, \tag{2.31}$$

is manifestly invariant under independent $SU(2)$ rotations of the left and right components

$$\begin{pmatrix} u_L \\ d_L \end{pmatrix} \to U_L \begin{pmatrix} u_L \\ d_L \end{pmatrix}, \qquad \begin{pmatrix} u_R \\ d_R \end{pmatrix} \to U_R \begin{pmatrix} u_R \\ d_R \end{pmatrix}, \tag{2.32}$$

where $U_R \in S(U)_L$ and $U_R \in SU(2)_R$. However, in the ground state of QCD observed in our universe, the quark billinears have a non zero expectation value

$$\langle \bar{u}u \rangle = \langle \bar{d}d \rangle = V^3 \simeq \Lambda^3_{\text{QCD}}. \tag{2.33}$$

Therefore the **chiral symmetry** $SU(2)_R \times SU(2)_R$ is spontaneously broken to the **vectorial group**, in which left and right components rotate the same way

$$SU(2)_R \times SU(2)_R = SU(2)_V. \tag{2.34}$$

The remaining symmetry under $SU(2)_V$ is known as **isospin**. The dynamics of spontaneous chiral symmetry breaking can be described in the linear sigma model

$$\mathscr{L}_{\text{eff}} \subset \left|\partial_\mu \Sigma\right|^2 + m^2 |\Sigma|^2 - \frac{\lambda}{4}|\Sigma|^4, \tag{2.35}$$

where Σ_{ij} is a set of fields, known as the **Goldstone matrix**, transforming as a **bi-fundamental** of $SU(2)_R \times SU(2)_R$

$$\Sigma \to U_L \Sigma U_R^\dagger. \tag{2.36}$$

We can check that Eq. (2.35) is minimized for

$$\langle \Sigma \rangle = \frac{v}{\sqrt{2}}\begin{pmatrix} 1 & 0 \\ 0 & 1 \end{pmatrix}, \qquad \text{with} \quad v = \frac{2m}{\sqrt{\lambda}}. \tag{2.37}$$

We can choose to parameterize the Goldstone matrix as

$$\Sigma(x) = \frac{v + \sigma(x)}{\sqrt{2}} \exp\left(2i \frac{\tau^a \pi^a(x)}{F_\pi}\right), \tag{2.38}$$

where the **pion decay constant** $F_\pi = v$ is chosen such that π^a are canonically normalized. The later can be measured from the pion decay and one finds $F_\pi \simeq 130\,\text{MeV}$ [109]. Under a chiral transformation $U_\text{L} = \exp(i\theta^a_\text{L}\tau^a)$ and $U_\text{R} = \exp(i\theta^a_\text{R}\tau^a)$, we can check that σ is invariant while

$$\pi^a \to \pi^a + \frac{F_\pi}{2}(\theta^a_\text{L} - \theta^a_\text{R}) - \frac{1}{2} f^{\text{abc}}(\theta^b_\text{L} + \theta^b_\text{R})\pi^c + \cdots, \tag{2.39}$$

where $f^{\text{abc}} = \varepsilon^{\text{abc}}$ are the structure constants of $SU(2)_\text{V}$. We can see that π^a transforms as an adjoint of $SU(2)_\text{V}$ but non-linearly—they **shift** at leading order—under the broken symmetry transformation $\theta^a_\text{L} \neq \theta^a_\text{R}$. Upon transforming the fundamental Lagrangian in Eq. (2.31) under $\theta^a_\text{L} = -\theta^a_\text{R} = \theta(x)$ where $\theta(x)$ depends on x, we obtain

$$\mathscr{L} \to \mathscr{L} + j^{\mu 5a}\partial_\mu\theta, \tag{2.40}$$

where $j^{\mu 5a}$ is the **conserved Noether current**

$$j^{\mu 5a} = \bar{Q}\gamma^\mu\gamma^5\tau^a Q, \qquad \text{with} \quad Q = \begin{pmatrix} u \\ d \end{pmatrix}. \tag{2.41}$$

Now upon applying the exact same transformation on the effective lagrangian $\mathscr{L}_\text{eff} = \frac{1}{2}(\partial\pi)^2 + \cdots$, and using Eq. (2.39), we obtain that the **chiral current** in Eq. (2.41), in the effective description matches to

$$j^{\mu 5a} = F_\pi \partial_\mu \pi. \tag{2.42}$$

By acting it on the meson state $|\pi\rangle$, we obtain

$$\langle 0| j^{\mu 5a} |\pi\rangle = i p^\mu F_\pi e^{-ipx}. \tag{2.43}$$

The conservation of the chiral current $\partial_\mu j^{\mu 5a} = 0$, implies that the mesons π are massless, so we are able to identify them as **Nambu-Goldstone bosons** of the chiral symmetry breaking. Next we want to use symmetries to write all the possible terms in the effective Lagrangian, in addition to those in Eq. (2.35). Since σ does not transform under any symmetry, we get rid of it by taking $m \to \infty$ and $\lambda \to \infty$, while holding F_π fixed

$$\frac{\sqrt{2}}{v}\Sigma(x) \to U(x) \equiv \exp\left[2i\frac{\tau^a\pi^a}{F_\pi}\right] = \exp\left[\frac{i}{F_\pi}\begin{pmatrix} \pi^0 & \sqrt{2}\pi^- \\ \sqrt{2}\pi^+ & -\pi^0 \end{pmatrix}\right], \tag{2.44}$$

where $\pi^0 = \pi^3$ and $\pi^\pm = \frac{1}{\sqrt{2}}\left(\pi^1 \pm i\pi^2\right)$. Such a theory with decoupled σ is known as **non-linear sigma model**. From this, we can write the most general effective Lagrangian invariant under $SU(2)_\mathrm{R} \times SU(2)_\mathrm{R}$, known as the **chiral lagrangian**

$$\mathscr{L}_\mathrm{eff} \subset \frac{F_\pi^2}{4}\mathrm{tr}\left[(D_\mu U)(D_\mu U)^\dagger\right] + L_1\mathrm{tr}\left[(D_\mu U)(D_\mu U)^\dagger\right]^2 \tag{2.45}$$

$$+L_2\mathrm{tr}\left[(D_\mu U)(D_\nu U)^\dagger\right]\mathrm{tr}\left[(D_\mu U)^\dagger(D_\nu U)\right] \tag{2.46}$$

$$+L_3\mathrm{tr}\left[(D_\mu U)(D_\mu U)^\dagger(D_\nu U)(D_\nu U)^\dagger\right] + \cdots \tag{2.47}$$

where the covariant derivative only contains electroweak gauge fields but not the gluons. The derivation of the theory of hadrons below the confining scale using symmetry of the fundamental lagrangian was pioneered by Callan, Coleman, Wess and Zumino (CCWZ) in 1969 [123, 124]. See [125–129] for reviews. In fact, the chiral symmetry is explicitly broken by the presence of the tiny up and down masses,

$$\mathscr{L} \subset \bar{q}M_q q, \qquad M_q = \begin{pmatrix} m_u & 0 \\ 0 & m_d \end{pmatrix} \tag{2.48}$$

and the mesons π are said to be **pseudo**-Nambu-Goldstone bosons. In order to compute the mass of the mesons, we can use the **spurion** trick. We trade the Lagrangian couplings responsible for the explicit breaking—here the mass matrix M_q—by fake fields which transform in such a way that the Lagrangian looks symmetric. We assume that M_q transforms under a bi-fundamental of the chiral transformation $M_q \to U_\mathrm{L} M_q U_\mathrm{R}$. The leading $SU(2)_\mathrm{R} \times SU(2)_\mathrm{R}$ invariant term which we can add is

$$\mathscr{L}_\mathrm{eff} \supset \frac{V^3}{2}\mathrm{tr}\left[M_q U + M_q^\dagger U^\dagger\right]$$

$$\supset V^3(m_u + m_d) - \frac{V^3}{2F_\pi^2}(m_u + m_d)(\pi_0^2 + \pi_1^2 + \pi_2^2) + \mathscr{O}(\pi^3). \tag{2.49}$$

The prefactor $V^3/2$ is fixed such that the vacuum energy in the effective lagrangian in Eq. (2.49) matches the vacuum energy in the fundamental lagrangian in Eq. (2.48). We conclude that the pion mass follows the relation

$$m_\pi^2 = \frac{V^3}{F_\pi^2}(m_u + m_d), \tag{2.50}$$

known as the **Gell-Mann-Oakes-Renner** relation [130]. Using $V \simeq 250$ MeV [131], $F_\pi \simeq 92$ MeV and $m_\pi = 135$ MeV, the sum of the up and down quark masses is predicted to be $m_u + m_d \simeq 10$ MeV. This is however only the leading order contribution to the pion mass and it overestimates the measured value for up and down quark masse, $m_u + m_d \simeq 6.83^{+0.97}_{-0.43}$ MeV, by more than 20% [109].

2.2.3 *Electroweak Symmetry Breaking*

The birth of the Glashow-Weinberg-Salam model:
Fermi, who died in 1954 did not get the chance to experience the major revolutions that his theory of weak interactions which he introduced in 1934 to explain beta decay, went through during the 60s. Before this, weak interactions were described using a **non-renormalizable** four-fermion interaction between left-handed currents, see e.g. in 1958 the formulation of the 'V-A' theory by Feynman and Gell-mann [132] or Sudarshan and Marshak [133].

A model unifying electromagnetism and weak interactions under the gauge group $SU(2)_L \times U(1)_Y$ was first proposed by Glashow in 1961 [4]. But the theory was not yet renormalizable.

In 1960, Nambu and Goldstone discovered, in the context of superconductivity, that whenever a symmetry is broken spontaneously, **massless modes** are generated [12, 13]. Shortly after, Goldstone, Salam and Weinberg proved the statement in the framework of Quantum Field Theory [14]. In 1964, Brout, Englert, Higgs, Guralnik, Hagen and Kibble provided a mechanism which gives a mass to a vector boson without violating gauge invariance [7–11]. Nowadays, this is known as the **Higgs mechanism** and it can be formulated as follows: if the local symmetry is spontaneously broken, the massless Nambu-Goldstone bosons become the longitudinal modes of the gauge fields which therefore acquire a mass.

Finally, this is in 1967 that the model of Glashow, unifying electromagnetism and weak interactions under the gauge group $\mathbf{SU(2)_L \times U(1)_Y}$ was improved by Weinberg and Salam who incorporated the spontaneous symmetry breaking into $\mathbf{U(1)_{em}}$ [5, 6]. The modern theory of Electroweak interactions was born. All three shared the Nobel prize in 1979. The $W^\pm$, Z^0 were discovered in 1983 with two corresponding Nobel prizes in 1984.

The renormalizability of such a massive gauge theory was proved in 1972 by 't Hooft and his adviser Veltman who developed the technique of dimensional regularization [15–17]. They received the Nobel prize in 1999.

Spontaneous symmetry breaking:
The third line of the SM Lagrangian in Eq. (2.15) contains the Higgs potential, which after minimization, is responsible, when $-\mu^2 < 0$, for a **vacuum expectation value** for the Higgs multiplet, which we can write, without loss of generality

$$\langle H \rangle = \begin{pmatrix} 0 \\ \frac{v}{\sqrt{2}} \end{pmatrix}, \tag{2.51}$$

with

$$v = \sqrt{\frac{-\mu^2}{\lambda}} \simeq 246 \text{ GeV}. \tag{2.52}$$

The expansion of the Higgs field around its vacuum background, gives

$$\langle H \rangle = \begin{pmatrix} \pi_1 + i\pi_2 \\ \frac{v+h}{\sqrt{2}} + i\pi_3 \end{pmatrix}, \tag{2.53}$$

where π_1, π_2 and π_3 are the $4 - 1 = 3$ **massless** modes predicted by the Nambu-Goldsone theorem and h is the **massive** excitation of the Higgs field, commonly known as the Higgs boson. First theorized in 1964 [7–11], the Higgs boson has been discovered at the Large Hadron Collider in 2012 at a mass of $m_{\mathrm{h}} \simeq 125$ GeV [134, 135]. Englert and Higgs get the Nobel prize in 2013.

From injecting Eq. (2.53) into Eq. (2.15), we read the Higgs mass

$$m_{\mathrm{h}} = \sqrt{-2\mu^2}. \tag{2.54}$$

Generation of the vector boson masses:
The fourth line of the SM Lagrangian in Eq. (2.16) contains the interactions with the Higgs boson responsible for the mass of the different fields. First, there is the kinetic term for the Higgs boson which contains the interactions with the vector bosons $W^\pm$ and Z^0 responsible for their mass. Second, there are the Yukawa terms responsible for the mass of the quarks and leptons. We have introduced

$$\tilde{H} = i\sigma_2 H^*, \tag{2.55}$$

where σ_2 is the second Pauli matrix. $\tilde{H}$ has the opposite hypercharge as H but the same transformation rules under $SU(2)_L$.[7] Upon assigning to the Higgs field its vacuum expectation value in Eq. (2.51), we obtain

$$\begin{aligned} |D_\mu H|^2 &= g_2^2 \frac{v^2}{8} (0\ 1) \begin{pmatrix} \frac{g_1}{g_2} B_\mu + W_\mu^3 & W_\mu^1 - iW_\mu^2 \\ W_\mu^1 + iW_\mu^2 & \frac{g_1}{g_2} B_\mu + W_\mu^3 \end{pmatrix} \begin{pmatrix} \frac{g_1}{g_2} B_\mu + W_\mu^3 & W_\mu^1 - iW_\mu^2 \\ W_\mu^1 + iW_\mu^2 & \frac{g_1}{g_2} B_\mu + W_\mu^3 \end{pmatrix} \begin{pmatrix} 0 \\ 1 \end{pmatrix} \\ &= g_2^2 \frac{v^2}{8} \left[(W_\mu^1)^1 + (W_\mu^2)^2 + \left(\frac{g_1}{g_2} B_\mu - W_\mu^3 \right)^2 \right]. \end{aligned} \tag{2.60}$$

[7] The field $\tilde{H} = i\sigma_2 H^*$ has opposite hypercharge as H, and so transforms the opposit way as H under $U(1)_Y$. However, since the fundamental representation of $SU(2)$ is pseudo-real (e.g. [136]),

$$\sigma_2 \left(e^{i\vec{\phi}.\vec{\sigma}}\right)^* \sigma_2 = e^{i\vec{\phi}.\vec{\sigma}}, \tag{2.56}$$

then $\tilde{H}$ transforms exactly the same way as H under $SU(2)_L$,

$$H \xrightarrow{SU(2)_L} e^{i\vec{\phi}.\vec{\sigma}} H, \tag{2.57}$$

$$\Leftrightarrow \quad H^* \xrightarrow{SU(2)_L} \left(e^{i\vec{\phi}.\vec{\sigma}}\right)^* H^*, \tag{2.58}$$

$$\Leftrightarrow \quad \tilde{H} \xrightarrow{SU(2)_L} -i\sigma_2 \left(e^{i\vec{\phi}.\vec{\sigma}}\right)^* \overbrace{i\sigma_2 i\sigma_2}^{=-1} H^* = e^{i\vec{\phi}.\vec{\sigma}} \tilde{H} \quad . \tag{2.59}$$

The diagonalization of the mass matrix after electroweak symmetry breaking defines the boson Z^0 and the photon A_μ

$$Z_\mu = W^3_\mu \cos\theta_w - B_\mu \sin\theta_w, \tag{2.61}$$

$$A_\mu = W^3_\mu \sin\theta_w + B_\mu \cos\theta_w, \tag{2.62}$$

where θ_w is the Weinberg angle

$$\sin\theta_w = \frac{g_1}{\sqrt{g_1^2 + g_2^2}} = (0.23122 \pm 0.00015)^{1/2}\,. \tag{2.63}$$

As it can be inferred from the next section, W^1 and W^2 do not have proper electric charge, instead we define the generators

$$\tau^\pm = \frac{1}{\sqrt{2}}(\tau^1 \pm i\tau^2), \tag{2.64}$$

with $[\tau^3,\ \tau^\pm] = \pm\tau^\pm$ such that the vector bosons

$$W^\pm_\mu = \frac{1}{\sqrt{2}}(W^1_\mu \mp i W^2_\mu), \tag{2.65}$$

have electromagnetic charge ± 1. The breaking of the electroweak gauge interactions to electromagnetism, $SU(2)_L \times U(1)_Y \to U(1)_{e.m.}$, also known as **Electroweak Symmetry Breaking** (EWSB) or Glashow-Weinberg-Salam model, is explicit from Eq. (2.60), where we can see that the photon is massless whereas the boson $W^\pm$ and Z^0 obtain the masses

$$m_W^2 = \frac{g_2^2}{2}v^2 \qquad \text{and} \qquad m_Z = \frac{g_1^2 + g_2^2}{2}v^2. \tag{2.66}$$

Generation of the Fermion masses:
Repeating the same steps for the Yukawa terms, the quarks and leptons get the masses

$$m_i = \frac{Y_i}{\sqrt{2}}v, \tag{2.67}$$

Fermion interactions:
Before EWSB, the interactions between the fermions and the electroweak bosons, contained in the fermion kinetic terms, in the second line of the SM Lagrangian, in Eq. (2.14), are

$$\mathscr{L} \supset g_2 W^1_\mu J^1_\mu + g_2 W^3_\mu J^2_\mu + g_2 W^3_\mu J^3_\mu + g_1 B_\mu J^Y_\mu \tag{2.68}$$

with

$$J^i_\mu = \sum_f \bar{\psi}^L_f \gamma^\mu \tau^i \psi^L_f, \quad \text{and} \quad J^Y_\mu = \sum_f Y^L_f \bar{\psi}^L_f \gamma^\mu \psi^L_f + \sum_f Y^R_f \bar{\psi}^R_f \gamma^\mu \psi^R_f. \tag{2.69}$$

After EWSB, they become

$$\mathscr{L} \supset g_2 W^+_\mu J^+_\mu + g_2 W^-_\mu J^-_\mu + \frac{e}{\sin\theta_w} Z_\mu J^Z_\mu + e A_\mu J^{EM}_\mu \tag{2.70}$$

with

$$J^Z_\mu = \frac{1}{\cos\theta_w}\left(J^3_\mu - \sin^2\theta_w J^{EW}_\mu\right), \tag{2.71}$$

$$J^{EM}_\mu = \sum_f Q_f \left(\bar{\psi}^L_f \gamma^\mu \psi^L_f + \bar{\psi}^R_f \gamma^\mu \psi^R_f\right), \tag{2.72}$$

$$J^+_\mu = \sum_f \bar{\psi}^L_f \gamma^\mu \tau^+ \psi^L_f, \tag{2.73}$$

$$J^+_\mu = \sum_f \bar{\psi}^L_f \gamma^\mu \tau^- \psi^L_f, \tag{2.74}$$

where we have defined the electric charge[8] Q_f and the electromagnetic coupling e

$$Q_f = T^3 + Y, \quad \text{and} \quad e = g_2 \sin\theta_w = g_1 \cos\theta_w. \tag{2.75}$$

The current J^Z_μ which has charge 0, is called the **neutral current**.

2.2.4 Weak CP Violation

A flavor mixing mass matrix:
The mass terms for the quarks can be written in a 3×3 matrix form

$$\mathscr{L} \supset -\frac{v}{\sqrt{2}} Y^d_{mn} \bar{d}_L d_R - \frac{v}{\sqrt{2}} Y^u_{mn} \bar{u}_L u_R + h.c. \tag{2.76}$$

$$\supset -\frac{v}{\sqrt{2}} \left[\bar{d}_L Y_d d_R + \bar{u}_L Y_u u_R\right] + h.c. \tag{2.77}$$

Hence the fermions from different flavors interact with each other in a complicated way, which calls for a diagonalization of the mass matrix. The Yukawa matrices Y are not hermitian but $YY^\dagger$ are. So it can be diagonalized with unitary matrices

[8] The first relation of Eq. (2.75) was already derived in the mid-50s, by Gell-Mann, Nakano and Nishijima [137, 138].

$$Y_d Y_d^\dagger = U_d M_d^2 U_d^\dagger, \quad \text{and} \quad Y_u Y_u^\dagger = U_u M_u^2 U_u^\dagger, \tag{2.78}$$

where $M_d = \frac{\sqrt{2}}{v}\text{diag}(m_d, m_s, m_b)$ and $M_u = \frac{\sqrt{2}}{v}\text{diag}(m_u, m_c, m_t)$. We can freely introduce other unitary matrices U_R and write

$$Y_d = U_L^d M_d \left(U_R^d\right)^\dagger, \quad \text{and} \quad Y_u = U_L^u M_u \left(U_R^u\right)^\dagger, \tag{2.79}$$

such that the Yukawa couplings read

$$\mathscr{L} \supset -\frac{v}{\sqrt{2}}\left[\bar{d}_L U_L^d M_d \left(U_R^d\right)^\dagger d_R + \bar{u}_L U_L^u M_u \left(U_R^u\right)^\dagger u_R\right]. \tag{2.80}$$

Mass eigenstates and the CKM matrix:
Now, we go to the mass basis, where the mass matrix is diagonal, by performing the unitary transformations

$$d_L \to U_L^d d_L, \qquad u_L \to U_L^u u_L, \tag{2.81}$$
$$d_R \to U_R^d d_R, \qquad u_R \to U_R^u u_R. \tag{2.82}$$

The matrices being unitary, all the terms of the Lagrangian which are flavor diagonal are preserved by the transformations above. Only the terms of interaction with the bosons $W^\pm$ which mix the flavors are not preserved and become

$$\mathscr{L} \supset \frac{g_2}{\sqrt{2}}\left[W_\mu^+ \bar{u}_L^i \gamma^\mu (V)^{ij} d_L^j + W_\mu^- \bar{d}_L^i \gamma^\mu \left(V^\dagger\right)^{ij} u_L^j\right]. \tag{2.83}$$

where $V = \left(U_L^u\right)^\dagger U_L^d$ is the **Cabiddo-Kobayashi-Maskawa (CKM) matrix**.

Leptons do not have such flavor mixing interactions in the SM since there are no neutrino mass terms. Hence, we are free to re-arrange ν_L in order to cancel the would-be $\left(U_L^\nu\right)^\dagger U_L^e$, analog of the CKM matrix in the lepton sector. However, such a matrix is needed to explain the neutrino oscillations, see Sect. 2.3.2.

Note that the CKM matrix is unitary. In 1970, Glashow, Iliopoulos and Maiani used the unitarity property of the CKM matrix to predict the existence of a fourth quark 'c' in order to account for the suppression of flavor-changing neutral currents [139]. The first charmed hadron, the $J/\psi : c\bar{c}$, was detected in 1974 at SLAC and BNL.

The origin of the *CP* violation:
The CKM matrix is a complex unitary 3×3 matrix and so has 9 real degrees of freedom. If it was a real $O(3)$ matrix it would have 3 real parameters. Hence a $U(3)$ matrix has 3 rotation angles and 6 complex phases. Actually, we can freely reparameterize the phases of the 6 quarks to set 5 phases to zero (nothing changes if all the rotations are the same, indeed this is due to baryon number conservation). So the physical parameters are 3 angles and 1 phase. Hence, we can parameterize the CKM matrix as follows [140]

$$
\begin{aligned}
V_{\mathrm{CKM}} &= \begin{pmatrix} 1 & 0 & 0 \\ 0 & c_{23} & s_{23} \\ 0 & -s_{23} & c_{23} \end{pmatrix} \begin{pmatrix} c_{13} & 0 & s_{13}\,e^{i\delta} \\ 0 & 1 & 0 \\ -s_{13}\,e^{i\delta} & 0 & c_{13} \end{pmatrix} \begin{pmatrix} c_{12} & s_{12} & 0 \\ -s_{12} & c_{12} & 0 \\ 0 & 0 & 1 \end{pmatrix} \\
&= \begin{pmatrix} c_{12}\,c_{13} & s_{12}\,c_{13} & s_{13}\,e^{-i\delta} \\ -s_{12}\,c_{23} - c_{12}\,s_{23}\,s_{13}\,e^{i\delta} & c_{12}\,c_{23} - s_{12}\,s_{23}\,s_{13}\,e^{i\delta} & s_{23}\,c_{13} \\ s_{12}\,s_{23} - c_{12}\,c_{23}\,s_{13}\,e^{i\delta} & -c_{12}\,s_{23} - s_{12}\,c_{23}\,s_{12}\,e^{i\delta} & c_{23}\,c_{12} \end{pmatrix}
\end{aligned} \tag{2.84}
$$

with $s_{ij} = \sin\theta_{ij}$ and $c_{ij} = \cos\theta_{ij}$. The complex phase δ is responsible for the **CP violation.**[9] However, this is just an arbitrary choice of normalization and it is useful to define basis-invariant quantity which encodes the CP violation in the electroweak quark sector. From Eq. (2.78), we can see that if $\left[Y_u Y_u^\dagger,\ Y_d Y_d^\dagger\right] = 0$, then we can find a unitary transformation $U = U_L^d = U_L^u$, acting on both down-type and up-type quarks, $d_L \to U d_L$ and $u_L \to U u_L$, which diagonalizes $Y_u Y_u^\dagger$ and $Y_d Y_d^\dagger$ simultaneously, and then the CKM matrix is trivial $V = \mathbb{1}$. Hence, the CP violation must be encoded in the commutator

$$
iC = \left[Y_u Y_u^\dagger,\ Y_d Y_d^\dagger\right] = \left[U_u M_u^2 U_u^\dagger,\ U_d M_d^2 U_d^\dagger\right] = U_u \left[M_u^2,\ V M_d^2\, V^\dagger\right] U_u^\dagger. \tag{2.85}
$$

The determinant of C is invariant under any unitary transformation of the quarks and can be computed with the Vandermond formula [143, 144]

$$
\det C = -16\, T\, B\, J, \tag{2.86}
$$

$$
\text{where} \quad T = (m_t^2 - m_u^2)(m_t^2 - m_c^2)(m_c^2 - m_u^2)/v^6, \tag{2.87}
$$

$$
B = (m_b^2 - m_d^2)(m_b^2 - m_s^2)(m_b^2 - m_s^2)(m_s^2 - m_d^2)/v^6, \tag{2.88}
$$

$$
J = s_{12}s_{23}s_{31}c_{12}c_{23}c_{31}^2\, \sin\delta, \tag{2.89}
$$

where J is known as the **Jarlskog invariant.**[10] CP is violated if and only if $\det C \neq 0$, otherwise the CKM matrix can be made real with any rephasing of the quark fields.

The entries of the CKM matrix have been measured from the decay of many hadrons [109]

$$
\begin{pmatrix} V_{ud} & V_{us} & V_{ub} \\ V_{cd} & V_{cs} & V_{cb} \\ V_{td} & V_{ts} & V_{tb} \end{pmatrix} = \begin{pmatrix} 0.97 \pm 0.0001 & 0.22 \pm 0.0004 & 0.0037 \pm 0.0001 \\ 0.22 \pm 0.0004 & 0.97 \pm 0.0001 & 0.042 \pm 0.0009 \\ 0.0090 \pm 0.0002 & 0.041 \pm 0.0007 & 0.999 \pm 0.00003 \end{pmatrix}. \tag{2.90}
$$

[9] The Lagrangian must be invariant under CPT [141, 142], where C, P and T stand for charge conjugation, mirror transformation and time reversal. So a CP-transformation is equivalent to a T-transformation. But T is anti-unitary (it sends i to $-i$) so any imaginary coefficient implies CP violation (assuming the field operators preserve T).

[10] The basis-invariant formulation of CP violation, $\det C$ in Eq. (2.86), has been introduced by Cecilia Jarlskog in 1985. She first introduced a formula containing linear masses [143] before correcting with squared mass two months later [144].

The corresponding magnitude for the mixing angled as well as the CP phase are [145]

$$\theta_{12} \simeq 13.02^\circ \pm 0.04^\circ, \quad \theta_{23} \simeq 2.36^\circ, \pm 0.08^\circ \quad \theta_{13} \simeq 0.20^\circ \pm 0.02^\circ, \quad \delta_{CP} \simeq 69^\circ \pm 5^\circ, \tag{2.91}$$

while the Jarlskog invariant is

$$J = (2.96 \pm 0.20) \times 10^{-5}. \tag{2.92}$$

Hence, even though the CP phase in the parameterization of Eq. (2.84) is close to 90°, the actual basis-invariant J is small. Also for some applications, e.g. baryogenesis discussed in Sect. 3.5.2, the physical quantity is $\det C \simeq 10^{-18}\, J$ [146–149], which is even smaller.

Explanation of the CP violation by the prediction of two new quarks:
Lee and Yang were the first to postulate back in 1956 that weak interactions violate parity.[11] P [150] They proposed an experiment which validates their prediction one year later [151, 152]. In 1963, Cabibbo introduced a mixing angle (the largest angle of the CKM matrix) to explain flavor-changing interactions [153]. In 1964, Cronin and Val Fitch provided evidence for CP violation in kaon decays [154]. In 1973, Kobayashi and Maskawa predicted the existence of a **third generation** of quarks in order to explain CP violation [22]. The first bottomium, $\Upsilon : b\bar{b}$, was detected in 1977 at Fermilab. The top was detected via $t\bar{t}$ pair production in 1995 at Fermilab. All these discoveries were rewarded with Nobel prizes.

2.2.5 *Anomaly Cancellation*

Let us consider a massless Dirac fermion coupled to a classical external $U(1)$ gauge field

$$\mathscr{L} = i\bar{\psi}\partial\!\!\!/\psi - eJ^\mu_V A_\mu, \qquad \text{with} \quad J^\mu_V = \bar{\psi}\gamma^\mu\psi. \tag{2.93}$$

The lagrangian looks invariant under the axial symmetry[12]

$$\psi \xrightarrow{U(1)_A} e^{i\gamma_5\theta_\psi}\psi. \tag{2.95}$$

The expectation value of the corresponding axial current $J^\mu_5 = \bar{\psi}\gamma^\mu\gamma^5\psi$ in the gauge field background can be expressed in the path integral formalism

---

[11] P violation in weak interactions comes from the fact that only left-handed fermions are charged under $SU(2)_L$.

[12] A chiral transformation rotates the left-handed and right-handed fields in an independent way

$$\psi_L \longrightarrow e^{i\alpha}\psi_L, \qquad \text{and} \qquad \psi_R \longrightarrow e^{i\beta}\psi_R. \tag{2.94}$$

The transformation $\psi \to e^{i\gamma_5\theta_\psi}$ is a particular chiral transformation called 'axial' transformation in which left-handed and right-handed fields rotate in the opposite way.

$$\langle A|J_5^\mu|A\rangle = \frac{\int \mathscr{D}\bar{\psi}\mathscr{D}\psi\; J_A^\mu\; e^{i\int d^4x(i\bar{\psi}\partial\!\!\!/\psi - eJ_V^\mu A_\mu)}}{\int \mathscr{D}\bar{\psi}\mathscr{D}\psi\; e^{i\int d^4x(i\bar{\psi}\partial\!\!\!/\psi - eJ_V^\mu A_\mu)}}. \tag{2.96}$$

Using perturbation theory, we get

$$\langle A|J_5^\mu|A\rangle = -ie\int d^4y\langle 0|T\left[J_5^\mu(x)J_V^\alpha(y)\right]|0\rangle A_\alpha(y) \tag{2.97}$$

$$-\frac{e^2}{2}\int d^4y_1 d^4y_2\langle 0|T\left[J_5^\mu(x)J_V^\alpha(y_1)J_V^\beta(y_2)\right]|0\rangle A_\alpha(y_1)A_\alpha(y_2), \tag{2.98}$$

where we used that $J_5^\mu(x)J_V^\alpha(y)$ is odd under parity in the first line. Hence, we see that $\langle A|J_5^\mu|A\rangle$ is proportional to $\langle 0|T\left[J_A^\mu(x)J_V^\alpha(y)J_V^\beta(0)\right]|0\rangle$ which via Feynman rules is given by a triangle diagram [61]. Careful computations realized in 1968 by Adler, Bell and [155, 156] show that

$$\langle A|\partial_\mu J_5^\mu|A\rangle = -\frac{e^2}{16\pi^2}\varepsilon^{\mu\nu\alpha\beta}F_{\mu\nu}F_{\alpha\beta}. \tag{2.99}$$

In absence of mass terms, the chiral transformation is an exact symmetry of the classical lagrangian. However, Eq. (2.99) shows that $U(1)_\mathrm{A}$ is broken by quantum effects in the presence of gauge fields. If the anomalous symmetry $U(1)_A$ is global it does not lead to any problem. However, if the anomalous symmetry $U(1)_A$ is gauged, then it can lead to inconsistencies (e.g. Lorentz violation). Therefore, we must check that the gauge groups in the SM are anomaly free. We can generalize the previous treatment to non-abelian symmetry groups in which case the Noether current reads

$$J_\mu^a = \sum_\psi \bar{\psi}_i\tau_{ij}^a\gamma^\mu\psi_j, \tag{2.100}$$

where τ_{ij}^a are the group generators. Now, each vertex of the triangle diagram picks up a group generator and we obtain

$$\partial_\alpha J_\alpha^a(x) = -\frac{A_{\alpha\beta\gamma}}{64\pi^2}\varepsilon^{\mu\nu\delta\sigma}F_{\mu\nu}^\beta F_{\delta\sigma}^\gamma, \tag{2.101}$$

where $A_{\alpha\beta\gamma}$ is a completely symmetry tensor

$$A_{\alpha\beta\gamma} = \mathrm{tr}\left[\tau_\alpha^L\;\{\tau_\beta^L\,\tau_\gamma^L\}\right] - \mathrm{tr}\left[\tau_\alpha^R\;\{\tau_\beta^R\,\tau_\gamma^R\}\right]. \tag{2.102}$$

Now, we can check that the $SU(2)_L^3$ anomaly vanishes since $\left\{\tau^a,\ \tau^b\right\} = \frac{1}{2}\delta^{ab}$, and so $A^{abc} = \delta^{bc}\mathrm{tr}[\tau^a] = 0$. The $SU(3)_c^3$ anomaly also cancels since right-handed and left-handed have the same color. All the anomalies of the type $SU(N)U(1)^2$ cancel since $\mathrm{tr}\,[\tau^a\,\{1,\,1\}] = 2\mathrm{tr}\,[\tau^a] = 0$. Therefore, the only potentially dangerous anomalies are the ones of $U(1)_Y$ in the background of $U(1)_Y$, $SU(2)_L$, $SU(3)_c$ and gravity (universal coupling). But the matter content of the SM is exactly such that the anomalies cancel[13]

$$U(1)_Y^3: \quad \left(2Y_L^3 - Y_e^3 - Y_\nu^3\right) + 3\left(2Y_Q^3 - Y_u^3 - Y_d^3\right) = 0, \tag{2.103}$$

$$SU(3)^2U(1)_Y: \quad 2Y_Q - Y_u - Y_d = 0, \tag{2.104}$$

$$SU(2)^2U(1)_Y: \quad 2Y_Q - Y_u - Y_d = 0, \tag{2.105}$$

$$\mathrm{Grav}^2U(1)_Y: \quad (2Y_L - Y_e - Y_\nu) + 3\left(2Y_Q - Y_u - Y_d\right) = 0. \tag{2.106}$$

Hence the SM is anomaly free: the marriage between quarks and leptons is a very peculiar one. Particularly, the electron has exactly the opposite charge as the proton. Actually, we can show that there is one more $U(1)$ symmetry which is anomaly free in the background of the four fundamental forces of Nature: $U(1)_{\mathrm{B-L}}$. In contrast, $SU(2)_L^2U(1)_B$ and $SU(2)_L^2U(1)_L$ are anomalous. As a consequence, $U(1)_B$ and $U(1)_L$ are broken by non-perturbative processes with $\Delta(B+L) \neq 0$, called sphalerons, which can play an important role for generating the observed asymmetry between matter and anti-matter, see Sect. 3.5.2 of Chap. 3.

The discovery of the chiral anomaly by ABJ in 1968 was a particularly important result at that time since it explained the decay of the neutral pion, $\pi_0 \to \gamma\gamma$, from the anomaly $U(1)_{\mathrm{e.m.}}^2U(1)_A$. Another triumph of chiral anomalies is the explanation by Weinberg and 't Hooft in 1975 [158, 159], of the heaviness of the η', with respect to the η which has the same flavor content, as resulting from the global anomaly $SU(3)_c^2U(1)_A$. In the next section, we discuss an other application of the $SU(3)_c^2U(1)_A$ anomaly.

2.2.6 *Strong CP Violation*

The SM Lagrangian contains terms in Eq. (2.14), of the form $\tilde{F}_{\mu\nu}F^{\mu\nu}$, which are total derivatives, violate CP and are generated by chiral transformations. The one involving the gluons of $SU(3)_c$ can have physical consequences.

Total derivative:
The quantity $\tilde{F}_{\mu\nu}$, called the Hodge-dual of the tensor strength, is defined by

$$\tilde{F}_{\mu\nu} \equiv \varepsilon_{\mu\nu\alpha\beta}F^{\alpha\beta}. \tag{2.107}$$

[13] Note that the mixed gauge-gravitational anomaly in Eq. (2.106) is automatically satisfied if the 3 other gauge anomalies, in Eqs. (2.103), (2.104) and (2.105), vanish and hypercharge is quantized. See [157] for an elegant proof which uses that Eqs. (2.103), (2.104) and (2.105) can be recast into the Diophantine equation $x^3 + y^3 = z^3$ and then applies Fermat's last theorem.

We can show that

$$\varepsilon^{\mu\nu\alpha\beta} F^a_{\mu\nu} F^a_{\alpha\beta} = \partial^\mu K_\mu, \qquad K_\mu = \varepsilon_{\mu\nu\alpha\beta} \left(A^a_{\text{nu}} F^a_{\alpha\beta} - \frac{g}{3} f^{abc} A^a_\nu A^b_\alpha A^c_\beta \right), \quad (2.108)$$

where K_μ is known as the Chern-Simons current. Total derivatives never contribute in perturbation theory since the corresponding Feynman rule contains a factor with the sum of all momenta going into the vertex minus the momenta going out, which is equal to zero. However, θ can still have physical consequences through non-perturbative effects.

***CP*-violation**:
Furthermore, we can check that the terms in Eq. (2.108) violate P, T and CP. This can be easily shown in the electromagnetism case where $F_{\mu\nu}\tilde{F}^{\mu\nu} = 4\vec{E}.\vec{B}$, the transformation rules of the electric and magnetic fields being $\vec{E} \xrightarrow{CP,T} \vec{E}$ and $\vec{B} \xrightarrow{CP,T} -\vec{B}$.

Generated by a chiral transformation:
Using the results from the previous section, e.g. Eq. (2.99), we now make the observation that the terms in Eq. (2.108) can be generated after performing a chiral rotation of the fermions

$$U(1)_A: \quad \int \mathscr{D}\bar{\psi}\mathscr{D}\psi \xrightarrow{\psi \to e^{i\gamma_5\theta}\psi} \int \mathscr{D}\bar{\psi}\mathscr{D}\psi \exp\left(i\theta \int \frac{g^2}{32\pi^2} \varepsilon^{\mu\nu\alpha\beta} F^a_{\mu\nu} F^a_{\alpha\beta}\right), \quad (2.109)$$

where $F^a_{\mu\nu}$ is the field strength for anything under which fermions are charged. Therefore, the terms $F\tilde{F}$ in the SM Lagrangian are in one-to-one connection with the anomaly of the chiral symmetry $U(1)_A$ in the background of the SM gauge fields: $SU(3)^2_c U(1)_A$, $SU(2)^2_L U(1)_A$ and $U(1)^2_Y U(1)_A$.

A physical parameter:
For $U(1)_Y$ which is abelian, the term $A^a_\nu A^b_\alpha A^c_\beta$ is absent in Eq. (2.108) and $\int d^4x \, F\tilde{F} = 0$.[14] For $SU(2)_L$, we can rotate the left-handed fermion fields to move the θ_2 angle in the mass matrix and then rotate the right-handed fields, which are uncharged and therefore not subject to the anomaly, to remove it completely.

However, for $SU(3)_c$, where both left-handed and right-handed quarks are colored, we can not rotate away θ_3. Instead, we can only move it back and forth between the anomaly term and the Yukawa matrix. From performing a chiral rotation to remove the Yukawa phase, we obtain

---

[14] For $\int d^4x \, F^2$ to converge, we must have $F_{\mu\nu} < \frac{1}{r^2}$. Naively, this would imply $A_\mu < \frac{1}{r}$. However, A_μ can scale like $A_\mu \sim \frac{1}{r}$ if it is pure gauge $U\partial_\mu U$ where U is the gauge matrix. For those fields we have $A_\nu F_{\rho\sigma} < \frac{1}{r^3}$ but $A_\nu A_\rho A_\sigma \sim \frac{1}{r^3}$. So we conclude that only the term $A_\nu A_\rho A_\sigma$ in Eq. (2.108) can contribute to $\int d^4x \, F\tilde{F}$. I thank Filippo Sala for pointing me this fact.

$$\mathscr{L}_\theta = \bar{\theta}\frac{g_s^2}{32\pi^2}F^a_{\mu\nu}\tilde{F}^{a\mu\nu}, \qquad \bar{\theta} \equiv \theta_3 - \theta_{\text{Yuk}}, \tag{2.110}$$

where θ_{Yuk} is the phase in the Yukawa matrix (if any)

$$\theta_{\text{Yuk}} \equiv \arg\det\left(\text{Y}_\text{d}\text{Y}_\text{u}\right). \tag{2.111}$$

Thus, $\bar{\theta}$ is a basis-independent measure of the strong CP violation and can be physical. Note, that θ_{Yuk} is not contained among the physical parameters of the CKM matrix in Sect. 2.2.4, because the counting $9 - 5 = 4$ ignores the quantum anomaly and assumes chiral transformations to be symmetries of the Lagrangian.

Where is the neutron EDM?:
Particularly, $\bar{\theta}$ generates an electron dipole moment (EDM) for the neutron [160–163]

$$d_N \simeq \left(2.5 \times 10^{-16}\ \text{e.cm}\right)\bar{\theta}. \tag{2.112}$$

The current bound on the neutron EDM is $d_N \lesssim 3 \times 10^{-26}$ e.cm [164–166], so that

$$\bar{\theta} \lesssim 10^{-10}. \tag{2.113}$$

The smallness of $\bar{\theta}$ without any symmetry protection, except CP, is known as the strong CP problem, see Sect. 2.3.4.

θ-vacua—dilute instanton gas approximation:
Another physical effect of $\bar{\theta}$ is that different values of the angle correspond to different vacua, called **θ-vacua**. The explicit dependence of the vacuum energy on the $\bar{\theta}$ parameter was first computed in the **dilute instanton gas approximation**, cf. original works pioneered by Callan, Dashen and Gross in 1976 [167–174] and reviews [175–177]. Let us start by discussing properties of the vacuum of $SU(N)$ gauge theories. In order to give a finite action, Yang-Mills fields must go to zero at infinity, up to a pure gauge configuration

$$A_\mu \to ig(x)\partial_\mu g(x)^{-1}, \qquad \text{where} \quad g(x) : S^3_\infty \to SU(N). \tag{2.114}$$

The set of maps $g(x)$ falls into an infinite number of distinct equivalence classes. Indeed, the third homotopy group of $SU(N)$ is non-trivial

$$\pi_3(SU(N)) = \mathbb{Z}, \qquad N \geq 2. \tag{2.115}$$

This means that the connected components of the space of finite-action solutions of Yang-Mills $SU(N)$ theories can be classified with an integer. This integer is known has the **winding number** or **Pontryagin number**

$$\nu(g) = \frac{1}{24\pi^2} \int_{S^3} d\theta_1 d\theta_2 d\theta_3 \varepsilon^{ijk} \text{tr}\left(g^{-1}\partial_i g g^{-1}\partial_j g g^{-1}\partial_k g\right) = \frac{1}{32\pi^2}\int d^4x\, F\tilde{F}, \tag{2.116}$$

where θ_1, θ_2 and θ_3 are three angles parameterizing S^3. The field configuration which interpolates between neighboring winding numbers are the configuration which have the lowest Euclidean action

$$\begin{aligned} S_{\text{E}} &= \frac{1}{4\,g^2}\int d^4x\, F_{\mu\nu}F^{\mu\nu} = \frac{1}{4\,g^2}\int d^4x \left[F\tilde{F} + \frac{1}{2}(F-\tilde{F})^2\right] \\ &= \frac{8\pi^2\nu}{g^2} + \frac{1}{8g^2}\int d^4x (F-\tilde{F})^2. \end{aligned} \tag{2.117}$$

They saturate the Bogomol'nyi bound [178] for $\nu = 1$ or $\nu = -1$,

$$F = \pm\tilde{F} \qquad \Longrightarrow \qquad S_0 = \frac{8\pi^2}{g^2}. \tag{2.118}$$

They have been computed by Belavin, Polyakov, Schwarz and Tyupkin in 1975 [179] and are known as **BPST instantons,**

$$A_\mu(x) = i f(r^2)\left(\frac{\tau + i\vec{x}\cdot\vec{\sigma}}{r}\right)\partial_\mu\left(\frac{\tau + i\vec{x}\cdot\vec{\sigma}}{r}\right)^{-1}, \qquad f(r^2) = \frac{r^2}{r^2+\rho^2}. \tag{2.119}$$

Since the instantons offer the possibility of quantum tunneling between the classical vacua $|n\rangle$ characterized by the winding number n of gauge fields at infinity, the quantum vacuum is a linear combination $|\text{vac}\rangle = \sum_{n=-\infty}^{+\infty} a_n|n\rangle$. We can easily show that the only linear combination which is invariant under the gauge transformation $|n\rangle \xrightarrow{g(x)} |n+1\rangle$ is the so-called θ**-vacuum**

$$|\theta\rangle = \sum_{n=-\infty}^{+\infty} e^{-i\theta}|n\rangle. \tag{2.120}$$

Due to tunneling, the infinite degeneracy between the energy of the classical vacua $|n\rangle$ is lifted. The energy of the ground state can be computed from the transition amplitude in the stationary limit

$$\langle\theta|e^{iHT}|\theta\rangle = \sum_i e^{-E_i T}\langle\theta|E_i\rangle\langle E_i|\theta\rangle \quad \xrightarrow{T\to\infty} \quad e^{-E_0 T}. \tag{2.121}$$

In quantum field theory, the transition amplitude can be computed in the Feynman path integral formulation

$$\langle\theta|e^{iHT}|\theta\rangle = N\int DA_\mu\, e^{-S_E}. \tag{2.122}$$

From Eq. (2.117), we can see that the path integral is dominated by the BPST instantons, whose action is minimal. They come as two types, the instantons $\nu = 1$ and the anti-instantons $\nu = -1$. Suppose we have a gas of n instantons with centers $x_1, x_2, \dots x_n$. We must integrate over the positions of all the individual instantons, which gives

$$\int_{-T/2}^{T/2} dt_1 \int_{-T/2}^{t_1} dt_2 \cdots \int_{-T/2}^{t_{n-1}} dt_n \int \prod_{i=1}^{n} d^3x_i = \frac{(TV)^n}{n!}. \tag{2.123}$$

In addition, we suppose that we have $\bar{n}$ anti-instantons. The path integral over all possible histories gives

$$\langle\theta|e^{iHT}|\theta\rangle = \sum_{n,\bar{n}} e^{i(n-\bar{n})\theta}\,\frac{(TVKe^{-S_0})^n}{n!}\,\frac{(TVKe^{-S_0})^{\bar{n}}}{\bar{n}!} \tag{2.124}$$

where S_0 is defined in Eq. (2.118) and where K is a prefactor accounting for the quantum fluctuations of the instanton size ρ, and which has been shown to contain an IR divergence by t'Hooft in 1976 [180], which he called **infrared embarrassment**. In Eq. (2.124), we have neglected the overlap between instantons and anti-instantons. This is valid as long as the instanton size ρ is larger than their separation. This is the so-called **dilute instanton gas approximation** [168, 175, 176]. We obtain the energy of the θ-vacuum

$$E_0(\theta)/V = Ke^{-S_0}\cos\theta. \tag{2.125}$$

Comparisons with lattice studies show that the dilute instanton gas approximation is valid only for $T \gtrsim 10^6$ GeV [181–183]. Lattice remains the only reliable tool to compute the axion potential for $\Lambda_{\rm QCD} \lesssim T \lesssim 10^6$ GeV.

θ-vacua—chiral lagrangian:
For $T \lesssim \Lambda_{\rm QCD}$, the axion potential can be computed rather precisely using the **Chiral lagrangian of QCD** [183, 184]. We already introduced the chiral Lagrangian of QCD in Eq. (2.45) of Sect. 2.2.2 in order to compute the mass of the pion. From Eq. (2.111), we see that the $\bar{\theta}$ parameter in the QCD Lagrangian in Eq. (2.110) can be moved to the quark mass matrix via the chiral transformation

$$\begin{pmatrix} u \\ d \end{pmatrix} \xrightarrow{U(1)_A} e^{i\gamma_5 Q_a \frac{\bar{\theta}}{2}} \begin{pmatrix} u \\ d \end{pmatrix}, \qquad \mathrm{tr}\,(Q_a) = 1. \tag{2.126}$$

The chiral Lagrangian of QCD in (2.45) becomes

$$\mathscr{L}_{\rm eff} \supset \frac{V^3}{2}\mathrm{tr}\left[M_a U + M_a^\dagger U^\dagger\right], \qquad V^3 = \frac{F_\pi^2 m_\pi^2}{m_u + m_d}, \tag{2.127}$$

with

$$U = e^{i\Pi/F_\pi}, \qquad \Pi = \begin{pmatrix} \pi^0 & \sqrt{2}\pi^+ \\ \sqrt{2}\pi^- & -\pi^0 \end{pmatrix}, \tag{2.128}$$

and

$$M_a = e^{i\gamma_5 Q_a \frac{\bar{\theta}}{2}} M_q e^{i\gamma_5 Q_a \frac{\bar{\theta}}{2}}, \qquad M_q = \begin{pmatrix} m_u & 0 \\ 0 & m_d \end{pmatrix}. \tag{2.129}$$

Keeping only the terms depending on $\bar{\theta}$, we obtain

$$V(\bar{\theta}) = -V^3 \left[m_u \cos\left(\frac{\pi^0}{F_\pi} - \frac{\bar{\theta}}{2}\right) + m_d \cos\left(\frac{\pi^0}{F_\pi} + \frac{\bar{\theta}}{2}\right) \right] \tag{2.130}$$

$$= -m_\pi^2 F_\pi^2 \sqrt{1 - \frac{4m_u m_d}{(m_u + m_d)^2} \sin^2\left(\frac{\bar{\theta}}{2}\right)} \cos\left(\frac{\pi^0}{F_\pi} - \phi_a\right), \tag{2.131}$$

where

$$\tan\phi_a \equiv \frac{m_u - m_d}{m_u + m_d} \tan\left(\frac{\bar{\theta}}{2}\right). \tag{2.132}$$

In order to minimize the potential, the vacuum expectation value of the meson field is equal to $\langle\pi^0\rangle = \phi_a F_\pi$. We obtain that the vacuum energy density is a function of the $\bar{\theta}$ parameter [183]

$$V(\bar{\theta}) = -m_\pi^2 f_\pi^2 \sqrt{1 - \frac{4m_u m_d}{(m_u + m_d)} \sin^2\left(\frac{\bar{\theta}}{2}\right)}, \tag{2.133}$$

where $m_\pi \simeq 135$ MeV and $F_\pi \simeq 92$ MeV. We can see that the vacuum with minimum energy is the CP-conserving vacuum $\bar{\theta} = 0$, in accordance with the Vafa-Witten theorem [185].

2.3 Open Problems

2.3.1 *Hierarchy Problem*

Renormalization:
The parameters in the SM Lagrangian density in Sect. 2.2.1, the gauge couplings g_i, the Yukawa couplings λ_i or the Higgs potential parameters μ^2 and λ, are called **bare** parameters. They are different from the **measured** or **renormalized** parameters due to the presence of quantum corrections. This is the essence of **renormalization**, initially developed at the end of the 40s by Tomonaga [187, 188], Schwinger [189–193] and Feynman [194–196] (Nobel prize 1965) but also Bethe [197], Pauli [198]

and Dyson [199, 200], in order to absorb the troublesome divergent integrals into a rescaling of the masses and coupling constants of the theory.[15] The quantum corrections are often UV divergent and then, they are sensitive to any unknown physics above the energy scale at which the SM has been tested experimentally. For instance, we can show that the quantum corrections to the fermion masses are logarithmically divergent (e.g. [60])

$$\delta m_{\psi_i} \propto \alpha\, m_{\psi_i} \log\left(\frac{\Lambda}{\mathrm{m}_{\psi_i}}\right), \tag{2.134}$$

where Λ is a UV cut-off parameterizing the UV divergence.

Technical naturalness:
The logarithmic dependence of the fermion mass on the UV scale, instead of a linear one $\delta m_{\psi_i} \propto \Lambda$, originates from symmetry consideration. Indeed, setting the fermion mass to zero introduces a new symmetry in the Lagrangian, the **chiral symmetry**, defined in footnote 12. Quantum corrections δm_{ψ_i} to the fermion masses must respect the possibility to enhance the symmetry group when we set $m_{\psi_i} = 0$, so they must be proportional to m_{ψ_i}. Therefore, from dimensional considerations, the renormalized mass can only depend logarithmically on the scale of new physics Λ. Whatever the scale of new physics Λ, the renormalized mass will only differ from the bare mass by tiny logarithmic corrections.

The quantum corrections to the photon mass are identically zero because of the $U(1)_{\mathrm{e.m.}}$ gauge symmetry. We say that the fermion and vector boson masses are **protected** from large quantum corrections by chiral and gauge symmetries respectively.[16]

The electroweak scale is not natural:
In contrast, setting to zero the Higgs mass m_{H} does not introduce any additional symmetry and the quantum corrections to the Higgs mass at 1-loop, see Fig. 2.2, are quadratically sensitive to the scale of new physics Λ, e.g. [204]

$$\delta m_{\mathrm{H}}^2 = \frac{3}{8\pi^2}\left(\frac{\Lambda}{v}\right)^2 \left(m_{\mathrm{H}}^2 + 2m_{\mathrm{W}}^2 + m_{\mathrm{Z}}^2 - 4m_{\mathrm{t}}^2\right). \tag{2.135}$$

The incompatibility between the quadratic sensitivity of the Higgs boson mass on the scale of new physics Λ and its measured value at $m_{\mathrm{H}} \simeq 125$ GeV, is called the **Naturalness** or **Hierarchy problem**.

---

[15] They applied it to the get predictions in Quantum Electrodynamics (QED), the theory of light, electrons and positrons. The approach of Tomonaga and Schwinger is based on operators whereas the approach of Feynman is based on diagrams. Nowadays, the value of the magnetic moment of the electron can be computed within the framework of QED up to 4 loops [201] and experimental measurements in Laboratoire Kastler-Brossel are able to agree with the theoretical predictions with a precision of one part out of 10^{14} [202].

[16] This is the idea being **technical naturalness**, introduced by 't Hooft [203]: an observed coupling constant can be much smaller than 1 if a symmetry is restored when the coupling is set to zero.

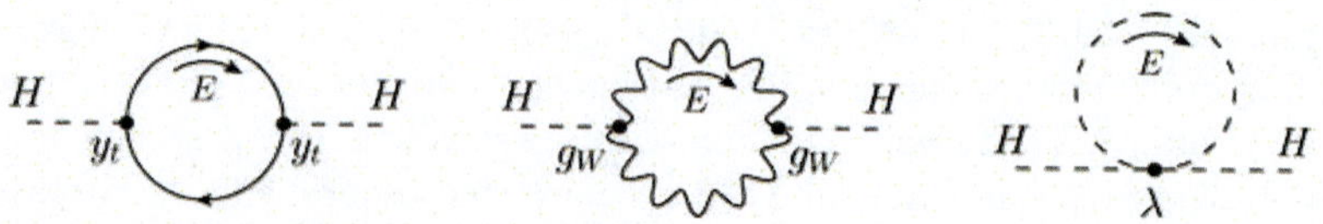

Fig. 2.2 1-loop corrections to the Higgs propagator from the top quark, the weak gauge bosons and the Higgs boson itself. Figure reproduced from [186]

The existence of physics Beyond the Standard Model (BSM) at an energy scale $\Lambda \gg m_{\mathrm{H}}$ is motivated by different open problems, the small neutrino mass problem, the strong CP problem, the flavor hierarchy problem, the dark matter puzzle or the baryonic asymmetry of the universe. Or at least **quantum gravity** around the **Planck scale** $\Lambda \simeq M_{\mathrm{pl}} \simeq 2.44 \times 10^{18}$ GeV, the scale where the strength of gravity becomes comparable to the other forces. In that last case, the hierarchy problem $M_{\mathrm{pl}} \gg m_{\mathrm{H}}$ can be rephrased as: why does the classical **gravity force** between two electrons

$$F_{\mathrm{Grav}} = \frac{G\, m_e^2}{r} \simeq \frac{m_{\mathrm{H}}^2}{M_{\mathrm{pl}}^2} \frac{Y_e^2}{r}, \tag{2.136}$$

is much smaller than the **Coulomb force**

$$F_{\mathrm{Coul}} = \frac{\alpha}{r}, \tag{2.137}$$

knowing that **quantum corrections** to m_{H}^2 are of order M_{pl}^2 ?

The Naturalness problem suggests the existence of a principle which protects the smallness of the Higgs mass relative to its would-be UV corrections. We cite below the most studied proposals to make the measured value of the Higgs mass natural.

Supersymmetry (SUSY):
In 1967, Coleman and Mandula proved that the largest symmetry group of quantum field theory is the direct product of some internal symmetries and the Poincaré group of spacetime isometries (4 transations, 3 rotations, 3 boosts) [205]. In 1971, Golfand and Likhtman found a loophole and extended the Poincaré group with a two-component weyl spinor operator which transforms **bosons into fermions**, and vice-versa, and whose anticommutator is a space-time translation [206]. If the universe is supersymmetric, each SM particle has a **super-partner** with equal mass but different spin, an idea first proposed by Pierre Fayet in 1977 [207]. In 1981, Dimopoulos and Georgi used SUSY to stabilize the EW scale [208]. Indeed, due to the relative sign between fermion loops and boson loops, the **quantum corrections cancel order by order** in perturbation. However, the non-observation in Nature of those super-particles (sfermions, gauginos, gravitino) implies that SUSY must be broken at low-energy in order to lift the mass degeneracy. After SUSY breaking the correction to the Higgs mass from the top t and stop $\tilde{t}$ is [209]

$$\delta m_{\rm H}^2 = \begin{cases} -\frac{3\lambda_t^2}{4\pi^2}\left(y_{\tilde{t}}^2 - y_t^2\right) = 0 & \text{SUSY is not broken,} \\ -\frac{3\lambda_t^2}{4\pi^2}\, m_{\tilde{t}}^2 \log\left[\frac{M_{SUSY}}{m_{\tilde{t}}}\right] & \text{SUSY is broken,} \end{cases} \quad (2.138)$$

where M_{SUSY} is the SUSY breaking scale and $m_{\tilde{t}}$ is the stop mass.

The discovery of the Higgs at 125 GeV and the searches on super-partners at LHC have already imposed the presence of a **fine-tuning** between radiative corrections of the order of 1 out of 100 for the minimal SUSY model [209, 210]. For instance, latest lower bounds on the stop mass are $m_{\tilde{t}} \gtrsim 1$ TeV [211]. Assuming the tree level Higgs mass $m_{\rm H}^2 \simeq m_{\rm Z}^2/2$ and plugging $M_{SUSY} \sim 10$ TeV and $m_{\tilde{t}} \sim 1$ TeV in Eq. (2.138), we find a needed fine-tuning of $\sim 2\%$ between the different radiative corrections in order to get the correct Higgs mass.

There are many alternatives to reduce the fine-tuning, e.g. next-to-minimal SUSY extensions (NMSSM) that involve an additional SM singlet [212, 213], colorless top partners if SUSY is 'folded' [214], a focus point on the SUSY RGE [215, 216], or top partners naturally lighter than others sparticle [217]. The last possibility can arise naturally if SUSY breaking is mediated through a gauged flavor symmetry [218]. Also, the missing transverse energy signal at LHC can be suppressed due to the sparticles spectrum being compressed [219], SUSY being 'Stealth' [220], or R-parity being violated [221]. For lecture notes/books, see e.g. [222–228]. For reviews, see e.g. [209, 229–234]. Particularly, see [209] for a review on experimental constraints on SUSY and its theoretical extensions.

In addition to provide a natural explanation for the stability of the EW scale, SUSY predicts gauge coupling unification [235] and natural Dark Matter candidates [236].

Low-scale SUSY also comes with its own difficulties. In its minimal version, it predicts large flavor and CP violation at odd with observations [237–244], and dramatic post-BBN entropy injection due to gravitino and moduli decay [245–252]. This motivated Arkani-Hamed and Dimopoulos to propose in 2004 that SUSY might be broken at a much higher scale than the TeV one, at the expense of a large fine-tuning [253–259]. A larger SUSY breaking scale preserves gauge coupling unification [253–255] and is compatible with the Higgs at 125 GeV [258]. Its realization in which it also explains dark matter and predicts signals at colliders is called **split SUSY** [253–255].

Composite Electroweak dynamics:
The QCD scale around 200 MeV arises from dimensional transmutation when the strong force confines, cf. Sect. 2.2.2. We could expect that the same occurs with the electroweak scale. This is the **Technicolor** idea introduced at the end of the 70s [203, 260–262]. We introduce additional massless fermions (technifermions) which are charged under an asymptotically free gauge theory (technicolor) which confines at the electroweak scale. Like in QCD, the global **chiral** symmetry of the technifermions is **spontaneously broken** down to the **vectorial** subgroup when the technifermions condense

$$SU(2)_L \times SU(2)_R \times U(1)_V \rightarrow SU(2)_V, \qquad (2.139)$$

where 2 is the number of techniflavors. $U(1)_V$ is the simultaneous vectorial rotation of all the technifermions at the same time. We did not include the axial rotation $U(1)_A$ which is anomalous. The SM gauge interactions $SU(2)_L \times U(1)_Y$ form a subgroup of the global chiral symmetry group in a way that it gets spontaneously broken to $SU(2)_L \times U(1)_Y \rightarrow U(1)_{\text{e.m.}}$ after chiral symmetry breaking. Particularly, three of the four **Nambu-Golstone bosons** generated in Eq. (2.139) become the longitudinal components of the SM vector bosons $W^{\pm}$ and Z, which then get massive. The SM fermions get their mass from bilinear operators coupling two SM fermions with an operator composed of technifields.

The **first** advantage of the Technicolor scenario is that there is **no need** to introduce the Higgs boson responsible for the hierarchy problem. A Higgs-like scalar can still be accommodated as one of the resonance of the strong sector. In that case, the Higgs boson is not sensitive to virtual effects above the compositeness scale, and its mass is natural.

The **second** advantage of Technicolor is that the electroweak scale is dynamically generated through **dimensional transmutation**, when the Technicolor gauge coupling runs toward a large value in the infrared, as in QCD, see Sect. 2.2.2. For reviews on Technicolor, see [263–266].

The discovery of the Higgs boson at 125 GeV, electroweak precision tests as well as the none-observation of exotic resonances at the LHC, have **excluded** most of the technicolor scenarios, e.g. see [267]. In order to conciliate the composite hypothesis with the LHC constraints, a solution is to introduce a natural little hierarchy between the compositeness scale f and the electroweak scale v. This is the design behind **Composite Higgs** (CH) scenarios which, for reasons which will become clear soon, are a sort of mix between the Technicolor idea and the SM Higgs mechanism. The composite sector of technicolor is enlarged with a global symmetry group G, which gets **spontaneously broken** to a subgroup H when the strong sector confines

$$G \rightarrow H. \qquad (2.140)$$

The Higgs field is defined as one of the Nambu-Goldstone bosons of the **coset** G/H. In contrast to Technicolor, the breaking pattern in Eq. (2.140) preserves the SM EW gauge group which is contained in the unbroken group

$$SU(2)_L \times U(1)_Y \subset H. \qquad (2.141)$$

We may ask then, where do the vector boson masses come from? In fact, the gauging of a subgroup of G only, as well as the embedding of SM fermions within large representations of the group G, introduce an **explicit breaking** of G. Therefore, loops of SM fermions and gauge fields generate a potential for the Higgs, which in turn can acquire a VEV. Consequently, the misalignment between the true vacuum configuration and the SM group $SU(2)_L \times U(1)_Y$, breaks the EW symmetry spontaneously. This is the **misalignement mechanism** introduced in 1983 by Georgi and

Kaplan [268–270]. The degree of misaligment can be measured by $\xi \equiv (v/f)^2$. The SM couplings are recovered in the 'alignement limit' $\xi \to 0$. In the case of minimal CH models, precision measurements of Higgs couplings to SM vector bosons and fermions impose a little hierarchy between the EW scale v of the compositeness scale f [271]

$$\xi \equiv (v/f)^2 \lesssim 0.1, \qquad \to \qquad f \gtrsim 800 \text{ GeV}. \tag{2.142}$$

In CH scenario, the SM fermions get their mass from operators which are **linear** in the SM fermions fields, and not bilinear as in the technicolor way. As a consequence, the physical fermions are linear combinations of elementary states and composite states, hence the name of **partial compositeness** proposed by D. B Kaplan in 1991 [272] . For reviews on CH, see [186, 231, 233, 273, 274].

Arrange ξ to be naturally small and compatible with Eq. (2.142), is the design of **Little Higgs**, proposed in 2001 [275, 276]. The gauge and fermion sector is extended such that the explicit breaking of G, responsible for the generation of the Higgs potential, arises only when **at least two** couplings are switched on simultaneously in the Lagrangian. As a consequence of this **collective breaking**, the Higgs potential is only generated at **two loops**. For reviews on Little Higgs, see [277, 278]. Another alternative to naturally lower the EW scale v with respect to the confining scale f is the **Twin Higgs** scenario [279, 280] where a second exact copy of the SM is introduced such that the quadratic dependence on the UV cut-off, in the Higgs potential, accidentally cancels. The Twin Higgs model is a well-known realization of **neutral naturalness**, an approach for solving the hierarchy problem with top partners which are entirely neutral under the SM gauge groups [281].

Large flat extra-dimensions:
In a model proposed in 1998 By Antoniadis, Arkani-Hamed, Dimopoulos and Dvali [282, 283], the SM is confined on a brane [284] whereas the gravity interactions are **diluted** due to their propagation in large compactified extra-dimensions. If the fundamental gravity scale in the full $4+n$ dimensional space is M_* and that the radii of the n extra dimensions is given by r, then the effective gravity scale M_{pl} in 4D is

$$M_{\text{pl}}^2 \sim M_*^{2+n} r^n. \tag{2.143}$$

Assuming that the fundamental scale is the electroweak scale $M_* \sim$ TeV and plugging the value $M_{\text{pl}} \simeq 2.44 \times 10^{18}$ GeV, we obtain the size of the extra-dimensions

$$r(n=1) \sim 10^{15} \text{ mm}, \qquad r(n=2) \sim 0.5 \text{ mm}, \tag{2.144}$$

$$r(n=3) \sim 10^{-6} \text{ mm}, \qquad r(n=4) \sim 10^{-8} \text{ mm}. \tag{2.145}$$

There is no hierarchy problem anymore since the cut-off is now M_*. Newton law has been tested up to a fraction of a millimeter using torsion balance experiment [285, 286] so the size of the extra-dimensions must be $r \lesssim 0.1$ mm. Hence $n = 1$ is excluded while $n = 2$ is under test. See [287–289] for reviews.

Warped extra-dimensions:
In 1999, Randall and Sundrum [290, 291] introduce a warped fifth dimension which **red-shifts** the Planck scale down to the electroweak scale. More precisely, the RS model is a slice of AdS_5 which is bounded by two branes, a UV-brane located at $y = 0$ and a IR-brane located at $y = y_{\text{IR}}$. The AdS_5 metric is given by

$$ds^2 = e^{-2ky}\eta_{\mu\nu}dx^\mu dx^\nu - dy^2, \tag{2.146}$$

where k is the AdS curvature scale and $\mu,\ \nu = 1...4$. Let's assume that the Higgs field is localized on the IR brane, such that the Higgs action reads

$$S \subset \int d^5x\sqrt{-g}\left[g^{\mu\nu}\partial_\mu H\partial_\nu H^\dagger - m_H^2\,|H|^2 - \lambda\,|H|^4\right]\delta\left(y - y_{\text{IR}}\right), \tag{2.147}$$

with the metric determinant $\sqrt{-g} = e^{-4ky}$. Integrating over the extra-dimension

$$S \subset \int d^4x\, e^{-4ky_{\text{IR}}}\left[e^{2ky_{\text{IR}}}\eta^{\mu\nu}\partial_\mu H\partial_\nu H^\dagger - m_H^2\,|H|^2 - \lambda\,|H|^4\right], \tag{2.148}$$

and rescaling $e^{-ky}H \to H$ to make the kinetic terms canonical, we obtain

$$S \subset \int d^4x\left[\eta^{\mu\nu}\partial_\mu H\partial_\nu H^\dagger - e^{-2ky_{\text{IR}}}m_H^2\,|H|^2 - \lambda\,|H|^4\right]. \tag{2.149}$$

The effective Higgs mass is **redshifted** $m_H \to e^{-ky_{\text{IR}}}m_H$ with a strength exponentially sensitive to the separation between the two branes. This solves the Hierarchy problem. For instance, if the fundamental Higgs mass in $5D$ is near the Planck scale, the effective Higgs mass on the brane can be around the electroweak scale for $k\,y_{\text{IR}} \simeq 35$. The stabilization of the position of the IR brane at $k\,y_{\text{IR}} \simeq 35$ is realized by promoting the branes inter-distance y_{IR} to a field having a non-trivial potential, this is the **Goldberger-Wise** mechanism [292, 293]. The gravity scales in 5D and 4D are related though $M_{\text{pl}}^2 = M_5^3/2k$, where we can safely choose $M_5 \sim k \sim M_{\text{pl}}$, in contrast to the proposal of large flat extra-dimensions. Warped geometry in 5D can be interpreted as **holographic dual** to Composite Higgs scenarios in 4D, e.g. see [294] for one of the first implementation. See reviews on warped extra-dimensions in [287, 288, 295, 296]. In Sect. 6.4.2 of Chap. 6, we introduce first-order cosmological phase transitions in the RS scenario as potential sources of large supercooling.

Cosmological naturalness:
The LHC has not discovered any new physics around the TeV scale as predicted by the three traditional solutions to the Hierarchy problem: SUSY, composite Higgs and extra-dimensions. Taking these results at face value together with the cosmological constant (CC) problem, might reveal the existence of fine tuning in Nature. This idea is already popular among string theorists, since the only consistent theory of quantum gravity predicts the existence of a vast landscape of values for the Higgs

mass m_h and for the CC [297, 298]. The question becomes **how can cosmology select a small Higgs mass and a small CC ?** The efforts in recent years which have been developed to address that question can be classified in three categories [299].

- **Dynamical** selection. The values of Higgs mass or CC are obtained as the result of a dynamical evolution, e.g. [300–313]
- **Statistical** selection. The Higgs mass or CC are linked to a scanning field whose quantum distributions peaks on the SM values, e.g. [311, 314–317]
- **Anthropic** selection. All values of the landscape are populated via eternal inflation, e.g. [318, 319].

We review some of the dynamical and anthropic selection ideas below. We study the case of the CC in more details in Sect. 3.5.1 of Chap. 3.

Cosmological relaxation:
The Higgs mass parameter is dynamical and set by a **rolling axion field** in the **early universe**. After the field reaches its equilibrium, the value of the electroweak scale has become independent of the UV cut-off. The first proposal of that type is the **relaxion** scenario (proposed in 2015 [307]), inspired from Abbot's idea for relaxing the Cosmological Constant (proposed in 1984 [300])

$$\mathscr{L} = g\Lambda^3\phi + \left(\Lambda^2 - g\Lambda\phi\right) h^2 + \Lambda_b^4 \left(\frac{\langle h\rangle}{v}\right)^n \cos\frac{\phi}{f}. \tag{2.150}$$

The first term is a **slope** for the scalar field ϕ where g is a constant. The second term is the Higgs mass. The last term are is a back-reaction term from Higgs-dependent **wiggles**, with potential barrier Λ_b and decay constant f. The scale Λ is the UV cut-off of the effective field theory. The dynamics is engineered such that the Higgs mass is set by the VEV of the relaxion ϕ and conversely, the size of the wiggles of ϕ are set by the VEV of the Higgs $\langle h\rangle$. Hence, when the Higgs mass turns negative, the Higgs gets a VEV and the relaxion ϕ gets a barrier. Shortly after the formation of the wiggles, the relaxion stops due to Hubble friction, and the Higgs gets a final tiny mass $m_h \ll \Lambda$. In order to have enough friction and time to relax, in the original proposal [307] the evolution must occur during inflation. Although see [320] for revisiting the assumption.

The parameters of the model need to fulfill some conditions

$$V(\phi) \lesssim V_{\rm inf} \quad \rightarrow \quad \frac{\Lambda^2}{M_{\rm pl}} \lesssim H_I, \qquad \text{inflaton potential dominates,} \tag{2.151}$$

$$H_I^{-1}\dot{\phi} \gtrsim H_I \quad \rightarrow \quad H_I \lesssim g^{1/3}\Lambda, \qquad \text{class. roll.} > \text{quant. fluct.,} \tag{2.152}$$

$$V' \simeq 0 \ \text{ exists} \quad \rightarrow \quad g\Lambda^3 \lesssim \frac{\Lambda_b^4}{f}, \qquad \text{barrier large enough,} \tag{2.153}$$

$$m_h \simeq 0 \ \text{ is reached} \quad \rightarrow \quad \Delta\phi \sim \frac{\Lambda}{g}, \qquad \text{field excursion,} \tag{2.154}$$

$$H_I \lesssim \Lambda_{\rm b}, \qquad \text{barrier is effective,} \tag{2.155}$$

where we used that the velocity of the rolling scalar field is set by the Hubble friction

$$\ddot{\phi} + 3H\dot{\phi} + V' = 0 \quad \rightarrow \quad \dot{\phi} \simeq \frac{g\Lambda^3}{H_I}, \qquad \text{slow-roll velocity.} \tag{2.156}$$

As a consequence, the original relaxion proposal suffers from several issues (see [320] for a visualization of the allowed parameter space)

1. From Eqs. (2.151), (2.152) and (2.153), we obtain $\Lambda \lesssim (\Lambda_b^4 M_{\rm pl}^3/f)^{1/6}$ where Λ_b must be 'natural' $\Lambda_b \lesssim$ TeV. Hence, assuming $f \gtrsim$ TeV, the **UV cut-off** Λ can not be larger than 10^{10} GeV. Therefore, the hierarchy problem re-appears if new physics couples to the Higgs above that scale.
2. From Eq. (2.153), the **linear coupling** g must be very small, e.g. smaller than $g \lesssim 10^{-30}$ for the largest UV cut-off [320]. Note however, that its smallness is technically natural since it is protected by a shift symmetry $\phi \rightarrow \phi + f$.
3. From Eq. (2.154), the relaxion has to perform huge **field excursion** $\Delta\phi \gg M_{\rm pl}$. This enters in conflict with the Swampland Distance Conjecture [321–323].
4. From Eqs. (2.154) and (2.156), we can compute the **number of e-folds** N_e

$$\Delta\phi \sim \dot{\phi}\frac{N_e}{H_I} \quad \rightarrow \quad N_e \sim \left(\frac{H_I}{g\Lambda}\right)^2, \qquad \text{number of e-folds.} \tag{2.157}$$

   The smallness of g implies that the whole relaxation process takes a huge number of e-folds $10^4 \lesssim N_e \lesssim 10^{60}$ [320].
5. If we assume the relaxion to be the **QCD axion**, then the potential barrier can be generated by QCD instantons, $\Lambda_b \simeq \Lambda_{\rm QCD}$ with $n = 1$ in Eq. (2.150), whenever the de Sitter temperature is not too large cf. Eq. (2.155). The final ϕ position is displaced from its CP-conserving minimum by the θ angle

$$\theta \sim \frac{g\Lambda^3 f}{\Lambda_b^4}, \qquad \text{displaced minimun.} \tag{2.158}$$

   From Eqs. (2.152), (2.158) and (2.155), we can check that the neutron EDM constraint $\theta \lesssim 10^{-10}$, together with the astrophysical constraints $f \gtrsim 10^9$ GeV, almost **exclude** the scenario where the relaxation is the QCD axion since the UV cut-off must be very low[17] $\Lambda \lesssim 30$ TeV [307].
6. The term $\Lambda_b^4(\langle h\rangle/v)^n \cos\phi/f$ is **unstable** under radiative corrections

$$\text{e.g.} \quad \varepsilon\Lambda_b^4 \cos\phi/f, \qquad \varepsilon\Lambda_b^3\, g\phi\, \cos\phi/f, \tag{2.159}$$

   where ε stands for the loop-suppression factors. Those terms should be included and since there are independent of the Higgs VEV, they spoil the relaxion mechanism, unless $\Lambda_b \lesssim v$, which implies new physics at the TeV scale. This is the

[17] Note that it can be relaxed to $\Lambda \lesssim 1000$ TeV if the classical rolling constraint in Eq. (2.152) is dropped. However, this leads to eternal inflation and therefore to a loss of predictivity.

coincident problem. Reference [324] proposes to solve the issue by introducing, in addition to the initial relaxion field ϕ which scans the Higgs mass, a second relaxion field σ which scans the potential barrier. This is the **double scanning mechanism**.

Interestingly, the relaxion can play the role of **Dark Matter** candidate [325, 326]. See [327] for a review of the experimental constraints on the relaxion.

References [320, 328, 329] demonstrate that the relaxion can also stop due to the dissipation resulting from **axion fragmentation**, instead of Hubble friction. Upon decomposing the field into the sum of the homogeneous background plus a perturbation

$$\phi(x,t) = \phi_0(t) + \delta\phi(x,t), \qquad \text{with } \delta\phi(x,t) = \int \frac{d^3k}{(2\pi)^3} a_k\, u_k(t)\, e^{ikx} + \text{h.c.}, \tag{2.160}$$

where $\left[a_k, a_k^\dagger\right] = (2\pi)^3\delta^3(k-k')$, then the equation of motion can be written as

$$\ddot{\phi} + 3H\dot{\phi} + V'(\phi) + \frac{1}{2}V'''(\phi)\int \frac{d^3k}{(2\pi)^3}|u_k|^2 = 0, \tag{2.161}$$

$$\ddot{u}_k + 3H\dot{u}_k + \left[\frac{k^2}{a^2} + V''(\phi)\right]u_k = 0. \tag{2.162}$$

In the relaxion scenario, the second equation reads

$$\ddot{u}_k + \left(k^2 - \frac{\Lambda_b^4}{f^2}\cos\frac{\dot{\phi}}{f}t\right)u_k = 0, \tag{2.163}$$

where it is implicitly assumed that Λ_b depends on the higgs vev $\langle h\rangle$. This is the well-known **Mathieu equation** [330]

$$\ddot{u}_k + (A_k - 2q\cos 2z)\,u_k = 0, \tag{2.164}$$

with $A_k = 4k^2/(\dot{\phi}/f)^2, q = 2\Lambda_b^4/\dot{\phi}^2$ and $z = \dot{\phi}/2f$. Its solution has an **exponential growth** when k is in one of the n **instability bands**, which in the **narrow-resonance** limit $q \ll A_k$, lies around $A_k \sim n$. The largest growth, $u_k \propto \exp qz/2$, arises in the first instability band $n = 1$, given by $1 - q \lesssim A_k \lesssim 1 + q$, and which can be rewritten as

$$\frac{\dot{\phi}^2}{4f^2} - \frac{\Lambda_b^4}{2f^2} \lesssim k^2 \lesssim \frac{\dot{\phi}^2}{4f^2} + \frac{\Lambda_b^4}{2f^2} \quad\rightarrow\quad \frac{\dot{\phi}}{2f} - \frac{\Lambda_b^4}{2f\dot{\phi}} \lesssim k \lesssim \frac{\dot{\phi}}{2f} + \frac{\Lambda_b^4}{2f\dot{\phi}}. \tag{2.165}$$

This alternative dissipation mechanism can help the relaxion to stop more efficiently. As a consequence, in some region of the parameter space, only a natural number of e-folds of inflation is required $N_e \sim 10^2$, the process can even occur after inflation.

The upper bound on g is relaxed, e.g. $g \sim 10^{-10}$, and the field excursion is never super-planckian. The price to pay is a smaller UV cut-off $\Lambda \lesssim 10^5$ GeV.

Anthropic solution:
Different studies suggest that the measured value of the weak scale $v \simeq 246$ GeV is consubstantial to our existence as observers.

- Proton decay arises from dimension 6 operators and yields the lifetime $\tau_p = M^4/m_p^5$. Assuming $M \simeq 10^{16}$ GeV, the weak scale must be smaller than $v \lesssim 400$ TeV in order for the proton to be stable over the universe age [331, 332].
- The pion, mediator of the nuclear force, has a mass proportional to the weak scale $m_\pi^2 \propto m_u + m_d \propto v$, see Eq. (2.50). An increase of the electroweak scale v by more than 60% would lead to the instability of complex nuclei, e.g. ^{16}O [333]. The same study finds that if the weak scale is lowered by 60%, the mass difference $m_d - m_u$ becomes too small to prevent the instability of hydrogen atoms $p + e \to n + \nu_e$. The bounds are revisited in [334]. Those bounds however suppose fixed Yukawa couplings.
- The increase of the weak scale v by one or two orders of magnitudes implies that the helium fraction reaches 100% which make halo cooling less efficient and stars[18] shorter-lived [334].
- If weak interactions whose strength is controlled by $G_F = 1/(\sqrt{2}v^2)$ were too small (large) then neutrinos produced in the final stages of stellar collapse might easily escape (get trapped) preventing supernova explosions, which are necessary to produce vital elements as oxygen. The gravitational free-fall time in neutron star is

$$\tau_{\text{fall}} \sim \frac{1}{\sqrt{G\rho}} \sim \frac{M_{\text{pl}}}{m_n^2}. \tag{2.166}$$

The neutrino free-streaming length is

$$l_\nu \sim \frac{1}{\sigma_{\text{weak}} n_n} \sim \frac{v^4}{T^5} \sim \frac{v^4}{m_n^5}. \tag{2.167}$$

The neutrino trapping time is the time it takes for them to diffuse over a distance equal to the star radius $R_s \sim M_{\text{pl}}/m_n^2$

$$\tau_{\text{trap}} \sim \frac{R_s^2}{l_\nu} \sim \frac{M_{\text{pl}}^2 m_n}{v^4}. \tag{2.168}$$

Core-collapse supernova explosion requires the neutrino to be trapped for a time τ_{trap} comparable to the gravitational time-scale τ_{fall}, which implies that the weak scale must be tuned to

$$v \sim c_0\, m_n^{3/4} M_{\text{pl}}^{1/4}, \qquad c_0 \sim 0.01, \tag{2.169}$$

within a factor of few [336].

[18] Note however the possibility of a universe containing stars, without weak interactions at all [335].

- The increase of the top mass by less than 5% might lead to an instability of our EW vacuum [337].[19]

Hence, the SM seems to sit inside a tiny **anthropic windows** outside which, a carbon-based life seems impossible. This motivates the possibility that the solution to the hierarchy problem is not microscopic but **environmental**.

The multiverse idea, based on eternal inflation and the string landscape, provides a theoretical framework to scan over the values of the observed physical parameters. The number of different vacua predicted in string theories is of googolplex order $10^{\text{few}\,100}$ [297, 298]. In 2000, Bousso, Polchinski and others [301, 302] propose to use the nucleation of membranes charged under four-form fields in order to scan over the different values of the cosmological constant within that landscape. The membrane nucleation is the exact $(3+1)D$ equivalently of the Schwinger process of e^+/e^- nucleation in $(1+1)D$ [338, 339]. A generalization of the Bousso-Polchinski approach to a scan over the values of the EW scale is straightforward upon the introduction of an operator—which is non-renormalizable in [303, 304] and renormalizable in [305, 306]—coupling the Higgs field with the four-form fields. More details on the relaxation à la Bousso-Polchinski can be found in Sect. 3.5.1 of Chap. 3 when discussing about the relaxation of the Cosmological Constant.

2.3.2 *Neutrino Oscillations*

The discovery of neutrino oscillations:
In 1968, Davis et al. [340] reported the first detection of **solar neutrinos**, using a tank of Chlorine and the charged-current interaction $\nu_e + n \to p + e$. The flux of electron neutrino ν_e coming from the sun is measured smaller than the prediction by a factor ~1/3. This was the famous **solar neutrino problem**. The same year, Pontecorvo [341] suggested that the electron neutrino produced by the sun are transformed in flight into other flavors to which the experiment was not sensitive to. This is the mechanism of **neutrino oscillation**. In 2001, the Super-Kamiokande [342] and Sudbury Neutrino Observatory [343, 344] were able to measure the total flux of neutrinos $\nu_e + \nu_\mu + \nu_\tau$ and found the missing $\sim 2/3$. Super-Kamiokande, which uses water tank, measures the total neutrino flux through the elastic interactions $\nu_{e,\mu,\tau} + e^- \to \nu_{e,\mu,\tau} + e^-$, while Sudbury Neutrino Observatory, which uses heavy water, can also detect neutral current interactions $\nu_{e,\mu,\tau} + n \to \nu_{e,\mu,\tau} + n$. The initial Davis experiments, which first detected solar neutrinos in 1968, received the Nobel prize in 2002, while the two experiments, Super-Kamiokande and Sudbury Neutrino Observatory, which confirmed the existence of neutrino oscillations and solved the solar neutrino problem, were co-awarded the Nobel Prize 2015.

[19] The analysis in [337] shows, first, that the EW vacuum is metastable with a lifetime much longer than the age of the the universe, and second, that the EW vacuum sits at the edge of a region of instability (+5% m_t). However, the analysis assumes the validity of the SM up to an energy scale of 10^{10} GeV. Hence, any new physics below that scale can potentially alter the statement.

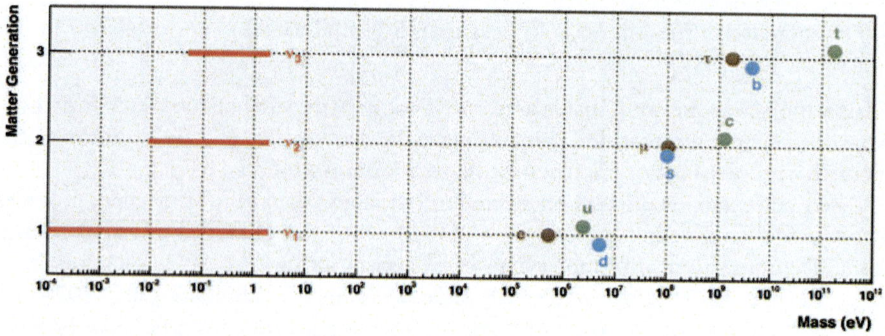

Fig. 2.3 Fermion masses in the SM. Figure reproduced from [346]

In fact, the first evidence for neutrino oscillations was already found a bit earlier in 1998 by Super-Kamiokande in the case of **atmospheric neutrinos**, generated by the decay $\pi^+ \to \mu^+ + \nu_\mu$ followed by $\mu^+ \to e^+ + \bar{\nu}_\mu + \nu_e$ [345]. Instead of measuring of factor ν_μ/ν_e of order 2, the collaboration measured ~ 1.2, showing the first evidence for neutrino oscillations.

At the theoretical level, neutrino oscillations rely on the existence of a mass matrix in the neutrino sector with two properties. First, it is non-diagonal and so contains mixing angles between the different neutrinos flavors. Second, it contains small mass differences Δm. The surprising features of the neutrino matrix are its small entries: the neutrino masses are incredibly small compared to other fermions, see Fig. 2.3. In what follows, we discuss how to generate a mass matrix for the neutrino sector, with small entries with respect to the electroweak scale.

The see-saw mechanism:
The smallness of the neutrino mass compared to the other fermions calls for an explanation. Actually, it is straightforward to account for the neutrino masses and oscillations by adding to the SM a dimension 5 operator, called the **Weinberg operator** [347]

$$\mathscr{L}_{\text{dim}-5} = -\lambda_{ij} \left(\bar{L}^i \tilde{H}\right) \left(\tilde{H} L_j\right)^\dagger, \tag{2.170}$$

where $\tilde{H}$ is defined in Eq. (2.55). After EWSB, the Weinberg operator provides to the neutrino sector, a non-diagonal mass matrix with flavor mixing and therefore accounts for the oscillations. However, even though the operator respects all the gauge symmetries in the SM, it is not renormalizable. In fact, this term is the low-energy description of a more complete theory which includes right-handed neutrinos ν_R^i, where i indexes the different right-handed neutrino families (generally one, two or three). If we assume that the right-handed neutrinos are neutral under both the weak and electromagnetic forces, then the most general renormalizable operators are

$$\mathscr{L}_{\nu,\,\text{mass}} = -Y^e_{ij} \bar{L}^i H e^j_R - Y^\nu_{ij} \bar{L}^i \tilde{H} \nu^j_R - i M_{\text{ij}} \left(\nu^i_R\right)^c \nu^j_R + h.c. \tag{2.171}$$

where $\nu_R^c = \nu_R^T \sigma_2$ denotes the charge conjugate. We can embed the Weyl spinors ν_R and ν_L into a Dirac notation[20]

$$\psi_L = \begin{pmatrix} \tilde{\nu}_R \\ \nu_R \end{pmatrix}, \quad \text{and} \quad \psi_R = \begin{pmatrix} \nu_L \\ \tilde{\nu}_L \end{pmatrix}, \tag{2.172}$$

where $\tilde{\nu}_{L,R} = i\sigma_2 \nu_{L,R}^*$. Then, the Dirac and Majorana mass terms can be simply written as

$$\mathscr{L}_{\nu,\,\text{mass}} = -m_{ij}\bar{\psi}_L^i \psi_R^j - \frac{M_{ij}}{2}\bar{\psi}_R^i \psi_R^j, \tag{2.173}$$

where

$$m_{ij} = \frac{Y_{ij}}{\sqrt{2}} v, \tag{2.174}$$

after EWSB. Assuming one flavor generation for simplicity, the mass eigenstates are linear combinations which diagonalize the matrix $\begin{pmatrix} 0 & m \\ m & M \end{pmatrix}$. In the limit $M \gg m$, we find

$$m_{\text{heavy}} \simeq M \quad \text{and} \quad m_{\text{light}} \simeq \frac{m^2}{M}. \tag{2.175}$$

Hence, the smallness of the neutrino mass results from the heaviness of the Majorana mass M, this is the **see-saw mechanism**, developed at the end of the 70s [348–352]. See [353, 354] for some reviews on the see-saw mechanism. The see-saw mechanism predicts the existence of right-handed neutrinos, called **sterile neutrinos**. The upper bound from neutrinoless double beta decay, presented later in Eq. (2.186), translates to a lower bound on the Majorana mass of sterile neutrinos

$$M \gtrsim 10^{14}\, Y_\nu^2. \tag{2.176}$$

Interestingly, such particle can explain the matter/anti-matter asymmetry through the **leptogenesis** mechanism, see Sect. 3.5.2 of Chap. 3. Unfortunately, for natural values of the Yukawa coupling Y_ν, the Majorana mass scale is beyond the reach of most of the experiments (colliders, telescopes, labs,..). On the other hand, Yukawa couplings below $\lesssim 10^{-6}$ and $\lesssim 10^{-10}$, predict sterile neutrinos in the collider energy range $\sim$ TeV, and in the keV range, respectively. The latter scenario can explain **Dark Matter** and can lead to astrophysical signatures [355, 356]. For reviews on sterile neutrinos, see [357–359].

[20] Dirac spinor are usually built with two independent left-handed and right-handed weyl fields $\psi = \begin{pmatrix} \psi_L \\ \psi_R \end{pmatrix}$, see Sect. 2.1.1.

Are neutrinos Majorana or Dirac fermions?
Upon taking the charge conjugate of Eq. (2.172), we find[21]

$$-i\gamma_2\psi^* = -i\begin{pmatrix}0 & \sigma_2\\ -\sigma_2 & 0\end{pmatrix}\begin{pmatrix}\nu_L\\ i\sigma_2\nu_L^*\end{pmatrix}^* = \begin{pmatrix}-\sigma_2\sigma_2^*\nu_L\\ i\sigma_2\nu_L^*\end{pmatrix} = \begin{pmatrix}\nu_L\\ i\sigma_2\nu_L^*\end{pmatrix}$$
$$= \psi, \tag{2.177}$$

and same for ν_R. Hence if neutrinos are majorana fermions, i.e. $M \neq 0$, they are their own anti-particle.[22] In contrast, if neutrinos are Dirac fermions, i.e. $M = 0$—if for instance the right-handed neutrino carries lepton number and the latter is conserved—then the right-handed and left-handed neutrinos become components of the same Dirac field $\psi = (\nu_L, \nu_R)$. Then charge conjugation gives

$$-i\gamma_2\psi^* = -i\begin{pmatrix}0 & \sigma_2\\ -\sigma_2 & 0\end{pmatrix}\begin{pmatrix}\nu_L\\ \nu_R\end{pmatrix}^* = \begin{pmatrix}-i\sigma_2\nu_R^*\\ i\sigma_2\nu_L^*\end{pmatrix} = \begin{pmatrix}\tilde{\nu}_R\\ \tilde{\nu}_L\end{pmatrix}$$
$$\neq \psi, \tag{2.178}$$

and neutrinos are different from their anti-particle. If neutrinos are Dirac fermions, then the tritium bound in Eq. (2.185) on the Dirac mass in Eq. (2.174), translates to a very tight upper bound on the Yukawa couplings $Y_\nu \lesssim 10^{-12}$. We can understand that this number seems unnatural and motivates alternative like the introduction of a large Majorana mass. But it is actually **technically natural** in the 't Hooft sense, see footote 16, since it is protected from large quantum corrections by the chiral symmety, see footnote 12.

The mixing angles and the PMNS matrix:
The diagonalization of the lepton Yukawa matrix in the presence of right-handed neutrinos modifies the flavor-mixing interaction terms with the $W^\pm$, in the same way as the quarks sector studied in Sect. 2.2.4

$$\mathcal{L} \supset \frac{g_2}{\sqrt{2}}\left[W_\mu^-\bar{e}_L^i\gamma^\mu\,(U)^{ij}\,\nu_L^j + W_\mu^+\bar{\nu}_L^i\gamma^\mu\left(U^\dagger\right)^{ij} e_L^j\right]. \tag{2.179}$$

The matrix U is the analog of the CKM matrix[23] in Eq. (2.171). It is called the **PMNS matrix** for Maki, Nakagawa and Sakata who introduced it in 1962 [362] in order to explain the neutrino oscillations predicted by Pontecorvo in 1957 [363]. As we did

[21] We have introduced the charge conjugation operation $\psi_c = -i\gamma_2\psi^*$. We can check that the Dirac equation $(i\partial\!\!\!/ - eA\!\!\!/ - m)\psi = 0$ becomes $(i\partial\!\!\!/ + eA\!\!\!/ - m)\psi_c = 0$.

[22] The proposition that fermions could be their own anti-particle was first theorized by Ettore Majorana, for the case of electron and positron, in 1937 [360], one year before its disparition under mysterious circumstances [361].

[23] The CKM is traditionally defined as $V = \left(U_L^u\right)^\dagger U_L^d$. A pure analog in the lepton sector, with respect to the $SU(2)_L$ charges, would be $U = \left(U_L^\nu\right)^\dagger U_L^e$. However, the PMNS matrix is traditionally defined as its complex conjugate, $U = \left(U_L^e\right)^\dagger U_L^\nu$.

for the CKM matrix, we can count how many physical parameters it contains. First of all, as a $U(3)$ matrix, it has 3 rotation angles and 6 complex phases. Notice that we can freely rotate the phases of the 3 charged-leptons, but because of the presence of the Majorana mass term, we can not rotate the phases of the neutrinos. So the number of physical parameters is 3 rotation angles and 3 phases. It can be written with an almost identical parameterization to Eq. (2.84)

$$U_{\text{PMNS}} = \begin{pmatrix} c_{12}\,c_{13} & s_{12}\,c_{13} & s_{13}\,e^{-i\delta} \\ -s_{12}\,c_{23} - c_{12}\,s_{23}\,s_{13}\,e^{i\delta} & c_{12}\,c_{23} - s_{12}\,s_{23}\,s_{13}\,e^{i\delta} & s_{23}\,c_{13} \\ s_{12}\,s_{23} - c_{12}\,c_{23}\,s_{13}\,e^{i\delta} & -c_{12}\,s_{23} - s_{12}\,c_{23}\,s_{12}\,e^{i\delta} & c_{23}\,c_{12} \end{pmatrix} P \tag{2.180}$$

where $P = \left(1,\, e^{i\phi_1},\, e^{i\phi_2}\right)$ accounts for the **two extra-phases** in the presence of Majorana masses.

Since two decades, measurements of neutrino oscillations have enabled the relatively precise determination of the neutrino mass differences and mixing angles. To the best of our knowledge, the physical parameters of the lepton sector are [109, 364]

$$m_e \simeq 0.511\ \text{MeV}, \qquad m_\mu \simeq 106\ \text{MeV}, \qquad m_\tau \simeq 1.78\ \text{GeV}, \tag{2.181}$$

$$\Delta m^2_{21} \simeq \left(7.39^{+0.21}_{-0.20}\right) \times 10^{-5}\ \text{eV}^2, \qquad \Delta m^2_{32} \simeq \begin{cases} +\left(2.525^{+0.033}_{-0.031}\right) \times 10^{-3}\ \text{eV}^2, & \text{NH} \\ -\left(2.512^{+0.034}_{-0.031}\right) \times 10^{-3}\ \text{eV}^2, & \text{IH} \end{cases}, \tag{2.182}$$

$$\theta_{12} = \left(33.82^{+0.78}_{-0.76}\right)^\circ, \quad \theta_{23} = \begin{cases} \left(49.7^{+0.9}_{-1.1}\right)^\circ, & \text{NH} \\ \left(49.7^{+0.9}_{-1.0}\right)^\circ, & \text{IH} \end{cases}, \quad \theta_{13} = \begin{cases} \left(8.61^{+0.12}_{-0.13}\right)^\circ, & \text{NH} \\ \left(8.65^{+0.12}_{-0.13}\right)^\circ, & \text{IH} \end{cases}, \tag{2.183}$$

$$\delta = \begin{cases} \left(215^{+40}_{-29}\right)^\circ, & \text{NH} \\ \left(284^{+27}_{-29}\right)^\circ, & \text{IH} \end{cases}, \tag{2.184}$$

where NH and IH denote normal ($m_{\nu_e} < m_{\nu_\mu} < m_{\nu_\tau}$) and inverted ($m_{\nu_\tau} < m_{\nu_e} < m_{\nu_\mu}$) hierarchy. The Majorana phases ϕ_1 and ϕ_2 have not yet been measured. Note that the neutrino oscillation dynamics is insensitive to the absolute neutrino mass which is then constrained by other experiments. For instance, KATRIN experiment [365] provides an upper bound on the neutrino mass from measuring the high-energy cut-off of the electron spectrum in tritium decay

$$m_\beta \equiv \sqrt{\sum_{k=1}^{3} |U_{ek}|^2\, m_k^2} \lesssim 1.1\ \text{eV}. \tag{2.185}$$

Other constraints on the absolute neutrino mass come from searches for neutrinoless double beta decay experiments like GERDA, KamLAND and EXO-200 (which however assumes that neutrino have Majorana masses) [366]

$$m_{ee} \equiv \left| \sum_{k=1}^{3} U_{ek}^2 m_k \right| \lesssim 0.2 - 0.4 \text{ eV}, \tag{2.186}$$

and cosmological probes (which rely on the ΛCDM model), CMB + BAO + Lyman-α [367]

$$\sum m_{\nu_j} \lesssim 0.12 \text{ eV}. \tag{2.187}$$

See [368] for an overview of the experimental prospects on neutrino physics and [369, 370] for their importance in cosmology.

2.3.3 *Flavor Hierarchy Problem*

The Flavor Puzzle:
In the previous section, Sect. 2.3.2, we have discussed the hierarchy between the neutrino masses and the masses of the other fermions. However, the matter sector suffers from two other (smaller) hierarchies which are sometimes called the **Flavor Puzzle**.

1. There is a hierarchy among the different entries of the CKM matrix of the electroweak interactions in the quark sector. By comparison, the PMNS matrix of the electroweak interactions in the lepton sector has a much weaker hierarchy, see Fig. 2.4. Particularly, the mixing angles of the CKM matrix, in Eq. (2.84), satisfy $\theta_{12} \gg \theta_{13}, \theta_{23}$, such that at first-order we can write it as

$$|V_{\text{CKM}}| = \begin{pmatrix} 1 & \lambda & \lambda^3 \\ \lambda & 1 & \lambda^2 \\ \lambda^3 & \lambda^2 & 1 \end{pmatrix}, \tag{2.188}$$

   with $\lambda = \sin\theta_{12} \simeq 0.22$ the Cabibbo angle.
2. There is a hierarchy among the masses of the different flavors of charged fermions, see Fig. 2.3. The mass ratios in the quark sector seem to follow the rules

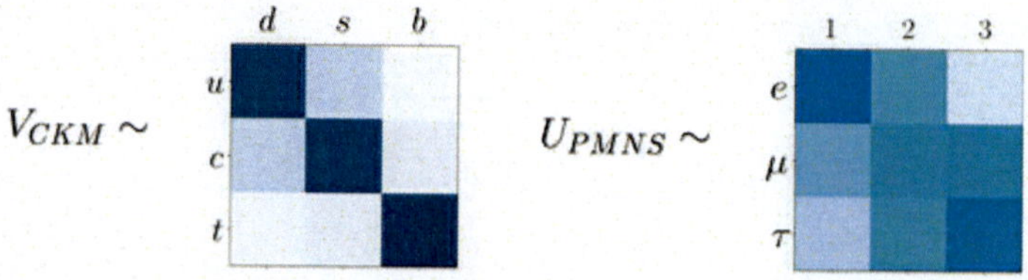

Fig. 2.4 Representation of the size of the mixing matrix elements. Lighter colors represent smaller absolute value with respect to darker ones. Figure reproduced from [371]

$$\frac{m_b}{m_t} \simeq \varepsilon^3, \qquad \frac{m_c}{m_t} \simeq \varepsilon^4, \qquad \frac{m_s}{m_t} \simeq \varepsilon^5, \qquad \frac{m_b}{m_t} \simeq \varepsilon^7, \qquad \frac{m_u}{m_t} \simeq \varepsilon^8, \tag{2.189}$$

with $\varepsilon \simeq \lambda$.

The Froggatt-Nielsen solution:
A popular explanation is the Froggatt-Nielsen mechanism introduced in 1978 [372] which introduces a new global symmetry $U(1)_{\rm FN}$ and a new field, the **flavon** S. Upon assigning $U(1)_{\rm FN}$ charges to the SM fermions and to the Higgs, we obtain

$$\mathcal{L}_{\rm FN} = -y^e_{ij}\left(\frac{S}{M_{\rm FN}}\right)^{n^e_{ij}} \bar{L}^i_{\rm L} H e^j_{\rm R} - y^d_{ij}\left(\frac{S}{M_{\rm FN}}\right)^{n^d_{ij}} \bar{Q}^i_{\rm L} H d^j_{\rm R} - y^u_{ij}\left(\frac{S}{M_{\rm FN}}\right)^{n^u_{ij}} \bar{Q}^i_{\rm L} \tilde{H} u^j_{\rm R} + h.c., \tag{2.190}$$

where $M_{\rm FN}$ is a heavy scale and the y parameters are numbers of order 1. The power exponents $n^{e,d,u}_{ij}$ depend on the $U(1)_{\rm PQ}$ charge assignment. The Yukawa matrix of the SM is generated when the flavon S acquires a VEV $\langle S\rangle$. The expansion parameter λ in the CKM and mass matrix is given by $\lambda = \langle S\rangle/M_*$. The different flavor hierarchies λ^n are controlled by the charges of the SM fermions under the new $U(1)_{\rm FN}$ symmetry. The non-renormalizable operators in Eq. (2.190) can be generated after integration of **messengers fermions** heavier than $M_{\rm FN}$. Generalization to other symmetry groups are possible, e.g. to $SU(3)$ [373] or to discrete symmetries [374, 375].

The partial-compositeness solution:
A Yukawa coupling hierarchy can also arise from wave-function localization in Randall-Sundrum-type 5D warped geometry, and their dual description in 4D strongly-coupled conformal Higgs sector [376–382]. In the 4D description, the flavors hierarchies are generated by the slow running of the Yukawa couplings, see partial compositeness in [186, 272] which replaces the old walking technicolor idea. See [383–388] for reviews on flavor physics.

2.3.4 *Strong CP Problem*

The non-detection of the electric dipole moment of the neutron provides a tight upper bound on the strong CP parameter, $\bar{\theta} \lesssim 10^{-10}$, cf. Eq. (2.113). This motivates the formulation of a principle responsible for its smallness. In the SM, CP violation by weak interactions generates quantum corrections to $\bar{\theta}$. However, $\bar{\theta}$ is not proportional solely to the CKM phase, but to the CKM phase times a flavor-neutral combination of Yukawa couplings, which implies a number of loops. This reduces the size of the quantum corrections and make the measured value of $\bar{\theta}$ not surprising from the point of view of the sole SM. However, it becomes a problem in the presence of any new physics which does not respect CP, and which is likely to induce large quantum corrections to $\bar{\theta}$. Possible solutions to the strong CP problem are listed below.[24]

The Nelson-Barr solution:
CP is asumed to be a fundamental symmetry of nature which is spontaneously broken by a scalar taking a complex vev [389–391]. For instance in [392], new fermionic states are introduced and engineered such that the mass matrix gives a zero $\bar{\theta} = \arg\det(\mathrm{Y_u Y_d}) = 0$ while generating a complex phase for the CKM matrix.

The Peccei-Quinn solution:
An additional anomalous $U(1)_{\mathrm{PQ}}$ symmetry is added to the SM [393–396]. If this $U(1)_{\mathrm{PQ}}$ is spontaneously broken, it will generate a pseudo-Nambu-Goldstone boson a, called the **axion**, which after a chiral rotation can be transferred to the $\bar{\theta}$ term. Hence, the $\bar{\theta}$ parameter in the vacuum energy in Eq. (2.133) is replaced by $\bar{\theta} \to \bar{\theta} + a/f_a$. Therefore, we have promoted the constant parameter $\bar{\theta}$ to a dynamical degree of freedom a which after minimizing its energy, given by Eq. (2.133), relaxes to the CP-preserving vacuum

$$\bar{\theta} + \frac{a}{f_a} = 0. \tag{2.191}$$

In fact, after QCD confinement the axion mass eigenstate does not exactly coincide with the $U(1)_{\mathrm{PQ}}$ Golstone boson but instead it couples to the SM mesons π^0 and η^0. This has great importance for phenomenology since it makes the axion mass eigenstate couple to photons, see e.g. [371] for the details. Upon expanding Eq. (2.133) to quadratic order, we get the well-known formula for the axion mass [396]

$$m_a^2 = \frac{\chi}{f_a^2}, \qquad \text{with} \;\; \chi = \frac{m_u m_d}{(m_u + m_d)^2} m_\pi^2 F_\pi^2 \simeq (78\ \mathrm{GeV})^4\,. \tag{2.192}$$

χ is known as the **susceptibility of the topological charge**. Higher-order corrections lead to $\chi = (75.5(5)\ \mathrm{GeV})^4$ [183]. In addition the axion is an excellent **Dark Matter** candidate [397–399] which continues to receive a lot of interest both theoretically and experimentally, see [400–403] for reviews.

The massless quark solution:
If one quark mass vanishes then we are free to perform a chiral rotation to cancel $\bar{\theta}$. In a sense, this last solution is similar to the axion solution. Indeed, by setting $m_u = 0$, we introduce an additional anomalous $U(1)$ symmetry, which is spontaneously broken when QCD confines. In that case, the axion a would exactly coincide with the η' [159]. This solution was proposed a long time ago but is no longer acceptable since lattice simulations have found a non-zero mass for the up quark [110].

Instead, we can introduce BSM massless fermions charged under both $SU(3)_c$ and a new confining force, called axicolor [404, 405], whose confining scale is much

---

[24] I am grateful for having benefited from the great expertise of Pablo Quílez on the strong CP problem and axion physics, to find answers to most of my silly questions.

larger than Λ_{QCD} [406–408]. The Goldstone boson which emerges when the chiral symmetry of the exotic fermions is spontaneously broken by axicolor confinement is called the **dynamical axion** [404, 405]. The axion shift symmetry can be preserved until operators of very high dimensions if the axicolored fermions are chiral [409, 410]. We say that we have an **accidental Peccei-Quinn symmetry**.

Axion quality problem:
One concern about any solutions to the strong CP problem is that the $U(1)_{\text{PQ}}$ being global, nothing prevent it to be violated by operators of dimension $n > 4$ [411–417]. For instance according to one of the swampland conjectures [321–323], any global symmetry is expected to be violated by **quantum gravity**. We can write

$$V_{\text{broken}}(\phi) = \varepsilon\, e^{i\delta} M^4 \left(\frac{\phi}{M}\right)^n + \text{h.c.} = 2\,\varepsilon\, M^4 \left(\frac{f_a}{M}\right)^n \cos\left(n\theta + \delta\right), \quad \phi = f_a e^{\theta/f_a}. \tag{2.193}$$

where M is a high energy scale and $\varepsilon e^{i\delta}$ a complex parameter. In order to not spoil the EDM bound in Eq. (2.113), the contribution of Eq. (2.193) to the axion mass Δm_a^2 must be smaller than 100 thousandth of the QCD contribution m_a

$$\Delta m_a^2 = 2\varepsilon n^2 M^2 \left(\frac{f_a}{M}\right)^{n-2} \lesssim 10^{-10} m_a^2. \tag{2.194}$$

In order to improve the quality of $U(1)_{\text{PQ}}$, we can either suppress the coupling constant ε, increase the dimension n of the $U(1)_{\text{PQ}}$-breaking operator or modify QCD to increase the axion mass m_a.

From quantum gravity contributions, the coupling constant ε is expected to be exponentially suppressed $\varepsilon \sim e^{-S}$ by the action S of the corresponding gravitational instanton. However, the order of magnitude of S depends on the setup. According to whether the axion decay constant f_a varies along the **wormhole** throat or not, the euclidean action is of order $S \sim M_{\text{pl}}/f_a$ [418–423] or $S \sim \log\left(M_{\text{pl}}/f_a\right)$ [423–425], respectively.

The minimal dimension n of the first $U(1)_{\text{PQ}}$-breaking operator proposals can be increased if $U(1)_{\text{PQ}}$ arises from gauge **discrete** symmetries [411, 415, 426, 429–435]. Such discrete gauge symmetries can emerge from spontaneous breaking of gauge symmetries in string theory frameworks [436]. Assuming $\varepsilon \sim 10^{-2}$, for the Peccei-Quinn solution to remain valid, we must forbid high-dimensional operators up to dimension $n \gtrsim 12$.

Another way to increase the order n of the $U(1)_{\text{PQ}}$-breaking operator is to make the axion composite [416, 417, 437–439, 442–446]. In that case, the axion is made of so many quantum fields that the minimal operator transforming under $U(1)_{\text{PQ}}$ which we can write is of high dimension. In that case, the PQ symmetry is not imposed on the Lagrangian but rather arises accidentally, due to the gauge symmetries and matter content of the theory. We say that we have an **accidental Peccei-Quinn symmetry**. In some sense, this solution is a realization of the massless quark solution described above. Lacking of massless quark in the SM, we introduce BSM massless fermions charged under both $SU(3)_c$ and a new confining force, called axicolor [437, 438], whose confining scale is much larger than Λ_{QCD} [406–408]. The Goldstone boson which emerges when the chiral symmetry of the exotic fermions is spontaneously broken by axicolor confinement is called the **dynamical axion** [437, 438]. The axion shift symmetry can be preserved until operators of very high dimensions if the axicolored fermions are chiral [445, 446].

We now comment on the **heavy** QCD axion solution to the quality problem. First of all, taking $M = M_{\rm pl}$, $\varepsilon = 1$, $n = 6$ and $f_a = 10^{10}$ GeV in Eq. (2.194), the $U(1)_{\rm PQ}$ symmetry is kept of good enough quality if the QCD axion is heavier than $m_a \gtrsim 3 \times 10^7$ GeV. A few years after t'Hooft pioneering paper on IR-dominated instantons in QCD, it was shown that the introduction of heavy colored particles would make the strong coupling constant α_s grow again at a higher scale Λ' [412, 448, 450, 451].[25] The axion mass would then receive a large contribution from UV instantons and could be as large as $(m_a f_a)^2 \sim (\Lambda')^4$ [451]. Other proposals are $SU(3)_c$ being the diagonal subgroup of an $SU(3)^N$ gauge group [452–454], mirror QCD [455–460], enlarged QCD $SU(3+N)$ [461–463], 5D instantons [464], and contributions from real [465] or virtual [466] monopoles. Heavy axion are further motivated by possible signatures in colliders [467–469] and gravitational wave interferometers [470, 471].

Note that also the Nelson-Barr solution suffers from a quality problem. In order for the mechanism to work, the fermionic mass matrix must have a special structure, with some zeroes here and there. High-dimensional operators are expected to give corrections to the Nelson-Barr mass matrix so that $\bar{\theta} = \arg\det\,(\mathrm{Y_u Y_d}) \neq 0$.

References

1. Einstein A (1905) Einstein's proposal of the photon conceptfia translation of the annalen der physik paper of 1905. Ann Physik 17:132
2. Dirac PA (1928) The quantum theory of the electron. Proc R Soc Lond A 117:610–624
3. Dirac P (1928) The quantum theory of electron 2. Proc R Soc Lond A A118:351
4. Glashow S (1961) Partial symmetries of weak interactions. Nucl Phys 22:579–588
5. Weinberg S (1967) A model of leptons. Phys Rev Lett 19:1264–1266
6. Salam A (1968) Weak and electromagnetic interactions. Conf Proc C 680519:367–377
7. Englert F, Brout R (1964) Broken symmetry and the mass of gauge vector mesons. Phys Rev Lett 13:321–323
8. Higgs PW (1964) Broken symmetries, massless particles and gauge fields. Phys Lett 12:132–133
9. Guralnik G, Hagen C, Kibble T (1964) Global conservation laws and massless particles. Phys Rev Lett 13:585–587
10. Higgs PW (1964) Broken symmetries and the masses of Gauge Bosons. Phys Rev Lett 13:508–509
11. Higgs PW (1966) Spontaneous symmetry breakdown without massless bosons. Phys Rev 145:1156–1163
12. Nambu Y (1960) Quasiparticles and gauge invariance in the theory of superconductivity. Phys Rev 117:648–663

[25] See $\alpha_s(\mu)$ in Fig. 1 of [451] which looks like a 'half-pipe' according to Iason Baldes.

13. Goldstone J (1961) Field theories with superconductor solutions. Nuovo Cim 19:154–164
14. Goldstone J, Salam A, Weinberg S (1962) Broken symmetries. Phys Rev 127:965–970
15. 't Hooft G (1971) Renormalizable lagrangians for massive yang-mills fields. Nucl Phys B 35:167–188
16. 't Hooft G (1971) Renormalization of massless yang-mills fields. Nucl Phys B 33: 173–199
17. 't Hooft G, Veltman M (1972) Regularization and renormalization of gauge fields. Nucl Phys B 44:189–213
18. Gross DJ, Wilczek F (1973) Ultraviolet behavior of nonabelian gauge theories. Phys Rev Lett 30:1343–1346
19. Politzer HD (1973) Reliable perturbative results for strong interactions? Phys Rev Lett 30:1346–1349
20. Fritzsch H, Gell-Mann M (1972) Current algebra: Quarks and what else? eConf C720906V2:135–165. arXiv:hep-ph/0208010
21. Fritzsch H, Gell-Mann M, Leutwyler H (1973) Advantages of the color octet gluon picture. Phys Lett B 47:365–368
22. Kobayashi M, Maskawa T (1973) CP violation in the renormalizable theory of weak interaction. Prog Theor Phys 49:652–657
23. Itzykson C, Zuber J (1980) Quantum field theory. In: International series in pure and applied physics. McGraw-Hill, New York
24. Huang K (1982) Quarks, leptons and gauge fields
25. Halzen F, Martin AD (1984) Quarks and leptons: an introductory course in modern particle physics
26. Cheng TP, Li LF (1984) Gauge theory of elementary particle physics. Oxford University Press, Oxford, UK
27. Cheng TP, Li LF (2000) Gauge theory of elementary particle physics: problems and solutions
28. Zinn-Justin J (2021) Quantum field theory and critical phenomena. In: International series of monographs on physics, vol 77. Oxford University Press, p 4
29. Brown LS (1994) Quantum field theory. Cambridge University Press, p 7
30. Kaku M (1993) Quantum field theory: a Modern introduction
31. Ryder LH (1986) Quantum field theory. Cambridge University Press, p 6
32. Donoghue JF, Golowich E, Holstein BR (2014) Dynamics of the standard model, vol 2. CUP. https://doi.org/10.1017/CBO9780511524370
33. Peskin ME, Schroeder DV (1995) An introduction to quantum field theory. Addison-Wesley, Reading, USA
34. Weinberg S (2005) The quantum theory of fields. Vol. 1: Foundations. Cambridge University Press, p 6
35. Weinberg S (2013) The quantum theory of fields. Vol. 2: Modern applications. Cambridge University Press, p 8
36. Huang K (1998) Quantum field theory: from operators to path integrals
37. Kiselev VG, Shnir YM, Tregubovich AY (2000) Introduction to quantum field theory
38. Atkinson D, Johnson PW (2002) Quantum field theory: a self-contained course, vol 2
39. Zee A (2003) Quantum field theory in a nutshell, p 11
40. Morii T, Lim CS, Mukherjee SN (2004) The physics of the standard model and beyond
41. Maggiore M (2005) A modern introduction to quantum field theory, vol 9
42. Nair VP (2005) Quantum field theory: a modern perspective
43. Burgess C, Moore G (2006) The standard model: a primer. Cambridge University Press, p 12
44. Cottingham WN, Greenwood DA (2007) An introduction to the standard model of particle physics. Cambridge University Press, p 4
45. Dine M (2016) Supersymmetry and string theory: beyond the standard model. Cambridge University Press, p 1
46. Srednicki M (2007) Quantum field theory. Cambridge University Press, p 1
47. Griffiths D (2008) Introduction to elementary particles
48. Bettini A (2008) Introduction to elementary particle physics
49. McMahon D (2009) Quantum field theory demystified: a self-teaching guide

50. Barnes KJ (2010) Group theory for the standard model of particle physics and beyond. Taylor & Francis. https://doi.org/10.1201/9781439895207
51. Langacker P (2010) The standard model and beyond
52. Mann RB (2010) An introduction to particle physics and the standard model. CRC Press, Boca Raton, FL
53. Nagashima Y (2010) Elementary particle physics. Vol. 1: quantum field theory and particles, p 11
54. Nagashima Y (2010) Elementary particle physics: foundations of the standard model, vol 2. Wiley-VCH, Weinheim
55. Nagashima Y (2014) Beyond the standard model of elementary particle physics. Wiley-VCH, Weinheim, USA
56. Boyarkin OM (2011) Advanced particle physics. Vol. 1: particles, fields, and quantum electrodynamics
57. Boyarkin OM (2011) Advanced particle physics: Vol. 2: the standard model and beyond. CRC Pr., Boca Raton, USA
58. Robinson M (2011) Symmetry and the standard model: mathematics and particle physics. Springer, New York, USA. https://doi.org/10.1007/978-1-4419-8267-4
59. Tully CG (2011) Elementary particle physics in a nutshell
60. Schwartz MD (2014) Quantum field theory and the standard model. Cambridge University Press, p 3
61. Alvarez-Gaume L, Vazquez-Mozo MA (2012) An invitation to quantum field theory, vol 839. https://doi.org/10.1007/978-3-642-23728-7
62. Shifman M (2012) Advanced topics in quantum field theory: a lecture course. Cambridge University Press, Cambridge, UK, p 2
63. Shifman M (2019) Quantum field theory II. WSP, p 2. https://doi.org/10.1142/10825
64. Banks T (2014) Modern quantum field theory: a concise introduction. Cambridge University Press, p 12
65. Lancaster T, Blundell SJ (2014) Quantum field theory for the gifted amateur. Oxford University Press
66. Manoukian EB (2016) Quantum field theory: Vol. 1: foundations and Abelian gauge theories. Graduate texts in physics. Springer, Cham. https://doi.org/10.1007/978-3-319-30939-2
67. Manoukian EB (2016) Quantum field theory: Vol. 2: introductions to quantum gravity, supersymmetry and string theory. Graduate texts in physics. Springer, Cham. https://doi.org/10.1007/978-3-319-33852-1
68. Kane G (2017) Modern elementary particle physics: explaining and extending the standard model, p 02. https://doi.org/10.1017/9781316691434
69. Fritzsch H (2017) Quantum field theory world scientific. https://doi.org/10.1142/10081
70. Hamilton MJ (2017) Mathematical gauge theory: with applications to the standard model of particle physics. Universitext. Springer International Publishing, Cham. https://doi.org/10.1007/978-3-319-68439-0
71. Vergados JD (2017) The standard model and beyond. World Scientific, Singapore. https://doi.org/10.1142/10669
72. Vergados JD (ed) (2017) Group and representation theory. WSP. https://doi.org/10.1142/10325
73. Altarelli G (2017) Collider physics within the standard model: a primer, vol 937. Springer, Berlin. https://doi.org/10.1007/978-3-319-51920-3
74. Coleman S (2018) Lectures of sidney coleman on quantum field theory. WSP, Hackensack, p 12. https://doi.org/10.1142/9371
75. Ecker G (2019) Particles, fields, quanta: from quantum mechanics to the standard model of particle physics. Springer, Cham. https://doi.org/10.1007/978-3-030-14479-1
76. Gelis F (2019) Quantum field theory. Cambridge University Press, p 7
77. Schmitz W (2019) Particles, fields and forces: a conceptual guide to quantum field theory and the standard model. The Frontiers Collection Springer. https://doi.org/10.1007/978-3-030-12878-4

78. Strickland M, Relativistic quantum field theory, vol 1
79. Strickland M, Relativistic quantum field theory, vol 2
80. Strickland M (2019) Relativistic quantum field theory, vol 3. IOP, p 11. https://doi.org/10.1088/2053-2571/ab3a99
81. Nair VP (2018) Concepts in particle physics. World Scientific, Singapore. https://doi.org/10.1142/10640
82. Peskin ME (2019) Concepts of elementary particle physics. Oxford Master Series in Physics. Oxford University Press, p 9. https://doi.org/10.1093/oso/9780198812180.001.0001
83. Larkoski AJ (2019) Elementary particle physics: an intuitive introduction. Cambridge University Press, p 6
84. Iliopoulos J, Tomaras TN (2021) Elementary particle physics. Oxford University Press, p 10
85. Donoghue JF, Sorbo L (2022) A prelude to quantum field theory. Princeton University Press, p 2
86. Hoddeson LH, Brown L, Riordan M, Dresden M (eds) (1997) The rise of the standard model: particle physics in the 1960s and 1970s. In: Proceedings, conference. Stanford, USA
87. Pickering A (1999) Constructing quarks: a sociological history of particle physics. Physics, history and sociology of science. University of Chicago Press
88. Baggott J, Baggott J (2011) The quantum story: a history in 40 moments. EBSCO Ebook Academic Collection, OUP Oxford
89. Farmelo G (2019) The universe speaks in numbers: how modern math reveals nature's deepest secrets. Basic Books, p 5
90. Frampton PH, Kim JE (2020) History of particle theory. World Scientific, p 7. https://doi.org/10.1142/11948
91. Schweber S (2020) QED and the men who made it: dyson, feynman, schwinger, and tomonaga. Princeton Series in Physics. Princeton University Press
92. Fierz M (1939) Über die relativistische Theorie kräftefreier Teilchen mit beliebigem Spin. Helvetica Physica Acta 12:3–37
93. Pauli W (1940) The connection between spin and statistics. Phys Rev 58:716–722
94. D'Ambrosio G, Giudice G, Isidori G, Strumia A (2002) Minimal flavor violation: An Effective field theory approach. Nucl Phys B 645:155–187. arXiv:hep-ph/0207036
95. Yukawa H (1935) On the interaction of elementary particles I. Proc Phys Math Soc Jpn 17:48–57
96. Gell-Mann M, The eightfold way: a theory of strong interaction symmetry
97. Ne'eman Y (1961) Derivation of strong interactions from a gauge invariance. Nucl Phys 26:222–229
98. Gell-Mann M (1964) A schematic model of baryons and mesons. Phys Lett 8:214–215
99. Gell-Mann M (2018) The eightfold way. CRC Press
100. Han M, Nambu Y (1965) Three triplet model with double SU(3) symmetry. Phys Rev 139:B1006–B1010
101. Bjorken J (1969) Asymptotic sum rules at infinite momentum. Phys Rev 179:1547–1553
102. Bjorken J, Paschos EA (1969) Inelastic electron proton and gamma proton scattering, and the structure of the nucleon. Phys Rev 185:1975–1982
103. Bloom ED et al (1969) High-energy inelastic e p scattering at 6-degrees and 10-degrees. Phys Rev Lett 23:930–934
104. Feynman RP (1969) Very high-energy collisions of hadrons. Phys Rev Lett 23:1415–1417
105. Yang C-N, Mills RL (1954) Conservation of isotopic spin and isotopic gauge invariance. Phys Rev 96:191–195
106. Barber D et al (1979) Discovery of three jet events and a test of quantum chromodynamics at PETRA energies. Phys Rev Lett 43:830
107. Caswell WE (1974) Asymptotic behavior of nonabelian gauge theories to two loop order. Phys Rev Lett 33:244
108. Greensite J (2011) An introduction to the confinement problem, vol 821. https://doi.org/10.1007/978-3-642-14382-3

109. Particle Data Group collaboration, Tanabashi M et al (2018) Review of particle physics. Phys Rev D98:030001
110. Flavour Lattice Averaging Group collaboration, Aoki S et al (2018) FLAG review 2019. Eur Phys J C 80:113. arXiv:1902.08191
111. Nambu Y (1969) Quark model and the factorization of the veneziano amplitude, talk presented at the international conference on symmetries and quark models, vol 2. Wayne University
112. Nambu Y (1979) QCD and the string model. Phys Lett B 80:372–376
113. Goto T (1971) Relativistic quantum mechanics of one-dimensional mechanical continuum and subsidiary condition of dual resonance model. Prog Theor Phys 46:1560–1569
114. Casher A, Kogut JB, Susskind L (1974) Vacuum polarization and the absence of free quarks. Phys Rev D 10:732–745
115. Artru X, Mennessier G (1974) String model and multiproduction. Nucl Phys B 70:93–115
116. Andersson B, Gustafson G, Peterson C (1979) A semiclassical model for quark jet fragmentation. Z Phys C 1:105
117. Andersson B, Gustafson G, Soderberg B (1983) A general model for jet fragmentation. Z Phys C 20:317
118. Andersson B, Gustafson G, Ingelman G, Sjostrand T (1983) Parton fragmentation and string dynamics. Phys Rep 97:31–145
119. Sjostrand T, Mrenna S, Skands PZ (2006) PYTHIA 6.4 physics and manual. JHEP 05:026. arXiv:hep-ph/0603175
120. Maldacena JM (1999) The Large N limit of superconformal field theories and supergravity. Int J Theor Phys 38:1113–1133. arXiv:hep-th/9711200
121. Petrov AA, Blechman AE (2016). Effective field theories WSP. https://doi.org/10.1142/8619
122. Burgess CP (2020) Introduction to effective field theory. Cambridge University Press, p 12. https://doi.org/10.1017/9781139048040
123. Coleman SR, Wess J, Zumino B (1969) Structure of phenomenological Lagrangians. 1. Phys Rev 177:2239–2247
124. Callan CG, Coleman SR, Wess J, Zumino B (1969) Structure of phenomenological Lagrangians 2. Phys Rev 177:2247–2250
125. Leutwyler H (1994) On the foundations of chiral perturbation theory. Annals Phys 235:165–203. arXiv:hep-ph/9311274
126. Ecker G (1995) Chiral perturbation theory. Prog Part Nucl Phys 35:1–80. arXiv:hep-ph/9501357
127. Pich A (1995) Chiral perturbation theory. Rep Prog Phys 58:563–610. arXiv:hep-ph/9502366
128. Scherer S (2003) Introduction to chiral perturbation theory. Adv Nucl Phys 27:277. arXiv:hep-ph/0210398
129. Scherer S, Schindler MR (2012) A primer for chiral perturbation theory, vol 830. https://doi.org/10.1007/978-3-642-19254-8
130. Gell-Mann M, Oakes RJ, Renner B (1968) Behavior of current divergences under SU(3) x SU(3). Phys Rev 175:2195–2199
131. Ioffe BL (2006) QCD at low energies. Prog Part Nucl Phys 56:232–277. arXiv:hep-ph/0502148
132. Feynman R, Gell-Mann M (1958) Theory of fermi interaction. Phys Rev 109:193–198
133. Sudarshan E, Marshak R (1958) Chirality invariance and the universal Fermi interaction. Phys Rev 109:1860
134. ATLAS collaboration, Aad G et al (2012) Observation of a new particle in the search for the standard model higgs boson with the ATLAS detector at the LHC. Phys Lett B716:1–29. arXiv:1207.7214
135. CMS collaboration, Chatrchyan S et al (2012) Observation of a new boson at a mass of 125 GeV with the CMS experiment at the LHC. Phys Lett B716:30–61. arXiv:1207.7235
136. Zee A (2016) Group theory in a nutshell for physicists. Princeton University Press, USA
137. Nakano T, Nishijima K (1953) Charge Independence for V-particles. Prog Theor Phys 10:581–582

138. Gell-Mann M (1956) The interpretation of the new particles as displaced charge multiplets. Nuovo Cim 4:848–866
139. Glashow S, Iliopoulos J, Maiani L (1970) Weak interactions with lepton-hadron symmetry. Phys Rev D 2:1285–1292
140. Chau L-L, Keung W-Y (1984) Comments on the parametrization of the Kobayashi-Maskawa matrix. Phys Rev Lett 53:1802
141. Streater R, Wightman A (1989) PCT, spin and statistics, and all that
142. Kostelecky V (1998) The Status of CPT. Physics beyond the standard model. In: Proceedings, 5th international WEIN symposium. Santa Fe, USA, pp 588–600, 10. arXiv:hep-ph/9810365
143. Jarlskog C (1985) Commutator of the quark mass matrices in the standard electroweak model and a measure of maximal CP violation. Phys Rev Lett 55:1039
144. Jarlskog C (1985) A basis independent formulation of the connection between quark mass matrices, CP violation and experiment. Z Phys C 29:491–497
145. Particle Data Group collaboration, Beringer J et al (2012) Review of particle physics (RPP). Phys Rev D 86:010001
146. Shaposhnikov M (1988) Structure of the high temperature gauge ground state and electroweak production of the baryon asymmetry. Nucl Phys B 299:797–817
147. Gavela M, Hernandez P, Orloff J, Pene O (1994) Standard model CP violation and baryon asymmetry. Mod Phys Lett A 9:795–810. arXiv:hep-ph/9312215
148. Konstandin T, Prokopec T, Schmidt MG (2004) Axial currents from CKM matrix CP violation and electroweak baryogenesis. Nucl Phys B 679:246–260. arXiv:hep-ph/0309291
149. Brauner T, Taanila O, Tranberg A, Vuorinen A (2012) Temperature dependence of standard model CP violation. Phys Rev Lett 108:041601. arXiv:1110.6818
150. Lee T, Yang C-N (1956) Question of parity conservation in weak interactions. Phys Rev 104:254–258
151. Wu C, Ambler E, Hayward R, Hoppes D, Hudson R (1957) Experimental test of parity conservation in β decay. Phys Rev 105:1413–1414
152. Garwin R, Lederman L, Weinrich M (1957) Observations of the failure of conservation of parity and charge conjugation in meson decays: the magnetic moment of the free muon. Phys Rev 105:1415–1417
153. Cabibbo N (1963) Unitary symmetry and leptonic decays. Phys Rev Lett 10:531–533
154. Christenson J, Cronin J, Fitch V, Turlay R (1964) Evidence for the 2π decay of the K_2^0 meson. Phys Rev Lett 13:138–140
155. Adler SL (1969) Axial vector vertex in spinor electrodynamics. Phys Rev 177:2426–2438
156. Bell J, Jackiw R (1969) A PCAC puzzle: $\pi^0 \to \gamma\gamma$ in the σ model. Nuovo Cim A 60:47–61
157. Lohitsiri N, Tong D (2020) Hypercharge quantisation and Fermat's last theorem. SciPost Phys 8:009 [arXiv:1907.00514]
158. Weinberg S (1975) The U(1) Problem. Phys Rev D 11:3583–3593
159. Hooft G (1976) Symmetry breaking through bell-jackiw anomalies. Phys Rev Lett 37:8–11
160. Crewther RJ, Di Vecchia P, Veneziano G, Witten E (1979) Chiral estimate of the electric dipole moment of the neutron in quantum chromodynamics. Phys Lett B 88:123
161. Pospelov M, Ritz A (2005) Electric dipole moments as probes of new physics. Annals Phys 318:119–169. arXiv:hep-ph/0504231
162. Dubbers D, Schmidt MG (2011) The neutron and its role in cosmology and particle physics. Rev Mod Phys 83:1111–1171. arXiv:1105.3694
163. Chupp T, Fierlinger P, Ramsey-Musolf M, Singh J (2019) Electric dipole moments of atoms, molecules, nuclei, and particles. Rev Mod Phys 91:015001. arXiv:1710.02504
164. Baker C et al (2006) An Improved experimental limit on the electric dipole moment of the neutron. Phys Rev Lett 97:131801 [arXiv:hep-ex/0602020]
165. Pendlebury JM et al (2015) Revised experimental upper limit on the electric dipole moment of the neutron. Phys Rev D 92:092003. arXiv:1509.04411
166. Abel C et al (2020) Measurement of the permanent electric dipole moment of the neutron. Phys Rev Lett 124:081803. arXiv:2001.11966

167. Callan CG Jr, Dashen RF, Gross DJ (1976) The structure of the gauge theory vacuum. Phys Lett B 63:334–340
168. Callan CG Jr, Dashen RF, Gross DJ (1978) Toward a theory of the strong interactions. Phys Rev D 17:2717
169. Callan CG Jr, Dashen RF, Gross DJ (1979) A theory of hadronic structure. Phys Rev D 19:1826
170. Bernard CW (1979) Gauge zero modes, instanton determinants, and QCD calculations. Phys Rev D 19:3013
171. Pisarski RD, Yaffe LG (1980) The density of instantons at finite temperature. Phys Lett B 97:110–112
172. Gross DJ, Pisarski RD, Yaffe LG (1981) QCD and instantons at finite temperature. Rev Mod Phys 53:43
173. Luscher M (1982) A semiclassical formula for the topological susceptibility in a finite space-time volume. Nucl Phys B 205:483–503
174. Morris TR, Ross DA, Sachrajda CT (1985) Higher order quantum corrections in the presence of an instanton background field. Nucl Phys B 255:115–148
175. Coleman SR (1979) The uses of instantons. Subnucl Ser 15:805
176. Vainshtein AI, Zakharov VI, Novikov VA, Shifman MA (1982) ABC's of instantons. Sov Phys Usp 25:195
177. Schäfer T, Shuryak EV (1998) Instantons in QCD. Rev Mod Phys 70:323–426. arXiv:hep-ph/9610451
178. Bogomolny EB (1976) Stability of classical solutions. Sov J Nucl Phys 24:449
179. Belavin AA, Polyakov AM, Schwartz AS, Tyupkin YS (1975) Pseudoparticle solutions of the yang-mills equations. Phys Lett B 59:85–87
180. 't Hooft G (1976) Computation of the quantum effects due to a four-dimensional pseudoparticle. Phys Rev D 14:3432–3450
181. Berkowitz E, Buchoff MI, Rinaldi E (2015) Lattice QCD input for axion cosmology. Phys Rev D 92:034507. arXiv:1505.07455
182. Borsanyi S, Dierigl M, Fodor Z, Katz SD, Mages SW, Nogradi D et al (2016) Axion cosmology, lattice QCD and the dilute instanton gas. Phys Lett B 752:175–181. arXiv:1508.06917
183. Grilli di Cortona G, Hardy E, Pardo Vega J, Villadoro G (2016) The QCD axion, precisely. JHEP 01: 034. arXiv:1511.02867
184. Di Vecchia P, Veneziano G (1980) Chiral dynamics in the large n limit. Nucl Phys B 171:253–272
185. Vafa C, Witten E (1984) Restrictions on symmetry breaking in vector-like gauge theories. Nucl Phys B 234:173–188
186. Panico G, Wulzer A (2016) The composite nambu-goldstone higgs. Lect Notes Phys 913:1–316. arXiv:1506.01961
187. Tomonaga S (1946) On a relativistically invariant formulation of the quantum theory of wave fields. Prog Theor Phys 1:27–42
188. Tomonaga S-I, Oppenheimer J (1948) On infinite field reactions in quantum field theory. Phys Rev 74:224–225
189. Schwinger JS (1948) On Quantum electrodynamics and the magnetic moment of the electron. Phys Rev 73:416–417
190. Schwinger JS (1948) Quantum electrodynamics. I A covariant formulation. Phys Rev 74:1439
191. Schwinger JS (1948) Quantum electrodynamics. 2. Vacuum polarization and selfenergy. Phys Rev 75:651
192. Schwinger JS (1949) Quantum electrodynamics. III: the electromagnetic properties of the electron: radiative corrections to scattering. Phys Rev 76:790–817
193. Schwinger JS (1951) On gauge invariance and vacuum polarization. Phys Rev 82:664–679
194. Feynman R (1949) The theory of positrons. Phys Rev 76:749–759
195. Feynman R (1949) Space—time approach to quantum electrodynamics. Phys Rev 76:769–789
196. Feynman R (1950) Mathematical formulation of the quantum theory of electromagnetic interaction. Phys Rev 80:440–457
197. Bethe H (1947) The electromagnetic shift of energy levels. Phys Rev 72:339–341

198. Pauli W, Villars F (1949) On the Invariant regularization in relativistic quantum theory. Rev Mod Phys 21:434–444
199. Dyson F (1949) The S matrix in quantum electrodynamics. Phys Rev 75:1736–1755
200. Dyson F (1949) The radiation theories of tomonaga, schwinger, and feynman. Phys Rev 75:486–502
201. Laporta S (2017) High-precision calculation of the 4-loop contribution to the electron g-2 in QED. Phys Lett B 772:232–238. arXiv:1704.06996
202. Bouchendira R, Clade P, Guellati-Khelifa S, Nez F, Biraben F (2011) New determination of the fine structure constant and test of the quantum electrodynamics. Phys Rev Lett 106:080801. arXiv:1012.3627
203. 't Hooft G (1980) Naturalness, chiral symmetry, and spontaneous chiral symmetry breaking. NATO Sci Ser B 59:135–157
204. Giudice GF (2013) Naturalness after LHC8. PoS EPS-HEP2013:163. arXiv:1307.7879
205. Coleman SR, Mandula J (1967) All possible symmetries of the S matrix. Phys Rev 159:1251–1256
206. Golfand Y, Likhtman E (1971) Extension of the algebra of poincare group generators and violation of p invariance. JETP Lett 13:323–326
207. Fayet P (1977) Spontaneously broken supersymmetric theories of weak, electromagnetic and strong interactions. Phys Lett B 69:489
208. Dimopoulos S, Georgi H (1981) Softly broken supersymmetry and SU(5). Nucl Phys B 193:150–162
209. Craig N (2013) The state of supersymmetry after run I of the LHC. In: Beyond the standard model after the first run of the LHC, p 9. arXiv:1309.0528
210. Hall LJ, Pinner D, Ruderman JT (2012) A natural SUSY higgs near 126 GeV. JHEP 04:131. arXiv:1112.2703
211. CMS collaboration, CMS Collaboration, Search for direct stop pair production in the dilepton final state at $\sqrt{s} = 13$ TeV
212. Ellwanger U, Hugonie C, Teixeira AM (2010) The next-to-minimal supersymmetric standard model. Phys Rep 496:1–77. arXiv:0910.1785
213. Ross GG, Schmidt-Hoberg K (2012) The fine-tuning of the generalised NMSSM. Nucl Phys B 862:710–719. arXiv:1108.1284
214. Burdman G, Chacko Z, Goh H-S, Harnik R (2007) Folded supersymmetry and the LEP paradox. JHEP 02:009. arXiv:hep-ph/0609152
215. Feng JL, Matchev KT, Moroi T (2000) Focus points and naturalness in supersymmetry. Phys Rev D 61:075005. arXiv:hep-ph/9909334
216. Feng JL, Sanford D (2012) A natural 125 GeV higgs boson in the MSSM from focus point supersymmetry with A-terms. Phys Rev D 86:055015. arXiv:1205.2372
217. Craig N, Dimopoulos S, Gherghetta T (2012) Split families unified. JHEP 04:116. arXiv:1203.0572
218. Craig N, McCullough M, Thaler J (2012) Flavor mediation delivers natural SUSY. JHEP 06:046. arXiv:1203.1622
219. LeCompte TJ, Martin SP (2011) Large hadron collider reach for supersymmetric models with compressed mass spectra. Phys Rev D 84:015004. arXiv:1105.4304
220. Fan J, Reece M, Ruderman JT (2011) Stealth supersymmetry. JHEP 11:012. arXiv:1105.5135
221. Csaki C, Grossman Y, Heidenreich B (2012) MFV SUSY: a natural theory for R-parity violation. Phys Rev D 85:095009. arXiv:1111.1239
222. Wess J, Bagger J (1992) Supersymmetry and supergravity. Princeton University Press, Princeton, NJ, USA
223. Martin SP (2010) A Supersymmetry primer, vol 21, pp 1–153. arXiv:hep-ph/9709356, https://doi.org/10.1142/9789812839657_0001
224. Weinberg S (2013) The quantum theory of fields. Vol. 3: supersymmetry. Cambridge University Press, p 6
225. Terning J (2006) Modern supersymmetry: Dynamics and duality, p 4. https://doi.org/10.1093/acprof:oso/9780198567639.001.0001

226. Binetruy P (2006) Supersymmetry: theory, experiment and cosmology
227. Aitchison I (2007) Supersymmetry in particle physics. An elementary introduction. Cambridge University Press, Cambridge. https://doi.org/10.1017/CBO9780511619250
228. Peskin ME (2008) Supersymmetry in elementary particle physics. In: Theoretical advanced study institute in elementary particle physics: exploring new frontiers using colliders and neutrinos, pp 609–704, 1. arXiv:0801.1928
229. Csáki C, Tanedo P (2015) Beyond the standard model. In: 2013 European school of high-energy physics, pp 169–268. arXiv:1602.04228, https://doi.org/10.5170/CERN-2015-004.169
230. McCullough M (2018) Lectures on physics beyond the standard model. In: 6th Tri-institute summer school on elementary particles, p 8
231. Wulzer A (2019) Behind the standard model. In: 2015 European school of high-energy physics, p 1. arXiv:1901.01017
232. Lee HM (2019) Lectures on physics beyond the standard model, p 7. arXiv:1907.12409
233. Rossia A (2019) New approaches to the hierarchy problem II: new approaches to the hierarchy problem. In: DESY workshop seminar 2019
234. Hebecker A, Lectures on naturalness, string landscape and multiverse. arXiv:2008.10625
235. Raby S (2006) Grand unified theories. In: 2nd world summit: physics beyond the standard model, p 8. arXiv:hep-ph/0608183
236. Jungman G, Kamionkowski M, Griest K (1996) Supersymmetric dark matter. Phys Rep 267:195–373. arXiv:hep-ph/9506380
237. Ellis JR, Nanopoulos DV (1982) Flavor changing neutral interactions in broken supersymmetric theories. Phys Lett B 110:44–48
238. Barbieri R, Gatto R (1982) Conservation laws for neutral currents in spontaneously broken supersymmetric theories. Phys Lett B 110:211
239. Campbell BA (1983) Supersymmetry and neutral flavor nonconservation. Phys Rev D 28:209–216
240. Duncan MJ (1983) Generalized cabibbo angles in supersymmetric gauge theories. Nucl Phys B 221:285
241. Gabbiani F, Masiero A (1989) FCNC in generalized supersymmetric theories. Nucl Phys B 322:235–254
242. Dimopoulos S, Hall LJ (1995) Electric dipole moments as a test of supersymmetric unification. Phys Lett B 344:185–192. arXiv:hep-ph/9411273
243. Barbieri R, Hall LJ (1994) Signals for supersymmetric unification. Phys Lett B 338:212–218. arXiv:hep-ph/9408406
244. Barbieri R, Hall LJ, Strumia A (1995) Hadronic flavor and CP violating signals of superunification. Nucl Phys B 449:437–461. arXiv:hep-ph/9504373
245. Weinberg S (1982) Cosmological constraints on the scale of supersymmetry breaking. Phys Rev Lett 48:1303
246. Coughlan GD, Holman R, Ramond P, Ross GG (1984) Supersymmetry and the entropy crisis. Phys Lett B 140:44–48
247. Ellis JR, Nanopoulos DV, Sarkar S (1985) The cosmology of decaying gravitinos. Nucl Phys B 259:175–188
248. Banks T, Kaplan DB, Nelson AE (1994) Cosmological implications of dynamical supersymmetry breaking. Phys Rev D 49:779–787. arXiv:hep-ph/9308292
249. Moroi T, Murayama H, Yamaguchi M (1993) Cosmological constraints on the light stable gravitino. Phys Lett B 303:289–294
250. Banks T, Berkooz M, Steinhardt PJ (1995) The Cosmological moduli problem, supersymmetry breaking, and stability in postinflationary cosmology. Phys Rev D 52:705–716. arXiv:hep-th/9501053
251. Kawasaki M, Moroi T (1995) Gravitino production in the inflationary universe and the effects on big bang nucleosynthesis. Prog Theor Phys 93:879–900. arXiv:hep-ph/9403364
252. Kawasaki M, Kohri K, Moroi T, Yotsuyanagi A (2008) Big-bang nucleosynthesis and gravitino. Phys Rev D 78:065011. arXiv:0804.3745

253. Arkani-Hamed N, Dimopoulos S, Giudice GF, Romanino A (2005) Aspects of split supersymmetry. Nucl Phys B 709:3–46. arXiv:hep-ph/0409232
254. Giudice GF, Romanino A (2004) Split supersymmetry. Nucl Phys B 699:65–89. arXiv:hep-ph/0406088
255. Arkani-Hamed N, Dimopoulos S, Giudice GF, Romanino A (2005) Aspects of split supersymmetry. Nucl Phys B 709:3–46. arXiv:hep-ph/0409232
256. Wells JD (2005) PeV-scale supersymmetry. Phys Rev D 71:015013. arXiv:hep-ph/0411041
257. Hall LJ, Nomura Y (2010) A finely-predicted higgs boson mass from a finely-tuned weak scale. JHEP 03:076. arXiv:0910.2235
258. Bagnaschi E, Giudice GF, Slavich P, Strumia A (2014) Higgs mass and unnatural supersymmetry. JHEP 09:092. arXiv:1407.4081
259. Baer H, Barger V, Salam S, Sengupta D (2022) Mini-review: expectations for supersymmetry from the string landscape. In: 2022 snowmass summer study, p 2. arXiv:2202.11578
260. Terazawa H, Akama K, Chikashige Y (1977) Unified model of the Nambu-Jona-Lasinio type for all elementary particle forces. Phys Rev D 15:480
261. Dimopoulos S, Susskind L, Mass without scalars
262. Farhi E, Susskind L (1981) Technicolor. Phys Rep 74:277
263. Chivukula R (2000) Lectures on technicolor and compositeness. Flavor physics for the millennium. In: Proceedings, theoretical advanced study institute in elementary particle physics, TASI 2000. Boulder, USA, pp 731–772, 6. arXiv:hep-ph/0011264
264. Lane K, Two lectures on technicolor. arXiv:hep-ph/0202255
265. Hill CT, Simmons EH (2003) Strong dynamics and electroweak symmetry breaking. Phys Rep 381:235–402. arXiv:hep-ph/0203079
266. Shrock R (2007) Some recent results on models of dynamical electroweak symmetry breaking. In: The origin of mass and strong coupling gauge theories. Proceedings, 5th International Workshop, SCGT'06, Nagoya, Japan, pp 227–241, 3. arXiv:hep-ph/0703050, https://doi.org/10.1142/9789812790750_0023
267. Arbey A, Cacciapaglia G, Cai H, Deandrea A, Le Corre S, Sannino F (2017) Fundamental composite electroweak dynamics: status at the LHC. Phys Rev D 95:015028. arXiv:1502.04718
268. Kaplan DB, Georgi H (1984) SU(2) x U(1) Breaking by vacuum misalignment. Phys Lett B 136:183–186
269. Kaplan DB, Georgi H, Dimopoulos S (1984) Composite higgs scalars. Phys Lett B 136:187–190
270. Dugan MJ, Georgi H, Kaplan DB (1985) Anatomy of a composite higgs model. Nucl Phys B 254:299–326
271. ATLAS collaboration, Aad G et al (2015) Constraints on new phenomena via Higgs boson couplings and invisible decays with the ATLAS detector. JHEP 11:206. arXiv:1509.00672
272. Kaplan DB (1991) Flavor at SSC energies: a New mechanism for dynamically generated fermion masses. Nucl Phys B 365:259–278
273. Contino R (2009) The higgs as a composite nambu-goldstone Boson. In: Physics of the large and the small, TASI 09, proceedings of the theoretical advanced study institute in elementary particle physics. Boulder, Colorado, USA, pp 235–306. arXiv:1005.4269, https://doi.org/10.1142/9789814327183_0005
274. Bellazzini B, Csáki C, Serra J (2014) Composite Higgses. Eur Phys J C 74:2766. arXiv:1401.2457
275. Arkani-Hamed N, Cohen AG, Georgi H (2001) Electroweak symmetry breaking from dimensional deconstruction. Phys Lett B 513:232–240. arXiv:hep-ph/0105239
276. Arkani-Hamed N, Cohen A, Katz E, Nelson A (2002) The littlest higgs. JHEP 07:034. arXiv:hep-ph/0206021
277. Perelstein M (2007) Little higgs models and their phenomenology. Prog Part Nucl Phys 58:247–291. arXiv:hep-ph/0512128
278. Schmaltz M, Tucker-Smith D (2005) Little Higgs review. Ann Rev Nucl Part Sci 55:229–270. arXiv:hep-ph/0502182

279. Chacko Z, Goh H-S, Harnik R (2006) The twin higgs: natural electroweak breaking from mirror symmetry. Phys Rev Lett 96:231802. arXiv:hep-ph/0506256
280. Batell B, Low M, Neil ET, Verhaaren CB (2022) Review of neutral naturalness. In: 2022 snowmass summer study, p 3. arXiv:2203.05531
281. Craig N, Knapen S, Longhi P (2015) Neutral naturalness from orbifold higgs models. Phys Rev Lett 114:061803. arXiv:1410.6808
282. Arkani-Hamed N, Dimopoulos S, Dvali G (1998) The Hierarchy problem and new dimensions at a millimeter. Phys Lett B 429:263–272. arXiv:hep-ph/9803315
283. Antoniadis I, Arkani-Hamed N, Dimopoulos S, Dvali G (1998) New dimensions at a millimeter to a Fermi and superstrings at a TeV. Phys Lett B 436:257–263. arXiv:hep-ph/9804398
284. Rubakov V, Shaposhnikov M (1983) Do we live inside a domain wall? Phys Lett B 125:136–138
285. Tan W-H et al (2020) Improvement for testing the gravitational inverse-square law at the submillimeter range. Phys Rev Lett 124:051301
286. Lee J, Adelberger E, Cook T, Fleischer S, Heckel B (2020) New test of the gravitational $1/r^2$ law at separations down to 52 μm. Phys Rev Lett 124:101101. arXiv:2002.11761
287. Rattazzi R (2003) Cargese lectures on extra-dimensions, particle physics and cosmology: the interface. In: Proceedings, NATO advanced study institute, school. Cargese, France, pp 461–517. arXiv:hep-ph/0607055
288. Csaki C (2004) TASI lectures on extra dimensions and branes, from fields to strings: circumnavigating theoretical physics. In: Ian Kogan memorial collection (3 volume set), pp 605–698, 4. arXiv:hep-ph/0404096
289. Cheng H-C (2011) Introduction to extra dimensions. In: Theoretical advanced study institute in elementary particle physics: physics of the large and the small, pp 125–162. arXiv:1003.1162, https://doi.org/10.1142/9789814327183_0003
290. Randall L, Sundrum R (1999) A Large mass hierarchy from a small extra dimension. Phys Rev Lett 83:3370–3373. arXiv:hep-ph/9905221
291. Randall L, Sundrum R (1999) An alternative to compactification. Phys Rev Lett 83:4690–4693. arXiv:hep-th/9906064
292. Goldberger WD, Wise MB (1999) Modulus stabilization with bulk fields. Phys Rev Lett 83:4922–4925. arXiv:hep-ph/9907447
293. Goldberger WD, Wise MB (2000) Phenomenology of a stabilized modulus. Phys Lett B 475:275–279. arXiv:hep-ph/9911457
294. Agashe K, Contino R, Pomarol A (2005) The minimal composite higgs model. Nucl Phys B 719:165–187. arXiv:hep-ph/0412089
295. Sundrum R (2005) Tasi 2004 lectures: to the fifth dimension and back. In: Theoretical advanced study institute in elementary particle physics: physics in D $\geqq$ 4, pp 585–630, 8. arXiv:hep-th/0508134
296. Gherghetta T (2011) A holographic view of beyond the standard model physics. In: Physics of the large and the small, TASI 09, proceedings of the Theoretical Advanced Study Institute in Elementary Particle Physics. Boulder, Colorado, USA, pp 165–232. arXiv:1008.2570, https://doi.org/10.1142/9789814327183_0004
297. Susskind L, The Anthropic landscape of string theory. arXiv:hep-th/0302219
298. Douglas MR (2003) The statistics of string/M theory vacua. JHEP 05:046. arXiv:hep-th/0303194
299. Teresi D (2021) GGI seminar—the weak scale as a trigger
300. Abbott L (1985) A mechanism for reducing the value of the cosmological constant. Phys Lett B 150:427–430
301. Bousso R, Polchinski J (2000) Quantization of four form fluxes and dynamical neutralization of the cosmological constant. JHEP 06:006. arXiv:hep-th/0004134
302. Feng JL, March-Russell J, Sethi S, Wilczek F (2001) Saltatory relaxation of the cosmological constant. Nucl Phys B 602:307–328. arXiv:hep-th/0005276
303. Dvali G, Vilenkin A (2004) Cosmic attractors and gauge hierarchy. Phys Rev D 70:063501. arXiv:hep-th/0304043

304. Dvali G (2006) Large hierarchies from attractor vacua. Phys Rev D 74:025018. arXiv:hep-th/0410286
305. Giudice G, Kehagias A, Riotto A (2019) The selfish Higgs. JHEP 10:199. arXiv:1907.05370
306. Kaloper N, Westphal A, A goldilocks higgs. arXiv:1907.05837
307. Graham PW, Kaplan DE, Rajendran S (2015) Cosmological Relaxation of the Electroweak Scale. Phys Rev Lett 115:221801. arXiv:1504.07551
308. Arkani-Hamed N, Cohen T, D'Agnolo RT, Hook A, Kim HD, Pinner D (2016) Solving the hierarchy problem at reheating with a large number of degrees of freedom. Phys Rev Lett 117:251801. arXiv:1607.06821
309. Bloch IM, Csáki C, Geller M, Volansky T, Crunching away the cosmological constant problem: dynamical selection of a small Λ. arXiv:1912.08840
310. Csáki C, D'Agnolo RT, Geller M, Ismail A, Crunching dilaton, hidden naturalness. arXiv: 2007.14396
311. Strumia A, Teresi D (2020) Relaxing the Higgs mass and its vacuum energy by living at the top of the potential. Phys Rev D 101:115002. arXiv:2002.02463
312. Tito D'Agnolo R, Teresi D, Sliding naturalness. arXiv:2106.04591
313. Tito D'Agnolo R, Teresi D, Sliding naturalness: cosmological selection of the weak scale. arXiv: 2109.13249
314. Dvali G (2006) A Vacuum accumulation solution to the strong CP problem. Phys Rev D 74:025019. arXiv:hep-th/0510053
315. Geller M, Hochberg Y, Kuflik E (2019) Inflating to the weak scale. Phys Rev Lett 122:191802. arXiv:1809.07338
316. Cheung C, Saraswat P, Mass hierarchy and vacuum energy. arXiv:1811.12390
317. Giudice GF, McCullough M, You T (2021) Self-organised localisation. JHEP 10:093. arXiv:2105.08617
318. Arvanitaki A, Dimopoulos S, Gorbenko V, Huang J, Van Tilburg K (2017) A small weak scale from a small cosmological constant. JHEP 05:071. arXiv:1609.06320
319. Arkani-Hamed N, Tito D'agnolo R, Kim HD, The weak scale as a trigger. arXiv:2012.04652
320. Fonseca N, Morgante E, Sato R, Servant G, Relaxion fluctuations (Self-stopping Relaxion) and overview of relaxion stopping mechanisms. arXiv:1911.08473
321. Brennan TD, Carta F, Vafa C (2017) The string landscape, the swampland, and the missing corner. PoS TASI2017:015. arXiv:1711.00864
322. Palti E (2019) The swampland: introduction and review. Fortsch Phys 67:1900037. arXiv:1903.06239
323. Graña M, Herráez A (2021) The swampland conjectures: a bridge from quantum gravity to particle physics. Universe 7:273. arXiv:2107.00087
324. Espinosa J, Grojean C, Panico G, Pomarol A, Pujolàs O, Servant G (2015) Cosmological higgs-axion interplay for a naturally small electroweak scale. Phys Rev Lett 115:251803. arXiv:1506.09217
325. Fonseca N, Morgante E (2019) Relaxion dark matter. Phys Rev D 100:055010. arXiv:1809.04534
326. Banerjee A, Kim H, Perez G (2019) Coherent relaxion dark matter. Phys Rev D 100:115026. arXiv:1810.01889
327. Banerjee A, Kim H, Matsedonskyi O, Perez G, Safronova MS, Probing the relaxed relaxion at the luminosity and precision frontiers. arXiv:2004.02899
328. Fonseca N, Morgante E, Sato R, Servant G (2020) Axion fragmentation. JHEP 04:010. arXiv:1911.08472
329. Morgante E, Ratzinger W, Sato R, Stefanek BA, Axion Fragmentation on the Lattice. arXiv: 2109.13823
330. Kofman L, Linde AD, Starobinsky AA (1997) Towards the theory of reheating after inflation. Phys Rev D 56:3258–3295. arXiv:hep-ph/9704452
331. Agrawal V, Barr SM, Donoghue JF, Seckel D (1998) Viable range of the mass scale of the standard model. Phys Rev D 57:5480–5492. arXiv:hep-ph/9707380

332. Agrawal V, Barr SM, Donoghue JF, Seckel D (1998) Anthropic considerations in multiple domain theories and the scale of electroweak symmetry breaking. Phys Rev Lett 80:1822–1825. arXiv:hep-ph/9801253
333. Damour T, Donoghue JF (2008) Constraints on the variability of quark masses from nuclear binding. Phys Rev D 78:014014. arXiv:0712.2968
334. Hall LJ, Pinner D, Ruderman JT (2014) The weak scale from BBN. JHEP 12:134. arXiv:1409.0551
335. Harnik R, Kribs GD, Perez G (2006) A Universe without weak interactions. Phys Rev D 74:035006. arXiv:hep-ph/0604027
336. D'Amico G, Strumia A, Urbano A, Xue W (2019) Direct anthropic bound on the weak scale from supernovæ explosions. Phys Rev D 100:083013. arXiv:1906.00986
337. Buttazzo D, Degrassi G, Giardino PP, Giudice GF, Sala F, Salvio A et al (2013) Investigating the near-criticality of the Higgs boson. JHEP 12:089. arXiv:1307.3536
338. Brown J, Teitelboim C (1987) Dynamical neutralization of the cosmological constant. Phys Lett B 195:177–182
339. Brown J, Teitelboim C (1988) Neutralization of the cosmological constant by membrane creation. Nucl Phys B 297:787–836
340. Davis R, Harmer DS, Hoffman KC (1968) Search for neutrinos from the sun. Phys Rev Lett 20:1205–1209
341. Pontecorvo B (1968) Neutrino experiments and the problem of conservation of leptonic charge. Sov Phys JETP 26:984–988
342. Super-Kamiokande collaboration Fukuda S et al (2001) Solar B-8 and hep neutrino measurements from 1258 days of Super-Kamiokande data. Phys Rev Lett 86:5651–5655. arXiv:hep-ex/0103032
343. SNO collaboration, Ahmad Q et al (2001) Measurement of the rate of $\nu_e + d \to p + p + e^-$ interactions produced by ^8B solar neutrinos at the Sudbury Neutrino Observator. Phys Rev Lett 87:071301. arXiv:nucl-ex/0106015
344. SNO collaboration, Ahmad Q et al (2002) Direct evidence for neutrino flavor transformation from neutral current interactions in the Sudbury Neutrino Observatory. Phys Rev Lett 89:011301. arXiv:nucl-ex/0204008
345. Super-Kamiokande collaboration, Fukuda Y et al (1998) Evidence for oscillation of atmospheric neutrinos. Phys Rev Lett 81:1562–1567. arXiv:hep-ex/9807003
346. Gomez-Cadenas J, Martin-Albo J, Mezzetto M, Monrabal F, Sorel M (2012) The Search for neutrinoless double beta decay. Riv Nuovo Cim 35:29–98. arXiv:1109.5515
347. Weinberg S (1979) Baryon and lepton nonconserving processes. Phys Rev Lett 43:1566–1570
348. Minkowski P (1977) $\mu \to e\gamma$ at a rate of one out of 10^9 muon decays? Phys Lett B 67:421–428
349. Ramond P (1979) The family group in grand unified theories. In: International symposium on fundamentals of quantum theory and quantum field theory, pp 265–280, 2. arXiv:hep-ph/9809459
350. Gell-Mann M, Ramond P, Slansky R (1979) Complex spinors and unified theories. Conf Proc C 790927:315–321. arXiv:1306.4669
351. Yanagida T (1979) Horizontal gauge symmetry and masses of neutrinos. Conf Proc C 7902131:95–99
352. Mohapatra RN, Senjanovic G (1980) Neutrino mass and spontaneous parity nonconservation. Phys Rev Lett 44:912
353. Drewes M (2013) The phenomenology of right handed neutrinos. Int J Mod Phys E 22:1330019. arXiv:1303.6912
354. King SF (2015) Models of neutrino mass, mixing and CP violation. J Phys G 42:123001. arXiv:1510.02091
355. Canetti L, Drewes M, Frossard T, Shaposhnikov M (2013) Dark matter, baryogenesis and neutrino oscillations from right handed neutrinos. Phys Rev D 87:093006. arXiv:1208.4607
356. Canetti L, Drewes M, Shaposhnikov M (2012) Matter and antimatter in the universe. New J Phys 14:095012. arXiv:1204.4186
357. Abazajian K et al, Light sterile neutrinos: a white paper. arXiv:1204.5379

358. Drewes M et al (2017) A white paper on keV sterile neutrino dark matter. JCAP 01:025. arXiv:1602.04816
359. Böser S, Buck C, Giunti C, Lesgourgues J, Ludhova L, Mertens S et al (2020) Status of light sterile neutrino searches. Prog Part Nucl Phys 111:103736. arXiv:1906.01739
360. Majorana E (1937) Teoria simmetrica dell'elettrone e del positrone, Il Nuovo Cimento (1924–1942) 14:171–184
361. Klein E (2000) Ettore majorana, le génial disparu, La Recherche 62–63
362. Maki Z, Nakagawa M, Sakata S (1962) Remarks on the unified model of elementary particles. Prog Theor Phys 28:870–880
363. Pontecorvo B (1957) Inverse beta processes and nonconservation of lepton charge. Zh Eksp Teor Fiz 34:247
364. Esteban I, Gonzalez-Garcia M, Hernandez-Cabezudo A, Maltoni M, Schwetz T (2019) Global analysis of three-flavour neutrino oscillations: synergies and tensions in the determination of θ_{23}, δ_{CP}, and the mass ordering. JHEP 01:106. arXiv:1811.05487
365. KATRIN collaboration, Aker M et al (2019) Improved upper limit on the neutrino mass from a direct kinematic method by KATRIN. Phys Rev Lett 123:221802. arXiv:1909.06048
366. Dolinski MJ, Poon AW, Rodejohann W (2019) Neutrinoless double-beta decay: status and prospects. Ann Rev Nucl Part Sci 69:219–251. arXiv:1902.04097
367. Yèche C, Palanque-Delabrouille N, Baur J, du Mas des Bourboux H (2017) Constraints on neutrino masses from Lyman-alpha forest power spectrum with BOSS and XQ-100. JCAP 06:047. arXiv:1702.03314
368. A. Argüelles C et al (2022) Snowmass white paper: beyond the standard model effects on neutrino flavor. In: 2022 snowmass summer study, p 3. arXiv:2203.10811
369. Lesgourgues J, Mangano G, Miele G, Pastor S (2013) Neutrino cosmology. Cambridge University Press, p 2
370. Abazajian KN et al, Synergy between cosmological and laboratory searches in neutrino physics: a white paper. arXiv:2203.07377
371. Quílez Lasanta P (2019) New dynamics in axions and flavor. PhD thesis, U. Autonoma, Madrid (main)
372. Froggatt C, Nielsen HB (1979) Hierarchy of quark masses, cabibbo angles and CP violation. Nucl Phys B 147:277–298
373. King S, Ross GG (2001) Fermion masses and mixing angles from SU(3) family symmetry. Phys Lett B 520:243–253. arXiv:hep-ph/0108112
374. Altarelli G, Feruglio F (2010) Discrete flavor symmetries and models of neutrino mixing. Rev Mod Phys 82:2701–2729. arXiv:1002.0211
375. King SF, Luhn C (2013) Neutrino mass and mixing with discrete symmetry. Rep Prog Phys 76:056201. arXiv:1301.1340
376. Grossman Y, Neubert M (2000) Neutrino masses and mixings in nonfactorizable geometry. Phys Lett B 474:361–371. arXiv:hep-ph/9912408
377. Gherghetta T, Pomarol A (2000) Bulk fields and supersymmetry in a slice of AdS. Nucl Phys B 586:141–162. arXiv:hep-ph/0003129
378. Huber SJ (2003) Flavor violation and warped geometry. Nucl Phys B 666:269–288. arXiv:hep-ph/0303183
379. Agashe K, Perez G, Soni A (2005) Flavor structure of warped extra dimension models. Phys Rev D 71:016002. arXiv:hep-ph/0408134
380. Csaki C, Falkowski A, Weiler A (2008) The flavor of the composite pseudo-goldstone higgs. JHEP 09:008. arXiv:0804.1954
381. Casagrande S, Goertz F, Haisch U, Neubert M, Pfoh T (2008) Flavor physics in the randall-sundrum model: I. Theoretical setup and electroweak precision tests. JHEP 10:094. arXiv:0807.4937
382. Bauer M, Casagrande S, Haisch U, Neubert M (2010) Flavor physics in the randall-sundrum model: II. Tree-level weak-interaction processes. JHEP 09:017. arXiv:0912.1625
383. Babu K (2008) TASI lectures on flavor physics. In: Proceedings of theoretical advanced study institute in elementary particle physics on the dawn of the LHC era (TASI 2008). Boulder. USA, pp 49–123. arXiv:0910.2948, https://doi.org/10.1142/9789812838360_0002

384. Isidori G, Nir Y, Perez G (2010) Flavor physics constraints for physics beyond the standard model. Ann Rev Nucl Part Sci 60:355. arXiv:1002.0900
385. Grinstein B (2016) Lectures on flavor physics and CP violation. In: Proceedings, 8th CERN–Latin-American school of high-energy physics (CLASHEP2015). Ibarra, Ecuador, pp 43–84. arXiv:1701.06916, https://doi.org/10.5170/CERN-2016-005.43
386. Grossman Y, Tanedo P (2016) Just a taste: lectures on flavor physics. In: Proceedings, theoretical advanced study institute in elementary particle physics : anticipating the next discoveries in particle physics (TASI 2016). Boulder, CO, USA, pp 109–295. arXiv:1711.03624, https://doi.org/10.1142/9789813233348_0004
387. Altmannshofer W, Zupan J (2022) Snowmass white paper: flavor model building. In: 2022 snowmass summer study, p 3. arXiv:2203.07726
388. Chauhan G, Dev PSB, Dziewit B, Flieger W, Gluza J, Grzanka K et al (2022) Discrete flavor symmetries and lepton masses and mixings. In: 2022 snowmass summer study, p 3. arXiv:2203.08105
389. Nelson AE (1984) Naturally weak CP violation. Phys Lett B 136:387–391
390. Barr SM (1984) Solving the strong CP problem without the peccei-quinn symmetry. Phys Rev Lett 53:329
391. Barr SM (1984) A natural class of nonpeccei-quinn models. Phys Rev D 30:1805
392. Bento L, Branco GC, Parada PA (1991) A minimal model with natural suppression of strong CP violation. Phys Lett B 267:95–99
393. Peccei RD, Quinn HR (1977) CP conservation in the presence of instantons. Phys Rev Lett 38:1440–1443
394. Peccei RD, Quinn HR (1977) Constraints imposed by CP conservation in the presence of instantons. Phys Rev D 16:1791–1797
395. Wilczek F (1978) Problem of strong P and T invariance in the presence of instantons. Phys Rev Lett 40:279–282
396. Weinberg S (1978) A new light Boson? Phys Rev Lett 40:223–226
397. Preskill J, Wise MB, Wilczek F (1983) Cosmology of the invisible axion. Phys Lett B 120:127–132
398. Abbott LF, Sikivie P (1983) A cosmological bound on the invisible axion. Phys Lett B 120:133–136
399. Dine M, Fischler W (1983) The not so harmless axion. Phys Lett B 120:137–141
400. Marsh DJE (2016) Axion cosmology. Phys Rep 643:1–79. arXiv:1510.07633
401. Hook A (2019) TASI lectures on the strong CP problem and axions. PoS TASI2018:004. arXiv:1812.02669
402. Irastorza IG, Redondo J (2018) New experimental approaches in the search for axion-like particles. Prog Part Nucl Phys 102:89–159. arXiv:1801.08127
403. Di Luzio L, Giannotti M, Nardi E, Visinelli L, The landscape of QCD axion models. arXiv:2003.01100
404. Kim JE (1985) A composite invisible axion. Phys Rev D 31:1733
405. Choi K, Kim JE (1985) Dynamical axion. Phys Rev D 32:1828
406. Weinberg S (1976) Implications of dynamical symmetry breaking. Phys Rev D 13:974–996
407. Susskind L (1979) Dynamics of spontaneous symmetry breaking in the weinberg-salam theory. Phys Rev D 20:2619–2625
408. Dimopoulos S, Preskill J (1982) Massless composites with massive constituents. Nucl Phys B 199:206–222
409. Redi M, Sato R (2016) Composite accidental axions. JHEP 05:104. arXiv:1602.05427
410. Gavela MB, Ibe M, Quilez P, Yanagida TT (2019) Automatic peccei-quinn symmetry. Eur Phys J C 79:542. arXiv:1812.08174
411. Georgi HM, Hall LJ, Wise MB (1981) Grand unified models with an automatic peccei-quinn symmetry. Nucl Phys B 192:409–416
412. Dine M, Seiberg N (1986) String theory and the strong CP problem. Nucl Phys B 273:109–124
413. Kamionkowski M, March-Russell J (1992) Planck scale physics and the Peccei-Quinn mechanism. Phys Lett B 282:137–141. arXiv:hep-th/9202003

414. Ghigna S, Lusignoli M, Roncadelli M (1992) Instability of the invisible axion. Phys Lett B 283:278–281
415. Holman R, Hsu SDH, Kephart TW, Kolb EW, Watkins R, Widrow LM (1992) Solutions to the strong CP problem in a world with gravity. Phys Lett B 282:132–136. arXiv:hep-ph/9203206
416. Barr SM, Seckel D (1992) Planck scale corrections to axion models. Phys Rev D 46:539–549
417. Dobrescu BA (1997) The Strong CP problem versus Planck scale physics. Phys Rev D 55:5826–5833. arXiv:hep-ph/9609221
418. Giddings SB, Strominger A (1988) Axion Induced Topology Change in Quantum Gravity and String Theory. Nucl Phys B 306: 890–907. https://doi.org/10.1016/0550-3213(88)90446-4
419. Lee K-M (1988) Wormholes and Goldstone Bosons. Phys Rev Lett 61:263–266. https://doi.org/10.1103/PhysRevLett.61.263
420. Bachlechner TC, Long C, McAllister L (2016) Planckian Axions and the Weak Gravity Conjecture. J High Energy Phys 01:091. https://doi.org/10.1007/JHEP01(2016)091. arXiv:1503.07853
421. Montero M, Uranga AM, Valenzuela I (2015) Transplanckian axions!? J High Energy Phys 08:032. https://doi.org/10.1007/JHEP08(2015)032. arXiv:1503.03886
422. Hebecker A, Mangat P, Theisen S, Witkowski LT (2017) Can Gravitational Instantons Really Constrain Axion Inflation? J High Energy Phys 02:097. https://doi.org/10.1007/JHEP02(2017)097. arXiv:1607.06814
423. Alonso R, Urbano A (2019) Wormholes and masses for Goldstone bosons. J High Energy Phys 02:136. https://doi.org/10.1007/JHEP02(2019)136. arXiv: 1706.07415
424. Abbott LF, Wise MB (1989) Wormholes and Global Symmetries. Nucl Phys B 325: 687–704. https://doi.org/10.1016/0550-3213(89)90503-8
425. Kallosh R, Linde AD, Linde DA, Susskind L (1995) Gravity and global symmetries. Phys Rev D 52:912–935. https://doi.org/10.1103/PhysRevD.52.912. arXiv:hep-ph/9502069
426. Kim JE (1981) Invisible Axion and Neutrino Oscillation in SU(11). Phys Rev D 24:3007. https://doi.org/10.1103/PhysRevD.24.3007
427. Georgi HM, Hall LJ, Wise MB (1981) Grand Unified Models With an Automatic Peccei-Quinn Symmetry. Nucl Phys B 192:409–416. https://doi.org/10.1016/0550-3213(81)90433-8
428. Holman R, Hsu SDH, Kephart TW, Kolb EW, Watkins R, Widrow LM (1992) Solutions to the strong CP problem in a world with gravity. Phys Lett B 282:132–136. https://doi.org/10.1103/10.1016/0370-2693(92)90491-L. arXiv: hep-ph/9203206
429. Babu KS, Gogoladze I, Wang K (2003) Stabilizing the axion by discrete gauge symmetries. Phys. Lett. B 560–222. https://doi.org/10.1016/S0370-2693(03)00411-8. arXiv: hep--ph/0212339
430. Dias AG, Pleitez V, Tonasse MD (2003) Naturally light invisible axion in models with large local discrete symmetries. Phys. Rev. D 67: 095008. https://doi.org/10.1103/PhysRevD.67.095008. arXiv: hep-ph/0211107
431. Harigaya K, Ibe M, Schmitz K, Yanagida TT (2013) Peccei-Quinn symmetry from a gauged discrete R symmetry. Phys. Rev. D 88: 075022. https://doi.org/10.1103/PhysRevD.88.075022. arXiv: 1308.1227
432. Dias AG, Machado ACB, Nishi CC, Ringwald A, Vaudrevange P (2014) The Quest for an Intermediate-Scale Accidental Axion and Further ALPs. JHEP 06:037. https://doi.org/10.1007/JHEP06(2014)037. arXiv: 1403.5760
433. Kim JE, Marsh DJE (2016) An ultralight pseudoscalar boson. Phys. Rev. D. 93: 025027 https://doi.org/10.1103/PhysRevD.93.025027. arXiv: 1510.01701
434. Di LL, Nardi E, Ubaldi L (2017) Accidental Peccei-Quinn symmetry protected to arbitrary order. Phys. Rev. D. 119: 011801. arXiv: 1704.01122, https://doi.org/10.1103/PhysRevLett.119.011801
435. Fukuda H, Ibe M, Suzuki M, Yanagida TT (2017) A "gauged" $U(1)$ Peccei–Quinn symmetry. Phys. Lett. B. 771: 327–331. arXiv: 1703.01112, https://doi.org/10.1016/j.physletb.2017.05.071

436. Butter D, Gaillard MK (2005) The Axion mass in modular invariant supergravity. Phys. Lett. B. 612: 304–310. arXiv: hep-th/0502100, https://doi.org/10.1016/j.physletb.2005.03.024
437. Kim JE (1985) A composite invisible axion. Phys. Rev. D. 31: 1733 https://doi.org/10.1103/PhysRevD.31.1733
438. Choi K, Kim JE (1985) Dynamical axion. Phys. Rev. D. 32: 1828 https://doi.org/10.1103/PhysRevD.32.1828
439. Randall L (1992) Composite axion models and Planck scale physics. Phys. Lett. B. 284: 77–80 https://doi.org/10.1016/0370-2693(92)91928-3
440. Barr SM, Seckel D (1992) Planck scale corrections to axion models. Phys. Rev. D. 46: 539–549 https://doi.org/10.1103/PhysRevD.46.539
441. Dobrescu BA (1997) The Strong CP problem versus Planck scale physics. Phys. Lett. B. 55: 5826–5833. arXiv: hep-ph/9609221, https://doi.org/10.1103/PhysRevD.55.5826
442. Fukuda H, Ibe M, Suzuki M, Yanagida, TT (2018) Gauged Peccei-Quinn symmetry — A case of simultaneous breaking of SUSY and PQ symmetry. JHEP. 07: 128. arXiv: 1803.00759, https://doi.org/10.1016/10.1007/JHEP07(2018)128
443. Ibe M, Suzuki M, Yanagida TT. (2018) $B - L$ as a Gauged Peccei-Quinn Symmetry. JHEP. 08: 049. arXiv: 1805.10029, https://doi.org/10.1016/10.1007/JHEP08(2018)049
444. Lillard B, Tait TMP (2018) A High Quality Composite Axion. JHEP. 11: 199.arXiv: 1811.03089, https://doi.org/10.1007/JHEP11(2018)199
445. Redi M, Sato R (2016) Composite Accidental Axions. JHEP. 05: 104. arXiv: 1602.05427, https://doi.org/10.1007/JHEP05(2016)104
446. Gavela MB, Ibe M, Quilez P, Yanagida, TT (2019) Automatic Peccei–Quinn symmetry. Eur. Phys. J. C. 79: 542.arXiv: 1812.08174, https://doi.org/10.1140/epjc/s10052-019-7046-3
447. Holdom B (1986) Two U(1)'s and Epsilon Charge Shifts. Phys Lett B 166: 196–198. https://doi.org/10.1016/0370-2693(86)91377-8
448. Holdom B (1982) Raising the Axion Mass. Nucl. Phys. B 208: 397–412. https://doi.org/10.1016/0550-3213(82)90228-0
449. Dine M, Seiberg, N (1986) String Theory and the Strong CP Problem. Nucl. Phys. B 273: 109–124. https://doi.org/10.1016/0550-3213(86)90043-X
450. Holdom B(1985) Strong QCD at High-energies and a Heavy Axion. Phys. Lett. B 154:316. https://doi.org/10.1016/0370-2693(85)90371-5
451. Flynn JM, Randall L (1987) A Computation of the Small Instanton Contribution to the Axion Potential. Nucl. Phys. B 293: 731–739. https://doi.org/10.1016/0550-3213(87)90089-7
452. Agrawal P, Howe K (2018) Factoring the Strong CP Problem. JHEP 12: 029. arXiv: hep--ph/1710.04213, https://doi.org/10.1007/JHEP12(2018)029
453. A Flavorful Factoring of the Strong CP Problem. JHEP 12: 035. arXiv:hep-ph/1712.05803, https://doi.org/10.1007/JHEP12(2018)035
454. Csáki C, Ruhdorfer M, Shirman Y (2020) UV Sensitivity of the Axion Mass from Instantons in Partially Broken Gauge Groups. JHEP 04: 031. arXiv:hep-ph/1912.02197, https://doi.org/10.1007/JHEP04(2020)031
455. Rubakov VA (1997) Grand unification and heavy axion. JETP Lett. 65: 621–624. arXiv: hep-ph/9703409, https://doi.org/10.1134/1.567390
456. Berezhiani Z, Gianfagna L, Giannotti M (2001) Strong CP problem and mirror world: The Weinberg-Wilczek axion revisited. Phys. Lett. B 500: 286–296. arXiv: hep-ph/0009290, https://doi.org/10.1016/S0370-2693(00)01392-7
457. Hook A (2015) Anomalous solutions to the strong CP problem. Phys. Rev. Lett. 114:141801. arXiv: hep-ph/1411.3325, https://doi.org/10.1103/PhysRevLett.114.141801
458. Fukuda H, Harigaya K, Ibe M, Yanagida TT (2015) Model of visible QCD axion. Phys. Rev. D 92: 015021. arXiv: hep-ph/1504.06084, https://doi.org/10.1103/PhysRevD.92.015021
459. Dimopoulos S, Hook A, Huang J, Marques-Tavares G (2016) A collider observable QCD axion. JHEP 11: 052. arXiv: hep-ph/1606.03097, https://doi.org/10.1007/JHEP11(2016)052
460. Hook A, Kumar S, Liu Z, Sundrum R (2020) High Quality QCD Axion and the LHC. Phys. Rev. Lett. 124: 221801. arXiv: hep-ph/1911.12364, https://doi.org/10.1103/PhysRevLett.124.221801

461. Dimopoulos S (1979) A Solution of the Strong CP Problem in Models With Scalars. Phys. Lett. B. 84: 435–439. https://doi.org/10.1016/0370-2693(79)91233-4
462. Gherghetta T, Nagata N, Shifman M (2016) A Visible QCD Axion from an Enlarged Color Group. Phys. Rev. D 93: 115010. https://doi.org/10.1103/PhysRevD.93.115010
463. Gherghetta T, Nguyen MD (2020) A Composite Higgs with a Heavy Composite Axion. JHEP 12: 094. arXiv: hep-ph/2007.10875, https://doi.org/10.1007/JHEP12(2020)094
464. Gherghetta T, Khoze VV, Pomarol A, Shirman Y (2020) The Axion Mass from 5D Small Instantons. JHEP 03: 063. arXiv: hep-ph/2001.05610, https://doi.org/10.1007/JHEP03(2020)063
465. Fischler W, Preskill J (1983) Dyon - Axion Dynamics. Phys. Lett. B 125: 165–170. https://doi.org/10.1016/0370-2693(83)91260-1
466. Fan J, Fraser K, Reece M, Stout J (2021) Axion Mass from Magnetic Monopole Loops. Phys. Rev. Lett. 127: 131602. arXiv: hep-ph/2105.09950, https://doi.org/10.1103/PhysRevLett.127.131602
467. Bauer M, Neubert M, Thamm A (2017) Collider Probes of Axion-Like Particles. JHEP 12:044. arXiv: hep-ph/1708.00443, https://doi.org/10.1007/JHEP12(2017)044
468. Bauer M, Heiles M, Neubert M, Thamm A (2019) Axion-Like Particles at Future Colliders. Eur. Phys. J. C 79:74. arXiv: hep-ph/1808.10323, https://doi.org/10.1140/epjc/s10052-019-6587-9
469. Cid VX, Mariotti A, Redigolo D, Sala F, Tobioka K (2019) New Axion Searches at Flavor Factories. JHEP 01:113. arXiv: hep-ph/1810.09452, https://doi.org/10.1007/JHEP01(2019)113
470. Zambujal FR, Notari A, Pujolàs O, Rompineve F (2022) High Quality QCD Axion at Gravitational Wave Observatories. Phys. Rev. Lett. 128: 141101. arXiv: hep-ph/2107.07542, https://doi.org/10.1103/PhysRevLett.128.141101
471. Ferreira RZ, Notari A, Pujolas O, Rompineve F (2022) Gravitational Waves from Domain Walls in Pulsar Timing Array Datasets. arXiv:2204.04228

Chapter 3
Standard Model of Cosmology

Humanity's understanding of the universe has evolved significantly over time. During the major part of History, Earth was regarded as the center of all things, with planets and stars orbiting around it. A profound shift in the understanding of the cosmos was made by Nicolaus Copernicus who suggested that Earth and the other planets in the solar system in fact orbited around the sun. The legend said he released the final version of his book in 1543 on the day of his death [1]. Despite waves of opposition, the heliocentric view of the solar system was finally accepted after the year 1687 when Isaac Newton formulated the law of gravitation to unify terrestrial and celestial mechanics [2]. An important step in the history of astronomy was made by Charles Messier with the release in 1781 of a catalog of 100 nebulas and star clusters visible from Paris [3]. During the next century, in 1864, the catalog got enlarged to 5000 objects including the ones visible from the south hemisphere, by Wiliams Hershell, his sister Carolina and his son John [4]. At that time the study of the universe was restricted to the observation and description of astrophysical objects. Any conjecture on the fundamental origin of those structures was speculative and metaphysical. Without any doubt, the greatest triumph was made in 1915 when Einstein formulated the theory of general relativity [5, 6]. Before Einstein, space and time were regarded as immutable coordinates introduced to describe the "real" physical objects. The universe itself was not part of the physical world, instead its role was restricted to be the container of the physical world. The masterstroke of Einstein was to understand that this container, space-time, is in fact a physical object subject to its own dynamics like any other. After 1915, it became possible to formulate theoretical models describing the past history of the universe. This task was conducted between 1922 and 1927 by Friedmann and Lemaitre, followed in 1935 by Robertson and Walker [7–16].

In parallel to the theoretical work, three important observations were made by astronomers. First, in 1912 Henrietta Leavitt discovered the relation between the luminosity and the period of Cepheid variables, providing a new way to measure the distance of faraway galaxies [17]. Second, the same year Vesto Slipher measured

Y. Gouttenoire, *Beyond the Standard Model Cocktail*, Springer Theses,
https://doi.org/10.1007/978-3-031-11862-3_3

galactic spectral redshifts which revealed anomalously large recessional velocities [18, 19]. Finally, in 1929 Edwin Hubble used the two previous discoveries to show that the universe is expanding according to Friedmann and Lemaitre prediction, and measured the constant ratio between the recession speed and the redshift, which is today known as the Hubble constant [20].

Upon reversing the arrow of time, the expansion measured by Hubble unavoidably predicts the universe to have started from a dense and hot state, this is the hot big-bang scenario. The term was coined in 1948 by a firm opponent of this theory, Fred Hoyle, who proposed the same year, in parallel of Bondi and Gold, that the universe is instead in a steady state [21, 22]. Fred Hoyle has never accepted the hot big-bang scenario but instead he has defended the steady state proposal until his death in 2001 [23–26].

However, arguments in favour of the big-bang theory became more and more difficult to discredit. In 1948, Alpher, Bethe, Gamow and Herman showed how a hot big-bang would create hydrogen, helium and heavier elements in the correct proportions to explain their abundance in the early universe [27–30]. In 1965, Robert Dicke, Jim Peebles and David Wilkinson at Princeton predicted that if the universe would have started with a hot big-bang, then the universe should be filled with microwave radiation [31]. The discovery by inadvertence of this cosmic microwave background, just 60 km away in Bell Labs the same year, by Arno Penzias and Robert Wilson [32], convinced most cosmologists that the hot big-bang model is the best explanation for the origin of the universe.

This new framework brings together the two most fundamental theories of Nature, General Relativity and Quantum Field Theory. In particular, the study of particle physics can be viewed as a proxy to study the early universe, and particle accelerators can be viewed as a place where the physical conditions of the early universe are recreated. The unification of electromagnetism and weak interactions by Glashow in 1961 [33], and Weinberg and Salam in 1967 [34, 35], implies the existence of a phase transition, taking place 10^{-12} s after the big-bang, during which the Higgs field acquires a vacuum expectation value. The discovery of asymptotic freedom by Gross, Wilczek and Politzer in 1973 [36, 37] implies that quarks were propagating freely before 10^{-7} s.

In 1980, an important breakthrough was made by Guth [38] who introduced the inflation paradigm to explain the flatness and homogeneity of the universe. Thanks to the works of, among others, Linde, Steinhardt, Starobinsky and Mukhanov [39–44], in the following years it became possible to relate the existence of the large scale structures of the universe to quantum fluctuations produced during an inflationary phase preceding the hot big-bang.

Since then, the interest in cosmology has blossomed due to the multiplication of the observational evidences for dark matter [45–50], and to the direct evidence in 1998 for the acceleration of the universe expansion from supernovae data [51, 52], but also thanks to impressive development in the science of telescopes. Today, cosmology has entered an era of precision. The Planck mission from 2009 to 2013, together with other probes such as Type-1a supernovae, Large Scale Structures (LSS), weak gravitational lensing and Lyman-α forests, have measured cosmological parameters like the abundance of dark matter or the slow-roll parameter below the percent level, or the sound horizon size below one part per 10^4. The 2nd Gaia data release in 2018, has measured the celestial position and apparent brightness for 1.7 billions of sources, with parallaxes and proper motions available for 1.3 billions of those [53]. The third gravitational-wave transcient catalogue released by LIGO-Virgo-Kagra Collaboration contains 90 compact binary merging events [54–56].

The near future of cosmology is really exciting. In the next decade, Planck successor Litebird in space [57–60] or CMB-S4 on ground [61–66] could detect primordial B-modes and measure the inflation scale, or reveal the existence of light relics contributing to N_{eff} [67]. Galaxy surveys like Euclid [68, 69], LSST [70–72] or SKA [73, 74] will measure the statistical distribution of LSS up to redshift $z \sim 3$ with such precision that theoretical uncertainties on non-linear scales will become the main obstacles [75]. 21-cm intensity mapping with SKA might permit to trace LSS back to the reionization era and even the cosmic dawn, up to redshift $z \sim 20$ [75–78]. One will be able to probe the cosmological properties of Dark Energy, Dark Matter [79, 80], or neutrinos with an unprecedented level of precision, possibly revealing deviation from ΛCDM. Planned multi-TeV cosmic-ray telescopes CTA [81–86], LHAASO [87] or KM3NeT [88, 89] will strengthen the constraints on WIMP dark matter. The other side of the mass spectrum, the so-called low energy frontier of particle physics [90, 91], will continue being probed by new generations of haloscope, helioscopes, light-shining-through the wall experiments [92–95], fifth force experiments [96], atomic clocks [97], and all sorts of astrophysical observables [98, 99]. In a bit more than a decade, GW interferometers LISA in space [100–106] or ET [107–110] and CE [111, 112] on ground will detect thousands of astrophysical sources, and will search for traces of gravitational waves of primordial origins which will open a new avenue of investigation of early universe cosmology.

In this chapter, we introduce the ΛCDM model (Sect. 3.1), the hot big-bang scenario (Sect. 3.2), the inflation paradigm (Sect. 3.3), and we discuss possible sources of primordial gravitational waves (Sect. 3.4). Next, we review the three major puzzles in cosmology, which are the cosmological constant problem (Sect. 3.5.1), the matter-antimatter asymmetry (Sect. 3.5.2) and the dark matter mystery (Sect. 3.5.3). Finally, we close the chapter by presenting three hot open problems which are the fragility of ΛCDM (Sect. 3.5.4), the Hubble tension (Sect. 3.5.5) and the 21 cm anomaly (Sect. 3.5.6).

For textbooks on general relativity and cosmology, we refer the reader to [113–156], and [157–186], respectively. For books on the historical development of general relativity and cosmology, we call attention to [162, 187–192]. We also mention various astrophysical complements, e.g. general astrophysics [193–201], galaxy formation [202–206] and dynamics [207, 208], stars formation [209–211, 211–213] and dynamics [214–217], black holes astrophysics [218–227], plasma astrophysics [228–233, 233–243], and astroparticles [244–258].

3.1 The ΛCDM Cosmological Model

3.1.1 A Homogeneous and Isotropic Expanding Universe

At large scales, the distribution of matter and radiation in the universe appears to be **homogeneous, isotropic** and in **expansion**. This suggests that the geometry of our universe is given by the Friedmann-Lemaitre-Robertson-Walker (FLRW) metric[1]

$$ds^2 = dt^2 - a^2(t)\left[\frac{dr^2}{1-kr^2} + r^2 d\theta^2 + r^2 \sin^2\theta\, d\phi^2\right], \tag{3.1}$$

where a is the scale factor and $k = -1,\ 0,\ +1$ for **open, flat** and **closed** universes. The evolution of the geometry of space-time is controlled by the distribution of energy-momentum within it, as stated by the Einstein equation of **General Relativity** (1915 [5, 6])

$$G_{\mu\nu} = \frac{1}{M_{\rm pl}^2} T_{\mu\nu}, \tag{3.2}$$

where $G_{\mu\nu}$ and $T_{\mu\nu}$ are the Einstein and energy-momentum tensors and $M_{\rm pl} \simeq 2.44 \times 10^{18}$ GeV is the reduced Planck mass.[2] The Einstein tensor can be built starting from the metric $g_{\mu\nu}$, by means of the Ricci tensor $R_{\mu\nu}$, the Ricci scalar R, the Riemann tensor $R^{\mu}_{\alpha\beta\gamma}$ and the Christoffel symbols $\Gamma^{\mu}_{\alpha\beta}$, e.g. [114, 124]

$$G_{\mu\nu} = R_{\mu\nu} - \frac{1}{2} g_{\mu\nu} R, \tag{3.3}$$

$$R_{\mu\nu} = R^{\alpha}_{\mu\alpha\nu}, \tag{3.4}$$

$$R = g^{\mu\nu} R_{\mu\nu}, \tag{3.5}$$

$$R^{\mu}_{\alpha\beta\gamma} = \partial_\beta \Gamma^{\mu}_{\alpha\gamma} - \partial_\gamma \Gamma^{\mu}_{\alpha\beta} + \Gamma^{\mu}_{\sigma\beta}\Gamma^{\sigma}_{\alpha\gamma} - \Gamma^{\mu}_{\sigma\gamma}\Gamma^{\sigma}_{\alpha\beta}, \tag{3.6}$$

$$\Gamma^{\mu}_{\alpha\beta} = \frac{1}{2} g^{\mu\sigma}\left(\partial_\alpha g_{\sigma\beta} + \partial_\beta g_{\alpha\sigma} - \partial_\sigma g_{\alpha\beta}\right). \tag{3.7}$$

From injecting the FLRW metric in Eq. (3.1) in the Einstein equation in Eq. (3.2), we find that the dynamics of the universe is governed by the **Friedmann equation** (1922),

$$H^2 + \frac{k}{a^2} = \frac{\rho}{3M_{\rm pl}^2}, \qquad H = \frac{\dot a}{a}, \tag{3.8}$$

where H is the Hubble constant and $\rho = \rho_M + \rho_R + \rho_\Lambda$ is the total energy density, sum of a matter, radiation and vacuum component. Assuming that the different fluids do not interact, the conservation of the energy-momentum of each individual component i in the FLRW universe $\partial_\mu T_{\mu\nu} = 0$ leads to the **continuity equation**

[1] The FLRW metric is a result of various works between 1922 and 1937 [7–16].

[2] The reduced Planck mass $M_{\rm pl}$ is related to the Newton gravitation constant G through $M_{\rm pl}^2 \equiv 1/8\pi G$.

$$\dot{\rho}_i + 3H(\rho_i + p_i) = 0. \tag{3.9}$$

It encodes how each fluid evolves in the expanding universe

$$\rho_i \propto a^{-3(\omega_i+1)}, \qquad \omega_i \equiv \frac{p_i}{\rho_i}, \tag{3.10}$$

with ω_i being the equation of state, $\omega_i = 1,\ \frac{1}{3},\ 0,\ -\frac{1}{3},\ -1$ for kination, radiation, matter, curvature and vacuum energy, respectively. From injecting Eq. (3.10) into the Friedmann equation in Eq. (3.8), we obtain

$$a(t) \propto t^{2/3(1+\omega)}. \tag{3.11}$$

We can define the fraction of energy density $\Omega_i = \rho_i/3M_{\rm pl}^2 H^2$, whose sum, Ω, is respectively equal to < 1, 1, > 1 for open, flat and closed universes

$$\Omega = \frac{\rho}{M_{\rm pl}^2 H^2} = \sum_i \Omega_i = 1 + \frac{k}{(a\,H)^2}. \tag{3.12}$$

3.1.2 *Energy Content of the Universe*

The observation of the Cosmic Microwave Background (CMB) with the telescope **Planck**, between 2009 and 2013, together with low-redshift probes (mainly the luminosities of Type Ia Supernova (SNe Ia) and the angular scale of Baryonic Acoustic Oscillations, have allowed the precise measurement of the energy content of the universe (per order of dominance) [259, 260]

$$\Omega_\Lambda = 0.685(7), \tag{3.13}$$

$$\Omega_{\rm DM} = 0.1200(12)\,h^{-2} = 0.265(7), \tag{3.14}$$

$$\Omega_b = 0.02237(15)\,h^{-2} = 0.0493(6), \tag{3.15}$$

$$\Omega_\nu = h^{-2}\sum m_{\nu_j}/93.14\,\text{eV} \quad < 0.003\ (\text{CMB} + \text{BAO}) \quad > 0.0012\ (\text{mixing}). \tag{3.16}$$

$$\Omega_\gamma = 2.473 \times 10^{-5}\,h^{-2} = 5.38(15) \times 10^{-5}, \qquad T_\gamma = 2.7255(6)\ \text{Kelvin}, \tag{3.17}$$

where Ω_Λ is the **Dark Energy** (vacuum energy), $\Omega_{\rm DM}$ is the **Dark Matter** (non-baryonic cold matter), Ω_b are the **baryons**, Ω_ν are the **neutrinos** and Ω_γ are the **photons**.

The Hubble parameter, which quantifies the expansion rate of the universe, is measured by Planck satellite to be [259, 260]

$$H_0 = 100\,h\ \text{km/s/Mpc} \qquad \text{with} \quad h = 0.674(5). \tag{3.18}$$

The Dark Energy is a fluid with **constant energy density** and **negative pressure** $\rho_\Lambda = -p_\Lambda =$ constant, responsible for the acceleration of the expansion of the universe, as confirmed in 1998 with SNe Ia measurements [261, 262]. The Dark Matter is a fluid which redshifts as **cold matter** $\rho_{\rm DM} \propto a^{-3}$.

Finally, the trigonometry in our universe appears to be close to **euclidean**, implying that the curvature is close to **flat** [259, 260][3]

$$\Omega_{\rm K} = 0.0007 \pm 0.0019. \tag{3.20}$$

3.2 The Hot Big-Bang Scenario

3.2.1 *Thermal Equilibrium*

The expansion of the universe implies that if we go backward in times the universe is contracting, and the temperature increases. The Einstein equation even predicts that the size of the universe $a(t)$, cf. Eq. (3.11), goes to zero when $t \to 0$, implying that the temperature becomes infinite. This is the **Hot Big-Bang Scenario.**[4]

At that time the interaction rate of particles Γ is larger than the expansion rate and the particles are at **thermal equilibrium**. Therefore we can describe the thermodynamics quantities of the plasma using **statistical mechanics**. The number density of particles with degrees of freedom g is given by

$$n = \frac{g}{(2\pi)^3} \int d^3p\, f(\vec{p}), \tag{3.21}$$

where $f(\vec{p})$ is the average number of fermions/bosons in the single-particle state of momentum $\vec{p}$

$$f(\vec{p}) = \frac{1}{e^{(E-\mu)/T} \pm 1}, \tag{3.22}$$

---

[3] Note however that the CMB fit leading to Eq. (3.20) has an anomaly in the lensing amplitude $A_{\rm L}$ at 2.8 σ [259]

$$A_{\rm L} = 1.180 \pm 0.065. \tag{3.19}$$

Interestingly, $A_{\rm L}$ and $\Omega_{\rm K}$ are degenerate and the resolution of the lensing anomaly, $A_{\rm L} \to 1$, is possible if our universe is closed at 3.4 standard deviations, $-0.007 > \Omega_{\rm K} > -0.095$ [263] (See also [264, 265]).

[4] In the standard big-bang scenario, Einstein equations predicts a cosmic singularity at $a \to 0$ and $T \to \infty$. This is the **Cosmic singularity problem**. Alternatively, in the **big-bounce scenario**, the initial singularity is replaced by a contracting phase in which the scale factor shrinks to a critical finite size and bounces. However, the inherent difficulty of the big-bounce scenario is the violation of the Null-Energy-Condition in Eq. (3.149). See [266–271] for reviews on bouncing cosmology.

with + for **Fermi-Dirac** (FD) statistics (1926) [272, 273] and—for **Bose-Einstein** (BE) statistics (1924) [274]. T is the photon temperature and μ is the chemical potential. The chemical equilibrium for $i + j \leftrightarrow k + l$ is realized when $\mu_i + \mu_j = \mu_k + \mu_l$. A given particle X has the opposite chemical potential as its antiparticle $\bar{X}$, $\mu_X = -\mu_{\bar{X}}$. Therefore, in a symmetric universe where $n_X = n_{\bar{X}}$, we have $\mu_X = 0$. Assuming it is the case, the **number density** at equilibrium is

$$n \simeq \begin{cases} c\, g\, \frac{\zeta(3)}{\pi^2}\, T^3, & T \gtrsim m, \\ g \left(\frac{m\,T}{2\pi}\right)^{3/2} e^{-m/T}, & T \lesssim m, \end{cases} \tag{3.23}$$

where $\zeta(3) \simeq 1.2$, $c = 1$ for BE, and $c = 3/4$ for FD.

The next important thermodynamic quantity is the **energy density** of particles at thermal equilibrium, which reads

$$\rho = \frac{g}{(2\pi)^3} \int d^3p\; E\; f(\vec{p}) \simeq \begin{cases} \frac{\pi^2}{30}\, g_{\rm eff}\, T^4, & T \gtrsim m, \\ m\, n, & T \lesssim m, \end{cases} \tag{3.24}$$

where $g_{\rm eff}$ is the effective number of relativistic degrees of freedom,

$$g_{\rm eff} = \sum_{i=b} g_i \left(\frac{T_i}{T}\right)^4 + \frac{7}{8} \sum_{i=f} g_i \left(\frac{T_i}{T}\right)^4 . \tag{3.25}$$

We have introduced an additional factor $(T_i/T)^4$ in order to account for relativistic species that are not in thermal equilibrium with the photons.

Similarly we define the **pressure**

$$P = \frac{g}{(2\pi)^3} \int d^3p\; \frac{p^2}{3E}\; f(\vec{p}). \tag{3.26}$$

which for radiation is simply $P = \rho/3$.

Finally, we define the **entropy density**[5]

[5] Upon assuming an adiabatic evolution, the entropy of the universe at the big-bang epoch was carried by the 410 photons and 340 neutrinos per cm^3 of the observable universe today. This is actually 10^{-122} times lower than the maximal entropy of the universe, which is given by the generalization of the Bekeinstein-Hawking formula $S = A^2/4$ for black hole horizons [275, 276] to cosmological horizons [277–280]. In comparison, a high-entropy big-bang universe would be born full of black holes. The fraction of the whole phase space in which our universe sits is one part in $e^{10^{122}}$, which is extremelly fine-tuned if one assumes a uniform distribution. This is the **entropy problem** raised by Roger Penrose in 1988 [281, 282]. Inflation, see Sect. 3.3, can not offer a solution if the distribution of high-entropy gravitational remnants from big-bang is scale-invariant as one could expect [281]. This leads to the question of whether 'the initial conditions of the universe could have originated from a Poincare recurrence' [283].

$$s = \frac{\rho + P}{T} = \frac{2\pi^2}{45} h_{\text{eff}}\, T^3, \tag{3.27}$$

with

$$h_{\text{eff}} = \sum_{i=b} g_i \left(\frac{T_i}{T}\right)^3 + \frac{7}{8}\sum_{i=f} g_i \left(\frac{T_i}{T}\right)^3. \tag{3.28}$$

If the evolution of the universe is **adiabatic**, the total entropy is conserved

$$\text{adiab. evolution :} \qquad h_{\text{eff}}\, T^3\, a^3 = \text{constant}. \tag{3.29}$$

Therefore, the scale factor evolves as $a \propto h_{\text{eff}}^{-1/3} T^{-1}$. It is sometimes convenient to introduce the **redshift parameter** z

$$1 + z(t) \equiv a_0/a(t), \tag{3.30}$$

which increases with the temperature. During a radiation-dominated evolution, the time-to-temperature relation is given by

$$t = 2.42\ \text{s}\ h_{\text{eff}}^{-1/2} \left(\frac{1\ \text{MeV}}{T}\right)^2. \tag{3.31}$$

3.2.2 *Beyond Thermal Equilibrium*

In an expanding universe, the thermal equilibrium condition is not always satisfied. A given particle χ (e.g. Dark Matter, baryons, neutrinos..) interacting with the hot SM plasma with a rate Γ, is at thermal equilibrium when

$$\Gamma \gtrsim H. \tag{3.32}$$

In the SM, the interactions between SM particles are mediated by the gauge bosons of $SU(3)_c \times SU(2)_L \times U(1)_Y$. In the early universe, when the temperature is larger than the mass of the gauge bosons, $T \gtrsim m_A$, the cross-section can be estimated by

$$\sigma \sim \alpha^2/T^2 \tag{3.33}$$

where α is the typical structure constant of the gauge interaction. The number of interaction per unit of time is given by

$$\Gamma = n_\chi \sigma v_{\text{rel}} \sim \alpha^2\, T. \tag{3.34}$$

It must be compared with the expansion rate of the universe $H \sim T^2/M_{\rm pl}$. The plasma of particles in the SM are interacting faster than the universe expands when $\Gamma \gtrsim H$ with

$$\frac{\Gamma}{H} \sim \left(\frac{\alpha}{1/30}\right)^2 \frac{10^{15}\ \text{GeV}}{T}. \tag{3.35}$$

Hence, the SM particles enter a thermal equilibrium state around $T \sim 10^{15}$ GeV.

If the universe was a hot plasma in the past, it is not the case anymore today. As a consequence of the universe expansion, the particles stop interacting. Very often,[6] this occurs when the temperature reaches the mass threshold, $T \lesssim m_\chi$, the thermal excitations of the plasma become too weak to afford for the production of χ, the production rate drops exponentially with the temperature and the particle χ leaves the thermal equilibrium. Consequently, the total number of particles becomes frozen $n_\chi\, a^3 =$ constant. It is common to introduce the comoving number density $Y_\chi = n_\chi/s$ which is conserved under Hubble expansion. The residual χ abundance can be computed in the **instantaneous freeze-out** approximation

$$\Gamma \simeq H \quad \rightarrow \quad Y_{\rm FO} \simeq \frac{H}{s\ \langle\sigma v_{\rm rel}\rangle} \simeq 0.75\, \frac{g_{\rm eff}^{1/2}}{h_{\rm eff}}\, \frac{x_{\rm FO}}{M_\chi\ M_{\rm pl}\ \langle\sigma v_{\rm rel}\rangle} \tag{3.36}$$

with the temperature of departure from thermal equilibrium given by

$$x_{\rm FO} \equiv \frac{M_\chi}{T_{\rm FO}} \simeq \text{Log}\left[0.192 \frac{g_\chi}{\sqrt{g_{\rm SM}^{\rm FO}}}\ \text{M}_{\rm pl}\ \text{M}_{\rm DM}\ \langle\sigma \text{v}_{\rm rel}\rangle\ \text{x}_{\rm FO}^{1/2}\right] \sim 30. \tag{3.37}$$

More details on thermal freeze-out are given in Sect. 4.1.2 of Chap. 4.

Baryons:
As an application, we propose to compute the abundance of baryons. The baryon abundance today is set when the proton/anti-proton annihilation process freezes-out, and we can assume the proton-antiproton annihilation cross-section to be [158]

$$\sigma_{p\bar{p}}\, v_{\rm rel} = \frac{c_1}{m_\pi^2}, \tag{3.38}$$

where c_1 is a constant of order 1 and $m_\pi \simeq 135$ MeV is the pion mass. From Eq. (3.37), we get $x_{\rm FO} \simeq 49$ where we have plugged $g_p = 4$, $g_{\rm SM}^{\rm FO} = 10.75$ and $c_1 = 1$. From Eq. (3.36) we find

$$\eta_{\rm B}^{\rm th} = \left.\frac{n_b}{n_\gamma}\right|_{\rm th} \simeq 1.8 \times 10^{-19} + \left.\frac{n_b - n_{\bar{b}}}{n_\gamma}\right|_{\rm prim}. \tag{3.39}$$

[6] If particles are very light, like the neutrinos, the freeze-out can occur at a temperature much larger than its mass, cf. paragraph on neutrinos of the present section.

where $\sim 10^{-19}$ is the expected baryon abundance today if $n_{\rm b} = n_{\bar{\rm b}}$, whereas $\left.\frac{n_b - n_{\bar{b}}}{n_\gamma}\right|_{\rm prim}$ is the primordial baryon asymmetry (which can not annihilate). The baryon-to-photon[7] ratio is precisely inferred from the comparison between the predictions for the abundance of light element produced during the Big-Bang Nucleosynthesis (BBN) and the measurement of these abundances [285]

$$\eta_{\rm B}^{\rm exp} = \left.\frac{n_b}{n_\gamma}\right|_{\rm exp} \simeq 6.5 \times 10^{-10}. \tag{3.41}$$

From comparing the expected baryon abundance assuming $n_b = n_{\bar{b}}$ in Eq. (3.39) and the observed abundance in Eq. (3.41), we understand that it is necessary to introduce a **primordial baryon asymmetry**, given by Eq. (3.41), in order to match the observations. This is problematic since the observed asymmetry $b/\bar{b}$ can not be explained in the SM of elementary particles. We give more details in Sect. 3.5.2.

Neutrinos:
For very light particles like the neutrinos, thermal freeze-out can occur at a temperature much above the neutrino mass. After that the temperature becomes smaller than the $W^\pm$ gauge boson mass, the cross-section of neutrino annihilation in Eq. (3.33) becomes $\sigma \sim G_F^2 T^2$ with $G_F^{-1} = \sqrt{2}\, v^2$, such that the neutrino decoupling occurs around 1 MeV

$$\frac{\Gamma_\nu}{H} \sim \left(\frac{T_{\nu,\,\rm dec}}{1\ \rm MeV}\right)^3. \tag{3.42}$$

The neutrino being relativistic at the time of the decoupling, their abundance today reads

$$\Omega_\nu h^2 = \frac{\sum m_\nu}{3M_{\rm pl}^2 (H_0/h)^2} Y_\nu^{\rm dec}\, s_0 \simeq \frac{\sum m_\nu}{94.1(93.1)\ \rm eV} \tag{3.43}$$

where $g_\nu = \frac{3}{4} \cdot 2$ (only one neutrino flavor), $h_{\rm eff}^{\rm dec} = 2 + \frac{7}{8}(4+6)$ (photons, e^+/e^- and neutrinos) and s_0 is the entropy today $s_0 = \frac{2\pi^2}{45}\left(2 + \frac{7}{8} \cdot 2 \cdot 3 \cdot \frac{4}{11}\right) T_0^3$. The factor $(T_\nu/T_\gamma)^3 = 4/11$ results from the $e+/e-$ annihilation around $\sim 511/30$ keV, which heats the photon bath but not the neutrino bath.[8] The number in bracket in Eq. (3.43) accounts for the correction due to the slight heating of the neutrino bath resulting from e^+/e^- annihilation [286]. Effectively, this is the same as replacing the number of neutrino species $N_\nu = 3$ by $N_{\nu,\,\rm eff} = 3.045$.

---

[7] Note however that the lepton asymmetry η_ν is only poorly constrained by BBN and is at most [284]

$$\eta_\nu^{\rm exp} \lesssim 0.1. \tag{3.40}$$

[8] The conservation of the entropy during the $e+/e-$ annihilation assuming that the neutrinos have already completely decoupled reads $(2 + \frac{7}{8} \cdot 4) T_\nu^3 = 2 T_\gamma^3$.

Photons:
The photon decouples from the thermal plasma when the rate of **Compton scattering**

$$e^- + \gamma \longleftrightarrow e^- + \gamma, \qquad \Gamma_\gamma \simeq n_e \sigma_T, \qquad \sigma_T \simeq 2 \times 10^{-3}\,\text{MeV}^{-2}, \tag{3.44}$$

becomes comparable with the expansion rate of the universe

$$\Gamma_\gamma(T_{\text{dec}}) \sim H(T_{\text{dec}}). \tag{3.45}$$

Compton scattering becomes inefficient when the electrons **recombine** with the protons to form hydrogen atoms. The electron density n_e can be computed from the chemical equilibrium condition

$$\mu_{e^-} + \mu_{p^+} = \mu_H + \mu_\gamma, \qquad \left(\frac{n_H}{n_e^2}\right)_{\text{eq}} = \left(\frac{2\pi}{m_e T}\right)^{3/2} e^{B_H/T} \qquad B_H \simeq 13.6\,\text{eV}. \tag{3.46}$$

From injecting Eq. (3.46) into Eq. (3.44) and then into Eq. (3.45) and using the numerical value of η_{B} in Eq. (3.41), we obtain the temperature at which the photon decouples from the hot plasma

$$T_{\text{dec}} \simeq 0.27\,\text{eV}, \qquad t_{\text{dec}} \simeq 380\,000\,\text{yrs} \tag{3.47}$$

Below that temperature, the universe becomes transparent and photons propagate freely.

The observation of these photons today constitutes the Cosmic Microwave Background (CMB). The CMB was predicted in 1965 by Robert Dicke, Jim Peebles and David Wilkinson at Princeton [31] and discovered inadvertently, just 60 km away in Bell Labs, the same year, by Arno Penzias and Robert Wilson [32]. Since then, the CMB has been measured with high precision by the Planck satellite from 2009 to 2013, cf. Fig. 3.1. It is isotropic at the 10^{-5} level and, for a given direction in the sky, it follows the spectrum of a black-body corresponding to a temperature $T_\gamma = 2.7255$ K.

3.3 Inflation

3.3.1 The Homogeneity Problem

On the one hand, the greatest distance from which an observer at the time of photon decoupling t_{dec} is able to receive signals is given by the **comoving particle horizon**

$$\chi_{\text{ph}}^{\text{dec}} = \int_0^{t_{\text{dec}}} \frac{dt}{a(t)}. \tag{3.48}$$

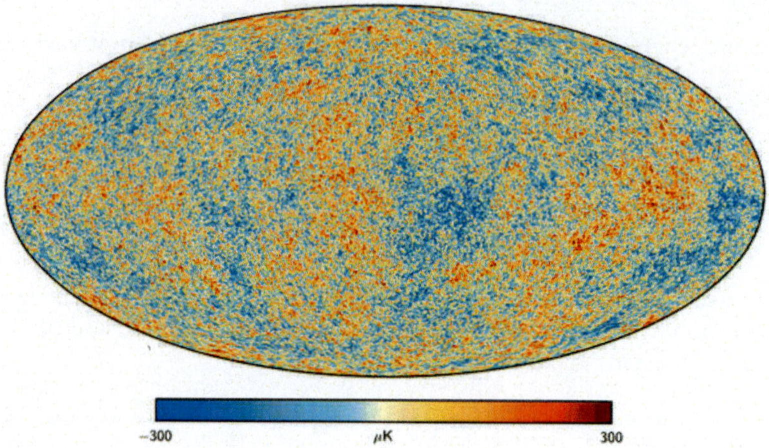

Fig. 3.1 Map of the temperature differences $\delta T(\theta, \phi)/T$ (at the 10^{-5} level) on the oldest known optical surface: the background sky when our universe first became transparent to light, 380 000 years after the big-bang. Credits: European Space Agency and Planck Collaboration

This gives the maximal size of the causal patch in the CMB we observed today. Actually, the true causal patch, called the **sound horizon**, is a bit smaller since the speed of sound is < 1 and close to $1/\sqrt{3}$.

On the other hand, the distance a photon has undergone from decoupling until today is $\int_{t_{\rm dec}}^{t_{\rm today}} \frac{dt}{a(t)}$. Therefore, the number of causal patches imprinted on the CMB sphere (or last-scattering-surface) we observe today, is equal to the area of the CMB sphere, divided by the area of an individual causal patch

$$\#_{\text{causal patches}} = \frac{4\pi}{\pi}\left(\frac{\int_{t_{\rm dec}}^{t_{\rm today}} \frac{dt}{a(t)}}{\int_0^{t_{\rm dec}} \frac{dt}{a(t)}}\right)^2 \simeq 10\,000 \tag{3.49}$$

where we computed $a(t)$ from integrating the Friedmann equation in Eq. (3.8) with the energy content measured by Planck, cf. Sect. 3.1.2 and from assuming a radiation-dominated universe before BBN. Hence, by looking at the celestial sphere, we should observed 10 000 causally independent Hubble patches with individual angular size 1°, the age of each patch being $t_{\rm dec}$. However, the temperature of the Cosmic Microwave Background appears to have a striking homogeneity of $\frac{\delta\rho}{\rho} \sim 10^{-5}$. This is the **homogeneity problem**.

3.3.2 The Flatness Problem

Another surprise is the observed flatness of the universe at $\Omega_K \lesssim 0.1\,\%$, cf. Eq. (3.20). From Eq. (3.12), we can see that the curvature fraction Ω_K should have increased with cosmic time like the square of the **comoving Hubble radius** $(a\,H)^{-1}$,

$$|\Omega_K(a)| = (a\,H)^{-2}, \tag{3.50}$$

implying that when the temperature was $T \sim$ TeV, the curvature would have been $\Omega_K^{\rm TeV} \lesssim 10^{-32}$. This, except if the geometry is rigorously flat $k = 0$, requires extreme fine-tuning. This is the **flatness problem**.

3.3.3 The Solution: Shrinking the Comoving Hubble Radius

The inflation idea:
The homogeneity and flatness problems motivated Alan Guth in 1980 [38] to postulate the existence of an early **inflation** period between $t_{\rm start}$ and $t_{\rm end}$, during which the scale factor increases exponentially fast

$$a(t) \propto e^{Ht} \qquad \longrightarrow \qquad \chi_{\rm ph}^{\rm dec} = \int_{t_{\rm start}}^{t_{\rm end}} \frac{dt}{a(t)} \simeq (a\,H)^{-1}_{t_{\rm start}}. \tag{3.51}$$

The comoving Hubble radius $(a\,H)^{-1}$ decreases and, for arbitrary long inflation, $a(t_{\rm start}) \to 0$, the comoving particle horizon $\chi_{\rm ph}^{\rm dec}$ can be made arbitrary large. As a consequence, the whole CMB sphere can be contained in a single inflated causal patch. The homogeneity problem is solved when the comoving Hubble radius $(a\,H)^{-1}$ during inflation has become smaller than the comoving Hubble radius today $(a_0\,H_0)^{-1}$. Also, the flatness problem is solved by stretching the curvature, see Eq. (3.50).

For pedagogical introductions to inflation, we mention the reviews [178, 287–293] and the reference textbooks [158, 163, 166, 167, 169, 172, 176].

Duration of inflation:
During inflation, a given wave number gets stretched as $k_{\rm phys} = k/a$, where k is the comoving wave number, until it leaves the causal patch when $k > a\,H$ and freezes. The number of e-folds of accelerated expansion between the time when the mode k exits the horizon and the end of inflation is

$$N_k \equiv \log \frac{a_{\rm end}}{a_k}, \qquad \text{with} \quad k = a_k\,H_k. \tag{3.52}$$

We introduce the comoving Hubble horizon today $a_0 H_0$, such that we can write (e.g. [294])

$$N_k = -\ln\left(\frac{k}{a_0 H_0}\right) + \ln\left(\frac{a_{\text{end}} H_{\text{end}}}{a_0 H_0}\right) - \ln\left(\frac{H_{\text{end}}}{H_k}\right). \tag{3.53}$$

We expect the last term to be of order 1, however the second term depends strongly on the evolution of the universe after the end of inflation. If we assume that the inflaton behaves like matter (if it oscillates in a quadratic potential), between the end of inflation and the end of the reheating phase, then

$$a_{\text{end}} = a_0 \left(\frac{H_0}{H_{\text{eq}}}\right)^{2/3} \left(\frac{H_{\text{eq}}}{H_{\text{reh}}}\right)^{1/2} \left(\frac{H_{\text{reh}}}{H_{\text{end}}}\right)^{2/3}, \tag{3.54}$$

where H_{reh} and H_{eq} are the Hubble parameters at the end of the reheating phase and at matter-radiation equality.[9] Plugging Eq. (3.54) into Eq. (3.53), we obtain

$$N_k \simeq 56 - \ln\left(\frac{k}{a_0 H_0}\right) + \frac{1}{3}\ln\left(\frac{T_{\text{reh}}}{10^9\ \text{GeV}}\right) + \frac{1}{3}\ln\left(\frac{H_{\text{end}}}{10^{13}\ \text{GeV}}\right) - \ln\left(\frac{H_{\text{end}}}{H_k}\right). \tag{3.55}$$

By setting $k \lesssim a_0 H_0$ we obtain the minimal number of e-folds of inflation for two given points within the Hubble patch today to be causally connected. Taking the lowest inflation scale such that sphaleron transitions are still active and successful baryogenesis is still possible, $T_{\text{reh}} \sim 1$ TeV (assuming instantaneous reheating), we obtain $N_{\min} \simeq 32$. If we consider the lowest temperature allowed by BBN, $T_{\text{reh}} \sim 10$ MeV, we obtain $N_{\min} \simeq 21$. Taking the maximal reheating temperature, $T_{\text{reh}} \sim 10^{16}$ GeV, compatible with the non-detection of fundamental B-modes in the CMB, see Eq. (3.73), we get $N_{\max} \simeq 62$.

3.3.4 *Slow-Roll Inflation*

The temperature anisotropies problem:
The observation of the CMB with the telescopes COBE (1989–1993), WMAP (2001–2010) and Planck (2009–2013) has revealed the presence of anisotropy of temperature at the level of $\delta T/T \sim 10^{-5}$. A theory which solves the horizon and the flatness problem must also be able to explain the presence of these temperature anisotropies. The temperature anisotropies in the CMB can be expanded under spherical harmonics as

$$\frac{\delta T}{T}(\theta, \phi) = \sum_{l=2}^{+\infty} \sum_{m=-l}^{+l} a_{lm} Y_{lm}(\theta, \phi), \tag{3.56}$$

[9] The Hubble factor at matter-equality can be computed as $H_{\text{eq}} \simeq H_0\, z_{\text{eq}}^{3/2}$ with $z_{\text{eq}} = 3402(26)$ [260].

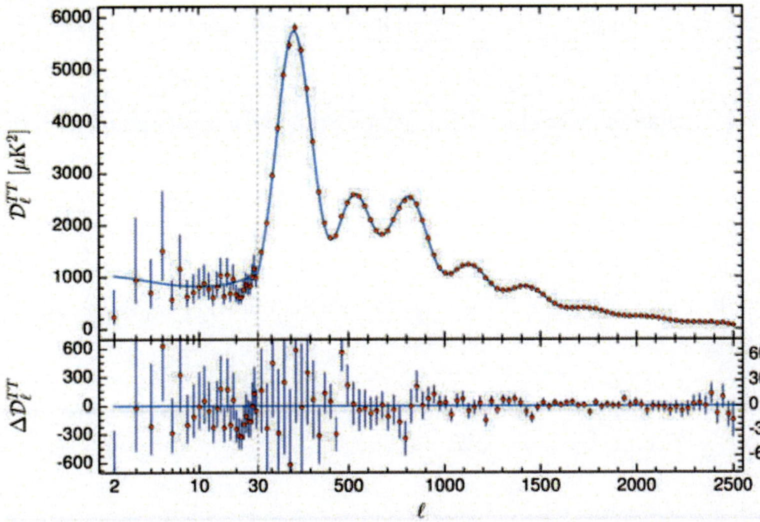

Fig. 3.2 Cosmic microwave background temperature power spectrum (CMB). The blue line shows the best fit to the Planck data [259]. The small (large) multipoles l correspond to the long (small) angular scales: the main acoustic peak at $l = 200$ corresponds to 1 degree on the sky whereas $l = 1800$ corresponds to 0.1 degree. Figure reproduced from [259]

where $Y_{lm}(\theta, \phi)$ are spherical harmonics. If the temperature fluctuations are assumed to be Gaussian, as it seems to be the case, the variance C_l of a_{lm}, given by

$$C_l \equiv \frac{1}{2l+1} \sum_{m=-l}^{+l} |a_{lm}|^2 , \tag{3.57}$$

contain all the information about the temperature anisotropies. The C_l have been very well measured by Planck, see Fig. 3.2.

The inflaton:
Guth's initial 1980 proposal for generating a fast expansion relied on the universe being stuck in a false vacuum of high density [38]. This idea was also proposed the same year by Kazanas [295]. However, Guth recognized himself that bubbles of true vacuum might never percolate and that the universe might remain stuck in the false vacuum forever [296, 297]. This led Linde, Albrecht and Steinhardt [39, 40] to propose the **slow roll** idea the next year. Right after, Starobinsky, Mukhanov and Chibisov showed that quantum fluctuations produced during slow roll generate a spectrum of inhomogeneities possibly accounting for the large scale structures in our universe [41–44].

The standard lore is to assume the universe to be dominated by a scalar field ϕ, the **inflaton**, which slow-roll down its potential $V(\phi)$ under the effect of Hubble friction, and to impose the parameter, e.g. [289]

$$\varepsilon \equiv -\frac{\dot{H}}{H^2} = -\frac{d\ln H}{dN} = \frac{\frac{1}{2}\dot{\phi}^2}{M_{\rm pl}^2 H^2}, \tag{3.58}$$

to be small. Deviation from slow-roll is encoded in the second parameter η defined through

$$\frac{d\ln\varepsilon}{dN} = 2(\varepsilon - \eta), \qquad \text{where } \eta \equiv -\frac{d\ln H_{,\phi}}{dN}. \tag{3.59}$$

The equation of motion for the slow-rolling scalar field reduces to $\dot{\phi} \simeq V'/3H$. In the slow-roll approximation $\varepsilon, |\eta| \ll 1$, the Hubble-slow-roll parameters ε and η can be related to the potential-slow-roll parameters

$$\varepsilon_V(\phi) \equiv \frac{M_{\rm pl}^2}{2}\left(\frac{V_{,\phi}}{V}\right)^2, \qquad \eta_V(\phi) \equiv M_{\rm pl}^2 \frac{V_{,\phi\phi}}{V}, \tag{3.60}$$

through

$$\varepsilon \simeq \varepsilon_V, \qquad \eta \simeq \eta_V - \varepsilon_V. \tag{3.61}$$

The number of e-folds of accelerated expansion between the time when the mode k exits the horizon at t_k and the end of inflation at t_e is

$$N_k = \int_{a_k}^{a_e} d\ln a = \int_{t_k}^{t_e} H(t)\,dt = \int_{\phi_k}^{\phi_e} \frac{H}{\dot{\phi}} d\phi = \int_{\phi_k}^{\phi_e} \frac{1}{\sqrt{2\varepsilon}} \frac{d\phi}{M_{\rm pl}}. \tag{3.62}$$

Primordial scalar perturbation:
Inflation ends when $\varepsilon(\phi_e) \approx 1$. However, quantum fluctuation of $\phi(t\,\vec{x}) = \bar{\phi}(t) + \delta\phi(t\,\vec{x})$ will make local difference in time when inflation ends, which leads to difference in local density $\delta\rho(t\,\vec{x})$, and therefore, in the CMB temperature $\delta T(t\,\vec{x})$. The inflaton fluctuations are equivalent to the ones of an harmonic oscillator in de Sitter space. In the spatially flat gauge, one finds

$$\Delta_{\delta\phi}^2(k) \equiv \langle 0|\delta\phi^2|0\rangle \simeq \left(\frac{H}{2\pi}\right)^2\Big|_{k=aH}. \tag{3.63}$$

In order to get ride of the gauge dependence, we usually introduce the gauge-invariant curvature perturbation

$$\Delta_{\mathcal{R}}^2(k) \equiv \frac{H}{\dot{\phi}}\Delta_{\delta\phi}^2 = \frac{1}{2\varepsilon}\frac{\Delta_{\delta\phi}^2}{M_{\rm pl}^2}, \tag{3.64}$$

such that the slow-roll prediction reads

$$\Delta_{\mathcal{R}}^2(k) \simeq \frac{1}{8\pi^2}\frac{1}{\varepsilon}\frac{H^2}{M_{\rm pl}^2}\Big|_{k=aH}. \tag{3.65}$$

Since during inflation, H is almost constant, the spectrum of primordial scalar perturbation is scale-invariant (power-law) with a slight tilt, usually expressed as

$$\Delta_R^2(k) \equiv A_s \left(\frac{k}{k_*}\right)^{n_s - 1}, \tag{3.66}$$

The Planck collaboration [259] has measured the amplitude of the scalar spectrum A_s and the tilt n_s around the pivot scale $k_* = 0.002\ \text{Mpc}^{-1}$, with a good precision

$$A_s = (2.101 \pm 0.034) \times 10^{-9}, \tag{3.67}$$

$$n_s = 0.9649 \pm 0.0042. \tag{3.68}$$

Primordial gravitational waves:
Independently of the inflaton dynamics, during the accelerated expansion, quantum fluctuations of the tensor modes h_{ij}, defined by

$$ds^2 = dt^2 - a^2(\delta_{ij} + h_{ij})dx^i dx^j, \tag{3.69}$$

are also generated. The power spectrum of these Gravitational Waves is found to be

$$\Delta_t^2(k) \equiv \langle 0|h_\times^2 + h_+^2|0\rangle \simeq \frac{2}{\pi^2}\frac{H^2}{M_{\text{pl}}^2}, \tag{3.70}$$

and again is expected to be scale-invariant and slightly tilted

$$\Delta_t^2(k) \equiv A_t \left(\frac{k}{k_*}\right)^{n_t - 1}. \tag{3.71}$$

See [298] for a review. The fundamental tensor modes are expected to source the so-called fundamental B-modes in the CMB. Their non-detection by neither by Planck 2018 or BICEP2/Keck sets an upper bound on the tensor-to-scalar ratio [299, 300]

$$r \equiv \frac{A_t}{A_s} \lesssim 0.055, \qquad (95\%\ \text{C.L.}) \tag{3.72}$$

which can be recast as an upper bound on the inflation scale

$$H_{\text{inf}} \lesssim 6 \times 10^{13}\ \text{GeV}. \tag{3.73}$$

The future experiment LiteBird [58, 301], planned to be launched within the next decade, is expected to have a sensitivity $r \lesssim 0.001$.

Slow-roll:
In the slow-roll approximation $\varepsilon_V, |\eta_V| \ll 1$, the scalar and tensor spectra are given by [289]

$$\Delta_s^2 \simeq \frac{1}{24\pi^2}\frac{V}{M_{\rm pl}^4}\frac{1}{\varepsilon_V}\bigg|_{k=aH}, \qquad \Delta_t^2 \simeq \frac{2}{3\pi^2}\frac{V}{M_{\rm pl}^4}\bigg|_{k=aH}. \tag{3.74}$$

The scalar and tensor spectral index are

$$n_s - 1 \simeq 2\eta_V - 6\varepsilon_V, \qquad n_t \simeq -2\varepsilon_V. \tag{3.75}$$

The tensor-to-scalar ratio is

$$r \simeq 16\varepsilon_V, \tag{3.76}$$

which implies the relation $r \simeq -8n_t$.

Inflation potentials:
Perhaps the simplest inflation model is $V(\phi) = \frac{1}{2}m^2\phi^2$. It belongs to the class of chaotic inflation models $V(\phi) \propto \phi^n$ introduced by Linde in 1983 [302]. It has slow-roll parameters [289]

$$\varepsilon_V = \eta_V = 2\left(\frac{M_{\rm pl}}{\phi}\right)^2. \tag{3.77}$$

The slow-roll conditions ε, $|\eta| \lesssim 1$ require super-Planckian values for the inflaton

$$\phi > \sqrt{2}M_{\rm pl}. \tag{3.78}$$

The possibility to generate inflation from a potential like $m^2\phi^2$ which is omnipresent in particle physics suggests that inflation is not only a peculiar phenomenon desirable for explaining the homogeneity and the flatness of the universe, but that *it is also a natural and inevitable consequence of the chaotic initial conditions in the very early universe* [302]. The number of inflation e-folding reads

$$N(\phi) = \frac{\phi^2}{4M_{\rm pl}^2} - \frac{1}{2}, \tag{3.79}$$

and fluctuations in the CMB are created at

$$\phi_{\rm cmb} = 2\sqrt{N_{\rm cmb}}M_{\rm pl} \sim 15M_{\rm pl}. \tag{3.80}$$

In order to agree with the observed scalar fluctuation

$$\Delta_s^2 = \frac{m^2}{M_{\rm pl}^2}\frac{N_{\rm cmb}^2}{3} \sim 10^{-9}, \tag{3.81}$$

we must fix $m \sim 10^{-6}M_{\rm pl}$. The scalar spectral index is

$$n_s = 1 + 2\eta_V - 6\varepsilon_V = 1 - \frac{2}{N_{\rm cmb}} \simeq 0.96, \tag{3.82}$$

and the tensor-to-scalar ratio is

$$r = 16\varepsilon_v = \frac{8}{N_{\text{cmb}}} \simeq 0.1. \tag{3.83}$$

For other power-law potentials $V(\phi) \propto \phi^n$, it generalizes to [300]

$$r = \frac{16n}{4N_{\text{cmb}} + n}. \tag{3.84}$$

The bound on r from Planck data in Eq. (3.72), excludes chaotic inflation models with $n \geq 2$. Other values $n = 4/3$, 1 and $2/3$, motivated by axion monodromy [303–305] are compatible with Planck.

Instead of chaotic models, constraints from Planck in the $r - n_s$ plane [300] favor a class of inflationary models with plateau-like inflaton potentials: natural (axionic) inflation [306, 307], symmetry breaking with non-minimal (quadratic) coupling [308–312], hilltop [313], R^2 [41], no-scale supergravity [314, 315], α-attractors [316–322]. For a review of the main inflation models and their constraints from Planck data, see [300, 323, 324].

It is worth to mention that in spite of the remarkable success of the inflation paradigm, the fact that Planck data favors plateau-like inflation models over 'more natural' chaotic inflation models has raised **criticisms** [325–330].

3.4 Gravitational Waves of Primordial Origin

Gravitational Waves (GW) have been predicted by Einstein in 1916 [331, 332], one year after publishing his work about the relatistic theory of gravitation [5, 6]). GW have been detected indirectly the first time by Hulse and Taylor in 1974 [333] and detected directly the first time by the LIGO collaboration on the 14 September of 2015 [334], one century after their prediction. The GW detected by LIGO were not of primordial origin but they have instead been produced 1.4 billions years ago when two $\sim$ 30 solar masses coalescing black holes merged into each other. Future experiments will have the sensitivity to probe potential GW backgrounds of primordial origin, and GW astronomy is opening a new era of exploration of high-energy physics in the early universe.

As discussed in Sect. 3.2.2, the universe only becomes transparent to light when the photons decouple from the electron/proton plasma at the time of recombination $e^- + p^+ \to H + \gamma$. So we can not use the electromagnetic probe to observe the state of the universe when it is younger than 380 000 yrs. However, GW decouple from the hot plasma much earlier. Indeed, assuming that the interaction rate of gravitons with SM is

$$\Gamma \sim GT^3, \tag{3.85}$$

where $G = 8\pi/M_{\rm pl}^2$ is the Newton constant, we obtain that gravitons decouple around

$$T \sim M_{\rm pl}. \tag{3.86}$$

Therefore, GW produced in the early universe propagate freely in the universe until today. Such primordial GW carry information about very high-energy physics and probe particle physics at scales that will never be probed by colliders. Current and future GW experiments, LIGO/VIRGO/KAGRA [335], LISA [106, 336], Einstein Telescope [107, 337], Cosmic Explorer [338], DECIGO, BBO [339], AION/MAGIS [340–342], AEDGE [343], constitute a new avenue of investigation in particles physics and cosmology. See [344] for a textbook and [106, 298, 345, 346] for reviews on primordial GW. See [347] for a review on the GW cosmological sources in the MHz and GHz range and the proposed experiments.

After having introduced GW as linearized solutions of Einstein equations in Sect. 3.4.1, we derive the power of emitted GW by a macroscopic object in Sect. 3.4.2, and use it to compute the GW spectrum produced by different cosmological sources in Sect. 3.4.3.

3.4.1 Linearized Wave Solutions of Einstein Equations

Gravitational-waves (GW) are small **tensor** perturbation around the **flat-space** metric [114, 124, 348]

$$g_{\mu\nu} = \eta_{\mu\nu} + h_{\mu\nu}, \qquad |h_{\mu\nu}| \ll 1. \tag{3.87}$$

General relativity is invariant under **any** coordinate transformations

$$x^\mu \to x'^\mu(x), \tag{3.88}$$

where $x'^\mu(x)$ is an arbitrary **smooth** function of x. In order to preserve the flat-space background in Eq. (3.87), we only perform tiny gauge transformations

$$x^\mu \to x'^\mu = x^\mu + \xi^\mu(x). \tag{3.89}$$

Using the transformation law of the metric

$$g_{\mu\nu}(x) \to g'_{\mu\nu}(x') = \frac{\partial x^\rho}{\partial x'^\mu}\frac{\partial x^\sigma}{\partial x'^\nu}\, g_{\rho\sigma}(x), \tag{3.90}$$

we find the transformation of $h_{\mu\nu}$ at **linearized** order

$$h_{\mu\nu}(x) \to h'_{\mu\nu}(x') = h_{\mu\nu}(x) - (\partial_\mu\xi_\nu + \partial_\nu\xi_\mu). \tag{3.91}$$

For convenient reasons, we introduce the variable

$$\bar{h}_{\mu\nu} = h_{\mu\nu} - \frac{1}{2}\eta_{\mu\nu}h, \tag{3.92}$$

with $h = \eta^{\mu\nu}h_{\mu\nu}$, which transforms as

$$\bar{h}_{\mu\nu}(x) \to \bar{h}'_{\mu\nu}(x') = \bar{h}_{\mu\nu}(x) - (\partial_\mu\xi_\nu + \partial_\nu\xi_\mu - \eta_{\mu\nu}\partial_\rho\xi^\rho). \tag{3.93}$$

The linearized equations of motion can be found after injecting Eq. (3.87) in the **Einstein equations** in Eq. (3.2)

$$\Box\bar{h}_{\mu\nu} + \eta_{\mu\nu}\partial^\rho\partial^\sigma\bar{h}_{\rho\sigma} - \partial^\rho\partial_\nu\bar{h}_{\mu\rho} - \partial^\rho\partial_\mu\bar{h}_{\nu\rho} = -\frac{16\pi G}{c^4}T_{\mu\nu}. \tag{3.94}$$

From imposing the **Lorentz gauge** (also called the Hilbert gauge, or the harmonic gauge, or the De Donder gauge)

$$\partial^\nu\bar{h}_{\mu\nu} = 0, \qquad \text{(Lorentz gauge)} \tag{3.95}$$

the equations of motion considerably simplify

$$\Box\bar{h}_{\mu\nu} = -\frac{16\pi G}{c^4}T_{\mu\nu}. \tag{3.96}$$

By looking at Eq. (3.93), we observe that even after imposing the Lorentz gauge in Eq. (3.95), we can still perform coordinates transformations $x^\mu \to x'^\mu = x^\mu + \xi^\mu(x)$ with

$$\Box\xi_\mu = 0. \tag{3.97}$$

The Lorentz gauge in Eq. (3.95) + the extra constraint in Eq. (3.97) imposes $4+4$ constraints on the 10 initial independent components of the symmetric 4×4 matrix $g_{\mu\nu}$, which therefore reduce to only 2 independents parameters. A convenient change of coordinates compatible with Eq. (3.97) is the one which imposes the tensor perturbation $h_{\mu\nu}$ to be transverse and traceless

$$h^{0i} = 0, \qquad h^i_i = 0, \qquad \text{(TT gauge)} \tag{3.98}$$

This defines the **transverse-traceless gauge** or TT gauge. Observe that $h^{0i} = 0$ together with the Lorentz gauge implies $\partial_0 h^{00} = 0$, which for non-stationary perturbations (GW), reduces to $h^{00} = 0$. Hence we can summarized the Lorentz + TT gauge as follows

$$h^{0\mu} = 0, \qquad h^i_i = 0, \qquad \partial^j h_{ij} = 0, \qquad \text{(Lorentz + TT gauge)}. \tag{3.99}$$

In what follows, we call TT gauge the association of the Lorentz gauge + the TT gauge. Note that only the spatial components are non-zero and $\bar{h}^{ij}_{\rm TT} = \bar{h}^{ij}$. Starting from a tensor perturbation with spatial component h^{ij} defined in an arbitrary gauge, we can construct the associated tensor h^{ij}_{TT} in the TT gauge by performing the **projection**

$$h^{ij}_{\rm TT} = \Lambda_{ij,kl}(\hat{\mathbf{k}})\, h^{kl}, \tag{3.100}$$

where

$$\Lambda_{ij,kl}(\hat{\mathbf{k}}) \equiv P^i_k P^j_l \; - \; \frac{1}{2} P^{ij} P_{kl}, \tag{3.101}$$

$$P_{ij}(\hat{\mathbf{k}}) \equiv \delta_{ij} \; - \; \hat{k}_i \hat{k}_j. \tag{3.102}$$

P_{ij} projects spatial tensor components onto the surface orthogonal to the spatial wavevector k_i while $\Lambda_{ij,kl}$ subtracts the trace. For any spatial vector v^i, we have the relations $k_i\, P^i_j\, v^j = 0$ and $P^i_k\, P^k_j\, v^j = P^i_j\, v^j$, which we can use to check that $h^{ij}_{\rm TT}$ satisfies Eqs. (3.99). For a plane wave $h^{ij} = A^{ij}\, e^{i\,(\omega t - k\,z)}$ propagating in the z direction, the GW amplitude in the TT gauge after applying the projection in Eq. (3.100), reads

$$h^{\mu\nu}_{\rm TT} = \begin{pmatrix} 0 & 0 & 0 & 0 \\ 0 & h_+ & h_\times & 0 \\ 0 & h_\times & -h_+ & 0 \\ 0 & 0 & 0 & 0 \end{pmatrix} \exp\, i\,(\omega t - k\,z), \tag{3.103}$$

with $h_\times = A^{12}$ and $h_+ = \frac{1}{2}(A^{11} - A^{22})$.

3.4.2 *Energy of Gravitational-Waves*

Energy-momentum tensor of gravitational-waves:
After having defined GW, we now discuss the definition of their energy-momentum tensor. Actually, the question does not make sense since there is no **local** notion of gravitational energy density in General Relativity. Indeed, we can always transform coordinates to a local **inertial** frame (a freely falling frame), at a point p where

$$g_{\mu\nu}\big|_p = \eta_{\mu\nu}, \tag{3.104}$$

$$\Gamma^\rho_{\mu\nu}\big|_p = 0. \tag{3.105}$$

However, we can show that GW act as a force between **two or more** nearby test masses. Hence, an energy-momentum tensor of GW can be defined after **averaging** over a space-time volume of size L larger than the GW wavelength $L \gg \lambda$. The energy-momentum sourced by GW $t_{\mu\nu}$ (the gravity of gravity) can be found by going to second order in perturbation

$$G^{(1)}_{\mu\nu} = -\frac{8\pi\, G}{c^4}\left(T_{\mu\nu} + t_{\mu\nu}\right) \tag{3.106}$$

with

$$t_{\mu\nu} \equiv \frac{c^4}{8\pi\, G}\left\langle G^{(2)}_{\mu\nu}\right\rangle, \tag{3.107}$$

where $G^{(1)}$ and $G^{(2)}$ are the Einstein tensors at first and second order in h^{ij}, and $\langle\cdots\rangle$ denotes the space-time average. Expansion at second order in the TT gauge yields [114, 124, 348]

$$t_{\mu\nu} \equiv \frac{c^4}{32\pi\, G}\left\langle \partial_\mu h^{\rm TT}_{ij}\, \partial_\nu h^{ij}_{\rm TT}\right\rangle. \tag{3.108}$$

Energy flux:
The outward-propagating GW carries away an energy flux

$$\frac{dE}{dA\, dt} = +c\, t^{00}, \tag{3.109}$$

$$= \frac{c^3}{32\pi\, G}\left\langle \dot{h}^{\rm TT}_{ij}\, \dot{h}^{\rm TT}_{ij}\right\rangle. \tag{3.110}$$

The quadrupole approximation:
The solution of the gravitational wave equation in Eq. (3.96), projected into the TT gauge as prescribed by Eq. (3.100), reads

$$\bar{h}^{\rm TT}_{ij}(t,\, \vec{x}) = \frac{4G}{c^4}\, \Lambda_{ij,\,kl}(\hat{k}) \int d^3x' \frac{1}{|\vec{x} - \vec{x}'|}\, T^{kl}\left(t - \frac{|\vec{x} - \vec{x}'|}{c},\, \vec{x}'\right) \tag{3.111}$$

$$= \frac{1}{r}\frac{4G}{c^4}\, \Lambda_{ij,\,kl}(\hat{k}) \int d^3x' \left(T_{kl}(t - \frac{r}{c},\, \vec{x}') + \frac{\vec{x}' \cdot \hat{k}}{c}\partial_0 T_{kl} + \cdots\right) \tag{3.112}$$

where we have assumed the distance to the source r to be larger than the source typical size $r \gg d$, and the motion of the source to be non-relativistic $v \ll c$ (we have kept the first term in the multipole expansion), such that we could write $\left|\vec{x} - \vec{x}'\right| \simeq r - \vec{x}' \cdot \hat{k}$, and perform an expansion in $\vec{x}' \cdot \hat{k}/c$. Now comes an interesting relation stemming from successive use of **integration by part** and **energy-momentum conservation** $\partial_\mu T^{\mu\nu} = 0$

$$\begin{aligned}\int d^3x\, T^{ij}(t,\, \vec{x}) &= -\int d^3x\, \partial_k\, T^{kj}(t,\, \vec{x})\, x^i \\ &= +\int d^3x\, \partial_0\, T^{0j}(t,\, \vec{x})\, x^i \\ &= -\frac{1}{2}\int d^3x\, \partial_0\partial_k\, T^{0k}(t,\, \vec{x})\, x^i\, x^j \\ &= +\frac{1}{2}\int d^3x\, \partial_0^2\, T^{00}(t,\, \vec{x})\, x^i\, x^j \end{aligned} \tag{3.113}$$

This allows to re-express the GW tensor component in Eq. (3.112) as a function of T^{00} only

$$\bar{h}_{ij}^{\mathrm{TT}}(t,\vec{x}) = \frac{1}{r}\frac{4G}{c^4}\,\Lambda_{ij,kl}(\hat{k})\int d^3x'\,\ddot{T}_{00}\left(t-\frac{r}{c},\,\vec{x}'\right). \tag{3.114}$$

Therefore, the GW radiated power in Eq. (3.110) at first-order in the multipole expansion (quadrupole) reads

$$\left(\frac{dP}{d\Omega}\right)_{\mathrm{quad}} = \frac{r^2c^3}{32\pi G}\left\langle \dot{h}_{ij}^{\mathrm{TT}}\dot{h}_{ij}^{\mathrm{TT}}\right\rangle \tag{3.115}$$

$$= \frac{G}{8\pi c^5}\,\Lambda_{ij,kl}(\hat{k})\left\langle \dddot{Q}_{ij}\,\dddot{Q}_{kl}\right\rangle, \tag{3.116}$$

where Q_{ij} is the quadrupole moment of the macroscopic object radiating the GW

$$Q_{ij} \equiv \frac{1}{c^2}\int d^3x\,T^{00}(t,\,\vec{x})\left(x^i\,x^j - \frac{1}{3}r^2\,\delta^{ij}\right) \tag{3.117}$$

which is the traceless version of Eq. (3.113). The angular dependence is only contained in $\Lambda_{ij,kl}(\hat{k})$. Using that

$$\int d\Omega\,\Lambda_{ij,kl}(\hat{k}) = \frac{2\pi}{15}\left(11\delta_{ik\delta_{jl}} - 4\delta_{ij}\delta_{kl} + \delta_{il}\delta_{jk}\right), \tag{3.118}$$

we find the **total radiated power** in the **quadrupole approximation**

$$P_{\mathrm{quad}} = \frac{G}{5c^5}\left\langle \dddot{Q}_{ij}\,\dddot{Q}_{kl}\right\rangle. \tag{3.119}$$

3.4.3 *Cosmological Signals*

Stochastic gravitational-waves background:
We are motivated by the possibility for future experiments to observe GW of primordial origin. They can be generated during inflation, during preheating, during a first-order phase transition or by a network of cosmic strings.. See [344] for a textbook and [298] for a review on primordial GW. Due to causality, a GW signal generated in the early universe must have a correlation length Δt_c smaller than the Hubble horizon at the time of production

$$\Delta t_c \lesssim H_*^{-1}. \tag{3.120}$$

The number of causal sources which we observe today can be estimated as the area of the sphere of events at time of GW production t_* divided by the area of an individual causal patch

$$\#_{\text{causal sources}} \sim \frac{4\pi}{\pi}\left(\frac{\int_{t_*}^{t_{\text{today}}} \frac{dt}{a(t)}}{\int_0^{t_*} \frac{dt}{a(t)}}\right)^2 \sim 10^{29}\left(\frac{g_*(T_*)}{106.75}\right)^{1/2}\left(\frac{T}{1\text{ TeV}}\right)^2, \tag{3.121}$$

where we computed $a(t)$ from integrating the Friedmann equation in Eq. (3.8) with the energy content measured by Planck, cf. Sect. 3.1.2 and from assuming a radiation-dominated universe before BBN. Therefore, a GW signal of primordial origin can only be observed today as a superposition of GW generated by an enormous number of causally independent sources. Individual sources can not be resolved but instead we can only observe a **Stochastic Gravitational-Wave Background (SGWB)**. For most of the cosmological sources, the SGWB is homogeneous, isotropic, gaussian and unpolarized such that we can write [298]

$$\langle h_r(\mathbf{k}, t) h_p^*(\mathbf{q}, t)\rangle = \frac{8\pi^5}{k^3}\delta^{(3)}(\mathbf{k}-\mathbf{q})\,\delta_{rp}\,h_c^2(k, t). \tag{3.122}$$

We now use the quadrupole formula in Eq. (3.119) in order to compute the SGWB generated by given cosmological sources in the early universe.

Model-independent:
The contribution to the total energy density of the universe today from GW generated by sources emitting during Δt in the early universe reads

$$\Omega_{\text{GW}} = \Omega_\gamma \frac{P_{\text{GW}}\,\Delta t\,H^3}{\rho_{\text{tot}}}, \tag{3.123}$$

where $\Omega_\gamma \simeq 4.2 \times 10^{-5}$ is the photon abundance today [260], P_{GW} is the radiated GW power per Hubble volume, H and ρ_{tot} are the Hubble scale and total energy density at the time of emission. We have used that the GW energy density ρ_{GW} redshifts as fast as the photon energy density ρ_γ. We have assumed GW emission to take place during radiation domination and we have assumed an adiabatic evolution. If the source is made of N macroscopic objects per Hubble volume, the power of GW emission can be estimated from the quadrupole formula in Eq. (3.119)

$$P_{\text{GW}} = NG\dddot{Q}^2, \qquad \text{with } \dddot{Q} = E\,L^2/\tau^3, \tag{3.124}$$

where E, L and τ are the characteristic energy, length and time scale of the dynamical objects. This can be recast in the simple form

$$\Omega_{\text{GW}} = \Omega_\gamma\,N\left(\frac{\rho_s}{\rho_{\text{tot}}}\right)^2 (L/\tau)^6\;(\Delta t\,H)\;(L\,H)^4\,. \tag{3.125}$$

where we have introduced the energy density of the source $\rho_s \equiv E/L^3$ and made use of the Friedmann equation $H^2 = \rho_{\text{tot}}/3M_{\text{pl}}^2$.

First-order phase transition:
Thanks to the universal formula in Eq. (3.125), we can recover some of the well-known SGWB of cosmological origin. For instance, the SGWB generated by **bubbles collision** (scalar field contribution) or **sound shell overlap** (sound wave contribution) during a first-order phase transition, see [101, 105, 349] and Chap. 6. The time scale of the bubble collision / sound shell overlap is given by $\tau \sim \beta^{-1}$ where β is the **time derivative of the tunneling rate**. The size of the bubbles/sound shells when they collide/overlap is given by $L \sim v_w/\beta^{-1}$ where v_w is the speed of the wall. The number of bubbles/sound shells per Hubble volume is $N \sim (\beta/v_w\, H)^3$. The fraction of kinetic energy density of the source at the time of the phase transition can be expressed in terms of the usual parameters α and κ

$$\frac{\rho_s\, L^2/\tau^2}{\rho_{\text{tot}}} = \frac{\kappa\, \alpha}{1+\alpha}, \tag{3.126}$$

where $\alpha \equiv \Delta V/\rho_{\text{tot}}$ is the vacuum energy (**latent heat**) fraction and $\kappa = (\rho_s\, L^2/\tau^2)\,/\,\Delta V$ is the **efficiency** of the energy transfer from vacuum energy to the GW source. The later is either the scalar field kinetic energy (scalar field contribution) or the fluid kinetic energy (sound wave contribution). Applying Eq. (3.125) to the case of 1stOPT gives

$$\Omega_{\text{GW}} \sim \Omega_\gamma \left(\frac{\kappa\, \alpha}{1+\alpha}\right)^2 v_w^3 \left(\frac{H}{\beta}\right)^2 (\Delta t\, \beta)\,, \tag{3.127}$$

where the duration of the emission is usually **short-lived** $\Delta t \sim \beta^{-1}$, except for sound-wave contributions of weak phase transitions (low α) which can be long-lived $\delta t \sim H^{-1}$ [350, 351]. The GW frequency today reads

$$f_0 = \frac{a_*}{a_0} f_* = 1.65 \times 10^{-5}\ \text{Hz} \left(\frac{T_*}{100\ \text{GeV}}\right) \left(\frac{g_{\text{eff},*}}{100}\right)^{1/6} \frac{f_*}{H_*}, \tag{3.128}$$

where the frequency at emission f_* is given by the inverse size of the bubbles/sound shells when they collide/overlap

$$f_* \sim \beta/v_w. \tag{3.129}$$

We have assumed instantaneous reheating after the time of GW emission t_*, see Eqs. (6.214) and (6.216) in Chap. 6 for the effect of the equation of state during reheating on the SGWB from 1stOPT. For $\beta/H = 100$ and $T_* = 100$ GeV, Eq. (3.128) becomes $f_0 \simeq 1.65$ mHz which corresponds to the sensitivity window of the future LISA[10] experiment [101, 105, 106, 349]. Hence, LISA is an observational window

---

[10] After an observation time of $T = 10$ years, LISA is expected to be able to detect SGWB with abundance $\Omega_{\text{GW}} = 10^{-11}$, with signal-to-noise ratio $S = 10$, in the range 48 μHz $\lesssim f_0 \lesssim$ 38 mHz, with a peak sensitiviy around 3.2 mHz, see Appendix 6.A.

on the electroweak scale for which the existence of new physics can be motivated by the hierarchy problem, cf. Sect. 2.3.1 in Chap. 2.

Inflation:
The **primordial** SGWB generated during inflation, cf. Sect. 3.3.4 and review [298], is not generated by macroscopic objects and the formula Eq. (3.125) should not by used. However, just as a **curiosity**, we can check that the correct expression for the inflationary SGWB is recovered with Eq. (3.125), if we assume that GW are generated by $N = 1$ quantum mode of Hubble size $L \sim H^{-1}$ with Hubble dynamical time scale $\tau \sim H^{-1}$, Hubble energy $E \sim H$, and emitting during one Hubble time $\Delta t \sim H^{-1}$. Indeed, we get

$$\Omega_{\rm GW} \sim \Omega_\gamma \left(\frac{H_{\rm inf}}{M_{\rm pl}}\right)^4, \tag{3.130}$$

which coincides with the well-known result [298]. The most precise probe of GW generated during inflation is the search for B-modes patterns in the CMB. As already mentioned in Eq. (3.73), the non-detection of primordial B-modes by neither Planck 2018 nor BICEP2/Keck implies $H_{\rm inf} \lesssim 6 \times 10^{13}$ GeV [299, 300].

Cosmic string (global):
The last major potential source of primordial SGWB are Cosmic Strings (CS). They are topological defects formed after spontaneous breaking of a $U(1)$ symmetry. CS can be either **global** or **local** according to whether the $U(1)$ symmetry is accompanied of a gauge boson or not. We devote Chap. 8 to the precise computation of the SGWB in the presence of a network of cosmic strings and Appendix 8.F for the case of global strings specifically. For global strings, the dominant GW emission results from the oscillations of $N = 1$ **Hubble-sized loops**, $L \sim \tau \sim H^{-1}$. Global loops are **short-lived**, due to the efficient **Goldstone bosons** emission. More precisely, the global loop lifetime is

$$\delta t \sim H^{-1} \log\left(\eta/H\right), \tag{3.131}$$

where the log term results from the IR divergence of the string tension due to the **long-range force** mediated by the massless mode and is of order 10^2. η is the typical energy scale in the string (scalar field VEV). The loop energy density goes like

$$\rho_s \sim \frac{\eta^2 \log\left(\eta/H\right)}{H^{-2}}, \tag{3.132}$$

where η is the radial mode VEV. Plugging the previous relations in Eq. (3.125) yields

$$\Omega_{\rm GW} \sim \Omega_\gamma \left(\frac{\eta}{M_{\rm pl}}\right)^4 \log^3\left(\eta/H\right). \tag{3.133}$$

See Appendix 8.F for a more precise derivation. Hence, the SGWB from global string receives a $\log^3 \simeq 10^6$ enhancement factor with respect to the inflationary spectrum, which bring the spectrum in the LISA band if $\eta > 5 \times 10^{14}$ GeV and the SKA band if $\eta > 10^{14}$ GeV. On the other hand, η receives an upper bound from the condition that the network should form after inflation and that the non-detection of B-modes translates to $V_{\rm inf} \lesssim 10^{16}$ GeV [299, 300]. Hence, the string scale which will be probed by LISA and SKA corresponds to the interval 10^{14} GeV $\lesssim \eta \lesssim 10^{16}$ GeV. However, in order to not overclose the universe the goldstone mass should be $m_a \lesssim 10^{-18}$ eV [352].

Cosmic string (local):
In the case where CS arise from the spontaneous breaking of a local symmetry, there is no massless mode and the strings are **long-lived**: if formed at time t_i, they decay much later, at time $\tilde{t} = t_i/G\mu$ where $\mu \sim \eta^2$ is the string tension. Hence, the dominant population of loops emitting at time $\tilde{t}$ were produced at a much earlier time, $t_i = G\mu\, \tilde{t}$, corresponding to one loop-lifetime ago.[11] Hence, the SGWB from local strings can be derived from the global case in Eq. (3.133), after replacing $N = 1$ by $N = (G\mu)^{-3/2}$ and removing the log divergence

$$\Omega_{\rm GW} \sim \Omega_\gamma \frac{\eta}{M_{\rm pl}} \tag{3.134}$$

The last expression being independent of the GW frequency at emission $f \sim L^{-1}$, the SWGB from CS is **flat** in frequency. This is not the case anymore in the presence of a non-standard cosmology. See Appendix 8.B for a thorough derivation of the SGWB from CS, in standard and non-standard cosmology.

We conclude the section by emphasizing that searches for primordial GW with current and future experiments offer a fantastic new field of exploration of particle physics and cosmology.

3.5 Open Problems

The cosmological model based on a flat universe containing today 69% of Dark Energy, 26% of Dark Matter, 5% baryons, a sub-leading neutrino background, photons at 2.72 K and inflationary initial conditions is called the **ΛCDM Standard Model of cosmology**. Thanks to the impressive work of many collaborations of experimentalists, in the last decade, the level of precision at which the parameters of the ΛCDM model are measured, has been considerably improved. However, big questions remain unanswered. What is the nature of the Dark Energy? The nature of Dark Matter ? What generates the asymmetry between matter and anti-matter? None of these observations can be consistently justified theoretically within the Standard Models of Cosmology and Elementary particles. In this section, we give short

[11] The loop-formation rate scales as $dn/dt \propto 1/t^4$. So the earlier the formation time, the larger the number of loops and the resulting GW signal. This explains why the dominant population of loops emitting at time $\tilde{t}$ are the loops produced the earliest as possible, which corresponds to $t_i = G\mu\, \tilde{t}$.

descriptions about these open questions. We also present the astrophysical small scale problems and the more recent anomalies in Hubble measurement and 21 cm signal.

Another open problem in cosmology (or astrophysics) which we only discuss briefly now is the **Lithium problem**: the theoretical prediction of ^{7}Li is a factor 3 above its determination in the atmosphere of metal-poor halo stars [353]. Unfortunately, CMB data can not allow to determine the ^{7}Li abundance [354]. The publication in 1999 of the NACRE compilation of nuclear reactions [355] (previous compilations were [356, 357]) allowed the precise determination of the abundance of light elements [358–364]. In 2013, an update compilation of nuclear rates [365] but also the release of Planck data [366] allowed to improve the precision of BBN predictions [367–370] and to confirm the ^{7}Li problem to an unprecedented level. Proposed solutions are modification of nuclear rates with resonances [371–374], decay of an exotic particle [375–382], stellar depletion [383–387], lithium diffusion in the post-recombination universe [388, 389], or variation of coupling constants [390–392]. For reviews on the ^{7}Li problem, see [393–397].

Another open problem which we briefly discuss now is the **origin of intergalactic magnetic fields**. Indeed, the non-observation by Fermi telescope of secondary $\gamma-$rays collinear to blazar-produced $\gamma-$rays suggest the presence of intergalactic magnetic fields with strength $B \gtrsim 3 \times 10^{-16}$ G [398]. Interestingly, the upcoming Cherenkov Telescope Array (CTA) may push the lower bound to $B \gtrsim 10^{-12}$ G [399], closer to the CMB upper bound at $B \lesssim 10^{-9}$ G [400] (or even $B \lesssim 10^{-11}$ G [401]). Primordial magnetic fields can be generated by particular inflation models or by a thermal first-order phase transition [402–404]. See [402, 403, 405–408] for reviews on cosmological magnetic fields.

3.5.1 Cosmological Constant Problem

The measurement of the dark energy:
Dark energy is a fluid with equation of state $\rho_\Lambda = -p_\Lambda =$ constant. Its precise abundance, $\Omega_\Lambda = 68.5 \pm 0.7\%$ of the total energy density of the universe, is inferred from two independent observations at low-redshift (when mixed with Planck measurements of the CMB)

- **Type Ia Supernova (SNe Ia)**: they are produced by the thermonuclear explosion of white dwarfs. Such an explosion arises when a white dwarf accretes gas from a companion star and reaches it Chandrasekhar limit around a mass of 1.44 $M_\odot$. Above that mass the electron degeneracy pressure in the core is insufficient to balance the gravitational self-attraction and the star collapses. The supernova explosion results from the outward-going shock wave generated by the bounce of the inner core. Since all SNe Ia involve a white dwarf with a well-defined mass 1.44 $M_\odot$, they have the same absolute luminosity and are qualified of **standard**

candles [409]. The measure of the luminosity-distance versus redshift allows to measure the expansion rate $H(z)$ and infer Ω_Λ.

- **Baryonic acoustic oscillations (BAO)**: As a consequence of the radiation pressure, an acoustic wave expands in the baryon-photon fluid from the moment of the Big-Bang until the time $t_{\rm dec}$ when the photons decouple. After the photons propagate freely, the sound horizon at decoupling stays imprinted in the baryon distribution and, after the non-linear evolution of the matter perturbations, at much later time, it appears as a little bump in the galaxy correlation function. The **sound horizon at recombination**, also known as the **BAO** scale, is well measured by Planck from the position of the first peak of the power spectrum of the temperature anisotropies in the CMB, cf. Fig. 3.2. The measurements of this **standard ruler** at different redshifts, using either galaxy redshift surveys or Lyman-α forests, allow to obtain the expansion parameter $H(z)$ and to measure Ω_Λ.

The cosmological constant:
The measurement of the presence of a fluid with equation of state $\omega = -1$ in the universe corresponds to add a constant Λ to the Einstein equations,

$$G_{\mu\nu} = \frac{1}{M_{\rm pl}^2}\left(T_{\mu\nu} + \Lambda\, g_{\mu\nu}\right), \qquad (3.135)$$

where $g_{\mu\nu}$ is the FLRW metric. This is the **Cosmological Constant (CC)**, first introduced by Einstein in 1917 as a mistake. From its appearance in Eq. (3.135), the CC coincides with the **vacuum energy** of our universe. The measured value of the constant in Planck unit is extremely small

$$\Lambda (= \rho_\Lambda = -p_\Lambda) = 3 M_{\rm pl}^2 H_0^2\, \Omega_\Lambda \simeq 5 \times 10^{-121}\, M_{\rm pl}^4 \simeq (1\ {\rm meV})^{1/4}\,. \qquad (3.136)$$

The reason why it constitutes 69% of the energy density of the universe in spite of being such a small number is that our universe is mainly made of vacuum.

The SM prediction:
In the SM of elementary particles, the vacuum energy comes from two contributions.

- **Zero-point energy**: In Quantum Field Theory, particles are quantum excitations of continuous fields that fill the whole space-time. A characteristic feature of quantum fields is that they constantly produce particles/anti-particles pairs which annihilate after a time $\lesssim \hbar/m$, in accordance with Heisenberg uncertainty principle. From the point of view of Feynman rules, the contribution of quantum fields to the vacuum energy are given by bubble diagrams.
- **Spontaneous symmetry breaking**: After each phase transition in the universe leading to spontaneous breaking of a symmetry (chiral symmetry when QCD confines, EW symmetry, supersymmetry, Peccei-Quinn symmetry, grand unified symmetry,...), a field operator gets a vev and the vacuum energy gets shifted.

Hence, the theoretical prediction of the vacuum energy is[12] [410]

$$\Lambda = \underbrace{\Lambda_{\rm bare}}_{\text{bare CC}} + \underbrace{\sum_i n_i \frac{m_i^4}{64\pi^2} \log \frac{m_i^2}{\mu^2}}_{\text{zero-point fluctuations}} + \rho_\Lambda^{\rm QCD} + \rho_\Lambda^{\rm EW} + \underbrace{\cdots}_{\text{other possible phase transitions}}, \tag{3.137}$$

where $n_i = (-1)^{2s_i} g_i$ where s_i and g_i are respectively the spin and number of degrees of freedom of the particle i. μ is the renormalization scale after dimensional regularization. The sum operates over all the fields in the universe. $\Lambda_{\rm bare}$ is the CC already present in the Lagrangian in the absence of the SM. Note that in the SM the negative contribution from the top/anti-top fluctuations largely dominates over the others. The vacuum energy of QCD is fixed by the gluon correlator [411], whereas the vacuum energy shift after EWPT is found from injecting Eq. (2.51) into Eq. (2.15), $\rho_\Lambda^{\rm EW} = m_h^2 v^2/8$. Hence, the prediction of the vacuum energy in the SM is

$$\Lambda = \underbrace{\Lambda_{\rm bare}}_{\text{bare CC}} + \underbrace{-(120\text{ GeV})^4}_{\text{zero-point fluctuations}} + \underbrace{(300\text{ MeV})^4}_{\rho_\Lambda^{\rm QCD}} + \underbrace{(105\text{ GeV})^4}_{\rho_\Lambda^{\rm EW}} + \underbrace{\cdots}_{\text{other possible phase transitions}}. \tag{3.138}$$

We can see that in order to compensate for the SM contribution $\rho_\Lambda^{\rm SM} = -(95\text{ GeV})^4$, we need to finely tune the bare constant $\rho_\Lambda^{\rm bare}$ at the level of 10^{-56}. Moreover, new physics at larger scales, e.g. GUT or Planck scale, could also contribute to the CC such that the needed fine-tuning could become even worst, e.g. 10^{-120}. This extreme fine-tuning is the **Cosmological Constant problem**. In addition, the CC appears to have started dominating the energy density of the universe only recently, around the cosmic age ~ 9 Gyrs. This is the **Coincidence problem** (there is even a triple coincidence $\rho_\Lambda \sim \rho_{\rm mat} \sim \rho_{\rm rad}$ [412]). Finally, another curious fact is the relation $\Lambda \sim \text{TeV}^2/M_{\rm pl}$, which is reminiscent of the **WIMP miracle**, which connects the measured Dark Matter abundance to the hierarchy problem, see Sect. 4.2.1.

Weinberg no-go theorem:
"The original symmetry of general covariance, which is always broken by the appearance of any given metric, cannot, without fine-tuning, be broken in such a way as to preserve the subgroup of space-time translations." Weinberg (1988 [413]).

In other words, a given universe with space-time metric $g_{\mu\nu}$, containing an adjusting scalar field ϕ, is **invariant under translations** (or under the whole Poincaré symmetry group) if $g_{\mu\nu}$ and ϕ are homogeneous and constant, which implies

1. $\frac{\partial \mathcal{L}}{\partial \phi} \approx 0 \quad \rightarrow \quad \frac{\partial V}{\partial \phi} \approx 0$: the adjusting field reaches a vev.

[12] Note that the naive estimation of the vacuum energy based on introducing a UV cut-off M on the momentum and leading to $\langle \rho \rangle = \frac{1}{2} \int_{|\vec{k}|<M} \frac{d^3\vec{k}}{(2\pi)^3}\, \omega(\vec{k}) \simeq \frac{M^4}{16\pi^2}$, violates Lorentz invariance and does not lead to the correct equation of state $\langle \rho \rangle = -\langle p \rangle$. Instead, one should use dimensional regularization. An interesting consequence is that the massless fields do not contribute to the vacuum energy. See [410] for more details.

2. $\frac{\partial \mathscr{L}}{\partial g_{\mu\nu}} \approx 0 \xrightarrow{\mathscr{L} \propto \sqrt{-g}\, V(\phi)} V(\phi) \approx 0$: the CC must vanish.

Any dynamical adjustment of the CC to a vanishing value, using the vev of a scalar field ϕ, needs to satisfy these two last conditions. Since the two conditions are **independent**, then their simultaneous enforcement requires **fine-tuning**. This is the **Weinberg no-go theorem** (1988 [413]). A possible resolution would be the existence of a symmetry such that the two conditions are not independent, but instead, they are related through

$$g_{\mu\nu} \frac{\partial \mathscr{L}}{\partial g_{\mu\nu}} = \kappa \frac{\partial \mathscr{L}}{\partial \phi}, \tag{3.139}$$

where κ is some parameter. We can check, using $\delta\sqrt{-g} = -\frac{1}{2}\sqrt{-g} g^{\alpha\beta} \delta g_{\alpha\beta}$, that this implies

$$\mathscr{L} = \sqrt{-g}\, e^{-2\kappa\phi} \mathscr{L}_0. \tag{3.140}$$

Then, the condition of Poincaré invariance can be fulfilled with a **unique** condition: $\phi \to \infty$. However, all masses in the Lagrangian, scale as $e^{-\phi}$, and would vanish concertedly with the CC. This corresponds to a scale-invariant universe which is not ours. Any proposed solution to the CC problem much find a may to evade the Weinberg no-go theorem. Either by cutting-off gravity at large scale, adding extra-dimensions, allowing the scalar field to vary in time, or introducing many scalar minima. See [410, 413–423] for some reviews on the Cosmological Constant problem and its attempted solutions.[13] We discuss some of them below.

Degravitate the CC in massive gravity:
The idea is to modify General Relativity at scales larger than $L \gtrsim 1/H_0$ such that the Newton constant $G_{\rm N}$ behaves as a **high-pass filter** [424, 425]

$$G_{\rm N}^{-1} G_{\mu\nu} = 8\pi T_{\mu\nu} \quad \to \quad G_{\rm N}^{-1} \left[1 + \left(\frac{m^2}{-\Box}\right)^{\kappa}\right] G_{\mu\nu} = 8\pi T_{\mu\nu}, \tag{3.141}$$

where $\frac{1}{2} < \kappa \lesssim 1$. Particularly, $\kappa = 1$ corresponds to massive gravity [426] whereas $\kappa < 1$ corresponds to having a continuum of massive gravitons. As a consequence, gravity is **screened** at large scales and the cosmology becomes insensitive to the vacuum energy. A prediction is a mass for the graviton around $H_0 \sim 10^{-33}$ eV. These effective field theories can find UV completions in higher dimensional models [427], see next paragraph. See [428–430] for reviews on modified gravity.

Self-tune the CC in 6D brane-world:
The claim is that our world could be localized on a (3+1)-brane in a $\gtrsim$ 6D space-time. The extra dimensions either have **infinite-volume** [431–433] (inspired from DGP model [434], they can be mapped to the case $\kappa \approx 1$ in Eq. (3.141) [427]), or **finite-volume** and **supersymmetric** [421, 435–437] (sub-milimeter, inspired from ADD model [438]). In those scenarios, the observed CC corresponds to the **brane tension**. With **more than two** extra-dimensions, interestingly the brane tension only warps the extra-dimensions but not the brane itself, similarly to a long cosmic string

[13] I thank all the participants of the **Workshop Seminar Series on Vacuum Energy** which has taken place in the DESY theory group between 28/10/2017 and 13/02/2018. An upcoming review written by the participants of the series is in preparation.

in 4D, which generates a deficit angle instead of curving itself under its tension. Therefore, an observer on the brane sees a flat curvature. See [429, 439] for reviews on brane-world gravity, and particularly [440] for recent applications of 6D brane models to the CC problem.

Self-adjust the CC with a scalar field:
The Weinberg no-go theorem assumes Poincaré invariance, not only at the level of the spacetime metric but also at the level of the self-adjusting scalar field. Hence, it may be interesting to drop the last assumption by allowing the scalar field to **vary in time**. We can even ask what is the most general covariant theory of gravity giving a viable self-adjusting mechanism? In order to address this question, in 2011, the authors of Ref. [441] have passed **Horndeski**'s theory of gravity, the most general covariant theory of gravity with second order equation of motion (first introduced in 1974 in [442] and revisited in [443]), through a **self-tuning filter**. They have obtained **Fab Four**, built on four operators, Einstein-Hilbert (George), Gauss-Bonnet (Ringo), double Hodge-dualized Riemann (Paul) and kinetic 'braiding' term (John)

$$\mathscr{L}_{\text{George}} = V_{\text{George}}(\phi)\, R, \tag{3.142}$$

$$\mathscr{L}_{\text{Ringo}} = V_{\text{Ringo}}(\phi)\left(R_{\mu\nu\alpha\beta}R^{\mu\nu\alpha\beta} - 4R_{\mu\nu}R^{\mu\nu} + R^2\right), \tag{3.143}$$

$$\mathscr{L}_{\text{Paul}} = V_{\text{Paul}}(\phi)\, \varepsilon^{\mu\nu\lambda\sigma}\varepsilon^{\alpha\beta\gamma\delta}R_{\lambda\sigma\gamma\delta}\, \nabla_\mu\phi\nabla_\nu\phi\nabla_\alpha\phi\nabla_\beta\phi, \tag{3.144}$$

$$\mathscr{L}_{\text{John}} = V_{\text{John}}(\phi)\left(R^{\mu\nu} - \frac{1}{2}g^{\mu\nu}R\right)\nabla_\mu\phi\nabla_\nu\phi, \tag{3.145}$$

A particular example is the k-essence scenario [444–446] where a negative pressure term arises from a non-linear kinetic term during matter-domination. The Fab Four theories admit a Minkowski vacuum for any value of the CC, even after an instantaneous change of the vacuum energy following a phase transition. However, the presence of light scalar field gives rise to **fifth force** constraints in the solar system such that we must employ a **screening mechanism**, e.g. Vainshtein mechanism (1972 [447]), where the fifth force is screened in local environment (at short length scales). Such a possibility is explored in Ref. [448] where the authors show for instance that Ringo in Eq. (3.143) should make himself tiny. They also demonstrate an apparent incompatibility between the requirement of having new effects in the IR, at horizon scale H_0^{-1}, and the requirement of maintaining perturbativity in the UV, down to mm, the shortest length scale at which the Newtonian law has been tested [449, 450].

Sequester the CC with global lagrange multipliers:
The Einstein equation are left intact locally but are modified **globally**, in the infinite wavelength limit. In addition to the vacuum energy of the SM, a CC **counterterm** Λ is introduced. The idea of **sequestering**, first exposed in 2013 in [451, 452], is to promote the CC counterterm Λ as well as the Planck mass M_{pl} to **global** dynamical degrees of freedom. Their equations of motion are global constraints which enforce the CC counterterm Λ to cancel the vacuum energy. Hence, the vacuum energy does not gravitate. A prediction is that the universe should be finite in space-time and therefore, should collapse in the future. Other cosmological consequences are

studied in [453]. However, we may still wonder what it means for a dynamical variable to be defined globally. In [454], a **local** version of sequestering is developed thanks to the introduction of auxiliary fields.

The weinberg window:
It can be shown that the values of the CC compatible with the existence of observers are very restricted and must be contained within the **Weinberg window**

$$-2 \times 10^{-119}\, M_{\rm pl}^4 \lesssim \Lambda \lesssim 5 \times 10^{-119}\, M_{\rm pl}^4. \tag{3.146}$$

The lower bound prevents the Universe to undergo a Big Crunch before galaxy formation, e.g. [455], while the upper bound forbids the possibility to have a CC domination before the galaxy formation (Weinberg 1988 [413]). This suggests that the CC may have an **anthropic origin**.

Dynamical relaxation of the CC on a landscape:
In 1984, Abbott [456] suggested a mechanism of dynamical adjustment of the CC based on a scalar field rolling down a slope with wiggles. This allows to relax the CC **step by step**, each time the scalar field **tunnels** from one minimum to the next one. In 1987, Brown and Teitelboim (BT) suggested to use 3-form fields $A^{\mu\nu\alpha}$ whose 4-form strenght tensor $F^{\mu\nu\alpha\beta}$ have the advantage to be **constant** in $(3+1)-D$, hence contributing to the CC [457, 458]. In this scenario, the CC is relaxed step by step each time a **membrane**—the $2D$ object charged under the 3-form gauge field—**nucleates**. The CC inside the membrane is decreased by one unit of membrane charge q. An effect analog to the Schwinger nucleation of e^-/e^+ pair in an electric field in $(1+1)-D$. However, the step-wise dynamical relaxation of the CC suffers from **two major flaws**.

- In order to be able to predict a CC within the Weinberg window in Eq. (3.146), the vacuum energy difference ε between two neighboring vacua in Abbott's model, or the 2-brane charge q in BT's model, must be smaller than $\lesssim 10^{-119}$ in Planck unit. This the **step-size problem**.
- The decay rate of the last vacua before entering inside the Weinberg window is ridiculously small. Hence, when the CC is finally adjusted to nearly zero, the universe has undergone an extremely long period of inflation and is nearly empty with no possibility of reheating. This is the **empty universe problem**.

In 2000, Bousso and Polchinski [419, 459] demonstrated that the two problems can be solved after introducing a **large number** N of four-form fields (such a large number of 4-form is moreover predicted in string compactification scenario). The parameter space of the N four-forms is a $N-$dimensional grid, whose inter-space is the membrane charge q, and in which the Weinberg window is represented by a shell. The trick is that, for a constant shell thickness, the number of grid points within the **Weinberg shell** grows extremely fast with the number of dimensions N. Hence, for $N \gtrsim 100$ we can get **more than one** vacuum configuration inside the Weinberg shell for membrane charge as large as $q \sim 1$ in Planck unit, which solves the step-size problem. The membrane charge being $q \sim 1$, the vacuum energy difference during the last step is also of order 1 in Planck unit. The conversion of this large vacuum energy into thermal energy leads to a hot big-bang and the empty universe problem is solved.

Upon introducing an operator (which is non-renormalizable in [460, 461] and renormalizable in [462, 463]) which couples the four-form fields with the Higgs field, the Bousso-Polchinski mechanism can be used to scan over the EW scale. Note that in that case, four-form fluxes act as a connection between the landscape of CCs and the landscape of EW scales. See also [464] where the authors discuss a possible connection between the CC and EW-scale landscapes in another context.

Now we discuss another possibility to solve the empty universe problem. Introduced in [465, 466], it consists to relax the CC to negative values, $-(\text{MeV})^4$, such that the universe crunches and undergoes a bounce, followed by a reheating phase. The inherent difficulty with bouncing cosmology scenarios is that Einstein equations predict an unavoidable singularity at $a \to 0$. From a combination between the Friedmann equation and the continuity equation, we can write

$$\frac{\ddot{a}}{a} - \left(\frac{\dot{a}}{a}\right)^2 = -\frac{1}{2M_{\text{pl}}^2}(\rho + p) + \frac{k}{a^2}. \tag{3.147}$$

From this, we can see that a way to allow for a bounce defined by

$$\text{Bounce}: \qquad \dot{a} = 0, \quad \ddot{a} > 0, \tag{3.148}$$

and therefore to avoid the singularity at $a \to 0$, is to introduce a fluid which violates the null energy condition (NEC)[14]

$$\text{NEC violation}: \qquad \rho + p < 0. \tag{3.149}$$

Assuming that the bounce can be realized, the negative CC $-(\text{meV})^4$ can be converted to a positive CC $+(\text{meV})^4$ through a phase transition [468]. More popular options for avoiding singularities rely on **beyond-Einstein** [270, 271, 329, 469–472] or **quantum gravity theories** [473]. See [266–271] for reviews on bouncing cosmology.

Eternal inflation and multiverse:
In scenarios where the CC is relaxed by vacuum tunneling, the quantum transition between two vacua is generally **much longer** that the Hubble time, which implies that there is at most one membrane of new vacuum per Hubble patch. Even if the

---

[14] From Eq. (3.147), we could think of producing a bounce using the positive curvature $k > 0$ of the universe, in light of the recent results [263]. The inherent difficulty is that the curvature term blueshift as $\propto a^{-2}$ too slowly compared to the other fluids, e.g. radiation $\propto a^{-4}$ or kination $\propto a^{-6}$, such that the curvature would hardly dominate the energy budget when $a \to 0$. Then the issue becomes how to convert, when the universe crunches, fast-blue-shifting fluids having $\omega > -1/3$ (e.g. matter, radiation, kination) with $\rho \propto a^{-3(1+\omega)}$, into slow-blue-shifting fluids $\omega < -1/3$ (e.g. dark energy [467]), such that the curvature $\omega = 1/3$ can take over when $a \to 0$. This is challenged by the second law of thermodynamics since the faster the blue-shift, the larger the entropy $s \propto (1 + \omega)$ (e.g. kination has the largest entropy while dark energy has a vanishing entropy).

just nucleated membrane of new vacuum expands at the speed of light, it is not fast enough to catch up with the expansion of the universe. Therefore, the old vacuum can not be entirely converted in the true vacuum. There is always some of the old vacuum left, which **inflates eternally**. We say that the membranes (or bubbles) do not **percolate** [297].[15] The consequence is the formation of a **multiverse** containing an infinite number of causal patches corresponding to different **membrane cascades**. In this picture, our universe would correspond to one of the membrane cascades among those ending up inside the Weinberg window in Eq. (3.146). However, the eternal inflation in a landscape suffers from **two main issues**, see also [474].

- The different patches in the multiverse are causally disconnected. Hence, the multiverse seems **not experimentally testable**. See however some attempts [475, 476].
- The multiverse scenario suffers from a loss of predictability. Not only the total number of distinct patches are infinite, but also the number of patches with the correct CC. *Anything that can happen will happen, and it will happen an infinite number of times* [477]. Hence, the probability that a given patch has a CC inside the Weinberg window can not be computed, $p = \frac{\infty}{\infty}$. This is the **measure problem**.

Note that the scenario of eternal inflation may be challenged [478] if the recent claim [263] that we live in a closed universe is confirmed.

In [479], eternal inflation and its intrinsic measure problem is evaded by **crunching** all the patches after some time using a **supercooled first-order phase transition** with AdS true vacuum. The patches with the largest CC are the ones with the temperature decreasing the fastest and which crunch the fastest. Conversely, regions with a small CC are long-lived.

In this section, we have outlined a set of creative ideas that have been put forward to solve the cosmological constant problem. However, it is fair to say that none of them is satisfactory. This is one of the biggest puzzles in high-energy physics and cosmology. We now move to the Baryon Asymmetry of the Universe (BAU).

3.5.2 *Matter-Anti-Matter Asymmetry*

When Dirac combined Quantum Mechanics and Special Relativity in 1928 in order to generalize the Schrodinger equation to relativistic particles [480, 481], he predicted the existence of **antimatter**. Dirac's prediction, followed with the discovery of the positron in 1932 by Anderson after investigation of cosmic rays in a cloud chamber [482, 483], is a major breakthrough in the history of science. Dirac received the Nobel prize the next year in 1933.

[15] However, in case of multiple bubble nucleation per Hubble volume, if the nucleation rate increases fast enough with time, percolation can occur during an inflating period. See Sect. 6.1.4 in Chap. 6 for a quantitative criterium.

However, our universe appears to have a large deficit of anti-matter with respect to matter. See e.g. [484] for a review. The small fraction of anti-protons and positrons in cosmic rays is well explained by secondary production processes. The existence of some region of the universe filled with anti-matter is ruled-out due to the non-observation of gamma-rays coming from matter/anti-matter annihilation, on scales ranging from the solar system, to the Milky-Way, to galaxy clusters, and even to distances comparable to the scale of the present horizon [485, 486]. Anyway, there are no known mechanisms to separate domains.

In Sect. 3.2.2, we explicitly showed that the baryon-to-light ratio n_b/n_γ, if one assumes a symmetric universe, would be much smaller than the observed ratio, hence demonstrating the need of a primordial baryon/anti-baryon asymmetry. Inflation is expected to dilute any pre-existing matter asymmetry, so the initial conditions are set to $B = L = 0$, where $L = L_e + L_\mu + L_\tau$, and we must provide a mechanism for generating $B \neq 0$.

In 1967, **Sakharov** gave three conditions that any theory seeking to explain the baryon asymmetry must satisfy [487]

- B violation: otherwise the baryon number is a conserved quantity.
- C and CP violation: otherwise for each process $\mathscr{P}$ generating B, there would exist a C (or CP) conjugate process $\bar{\mathscr{P}}$ which would generate $-B$ with the same probability.
- Out of equilibrium: otherwise the system is time-invariant and $\frac{d}{dt}B(t) = 0$.

In Sect. 2.2.1 of Chap. 2, we discussed that the baryonic and leptonic charges B and L were a priori conserved in the SM at tree level. But actually, B and L are violated by non-perturbative processes called **sphalerons.**[16] This results from the chirality of the weak interactions (only left-handed fermions are charged) which leads to the anomalies $SU(2)_L^2U(1)_B$ and $SU(2)_L^2U(1)_L$, see Sect. 2.2.5 of Chap. 2. As a consequence, we can write

$$\partial_\mu j_B^\mu = \partial_\mu j_L^\mu = \frac{n_f}{32\pi^2}\left(g_2^2\, W^a_{\mu\nu}\tilde{W}^{a\mu\nu} - g_1^2 B_{\mu\nu}B^{\mu\nu}\right), \tag{3.150}$$

where j_B^μ and j_L^μ are the B and L Noether currents. Note that only $B + L$ is violated while $B - L$ is conserved. 't Hooft was the first to introduce $B + L$ violation by non-perturbative effects in 1976 [489]. In his paper, he showed that the tunneling amplitude is extremely suppressed at zero temperature $\mathscr{A} \sim e^{-8\pi^2/g_2^2} \sim e^{-170}$. In 1978, Dimopoulos and Susskind proposed that such processess are active at high temperature [490] and a number of subsequent papers investigated this effect [488, 491–495]. Particularly, in 1985, Kuzmin, Rubakov, and Shaposhnikov [493] computed that sphaleron transitions are at thermal equilibrium in the early universe above 130 GeV (see [496] for recent lattice simulations). More particularly, the sphaleron rate per unit of volume goes like

[16] The world 'sphalerons' which means 'ready to fall' in Greek, was proposed by Klinkhamer and Manton in 1984 [488].

$$\Gamma_{\rm sph} \sim \left(\frac{E_{\rm sph}}{T}\right)^3 \left(\frac{m_W}{T}\right) T^4\, e^{-E_{\rm sph}/T} \qquad T \lesssim 130\ \text{GeV}, \tag{3.151}$$

$$\Gamma_{\rm sph} \sim 30\,\alpha_w^5\, T^4, \qquad T \gtrsim 130\ \text{GeV}, \tag{3.152}$$

where $\alpha_w = g_2^2/4\pi \sim 1/30$ and $E_{\rm sph} = 8\pi v/g_2 \simeq 10$ TeV is the height of the energy barrier between two neighboring values of $B + L$. Hence, above $T \gtrsim 130$ GeV, any $B + L \neq 0$ will be quickly relaxed to $B + L \simeq 0$, while $B - L \neq 0$ will remain conserved, such that the final baryon asymmetry will be $B_{\rm eq} \simeq -L_{\rm eq}/2$. More precisely, thermal equilibrium of various processes[17] in the SM implies [498]

$$B_{\rm eq} = \frac{8N_f + 4}{22N_f + 13}(B_{\rm eq} - L_{\rm eq}), \tag{3.153}$$

where $N_f = 3$ is the number of flavors.

We list below the main proposals for baryogenesis.[18]

Baryogenesis in Grand Unified Theories (GUT):
Models which unify strong and electroweak interactions, the simplest being $SU(5) \to SU(3)_c \times SU(2)_L$ proposed in 1974 by Georgi and Glashow [500],[19] put leptons and hadrons in the same multiplet, unavoidably leading to baryon number violation. Hence, a baryon asymmetry can be created by the **decay** of a heavy particle X with mass comparable to the GUT scale $M_X \sim 10^{16}$. When the decay occurs **out of equilibrium**, namely when $\Gamma_X \lesssim H$ for $T \sim M_X$, the baryon asymmetry is given by

$$n_{\rm B}/s \sim \frac{\kappa}{g_*} \sum_f B_f \frac{\Gamma(X \to f) - \Gamma(\bar{X} \to \bar{f})}{\Gamma_X}, \tag{3.154}$$

where the sum is performed over all the final state f of the X decay. f has baryon number B_f. The total decay rate of X and $\bar{X}$ are equal, $\Gamma_X = \Gamma_{\bar{X}}$, because of CPT invariance and unitary. κ accounts for the possible partial wash-out due to inverse-decay if the out-of-equilibrium conditions are not satisfied

$$\kappa \sim \begin{cases} 1 & \text{if } \Gamma_X \lesssim H\Big|_{T \simeq M_X}, \\ \frac{H}{\Gamma_X}\Big|_{T \simeq M_X} & \text{if } \Gamma_X \gtrsim H\Big|_{T \simeq M_X}. \end{cases} \tag{3.155}$$

---

[17] Sphalerons transitions only acts on left-handed fermions which explains why $B_{\rm eq}$ is not exactly equal to $-L_{\rm eq}/2$. In addition to be set by $SU(2)$ EW instantons, the baryon number at equilibrium in Eq. (3.153), also relies on $SU(3)$ instantons which equilibrate left-handed with right-handed quarks, hypercharge-conserving processes which equilibrate Y, and on Yukawa interactions which relate the population of left-handed with right-handed fermions [497].

[18] Recent years have seen the development of a variety of new theoretical ideas for explaining the baryon asymmetry of the universe. They can be very different from the mainstream mechanisms discussed in this book, and we refer to [499] for a review.

[19] The GUT based on $SO(10)$ was proposed the same year by Fritzsch and Minkowski [501].

Particularly, if X decays into $f = a,\ b$, then we obtain

$$n_{\mathrm{B}}/s \sim \frac{\kappa}{g_*}(B_a - B_b)\,\frac{\Gamma(X \to a) - \Gamma(\bar{X} \to \bar{a})}{\Gamma_X}, \tag{3.156}$$

where we have used $\Gamma_X = \Gamma(X \to b) + \Gamma(X \to a)$ and idem for $\bar{X}$. We can explicitly see that we need **B-violation**, $B_a \neq B_b$, as well as **CP-violation**, $\Gamma(X \to a) \neq \Gamma(\bar{X} \to \bar{a})$, satisfied together to produce a baryon asymmetry.

However in practice, there are some difficulties with GUT baryogenesis. Any **entropy production** below the GUT scale, due to inflation or the decay of heavy relic, would **dilute** the baryon asymmetry. Particularly, the upper bound on the inflationary vacuum energy density, from B-modes non-obervation in the CMB, is very close to the GUT scale, $V_{\mathrm{inf}}^{1/4} = (1.88 \times 10^{16})(r/0.10)^{1/4}$, with the scalar-to-tensor ratio r constrained to $r \lesssim 0.055$ [259]. Another flaw is that minimal GUT models conserve $B - L$ such that any B-asymmetry produced at the GUT scale is subsequently washed-out by sphaleron transitions. Note also that minimal GUT models, like the simple $SU(5)$, are constrained by the lower bound on the proton decay, $\tau_p/BR(p \to e^+\pi^0) > 1.67 \times 10^{34}$ years [260, 502]. See some of the pioneering articles in 1978 [490, 503–508] and some reviews [509–511].

Affleck-dine mechanism:
A non-zero baryon number can be generated dynamically by the **coherent oscillation** of a scalar field ϕ charged under baryon number, in a potential $V(\phi)$ containing correction terms which explicitly break baryon number. The baryon asymmetry is given by the current

$$n_{\mathrm{B}} = \mathrm{Im}\left[\phi^\dagger \dot{\phi}\right], \tag{3.157}$$

and its evolution follows from the Klein-Gordon equation for ϕ

$$\dot{n}_{\mathrm{B}} + 3Hn_{\mathrm{B}} = \mathrm{Im}\left[\phi \frac{\partial V}{\partial \phi}\right]. \tag{3.158}$$

Explicit breaking of baryon baryon number can be provided by $V(\phi) \supset \frac{\phi^n}{\Lambda^{n-4}} + \mathrm{h.c.}$ with n odd. The baryon number is transmitted to the SM when the scalar condensate decays . See the seminal article in 1984 [512], a review [510] and a more recent study involving the QCD axion [513–515] .

Spontaneous baryogenesis:
In 1987, Cohen and Kaplan (and later Nelson) [516, 517] proposed a minimal scenario for generating a baryon asymmetry simply based on

$$\mathscr{L} \supset V(\phi) + \frac{1}{f}\partial_\mu \phi\, j_{\mathrm{B}}^\mu + \mathcal{O}_{\not{B}}. \tag{3.159}$$

The first term induces a **motion of the scalar field** $\dot{\phi} \neq 0$, which thanks to the second term, generates an energy difference between baryons and anti-baryons by an amount

$2\dot{\phi}/f$

$$\Delta\mathscr{H} = \frac{\dot{\phi}}{f}\left(n_b - n_{\bar{b}}\right) = -\mu\,\Delta B, \qquad \text{(assuming } \partial_i\phi = 0\text{)}. \tag{3.160}$$

This is equivalent to give b and $\bar{b}$ a non-vanishing opposite **chemical potential** $\mu_{\bar{b}} = -\mu_b = \dot{\phi}/f$. If additionally we switch on some B-violation operator $\mathscr{O}_{\not{B}}$, then we would start producing a net B-number $\Delta B \neq 0$. In the presence of $\dot{\phi} \neq 0$, the CPT symmetry is spontaneously broken which is another Sakharov condition. If the B-violating interactions freeze-out, at $T_{\rm fo}$, before ϕ stops rolling, then a residual baryon asymmetry remains

$$n_{\rm B}/s \sim \frac{\dot{\phi}}{f\,g_*\,T_{\rm fo}}. \tag{3.161}$$

Spontaneous baryogenesis can occur during axion inflation either from the operator $\frac{\phi}{f}F\tilde{F}$ or $\frac{\partial_\mu\phi}{f}\bar{\psi}\gamma^\mu\gamma_5\psi$ [518–520]. These two operators are equivalent up to a field rotation, and [521] have explicitly checked that the transport equations for the baryonic charge $n_{\rm B}$ are independent of this choice of representation. See [522] for a realization of spontaneous baryogenesis with the relaxion.

Spontaneous breaking of CPT has also been proposed to be generated by the expansion of the universe through the operator [523]

$$\sqrt{-g}\frac{(\partial_\mu\mathscr{R})}{M_{\rm pl}}J^\mu_{\rm B}, \tag{3.162}$$

where $\mathscr{R}$ is the Ricci scalar curvature whose time derivative

$$\dot{\mathscr{R}} = \sqrt{3}(1-3\omega)(1+\omega)\frac{\rho^{3/2}}{M_{\rm pl}^2}, \tag{3.163}$$

is different from zero when the equation of state ω departs from pure-radiation or pure-inflation. This mechanism is called **gravitational baryogenesis**. Interestingly, the needed source of B-violation $\mathscr{O}_{\not{B}}$ can be generated by the Hawking radiation from black holes [524].

Leptogenesis:
This is the successor of GUT baryogenesis, the role of the heavy decaying particle is played by a heavy right-handed neutrino N_I which decays into a left-handed neutrino and a Higgs. Right-handed neutrino are motivated for explaining the neutrino oscillations in the seesaw model, see Sect. 2.3.2 of Chap. 2. Since right-handed neutrinos do not carry lepton number, their decay violates the L number. The decay of N_I must occur out of equilibrium, which implies $\Gamma_{N_I} \lesssim H$ when $T \lesssim M_{N_I}$. Assuming it is the case, then the L asymmetry is given by

$$\frac{n_L}{s} \simeq \frac{\kappa}{g_*} \sum_\alpha \frac{\Gamma(N_I \to l_\alpha\, h^\dagger) - \Gamma(N_I \to \bar{l}_\alpha\, h)}{\Gamma(N_I \to l_\alpha\, h^\dagger) + \Gamma(N_I \to \bar{l}_\alpha\, h)}, \tag{3.164}$$

where the sum runs over SM leptons. The needed CP violation, $\Gamma(N_I \to l_\alpha\, h^\dagger) \neq \Gamma(N_I \to \bar{l}_\alpha\, h)$, is introduced via complex phases in the neutrino mass matrix. It vanishes at tree level and the leading contribution comes from an interference between the tree-level and one-loop diagrams. κ is an efficiency factor which takes into account an eventual wash-out due to inverse decay and other L-changing scattering processes. It is given by a similar expression as Eq. (3.155). Then, the L number is subsequently converted into B number via sphaleron transitions, see Eq. (3.153).

Leptogenesis was introduced in 1986 by Fukugita and Yanagida the year after sphalerons were shown to reach thermal equilibrium in the early universe [525]. See some reviews [497, 526–531].

At first, it was asserted that successful leptogenesis requires the right-handed neutrinos to be heavier than $m_{\nu_R} \gtrsim 10^9$ GeV [532]. But then, it was shown that it can be relaxed to $m_{\nu_R} \gtrsim 1$ TeV using nearly-degenerate right-handed neutrino masses [533]. It has even been reduced to the keV/GeV range by Shaposhnikov et al. [534] using CP-violating right-handed neutrino oscillations [535]. In the latter model, called Neutrino Minimal Standard Model, the KeV sterile neutrino can be a Dark Matter candidate [484, 536]. See [537–539] for reviews on sterile neutrinos.

Electroweak baryogenesis:
Baryon creation can occur in the vicinity of expanding bubble walls during an electroweak first-order phase transition. When the interactions between particles in the plasma and the Higgs bubble violate CP, a CP asymmetry form in the symmetric phase, outside the bubble, which sphalerons convert to B-asymmetry. Then, due to the wall motion, the baryons diffuse into the broken phase, inside the bubble, where the sphaleron rate is exponentially suppressed by $\langle\phi\rangle / T$. The baryon asymmetry is given by

$$n_B/s \sim \frac{1}{v_w\, g_*\, T_{\text{nuc}}^4} \int_{-\infty}^{\infty} dz\, \Gamma_{\text{sph}}\, \mu_L(z)\, e^{\,c \int_{-\infty}^{z} d\tilde{z}\, \Gamma_{\text{sph}}/T_{\text{nuc}}^3}, \tag{3.165}$$

where z is the radius coordinate along the bubble wall profile (broken phase in $z = -\infty$, wall in $z = 0$ and symmetric phase in $z = +\infty$) and v_w is the bubble wall velocity. $\mu_L(z)$ is the local chemical potential of the left-handed fermions in the plasma. It can be computed by solving a diffusion equation [540–542]. Γ_{sph} is the sphaleron rate given by Eqs. (3.151) and (3.152). In order to prevent the sphaleron wash-out inside the bubble (due to the exponential term), the nucleation temperature T_{nuc} of the phase transition must satisfy $v/T_{\text{nuc}} \gtrsim 1$. For the same reason, the reheating temperature[20] must satisfy $v/T_{\text{RH}} \gtrsim 1$. If these last two conditions are satisfied, we obtain

$$n_B/s \sim \frac{\Gamma_{\text{sph}}\, \mu_L\, L_w}{v_w\, g_*\, T_{\text{nuc}}^4}, \tag{3.166}$$

[20] For clarity, we omitted the dilution factor $(T_{\text{nuc}}/T_{\text{RH}})^3$ in Eq. (3.165).

where L_w is the typical wall thickness.

In the SM, the electroweak phase transition is a **cross-over** and needs BSM physics to make it **first-order**, e.g. supersymmeric extensions [543], higher dimensional operator [544], a second scalar [545, 546] or varying Yukawa [547].

Also the CP violation from the **CKM matrix** appears to be to small for successful baryogenesis, since it appears at the 12th order in Yukawa couplings, see Sect. 2.2.4 in Chap. 2 or [548–551]. It is then necessary to introduce new sources of CP violation, e.g. supersymmeric extensions [552, 553], two-higgs doublet model [554, 555], strong CP violation [556, 557], varying Yukawa [558–561] or Composite Higgs scenario [562–564].

It is commonly assumed [565] that successful baryogenesis requires the speed of the wall to be within the range $10^{-3} \lesssim v_w < 1/\sqrt{3}$, where $1/\sqrt{3}$ is the speed of sound in a relativistic plasma. However, loopholes to the claim that successfull baryogenesis requires subsonic wall have been proposed [566–571]. This would offer the possibility to test electroweak baryogenesis models with GW [567, 572]. Another proposal based on particle production at the wall boundary allows baryogenesis via relativistic bubble walls $v_w \simeq 1$ [573, 574].

See the first implementations of electroweak baryogenesis in 1990 by Cohen, Kaplan and Nelson [575–577], based on the proposal of Kuzmin, Rubakov, and Shaposhnikov in 1985 [493], and some reviews [509, 511, 530, 531, 578–580].

3.5.3 *Dark Matter Puzzle*

So far, DM has only been observed indirectly through its **gravitational** effects at the scale of individual galaxies, clusters of galaxies, or at larger scales. For reviews on DM history, observations, models and searches, we refer to [581–606]. For textbooks, we refer to [45–50, 607].

Galactic scale:
The most famous evidence of dark matter is probably the disagreement between the mass distribution $M_{\text{visible}}(R)$ of spiral galaxies based on the observed **light distribution**, as we can see in Fig. 3.3, assuming a constant mass-to-light ratio, and the mass distribution infered from the **circular velocity** V_{circ} of stars and gas. The discordance arises in the outer region and can be accounted for by invoking the presence of a **spherical halo of dark matter** which extends far beyond the visible galaxy. The discrepancy was first pointed out by Vera Rubin, Kent Ford and Norbert Thonnard in 1978 [609, 610] and we refer to [611] for an old good review written at that epoch. From Newton's law of gravitation, the circular velocity follows

$$V_{\text{circ}}^2 = \frac{GM_{\text{halo}}}{R} + \frac{GM_{\text{visible}}}{R} + \frac{GM_{\text{gas}}}{R}, \tag{3.167}$$

where V_{circ} is measured from the Doppler shift of the H_α lines for the ionized hydrogen in the inner part, and from the Doppler shift of the 21-cm lines for the neutral

Fig. 3.3 Overall view of Andromeda galaxy (M31) taken by Hubble telescope. The visible matter only represents ∼20% of the total mass orbiting around the galaxy, the remaining ∼80% being under the form of Dark Matter

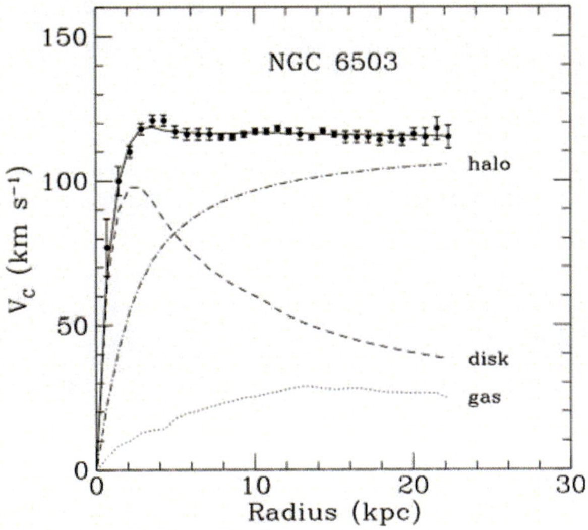

Fig. 3.4 The rotation curves of a galaxy (NGC 6503 here) is well-fitted (solid) upon introducing a spherical dark-halo (dot-dashed) which adds up to the baryonic components measured optically (dashed) and the gas component measured with the 21 cm line of neutral hydrogen (dotted), cf. Eq. (3.167). The fitting parameters are the mass-to-light ratio of the disc, the halo core radius, and the halo asymptotic circular velocity. Figure reproduced from [608]

hydrogen in the outer part, respectively. $M_{\text{visible}}(R)$ and $M_{\text{gas}}(R)$ are infered from the light distribution observed optically and with 21 cm. Assuming a given mass-to-light ratio of the disc, M_{halo} is the spherical dark matter halo needed to resolve the inconsistency. A rotation curve and its appropriate fit assuming the presence of a spherical dark matter halo is shown in Fig. 3.4. Without dark matter halo, we would expect the measured rotation curve to coincide with the one infered from the baryonic mass, namely disk + gas lines, where in that case $V_{\text{circ}} \propto 1/R$. Instead, the measured rotation curve does not decrease but quickly flattens, hinting for a conspiracy between baryonic matter and dark matter at intermediate radius and for $M_{\rm halo} \propto R \ \rightarrow \ \rho_{\rm halo} \propto 1/R^2$ at large radius.

Inter-galactic scale: first dark matter evidence:
Historically, the first proposal for the existence of dark matter goes back to Fritz Zwicky in 1933 and 1937 [613, 614] who found a discrepancy between the **optical mass** of the Coma cluster infered from the measurement of its visible light emission and the **dynamical mass** measured from the average dispersion velocity of its constituent galaxies. More precisely, the **Virial theorem** allows to write [614]

$$\langle V_{\rm s}^2 \rangle = \frac{5GM}{3R}, \tag{3.168}$$

where $\langle V_{\rm s}^2 \rangle$ is the average quadratic velocity along the line-of-sight, R and M are the radius and total mass of the Coma cluster. After taking $R \simeq 6.5$ Mpc and $\langle V_{\rm s}^2 \rangle \simeq (700\ \mathrm{km/s})^2$, Zwicky computed the total mass $M \simeq 4.5 \times 10^{13}\ M_{\odot}$, which after division by the measured total luminosity $L = 8.5 \times 10^{10}\ L_{\odot}$, led him to the **mass-to-light ratio** of the Coma cluster

$$M/L \sim 500, \tag{3.169}$$

which is two orders of magnitude above the solar neighborhood value. Hence, Zwicky claimed that most of the matter constituting the Coma cluster is **dark**. His result was not taken seriously during his time, and unfortunately, Zwicky died in 1974, few years before the seminal work from Rubin et al. got published. More precise analysis have later confirmed Zwicky's finding, e.g. [615] which finds $M/L \simeq 351$.[21]

Inter-galactic scale: structure formation (analytical):
Nowadays, the mass of a cluster can be computed by at least three different methods [616]

- Measurement of the dispersion velocities of individual galaxies as initiated by Zwicky.
- Measurement of the temperature of the gas, which can be related to the gravitational potential assuming hydrostatic equilibrium [586].
- Measurement of the gravitational mass through lensing effects, cf. Fig. 3.6.

By taking the mean value of the three independent measurements explicited above, we can minimize the uncertainties. Then, we can precisely measure the number of clusters, for each mass range (an histogram), as a function of the redshift. On the theory side, this number of condensed objects for each mass range, as a function of the redshift, can be predicted, assuming a given dark matter abundance, in the Press-Schlechter formalism [617]. The comparison between observations and analytical

---

[21] The reason for such a large mass-to-light ratio is not only attributed to dark matter but also to the presence of a lot of gas. More precisely, the study in [615] finds in the Coma cluster only 2% of stellar mass against 13% of gas and 85% of dark matter.

models allows to measure the dark matter abundance at the inter-galactic scale, e.g. $\Omega_{DM} = 0.22^{+0.25}_{-0.1}$ [616] (study performed in 1998).

Inter-galactic scale: structure formation (numerical):
In the last decade [618], the major improvement of the computational power has permitted **N-body simulations** of large portions of universe, e.g. the 2014 Illustris project [612] which covers a volume (106.5 Mpc)^3 with mass resolution as low as 10^6 $M_\odot$ and includes baryonic effects as stellar feeback or supermassive black hole growth, or the 2015 APOSTLE project [619] which simulates the Local group environment. As shown in Fig. 3.5, the prediction by Illustris of the star formation rate as a function of the redshift is in good agreement with observations. This illustrates the success of the ΛCDM cosmological model where 26% of the energy density of the universe is contained under the form of cold dark matter and 5% under the form of baryons.

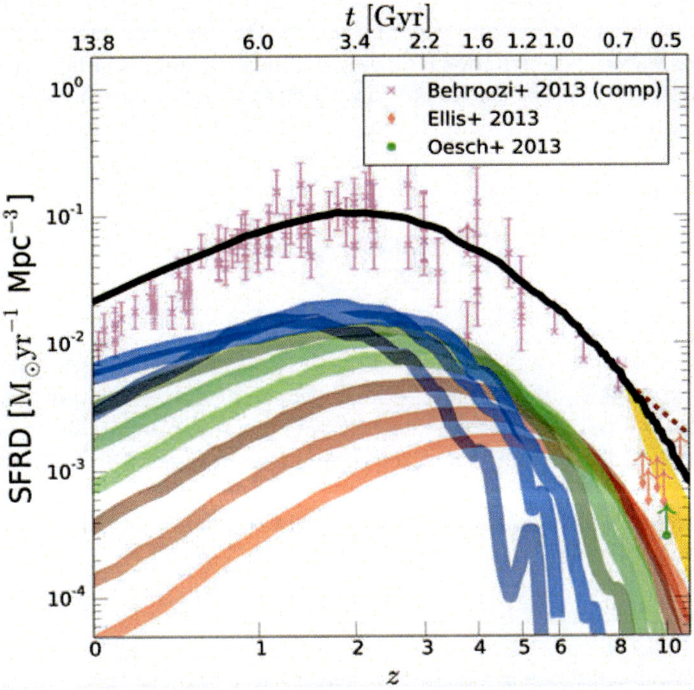

Fig. 3.5 Success of ΛCDM numerical simulations [612] in predicting the Star Formation Rate Density (SFRD), compared to observations (Behroozi et al. 2013a, Ellis et al. 2013, Oesch et al. 2013). Star formation depends on the details of structure growth which is itself sensitive to the dark matter fraction in the universe. The black solid line shows the total SFRD whereas colored thick lines show the contributions from galaxies of different stellar masses ranging from low mass (red lines: $M_* = 10^7$, $10^{7.5}$, 10^8 $M_\odot$), to intermediate mass (green lines: $M_* = 10^{8.5}$, 10^9, $10^{9.5}$ $M_\odot$), to the most massive systems (blue lines: $M_* = 10^{10}$, $10^{10.5}$, 10^{11} $M_\odot$)

Inter-galactic scale: gravitational lensing:
By studying the photon trajectory in Schwarzschild geometry, we find that the **deflection angle** $\Delta\phi$ of a light ray passing nearby an object of mass M with an impact parameter b is [124]

$$\Delta\phi = \frac{4GM}{b}. \tag{3.170}$$

When considering full images, bending of light leads to **gravitational lensing** which can be used to detect the presence of Dark Matter. By analysing the distortion of the light emitted by the background, we can map the mass distribution of the dark matter in the foreground. The lensing is said **strong**, when we observe multiple images of the same physical source. If the foreground object responsible for the distorsion is spherically symmetric, we can even observed **Einstein rings** [620]. In the **weak** lensing regime, we can observe magnification and shear.

Well-known examples of weak lensing are the **bullet cluster** and the **baby bullet** shown in Fig. 3.6. On those images, we can see two clusters after they have collided. On the one hand, the gas traced by its X-ray emission in purple, is found mostly on the collision site at the center of the image. On the other hand, the two dark matter halos traced by gravitational lensing in blue are found at large distance from the collision site, as so for the galaxies in yellow. We deduce that in contrast to the gas which has strongly interacted during the collision, the dark matter halos have passed through each other. Hence, besides revealing the existence of dark matter, the analysis provides an upper bound on the self-interaction cross-section of dark matter [621–623]

$$\frac{\sigma}{M_{\text{DM}}} \lesssim 1\ \text{cm}^{2/\text{g}}. \tag{3.171}$$

Cosmological scale:
The observation of the temperature anisotropies imprinted in the **Cosmic Microwave Background (CMB)** , the furthest photons we can observe in the universe, by the satellite Planck from 2009 to 2013, see Figs. 3.1 and 3.2, provides the most precise measurement of Dark Matter abundance. As shown in Fig. 3.7, the amplitudes and positions of the peaks of the temperature power spectrum depend crucially on the DM abundance. In the Planck 2018 paper [259], we read the dark matter and baryon density in the universe

$$\Omega_{\text{DM}}h^2 = 0.1200 \pm 0.0011, \quad \Omega_{\text{b}}h^2 = 0.02237 \pm 0.00015, \tag{3.172}$$

where $h = 0.674 \pm 0.005$ is the Hubble constant in unit of 100 km/s/Mpc. In order to further reduce the uncertainties, it is useful to combine the CMB analysis with the measurement of the scale of **Baryonic Acoustic Oscillations** (which is around 147 Mpc today) [259]

$$\Omega_{\text{DM}}h^2 = 0.11911 \pm 0.00091, \quad \Omega_{\text{b}}h^2 = 0.02242 \pm 0.00014. \tag{3.173}$$

Fig. 3.6 The 'bullet cluster' and the 'baby bullet'. The background pictures show the location of galaxies in yellow. On top, the hot, intra-cluster gas in purple is traced from its X-ray emission whereas the total mass in blue is reconstructed from gravitational lensing measurement. It is found that there is far more mass than that of the galaxies, providing strong evidence for the presence of dark matter. Moreover, the coincident location of dark matter at the same place as the galaxies implies that dark matter has a similarly small interaction cross-section. Figure reproduced from [620]. For pictures with more quantitative informations, e.g. the gravitation lensing contours, see [624]. Figure credit: Left: X-ray: NASA/CXC/CfA/ M.Markevitch et al.; Lensing Map: NASA/STScI; ESO WFI; Magellan/U.Arizona/ D.Clowe et al. Optical image: NASA/STScI; Magellan/U.Arizona/D.Clowe et al.; Right: NASA/ESA/M.Bradac et al.

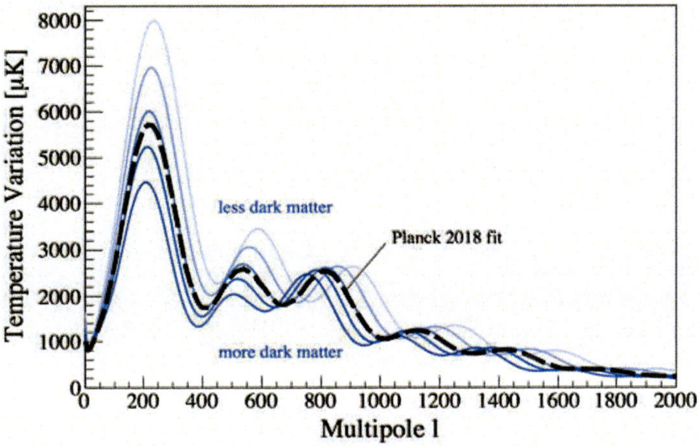

Fig. 3.7 Power spectrum of the temperature anisotropies in the Cosmic Microwave Background (CMB), cf. Eq. (3.57), for a dark matter density contribution Ω_{DM} varying between 0.11 and 0.43 (blue lines). The dotted black lines show the best fit to the Planck data [259]. The small (large) multipoles l correspond to the long (small) angular scales: the main acoustic peak at $l = 200$ corresponds to 1 degree on the sky whereas $l = 1800$ corresponds to 0.1 degree. Figure reproduced from [625]

Fig. 3.8 Most studied DM candidates. Figure reproduced from [605]

Dark matter candidates:
There are very numerous DM candidates from masses as small as 10^{-20} eV for the ultra-light bosons to masses as large as $10^{-11}\ M_\odot = 10^{46}$ eV for primordial black holes (PBH). Figure 3.8 shows the most studied DM candidates [605, 635]. We discuss some of them below.

Thermal DM:
Thermal Dark Matter is motivated by the simplicity of its production mechanism which makes the final abundance only depend on the annihilation cross-section, and by its ubiquity in theories aiming at solving the electroweak hierarchy problem, see Sect. 2.3.1 in Chap. 2. In view of its importance, we dedicate the whole Chap. 4 to thermal DM.

QCD axion:
A second well-studied DM candidate is motivated by the strong CP problem, see Sect. 2.3.4 in Chap. 2. Future experiments aiming at searching for the QCD axion or more generally, for any Axion-Like-Particle, seem promising, e.g. Casper, Abracadabra, ALPS II, the IAXO helioscope and a series of haloscopes, see Fig. 3.9. We

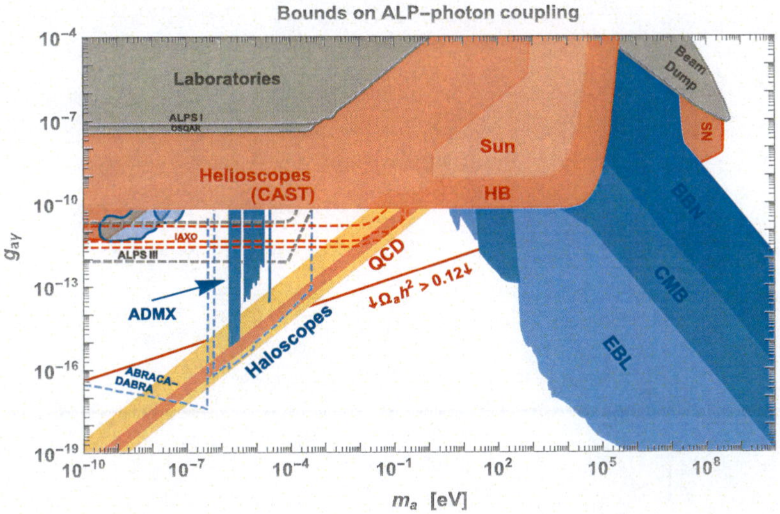

Fig. 3.9 Current constraints and prospects on the coupling $g_{a\gamma\gamma}$ [GeV] between the photon and a CP-odd axion-like particle $\mathscr{L} \supset g_{a\gamma\gamma} a\tilde{F}F$ [95]. The red line labeled 'QCD' shows where the Peccei-Quinn QCD axion is supposed to lie, based on the $m_a - f_a$ relation in Eq. (2.192), $m_a = (75.5\ \mathrm{GeV})_{\mathrm{a}}^{2/\mathrm{f}}$, and $g_{a\gamma\gamma} = \frac{\alpha_{\mathrm{e.m.}}}{2\pi f_a}(1.92 - \frac{\mathrm{E}}{\mathrm{N}})$ where 1.92 is induced by the mixing between the Peccei-Quinn axion and the SM π^0 and η^0 [626], while E/N depends on the UV realization. E.g. in KSVZ model, E and N are the anomaly coefficients of heavy Peccei-Quinn fermions in the background of $U(1)_{\mathrm{e.m.}}$ and $SU(3)_c$. The second red line shows the region where the axion abundance overcloses the universe, assuming standard misalignement mechanism and that the Peccei-Quinn symmetry is broken after inflation

refer the reader to [636–638] for original articles and [92–95, 639–647] for reviews on the QCD axion and ALPs. See [648] for a review of the open questions in axion theory.

Primordial black holes:
They are black holes formed during the early universe whose abundance is a substantial fraction of DM. Figure 3.10 shows the current state of the phenomenological constraints on the fraction of DM contained in PBHs. The mass window [10^{17} g, 10^{23} g] is compatible with 100% of DM being under the form of PBHs. We call the attention to [627–629, 649, 650] for reviews.

We now discuss astrophysical puzzles in galaxy dynamics which challenge ΛCDM model. We discuss 5 types of DM candidates whose properties have been shown to possibly resolve some of those issues: self-interacting DM, ultra-light DM, MOND, superfluid DM and baryon-interacting DM.

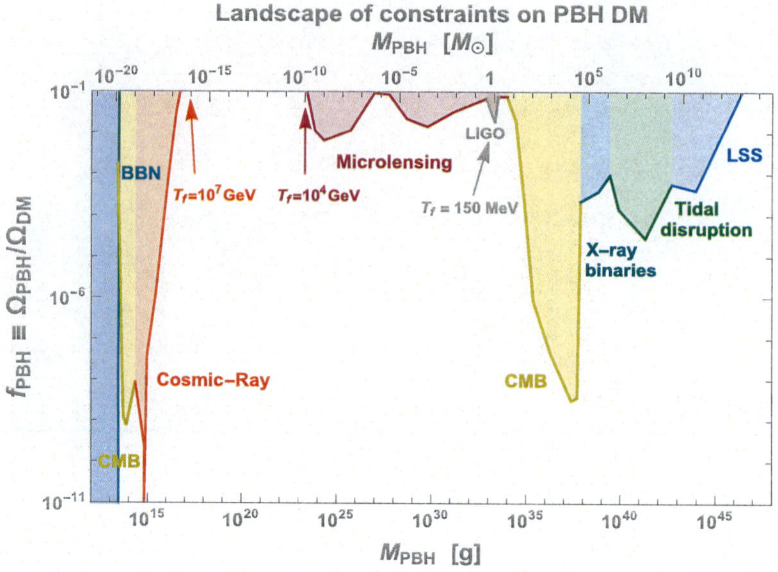

Fig. 3.10 Constraints on the PBH fraction of DM [627–629]. The scenario where 100% of DM is made of PBH with a mass within $[10^{17}\ \text{g},\ 10^{23}\ \text{g}]$ is compatible with experiments. The mechanism producing those BHs would have to take place within the temperature range $[10^4\ \text{GeV},\ 10^7\ \text{GeV}]$, where we assumed the PBH mass to be the energy contained within the acoustic horizon $c_s H^{-1}$ at the time of their formation, with $c_s = 1/\sqrt{3}$. There are claims that BH observed in LIGO/VIRGO data might be of primordial origins [630–634]. The mechanism responsible for their formation would have to take place around the QCD scale

3.5.4 *The Fragility of ΛCDM*

ΛCDM N-body simulations predict DM halos to follow the Navarro-Frenk-White density profile [651] (1995)

$$\rho_{\text{NFW}}(r) = \frac{\rho_0}{\frac{r}{r_0}\left(1+\frac{r}{r_0}\right)^2} \tag{3.174}$$

Typical parameters are e.g. $\rho_0 = 24.42\ \text{GeV/cm}^3$ and $r_0 = 0.184$ kpc in order for the local DM density at the sun position $r_\odot = 8.33$ kpc to be $\rho_\odot = 0.3\ \text{GeV/cm}^3$ and for the total DM mass contained within 60 kpc to be $M_{60} \equiv 4.7 \times 10^{11}\ M_\odot$ [652]. However, ΛCDM N-body simulations disagrees with observations on some points which we review below. See [653–659] for reviews on the small-scale problems. See also the rather rich introduction in [660].

The cusp versus core problem:
The presence of a cusp in NFW (DM profile peaked as $\rho_{DM} \sim 1/r$ at the center) contradicts the observation of a core in dwarf galaxies ($\rho_{DM} \sim r^0$ at the center).

More precisely, it has been found that the star rotation velocity in the inner part of the galaxy shows a solid-body behaviour (rises linearly with the radius) hence indicating the presence of a central core in the DM distribution [661–670], although the possibility to distinguish, in the data, a core from a cusp is still debated [671–673]. Baryonic physics have been proposed to resolve the conflict: the center density of DM halos can be flattened by repeated gas outflows due to supernova feedback [674–677] or dynamical friction [678–681]. On the one hand, energy transfer to DM from gas outflows grows with the stellar mass, implying a minimum stellar-to-halo mass for the center halo density to be flattened [677, 682]. One the other hand, effects of the gas outflow are overcome by the enlarged gravitational potential at larger stellar mass, implying also a maximum stellar-to-halo mass for the center halo density to be flattened [683, 684]. The dependence of the core creation with the mass in stars has been verified later by hydrodynamical simulations, e.g. NIHAO [685] and FIRE-2 [686, 687], see however [688] who find cores for all stellar masses or [619] who find no cores at all.

The missing satellite problem:
High-resolution hydrodynamical simulations predict a mass spectrum of DM halos which is self-similar and grows as a power law up to the numerical cut-off, hence predicting thousands of visible MW satellites [689–692]. However, only $\sim$50 satellites in the local group (MW+M31) have been observed so far [693–695]. The missing satellite problem might be explained by inhibition of galaxy formation in low-mass halo in the presence of photoionizing background due to stars and quasars [696–699], or by ram pressure stripping and tidal effects due to interaction with central hosts [700, 701]. Hence, DM halos become increasingly inefficient at making galaxies at low masses.

The too-big-to-fail (TBTF) problem:
The missing satellite problem can be solved by assigning the known MW satellites to the largest DM halos by abundance matching and by attributing the lack of observed satellites to the inefficiency of smaller subhalos to form galaxies. The problem is that the largest DM halos predicted in simulations are too dense compared to the brightest observed satellites. More precisely, ΛCDM N-body simulations predict at least 10 subhalos with $V_{\rm max} > 25$ km/s in the satellite population of the MW, and we don't observe any of them [702, 703]. Note the claim of [704] that TBTF is not a statement of high statistical significance. Later, the TBTF problem has been found in Andromeda (M31) [705] as well as in satellites 'in the field', beyond the virial radius of the MW and Andromeda [706]. The existence of TBTF beyond the influence of MW and M31 implies that environment effects, e.g. ram pressure stripping and tidal interactions, which are absent for field satellites, are unlikely to explain the TBTF problem [707]. Another proposal is to lower the center density of DM halo using repeated gas outflows due to supernova feedback [674–677] or dynamical friction [678–681]. Evidences of gas outflows include the observation of blue-shifted absorption lines [708, 709] or the observation of bulgeless galaxy [710]. High-resolution hydrodynamics simulations of the Local Group including star formation and stellar/supernova feedback, e.g. APOSTLE [619] or FIRE-2 [687,

711], find no evidence for missing satellite and TBTF problem, which confirms the results from earlier simulations [677, 712, 713] and seems to clear up the conflict whether SN feedback is likely to inject enough energy into the DM halo [714] or not [715, 716]. The crucial feature of baryonic feedback is that energy injection in the DM halo is additive over multiple SN burst events occurring during an extended period of time [654].

However, the existence of TBTF problem for satellites in the field [706], where external feedback (environment effects) is absent, and below the critical stellar mass $10^6\ M_\odot$, where internal feedback (e.g. SN feedback) is inefficient, remains to be explained. A possibility is the ram pressure induced by the flow of gas in filaments and pancakes of the DM cosmic web [717]. Another possibility is the presence of biases in the measurement of the rotation curves of ultra-faint dwarfs [718]. Whereas the TBTF problem can be interpreted as a new formulation of the missing satellite problem in light of the abundance matching between brightest dwarfs satellites and densest DM sub-halos, the cusp vs core problem can be uncorrelated, e.g. [619] which solves the TBTF problem with only cuspy profiles.

The diversity problem:
Rotation curves are observed to have a large scatter at small radius for a fixed value of the velocity at large radius $V_{\rm max}$ [719–722]. DM self-interactions are believed to lead to the formation of a core whose size increases with the strength of the self-interaction cross-section. The size of the core is also tight to the baryon density: an increase of the baryon density enlarges the gravitational well and then decreases the core size. For a given halo mass ($V_{\rm max}$), different baryon contents corresponding to different galaxy formation histories lead to different inner rotation curves. This is how the diversity problem is addressed in [723]. The diversity problem is also well addressed with DM self-interactions replaced by baryonic feedback in [724].

Proposed solution: self-interacting DM:
In the text above, we have reviewed solutions to the small-scale problems involving baryonic physics. An alternative approach, proposed by Spergel & Steinhardt in 1999 [725], is to go beyond the assumption that DM is collisionless and instead to postulate that DM is self-interacting. The inclusion of DM collisions in N-body simulations, e.g. [726–738], favors a self-interaction cross-section of order $\sigma/m \sim 1\ \mathrm{cm}^2/g$. Notice that it is much larger—about 15 orders of magnitude—than the weak-scale cross section expected for a usual WIMP DM candidate and is close to the upper bound from bullet cluster observation, cf. Eq. (3.171).

The presence of collisions generates a cosmological pressure p for the DM fluid, which can be related to the DM energy density ρ and temperature T using the ideal gas law

$$p = \frac{k_{\rm B} T}{m} \rho, \tag{3.175}$$

where m is the DM mass and $k_{\rm B}$ is the Boltzmann constant. The Navier-Stokes equation at equilibrium reads

$$\frac{dp}{dr} = \frac{k_B T}{m}\frac{d\rho}{dr} = -\rho\frac{d\phi}{dr}, \tag{3.176}$$

Everywhere collisions are fast enough, we expect heat flow to homogenize the temperature $T = \text{cst}$. This is the so-called **isothermal** limit for which the previous equation becomes

$$\frac{d\ln\rho}{dr} = -\frac{d(\phi/\sigma_v^2)}{dr}, \tag{3.177}$$

where $\sigma_v^2 \equiv \frac{k_B T}{m}$ is the mean velocity dispersion squared. The later equation can be trivially integrated to give the density profile of an isothermal self-gravitating sphere [208].

$$\rho_{\text{iso}}(r) = \rho_0 \exp\left(\frac{\phi(0) - \phi(r)}{\sigma_v^2}\right). \tag{3.178}$$

Gauss's theorem states that at the center of the galaxy the gravitational potential must reach a constant $\phi(r) \to \phi_0$. From Eq. (3.178), it implies that the density must reach a plateau ρ_0. Hence, while ΛCDM N-body simulations find the density profile to be peaked at the center $\rho \propto r^{-1}$, cf. Eq. (3.174), the presence of self-interaction leads to a cored profile. The radius r_0 at which the density flattens, also called the radius of the core, can be simply estimated by dimensional analysis from Eq. (3.177) together with Poisson's equation $\nabla^2\phi = 4\pi G\rho$

$$r_0 \simeq \sqrt{\frac{\sigma^2}{2\pi G\rho_0}}. \tag{3.179}$$

Far from the galactic center where the DM density is weaker, we expect the self-interaction to be negligible and the NFW profile in Eq. (3.174) to be a good approximation. However, below the radius r_1 where the number of self-interactions per DM particles is larger than 1

$$\frac{\rho(r_1)}{m}\langle\sigma v\rangle\, t_{\text{age}} \equiv 1, \tag{3.180}$$

the collisionless NFW profile should be replaced by the isothermal profile in Eq. (3.178). The inner region where self-interactions dominates is also the region where the gravitational potential is dominated by baryons, hence leading to a correlation between the effective stellar radius and the SIDM core size [739]. This analytical model allows to address the cusp vs core problem [739], the diversity problem [740–742], and the regularity problem [723].

The simplest and first particle physics model of self-interacting DM, proposed in 2000, is a quartic scalar $\lambda\phi^4$ [743, 744]. However it is now excluded by invisible constraints of the Higgs boson which leads to DM overabundance [745]. Instead the abundance could be depleted further with $3 \to 2$ scattering [746–751]. Unfortunately a model with a quartic scalar produces a constant cross-section over all scales, which is disfavored by astrophysical data from galaxies to clusters [622, 752]. Indeed while

$\sigma/m \sim 1\ \mathrm{cm^{2}/g}$ is favored in dwarf galaxies, it must be less than $\sigma/m \sim 0.1\ \mathrm{cm^{2}/g}$ on cluster scales [622, 752].

A non-trivial velocity-dependence of the self-scattering cross-section can be obtained with Yukawa interactions [753–757]. However these models with light mediators face strong bounds from direct and indirect probes of dark matter [758–760], except if DM has a large asymmetric component [761]. Other proposals for generating velocity-dependant scattering properties are models involving atomic interactions [762, 763], a Breit-Wigner resonance [764, 765], or strongly-coupled interactions [766–769].

Another solution is to make DM warmer in order to damp structure formation at small scales. Warm DM was first proposed in 2000 [770–772]. However, the resolution of small scale problems which suggests $m_{\mathrm{WDM}} \sim 2$ keV [773–776], is disfavored by Lyman-α forest observations which imposes $m_{\mathrm{WDM}} \gtrsim 3$ keV [777, 778]. Other proposals for damping structure formation at small scales using DM heating is delayed DM kinetic decoupling [753, 779–786], dark matter - radiation interactions [787, 788], or the presence of a phase of DM cannibalism [746–751].

We refer to [658] for a review on the particle solution to the small scale crisis.

Proposed solution: ultra-light DM:
They are DM candidates with a mass smaller than $m \lesssim 5$ eV. In the galaxy, their particle number per de Broglie wavelength $\lambda_{\mathrm{dB}} = 1/mv$ is much larger than 1

$$n_{\mathrm{DM}}\lambda_{\mathrm{dB}}^{3} = \frac{\rho_{\mathrm{DM}}}{m}\lambda_{\mathrm{dB}}^{3} = 10\left(\frac{\rho_{\mathrm{DM}}}{0.3\ \mathrm{GeV/cm^{3}}}\right)\left(\frac{5\ \mathrm{eV}}{m}\right)^{4}\left(\frac{220\ \mathrm{km/s}}{v}\right)^{3} \gg 1. \tag{3.181}$$

Obviously, Fermi-Dirac statistic forbids ULDM to be a fermion, as already stated by Tremaine and Gunn in 1979 [789].[22] For a large occupation number of a phase-space cell, the particle description breaks down and must be replaced by a wave description. A first possibility is the Schrodinger-Poisson system of equations

$$i\dot{\psi} = -\frac{1}{2m}\nabla^{2}\psi + m\Phi\psi - \frac{3}{2}iH\psi + g_{2}|\psi|^{2}\psi + g_{3}|\psi|^{4}\psi + \cdot \tag{3.182}$$

$$\nabla^{2}\Phi = 4\pi G(\rho - \overline{\rho}). \tag{3.183}$$

where Φ is the gravitational potential and the g_i are the coupling constants of i-body self-interactions. In the literature, the case $g_i = 0$ is called Fuzzy DM (FDM) while models with $g_i \neq 1$ are called self-interacting fuzzy DM (SIFDM). In the long-wavelength limit, the field theory can be replaced by a set of hydrodynamical-like equations using the Madelung decomposition [791]

$$\psi \equiv \sqrt{\frac{\rho}{m}}e^{i\theta}, \quad \mathbf{v} = \frac{1}{am}\nabla\theta = \frac{1}{2ima}\left(\frac{1}{\psi}\nabla\psi - \frac{1}{\psi^{*}}\nabla\psi^{*}\right). \tag{3.184}$$

[22] Careful analysis of the phase space distribution of DM in Dwarf galaxies implies the lower bound on the mass of fermion DM to be $m_{\mathrm{DM}} \gtrsim 1$ keV [790].

The equations of motion for ψ become

$$\dot{\rho} + 3H\rho + \frac{1}{a}\nabla \cdot (\rho \mathbf{v}) = 0, \tag{3.185}$$

$$\rho\dot{\mathbf{v}} + \rho H\mathbf{v} + \frac{\rho}{a}(\mathbf{v}\cdot\nabla)\mathbf{v} = -\frac{\rho}{a}\nabla\Phi + \nabla\left(P_{\rm int} + P_{\rm QG}\right), \tag{3.186}$$

with

$$P_{\rm int} = \sum K_j \rho^{(j+1)/j} = g_2 n^2 + g_3 n^3 + \cdots, \tag{3.187}$$

$$\nabla P_{\rm QG} = \frac{1}{2a^3 m^2}\nabla\left(\frac{\nabla^2\sqrt{\rho}}{\sqrt{\rho}}\right). \tag{3.188}$$

$P_{\rm int}$ is the pressure due to self-interactions with j being the polytropic index. $P \propto n^2$ is generated from 2-body interactions, $P \propto n^3$ from 3-body interactions [792], and so on. The term $P_{\rm QG}$ which depends on the curvature of the amplitude of the wave function is the quantum pressure. It plays the role of a repulsive interaction which can be understood as arising from Heisenberg uncertainty principle. Interests for Fuzzy DM started in 2000 after that Hu, Barkana and Gruzinov [793] proposed to use the quantum pressure $P_{\rm QG}$ to address the incompatibility between astrophysics observation and ΛCDM cosmology at small scales. Indeed, the impact of $P_{\rm QG}$ is to smoothen structures below the de Broglie wavelength

$$\lambda_{\rm dB} = 2\ {\rm kpc}\left(\frac{10^{-22}\ {\rm eV}}{m}\right)\left(\frac{10\ {\rm km/s}}{v}\right). \tag{3.189}$$

However, $m \sim 10^{-22}$ eV is in tension with Lyman-α forest observations which imply $m \gtrsim 2\times 10^{-22}$ eV [794], and raise doubts on the possibility for fuzzy DM to address small scale problems. See also the constraints coming from CMB [795] and LSS [796]. For more details, we refer the reader to existing reviews on astrophysical probes of ULDM [797–802]. Even if the resolution of small scale problems by ULDM becomes completely ruled out in the near future, ULDM remains an interesting scenario due to the ubiquity of ultra light scalars in the string landscape, with possibly axion-like or dilaton-like coupling to SM [98, 803–806]. When composing a significant fraction of DM they can be responsible for variation of coupling constants and give signatures in atomic clocks [97, 807–814], GW interferometers [815–818], resonant mass detectors [819], or BBN physics [820, 821]. Even when they do not constitute DM, the existence of ultra light scalars can be probed by fifth force experiments [96, 822–825] and black hole superradiance [814, 826–830].

The regularity problem:
We now discuss another astrophysical deviation from ΛCDM whose nature is a bit different from the cusp-core, satellite, too-big-to-fail and diversity problems at small scales < kpc. The regularity problem arises at the scale of the whole DM halo, since it deals with the asymptotic region of the rotation curves. Analog to

Kepler's laws for planets, galaxies have shown to have their own rotation laws. Their stricking feature is their intrinsic small scatter. It seems that the diversity problem, i.e. large scatter of the raising part of the rotation curves, is correlated with the baryon distribution in such a way that tight galactic laws, with a small scatter, emerge. They can be formulated **globally, centrally** or **locally** [831].

- **Global law**: a tight relation between the maximal rotation velocity $V_{\rm max}$ and the total baryonic mass $M_{\rm bar}$

$$V_{\rm max}^4 \simeq g_0\, G\, M_{\rm bar}. \tag{3.190}$$

It has first been known as a tight correlation between the maximum circular velocity and the stellar mass [832] before being generalized to the total baryonic mass [833], and so is called the baryonic Tully-Fisher relation (BTFR). Introducing the factors $f_{\rm V} \equiv V_{\rm max}/V_{\rm vir}$ and $f_{\rm b} \equiv M_{\rm bar}/M_{\rm vir}$, ΛCDM cosmology suggests the relation [834]

$$V_{\rm max}^3 \propto f_{\rm V}^3 f_{\rm b}^{-1} M_{\rm bar}, \tag{3.191}$$

with a large scatter [835]. Consistency with the data requires $f_{\rm V}^3 f_{\rm b}^{-1} \propto V_{\rm max}^{-1}$. Whereas f_V remains close to 1, the baryon content $f_{\rm b}$ can decrease with the baryonic feedback ε, e.g. $f_{\rm b} \propto \varepsilon^{-1}$. We expect baryonic feedback to increase with the stellar mass, hence with $V_{\rm max}$, which is in contradiction with the requirement from the BTFR. In spite of this apparent contradiction, they are many attempts of recovering the BTFR in ΛCDM either with hydrodynamical simulations [836, 837] or analytical modelisations [723, 838–840], where it is claimed that the scatter from ΛCDM estimation may be overestimated.

- **Central law**: a tight relation between the central surface stellar density

$$\Sigma_b(0) \equiv \int_{-\infty}^{+\infty} dz\, \rho_b(\vec{x}), \tag{3.192}$$

where z is the coordinate transverse to the disc, and the dynamical mass [841]

$$\Sigma_{\rm dyn}(0) \equiv \frac{1}{2\pi G} \int_0^{\infty} \frac{V^2(R)}{R^2} dR = \int_{-\infty}^{+\infty} dz\, \rho(\vec{x}), \tag{3.193}$$

High-surface density galaxies, dominated by baryon, follow the 1:1 line whereas low-surface galaxies, dominated by DM, deviate from the 1:1 line with a small scatter, showing that DM and baryon are correlated

$$\Sigma_{\rm dyn}(0) = \begin{cases} \Sigma_b(0), & \Sigma_b(0) > \frac{a_0}{G}, \\ \sqrt{\frac{2}{\pi}\frac{a_0}{G}\Sigma_b(0)}, & \Sigma_b(0) < \frac{a_0}{G}, \end{cases} \tag{3.194}$$

where the critical acceleration scale a_0 is measured to be $a_0 \simeq 1.2 \times 10^{-8}$ m^2/s. The central law is called the central surface density relation (CSDR).

- **Local law**: a tight relation between the radial acceleration traced by rotation curves

$$g_{\rm obs} \equiv \frac{v^2}{R} = \frac{GM(R)}{R}, \tag{3.195}$$

and that predicted by the observed distribution of baryons

$$g_{\rm bar} \equiv \frac{GM_b(R)}{R}. \tag{3.196}$$

Large accelerations are baryon dominated and follow the 1:1 line [842]. However below the critical acceleration scale $a_0 \simeq 1.2 \times 10^{-8}$ m^2/s, the observed acceleration $g_{\rm obs}$ scales as the square root of the baryon gravitational potential $g_{\rm bar}$

$$g_{\rm obs} = \begin{cases} g_{\rm bar}, & g_{\rm bar} > a_0, \\ \sqrt{g_{\rm bar} a_0}, & g_{\rm bar} < a_0. \end{cases} \tag{3.197}$$

Again the small scatter shows a tight correlation between the DM and the baryons. This local relation is called the mass discrepancy-acceleration relation (MDAR) or Milgrom's relation. It is a powerful statement about how DM is distributed in galaxies: in regions where baryons dominate, the theory behaves like Newtonian theory, and in regions where the DM dominates, the DM mass is uniquely determined by the baryonic distribution. Also surprising is the numerical coincidence between the critical acceleration scale a_0 and the current expansion scale of the universe H_0

$$a_0 \simeq 1.2 \times 10^{-8}\ \mathrm{m}^2/s \simeq H_0/6. \tag{3.198}$$

Proposed solution: MOND:
Probably the simplest explanation for the MDAR in Eq. (3.197) (and then BTFR in Eq. (3.190) follows trivially) is given by Modified Newtonian Dynamics (MOND) proposed by Milgrom in 1983 [843–845]

$$\mu\left(\frac{g}{a_0}\right)\vec{g} = \vec{g}_{\rm N}, \tag{3.199}$$

where $\vec{g} \equiv d^2\vec{x}/dt^2$, $g_{\rm N} \equiv GM/r^2$ and

$$\mu(x) = \begin{cases} 1, & x \gg 1, \\ x, & x \ll 1. \end{cases} \tag{3.200}$$

A first Lagrangian formation, called AQUAL, is proposed the following year by Bekenstein and Milgrom [846], which we now review. Newtonian gravity can be derived from the non-relativistic action

$$S_{\rm N} = \int d^3x\, dt \left[\frac{\rho \vec{v}^2}{2} - \rho\Phi_{\rm N} - \frac{|\vec{\nabla}\Phi_{\rm N}|^2}{8\pi G}\right]. \tag{3.201}$$

Varying $S_{\rm N}$ with respect to space coordinates $\vec{x}$ and with respect to potential $\Phi_{\rm N}$ yields the usual equation of motion and Poisson's equation

$$\frac{d^2\vec{x}}{dt^2} = -\vec{\nabla}\Phi_{\rm N}, \qquad \text{and} \qquad \vec{\nabla}^2\Phi_{\rm N} = 4\pi G\rho. \tag{3.202}$$

Milgrom's law in Eq. (3.199) can be recovered after replacing the last term of the Newtonian Lagrangian in Eq. (3.201) by some function $F(z)$

$$S_{\rm BM,grav} = -\frac{a_0}{8\pi G}\int d^3x\, dt\, F\left(\frac{|\nabla\Phi|^2}{a_0^2}\right). \tag{3.203}$$

This is the original Aquadratic Lagrangian (AQUAL) of Bekenstein and Milgrom [846]. Varying with respect to Φ leads to a generalization of Poisson's equation

$$\vec{\nabla}\cdot\left[\mu\left(\frac{|\vec{\nabla}\Phi|}{a_0}\right)\vec{\nabla}\Phi\right] = 4\pi G\rho, \tag{3.204}$$

where $\mu(x) = F'(z)$ and $z = x^2$. Agreement with Milgrom's law in Eq. (3.199) imposes

$$F(z) = \begin{cases} z, & x \gg 1, \\ \frac{2}{3}z^{3/2}, & x \ll 1. \end{cases} \tag{3.205}$$

In the low gravity regime, the AQUAL Lagrangian reads

$$|\vec{\nabla}\Phi| \ll a_0 \qquad \Longrightarrow \qquad S_{\rm BM} = -\int d^3x\, dt \left[\rho\Phi + \frac{1}{12\pi G a_0}\left(|\vec{\nabla}\Phi|^2\right)^{3/2}\right], \tag{3.206}$$

Later, other Lagrangian formulations of Milgrom's law have been proposed, either non-relatistic: QMOND [847], or relativistic: TeVeS [848] and BMOND [849]. MOND can be interpreted as a DM fluid which has a dipole interaction with gravity [850, 851]. Another MOND proposal is to assume that spacetime and gravity emerge together from the entanglement structure of an underlying microscopic theory [852–854]. However, MOND without DM has been challenged by measurements of gravitational lensing (e.g. bullet cluster in Fig. 3.6) and CMB anisotropies, in particular by the amplitude of the third peak of the CMB power spectrum [855, 856], even though progress is still on going, e.g. [857]. Lower bounds on the speed of gravitational waves following neutron star merger events have also brought additional constraints on MOND theories [858].

Proposed solution: superfluid DM:
From the particle theory side, the Lagrangian in Eq. (3.206) with a fractional power can be obtained by a complex scalar field dominated by a 3-body interactions

$$\mathcal{L} = -|\partial_\mu \Phi|^2 - m^2|\Phi|^2 - \frac{\lambda}{3}|\Phi|^6. \tag{3.207}$$

This theory is invariant under a global $U(1)$ symmetry with associated conservation of number of particles. We replace $\psi = \sqrt{2m}\,\Phi\, e^{imt}$ such that the Lagrangian in the non-relativistic limit becomes

$$\mathcal{L} = \frac{i}{2}\left(\psi\partial_t\psi^* - \psi^*\partial_t\psi\right) - \frac{|\vec{\nabla}\psi|^2}{2m} - \frac{\lambda}{24m^3}|\psi|^6, \tag{3.208}$$

whose equation of motion is a non-linear Schrodinger equation

$$-i\partial_t\psi + \frac{\nabla^2\psi}{2m} - \frac{\lambda}{8m^3}|\psi|^4\psi = 0. \tag{3.209}$$

The homogeneous solution describes a Bose-Einstein condensate at zero temperature

$$\psi_0 = \sqrt{2m}\,v\,e^{i\mu t}, \qquad \mu \equiv \frac{\lambda v^4}{2m}, \tag{3.210}$$

where μ is the chemical potential. The solution spontaneously breaks the $U(1)$ symmetry because it is associated with a finite number density n, given by the time component of the Noether current $J^\mu = \frac{1}{m}\mathrm{Im}\,(\psi^*\partial^\mu\psi)$

$$n = \frac{1}{m}\mathrm{Im}\,(\psi^*\dot{\psi}) = 2\mu v^2 = \left(\frac{8\mu m^2}{\lambda}\right)^{1/2}. \tag{3.211}$$

Perturbations around the homogeneous solution read

$$\psi = \sqrt{2m}(v+\rho)e^{i(\mu t+\phi)}, \tag{3.212}$$

here ρ is the perturbation of the order parameter, while ϕ is the Goldstone boson. Substituting into Eq. (3.208) gives

$$\mathcal{L} = -(\vec{\nabla}\rho)^2 + 2m(v+\rho)^2\left[\mu + \dot{\phi} - \frac{(\vec{\nabla}\phi)^2}{2m}\right] - \frac{\lambda}{3}(v+\rho)^6. \tag{3.213}$$

The dispersion relation of the linearized equation of motion is

$$\omega^2 = c_s^2 k^2 + \mathcal{O}(k^4), \qquad c_s^2 = \frac{\lambda v^4}{m^2} = \frac{2\mu}{m}. \tag{3.214}$$

The linear dispersion of phonons ϕ leads to the existence of superfluidity via the Landau criterion [859]: an impurity moving at a subsonic velocity $v < c_s$ moves without friction.[23] Integrating out the heavy mode ρ, we obtain an EFT for the phonons [860]

$$\mathcal{L} = \frac{4}{3} m \left(\mu + \dot{\phi} - \frac{(\vec{\nabla}\phi)^2}{2m}\right) \left(\frac{2m}{\lambda}\left[\mu + \dot{\phi} - \frac{(\vec{\nabla}\phi)^2}{2m}\right]\right)^{1/2}, \tag{3.217}$$

Upon adding a baryon-phonon interaction

$$\mathcal{L}_{\text{int}} = -\frac{\Lambda}{M_{\text{pl}}} \phi \rho_b, \tag{3.218}$$

in the limit $\frac{(\vec{\nabla}\phi)^2}{2m} \gg \dot{\phi},\ \mu$ we obtain the Bekenstein-Milgrom Lagrangian Eq. (3.206) with the fractional $3/2$ power and the correct sign in front of the kinetic term if

$$\lambda < 0. \tag{3.219}$$

However because of Eq. (3.214), $\lambda < 0$ leads to an imaginary frequency and to a destruction of the condensate. More precisely, in the small perturbation regime the Lagrangian in Eq (3.217) becomes

$$\mathcal{L}_{\text{int}} = \frac{n}{2}\left(\frac{4}{3}\mu + 2\dot{\phi} + \frac{1}{2\mu}\left(\dot{\phi}\right)^2 - \frac{1}{m}\left(\vec{\nabla}\phi\right)^2\right). \tag{3.220}$$

From Eq. (3.210), $\lambda < 0$ leads to $\mu < 0$ which implies a negative pressure (constant term in Eq. (3.217)) and to a ghost-like kinetic term. Negative λ corresponds to an attractive interaction which makes the BEC unstable against collapse. It is therefore needed to add repulsive interactions to stabilise the BEC. We refer to [860, 861] for more details on this point. Superfluid DM is shown to provide successful model

---

[23] Suppose an impurity of mass M moving with initial velocity $\vec{v}_i$ which produces a phonon of momentum $\vec{k}$ and acquires the final velocity $\vec{v}_f$ after scattering. Energy-momentum conservation reads

$$\frac{Mv_i^2}{2} = \frac{Mv_f^2}{2} + \omega, \qquad \text{and} \qquad M\vec{v}_i = M\vec{v}_f + \vec{k}, \tag{3.215}$$

The 2nd equation leads to $\vec{v}_f = \vec{v}_i - \vec{k}/M$, which after injection in the 1st equation leads to the condition that phonon emission can only occur if the initial impurity velocity is supersonic

$$v_i = \frac{c_s + k/2M}{\cos\theta} \geq c_s. \tag{3.216}$$

of galaxy clusters and galaxy rotation curves [862, 863]. The dissipationless nature of subsonic motion can be used to explain existing astrophysical puzzles [864]. Dynamics explaining dark energy can be added [865, 866]. Superfluid DM has also received criticisms [867–869]. For more details, we refer the reader to the reviews [801, 870].

Proposed solution: environment-dependent baryon-DM interactions:
Another proposal to explain BTFR and MDAR is that baryons and DM interact together with a cross-section scaling as the inverse of the DM density and squared velocity [871–873]

$$\sigma_{\text{int}} \propto \frac{1}{\rho v^2}. \tag{3.221}$$

We now propose to quickly review the reasoning. The phase space density $f(t, \vec{r}, \vec{v})$ of the DM component obeys the Boltzmann transport equation

$$\frac{\partial f}{\partial t} + \vec{v} \cdot \frac{\partial f}{\partial \vec{r}} + \vec{g} \cdot \frac{\partial f}{\partial \vec{v}} = I\left[f,\ f_b\right], \tag{3.222}$$

where $I\left[f,\ f_b\right]$ accounts for interactions between DM particles and baryons. The 0th, 1st and 2nd velocity moments of the Boltzmann equation, known as the continuity, Jeans and heat equations, read

$$\frac{\partial \rho}{\partial t} + \vec{\nabla} \cdot (\rho \vec{u}) = 0, \tag{3.223}$$

$$\rho\left(\frac{\partial}{\partial t} + \vec{u} \cdot \vec{\nabla}\right) u^i + \partial_j P^{ij} = \rho \vec{g}, \tag{3.224}$$

$$\frac{3}{2}\left(\frac{\partial}{\partial t} + \vec{u} \cdot \vec{\nabla}\right) T + \frac{m}{\rho} P^{ij} \partial_i u_j + \frac{m}{\rho} \vec{\nabla} \cdot \vec{q} = \dot{\varepsilon}, \tag{3.225}$$

where $\vec{u} \equiv \langle \vec{v} \rangle$ is the bulk velocity,[24] $P^{ij} \equiv \rho \left\langle \left(v^i - u^i\right)\left(v^j - u^j\right)\right\rangle$ is the pressure tensor, $T \equiv \frac{m}{3}\left\langle |\vec{v} - \vec{u}|^2 \right\rangle$ is the local DM temperature, and $\vec{q} \equiv \frac{1}{2}\rho \left\langle (\vec{v} - \vec{u})|\vec{v} - \vec{u}|^2 \right\rangle$ is the heat flux. The local heat rate $\dot{\varepsilon}$ is due to interaction with baryons. For the sake of simplicity, we assume that the DM bulk velocity is zero and that the DM distribution is described by a Maxwell-Boltzmann distribution,

$$\vec{u} \simeq 0, \qquad f(\vec{r}, \vec{v}) \simeq f_{\text{MB}}(\vec{r}, \vec{v}). \tag{3.226}$$

This implies $P^{ij} = \rho v^2 \delta^{ij}$ and $v = \sqrt{T/M}$. The heat flux can be obtained by perturbing the M-B distribution

$$\vec{q} \simeq -\kappa \vec{\nabla} T, \qquad \text{with} \quad \kappa = \frac{3}{2} n \frac{l^2}{t_{\text{relax}}}, \tag{3.227}$$

---

[24] We define $\langle A \rangle = \frac{1}{n} \int d^3 v\, f(t, \vec{r}, \vec{v}) A$.

where κ is the thermal conductivity with l the system size and $t_{\rm relax}$ the time scale for DM particle to lose memory of its initial velocity. For dilute system like galaxies, the relaxation time is set by the dynamical time or Jeans time, $t_{\rm relax} \sim t_{\rm dyn} \sim 1/\sqrt{G\rho}$, so we have $l \sim v\, t_{\rm dyn} \sim r$ and

$$\kappa \sim n\, r\, v. \tag{3.228}$$

The heating rate can be written

$$\dot{\varepsilon} = -\Gamma_{\rm int}\varepsilon, \qquad \Gamma_{\rm int} = n_b \sigma_{\rm int} v, \qquad \varepsilon \simeq \frac{m m_b}{m + m_b} v^2, \tag{3.229}$$

where $\sigma_{\rm int}$ is the DM-baryon scattering cross-section and ε is the exchanged energy per scattering. We now show that the following ansatz reproduces the TBFR [871–873]

$$\sigma_{\rm int} = \frac{C a_0 m_b}{n \varepsilon} \simeq \frac{m + m_b}{m} \frac{C a_0}{n v^2}, \tag{3.230}$$

with $a_0 \simeq 1.2 \times 10^{-8}$ m/s^2 and $C = \mathcal{O}(1)$. Poisson and approximated Jeans and heat equations read

$$\vec{\nabla} \cdot \vec{g} = -4\pi G (\rho + \rho_b) \quad \Longrightarrow \quad g = -\frac{G\,(M + M_b)}{r^2}, \tag{3.231}$$

$$\frac{1}{\rho} \partial \left(\rho v^2\right) = g, \tag{3.232}$$

$$\vec{\nabla} \cdot \left(\frac{\rho r v}{a_0} \vec{\nabla}\left(v^2\right) \right) = C v \rho_b. \tag{3.233}$$

The heat equation in spherical coordinate gives

$$\frac{1}{r^2} \frac{d}{dr} \left(\rho v r^3 \frac{dv^2}{dr} \right) = C a_0 v \rho_b, \qquad \Longrightarrow \qquad \rho r \frac{dv^2}{dr} \sim C a_0 \frac{M_b}{4\pi r^2}. \tag{3.234}$$

The Jeans equation in DM-dominated region $M(r) \gg M_b(r)$ reads

$$\rho(r) \sim \frac{v^2(r)}{2\pi G r^2}, \tag{3.235}$$

which after substituting in the previous equation gives the BTFR, cf. Eq. (3.190)

$$v^4(r) \sim C a_0 G M_b(r), \tag{3.236}$$

up to a log factor which we neglected. We refer the reader to the original articles [871–873] for more details. These are exploratory works and further studies are needed to find a particle physics model giving rise to the master DM-baryon scattering cross-section in Eq. (3.230). A proposal is superfluid DM. Indeed, upon canonicalizing the gradient term of the linearized lagrangian in Eq. (3.220), the baryon-phonon interaction in Eq. (3.218) becomes

$$\mathscr{L}_{\text{int}} = -\frac{\Lambda}{M_{\text{pl}}}\phi\rho_b \qquad \Longrightarrow \qquad \mathscr{L}_{\text{int}}^{\text{canon.}} = -\frac{2m}{n}\frac{\Lambda}{M_{\text{pl}}}\tilde{\phi}\rho_b, \tag{3.237}$$

which has the $1/n$ behavior of the master relation in Eq. (3.230).

In the remaining part of this chapter, we discuss two other cosmological (or astrophysical) problems which should not be put on the same footing as the previous ones and may end up being attributed to experimental issues.

3.5.5 *The Hubble Tension*

The H_0 tension results from the $\sim 5\,\sigma$ discordance between the measurements of the Hubble parameter today H_0 using late time observables and the measurements using early time observables.

1. **Late time**: The expansion parameter is inferred from the luminosity-distance versus redshift relation of Cepheid-calibrated[25] **Supernova Ia**, also known as the local distance ladder. Particularly, the SHOES collaboration has determined [874]

$$H_0 = 74.03 \pm 1.42 \text{ km/s/Mpc}. \tag{3.238}$$

Another solid low-redshift observable is the **time-delays of strongly-lensed quasars** such as measured by H0LiCOW [875]

$$H_0 = 73.3^{+1.7}_{-1.8} \text{ km/s/Mpc}. \tag{3.239}$$

2. **Early time**: the high-precision measurement of the **CMB**, by Planck leads to [259]

$$H_0 = 67.3 \pm 0.6 \text{ km/s/Mpc}. \tag{3.240}$$

This is supplemented at high redshift by the observation of the **Baryonic Acoustic Oscillations** in galaxy redshift surveys and Lyman-α forests (also known as inverse distance ladder) calibrated against **Big-Bang Nucleosynthesis** (deuterium + helium abundance) [876] (see also [877])

$$H_0 = 68.3^{+1.1}_{-1.2} \text{ km/s/Mpc}. \tag{3.241}$$

[25] Cepheids are stars whose brightness oscillates with a period precisely correlated with its amplitude through the period-luminosity relation (PLR). They are more accurate standard candles than SNe Ia, but since they are less bright, they can not be observed at high redshift in contrast to SNe Ia. They are useful for calibrating the SNe Ia.

The robustness of the Hubble tension lies on the fact that it is verified by different independent observables, see [878] for a review on the different H_0 measurements. Therefore, proposed solutions which focus on only one probe, like dimming the supernovas [879] or introducing a fifth force which increases the gravity on the Cepheids in order to change the luminosity-profile relation and bias the SNe Ia measurement, are difficult to reconcile with other late-time probes.

The Hubble tension can be recast as a mismatch between H_0 at low redshift, and the sound horizon at high redshift [880], see however [881]. Early time observables, BAO and CMB, measure the Hubble parameter[26] H_0 through the **sound horizon angle** [876]

$$\theta_s(z)\Big|_{\text{obs}} \equiv \frac{r_s}{D_A} = \frac{\int_{z_D}^{\infty} d\tilde{z}\; c_s(\tilde{z})\; H(\tilde{z})^{-1}}{\int_0^z d\tilde{z}\; H(\tilde{z})^{-1}} \simeq \frac{\int_{z_D}^{\infty} d\tilde{z}\; c_s(\tilde{z}) \left[\Omega_r/\Omega_m(1+\tilde{z})^4) + (1+\tilde{z})^3\right]^{-1/2}}{\int_0^z d\tilde{z}\; \left[\Omega_\Lambda/\Omega_m + (1+\tilde{z})^3\right]^{-1/2}} \tag{3.242}$$

where $c_s(z) = 1/\sqrt{3(1+3\rho_b/4\rho_r)}$ is the speed of sound. So the sound horizon is sensitive to the ratios Ω_r/Ω_m and Ω_Λ/Ω_m. Hence, by increasing N_{eff}, we must increase Ω_m while keeping Ω_Λ/Ω_m constant which increases H_0 and solves the discrepancy between Planck and SNe Ia. But, this enters in tension with BAO+BBN [876]. A possible solution is to increase N_{eff} between BBN and CMB [880, 882–884]. Moreover, just by looking at the CMB fit alone, it can be shown that a period of non-standard cosmology hoping for solving the Hubble tension must occur one or two decades of scale factor evolution prior to recombination [885, 886]. Nonetheless, to increase N_{eff} has also effects on the CMB at the perturbation level (like Silk damping and neutrino drag at high l) which must be counterbalanced with other ingredients. Particularly, the CMB is sensitive to the so-called **damping angle**

$$\theta_D \simeq \frac{r_D}{\int_0^z d\tilde{z}\; H(\tilde{z})^{-1}} \qquad \text{where} \qquad r_D^2 \propto \int_{z_D}^{\infty} \frac{dz}{H(z)} \frac{1}{a\,\sigma_T\, n_e}. \tag{3.243}$$

While the sound horizon scale r_s in Eq. (3.242) was linear in H_D, the damping scale r_D, which is the variance of a **random walk**, is a **quadratic** object. As a consequence, the decrease of 7% of r_s needed to solve the Hubble tension (as well as the decrease of D_A by the same amount in order to keep θ_s constant in Eq. (3.242)), implies a decrease of 3.5% of r_D and therefore an increase of 3.5% of θ_D. This is the θ_D/θ_s **problem** [887].

A proposed solution is to increase N_{eff} with new species which also have self-interactions in order to reduce their free-streaming length, e.g. self-interacting neutrinos [888], self-interacting dark radiation [889],

An other popular solution [890–894] is a **frozen scalar field** behaving as dark energy, reaching 10% of the total energy density at redshift $z \sim 3000$, and which subsequently **dilutes faster than matter** when oscillating in its potential, and which

[26] The location of the first peak of the CMB power spectrum is given by $l \simeq \pi/\theta(z)$.

could solve the Hubble tension at the background and perturbation level. This solution may be challenged by LSS contraints on the matter power spectrum which break the degeneracy between the sound horizon r_s and the Hubble constant H_0 [895–897], or not [898–900].

They are many other proposals, e.g. primordial magnetic fields [901], early AdS phase [902], time-variation of fine-strucutre constant and electron mass [903, 904], Newton constant $G_{\rm N}$ [905–910], so instead of citing them all we refer to the reviews [887, 911, 912].

In the forthcoming years, we expect others probes of H_0 to enter the same precision level as CMB or SNe Ia. This is the case of **Tip of Red Giant Branch (TRGB)** measurement [913]

$$H_0 = 69.8 \pm 1.7 \text{ km/s/Mpc}, \tag{3.244}$$

which currently, appears to sit halfway between CMB and SNe Ia.

Another interesting promising probe is the detection of gravitational waves emited by **binary neutron stars (BNS) mergers**, also known as **standard sirens**. It is claimed that a sample of 50 BNS, detectable by LIGO and VIRGO within a decade would be able to arbitrate the H_0 tension between CMB and local distance ladder [914, 915].

Finally, another exciting possibility for measuring the current expansion rate of the universe is the direct observation of the time evolution of the cosmological redshift of distant sources due to the accelelation of the universe expansion, also known as **redshift drift** [916–918]. Even though the level of spectral drifts is of the order of $10^{-10}/10$ yr, ultra stable and ultra precise spectroscopy coupled with ultra large telescopes like ELT or SKA would be able to detect them [919–921].

3.5.6 The 21-cm Anomaly

The 21 cm line (1420 MHz) arises from the hyperfine splitting of the $1S$ ground state of hydrogen due to the interaction between the magnetic moments of the proton and the electron. The spin temperature $T_{\rm S}$ determines the relative abundance of the triplet state relative to the singlet

$$\frac{n_1}{n_0} = \frac{3}{1} \exp\left(-\frac{E_{21}}{T_{\rm S}}\right), \tag{3.245}$$

with $E_{21} = 2\pi/\lambda_{21} = 0.068$ K.

By passing through the hydrogen clouds, the CMB photons may subtract or add energy to it by absorption or emission. The resulting intensity at a given frequency ν and a given redshift z is found after integrating the radiative transfer equation [922]

$$I_{\rm CMB}(\tau) = I_{\rm H}\left(1 - e^{-\tau}\right) + I_{\rm CMB}(0)e^{-\tau}, \tag{3.246}$$

where $I_{\rm CMB}(0)$ is the black body CMB intensity in a transparent universe, $I_{\rm H}$ is the intensity emitted by the medium considered as a black body with temperature $T_{\rm S}$, and τ is the optical depth, found after integrating the inverse mean free path over the line of sight [923]

$$\tau = \int_{\rm l.o.s} ds \frac{h\nu}{4\pi}\left[n_0 B_{01} - n_1 B_{10}\right] \simeq \frac{3\lambda_{21}^2 n_H A_{10}}{16 T_{\rm S}\, H(z)}, \qquad A_{10} \simeq 1.85 \times 10^{-15}\ {\rm s}^{-1}. \tag{3.247}$$

We have used the relation between the Einstein coefficients, $g_1 B_{10} = g_0 B_{01}$ and $\frac{2h\nu_{21}^3}{c^3} B_{10} = A_{10}$, and approximated the neutral hydrogen column length by the Hubble horizon $\int ds \simeq H(z)^{-1}$. We recall that Eq. (3.246) is evaluated for a particular frequency ν and a particular redshift z. We assume the Rayleigh-Jeans limit, $\nu \ll T_{\rm S}, T_{\rm CMB}$, such that we can approximate $I(\nu, T, z) \simeq \frac{2\nu^2}{c^2} k_B T(z)$. The contrast temperature observed today, between the spin temperature $T_{\rm S}$ and the CMB temperature $T_{\rm CMB}$,[27] also known as the differential **brightness temperature**, reads

$$\delta T_b = \frac{T_{\rm S} - T_{\rm CMB}}{1+z}\left(1 - e^{-\tau_\nu}\right) \tag{3.248}$$

$$\simeq 27\ {\rm mK}\ x_H (1+\delta_b)\left(\frac{\Omega_b h^2}{0.022}\right)\left(\frac{0.14}{\Omega_m h^2}\frac{1+z}{10}\right)^{1/2} \frac{T_{\rm S} - T_{\rm CMB}}{T_{\rm S}}, \tag{3.249}$$

where x_H is the neutral-to-ionized hydrogen fraction and δ_b is a density fluctuation parameter (important for 21 cm forests).

The spin temperature results from three competing processes

$$T_{\rm S}^{-1} = \frac{T_{\rm CMB}^{-1} + x_c T_{\rm K}^{-1} + x_\alpha T_{\rm C}^{-1}}{1 + x_c + x_\alpha}, \tag{3.250}$$

1. The absorption/emission of 21 cm photons from/to the CMB at temperature $T_{\rm CMB}$.
2. The collision with other hydrogen atoms, electrons and protons which have a kinetic temperature $T_{\rm K}$.
3. The resonant scattering of Lyman-α photons with color temperature $T_{\rm C}$ which pump the 21 cm transition through an intermediate state (the 1st excited hydrogen level).

The evolution of the 21 cm cosmic hydrogen signal δT_b with redshift is summarized on Fig. 3.11.

[27] In the context of 21 cm forest astronomy, the brightness temperature of the 21 cm line is defined relative to bright radio sources, e.g. active galactic nuclei or gamma-ray bursts, instead of relative to the CMB. 21 cm forests are a promising probe of the Epoch of Reionization and might be detected with future experiments SKA [73–78].

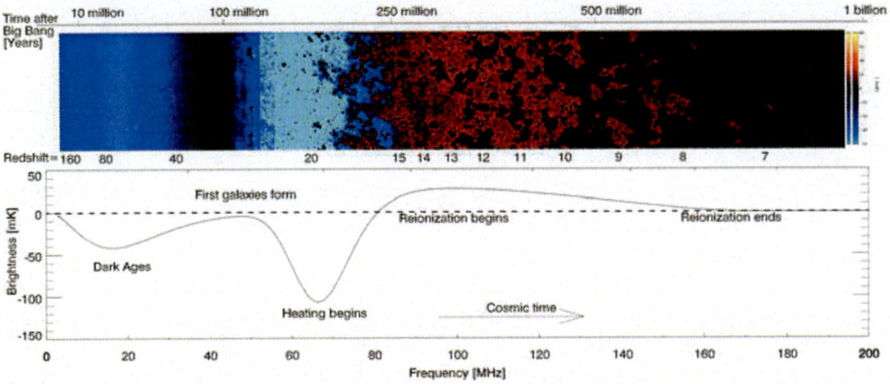

Fig. 3.11 The 21 cm cosmic hydrogen signal $\delta T_b \propto T_S - T_{CMB}$. **Top**: The fluctuation of the brightness temperature δT_b. **Bottom**: The sky-averaged signal $\delta \bar{T}_b$. Figure reproduced from [76]

- **200** $\lesssim$ **z** $\lesssim$ **1100**: The Compton scattering between CMB photons and residual electrons allows $T_K = T_{CMB}$. Moreover, the high gas density leads to $T_S = T_K$ and there is no 21 cm signal $\delta T_b = 0$.
- **40** $\lesssim$ **z** $\lesssim$ **200**: The gas cools adiabatically $T_K \propto (1+z)^2$ leading to $T_K \lesssim T_{CMB}$ while collisional coupling sets $T_S < T_{CMB}$, leading to a first absorption signal $\delta T_b < 0$.
- **25** $\lesssim$ **z** $\lesssim$ **40**: The spin temperature decouples from the gas, whose density decreases due to Hubble expansion and $T_S \sim T_{CMB}$. The 21 cm signal switches off $\delta T_b \approx 0$.
- **15** $\lesssim$ **z** $\lesssim$ **25**: As the first stars form, they emit Lyman-α photons such that $T_S \simeq T_C$ with $T_C \simeq T_K$. This results in an absorption depth $\delta T_b \lesssim 0$.
- **8** $\lesssim$ **z** $\lesssim$ **15**: Continuated heating drives $T_K > T_{CMB}$ giving rise to an emission signal $\delta T_b > 0$.
- **z** $\lesssim$ **8**: A significant fraction of the hydrogen atoms become ionized ($x_H \to 0$ in Eq. (3.249)) such that $\delta \bar{T}_b \to 0$.

For reviews on 21 cm cosmology, see [76, 923–925]. In 2018, upon analysing EDGES data, the authors of [926] have claimed to have found a 21 cm absorption depth in the CMB,[28] due to the Epoch of Reionization, peaked at the frequency $(21\,\text{cm})^{-1}/(1+z)$ with $z \simeq 17$

$$\delta T_b\big|_{\text{EDGES}} = -0.500^{+200}_{-500}\ \text{mK} \qquad (99\%\ \text{C.L.}). \tag{3.251}$$

However, the ΛCDM cosmology predicts a baryon temperature $T_K \simeq 7$ K and a ionized fraction $x_H \simeq 1$ [927, 928], which from imposing $T_S \gtrsim T_K$ in Eq. (3.249), leads to

$$\delta T_b|_{\Lambda}\text{CDM} = -0.22\ \text{mK}. \tag{3.252}$$

[28] The EDGES absorption depth has been observed at 78 MHz which corresponds to the low energy tail of the CMB, whose peak frequency is much larger: 160 GHz (Wien's law).

This missing factor two between the observed value and the theoretical prediction is the **21 cm anomaly**. We now briefly discuss the proposed solutions.

1. A first explanation is a lower baryon temperature than expected from ΛCDM. From Eqs. (3.249) and (3.251), we deduce the baryon temperature inferred by EDGES result
$$T_{\text{baryon}}\big|_{\text{EDGES}} \lesssim T_S\big|_{\text{EDGES}} \simeq 3.3^{+2}_{-1.6}\ \text{K}, \tag{3.253}$$
such that
$$\Delta T_{\text{baryon}} = T_{\text{baryon}}\big|_{\text{EDGES}} - T_{\text{baryon}}\big|_{\Lambda}\text{CDM} \lesssim -3.7^{+2}_{-1.6}\ \text{K}, \tag{3.254}$$
with $T_{\text{baryon}}\big|_{\Lambda\text{CDM}} \simeq 7$ K [927, 928]. Hence, the 21 cm anomaly predicts the baryons at redshift $z = 17$ to be at least 1.7 K cooler (at 99% $C.L.$) than in the ΛCDM scenario. Such a cooling can arise in the presence of long range interactions between baryons and DM when the latter is millicharged [929–931] (note however [932]). However, in [933–937], it was shown that this possibility is severly constrained by colliders, CMB, BBN, cooling of SN1987A, stars and DM overclosure. This possibility can be revived if a small fraction of DM is milicharged under both SM and dominant DM [938] or if the milicharged DM forms composites blobs with higher millicharge [939]. Also, upon asking minimal baryon heating due to DM annihilation, generic upper bound on DM annihilation cross-section can be obtained [940, 941], which allows to exclude models of thermal WIMP DM lighter than $M_{\text{DM}} \lesssim 10$ GeV. A lower baryon temperature can also be obtained with the presence of primordial magnetic fields [942].
2. The second possibility is to inject photons in the Rayleigh-Jeans tail of the CMB via invoking additional radio sources [943], e.g. black holes population [944] or hidden dark photons from DM decay, oscillating into photons [945].

In the meantime, the proper interpretation of the data has raised doubts. Particularly, the authors of [946] find that the conclusion depends on the modelization of the foreground, while the authors of [947] suggest the absorption depth to be due to the presence of an unsuspected resonant cavity in the EDGES antenna. The interpretation of EDGES anomaly in terms of systematic effects receives further support in December 2021 after that another radiometer, SARAS 3 [948], did not detect any sign of the profile found by EDGES [926] in their data.

We conclude the chapter by saying that the measurement of the cosmological parameters by Planck satellite has considerably transformed the field of Cosmology into a science of precision. However, the big questions remains unresolved. The future of the field is promising thanks to planned experiments like GW interferometers, multi-TeV telescopes, direct-detection experiments, observation of 21-cm forests, B-modes searches... We know move to the chapter on thermal Dark Matter.

References

1. Copernicus N (2010) On the revolutions of heavenly spheres. Prometheus Books
2. Newton I (1687) Philosophiæ Naturalis Principia Mathematica. England
3. Messier C (1781) Catalogue des nébuleuses et des amas d'étoiles (catalog of nebulae and star clusters). Connoissance des Temps ou des Mouvements Célestes 227–267
4. Herschel JFW (1864) Catalogue of nebulae and clusters of stars. Philoso Trans R Soc Lond 1–137
5. Einstein A (1915) The field equations of gravitation. Sitzungsber Preuss Akad Wiss Berlin (Math Phys) 1915:844–847
6. Einstein A (1916) The foundation of the general theory of relativity. Annalen Phys 49:769–822
7. Friedmann A (1922) 125. on the curvature of space. Zeitschrift für Physik 10:377–386
8. Friedmann A (1924) On the possibility of a world with constant negative curvature of space. Z Phys 21:326–332
9. Lemaitre G (2013) A homogeneous universe of constant mass and growing radius accounting for the radial velocity of extragalactic nebulae. Gen Rel Grav 45:1635–1646
10. Lemaitre G (1931) A homogeneous universe of constant mass and increasing radius accounting for the radial velocity of extra-galactic nebulae. Mon Not R Astron Soc 91:483–490
11. Lemaitre G (1997) The expanding universe. Gen Rel Grav 29:641–680
12. Robertson H (1935) Kinematics and world-structure. Astrophys J 82:284–301
13. Robertson H (1935) Kinematics and world-structure. 2. Astrophys J 83:187–201
14. Robertson H (1936) Kinematics and world-structure. 3. Astrophys J 83:257–271
15. Walker A (1935) On riemanntan spaces with spherical symmetry about a line, and the conditions for isotropy in genj relativity. Q J Math 81–93
16. Walker AG (1937) On milne's theory of world-structure. Proc Lond Math Soc 2:90–127
17. Leavitt HS, Pickering EC (1912) Periods of 25 variable stars in the small magellanic cloud. Harvard Obs Circ 173:1–3
18. Slipher VM (1913) The radial velocity of the andromeda nebula. Lowell Obs Bull 2:56–57
19. Slipher VM (1915) Spectrographic observations of nebulae. Popular Astron 23:21–24
20. Hubble E (1929) A relation between distance and radial velocity among extra-galactic nebulae. Proc Nat Acad Sci 15:168–173
21. Hoyle F (1948) A new model for the expanding universe. Mon Not R Astron Soc 108:372–382
22. Bondi H, Gold T (1948) The steady-state theory of the expanding universe. Mon Not R Astron Soc 108:252
23. Hoyle F, Burbidge G, Narlikar JV (1994) Astrophysical deductions from the quasisteady state cosmology. Mon Not R Astron Soc 267:1007
24. Hoyle F, Burbidge G, Narlikar JV (1994) Further astrophysical quantities expected in a quasisteady state universe. Astron Astrophys 289:729–739
25. Hoyle F, Burbidge G, Narlikar JV (1995) The Basic theory underlying the quasisteady state cosmology. Proc R Soc Lond A 448:191
26. Narlikar JV, Vishwakarma RG, Hajian A, Souradeep T, Burbidge G, Hoyle F (2003) Inhomogeneities in the microwave background radiation interpreted within the framework of the quasisteady state cosmology. Astrophys J 585:1–11. arXiv:astro-ph/0211036
27. Alpher RA, Bethe H, Gamow G (1948) The origin of chemical elements. Phys Rev 73:803–804

28. Alpher RA, Herman R (1948) Evolution of the universe. Nature 162:774–775
29. Alpher RA (1948) A neutron-capture theory of the formation and relative abundance of the elements. Phys Rev 74:1577–1589
30. Alpher RA, Herman RC (1948) On the relative abundance of the elements. Phys Rev 74:1737–1742
31. Dicke R, Peebles P, Roll P, Wilkinson D (1965) Cosmic black-body radiation. Astrophys J 142:414–419
32. Penzias AA, Wilson RW (1965) A measurement of excess antenna temperature at 4080-Mc/s. Astrophys J 142:419–421
33. Glashow S (1961) Partial symmetries of weak interactions. Nucl Phys 22:579–588
34. Weinberg S (1967) A model of leptons. Phys Rev Lett 19:1264–1266
35. Salam A (1968) Weak and electromagnetic interactions. Conf Proc C 680519:367–377
36. Gross DJ, Wilczek F (1973) Ultraviolet behavior of nonabelian gauge theories. Phys Rev Lett 30:1343–1346
37. Politzer HD (1973) Reliable perturbative results for strong interactions? Phys Rev Lett. 30:1346–1349
38. Guth AH (1981) The inflationary universe: a possible solution to the horizon and flatness problems. Phys Rev D 23:347–356
39. Linde AD (1982) A new inflationary universe scenario: a possible solution of the horizon, flatness, homogeneity, isotropy and primordial monopole problems. Phys Lett B 108:389–393
40. Albrecht A, Steinhardt PJ (1982) Cosmology for grand unified theories with radiatively induced symmetry breaking. Phys Rev Lett 48:1220–1223
41. Starobinsky AA (1980) A new type of isotropic cosmological models without singularity. Phys Lett B 91:99–102
42. Starobinsky AA (1982) Dynamics of phase transition in the new inflationary universe scenario and generation of perturbations. Phys Lett B 117:175–178
43. Mukhanov VF, Chibisov GV (1981) Quantum fluctuations and a nonsingular universe. JETP Lett 33:532–535
44. Mukhanov VF, Feldman HA, Brandenberger RH (1992) Theory of cosmological perturbations. Part 1. Classical perturbations. Part 2. Quantum theory of perturbations. Part 3. Extensions. Phys Rep 215:203–333
45. Silk J et al (2010) Particle dark matter: observations, models and searches. Cambridge University Press, Cambridge. https://doi.org/10.1017/CBO9780511770739
46. Einasto J (2013) Dark matter and cosmic web story. World Sci. https://doi.org/10.1142/8928
47. Majumdar D (2014) Dark matter: an introduction. Taylor & Francis Group, Boca Raton, p 8
48. Vavilova IB, Bolotin YL, Boyarsky AM, Danevich FA, Kobychev VV, Tretyak VI et al (2015) Dark matter: observational manifestation and experimental searches, vol 3. Dark energy and dark matter in the Universe, Akademperiodyka, Kyiv, Ukraine
49. Profumo S (2017) An introduction to particle dark matter. World Sci. https://doi.org/10.1142/q0001
50. Essig R, Feng J, Zurek K (eds) (2019) Proceedings, of a simons symposium: illuminating dark matter. Kruen, Germany, Springer. https://doi.org/10.1007/978-3-030-31593-1
51. Supernova Search Team collaboration, Riess AG et al (1988) Observational evidence from supernovae for an accelerating universe and a cosmological constant. Astron J 116:1009–1038. arXiv:astro-ph/9805201
52. Supernova Cosmology Project collaboration, Perlmutter S et al (1999) Measurements of Ω and Λ from 42 high redshift supernovae. Astrophys J 517:565–586. arXiv:astro-ph/9812133
53. Gaia collaboration, Brown AGA et al (2018) Gaia data release 2: Lof the contents and survey properties. Astron Astrophys 616:A1. arXiv:1804.09365
54. LIGO Scientific, VIRGO, KAGRA collaboration, Abbott R et al, The population of merging compact binaries inferred using gravitational waves through GWTC-3. arXiv:2111.03634
55. LIGO Scientific, VIRGO, KAGRA collaboration, Abbott R et al, Constraints on the cosmic expansion history from GWTC-3. arXiv:2111.03604

56. LIGO Scientific, VIRGO, KAGRA collaboration, Abbott R et al, Tests of general relativity with GWTC-3. arXiv:2112.06861
57. Suzuki A et al (2018) The LiteBIRD satellite mission—sub-kelvin instrument. J Low Temp Phys 193:1048–1056. arXiv:1801.06987
58. Hazumi M et al (2019) LiteBIRD: A satellite for the studies of b-mode polarization and inflation from cosmic background radiation detection. J Low Temp Phys 194:443–452
59. Sugai H et al, Updated design of the CMB polarization experiment satellite LiteBIRD. J Low Temp Phys 199:1107–1117. arXiv:2001.01724
60. LiteBIRD collaboration, Hazumi M et al (2020) LiteBIRD: JAXA's new strategic L-class mission for all-sky surveys of cosmic microwave background polarization Proc SPIE Int Soc Opt Eng 11443:114432F. arXiv:2101.12449
61. CMB-S4 collaboration, Abazajian KN et al, CMB-S4 science book, 1st edn. arXiv:1610.02743
62. CMB-S4 collaboration, Abitbol MH et al, CMB-S4 technology book, 1st edn. arXiv:1706.02464
63. Abazajian K et al, CMB-S4 science case, reference design, and project plan. arXiv:1907.04473
64. Abazajian K et al (2019) CMB-S4 decadal survey APC white paper. Bull Am Astron Soc 51:209. arXiv:1908.01062
65. CMB-S4 collaboration, Abazajian K et al (2022) CMB-S4: forecasting constraints on primordial gravitational waves. Astrophys J 926:54. arXiv:2008.12619
66. Abazajian K et al (2022) Snowmass 2021 CMB-S4 white paper. In: 2022 snowmass summer study, p 3. arXiv:2203.08024
67. Dvorkin C et al (2022) The physics of light relics. In: 2022 snowmass summer study, p 3. arXiv:2203.07943
68. EUCLID collaboration, Laureijs R et al, Euclid definition study report. arXiv:1110.3193
69. Amendola L et al (2018) Cosmology and fundamental physics with the Euclid satellite. Living Rev Rel 21:2. arXiv:1606.00180
70. LSST Science, LSST Project collaboration, Abell PA et al, LSST Science Book, Version 2.0. arXiv:0912.0201
71. LSST Dark Energy Science collaboration, Mandelbaum R et al, The LSST Dark Energy Science Collaboration (DESC) science requirements document. arXiv:1809.01669
72. Ferraro S, Sailer N, Slosar A, White M, Snowmass2021 cosmic frontier white paper: cosmology and fundamental physics from the three-dimensional large scale structure. arXiv:2203.07506
73. Santos MG et al (2015) Cosmology from a SKA HI intensity mapping survey. PoS AASKA14:019. arXiv:1501.03989
74. SKA Cosmology SWG collaboration, Maartens R, Abdalla FB, Jarvis M, Santos MG (2015) Overview of cosmology with the SKA. PoS AASKA14:016. arXiv:1501.04076
75. Sprenger T, Archidiacono M, Brinckmann T, Clesse S, Lesgourgues J (2019) Cosmology in the era of euclid and the square kilometre array. JCAP 02:047. arXiv:1801.08331
76. Pritchard JR, Loeb A (2012) 21-cm cosmology. Rep Prog Phys 75:086901. arXiv:1109.6012
77. Cosmology-SWG, EoR/CD-SWG collaboration, Pritchard J et al (2015) Cosmology from EoR/Cosmic Dawn with the SKA. PoS AASKA14:012. arXiv:1501.04291
78. Snowmass 2021 Cosmic Frontier 5 Topical Group collaboration, Liu A, Newburgh L, Saliwanchik B, Slosar A, Snowmass2021 cosmic frontier white paper: 21cm radiation as a probe of physics across cosmic ages. arXiv:2203.07864
79. Dvorkin C et al (2022) Dark matter physics from the CMB-S4 experiment. In: 2022 snowmass summer study, p 3. arXiv:2203.07064
80. Chakrabarti S et al (2022) Snowmass2021 cosmic frontier white paper: observational facilities to study dark matter. In: 2022 snowmass summer study, p 3. arXiv:2203.06200
81. CTA Consortium collaboration, Actis M et al (2011) Design concepts for the Cherenkov Telescope Array CTA: An advanced facility for ground-based high-energy gamma-ray astronomy. Exper Astron 32:193–316. arXiv:1008.3703
82. CTA Consortium collaboration, Acharya BS et al (2013) Introducing the CTA concept. Astropart Phys 43:3–18

83. Silverwood H, Weniger C, Scott P, Bertone G (2015) A realistic assessment of the CTA sensitivity to dark matter annihilation. JCAP 03:055 [arXiv:1408.4131]
84. Pierre M, Siegal-Gaskins JM, Scott P (2014) Sensitivity of CTA to dark matter signals from the Galactic Center. JCAP 06:024 [arXiv:1401.7330]
85. Ibarra A, Lamperstorfer AS, López-Gehler S, Pato M, Bertone G (2015) On the sensitivity of CTA to gamma-ray boxes from multi-TeV dark matter. JCAP 09:048. arXiv:1503.06797
86. Duangchan C et al, CTA Sensitivity on TeV scale dark matter models with complementary limits from direct detection. arXiv:2202.07321
87. Addazi A et al (2022) The Large High Altitude Air Shower Observatory (LHAASO) science book (2021 Edition). Chin Phys C 46:035001–035007. arXiv:1905.02773
88. KM3Net collaboration, Adrian-Martinez S et al (2016) Letter of intent for KM3NeT 2.0. J Phys G 43:084001. arXiv:1601.07459
89. KM3NeT collaboration, Lopez-Coto D, Navas S, Zornoza JD (2022) Indirect dark matter searches with the ANTARES and KM3NeT neutrino telescopes. PoS EPS-HEP2021:174
90. Jaeckel J, Ringwald A (2010) The low-energy frontier of particle physics. Ann Rev Nucl. Part Sci 60:405–437. arXiv:1002.0329
91. Essig R et al (2013) Working group report: new light weakly coupled particles. In: Community summer study 2013: snowmass on the Mississippi, p 10. arXiv:1311.0029
92. Marsh DJE (2016) Axion cosmology. Phys Rep 643:1–79. arXiv:1510.07633
93. Hook A (2019) TASI lectures on the strong CP problem and axions. PoS TASI2018:004. arXiv:1812.02669
94. Irastorza IG, Redondo J (2018) New experimental approaches in the search for axion-like particles. Prog Part Nucl Phys 102:89–159. arXiv:1801.08127
95. Di Luzio L, Giannotti M, Nardi E, Visinelli L, The landscape of QCD axion models. arXiv:2003.01100
96. Bergé J, Brax P, Métris G, Pernot-Borràs M, Touboul P, Uzan J-P (2018) MICROSCOPE mission: first constraints on the violation of the weak equivalence principle by a light scalar dilaton. Phys Rev Lett 120:141101. arXiv:1712.00483
97. Arvanitaki A, Huang J, Van Tilburg K (2015) Searching for dilaton dark matter with atomic clocks. Phys Rev D 91:015015. arXiv:1405.2925
98. Arvanitaki A, Dimopoulos S, Dubovsky S, Kaloper N, March-Russell J (2010) String Axiverse. Phys Rev D 81:123530. arXiv:0905.4720
99. Caputo A, Millar AJ, O'Hare CAJ, Vitagliano E (2021) Dark photon limits: a handbook. Phys Rev D 104:095029. arXiv:2105.04565
100. Klein A et al (2016) Science with the space-based interferometer eLISA: supermassive black hole binaries. Phys Rev D 93:024003. arXiv:1511.05581
101. Caprini C et al (2016) Science with the space-based interferometer eLISA. II: Gravitational waves from cosmological phase transitions. JCAP 04:001. arXiv:1512.06239
102. Tamanini N, Caprini C, Barausse E, Sesana A, Klein A, Petiteau A (2016) Science with the space-based interferometer eLISA. III: probing the expansion of the Universe using gravitational wave standard sirens. JCAP 04:002. arXiv:1601.07112
103. Bartolo N et al (2016) Science with the space-based interferometer LISA. IV: Probing inflation with gravitational waves. JCAP 12:026. arXiv:1610.06481
104. Babak S, Gair J, Sesana A, Barausse E, Sopuerta CF, Berry CPL et al (2017) Science with the space-based interferometer LISA. V: Extreme mass-ratio inspirals. Phys Rev D 95:103012. arXiv:1703.09722
105. Caprini C et al (2020) Detecting gravitational waves from cosmological phase transitions with LISA: an update. JCAP 03:024. arXiv:1910.13125
106. LISA Cosmology Working Group collaboration, Auclair P et al, Cosmology with the laser interferometer space antenna. arXiv:2204.05434
107. Punturo M et al (2010) The Einstein telescope: a third-generation gravitational wave observatory. Class Quant Grav 27:194002
108. Sathyaprakash B et al (2012) Scientific objectives of Einstein telescope. Class Quant Grav 29:124013. arXiv:1206.0331

109. Regimbau T et al (2012) A mock data challenge for the Einstein gravitational-wave telescope. Phys Rev D 86:122001. arXiv:1201.3563
110. Maggiore M et al (2020) Science case for the Einstein telescope. JCAP 03:050. arXiv:1912.02622
111. Reitze D et al (2019) Cosmic explorer: the U.S. contribution to gravitational-wave astronomy beyond LIGO. Bull Am Astron Soc 51:035. arXiv:1907.04833
112. Evans M et al, A horizon study for cosmic explorer: science, observatories, and community. arXiv:2109.09882
113. Weinberg S (1972) Gravitation and cosmology: principles and applications of the general theory of relativity. Wiley, New York
114. Misner CW, Thorne KS, Wheeler JA (1973) Gravitation. W. H, Freeman, San Francisco
115. Birrell ND, Davies PCW (1984) Quantum fields in curved space. Cambridge Monographs on Mathematical Physics. Cambridge University Press, Cambridge, UK, P 2. https://doi.org/10.1017/CBO9780511622632
116. Wald RM (1984) General relativity. Chicago University Pr., Chicago, USA. https://doi.org/10.7208/chicago/9780226870373.001.0001
117. Stewart JM (1994) Advanced general relativity. Cambridge Monographs on Mathematical Physics. Cambridge University Press, p 4. https://doi.org/10.1017/CBO9780511608179
118. Schutz BF (2003) Gravity from the ground up
119. Hartle JB (2003) An introduction to Einstein's general relativity
120. Carroll SM (2019) Spacetime and geometry. Cambridge University Press, p 7
121. Stephani H, Kramer D, MacCallum MAH, Hoenselaers C, Herlt E (2003) Exact solutions of Einstein's field equations. Cambridge Monographs on Mathematical Physics. Cambridge University Press, Cambridge. https://doi.org/10.1017/CBO9780511535185
122. Stephani H (2004) Relativity: an introduction to special and general relativity
123. Cheng TP (2010) Relativity, gravitation, and cosmology: a basic introduction. Oxford University, Pr, Oxford, UK
124. Hobson M, Efstathiou G, Lasenby A (2006) General relativity: an introduction for physicists, p 10
125. Rindler W (2006) Relativity: special, general, and cosmological, p 11
126. Hoyng P (2006) Relativistic astrophysics and cosmology: a primer
127. Plebanski J, Krasinski A (2006) An introduction to general relativity and cosmology
128. Groen O, Hervik S (2007) Einstein's general theory of relativity: with modern applications in cosmology
129. Woodhouse NMJ (2007) General relativity, p 2
130. Mukhanov V, Winitzki S (2007) Introduction to quantum effects in gravity. Cambridge University Press, p 6
131. Chow TL (2008) Gravity, black holes, and the very early universe: an introduction to general relativity and cosmology
132. Parker LE, Toms D (2009) Quantum field theory in curved spacetime: quantized field and gravity. Cambridge Monographs on Mathematical Physics. Cambridge University Press, p 8. https://doi.org/10.1017/CBO9780511813924
133. Griffiths JB, Podolsky J (2009) Exact space-times in Einstein's general relativity. Cambridge Monographs on Mathematical Physics. Cambridge University Press, Cambridge. https://doi.org/10.1017/CBO9780511635397
134. Ryder L (2009) Introduction to general relativity
135. Taylor E, Wheeler J, Bertschinger E (2010) Exploring black holes: introduction to general relativity. Addison-Wesley
136. Lambourne RJA (2010) Relativity, gravitation and cosmology
137. Baumgarte TW, Shapiro SL (2010) Numerical relativity: solving Einstein's equations on the computer. Cambridge University Press. https://doi.org/10.1017/CBO9781139193344
138. Baumgarte TW, Shapiro SL (2021) Numerical relativity: starting from scratch. Cambridge University Press, p 2. https://doi.org/10.1017/9781108933445

139. Straumann N (2013) General relativity. Graduate texts in physics. Springer, Dordrecht. https://doi.org/10.1007/978-94-007-5410-2
140. Zee A (2013) Einstein gravity in a nutshell. Princeton University Press, New Jersey, p 5
141. Gourgoulhon E (2013) Special relativity in general frames. Graduate texts in physics. Springer, Berlin. https://doi.org/10.1007/978-3-642-37276-6
142. Choquet-Bruhat Y (2014) Introduction to general relativity, black holes, and cosmology. Oxford University Press, p 11
143. Ashtekar A, Berger BK, Isenberg J, MacCallum M (2015) General relativity and gravitation: a centennial perspective. Cambridge University Press, p 7
144. Williams C (ed) (2015) Advances in general relativity research. Nova
145. Böhmer CG (2016) Introduction to general relativity and cosmology. Essential textbooks in physics. World Scientific, p 12. https://doi.org/10.1142/q0034
146. Deruelle N, Uzan J-P (2018) Relativity in modern physics. Oxford graduate texts. Oxford University Press, p 8
147. Bambi C (2018) Introduction to general relativity. Undergraduate lecture notes in physics. Springer, Singapore. https://doi.org/10.1007/978-981-13-1090-4
148. Hall MJW, General relativity: an introduction to black holes, gravitational waves, and cosmology
149. Carlip S (2019) General relativity. Oxford University Press, p 1
150. Compère G (2019) Advanced lectures on general relativity, vol 952. Springer, Cham, Cham, Switzerland, p 2. https://doi.org/10.1007/978-3-030-04260-8
151. Guidry M (2019) Modern general relativity. Cambridge University Press, p 2. https://doi.org/10.1017/9781108181938
152. Soffel MH, Han W-B (2019) Applied general relativity. Springer, Berlin. https://doi.org/10.1007/978-3-030-19673-8
153. Mayerson DR, Charles AM, Golec JE (2019) Relativity: a journey through warped space and time. Springer, Berlin. https://doi.org/10.1007/978-3-030-18914-3
154. Krasnov K (2020) Formulations of general relativity. Cambridge monographs on mathematical physics. Cambridge University Press, p 11. https://doi.org/10.1017/9781108674652
155. Ferrari V, Gualtieri L, Pani P (2020) General relativity and its applications. CRC Press, Taylor & Francis Group
156. Rovelli C (2021) General relativity: the essentials. Cambridge University Press. https://doi.org/10.1017/9781009031806
157. Hawking SW, Ellis GFR (2011) The large scale structure of space-time. Cambridge monographs on mathematical physics. Cambridge University Press, p 2. https://doi.org/10.1017/CBO9780511524646
158. Kolb EW, Turner MS (1990) The early universe. Front Phys 69:1–547
159. Linde AD (1990) Particle physics and inflationary cosmology, vol 5
160. Peebles PJE (1994) Principles of physical cosmology
161. Roos M (1994) Introduction to cosmology
162. Hawley JF, Holcomb KA (1998) Foundations of modern cosmology
163. Liddle AR, Lyth DH (2000) Cosmological inflation and large scale structure, p 9
164. Bonometto S, Gorini V, Moschella U (eds) (2002) Modern cosmology
165. Narlikar JV (2002) Introduction to cosmology, 3rd edn. Cambridge University Press, Cambridge
166. Dodelson S (2003) Modern cosmology. Academic, Amsterdam
167. Mukhanov V (2005) Physical foundations of cosmology. Cambridge University Press, Oxford
168. Liebscher DE (2005) Cosmology. Springer Tracts Mod Phys 210:1–294
169. Weinberg S (2008) Cosmology, p 9
170. Belusevic R (2008) Relativity, astrophysics, and cosmology, vol 1
171. Belusevic R (2008) Relativity, astrophysics, and cosmology, vol 2
172. Lyth DH, Liddle AR (2009) The primordial density perturbation: cosmology, inflation and the origin of structure

173. Montani G, Battisti MV, Benini R, Imponente G (2009) Primordial cosmology. World Scientific, Singapore
174. Goodstein D (ed) (2012) Adventures in cosmology. https://doi.org/10.1142/7820
175. Lachièze-Rey M (2012) Theoretical and observational cosmology. Springer, Netherlands, Nato Science Series C
176. Peter P, Uzan J-P (2013) Cosmology. Oxford graduate texts. Oxford University Press, p 2
177. Lesgourgues J, Mangano G, Miele G, Pastor S (2013) Neutrino cosmology. Cambridge University Press, p 2
178. Baumann D, McAllister L (2015) Inflation and string theory. Cambridge monographs on mathematical physics. Cambridge University Press, p 5. https://doi.org/10.1017/CBO9781316105733
179. Lambourne RJA, Serjeant S, Jones MH (eds) (2015) An introduction to galaxies and cosmology. Cambridge University Press, p 1
180. Bambi C, Dolgov AD (2015) Introduction to particle cosmology. Unitext for physics Springer, Berllin. https://doi.org/10.1007/978-3-662-48078-6
181. Heacox WD (2015) The expanding universe a primer on relativistic cosmology. Cambridge University Press, p 11
182. Jones BJT (2017) Precision cosmology. Cambridge University Press, p 4
183. Lyth D (2017) Cosmology for physicists. CRC Press, Boca Raton
184. Saunders S, Silk J, Barrow JD, Chamcham K (eds) (2017) The philosophy of cosmology. Cambridge University Press, p 4. https://doi.org/10.1017/9781316535783
185. Fazio GG, Barkana R, Tsujikawa S, Kim JE (2018). The encyclopedia of cosmology world scientific. https://doi.org/10.1142/9496
186. Baumann D (2022) Cosmology. Cambridge University Press
187. Friedmann A, Lemaître G, Grib A (1997) Essais de cosmologie. Points Sciences, Seuil
188. Renn J, Renn J, Schemmel M, Smeenk C, Martin C, Divarci L (eds) (2007) The genesis of general relativity: sources and interpretations, vol 250 of Boston studies in the philosophy and history of science. Springer, Berlin. https://doi.org/10.1007/978-1-4020-4000-9
189. Kragh EBH, Longair M (2019) The Oxford handbook of the history of modern cosmology. Oxford Handbooks. Oxford University Press, p 3
190. Kolata JJ (2020) Elementary cosmology, 2nd edn. IOP, p 12
191. Peebles PJE, Cosmology's Century: an inside history of our modern understanding of the universe. Princeton University Press
192. Kragh H (2021) Cosmology and controversy: the historical development of two theories of the universe. Princeton University Press
193. Ginzburg V, Haar D (2013) Theoretical physics and astrophysics. International series on nuclear energy. Elsevier Science
194. Padmanabhan P (2000) Theoretical astrophysics. Vol. 1: astrophysical processes
195. Padmanabhan T (2000) Theoretical astrophysics: volume 2, stars and stellar systems, vol 2. Cambridge University Press
196. Padmanabhan T (2002) Theoretical astrophysics: volume 3, galaxies and cosmology. Cambridge University Press
197. Duric N (2004) Advanced astrophysics. Cambridge University Press
198. Makarov I (2014) Introduction to theoretical astrophysics. Createspace Independent Pub
199. Carroll BW, Ostlie DA (2017) An introduction to modern astrophysics. Cambridge University Press
200. Longair MS (2010) High energy astrophysics. Cambridge university press
201. Weinberg S (2019) Lectures on astrophysics. Cambridge University Press
202. Chabrier G (2011) Structure formation in astrophysics. Contemporary astrophysics. Cambridge University Press, Cambridge
203. Lagos C (2013) The physics of galaxy formation. Springer International Publishing, Springer Theses
204. Fox A, Davé R (2017) Gas accretion onto galaxies. Astrophysics and space science library. Springer International Publishing

205. Mo H, van den Bosch F, White S (2010) Galaxy formation and evolution. Galaxy Formation and Evolution. Cambridge University Press
206. Cimatti A, Fraternali F, Nipoti C (2019) Introduction to galaxy formation and evolution: from primordial gas to present-day galaxies. Cambridge University Press
207. Longair MS (2007) Galaxy formation. Springer Science & Business Media
208. Binney J, Tremaine S (2011) Galactic dynamics, vol 13. Princeton university Press
209. Hartmann L (2000) Accretion processes in star formation, vol 32. Cambridge University Press
210. Stahler S, Palla F (2008) The formation of stars. Wiley, New York
211. Ward-Thompson D, Whitworth A (2011) An introduction to star formation. Cambridge University Press
212. Bodenheimer P (2011) Principles of star formation. Springer Science & Business Media
213. Krumholz M (2017) Star formation. World scientific series in astrophysics. World Scientific Publishing Company
214. Raffelt GG (1996) Stars as laboratories for fundamental physics: the astrophysics of neutrinos, axions, and other weakly interacting particles, p 5
215. Salaris M, Cassisi S (2005) Evolution of stars and stellar populations. Wiley, New York
216. Schwarzschild M (2015) Structure and evolution of stars. Princeton University Press
217. Iliadis C (2015) Nuclear physics of stars. Wiley, New York
218. Melia F (2007) The galactic supermassive black hole. Princeton paperbacks. Princeton University Press, Princeton
219. Frolov V, Zelnikov A (2011) Introduction to black hole physics. OUP Oxford
220. Meier D (2012) Black hole astrophysics: the engine paradigm. Springer praxis books. Springer, Berlin
221. Romero G, Vila G (2013) Introduction to black hole astrophysics. Lecture Notes in physics. Springer, Berlin
222. Haardt F, Gorini V, Moschella U, Treves A, Colpi M (2015) Astrophysical black holes. Lecture notes in physics. Springer International Publishing
223. Grenzebach A (2016) The shadow of black holes: an analytic description. SpringerBriefs in physics. Springer International Publishing
224. Falanga M, Belloni T, Casella P, Gilfanov M, Jonker P, King A (2016) The physics of accretion onto black holes. Space sciences series of ISSI. Springer, New York
225. Di Matteo T, King A, Cornish N, Walter R, Jetzer P, Mayer L et al (2019) Black hole formation and growth: saas-fee advanced course 48. Swiss society for astrophysics and astronomy. Saas-fee advanced course. Springer, Berlin
226. Latif M, Schleicher D (2019) Formation of the first black holes. World Scientific Publishing Company
227. Brito R, Cardoso V, Pani P (2020) Superradiance: new frontiers in black hole physics. Lecture notes in physics. Springer International Publishing
228. Tucker W (1975) Radiation processes in astrophysics. The harvard books on astronomy
229. Rybicki GB, Lightman AP (1991) Radiative processes in astrophysics. Wiley, New York
230. Melrose DB (1980) Plasma astrophysics, vol 1. CRC Press
231. Draine b (2011) Physics of the Interstellar and Intergalactic Medium. Princeton series in astrophysics. Princeton University Press
232. LeBlanc F (2011) An introduction to stellar astrophysics. Wiley, New York
233. Somov BV (2012) Plasma astrophysics, part II: reconnection and flares, vol 392. Springer Science & Business Media
234. Zheleznyakov VV (2012) Radiation in astrophysical plasmas, vol 204. Springer Science & Business Media
235. Benz AO (2012) Plasma astrophysics: kinetic processes in solar and stellar coronae, vol 184. Springer Science & Business Media
236. Goossens M (2012) An introduction to plasma astrophysics and magnetohydrodynamics. Astrophysics and space science library. Springer, Netherlands
237. Ghisellini G (2013) Radiative processes in high energy astrophysics, vol 873. Springer
238. Chiuderi C, Velli M (2014) Basics of plasma astrophysics. Unitext for physics. Springer Milan

239. Williams A (2018) Plasma astrophysics. CreateSpace Independent Publishing Platform
240. Tajima T, Shibata K (2018) Plasma astrophysics. CRC Press
241. Goedbloed H, Keppens R, Poedts S (2019) Magnetohydrodynamics of laboratory and astrophysical plasmas. Cambridge University Press
242. Kulsrud RM (2020) Plasma physics for astrophysics. Princeton University Press
243. Williams J (2021) Introduction to the interstellar medium. Cambridge University Press
244. Kim CW, Pevsner A, Neutrinos in physics and astrophysics
245. Klapdor-Kleingrothaus HV, Zuber K (1997) Particle astrophysics. CRC Press
246. Grieder PK (2001) Cosmic rays at Earth. Elsevier
247. Miroshnichenko L, Solar cosmic rays
248. Zuber K (2003) Neutrino physics. Taylor & Francis
249. Grupen C, Cowan G, Eidelman S, Eidelman G, Stroh T (2005) Astroparticle physics. Springer
250. Sarkar U (2007) Particle and astroparticle physics. Series in high energy physics, cosmology and gravitation. CRC Press
251. Ginzburg V, The origin of cosmic rays
252. Valle J, Romao J (2015) Neutrinos in high energy and astroparticle physics. Wiley
253. Gaisser TK, Engel R, Resconi E (2016) Cosmic rays and particle physics. Cambridge University Press
254. Sigl G (2016) Astroparticle physics: theory and phenomenology. Atlantis studies in astroparticle physics and cosmology. Atlantis Press
255. Duffy R (2017) Astroparticle physics: theory and phenomenology. CreateSpace independent publishing platform
256. De Angelis A, Pimenta M (2015) Introduction to particle and astroparticle physics: questions to the universe. undergraduate lecture notes in physics. Springer Milan
257. De Angelis A, Pimenta M (2018) Introduction to particle and astroparticle physics: multimessenger astronomy and its particle physics foundations. Undergraduate lecture notes in physics. Springer International Publishing
258. De Angelis A, Pimenta M, Conceição R (2021) Particle and astroparticle physics: problems and solutions. Undergraduate lecture notes in physics. Springer International Publishing
259. Planck collaboration, Aghanim N et al, Planck 2018 results. VI. Cosmological parameters. arXiv:1807.06209
260. Particle Data Group collaboration, Tanabashi M et al (2018) Review of particle physics. Phys Rev D98:030001
261. Supernova Search Team collaboration, Riess AG et al (1998) Observational evidence from supernovae for an accelerating universe and a cosmological constant. Astron J 116:1009–1038. arXiv:astro-ph/9805201
262. Supernova Cosmology Project collaboration, Perlmutter S et al (1999) Measurements of Ω and Λ from 42 high redshift supernovae. Astrophys J 517:565–586. arXiv:astro-ph/9812133
263. Di Valentino E, Melchiorri A, Silk J (2019) Planck evidence for a closed Universe and a possible crisis for cosmology. Nat Astron 4:196–203. arXiv:1911.02087
264. Park C-G, Ratra B (2019) Using the tilted flat-Λ CDM and the untilted non-flat Λ CDM inflation models to measure cosmological parameters from a compilation of observational data. Astrophys J 882:158. arXiv:1801.00213
265. Handley W, Curvature tension: evidence for a closed universe. arXiv:1908.09139
266. Lehners J-L (2008) Ekpyrotic and cyclic cosmology. Phys Rep 465:223–263. arXiv:0806.1245
267. Cai Y-F (2014) Exploring bouncing cosmologies with cosmological surveys. Sci China Phys Mech Astron 57:1414–1430. arXiv:1405.1369
268. Battefeld D, Peter P (2015) A critical review of classical bouncing cosmologies. Phys Rep 571:1–66. arXiv:1406.2790
269. Brandenberger R, Peter P (2017) Bouncing cosmologies: progress and problems. Found Phys 47:797–850. arXiv:1603.05834
270. Ijjas A, Steinhardt PJ (2018) Bouncing Cosmology made simple. Class Quant Grav 35:135004. arXiv:1803.01961

271. Mironov S, Rubakov V, Volkova V, Cosmological scenarios with bounce and Genesis in Horndeski theory and beyond: an essay in honor of I.M. Khalatnikov on the occasion of his 100th birthday. arXiv:1906.12139
272. Zannoni A (1999) On the quantization of the monoatomic ideal gas. cond-mat/9912229
273. Dirac PAM (1926) On the theory of quantum mechanics Proc R Soc Lond Ser Contain Pap Math Phys Character 112:661–677
274. Einstein A (1924) Quantentheorie des einatomigen idealen gases. S B Preuss Akad Wiss phys-math Klasse
275. Bekenstein JD (1972) Black holes and the second law. Lett Nuovo Cim 4:737–740
276. Hawking SW (1975) Particle creation by black holes. Commun Math Phys 43:199–220
277. Fischler W, Susskind L, Holography and cosmology. arXiv:hep-th/9806039
278. Bousso R (1999) A Covariant entropy conjecture. JHEP 07:004. arXiv:hep-th/9905177
279. Spradlin M, Strominger A, Volovich A (2001) Les Houches lectures on de Sitter space. In: Les Houches summer school: Session 76: Euro summer school on unity of fundamental physics: gravity, gauge theory and strings, pp 423–453, 10. arXiv:hep-th/0110007
280. Bousso R (2002) Adventures in de Sitter space. In: Workshop on conference on the future of theoretical physics and cosmology in honor of Steven Hawking's 60th Birthday, pp 539–569, 5. arXiv:hep-th/0205177
281. Penrose R (1989) Difficulties with inflationary cosmology. Annals N Y Acad Sci 571:249–264
282. Barnes LA, Lewis GF, Under an Iron Sky: On the Entropy at the Start of the Universe. arXiv:2012.06975
283. Dyson L, Kleban M, Susskind L (2002) Disturbing implications of a cosmological constant. JHEP 10:011. arXiv:hep-th/0208013
284. Mangano G, Miele G, Pastor S, Pisanti O, Sarikas S (2012) Updated BBN bounds on the cosmological lepton asymmetry for non-zero θ_{13}. Phys Lett B 708:1–5. arXiv:1110.4335
285. Fields BD, Molaro P, Sarkar S (2014) Big-bang nucleosynthesis. Chin Phys C 38:339–344. arXiv:1412.1408
286. Lesgourgues J, Mangano G, Miele G, Pastor S (2013) Neutrino cosmology. Cambridge University Press, p 2
287. Riotto A (2003) Inflation and the theory of cosmological perturbations. ICTP Lect Notes Ser 14:317–413. arXiv:hep-ph/0210162
288. Linde AD (1990) Particle physics and inflationary cosmology, vol 5
289. Baumann D (2011) Inflation. In: Theoretical advanced study institute in elementary particle physics: physics of the large and the small, pp 523–686. arXiv:0907.5424, https://doi.org/10.1142/9789814327183_0010
290. Langlois D (2010) Lectures on inflation and cosmological perturbations. Lect Notes Phys 800:1–57. arXiv:1001.5259
291. Senatore L (2013) TASI 2012 Lectures on Inflation. In: Theoretical advanced study institute in elementary particle physics: searching for new physics at small and large scales, pp 221–302. https://doi.org/10.1142/9789814525220_0006
292. Burgess CP, Intro to effective field theories and inflation. arXiv:1711.10592
293. A. Vázquez J, Padilla LE, Matos T, Inflationary cosmology: from theory to observations. arXiv:1810.09934
294. Liddle AR, Lyth DH (1993) The Cold dark matter density perturbation. Phys Rep 231:1–105. arXiv:astro-ph/9303019
295. Kazanas D (1980) Dynamics of the universe and spontaneous symmetry breaking. Astrophys J Lett 241:L59–L63
296. Guth AH, Weinberg EJ (1981) Cosmological consequences of a first order phase transition in the SU(5) grand unified model. Phys Rev D 23:876
297. Guth AH, Weinberg EJ (1983) Could the universe have recovered from a slow first order phase transition? Nucl Phys B 212:321–364
298. Caprini C, Figueroa DG (2018) Cosmological backgrounds of gravitational waves. Class Quant Grav 35:163001. arXiv:1801.04268

299. BICEP2, Keck Array collaboration, Ade PAR et al (2018) BICEP2/Keck Array x: constraints on primordial gravitational waves using planck, WMAP, and New BICEP2/Keck observations through the 2015 season. Phys Rev Lett 121:221301. arXiv:1810.05216
300. Planck collaboration, Akrami Y et al (2020) Planck 2018 results. X. Constraints on inflation. Astron Astrophys 641:A10. arXiv:1807.06211
301. Matsumura T et al (2014) Mission design of LiteBIRD. J Low Temp Phys 176:733. arXiv:1311.2847
302. Linde AD (1983) Chaotic inflation. Phys Lett B 129:177–181
303. Silverstein E, Westphal A (2008) Monodromy in the CMB: gravity waves and string inflation. Phys Rev D 78:106003. arXiv:0803.3085
304. McAllister L, Silverstein E, Westphal A (2010) gravity waves and linear inflation from axion monodromy. Phys Rev D 82:046003. arXiv:0808.0706
305. McAllister L, Silverstein E, Westphal A, Wrase T (2014) The powers of monodromy. JHEP 09:123. arXiv:1405.3652
306. Freese K, Frieman JA, Olinto AV (1990) Natural inflation with pseudo—Nambu-Goldstone bosons. Phys Rev Lett 65:3233–3236
307. Adams FC, Bond JR, Freese K, Frieman JA, Olinto AV (1993) Natural inflation: particle physics models, power law spectra for large scale structure, and constraints from COBE. Phys Rev D 47:426–455. arXiv:hep-ph/9207245
308. Salopek DS, Bond JR, Bardeen JM (1989) Designing density fluctuation spectra in inflation. Phys Rev D 40:1753
309. Fakir R, Unruh WG (1990) Improvement on cosmological chaotic inflation through nonminimal coupling. Phys Rev D 41:1783–1791
310. Bezrukov FL, Shaposhnikov M (2008) The standard model higgs boson as the inflaton. Phys Lett B 659:703–706. arXiv:0710.3755
311. Kaiser DI, Sfakianakis EI (2014) Multifield inflation after planck: the case for nonminimal couplings. Phys Rev Lett 112:011302. arXiv:1304.0363
312. Cheong DY, Lee SM, Park SC, Progress in higgs inflation. arXiv:2103.00177
313. Boubekeur L, Lyth DH (2005) Hilltop inflation. JCAP 07:010. arXiv:hep-ph/0502047
314. Ellis J, Nanopoulos DV, Olive KA (2013) No-scale supergravity realization of the starobinsky model of inflation. Phys Rev Lett 111:111301. arXiv:1305.1247
315. Ellis J, Nanopoulos DV, Olive KA (2013) Starobinsky-like inflationary models as avatars of no-scale supergravity. JCAP 10:009. arXiv:1307.3537
316. Kallosh R, Linde A (2013) Universality class in conformal inflation. JCAP 07:002. arXiv:1306.5220
317. Kallosh R, Linde A (2013) Superconformal generalizations of the Starobinsky model. JCAP 06:028. arXiv:1306.3214
318. Ferrara S, Kallosh R, Linde A, Porrati M (2013) Minimal supergravity models of inflation. Phys Rev D 88:085038. arXiv:1307.7696
319. Kallosh R, Linde A (2013) Non-minimal inflationary attractors. JCAP 10:033. arXiv:1307.7938
320. Kallosh R, Linde A, Roest D (2013) Superconformal inflationary α-attractors. JHEP 11:198. arXiv:1311.0472
321. Kallosh R, Linde A (2013) Multi-field conformal cosmological attractors. JCAP 12:006. arXiv:1309.2015
322. Kallosh R, Linde A, Roest D (2014) Universal attractor for inflation at strong coupling. Phys Rev Lett 112:011303. arXiv:1310.3950
323. Martin J, Ringeval C, Vennin V (2014) Encyclopædia Inflationaris. Phys Dark Univ 5–6:75–235. arXiv:1303.3787
324. Martin J, Ringeval C, Trotta R, Vennin V (2014) The best inflationary models after planck. JCAP 03:039. arXiv:1312.3529
325. Ijjas A, Steinhardt PJ, Loeb A (2013) Inflationary paradigm in trouble after Planck. Phys Lett B 723:261–266. arXiv:1304.2785

326. Guth AH, Kaiser DI, Nomura Y (2013) Inflationary paradigm after Planck. Phys Lett B 733:112–119. arXiv:1312.7619
327. Linde A (2015)Inflationary cosmology after planck 2013. In: 100e Ecole d'Ete de Physique: post-planck cosmology, pp 231–316. arXiv:1402.0526, https://doi.org/10.1093/acprof:oso/9780198728856.003.0006
328. Ijjas A, Steinhardt PJ, Loeb A (2014) Inflationary schism. Phys Lett B 736:142–146. arXiv:1402.6980
329. Ijjas A, Steinhardt PJ (2016) Implications of Planck 2015 for inflationary, ekpyrotic and anamorphic bouncing cosmologies. Class Quant Grav 33:044001. arXiv:1512.09010
330. Chowdhury D, Martin J, Ringeval C, Vennin V (2019) Assessing the scientific status of inflation after Planck. Phys Rev D 100:083537. arXiv:1902.03951
331. Einstein A (1916) Approximative integration of the field equations of gravitation. Sitzungsber Preuss Akad Wiss Berlin (Math Phys) 1916:688–696
332. Einstein A (1918) Über Gravitationswellen. Sitzungsber Preuss Akad Wiss Berlin (Math Phys) 1918:154–167
333. Hulse R, Taylor J (1975) Discovery of a pulsar in a binary system. Astrophys J Lett 195:L51–L53
334. LIGO Scientific, Virgo collaboration, Abbott BP et al (2016) GW150914: first results from the search for binary black hole coalescence with Advanced LIGO. Phys Rev D93:122003. arXiv:1602.03839
335. LIGO Scientific, VIRGO collaboration, Aasi J et al (2015) Characterization of the LIGO detectors during their sixth science run. Class Quant Grav 32:115012. arXiv:1410.7764
336. LISA collaboration, AudleyH et al, Laser interferometer space antenna. arXiv:1702.00786
337. Hild S et al (2011) Sensitivity studies for third-generation gravitational wave observatories. Class Quant Grav 28:094013. arXiv:1012.0908
338. LIGO Scientific collaboration, Abbott BP et al (2017) Exploring the sensitivity of next generation gravitational wave detectors. Class Quant Grav 34:044001. arXiv:1607.08697
339. Yagi K, Seto N (2011) Detector configuration of DECIGO/BBO and identification of cosmological neutron-star binaries. Phys Rev D 83:044011. arXiv:1101.3940
340. Graham PW, Hogan JM, Kasevich MA, Rajendran S (2016) Resonant mode for gravitational wave detectors based on atom interferometry. Phys Rev D 94:104022. arXiv:1606.01860
341. MAGIS collaboration, Graham PW, Hogan JM, Kasevich MA, Rajendran S, Romani RW, Mid-band gravitational wave detection with precision atomic sensors. arXiv:1711.02225
342. Badurina L et al (2020) AION: an atom interferometer observatory and network. JCAP 05:011. arXiv:1911.11755
343. AEDGE collaboration, El-Neaj YA et al (2020) AEDGE: atomic experiment for dark matter and gravity exploration in space. EPJ Quant Technol 7:6. arXiv:1908.00802
344. Maggiore M (2018) Gravitational waves. Vol. 2: astrophysics and cosmology. Oxford University Press, p 3
345. Kuroyanagi S, Chiba T, Takahashi T (2018) Probing the universe through the stochastic gravitational wave background. JCAP 11:038. arXiv:1807.00786
346. Caldwell R et al, Detection of early-universe gravitational wave signatures and fundamental physics. arXiv:2203.07972
347. Aggarwal N et al, Challenges and opportunities of gravitational wave searches at MHz to GHz frequencies. arXiv:2011.12414
348. Maggiore M (2007) Gravitational waves. Vol. 1: theory and experiments. Oxford master series in physics. Oxford University Press
349. Grojean C, Servant G (2007) Gravitational waves from phase transitions at the electroweak scale and beyond. Phys Rev D 75:043507. arXiv:hep-ph/0607107
350. Caprini C, Durrer R, Servant G (2009) The stochastic gravitational wave background from turbulence and magnetic fields generated by a first-order phase transition. JCAP 0912:024. arXiv:0909.0622
351. Ellis J, Lewicki M, No JM (2020) Gravitational waves from first-order cosmological phase transitions: lifetime of the sound wave source. JCAP 07:050. arXiv:2003.07360

352. Gorghetto M, Hardy E, Nicolaescu H (2021) Observing invisible axions with gravitational waves. JCAP 06:034. arXiv:2101.11007
353. Lind K, Melendez J, Asplund M, Collet R, Magic Z (2013) The lithium isotopic ratio in very metal-poor stars. Astron Astrophys 554:A96. arXiv:1305.6564
354. Switzer ER, Hirata CM (2005) Ionizing radiation from hydrogen recombination strongly suppresses the lithium scattering signature in the CMB. Phys Rev D 72:083002. arXiv:astro-ph/0507106
355. Angulo C et al (1999) A compilation of charged-particle induced thermonuclear reaction rates. Nucl Phys A 656:3–183
356. Fowler WA, Caughlan GR, Zimmerman BA (1967) Thermonuclear reaction rates. Ann Rev Astron Astrophys 5:525–570
357. Fowler WA, Caughlan GR, Zimmerman BA (1975) Thermonuclear reaction rates. 2. Ann Rev Astron Astrophys 13:69–112
358. Cyburt RH, Fields BD, Olive KA (2001) The NACRE thermonuclear reaction compilation and big bang nucleosynthesis. New Astron 6:215–238. arXiv:astro-ph/0102179
359. Coc A, Vangioni-Flam E, Descouvemont P, Adahchour A, Angulo C (2004) Updated Big Bang nucleosynthesis confronted to WMAP observations and to the abundance of light elements. Astrophys J 600:544–552. arXiv:astro-ph/0309480
360. Cuoco A, Iocco F, Mangano G, Miele G, Pisanti O, Serpico PD (2004) Present status of primordial nucleosynthesis after WMAP: results from a new BBN code. Int J Mod Phys A 19:4431–4454. arXiv:astro-ph/0307213
361. Descouvemont P, Adahchour A, Angulo C, Coc A, Vangioni-Flam E (2004) Compilation and R-matrix analysis of Big Bang nuclear reaction rates. Atom Data Nucl Data Tabl 88:203–236. arXiv:astro-ph/0407101
362. Cyburt RH (2004) Primordial nucleosynthesis for the new cosmology: determining uncertainties and examining concordance. Phys Rev D 70:023505. arXiv:astro-ph/0401091
363. Cyburt RH, Fields BD, Olive KA (2008) An Update on the big bang nucleosynthesis prediction for Li-7: The problem worsens. JCAP 11:012 [arXiv:0808.2818]
364. Coc A, Goriely S, Xu Y, Saimpert M, Vangioni E (2012) Standard big-bang nucleosynthesis up to CNO with an improved extended nuclear network. Astrophys J 744:158. arXiv:1107.1117
365. Xu Y, Takahashi K, Goriely S, Arnould M, Ohta M, Utsunomiya H (2013) NACRE II: an update of the NACRE compilation of charged-particle-induced thermonuclear reaction rates for nuclei with mass number $A<16$. Nucl Phys A 918:61–169. arXiv:1310.7099
366. Planck collaboration, Ade PAR et al (2014) Planck 2013 results. XVI. Cosmological parameters. Astron Astrophys 571:A16. arXiv:1303.5076
367. Coc A, Uzan J-P, Vangioni E (2014) Standard big bang nucleosynthesis and primordial CNO Abundances after Planck. JCAP 10:050. arXiv:1403.6694
368. Cyburt RH, Fields BD, Olive KA, Yeh T-H (2016) Big bang nucleosynthesis: 2015. Rev Mod Phys 88:015004. arXiv:1505.01076
369. Pitrou C, Coc A, Uzan J-P, Vangioni E (2018) Precision big bang nucleosynthesis with improved Helium-4 predictions. Phys Rep 754:1–66. arXiv:1801.08023
370. Fields BD, Olive KA, Yeh T-H, Young C (2020) Big-bang nucleosynthesis after planck. JCAP 03:010. arXiv:1912.01132
371. Cyburt RH, Pospelov M (2012) Resonant enhancement of nuclear reactions as a possible solution to the cosmological lithium problem. Int J Mod Phys E 21:1250004. arXiv:0906.4373
372. Boyd RN, Brune CR, Fuller GM, Smith CJ (2010) New nuclear physics for big bang nucleosynthesis. Phys Rev D 82:105005. arXiv:1008.0848
373. Chakraborty N, Fields BD, Olive KA (2011) Resonant destruction as a possible solution to the cosmological lithium problem. Phys Rev D 83:063006. arXiv:1011.0722
374. Broggini C, Canton L, Fiorentini G, Villante FL (2012) The cosmological 7Li problem from a nuclear physics perspective. JCAP 06:030. arXiv:1202.5232
375. Pospelov M (2007) Particle physics catalysis of thermal big bang nucleosynthesis. Phys Rev Lett 98:231301. arXiv:hep-ph/0605215

376. Jedamzik K, Pospelov M (2009) Big bang nucleosynthesis and particle dark matter. New J Phys 11:105028. arXiv:0906.2087
377. Cyburt RH, Ellis J, Fields BD, Luo F, Olive KA, Spanos VC (2010) Nuclear reaction uncertainties. Massive gravitino decays and the cosmological lithium problem. JCAP 10:032. arXiv:1007.4173
378. Cyburt RH, Ellis J, Fields BD, Luo F, Olive KA, Spanos VC (2012) Metastable charged sparticles and the cosmological Li7 problem. JCAP 12:037. arXiv:1209.1347
379. Cyburt RH, Ellis J, Fields BD, Luo F, Olive KA, Spanos VC (2013) gravitino decays and the cosmological lithium problem in light of the LHC higgs and supersymmetry searches. JCAP 05:014. arXiv:1303.0574
380. Poulin V, Serpico PD (2015) Loophole to the universal photon spectrum in electromagnetic cascades and application to the cosmological lithium problem. Phys Rev Lett 114:091101. arXiv:1502.01250
381. Goudelis A, Pospelov M, Pradler J (2016) Light particle solution to the cosmic lithium problem. Phys Rev Lett 116:211303. arXiv:1510.08858
382. Depta PF, Hufnagel M, Schmidt-Hoberg K, Updated BBN constraints on electromagnetic decays of MeV-scale particles. arXiv:2011.06519
383. Vauclair S, Charbonnel C (1998) Element segregation in low metallicity stars and the primordial lithium abundance. Astrophys J 502:372. arXiv:astro-ph/9802315
384. Pinsonneault MH, Walker TP, Steigman G, Narayanan VK (2002) Halo star lithium depletion. Astrophys J 527:180–198. arXiv:astro-ph/9803073
385. Pinsonneault MH, Steigman G, Walker TP, Narayanans VK (2002) Stellar mixing and the primordial lithium abundance. Astrophys J 574:398–411. arXiv:astro-ph/0105439
386. Richard O, Michaud G, Richer J (2005) Implications of WMAP observations on Li abundance and stellar evolution models. Astrophys J 619:538–548. arXiv:astro-ph/0409672
387. Korn AJ, Grundahl F, Richard O, Barklem PS, Mashonkina L, Collet R et al (2006) A probable stellar solution to the cosmological lithium discrepancy. Nature 442:657–659. arXiv:astro-ph/0608201
388. Pospelov M, Afshordi N, Lithium diffusion in the post-recombination universe and spatial variation of [Li/H]. arXiv:1208.0793
389. Kusakabe M, Kawasaki M (2015) Chemical separation of primordial Li^+ during structure formation caused by nanogauss magnetic field. Mon Not R Astron Soc 446:1597–1624. arXiv:1404.3485
390. Coc A, Nunes NJ, Olive KA, Uzan J-P, Vangioni E (2007) Coupled variations of fundamental couplings and primordial nucleosynthesis. Phys Rev D 76:023511. arXiv:astro-ph/0610733
391. Berengut JC, Flambaum VV, Dmitriev VF (2010) Effect of quark-mass variation on big bang nucleosynthesis. Phys Lett B 683:114–118. arXiv:0907.2288
392. Coc A, Descouvemont P, Olive KA, Uzan J-P, Vangioni E (2012) The variation of fundamental constants and the role of A = 5 and A = 8 nuclei on primordial nucleosynthesis. Phys Rev D 86:043529. arXiv:1206.1139
393. Iocco F, Mangano G, Miele G, Pisanti O, Serpico PD (2009) Primordial nucleosynthesis: from precision cosmology to fundamental physics. Phys Rep 472:1–76. arXiv:0809.0631
394. Pospelov M, Pradler J (2010) Big bang nucleosynthesis as a probe of new physics. Ann Rev Nucl Part Sci 60:539–568. arXiv:1011.1054
395. Spite M, Spite F, Bonifacio P (2012) The cosmic lithium problem: an observer's perspective. Mem Soc Astron Ital Suppl 22:9. arXiv:1208.1190
396. Iocco F (2012) The lithium problem, a phenomenologist's perspective. Mem Soc Astron Ital Suppl 22:19. arXiv:1206.2396
397. Mathews GJ, Kedia A, Sasankan N, Kusakabe M, Luo Y, Kajino T et al (2020) Cosmological solutions to the lithium problem. JPS Conf Proc 31:011033. arXiv:1909.01245
398. Neronov A, Vovk I (2010) Evidence for strong extragalactic magnetic fields from Fermi observations of TeV blazars. Science 328:73–75. arXiv:1006.3504
399. Korochkin A, Kalashev O, Neronov A, Semikoz D, Sensitivity reach of gamma-ray measurements for strong cosmological magnetic fields. arXiv:2007.14331

400. Planck collaboration, Ade P et al (2016) Planck 2015 results. XIX. Constraints on primordial magnetic fields. Astron Astrophys 594:A19. arXiv:1502.01594
401. Jedamzik K, Saveliev A (2019) Stringent limit on primordial magnetic fields from the cosmic microwave background radiation. Phys Rev Lett 123:021301. arXiv:1804.06115
402. Durrer R, Neronov A (2013) Cosmological magnetic fields: their generation, evolution and observation. Astron Astrophys Rev 21:62. arXiv:1303.7121
403. Subramanian K (2016) The origin, evolution and signatures of primordial magnetic fields. Rep Prog Phys 79:076901. arXiv:1504.02311
404. Neronov A, Roper Pol A, Caprini C, Semikoz D, NANOGrav signal from MHD turbulence at QCD phase transition in the early universe. arXiv:2009.14174
405. Grasso D, Rubinstein HR (2001) Magnetic fields in the early universe. Phys Rep 348:163–266. arXiv:astro-ph/0009061
406. Widrow LM (2002) Origin of galactic and extragalactic magnetic fields. Rev Mod Phys 74:775–823. arXiv:astro-ph/0207240
407. Vallée JP (2004) Cosmic magnetic fields—as observed in the Universe, in galactic dynamos, and in the Milky Way. New Astron Rev 48:763–841
408. Vachaspati T, Progress on cosmological magnetic fields. arXiv:2010.10525
409. Phillips M (1993) The absolute magnitudes of type IA supernovae. Astrophys J 413:L105–L108
410. Martin J (2012) Everything you always wanted to know about the cosmological constant problem (But Were Afraid To Ask). Comptes Rendus Physique 13:566–665. arXiv:1205.3365
411. Narison S (2012) Gluon Condensates and precise $\overline{m}_{c,b}$ from QCD-moments and their ratios to order α_s^3 and $<G^4>$. Phys Lett B 706:412–422. arXiv:1105.2922
412. Arkani-Hamed N, Hall LJ, Kolda CF, Murayama H (2000) A new perspective on cosmic coincidence problems. Phys Rev Lett 85:4434–4437. arXiv:astro-ph/0005111
413. Weinberg S (1989) The cosmological constant problem. Rev Mod Phys 61:1–23
414. Carroll SM (2001) The cosmological constant. Living Rev Rel 4:1. arXiv:astro-ph/0004075
415. Straumann N (2002) The History of the cosmological constant problem. In: 18th IAP colloquium on the nature of dark energy: observational and theoretical results on the accelerating universe, p 8. arXiv:gr-qc/0208027
416. Copeland EJ, Sami M, Tsujikawa S (2006) Dynamics of dark energy. Int J Mod Phys D 15:1753–1936. arXiv:hep-th/0603057
417. Nobbenhuis S (2006) The cosmological constant problem, an inspiration for new physics. PhD thesis, Utrecht U. arXiv:gr-qc/0609011
418. Bousso R (2008) TASI lectures on the cosmological constant. Gen Rel Grav 40:607–637. arXiv:0708.4231
419. Bousso R (2011) The cosmological constant problem, dark energy, and the landscape of string theory. Pontif Acad Sci Scr Varia 119:129–151. arXiv:1203.0307
420. Sola J (2013) Cosmological constant and vacuum energy: old and new ideas. J Phys Conf Ser 453:012015. arXiv:1306.1527
421. Burgess C (2015) The cosmological constant problem: why it's hard to get dark energy from micro-physics. In: 100e Ecole d'Ete de Physique: post-planck cosmology, pp 149–197. arXiv:1309.4133, https://doi.org/10.1093/acprof:oso/9780198728856.003.0004
422. Padilla A, Lectures on the cosmological constant problem. arXiv:1502.05296
423. Brax P (2018) What makes the Universe accelerate? a review on what dark energy could be and how to test it. Rep Prog Phys 81:016902
424. Arkani-Hamed N, Dimopoulos S, Dvali G, Gabadadze G, Nonlocal modification of gravity and the cosmological constant problem. arXiv:hep-th/0209227
425. Dvali G, Hofmann S, Khoury J (2007) Degravitation of the cosmological constant and graviton width. Phys Rev D 76:084006. arXiv:hep-th/0703027
426. de Rham C, Gabadadze G, Heisenberg L, Pirtskhalava D (2011) Cosmic acceleration and the helicity-0 graviton. Phys Rev D 83:103516. arXiv:1010.1780
427. de Rham C, Hofmann S, Khoury J, Tolley AJ (2008) Cascading gravity and degravitation. JCAP 02:011. arXiv:0712.2821

428. Hinterbichler K (2012) Theoretical aspects of massive gravity. Rev Mod Phys 84:671–710. arXiv:1105.3735
429. Clifton T, Ferreira PG, Padilla A, Skordis C (2012) Modified gravity and cosmology. Phys Rep 513:1–189. arXiv:1106.2476
430. Joyce A, Jain B, Khoury J, Trodden M (2015) Beyond the cosmological standard model. Phys Rep 568:1–98. arXiv:1407.0059
431. Dvali G, Gabadadze G (2001) Gravity on a brane in infinite volume extra space. Phys Rev D 63:065007. arXiv:hep-th/0008054
432. Dvali G, Gabadadze G, Shifman M (2003) Diluting cosmological constant in infinite volume extra dimensions. Phys Rev D 67:044020. arXiv:hep-th/0202174
433. Niedermann F, Schneider R, Hofmann S, Khoury J (2015) Universe as a cosmic string. Phys Rev D 91:024002. arXiv:1410.0700
434. Dvali G, Gabadadze G, Porrati M (2000) 4-D gravity on a brane in 5-D Minkowski space. Phys Lett B 485:208–214. arXiv:hep-th/0005016
435. Aghababaie Y, Burgess C, Parameswaran S, Quevedo F (2004) Towards a naturally small cosmological constant from branes in 6-D supergravity. Nucl Phys B 680:389–414. arXiv:hep-th/0304256
436. Burgess C (2004) Towards a natural theory of dark energy: supersymmetric large extra dimensions. AIP Conf Proc 743:417–449. arXiv:hep-th/0411140
437. Niedermann F, Schneider R (2016) Fine-tuning with brane-localized flux in 6D supergravity. JHEP 02:025. arXiv:1508.01124
438. Arkani-Hamed N, Dimopoulos S, Dvali G (1998) The Hierarchy problem and new dimensions at a millimeter. Phys Lett B 429:263–272. arXiv:hep-ph/9803315
439. Maartens R, Koyama K (2010) Brane-world gravity. Living Rev Rel 13:5. arXiv:1004.3962
440. Niedermann F (2016) Natural braneworlds in six dimensions and the cosmological constant problem. PhD thesis, Munich U., p 3
441. Charmousis C, Copeland EJ, Padilla A, Saffin PM (2012) General second order scalar-tensor theory, self tuning, and the Fab Four. Phys Rev Lett 108:051101. arXiv:1106.2000
442. Horndeski GW (1974) Second-order scalar-tensor field equations in a four-dimensional space. Int J Theor Phys 10:363–384
443. Deffayet C, Gao X, Steer D, Zahariade G (2011) From k-essence to generalised Galileons. Phys Rev D 84:064039. arXiv:1103.3260
444. Armendariz-Picon C, Mukhanov VF, Steinhardt PJ (2000) A dynamical solution to the problem of a small cosmological constant and late time cosmic acceleration. Phys Rev Lett 85:4438–4441. arXiv:astro-ph/0004134
445. Armendariz-Picon C, Mukhanov VF, Steinhardt PJ (2001) Essentials of k essence. Phys Rev D 63:103510. arXiv:astro-ph/0006373
446. Erickson JK, Caldwell R, Steinhardt PJ, Armendariz-Picon C, Mukhanov VF (2002) Measuring the speed of sound of quintessence. Phys Rev Lett 88:121301. arXiv:astro-ph/0112438
447. Vainshtein A (1972) To the problem of nonvanishing gravitation mass. Phys Lett B 39:393–394
448. Kaloper N, Sandora M (2014) Spherical cows in the sky with fab four. JCAP 05:028. arXiv:1310.5058
449. Tan W-H et al (2020) Improvement for testing the gravitational inverse-square law at the submillimeter range. Phys Rev Lett 124:051301
450. Lee J, Adelberger E, Cook T, Fleischer S, Heckel B (2020) New test of the gravitational $1/r^2$ law at separations down to 52 μ m. Phys Rev Lett 124:101101. arXiv:2002.11761
451. Kaloper N, Padilla A (2014) Sequestering the standard model vacuum energy. Phys Rev Lett 112:091304. arXiv:1309.6562
452. Kaloper N, Padilla A (2014) Vacuum energy sequestering: the framework and its cosmological consequences. Phys Rev D 90:084023. arXiv:1406.0711
453. Coltman B, Li Y, Padilla A (2019) Cosmological consequences of Omnia Sequestra. JCAP 06:017. arXiv:1903.02829

454. Kaloper N, Padilla A, Stefanyszyn D, Zahariade G (2016) Manifestly local theory of vacuum energy sequestering. Phys Rev Lett 116:051302. arXiv:1505.01492
455. Kallosh R, Linde AD (2003) M theory, cosmological constant and anthropic principle. Phys Rev D 67:023510. arXiv:hep-th/0208157
456. Abbott L (1985) A mechanism for reducing the value of the cosmological constant. Phys Lett B 150:427–430
457. Brown J, Teitelboim C (1987) Dynamical neutralization of the cosmological constant. Phys Lett B 195:177–182
458. Brown J, Teitelboim C (1988) Neutralization of the cosmological constant by membrane creation. Nucl Phys B 297:787–836
459. Bousso R, Polchinski J (2000) Quantization of four form fluxes and dynamical neutralization of the cosmological constant. JHEP 06:006. arXiv:hep-th/0004134
460. Dvali G, Vilenkin A (2004) Cosmic attractors and gauge hierarchy. Phys Rev D 70:063501. arXiv:hep-th/0304043
461. Dvali G (2006) Large hierarchies from attractor vacua. Phys Rev D 74:025018. arXiv:hep-th/0410286
462. Giudice G, Kehagias A, Riotto A (2019) The selfish higgs. JHEP 10:199. arXiv:1907.05370
463. Kaloper N, Westphal A, A goldilocks higgs. arXiv:1907.05837
464. Arkani-Hamed N, Tito D'agnolo R, Kim HD, The weak scale as a trigger. arXiv:2012.04652
465. Steinhardt PJ, Turok N (2006) Why the cosmological constant is small and positive. Science 312:1180–1182. arXiv:astro-ph/0605173
466. Graham PW, Kaplan DE, Rajendran S (2018) Born again universe. Phys Rev D 97:044003. arXiv:1709.01999
467. Barrau A, A pure general relativistic non-singular bouncing origin for the Universe. arXiv:2005.04693
468. Graham PW, Kaplan DE, Rajendran S (2019) Relaxation of the cosmological constant. Phys Rev D 100:015048. arXiv:1902.06793
469. Ijjas A, Steinhardt PJ (2016) Classically stable nonsingular cosmological bounces. Phys Rev Lett 117:121304. arXiv:1606.08880
470. Ijjas A, Steinhardt PJ (2019) A new kind of cyclic universe. Phys Lett B 795:666–672. arXiv:1904.08022
471. Rubakov VA (2014) The Null Energy Condition and its violation. Usp Fiz Nauk 184:137–152. arXiv:1401.4024
472. Sloan D, Dimopoulos K, Karamitsos S (2020) T-model inflation and bouncing cosmology. Phys Rev D 101:043521. arXiv:1912.00090
473. Bojowald M (2007) Singularities and quantum gravity. AIP Conf Proc 910:294–333. arXiv:gr-qc/0702144
474. Freivogel B (2011) Making predictions in the multiverse. Class Quant Grav 28:204007. arXiv:1105.0244
475. Hook A, Huang J (2019) Searches for other vacua. Part I. Bubbles in our universe. JHEP 08:148. arXiv:1904.00020
476. Hook A, Huang J, Racco D (2020) Searches for other vacua. Part II. A new Higgstory at the cosmological collider. JHEP 01:105. arXiv:1907.10624
477. Guth AH (2007) Eternal inflation and its implications. J Phys A 40:6811–6826. arXiv:hep-th/0702178
478. Guth AH, Nomura Y (2012) What can the observation of nonzero curvature tell us? Phys Rev D 86:023534. arXiv:1203.6876
479. Bloch IM, Csáki C, Geller M, Volansky T, Crunching away the cosmological constant problem: dynamical selection of a small Λ. arXiv:1912.08840
480. Dirac PA (1928) The quantum theory of the electron. Proc R Soc Lond A A 117:610–624
481. Dirac P (1928) The quantum theory of electron. 2. Proc R Soc Lond A A118:351
482. Anderson CD (1932) The apparent existence of easily deflectable positives. Science 76:238–239
483. Anderson C (1933) The positive electron. Phys Rev 43:491–494

484. Canetti L, Drewes M, Shaposhnikov M (2012) Matter and antimatter in the universe. New J Phys 14:095012. arXiv:1204.4186
485. Cohen AG, De Rujula A, Glashow S (1998) A matter—antimatter universe? Astrophys J 495:539–549. arXiv:astro-ph/9707087
486. Steigman G (2008) When clusters collide: constraints on antimatter on the largest scales. JCAP 10:001. arXiv:0808.1122
487. Sakharov A (1991) Violation of CP Invariance, C asymmetry, and baryon asymmetry of the universe. Sov Phys Usp 34:392–393
488. Klinkhamer FR, Manton N (1984) A saddle point solution in the weinberg-salam theory. Phys Rev D 30:2212
489. 't Hooft G (1976) Symmetry breaking through Bell-Jackiw anomalies. Phys Rev Lett 37:8–11
490. Dimopoulos S, Susskind L (1978) On the baryon number of the universe. Phys Rev D 18:4500–4509
491. Christ N (1980) Conservation law violation at high-energy by anomalies. Phys Rev D 21:1591
492. Manton N (1983) Topology in the weinberg-salam theory. Phys Rev D 28:2019
493. Kuzmin V, Rubakov V, Shaposhnikov M (1985) On the anomalous electroweak baryon number nonconservation in the early universe. Phys Lett B 155:36
494. Arnold PB, McLerran LD (1987) Sphalerons, small fluctuations and baryon number violation in electroweak theory. Phys Rev D 36:581
495. Arnold PB, McLerran LD (1988) The sphaleron strikes back. Phys Rev D 37:1020
496. D'Onofrio M, Rummukainen K, Tranberg A (2014) Sphaleron rate in the minimal standard model. Phys Rev Lett 113:141602. arXiv:1404.3565
497. Buchmuller W, Peccei R, Yanagida T (2005) Leptogenesis as the origin of matter. Ann Rev Nucl Part Sci 55:311–355. arXiv:hep-ph/0502169
498. Khlebnikov S, Shaposhnikov M (1988) The statistical theory of anomalous fermion number nonconservation. Nucl Phys B 308:885–912
499. Elor G et al (2022) New ideas in baryogenesis: a snowmass white paper. In: 2022 snowmass summer study, p 3. arXiv:2203.05010
500. Georgi H, Glashow S (1974) Unity of all elementary particle forces. Phys Rev Lett 32:438–441
501. Fritzsch H, Minkowski P (1975) Unified interactions of leptons and hadrons. Annals Phys 93:193–266
502. Super-Kamiokande collaboration, Abe K et al (2017) Search for proton decay via $p \to e^+\pi^0$ and $p \to \mu^+\pi^0$ in 0.31 megaton· years exposure of the Super-Kamiokande water Cherenkov detector. Phys Rev D 95:012004. arXiv:1610.03597
503. Yoshimura M (1978) Unified gauge theories and the baryon number of the universe. Phys Rev Lett 41:281–284
504. Ellis JR, Gaillard MK, Nanopoulos DV (1979) Baryon number generation in grand unified theories. Phys Lett B 80:360
505. Toussaint D, Treiman S, Wilczek F, Zee A (1979) Matter—antimatter accounting, thermodynamics, and black hole radiation. Phys Rev D 19:1036–1045
506. Weinberg S (1979) Baryon and lepton nonconserving processes. Phys Rev Lett 43:1566–1570
507. Kolb EW, Wolfram S (1980) Baryon number generation in the early universe. Nucl Phys B 172:224
508. Fry JN, Olive KA, Turner MS (1980) Evolution of cosmological baryon asymmetries. Phys Rev D 22:2953
509. Riotto A (1998) Theories of baryogenesis. In: Proceedings, summer school in high-energy physics and cosmology. Trieste, Italy, pp 326–436, 7. arXiv:hep-ph/9807454
510. Dine M, Kusenko A (2003) The origin of the matter—antimatter asymmetry. Rev Mod Phys 76:1. arXiv:hep-ph/0303065
511. Cline JM (2006) Baryogenesis. In: Les Houches summer school - session 86: particle physics and cosmology: the fabric of spacetime, p 9. arXiv:hep-ph/0609145
512. Affleck I, Dine M (1985) A new mechanism for baryogenesis. Nucl Phys B 249:361–380
513. Harigaya K (2019) Nambu-goldstone affleck-dine baryogenesis. JHEP 08:085.arXiv:1906.05286

514. Co RT, Harigaya K (2020) Axiogenesis. Phys Rev Lett 124:111602. arXiv:1910.02080
515. Gouttenoire Y, Servant G, Simakachorn P, Kination cosmology from scalar fields and gravitational-wave signatures. arXiv:2111.01150
516. Cohen AG, Kaplan DB (1987) Thermodynamic generation of the baryon asymmetry. Phys Lett B 199:251–258
517. Cohen AG, Kaplan D, Nelson A (1991) Spontaneous baryogenesis at the weak phase transition. Phys Lett B 263:86–92
518. Adshead P, Pearce L, Peloso M, Roberts MA, Sorbo L (2018) Phenomenology of fermion production during axion inflation. JCAP 06:020. arXiv:1803.04501
519. Domcke V, Mukaida K (2018) Gauge field and fermion production during axion inflation. JCAP 11:020. arXiv:1806.08769
520. Domcke V, von Harling B, Morgante E, Mukaida K (2019) Baryogenesis from axion inflation. JCAP 10:032. arXiv:1905.13318
521. Domcke V, Ema Y, Mukaida K, Yamada M, Spontaneous baryogenesis from axions with generic couplings. arXiv:2006.03148
522. Abel S, Gupta R, Scholtz J (2019) Out-of-the-box baryogenesis during relaxation. Phys Rev D 100:015034. arXiv:1810.05153
523. Davoudiasl H, Kitano R, Kribs GD, Murayama H, Steinhardt PJ (2004) Gravitational baryogenesis. Phys Rev Lett. 93:201301. arXiv:hep-ph/0403019
524. Hook A (2014) Baryogenesis from hawking radiation. Phys Rev D 90:083535. arXiv:1404.0113
525. Fukugita M, Yanagida T (1986) Baryogenesis without grand unification. Phys Lett B 174:45–47
526. Chen M-C (2007) TASI 2006 lectures on leptogenesis. In: Proceedings of theoretical advanced study institute in elementary particle physics : exploring new frontiers using colliders and neutrinos (TASI 2006). Boulder, Colorado, pp 123–176, p 3. arXiv:hep-ph/0703087
527. Davidson S, Nardi E, Nir Y (2008) Leptogenesis. Phys Rep 466:105–177. arXiv:0802.2962
528. Di Bari P (2012) An introduction to leptogenesis and neutrino properties. Contemp Phys 53:315–338. arXiv:1206.3168
529. Fong CS, Nardi E, Riotto A (2012) Leptogenesis in the universe. Adv High Energy Phys 2012:158303. arXiv:1301.3062
530. Garbrecht B (2020) Why is there more matter than antimatter? Calculational methods for leptogenesis and electroweak baryogenesis. Prog Part Nucl Phys 110:103727. arXiv:1812.02651
531. Bodeker D, Buchmuller W, Baryogenesis from the weak scale to the GUT scale. arXiv:2009.07294
532. Davidson S, Ibarra A (2002) A Lower bound on the right-handed neutrino mass from leptogenesis. Phys Lett B 535:25–32. arXiv:hep-ph/0202239
533. Pilaftsis A, Underwood TE (2004) Resonant leptogenesis. Nucl Phys B 692:303–345. arXiv:hep-ph/0309342
534. Asaka T, Shaposhnikov M (2005) The ν MSM, dark matter and baryon asymmetry of the universe. Phys Lett B 620:17–26. arXiv:hep-ph/0505013
535. Akhmedov EK, Rubakov V, Smirnov A (1998) Baryogenesis via neutrino oscillations. Phys Rev Lett 81:1359–1362. arXiv:hep-ph/9803255
536. Canetti L, Drewes M, Frossard T, Shaposhnikov M (2013) Dark matter, baryogenesis and neutrino oscillations from right handed neutrinos. Phys Rev D 87:093006. arXiv:1208.4607
537. Abazajian K et al, Light sterile neutrinos: a white paper. arXiv:1204.5379
538. Drewes M et al (2017) A white paper on kev sterile neutrino dark matter. JCAP 01:025. arXiv:1602.04816
539. Böser S, Buck C, Giunti C, Lesgourgues J, Ludhova L, Mertens S et al (2020) Status of light sterile neutrino searches. Prog Part Nucl Phys 111:103736. arXiv:1906.01739
540. Prokopec T, Schmidt MG, Weinstock S (2004) Transport equations for chiral fermions to order h bar and electroweak baryogenesis. Part 1. Annals Phys 314:208–265. arXiv:hep-ph/0312110
541. Prokopec T, Schmidt MG, Weinstock S (2004) Transport equations for chiral fermions to order h-bar and electroweak baryogenesis. Part II, Annals Phys 314:267–320. arXiv:hep-ph/0406140

542. Konstandin T (2013) Quantum transport and electroweak baryogenesis. Phys Usp 56:747–771. arXiv:1302.6713
543. Cline JM, Joyce M, Kainulainen K (2000) Supersymmetric electroweak baryogenesis. JHEP 07:018. arXiv:hep-ph/0006119
544. Grojean C, Servant G, Wells JD (2005) First-order electroweak phase transition in the standard model with a low cutoff. Phys Rev D 71:036001. arXiv:hep-ph/0407019
545. Espinosa JR, Konstandin T, Riva F (2012) Strong electroweak phase transitions in the standard model with a singlet. Nucl Phys B 854:592–630. arXiv:1107.5441
546. Cline JM, Kainulainen K (2013) Electroweak baryogenesis and dark matter from a singlet Higgs. JCAP 01:012. arXiv:1210.4196
547. Baldes I, Konstandin T, Servant G (2018) A first-order electroweak phase transition from varying Yukawas. Phys Lett B 786:373–377. arXiv:1604.04526
548. Shaposhnikov M (1988) Structure of the high temperature gauge ground state and electroweak production of the baryon asymmetry. Nucl Phys B 299:797–817
549. Gavela M, Hernandez P, Orloff J, Pene O (1994) Standard model CP violation and baryon asymmetry. Mod Phys Lett A 9:795–810. arXiv:hep-ph/9312215
550. Konstandin T, Prokopec T, Schmidt MG (2004) Axial currents from CKM matrix CP violation and electroweak baryogenesis. Nucl Phys B 679:246–260. arXiv:hep-ph/0309291
551. Brauner T, Taanila O, Tranberg A, Vuorinen A (2012) Temperature dependence of standard model CP violation. Phys Rev Lett 108:041601. arXiv:1110.6818
552. Cirigliano V, Li Y, Profumo S, Ramsey-Musolf MJ (2010) MSSM baryogenesis and electric dipole moments: an update on the phenomenology. JHEP 01:002. arXiv:0910.4589
553. Kozaczuk J, Profumo S, Ramsey-Musolf MJ, Wainwright CL (2012) Supersymmetric electroweak baryogenesis via resonant sfermion sources. Phys Rev D 86:096001. arXiv:1206.4100
554. Dorsch G, Huber S, Konstandin T, No J (2017) A second higgs doublet in the early universe: baryogenesis and gravitational waves. JCAP 05:052. arXiv:1611.05874
555. Alanne T, Kainulainen K, Tuominen K, Vaskonen V (2016) Baryogenesis in the two doublet and inert singlet extension of the standard model. JCAP 08:057. arXiv:1607.03303
556. Kuzmin V, Shaposhnikov M, Tkachev I (1992) Strong CP violation, electroweak baryogenesis, and axionic dark matter. Phys Rev D 45:466–475
557. Servant G (2014) Baryogenesis from strong CP violation and the QCD axion. Phys Rev Lett 113:171803. arXiv:1407.0030
558. Baldes I, Konstandin T, Servant G (2016) Flavor cosmology: dynamical yukawas in the Froggatt-Nielsen mechanism. JHEP 12:073. arXiv:1608.03254
559. von Harling B, Servant G (2017) Cosmological evolution of Yukawa couplings: the 5D perspective. JHEP 05:077. arXiv:1612.02447
560. Bruggisser S, Konstandin T, Servant G (2017) CP-violation for electroweak baryogenesis from dynamical CKM matrix. JCAP 11:034. arXiv:1706.08534
561. Servant G (2018) The serendipity of electroweak baryogenesis. Philso Trans R Soc Lond A 376:20170124. arXiv:1807.11507
562. Espinosa JR, Gripaios B, Konstandin T, Riva F (2012) Electroweak baryogenesis in non-minimal composite higgs models. JCAP 01:012. arXiv:1110.2876
563. Bruggisser S, Von Harling B, Matsedonskyi O, Servant G (2018) Baryon asymmetry from a composite higgs Boson. Phys Rev Lett 121:131801. arXiv:1803.08546
564. Bruggisser S, Von Harling B, Matsedonskyi O, Servant G (2018) Electroweak phase transition and baryogenesis in composite higgs models. JHEP 12:099. arXiv:1804.07314
565. Fromme L, Huber SJ (2007) Top transport in electroweak baryogenesis. JHEP 03:049. arXiv:hep-ph/0604159
566. Caprini C, No JM (2012) Supersonic electroweak baryogenesis: achieving baryogenesis for fast bubble walls. JCAP 01:031. arXiv:1111.1726
567. No JM (2011) Large gravitational wave background signals in electroweak baryogenesis scenarios. Phys Rev D 84:124025. arXiv:1103.2159
568. Cline JM, Kainulainen K (2020) Electroweak baryogenesis at high bubble wall velocities. Phys Rev D 101:063525. arXiv:2001.00568

569. Laurent B, Cline JM, Fluid equations for fast-moving electroweak bubble walls. arXiv:2007.10935
570. Dorsch GC, Huber SJ, Konstandin T, On the wall velocity dependence of electroweak baryogenesis. arXiv:2106.06547
571. De Curtis S, Rose LD, Guiggiani A, Muyor AG, Panico G, Bubble wall dynamics at the electroweak phase transition. arXiv:2201.08220
572. Cline JM, Friedlander A, He D-M, Kainulainen K, Laurent B, Tucker-Smith D, Baryogenesis and gravity waves from a UV-completed electroweak phase transition. arXiv:2102.12490
573. Azatov A, Vanvlasselaer M, Yin W (2021) Baryogenesis via relativistic bubble walls. JHEP 10:043. arXiv:2106.14913
574. Baldes I, Blasi S, Mariotti A, Sevrin A, Turbang K, Baryogenesis via relativistic bubble expansion. arXiv:2106.15602
575. Cohen AG, Kaplan DB, Nelson AE (1990) Weak scale baryogenesis. Phys Lett B 245:561–564
576. Cohen AG, Kaplan DB, Nelson AE (1991) Baryogenesis at the weak phase transition. Nucl Phys B 349:727–742
577. Cohen AG, Kaplan D, Nelson A (1993) Progress in electroweak baryogenesis. Ann Rev Nucl Part Sc 43:27–70. arXiv:hep-ph/9302210
578. Riotto A, Trodden M (1999) Recent progress in baryogenesis. Ann Rev Nucl Part Sci 49:35–75. arXiv:hep-ph/9901362
579. Cline JM (2019) TASI lectures on early universe cosmology: inflation, baryogenesis and dark matter. PoS TASI2018:001. arXiv:1807.08749
580. Morrissey DE, Ramsey-Musolf MJ (2012) Electroweak baryogenesis. New J Phys 14:125003. arXiv:1206.2942
581. Jungman G, Kamionkowski M, Griest K (1996) Supersymmetric dark matter. Phys Rep 267:195–373. arXiv:hep-ph/9506380
582. Bergström L (2000) Nonbaryonic dark matter: observational evidence and detection methods. Rep Prog Phys 63:793. arXiv:hep-ph/0002126
583. Bergstrom L (2009) Dark matter candidates. New J Phys 11:105006. arXiv:0903.4849
584. Bergstrom L (2012) Dark matter evidence, particle physics candidates and detection methods. Annalen Phys 524:479–496. arXiv:1205.4882]
585. Olive KA (2003) TASI lectures on dark matter. In: Theoretical advanced study institute in elementary particle physics (TASI 2002): particle physics and cosmology: the quest for physics beyond the standard model(s), pp 797–851, 1. arXiv:astro-ph/0301505
586. Bertone G, Hooper D, Silk J (2005) Particle dark matter: evidence, candidates and constraints. Phys Rep 405:279–390. arXiv:hep-ph/0404175
587. Murayama H (2007) Physics beyond the standard model and dark matter. In: Les Houches summer school—session 86: particle physics and cosmology: the fabric of spacetime, p 4. arXiv:0704.2276
588. Steffen FD (2009) Dark matter candidates—axions, neutralinos, gravitinos, and axinos. Eur Phys J C 59:557–588. arXiv:0811.3347
589. Hooper D (2010) Particle dark matter. In: Theoretical advanced study institute in elementary particle physics: the dawn of the LHC Era, pp 709–764. arXiv:0901.4090, https://doi.org/10.1142/9789812838360_0014
590. Feng JL (2010) Dark matter candidates from particle physics and methods of detection. Ann Rev Astron Astrophys 48:495–545. arXiv:1003.0904
591. Roos M, Dark matter: the evidence from astronomy, astrophysics and cosmology. arXiv:1001.0316
592. Bertone G (2010) The moment of truth for WIMP dark matter. Nature 468:389–393. arXiv:1011.3532
593. Schnee RW (2011) Introduction to dark matter experiments. In: Theoretical advanced study institute in elementary particle physics: physics of the large and the small, pp 775–829. arXiv:1101.5205, https://doi.org/10.1142/9789814327183_0014
594. Peter AHG, Dark matter: a brief review. arXiv:1201.3942

595. Zurek KM (2014) Asymmetric dark matter: theories, signatures, and constraints. Phys Rep 537:91–121. arXiv:1308.0338
596. Profumo S (2013) Astrophysical probes of dark matter. In: Theoretical advanced study institute in elementary particle physics: searching for new physics at small and large scales, pp 143–189. arXiv:1301.0952, https://doi.org/10.1142/9789814525220_0004
597. Lisanti M (2017) Lectures on dark matter physics. In: Theoretical advanced study institute in elementary particle physics: new frontiers in fields and strings, pp 399–446. arXiv:1603.03797, https://doi.org/10.1142/9789813149441_0007
598. Alexander J et al, Dark sectors 2016 workshop: community report, p 8. arXiv:1608.08632
599. Bauer M, Plehn T (2019) Yet another introduction to dark matter: the particle physics approach, vol. 959 of lecture notes in physics. Springer. https://doi.org/10.1007/978-3-030-16234-4
600. Bertone G, Hooper D (2018) History of dark matter. Rev Mod Pys 90:045002. arXiv:1605.04909
601. Freese K (2017) Status of dark matter in the universe. Int J Mod Phys 1:325–355. arXiv:1701.01840
602. Roszkowski L, Sessolo EM, Trojanowski S (2018) WIMP dark matter candidates and searches–current status and future prospects. Rep Prog Phys 81:066201. arXiv:1707.06277
603. Arcadi G, Dutra M, Ghosh P, Lindner M, Mambrini Y, Pierre M et al, The waning of the wimp? a review of models, searches, and constraints. arXiv:1703.07364
604. Battaglieri M et al (2017) US cosmic visions: new ideas in dark matter 2017: community report. In: U.S. cosmic V ideas in dark matter, p 7. arXiv:1707.04591
605. Bertone G, Tait TMP (2018) A new era in the search for dark matter. Nature 562:51–56. arXiv:1810.01668
606. Blanco C, Escudero M, Hooper D, Witte SJ (2019) Zediated WIMPs: dead, dying, or soon to be detected? JCAP 11:024. arXiv:1907.05893
607. Mambrini Y (2021) Particles in the dark universe: a student's guide to particle physics and cosmology. Springer International Publishing
608. Begeman K, Broeils A, Sanders R (1991) Extended rotation curves of spiral galaxies: Dark haloes and modified dynamics. Mon Not R Astron Soc 249:523
609. Rubin VC, Ford J, Kent W, Thonnard N (1978) Extended rotation curves of high-luminosity spiral galaxies. IV. Systematic dynamical properties, Sa through Sc. Astrophys J 225:L107–L111
610. Rubin VC, Ford Jr WK, Thonnard N (1980) Rotational properties of 21 SC galaxies with a large range of luminosities and radii, from NGC 4605 /R = 4kpc/ to UGC 2885 /R = 122 kpc/. Astrophys J 238:471
611. Albada T, Sancisi R, Petrou M, Tayler R (1986) Dark matter in spiral galaxies [and Discussion]. Philso Trans R Soc Lond A 320:447–464
612. Vogelsberger M, Genel S, Springel V, Torrey P, Sijacki D, Xu D et al (2014) Introducing the illustris project: simulating the coevolution of dark and visible matter in the universe. Mon Not R Astron Soc 444:1518–1547. arXiv:1405.2921
613. Zwicky F (1933) Die Rotverschiebung von extragalaktischen Nebeln. Helv Phys Acta 6:110–127
614. Zwicky F (1937) On the masses of nebulae and of clusters of nebulae. Astrophys J 86:217–246
615. Lokas EL, Mamon GA (2003) Dark matter distribution in the coma cluster from galaxy kinematics: breaking the mass—anisotropy degeneracy. Mon Not R Astron Soc 343:401. arXiv:astro-ph/0302461
616. Bahcall NA, Fan X-H (1998) The most massive distant clusters: determining omega and sigma_8. Astrophys J 504:1. arXiv:astro-ph/9803277
617. Press WH, Schechter P (1974) Formation of galaxies and clusters of galaxies by selfsimilar gravitational condensation. Astrophys J 187:425–438
618. Banerjee A et al (2022) Snowmass2021 cosmic frontier white paper: cosmological simulations for dark matter physics. In: 2022 snowmass summer study, p 3. arXiv:2203.07049
619. Sawala T et al (2016) The APOSTLE simulations: solutions to the Local Group's cosmic puzzles. Mon Not R Astron Soc 457:1931–1943. arXiv:1511.01098

620. Massey R, Kitching T, Richard J (2010) The dark matter of gravitational lensing. Rep Prog Phys 73:086901. arXiv:1001.1739
621. Kahlhoefer F, Schmidt-Hoberg K, Frandsen MT, Sarkar S (2014) Colliding clusters and dark matter self-interactions. Mon Not R Astron Soc 437:2865–2881. arXiv:1308.3419
622. Harvey D, Massey R, Kitching T, Taylor A, Tittley E (2015) The non-gravitational interactions of dark matter in colliding galaxy clusters. Science 347:1462–1465. arXiv:1503.07675
623. Robertson A, Massey R, Eke V (2017) What does the bullet cluster tell us about self-interacting dark matter?. Mon Not R Astron Soc 465:569–587. arXiv:1605.04307
624. Clowe D, Bradac M, Gonzalez AH, Markevitch M, Randall SW, Jones C et al (2006) A direct empirical proof of the existence of dark matter. Astrophys J Lett 648:L109–L113. arXiv:astro-ph/0608407
625. Schumann M (2019) Direct detection of WIMP dark matter: concepts and status. J Phys G 46:103003. arXiv:1903.03026
626. Quílez Lasanta P (2019) New dynamics in axions and flavor. PhD thesis, U. Autonoma, Madrid (main)
627. Carr B, Kohri K, Sendouda Y, Yokoyama J, Constraints on primordial black holes. arXiv:2002.12778
628. Green AM, Kavanagh BJ, Primordial black holes as a dark matter candidate. arXiv:2007.10722
629. Carr B, Kuhnel F, Primordial black holes as dark matter: recent developments. arXiv:2006.02838
630. De Luca V, Franciolini G, Pani P, Riotto A (2020) Primordial black holes confront LIGO/virgo data: current situation. JCAP 06:044. arXiv:2005.05641
631. De Luca V, Desjacques V, Franciolini G, Pani P, Riotto A (2021) GW190521 mass gap event and the primordial black hole scenario. Phys Rev Lett 126:051101. arXiv:2009.01728
632. Bhagwat S, De Luca V, Franciolini G, Pani P, Riotto A (2021) The importance of priors on LIGO-Virgo parameter estimation: the case of primordial black holes. JCAP 01:037. arXiv:2008.12320
633. De Luca V, Franciolini G, Pani P, Riotto A (2021) Bayesian evidence for both astrophysical and primordial black holes: mapping the GWTC-2 catalog to third-generation detectors. JCAP 05:003. arXiv:2102.03809
634. Franciolini G, Baibhav V, De Luca V, Ng KKY, Wong KWK, Berti E et al, Quantifying the evidence for primordial black holes in LIGO/Virgo gravitational-wave data. arXiv:2105.03349
635. Baer H, Choi K-Y, Kim JE, Roszkowski L (2015) Dark matter production in the early Universe: beyond the thermal WIMP paradigm. Phys Rep 555:1–60. arXiv:1407.0017
636. Preskill J, Wise MB, Wilczek F (1983) Cosmology of the invisible axion. Phys Lett B 120:127–132
637. Abbott LF, Sikivie P (1983) A cosmological bound on the invisible axion. Phys Lett B 120:133–136
638. Dine M, Fischler W (1983) The not so harmless axion. Phys Lett B 120:137–141
639. Sikivie P (2008) Axion cosmology. Lect Notes Phys 741:19–50. arXiv:astro-ph/0610440
640. Wantz O, Shellard EPS (2010) Axion cosmology revisited. Phys Rev D 82:123508. arXiv:0910.1066
641. Kim JE, Carosi G (2010) Axions and the strong CP problem. Rev Mod Phys 82:557–602. arXiv:0807.3125
642. Arias P, Cadamuro D, Goodsell M, Jaeckel J, Redondo J, Ringwald A (2012) WISPy cold dark matter. JCAP 06:013. arXiv:1201.5902
643. Kawasaki M, Nakayama K (2013) Axions: theory and cosmological role. Ann Rev Nucl Part Sci 63:69–95. arXiv:1301.1123
644. Ringwald A (2014) Axions and axion-like particles. In: 49th Rencontres de Moriond on electroweak interactions and unified theories, pp 223–230. arXiv:1407.0546
645. Graham PW, Irastorza IG, Lamoreaux SK, Lindner A, van Bibber KA (2015) Experimental searches for the axion and axion-like particles. Ann Rev Nucl Part Sci 65:485–514. arXiv:1602.00039
646. Sikivie P (2021) Invisible axion search methods. Rev Mod Phys 93:015004. arXiv:2003.02206

647. Adams CB et al (2022) Axion dark matter. In: 2022 Snowmass summer study, p 3. arXiv:2203.14923
648. Agrawal P, Berghaus KV, Fan J, Hook A, Marques-Tavares G, Rudelius T (2022) Some open questions in axion theory. In: 2022 snowmass summer study, p 3. arXiv:2203.08026
649. García-Bellido J (2017) Massive primordial black holes as dark matter and their detection with gravitational waves. J Phys Conf Ser 840:012032. arXiv:1702.08275
650. Sasaki M, Suyama T, Tanaka T, Yokoyama S (2018) Primordial black holes–perspectives in gravitational wave astronomy. Class Quant Grav 35:063001. arXiv:1801.05235
651. Navarro JF, Frenk CS, White SDM (1996) The structure of cold dark matter halos. Astrophys J 462:563–575. arXiv:astro-ph/9508025
652. Cirelli M, Corcella G, Hektor A, Hutsi G, Kadastik M, Panci P et al (2011) PPPC 4 DM ID: a poor particle physicist cookbook for dark matter indirect detection. JCAP 03:051. arXiv:1012.4515
653. Famaey B, McGaugh S (2012) Modified Newtonian Dynamics (MOND): observational phenomenology and relativistic extensions. Living Rev Rel 15:10. arXiv:1112.3960
654. Pontzen A, Governato F (2014) Cold dark matter heats up. Nature 506:171–178 [arXiv:1402.1764]
655. Del Popolo A, Le Delliou M (2017) Small scale problems of the Λ CDM model: a short review. Galaxies 5:17. arXiv:1606.07790
656. Bullock JS, Boylan-Kolchin M (2017) Small-scale challenges to the Λ CDM paradigm. Ann Rev Astron Astrophys 55:343–387. arXiv:1707.04256
657. Buckley MR, Peter AHG (2018) Gravitational probes of dark matter physics. Phys Rep 761:1–60. arXiv:1712.06615
658. Tulin S, Yu H-B (2018) Dark matter self-interactions and small scale structure. Phys Rep 730:1–57. arXiv:1705.02358
659. Bechtol K et al (2022) Snowmass2021 cosmic frontier white paper: dark matter physics from halo measurements. In: 2022 snowmass summer study, p 3. arXiv:2203.07354
660. Fitts A et al, Dwarf galaxies in CDM, WDM, and SIDM: disentangling baryons and dark matter physics. arXiv:1811.11791
661. Moore B (1994) Evidence against dissipationless dark matter from observations of galaxy haloes. Nature 370:629
662. Flores RA, Primack JR (1994) Observational and theoretical constraints on singular dark matter halos. Astrophys J 427:L1-4. arXiv:astro-ph/9402004
663. Battaglia G, Helmi A, Tolstoy E, Irwin M, Hill V, Jablonka P (2008) The kinematic status and mass content of the Sculptor dwarf spheroidal galaxy. Astrophys J 681:L13. arXiv:0802.4220
664. de Blok WJG, Walter F, Brinks E, Trachternach C, Oh S-H, Kennicutt RC Jr (2008) High-resolution rotation curves and galaxy mass models from THINGS. Astron J 136:2648–2719. arXiv:0810.2100
665. Walker MG, Penarrubia J (2011) A method for measuring (Slopes of) the mass profiles of dwarf spheroidal galaxies. Astrophys J 742:20. arXiv:1108.2404
666. Amorisco NC, Evans NW (2012) Dark matter cores and cusps: the case of multiple stellar populations in dwarf spheroidals. Mon Not R Astron Soc 419:184–196. arXiv:1106.1062
667. Amorisco NC, Agnello A, Evans NW (2013) The core size of the Fornax dwarf Spheroidal. Mon Not R Astron Soc 429:L89–L93. arXiv:1210.3157
668. Agnello A, Evans NW (2012) A virial core in the sculptor dwarf spheroidal galaxy. Astrophys J 754:L39. arXiv:1205.6673
669. Adams JJ et al (2014) Dwarf galaxy dark matter density profiles inferred from stellar and gas kinematics. Astrophys J 789:63. arXiv:1405.4854
670. Oh S-H et al (2015) High-resolution mass models of dwarf galaxies from little things. Astron J 149:180. arXiv:1502.01281
671. Frenk C, See talk by Carlos Frenk at Colmology 2018 in Dubrovnik
672. Pineda JCB, Hayward CC, Springel V, Mendes de Oliveira C (2016) Rotation curve fitting and its fatal attraction to cores in realistically simulated galaxy observations. Mon Not R Astron Soc 466(11):63–87. http://oup.prod.sis.lan/mnras/article-pdf/466/1/63/10865036/stw3004.pdf

673. Oman KA, Marasco A, Navarro JF, Frenk CS, Schaye J, Benítez-Llambay A (2019) Non-circular motions and the diversity of dwarf galaxy rotation curves. Mon Not R Astron Soc 482:821–847. arXiv:1706.07478
674. Navarro JF, Eke VR, Frenk CS (1996) The cores of dwarf galaxy halos. Mon Not R Astron Soc 283:L72–L78. arXiv:astro-ph/9610187
675. Read JI, Gilmore G (2005) Mass loss from dwarf spheroidal galaxies: the Origins of shallow dark matter cores and exponential surface brightness profiles. Mon Not R Astron Soc 356:107–124. arXiv:astro-ph/0409565
676. Pontzen A, Governato F (2012) How supernova feedback turns dark matter cusps into cores. Mon Not R Astron Soc 421:3464. arXiv:1106.0499
677. Governato F, Zolotov A, Pontzen A, Christensen C, Oh SH, Brooks AM et al (2012) Cuspy no more: how outflows affect the central dark matter and baryon distribution in lambda CDM galaxies. Mon Not R Astron Soc 422:1231–1240. arXiv:1202.0554
678. White SDM (1976) Dynamical friction in spherical clusters. mnras 174:19–28
679. El-Zant A, Shlosman I, Hoffman Y (2001) Dark halos: the flattening of the density cusp by dynamical friction. Astrophys J 560:636. arXiv:astro-ph/0103386
680. Tonini C, Lapi A (2006) Angular momentum transfer in dark matter halos: erasing the cusp. Astrophys J 649:591–598. arXiv:astro-ph/0603051
681. Romano-Diaz E, Shlosman I, Hoffman Y, Heller C (2008) Erasing dark matter cusps in cosmological galactic halos with baryons. Astrophys J 685:L105. arXiv:0808.0195
682. Penarrubia J, Pontzen A, Walker MG, Koposov SE (2012) The coupling between the core/cusp and missing satellite problems. Astrophys J 759:L42. arXiv:1207.2772
683. Di Cintio A, Brook CB, Macciò AV, Stinson GS, Knebe A, Dutton AA et al (2014) The dependence of dark matter profiles on the stellar-to-halo mass ratio: a prediction for cusps versus cores. Mon Not R Astron Soc 437:415–423. arXiv:1306.0898
684. Di Cintio A, Brook CB, Dutton AA, Macciò AV, Stinson GS, Knebe A (2014) A mass-dependent density profile for dark matter haloes including the influence of galaxy formation. Mon Not R Astron Soc 441:2986–2995. arXiv:1404.5959
685. Tollet E et al (2016) NIHAO—IV: core creation and destruction in dark matter density profiles across cosmic time. Mon Not R Astron Soc 456:3542–3552. arXiv:1507.03590
686. Fitts A et al (2017) Fire in the field: simulating the threshold of galaxy formation. Mon Not R Astron Soc 471:3547–3562. arXiv:1611.02281
687. Hopkins PF et al (2018) Fire-2 simulations: physics versus numerics in galaxy formation. Mon Not R Astron Soc 480:800–863. arXiv:1702.06148
688. Read JI, Agertz O, Collins MLM (2016) Dark matter cores all the way down. Mon Not R Astron Soc 459:2573–2590. arXiv:1508.04143
689. Springel V, Wang J, Vogelsberger M, Ludlow A, Jenkins A, Helmi A et al (2008) The aquarius project: the subhalos of galactic halos. Mon Not R Astron Soc 391:1685–1711. arXiv:0809.0898
690. Stadel J, Potter D, Moore B, Diemand J, Madau P, Zemp M et al (2009) Quantifying the heart of darkness with GHALO—a multi-billion particle simulation of our galactic halo. Mon Not R Astron Soc 398:L21–L25. arXiv:0808.2981
691. Garrison-Kimmel S, Boylan-Kolchin M, Bullock J, Lee K (2014) ELVIS: exploring the local volume in simulations. Mon Not R Astron Soc 438:2578–2596. arXiv:1310.6746
692. Griffen BF, Ji AP, Dooley GA, Gómez FA, Vogelsberger M, O'Shea BW et al (2016) The caterpillar project: a large suite of Milky way sized halos. Astrophys J 818:10. arXiv:1509.01255
693. Tollerud EJ, Bullock JS, Strigari LE, Willman B (2008) Hundreds of Milky way satellites? luminosity bias in the satellite luminosity function. Astrophys J 688:277–289. arXiv:0806.4381
694. Hargis JR, Willman B, Peter AHG (2014) Too many, too few, or just right? the predicted number and distribution of milky way dwarf galaxies. Astrophys J 795:L13. arXiv:1407.4470
695. Newton O, Cautun M, Jenkins A, Frenk CS, Helly J (2018) The total satellite population of the Milky Way. Mon Not R Astron Soc 479:2853–2870. arXiv:1708.04247

696. Bullock JS, Kravtsov AV, Weinberg DH (2000) Reionization and the abundance of galactic satellites. Astrophys J 539:517. arXiv:astro-ph/0002214
697. Benson AJ, Lacey CG, Baugh CM, Cole S, Frenk CS (2002) The effects of photoionization on galaxy formation. 1. Model and results at z = 0. Mon Not R Astron Soc 333:156. arXiv:astro-ph/0108217
698. Somerville RS, Bullock JS, Livio M (2003) The epoch of reionization in models with reduced small-scale power. Astrophys J 593:616–621
699. Okamoto T, Gao L, Theuns T (2008) Mass loss of galaxies due to an ultraviolet background. Mon Not R Astron Soc 390(10):920–928. http://oup.prod.sis.lan/mnras/article-pdf/390/3/920/18427117/mnras0390-0920.pdf
700. Zolotov A, Brooks AM, Willman B, Governato F, Pontzen A, Christensen C et al (2012) Baryons matter: why luminous satellite galaxies have reduced central masses. Astrophys J 761:71. arXiv:1207.0007
701. Arraki KS, Klypin A, More S, Trujillo-Gomez S (2014) Effects of baryon removal on the structure of dwarf spheroidal galaxies. Mon Not R Astron Soc 438:1466–1482. arXiv:1212.6651
702. Boylan-Kolchin M, Bullock JS, Kaplinghat M (2011) Too big to fail? the puzzling darkness of massive Milky Way subhaloes. Mon Not R Astron Soc 415:L40. arXiv:1103.0007
703. Boylan-Kolchin M, Bullock JS, Kaplinghat M (2012) The Milky Way's bright satellites as an apparent failure of LCDM. Mon Not R Astron Soc 422:1203–1218. arXiv:1111.2048
704. Purcell CW, Zentner AR (2012) Bailing out the milky way: variation in the properties of massive dwarfs among galaxy-sized systems. JCAP 1212:007. arXiv:1208.4602
705. Tollerud EJ, Boylan-Kolchin M, Bullock JS (2014) M31 satellite masses compared to LCDM subhaloes. Mon Not R Astron Soc 440:3511–3519. arXiv:1403.6469
706. Kirby EN, Bullock JS, Boylan-Kolchin M, Kaplinghat M, Cohen JG (2014) The dynamics of isolated local group galaxies. Mon Not R Astron Soc 439:1015–1027. arXiv:1401.1208
707. Garrison-Kimmel S, Boylan-Kolchin M, Bullock JS, Kirby EN (2014) Too big to fail in the local group. Mon Not R Astron Soc 444:222–236. arXiv:1404.5313
708. Heckman TM, Lehnert MD, Strickland DK, Armus L (2000) Absorption-line probes of gas and dust in galactic superwinds. Astrophys J Suppl 129:493–516. arXiv:astro-ph/0002526
709. Martin CL, Shapley AE, Coil AL, Kornei KA, Bundy K, Weiner BJ et al (2012) Demographics and physical properties of gas out/inflows at 0.4 < z < 1.4. Astrophys J 760:127. arXiv:1206.5552
710. Governato F et al (2010) At the heart of the matter: the origin of bulgeless dwarf galaxies and dark matter cores. Nature 463:203–206. arXiv:0911.2237
711. Garrison-Kimmel S, Hopkins PF, Wetzel A, Bullock JS, Boylan-Kolchin M, Kereš D et al (2019) The local group on fire: dwarf galaxy populations across a suite of hydrodynamic simulations. mnras 487:1380–1399. arXiv:1806.04143
712. Brooks AM, Kuhlen M, Zolotov A, Hooper D (2013) A baryonic solution to the missing satellites problem. Astrophys J 765:22. arXiv:1209.5394
713. Brooks AM, Zolotov A (2014) Why baryons matter: the kinematics of dwarf spheroidal satellites. Astrophys J 786:87. arXiv:1207.2468
714. Amorisco NC, Zavala J, de Boer TJL (2014) Dark matter cores in the Fornax and Sculptor dwarf galaxies: joining halo assembly and detailed star formation histories. Astrophys J 782:L39. arXiv:1309.5958
715. Gnedin OY, Zhao H (2002) Maximum feedback and dark matter profiles of dwarf galaxies. Mon Not R Astron Soc 333:299. arXiv:astro-ph/0108108
716. Garrison-Kimmel S, Rocha M, Boylan-Kolchin M, Bullock J, Lally J (2013) Can feedback solve the too big to fail problem? Mon Not R Astron Soc 433:3539. arXiv:1301.3137
717. Benitez-Llambay A, Navarro JF, Abadi MG, Gottloeber S, Yepes G, Hoffman Y et al (2013) Dwarf galaxies and the cosmic web. Astrophys J 763:L41. arXiv:1211.0536
718. Papastergis E, Ponomareva AA (2017) Testing core creation in hydrodynamical simulations using the hi kinematics of field dwarfs. Astron Astrophys 601:A1
719. Burkert A (1996) The Structure of dark matter halos in dwarf galaxies. IAU Symp 171:175. arXiv:astro-ph/9504041

720. Persic M, Salucci P, Stel F (1996) The universal rotation curve of spiral galaxies: 1. The dark matter connection. Mon Not R Astron Soc 281:27. arXiv:astro-ph/9506004
721. Navarro JF, Ludlow A, Springel V, Wang J, Vogelsberger M, White SDM et al (2010) The diversity and similarity of cold dark matter halos. Mon Not R Astron Soc 402:21. arXiv:0810.1522
722. Oman KA et al (2015) The unexpected diversity of dwarf galaxy rotation curves. Mon Not R Astron Soc 452:3650–3665. arXiv:1504.01437
723. Ren T, Kwa A, Kaplinghat M, Yu H-B (2019) Reconciling the diversity and uniformity of galactic rotation curves with self-interacting dark matter. Phys Rev X 9:031020. arXiv:1808.05695
724. Katz H, Lelli F, McGaugh SS, Di Cintio A, Brook CB, Schombert JM (2017) Testing feedback-modified dark matter haloes with galaxy rotation curves: estimation of halo parameters and consistency with Λ CDM scaling relations. Mon Not R Astron Soc 466:1648–1668. arXiv:1605.05971
725. Spergel DN, Steinhardt PJ (2000) Observational evidence for selfinteracting cold dark matter. Phys Rev Lett 84:3760–3763. arXiv:astro-ph/9909386
726. Rocha M, Peter AHG, Bullock JS, Kaplinghat M, Garrison-Kimmel S, Onorbe J et al (2013) cosmological simulations with self-interacting dark matter I: constant density cores and substructure. Mon Not R Astron Soc 430:81–104. arXiv:1208.3025
727. Peter AHG, Rocha M, Bullock JS, Kaplinghat M (2013) cosmological simulations with self-interacting dark matter II: halo shapes vs. observations. Mon Not R Astron Soc 430:105. arXiv:1208.3026
728. Vogelsberger M, Zavala J, Loeb A (2012) Subhaloes in self-interacting galactic dark matter haloes. Mon Not R Astron Soc 423:3740. arXiv:1201.5892
729. Zavala J, Vogelsberger M, Walker MG (2013) Constraining self-interacting dark matter with the milky way's dwarf spheroidals. Mon Not R Astron Soc 431:L20–L24. arXiv:1211.6426
730. Elbert OD, Bullock JS, Garrison-Kimmel S, Rocha M, Oñorbe J, Peter AHG (2015) Core formation in dwarf haloes with self-interacting dark matter: no fine-tuning necessary. Mon Not R Astron Soc 453:29–37. arXiv:1412.1477
731. Vogelsberger M, Zavala J, Simpson C, Jenkins A (2014) Dwarf galaxies in CDM and SIDM with baryons: observational probes of the nature of dark matter. Mon Not R Astron Soc 444:3684–3698. arXiv:1405.5216
732. Fry AB, Governato F, Pontzen A, Quinn T, Tremmel M, Anderson L et al (2015) All about baryons: revisiting SIDM predictions at small halo masses. Mon Not R Astron Soc 452:1468–1479. arXiv:1501.00497
733. Dooley GA, Peter AHG, Vogelsberger M, Zavala J, Frebel A (2016) Enhanced tidal stripping of satellites in the galactic halo from dark matter self-interactions. Mon Not R Astron Soc 461:710–727. arXiv:1603.08919
734. Robertson A, Massey R, Eke V (2017) Cosmic particle colliders: simulations of self-interacting dark matter with anisotropic scattering. Mon Not R Astron Soc 467:4719–4730. arXiv:1612.03906
735. Elbert OD, Bullock JS, Kaplinghat M, Garrison-Kimmel S, Graus AS, Rocha M (2018) A testable conspiracy: simulating baryonic effects on self-interacting dark matter halos. Astrophys J 853:109. arXiv:1609.08626
736. Kummer J, Kahlhoefer F, Schmidt-Hoberg K (2018) Effective description of dark matter self-interactions in small dark matter haloes. Mon Not R Astron Soc 474:388–399. arXiv:1706.04794
737. Kummer J, Brüggen M, Dolag K, Kahlhoefer F, Schmidt-Hoberg K (2019) Simulations of core formation for frequent dark matter self-interactions. Mon Not R Astron Soc 487:354–363. arXiv:1902.02330
738. Robles VH, Kelley T, Bullock JS, Kaplinghat M, The Milky Way's halo and subhalos in self-interacting dark matter. arXiv:1903.01469
739. Kaplinghat M, Keeley RE, Linden T, Yu H-B (2014) Tying dark matter to baryons with self-interactions. Phys Rev Lett 113:021302. arXiv:1311.6524

740. Kamada A, Kaplinghat M, Pace AB, Yu H-B (2017) How the self-interacting dark matter model explains the diverse galactic rotation curves. Phys Rev Lett 119:111102. arXiv:1611.02716
741. Creasey P, Sameie O, Sales LV, Yu H-B, Vogelsberger M, Zavala J (2017) Spreading out and staying sharp—creating diverse rotation curves via baryonic and self-interaction effects. Mon Not R Astron Soc 468:2283–2295. arXiv:1612.03903
742. Kahlhoefer F, Kaplinghat M, Slatyer TR, Wu C-L, Diversity in density profiles of self-interacting dark matter satellite halos. arXiv:1904.10539
743. Bento MC, Bertolami O, Rosenfeld R, Teodoro L (2000) Selfinteracting dark matter and invisibly decaying Higgs. Phys Rev D 62:041302. arXiv:astro-ph/0003350
744. McDonald J (2002) Thermally generated gauge singlet scalars as selfinteracting dark matter. Phys Rev Lett 88:091304. arXiv:hep-ph/0106249
745. CMS collaboration, Khachatryan v et al (2017) Searches for invisible decays of the Higgs boson in pp collisions at $\sqrt{s} = 7$, 8, and 13 TeV. JHEP 02:135. arXiv:1610.09218
746. Carlson ED, Machacek ME, Hall LJ (1992) Self-interacting dark matter. Astrophys J 398:43–52
747. de Laix AA, Scherrer RJ, Schaefer RK (1995) Constraints of selfinteracting dark matter. Astrophys J 452:495. arXiv:astro-ph/9502087
748. Hochberg Y, Kuflik E, Murayama H, Volansky T, Wacker JG (2015) Model for thermal relic dark matter of strongly interacting massive particles. Phys Rev Lett 115:021301. arXiv:1411.3727
749. Bernal N, Chu X, Garcia-Cely C, Hambye T, Zaldivar B (2016) Production regimes for self-interacting dark matter. JCAP 03:018. arXiv:1510.08063
750. Pappadopulo D, Ruderman JT, Trevisan G (2016) Dark matter freeze-out in a nonrelativistic sector. Phys Rev D 94:035005. arXiv:1602.04219
751. Farina M, Pappadopulo D, Ruderman JT, Trevisan G (2016) Phases of cannibal dark matter. JHEP 12:039. arXiv:1607.03108
752. Kaplinghat M, Tulin S, Yu H-B (2016) Dark matter halos as particle colliders: unified solution to small-scale structure puzzles from dwarfs to clusters. Phys Rev Lett 116:041302. arXiv:1508.03339
753. Feng JL, Kaplinghat M, Yu H-B (2010) Halo shape and relic density exclusions of sommerfeld-enhanced dark matter explanations of cosmic ray excesses. Phys Rev Lett 104:151301. arXiv:0911.0422
754. Buckley MR, Fox PJ (2010) Dark matter self-interactions and light force carriers. Phys Rev D 81:083522. arXiv:0911.3898
755. Tulin S, Yu H-B, Zurek KM (2013) Resonant dark forces and small scale structure. Phys Rev Lett 110:111301. arXiv:1210.0900
756. Tulin S, Yu H-B, Zurek KM (2013) Beyond collisionless dark matter: particle physics dynamics for dark matter halo structure. Phys Rev D 87:115007. arXiv:1302.3898
757. Schutz K, Slatyer TR (2015) Self-scattering for dark matter with an excited state. JCAP 01:021. arXiv:1409.2867
758. Del Nobile E, Kaplinghat M, Yu H-B (2015) Direct detection signatures of self-interacting dark matter with a light mediator. JCAP 10:055. arXiv:1507.04007
759. Bringmann T, Kahlhoefer F, Schmidt-Hoberg K, Walia P (2017) Strong constraints on self-interacting dark matter with light mediators. Phys Rev Lett 118:141802. arXiv:1612.00845
760. Kahlhoefer F, Schmidt-Hoberg K, Wild S (2017) Dark matter self-interactions from a general spin-0 mediator. JCAP 08:003. arXiv:1704.02149
761. Baldes I, Cirelli M, Panci P, Petraki K, Sala F, Taoso M (2018) Asymmetric dark matter: residual annihilations and self-interactions. SciPost Phys 4:041. arXiv:1712.07489
762. Cline JM, Liu Z, Moore G, Xue W (2014) Scattering properties of dark atoms and molecules. Phys Rev D 89:043514. arXiv:1311.6468
763. Boddy KK, Kaplinghat M, Kwa A, Peter AHG (2016) Hidden sector hydrogen as dark matter: small-scale structure formation predictions and the importance of hyperfine interactions. Phys Rev D 94:123017. arXiv:1609.03592

764. Chu X, Garcia-Cely C, Murayama H (2019) Velocity dependence from resonant self-interacting dark matter. Phys Rev Lett 122:071103. arXiv:1810.04709
765. Chu X, Garcia-Cely C, Murayama H (2020) A practical and consistent parametrization of dark matter self-interactions. JCAP 06:043. arXiv:1908.06067
766. Boddy KK, Feng JL, Kaplinghat M, Tait TMP (2014) Self-interacting dark matter from a non-abelian hidden sector. Phys Rev D 89:115017. arXiv:1402.3629
767. Soni A, Zhang Y (2016) Hidden SU(N) Glueball dark matter. Phys Rev D 93:115025. arXiv:1602.00714
768. Prilepina V, Tsai Y (2017) Reconciling large and small-scale structure in twin higgs models. JHEP 09:033. arXiv:1611.05879
769. Kribs GD, Neil ET (2016) Review of strongly-coupled composite dark matter models and lattice simulations. Int J Mod Phys A 31:1643004. arXiv:1604.04627
770. Bode P, Ostriker JP, Turok N (2001) Halo formation in warm dark matter models. Astrophys J 556:93–107. arXiv:astro-ph/0010389
771. Dalcanton JJ, Hogan CJ (2001) Halo cores and phase space densities: Observational constraints on dark matter physics and structure formation. Astrophys J 561:35–45. arXiv:astro-ph/0004381
772. Zentner AR, Bullock JS (2003) Halo substructure and the power spectrum. Astrophys J 598:49. arXiv:astro-ph/0304292
773. Smith RE, Markovic K (2011) Testing the warm dark matter paradigm with large-scale structures. Phys Rev D 84:063507. arXiv:1103.2134
774. Lovell MR, Eke V, Frenk CS, Gao L, Jenkins A, Theuns T et al (2012) The haloes of bright satellite galaxies in a warm dark matter universe. Mon Not R Astron Soc 420:2318–2324. arXiv:1104.2929
775. Anderhalden D, Schneider A, Maccio AV, Diemand J, Bertone G (2013) Hints on the nature of dark matter from the properties of milky way satellites. JCAP 1303:014. arXiv:1212.2967
776. Lovell MR, Frenk CS, Eke VR, Jenkins A, Gao L, Theuns T (2014) The properties of warm dark matter haloes. Mon Not R Astron Soc 439:300–317. arXiv:1308.1399
777. Viel M, Becker GD, Bolton JS, Haehnelt MG (2013) Warm dark matter as a solution to the small scale crisis: new constraints from high redshift Lyman-α forest data. Phys Rev D 88:043502. arXiv:1306.2314
778. Schneider A, Anderhalden D, Maccio A, Diemand J (2014) Warm dark matter does not do better than cold dark matter in solving small-scale inconsistencies. Mon Not R Astron Soc 441:6. arXiv:1309.5960
779. van den Aarssen LG, Bringmann T, Pfrommer C (2012) Is dark matter with long-range interactions a solution to all small-scale problems of \ Lambda CDM cosmology? Phys Rev Lett 109:231301. arXiv:1205.5809
780. Ahlgren B, Ohlsson T, Zhou S (2013) Comment on is dark matter with long-range interactions a solution to all small-scale problems of Λ cold dark matter cosmology?. Phys Rev Lett 111:199001. arXiv:1309.0991
781. Laha R, Dasgupta B, Beacom JF (2014) Constraints on new neutrino interactions via light abelian vector Bosons. Phys Rev D 89:093025. arXiv:1304.3460
782. Bertoni B, Ipek S, McKeen D, Nelson AE (2015) Constraints and consequences of reducing small scale structure via large dark matter-neutrino interactions. JHEP 04:170. arXiv:1412.3113
783. Chu X, Dasgupta B (2014) Dark radiation alleviates problems with dark matter Halos. Phys Rev Lett 113:161301. arXiv:1404.6127
784. Buckley MR, Zavala J, Cyr-Racine F-Y, Sigurdson K, Vogelsberger M (2014) Scattering, damping, and acoustic oscillations: simulating the structure of dark matter halos with relativistic force carriers. Phys Rev D 90:043524. arXiv:1405.2075
785. Tang Y (2016) Interacting scalar radiation and dark matter in cosmology. Phys Lett B 757:387–392. arXiv:1603.00165
786. Agrawal P, Cyr-Racine F-Y, Randall L, Scholtz J (2017) Dark catalysis. JCAP 08:021. arXiv:1702.05482

787. Boehm C, Schewtschenko JA, Wilkinson RJ, Baugh CM, Pascoli S (2014) Using the milky way satellites to study interactions between cold dark matter and radiation. Mon Not R Astron Soc 445:L31–L35 [arXiv:1404.7012]
788. Schewtschenko JA, Baugh CM, Wilkinson RJ, Bœhm C, Pascoli S, Sawala T (2016) Dark matter–radiation interactions: the structure of Milky Way satellite galaxies. Mon Not R Astron Soc 461:2282–2287. arXiv:1512.06774
789. Tremaine S, Gunn JE (1979) Dynamical role of light neutral leptons in cosmology. Phys Rev Lett 42:407–410
790. Boyarsky A, Ruchayskiy O, Iakubovskyi D (2009) A lower bound on the mass of dark matter particles. JCAP 03:005. arXiv:0808.3902
791. Madelung E (1927) Quantum theory in hydrodynamical form. z Phys 40
792. Sharma A, Khoury J, Lubensky T (2019) the equation of state of dark matter superfluids. JCAP 05:054. arXiv:1809.08286
793. Hu W, Barkana R, Gruzinov A (2000) Cold and fuzzy dark matter. Phys Rev Lett 85:1158–1161. arXiv:astro-ph/0003365
794. Rogers KK, Peiris HV (2021) Strong bound on canonical ultralight axion dark matter from the lyman-alpha forest. Phys Rev Lett 126:071302. arXiv:2007.12705
795. Hlozek R, Grin D, Marsh DJE, Ferreira PG (2015) A search for ultralight axions using precision cosmological data. Phys Rev D 91:103512. arXiv:1410.2896
796. Laguë A, Bond JR, Hložek R, Rogers KK, Marsh DJE, Grin D (2022) Constraining ultralight axions with galaxy surveys. JCAP 01:049. arXiv:2104.07802
797. Suárez A, Robles VH, Matos T (2014) A review on the scalar field/bose-Einstein condensate dark matter model. Astrophys Space Sci Proc 38:107–142. arXiv:1302.0903
798. Hui L, Ostriker JP, Tremaine S, Witten E (2017) Ultralight scalars as cosmological dark matter. Phys Rev D 95:043541. arXiv:1610.08297
799. Ureña López LA (2019) Brief review on scalar field dark matter models. Front Astron Space Sci 6:47
800. Niemeyer JC, Small-scale structure of fuzzy and axion-like dark matter. arXiv:1912.07064
801. Ferreira EGM (2021) Ultra-light dark matter. Astron Astrophys Rev 29:7. arXiv:2005.03254
802. Marsh DJE, Hoof S, Astrophysical searches and constraints on ultralight bosonic dark matter. arXiv:2106.08797
803. Cicoli M, Goodsell M, Ringwald A (2012) The type IIB string axiverse and its low-energy phenomenology. JHEP 10:146. arXiv:1206.0819
804. Mehta VM, Demirtas M, Long C, Marsh DJE, Mcallister L, Stott MJ, Superradiance exclusions in the landscape of type IIB string theory. arXiv:2011.08693
805. Mehta VM, Demirtas M, Long C, Marsh DJE, McAllister L, Stott MJ (2021) Superradiance in string theory. JCAP 07:033. arXiv:2103.06812
806. Demirtas M, Gendler N, Long C, McAllister L, Moritz J, PQ axiverse. arXiv:2112.04503
807. Flambaum VV, Dzuba VA (2009) Search for variation of the fundamental constants in atomic, molecular and nuclear spectra. Can J Phys 87:25–33. arXiv:0805.0462
808. Ludlow AD, Boyd MM, Ye J, Peik E, Schmidt PO (2014) Optical atomic clocks. https://doi.org/10.48550/ARXIV.1407.3493
809. Van Tilburg K, Leefer N, Bougas L, Budker D (2015) Search for ultralight scalar dark matter with atomic spectroscopy. Phys Rev Lett 115:011802. arXiv:1503.06886
810. Hees A, Guéna J, Abgrall M, Bize S, Wolf P (2016) Searching for an oscillating massive scalar field as a dark matter candidate using atomic hyperfine frequency comparisons. Phys Rev Lett 117:061301. arXiv:1604.08514
811. Stadnik YV, Flambaum VV (2016) Improved limits on interactions of low-mass spin-0 dark matter from atomic clock spectroscopy. Phys Rev A 94:022111. arXiv:1605.04028
812. Flambaum VV, Tran Tan HB, Samsonov IB, Stadnik YV, Budker D (2018) Resonant detection and production of axions with atoms. Int J Mod Phys A 33:1844030
813. Kennedy CJ, Oelker E, Robinson JM, Bothwell T, Kedar D, Milner WR et al (2020) Precision metrology meets cosmology: improved constraints on ultralight dark matter from atom-cavity frequency comparisons. Phys Rev Lett 125:201302. arXiv:2008.08773

814. Berti E et al (2022) Dark matter in extreme astrophysical environments. In: 2022 snowmass summer study, p 3. arXiv:2203.07984
815. Stadnik YV, Flambaum VV (2015) Searching for dark matter and variation of fundamental constants with laser and maser interferometry. Phys Rev Lett 114:161301. arXiv:1412.7801
816. Stadnik YV, Flambaum VV (2016) Enhanced effects of variation of the fundamental constants in laser interferometers and application to dark matter detection. Phys Rev A 93:063630. arXiv:1511.00447
817. Arvanitaki A, Graham PW, Hogan JM, Rajendran S, Van Tilburg K (2018) Search for light scalar dark matter with atomic gravitational wave detectors. Phys Rev D 97:075020. arXiv:1606.04541
818. Grote H, Stadnik YV (2019) Novel signatures of dark matter in laser-interferometric gravitational-wave detectors. Phys Rev Res 1:033187. arXiv:1906.06193
819. Arvanitaki A, Dimopoulos S, Van Tilburg K (2016) Sound of dark matter: searching for light scalars with resonant-mass detectors. Phys Rev Lett 116:031102. arXiv:1508.01798
820. Stadnik YV, Flambaum VV (2015) Can dark matter induce cosmological evolution of the fundamental constants of Nature? Phys Rev Lett 115:201301. arXiv:1503.08540
821. Sibiryakov S, Sørensen P, Yu T-T (2020) BBN constraints on universally-coupled ultralight scalar dark matter. JHEP 12:075. arXiv:2006.04820
822. Schlamminger S, Choi KY, Wagner TA, Gundlach JH, Adelberger EG (2008) Test of the equivalence principle using a rotating torsion balance. Phys Rev Lett 100:041101. arXiv:0712.0607
823. Piazza F, Pospelov M (2010) Sub-eV scalar dark matter through the super-renormalizable Higgs portal. Phys Rev D 82:043533. arXiv:1003.2313
824. Graham PW, Kaplan DE, Mardon J, Rajendran S, Terrano WA (2016) Dark matter direct detection with accelerometers. Phys Rev D 93:075029. arXiv:1512.06165
825. Hees A, Minazzoli O, Savalle E, Stadnik YV, Wolf P (2018) Violation of the equivalence principle from light scalar dark matter. Phys Rev D 98:064051. arXiv:1807.04512
826. Arvanitaki A, Dubovsky S (2011) Exploring the string axiverse with precision black hole physics. Phys Rev D 83:044026. arXiv:1004.3558
827. Arvanitaki A, Baryakhtar M, Huang X (2015) Discovering the QCD axion with black holes and gravitational waves. Phys Rev D 91:084011. arXiv:1411.2263
828. Brito R, Ghosh S, Barausse E, Berti E, Cardoso V, Dvorkin I et al (2017) Gravitational wave searches for ultralight bosons with LIGO and LISA. Phys Rev D 96:064050. arXiv:1706.06311
829. Brito R, Ghosh S, Barausse E, Berti E, Cardoso V, Dvorkin I et al (2017) Stochastic and resolvable gravitational waves from ultralight bosons. Phys Rev Lett 119:131101. arXiv:1706.05097
830. Stott MJ, Marsh DJE (2018) Black hole spin constraints on the mass spectrum and number of axionlike fields. Phys Rev D 98:083006. arXiv:1805.02016
831. Lelli F (2017) See talk by F. Lelli at LCDM, Modified Gravity or new Dark Matter models? - 2017 - LPTHE
832. Tully RB, Fisher JR (1977) A New method of determining distances to galaxies. Astron Astrophys 54:661–673
833. McGaugh SS, Schombert JM, Bothun GD, de Blok WJG (2000) The baryonic tully-fisher relation. Astrophys J 533:L99–L102. arXiv:astro-ph/0003001
834. McGaugh S (2012) The baryonic tully-fisher relation of gas rich galaxies as a test of LCDM and MOND. Astron J 143:40. arXiv:1107.2934
835. Lelli F, McGaugh SS, Schombert JM (2016) The small scatter of the baryonic tully-fisher relation. Astrophys J Lett 816:L14. arXiv:1512.04543
836. Keller BW, Wadsley JW (2017) Lambda CDM is Consistent with SPARC Radial Acceleration Relation. Astrophys J 835:L17. arXiv:1610.06183
837. Ludlow AD et al (2017) Mass-discrepancy acceleration relation: a natural outcome of galaxy formation in cold dark matter halos. Phys Rev Lett 118:161103. arXiv:1610.07663
838. Di Cintio A, Lelli F (2016) The mass discrepancy acceleration relation in a Λ CDM context. Mon Not R Astron Soc 456:L127–L131. arXiv:1511.06616
839. Desmond H (2017) A statistical investigation of the mass discrepancy–acceleration relation. Mon Not R Astron Soc 464:4160–4175. arXiv:1607.01800

840. Navarro JF, Benítez-Llambay A, Fattahi A, Frenk CS, Ludlow AD, Oman KA et al (2017) The origin of the mass discrepancy-acceleration relation in Λ CDM. Mon Not R Astron Soc 471:1841–1848. arXiv:1612.06329
841. Lelli F, McGaugh SS, Schombert JM, Pawlowski MS (2016) The relation between stellar and dynamical surface densities in the central regions of disk galaxies. Astrophys J Lett 827:L19. arXiv:1607.02145
842. McGaugh S, Lelli F, Schombert J (2016) Radial acceleration relation in rotationally supported galaxies. Phys Rev Lett. 117:201101. arXiv:1609.05917
843. Milgrom M (1983) A Modification of the Newtonian dynamics as a possible alternative to the hidden mass hypothesis. Astrophys J 270:365–370
844. Milgrom M (1983) A modification of the newtonian dynamics: implications for galaxies. Astrophys J 270:371–383
845. Sanders RH, McGaugh SS (2002) Modified Newtonian dynamics as an alternative to dark matter. Ann Rev Astron Astrophys 40:263–317. arXiv:astro-ph/0204521
846. Bekenstein J, Milgrom M (1984) Does the missing mass problem signal the breakdown of Newtonian gravity? Astrophys J 286:7–14
847. Milgrom M (2010) Quasi-linear formulation of MOND. Mon Not R Astron Soc 403:886. arXiv:0911.5464
848. Bekenstein JD (2004) Relativistic gravitation theory for the MOND paradigm. Phys Rev D 70:083509. arXiv:astro-ph/0403694
849. Milgrom M (2009) Bimetric MOND gravity. Phys Rev D 80:123536. arXiv:0912.0790
850. Blanchet L (2007) Gravitational polarization and the phenomenology of MOND. Class Quant Grav 24:3529–3540. arXiv:astro-ph/0605637
851. Blanchet L, Le Tiec A (2009) Dipolar dark matter and dark energy. Phys Rev D 80:023524. arXiv:0901.3114
852. Verlinde EP (2011) On the origin of gravity and the laws of newton. JHEP 04:029. arXiv:1001.0785
853. Verlinde EP (2017) Emergent gravity and the dark universe. SciPost Phys 2:016. arXiv:1611.02269
854. Lelli F, McGaugh SS, Schombert JM (2017) Testing Verlinde's emergent gravity with the radial acceleration relation. Mon Not R Astron Soc 468:L68–L71. arXiv:1702.04355
855. Skordis C, Mota DF, Ferreira PG, Boehm C (2006) Large scale structure in bekenstein's theory of relativistic modified newtonian dynamics. Phys Rev Lett 96:011301. arXiv:astro-ph/0505519
856. Skordis C (2009) The tensor-vector-scalar theory and its cosmology. Class Quant Grav 26:143001. arXiv:0903.3602
857. Skordis C, Zlosnik T (2021) New relativistic theory for modified newtonian dynamics. Phys Rev Lett 127:161302. arXiv:2007.00082
858. Skordis C, Złośnik T (2019) Gravitational alternatives to dark matter with tensor mode speed equaling the speed of light. Phys Rev D 100:104013. arXiv:1905.09465
859. Landau LD (1941) The theory of superfuidity of helium II. J Phys (USSR) 5:71–100
860. Berezhiani L, Khoury J (2015) Theory of dark matter superfluidity. Phys Rev D 92:103510. arXiv:1507.01019
861. Khoury J (2016) Another path for the emergence of modified galactic dynamics from dark matter superfluidity. Phys Rev D 93:103533. arXiv:1602.05961
862. Hodson A, Zhao H, Khoury J, Famaey B (2017) Galaxy clusters in the context of superfluid dark matter. Astron Astrophys 607:A108. arXiv:1611.05876
863. Berezhiani L, Famaey B, Khoury J (2018) Phenomenological consequences of superfluid dark matter with baryon-phonon coupling. JCAP 09:021. arXiv:1711.05748
864. Berezhiani L, Elder B, Khoury J (2019) Dynamical friction in superfluids. JCAP 10:074. arXiv:1905.09297
865. Khoury J, Sakstein J, Solomon AR (2018) Superfluids and the cosmological constant problem. JCAP 08:024. arXiv:1805.05937

866. Ferreira EGM, Franzmann G, Khoury J, Brandenberger R (2019) Unified Superfluid Dark Sector. JCAP 08:027. arXiv:1810.09474
867. Lisanti M, Moschella M, Outmezguine NJ, Slone O (2019) Testing dark matter and modifications to gravity using local milky way observables. Phys Rev D 100:083009. arXiv:1812.08169
868. Lisanti M, Moschella M, Outmezguine NJ, Slone O, The inconsistency of superfluid dark matter with milky way dynamics. arXiv:1911.12365
869. Mistele T, McGaugh S, Hossenfelder S, Galactic mass-to-light ratios with superfluid dark matter. arXiv:2201.07282
870. Khoury J (2021) Dark matter superfluidity. In: Les Houches summer school on Dark Matter, p 9. arXiv:2109.10928, https://doi.org/10.21468/SciPostPhysLectNotes.42
871. Famaey B, Khoury J, Penco R (2018) Emergence of the mass discrepancy-acceleration relation from dark matter-baryon interactions. JCAP 1803:038. arXiv:1712.01316
872. Khoury J (2019) See talk by J. Khoury at novel ideas of dark matter
873. Famaey B, Khoury J, Penco R, Sharma A (2020) Baryon-interacting dark matter: heating dark matter and the emergence of galaxy scaling relations. JCAP 06:025. arXiv:1912.07626
874. Riess AG, Casertano S, Yuan W, Macri LM, Scolnic D (2019) Large magellanic cloud cepheid standards provide a 1% foundation for the determination of the hubble constant and stronger evidence for physics beyond Λ CDM. Astrophys J 876:85. arXiv:1903.07603
875. Wong KC et al, H0LiCOW XIII. A 2.4% measurement of H_0 from lensed quasars: 5.3σ tension between early and late-Universe probes. arXiv:1907.04869
876. Schöneberg N, Lesgourgues J, Hooper DC (2019) The BAO+BBN take on the Hubble tension. JCAP 10:029. arXiv:1907.11594
877. DES collaboration, Abbott T et al (2018) Dark energy survey year 1 results: a precise H0 estimate from DES Y1, BAO, and D/H Data. Mon Not R Astron Soc 480:3879–3888. arXiv:1711.00403
878. Verde L, Treu T, Riess A (2019) Tensions between the early and the late universe, p 7. arXiv:1907.10625, https://doi.org/10.1038/s41550-019-0902-0
879. Csaki C, Kaloper N, Terning J (2002) Dimming supernovae without cosmic acceleration. Phys Rev Lett 88:161302. arXiv:hep-ph/0111311
880. Bernal JL, Verde L, Riess AG (2016) The trouble with H_0. JCAP 10:019. arXiv:1607.05617
881. Jedamzik K, Pogosian L, Zhao G-B, Why reducing the cosmic sound horizon can not fully resolve the Hubble tension. arXiv:2010.04158
882. Binder T, Gustafsson M, Kamada A, Sandner SMR, Wiesner M, Reannihilation of self-interacting dark matter. arXiv:1712.01246
883. Bringmann T, Kahlhoefer F, Schmidt-Hoberg K, Walia P (2018) Converting nonrelativistic dark matter to radiation. Phys Rev D 98:023543. arXiv:1803.03644
884. Hooper D, Krnjaic G, McDermott SD (2019) Dark radiation and superheavy dark matter from black hole domination. JHEP 08:001. arXiv:1905.01301
885. Hojjati A, Linder EV, Samsing J (2013) New constraints on the early expansion history of the universe. Phys Rev Lett 111:041301. arXiv:1304.3724
886. Aylor K, Joy M, Knox L, Millea M, Raghunathan S, Wu WK (2019) Sounds discordant: classical distance ladder & Λ CDM -based determinations of the cosmological sound horizon. Astrophys J 874:4. arXiv:1811.00537
887. Knox L, Millea M (2020) Hubble constant hunter's guide. Phys Rev D 101:043533. arXiv:1908.03663
888. Kreisch CD, Cyr-Racine F-Y, Doré O, The neutrino puzzle: anomalies, interactions, and cosmological tensions. arXiv:1902.00534
889. Aloni D, Berlin A, Joseph M, Schmaltz M, Weiner N, A step in understanding the hubble tension. arXiv:2111.00014
890. Poulin V, Smith TL, Karwal T, Kamionkowski M (2019) Early Dark Energy Can Resolve The Hubble Tension. Phys Rev Lett. 122:221301. arXiv:1811.04083
891. Agrawal P, Cyr-Racine F-Y, Pinner D, Randall L, Rock 'n' roll solutions to the hubble tension. arXiv:1904.01016

892. Lin M-X, Benevento G, Hu W, Raveri M (2019) Acoustic dark energy: potential conversion of the hubble tension. Phys Rev D 100:063542. arXiv:1905.12618
893. Smith TL, Poulin V, Amin MA (2020) Oscillating scalar fields and the Hubble tension: a resolution with novel signatures. Phys. Rev. D 101:063523 [arXiv:1908.06995]
894. Berghaus KV, Karwal T (2020) Thermal Friction as a Solution to the Hubble Tension. Phys. Rev. D 101:083537 [arXiv:1911.06281]
895. Hill JC, McDonough E, Toomey MW, Alexander S, Early dark energy does not restore cosmological concordance. arXiv:2003.07355
896. D'Amico G, Senatore L, Zhang P, Zheng H, The hubble tension in light of the full-shape analysis of large-scale structure data. arXiv:2006.12420
897. Ivanov MM, McDonough E, Hill JC, Simonović M, Toomey MW, Alexander S et al, Constraining early dark energy with large-scale structure. arXiv:2006.11235
898. Niedermann F, Sloth MS, New early dark energy is compatible with current LSS data. arXiv:2009.00006
899. Smith TL, Poulin V, Bernal JL, Boddy KK, Kamionkowski M, Murgia R, Early dark energy is not excluded by current large-scale structure data. arXiv:2009.10740
900. Chudaykin A, Gorbunov D, Nedelko N, Exploring early dark energy solution to the hubble tension with planck and SPTPol data. arXiv:2011.04682
901. Jedamzik K, Pogosian L (2020) Relieving the Hubble tension with primordial magnetic fields. Phys Rev Lett 125:181302. arXiv:2004.09487
902. Ye G, Piao Y-S (2020) Is the hubble tension a hint of AdS phase around recombination? Phys Rev D 101:083507. arXiv:2001.02451
903. Hart L, Chluba J (2018) New constraints on time-dependent variations of fundamental constants using Planck data. Mon Not R Astron Soc 474:1850–1861. arXiv:1705.03925
904. Hart L, Chluba J, Varying fundamental constants principal component analysis: additional hints about the Hubble tension. arXiv:2107.12465
905. Lin M-X, Raveri M, Hu W (2019) Phenomenology of modified gravity at recombination. Phys Rev D 99:043514. arXiv:1810.02333
906. Ballesteros G, Notari A, Rompineve F (2020) The H_0 tension: ΔG_N vs. $\Delta N_{\rm eff}$, JCAP 11:024. arXiv:2004.05049
907. Braglia M, Ballardini M, Emond WT, Finelli F, Gumrukcuoglu AE, Koyama K et al (2020) Larger value for H_0 by an evolving gravitational constant. Phys Rev D 102:023529. arXiv:2004.11161
908. Braglia M, Ballardini M, Finelli F, Koyama K (2021) Early modified gravity in light of the H_0 tension and LSS data. Phys Rev D 103:043528. arXiv:2011.12934
909. Ballardini M, Braglia M, Finelli F, Paoletti D, Starobinsky AA, Umiltà C (2020) Scalar-tensor theories of gravity, neutrino physics, and the H_0 tension. JCAP 10:044. arXiv:2004.14349
910. Abadi T, Kovetz ED, Can conformally coupled modified gravity solve the hubble tension?. arXiv:2011.13853
911. Di Valentino E, Mena O, Pan S, Visinelli L, Yang W, Melchiorri A et al (2021) In the realm of the Hubble tension–a review of solutions. Class Quant Grav 38:153001. arXiv:2103.01183
912. Schöneberg N, Franco Abellán G, Pérez Sánchez A, Witte SJ, Poulin V, Lesgourgues J, The H_0 olympics: a fair ranking of proposed models. arXiv:2107.10291
913. Freedman WL et al, The carnegie-chicago hubble program. VIII. An independent determination of the hubble constant based on the tip of the red giant branch. arXiv:1907.05922
914. Chen H-Y, Fishbach M, Holz DE (2018) A two per cent Hubble constant measurement from standard sirens within five years. Nature 562:545–547. arXiv:1712.06531
915. Feeney SM, Peiris HV, Williamson AR, Nissanke SM, Mortlock DJ, Alsing J et al (2019) Prospects for resolving the Hubble constant tension with standard sirens. Phys Rev Lett 122:061105. arXiv:1802.03404
916. Sandage A (1962) The change of redshift and apparent luminosity of galaxies due to the deceleration of selected expanding universes. Astrophys J 136:319
917. McVittie G (1962) Appendix to the change of redshift and apparent luminosity of galaxies due to the deceleration of selected expanding universes. Astrophys J 136:334

918. Loeb A (1998) Direct measurement of cosmological parameters from the cosmic deceleration of extragalactic objects. Astrophys J Lett 499:L111–L114. arXiv:astro-ph/9802122
919. Balbi A, Quercellini C (2007) The time evolution of cosmological redshift as a test of dark energy. Mon Not R Astron Soc 382:1623–1629. arXiv:0704.2350
920. Liske J et al (2008) Cosmic dynamics in the era of extremely large telescopes. Mon Not R Astron Soc 386:1192–1218. arXiv:0802.1532
921. Bolejko K, Wang C, Lewis GF, Direct detection of the cosmic expansion: the redshift drift and the flux drift. arXiv:1907.04495
922. Rybicki GB (2004). Radiative processes in astrophysics Wiley-VCH. https://doi.org/10.1002/9783527618170
923. Madau P, Meiksin A, Rees MJ (1997) 21-CM tomography of the intergalactic medium at high redshift. Astrophys J 475:429. arXiv:astro-ph/9608010
924. Furlanetto S, Oh S, Briggs F (2006) Cosmology at low frequencies: the 21 cm transition and the high-redshift universe. Phys Rep 433:181–301. arXiv:astro-ph/0608032
925. Furlanetto SR, The fundamentals of the 21-cm line. arXiv:1909.13740
926. Bowman JD, Rogers AEE, Monsalve RA, Mozdzen TJ, Mahesh N (2018) An absorption profile centred at 78 megahertz in the sky-averaged spectrum. Nature 555:67–70
927. Chluba J, Thomas R (2011) Towards a complete treatment of the cosmological recombination problem. Mon Not R Astron Soc 412:748. arXiv:1010.3631
928. Ali-Haimoud Y, Hirata CM (2011) HyRec: a fast and highly accurate primordial hydrogen and helium recombination code. Phys Rev D 83:043513. arXiv:1011.3758
929. Tashiro H, Kadota K, Silk J (2014) Effects of dark matter-baryon scattering on redshifted 21 cm signals. Phys Rev D 90:083522. arXiv:1408.2571
930. Muñoz JB, Loeb A (2018) A small amount of mini-charged dark matter could cool the baryons in the early Universe. Nature 557:684. arXiv:1802.10094
931. Barkana R (2018) Possible interaction between baryons and dark-matter particles revealed by the first stars. Nature 555:71–74. arXiv:1803.06698
932. Muñoz JB, Kovetz ED, Ali-Haïmoud Y (2015) Heating of baryons due to scattering with dark matter during the dark ages. Phys Rev D 92:083528. arXiv:1509.00029
933. Barkana R, Outmezguine NJ, Redigolo D, Volansky T (2018) Strong constraints on light dark matter interpretation of the EDGES signal. Phys Rev D 98:103005. arXiv:1803.03091
934. Berlin A, Hooper D, Krnjaic G, McDermott SD (2018) Severely constraining dark matter interpretations of the 21-cm anomaly. Phys Rev Lett 121:011102. arXiv:1803.02804
935. Kovetz ED, Poulin V, Gluscevic V, Boddy KK, Barkana R, Kamionkowski M (2018) Tighter limits on dark matter explanations of the anomalous EDGES 21 cm signal. Phys Rev D 98:103529. arXiv:1807.11482
936. Falkowski A, Petraki K, 21cm absorption signal from charge sequestration. arXiv:1803.10096
937. Creque-Sarbinowski C, Ji L, Kovetz ED, Kamionkowski M (2019) Direct millicharged dark matter cannot explain the EDGES signal. Phys Rev D 100:023528. arXiv:1903.09154
938. Liu H, Outmezguine NJ, Redigolo D, Volansky T, Reviving millicharged dark matter for 21-cm cosmology. arXiv:1908.06986
939. Mathur A, Rajendran S, Ramani H, A composite solution to the EDGES anomaly. arXiv:2102.11284
940. D'Amico G, Panci P, Strumia A, Bounds on dark matter annihilations from 21 cm data. arXiv:1803.03629
941. Liu H, Slatyer TR, Too hot, too cold or just right? implications of a 21-cm signal for dark matter annihilation and decay. arXiv:1803.09739
942. Natwariya PK, Bhatt JR (2020) EDGES signal in the presence of magnetic fields. Mon Not R Astron Soc 497:L35–L39. arXiv:2001.00194
943. Feng C, Holder G (2018) Enhanced global signal of neutral hydrogen due to excess radiation at cosmic dawn. Astrophys J Lett 858:L17. arXiv:1802.07432
944. Ewall-Wice A, Chang T-C, Lazio J, Dore O, Seiffert M, Monsalve R (2018) Modeling the Radio Background from the First Black Holes at Cosmic Dawn: Implications for the 21 cm Absorption Amplitude. Astrophys J 868:63. arXiv:1803.01815

945. Pospelov M, Pradler J, Ruderman JT, Urbano A (2018) Room for new physics in the rayleigh-jeans tail of the cosmic microwave background. Phys Rev Lett 121:031103. arXiv:1803.07048
946. Hills R, Kulkarni G, Meerburg PD, Puchwein E (2018) Concerns about modelling of the EDGES data. Nature 564:E32–E34. arXiv:1805.01421
947. Bradley RF, Tauscher K, Rapetti D, Burns JO (2019) A Ground Plane Artifact that Induces an Absorption Profile in Averaged Spectra from Global 21-cm Measurements—with Possible Application to EDGES. Astrophys J 874:153. arXiv:1810.09015
948. Singh S, Subrahmanyan JNTR, Shankar NU, Girish BS, Raghunathan A et al, On the detection of a cosmic dawn signal in the radio background. arXiv:2112.06778

Chapter 4
Thermal Dark Matter

Since there are zillions of Dark Matter models and thus different types of production mechanisms characterized by different masses and couplings, see Fig. 3.8, instead of listing the different possible candidates, we prefer to focus on one of the best motivated and most studied scenario: the one in which Dark Matter is thermally produced. For reviews on DM history, observations, models and searches, we refer to [1–26]. For textbooks, we call attention to [27–33]. The results of the present chapter, in particular Sect. 4.3 which deals with Sommerfeld effects and the unitarity limit, are crucial for Chap. 5 on Homeopathic Dark Matter. All the figures in the chapter are mine (Figs. 4.1, 4.2, 4.3, 4.4, 4.5 and 4.6).

4.1 Production Mechanism

4.1.1 The Boltzmann Equation

The production rate of DM:
The numerous, compelling evidences for the presence of dark matter in our universe raise the question of its nature and of its production mechanism. Probably the most straightforward assumption is that DM is a particle which was at thermal equilibrium with the SM in the early universe. An example of quantum process involving both DM and SM is a pair of SM particles which annihilate into a pair of DM and anti-DM particles.[1] Due to this interaction, DM can be thermally produced out of the primordial SM plasma with a rate per unit of volume

[1] At least at first approximation, the SM should be neutral with respect to the DM quantum charge (if any), hence the SM can only produce $DM\bar{DM}$ but not $DMDM$ or $\bar{DM}\bar{DM}$. with a cross-section $\sigma_{\text{SMSM}\to DM\bar{DM}}$

Y. Gouttenoire, *Beyond the Standard Model Cocktail*, Springer Theses,
https://doi.org/10.1007/978-3-031-11862-3_4

$$\gamma_{\mathrm{SMSM}\to\mathrm{DM}\overline{\mathrm{DM}}} = n_{\mathrm{SM}}^2 \langle \sigma_{\mathrm{SMSM}\to\mathrm{DM}\overline{\mathrm{DM}}} v_{\mathrm{rel}} \rangle , \tag{4.1}$$

where v_{rel} is the relative velocity between the incoming particles and $\langle \, . \, \rangle$ denotes the average over the thermal distributions of the incoming particles

$$\langle \sigma \, v_{\mathrm{rel}} \rangle = \frac{\int \sigma \, v_{\mathrm{rel}} \, e^{-E_1/T} e^{-E_2/T} d^3 p_1 \, d^3 p_2}{\int e^{-E_1/T} e^{-E_2/T} d^3 p_1 \, d^3 p_2} . \tag{4.2}$$

The destruction rate of DM:
DM production is not the only process happening, but instead it competes with the inverse process where DM annihilates into SM and whose rate per unit of volume is

$$\gamma_{\mathrm{DM}\overline{\mathrm{DM}}\to\mathrm{SMSM}} = n_{\mathrm{DM}} n_{\overline{\mathrm{DM}}} \langle \sigma_{\mathrm{DM}\overline{\mathrm{DM}}\to\mathrm{SMSM}} \, v_{\mathrm{rel}} \rangle . \tag{4.3}$$

The change per unit of time in the number of DM particles in a comoving volume is simply the production rate minus the destruction rate

$$\frac{1}{a^3} \frac{d(n_{\mathrm{DM}}^3 a^3)}{dt} = \gamma_{\mathrm{SMSM}\to\mathrm{DM}\overline{\mathrm{DM}}} - \gamma_{\mathrm{DM}\overline{\mathrm{DM}}\to\mathrm{SMSM}} , \tag{4.4}$$

where a is the scale factor of the expanding universe. The SM number density can be safely fixed to their thermal equilibrium value

$$n_{\mathrm{SM}} = n_{\mathrm{SM}}^{\mathrm{eq}} . \tag{4.5}$$

At this stage, it is useful to note that DM is at thermal equilibium when its destruction rate is equal to its production rate, hence leading to a relation between the thermally averaged cross-sections of the direct and inverse process (often called the 'detailed balance condition')

$$\left(n_{\mathrm{SM}}^{\mathrm{eq}} \right)^2 \langle \sigma_{\mathrm{SMSM}\to\mathrm{DM}\overline{\mathrm{DM}}} \, v_{\mathrm{rel}} \rangle = n_{\mathrm{DM}}^{\mathrm{eq}} n_{\overline{\mathrm{DM}}}^{\mathrm{eq}} \langle \sigma_{\mathrm{DM}\overline{\mathrm{DM}}\to\mathrm{SMSM}} \, v_{\mathrm{rel}} \rangle . \tag{4.6}$$

Upon injecting the equilibrium condition (4.6) in the the out-of-equilibrium equation (4.4), we obtain

$$\frac{1}{a^3} \frac{d(n_{\mathrm{DM}} a^3)}{dt} = - \langle \sigma \, v_{\mathrm{rel}} \rangle \left(n_{\mathrm{DM}} n_{\overline{\mathrm{DM}}} - n_{\mathrm{DM}}^{\mathrm{eq}} n_{\overline{\mathrm{DM}}}^{\mathrm{eq}} \right) , \tag{4.7}$$

where $\langle \sigma \, v_{\mathrm{rel}} \rangle$ is short for $\langle \sigma_{\mathrm{DM}\overline{\mathrm{DM}}\to\mathrm{SMSM}} \, v_{\mathrm{rel}} \rangle$.

Two convenient changes of variable:
If we assume that the universe follows an adiabatic expansion, cf. Eq. (3.29) in Chap. 3, then the SM entropy per unit of comoving volume is conserved

$$\frac{d(s a^3)}{dt} = 0 , \tag{4.8}$$

where s is the entropy density. Therefore, it is convenient to trade the variable $n_{\rm DM}a^3$ for $Y_{\rm DM} \equiv n_{\rm DM}/s$ with

$$\frac{1}{a^3}\frac{d(n_{\rm DM}^3 a^3)}{dt} = s\frac{dY_{\rm DM}}{dt}. \tag{4.9}$$

A second advantageous choice is to replace the cosmic time t by the inverse temperature $x \equiv M_{\rm DM}/T$ using $Hdt = da/a$ and

$$d(h_{\rm eff}\,T^3\,a^3) = 0 \quad\rightarrow\quad \frac{da}{a} = \left(1 + \frac{T}{3h_{\rm eff}}\frac{dh_{\rm eff}}{dT}\right)\frac{dx}{x}. \tag{4.10}$$

Finally, we obtain the so-called Boltzmann equation for annihilating DM

$$\frac{dY_{\rm DM}}{dx} = -\frac{s\langle\sigma\, v_{\rm rel}\rangle}{x\,H}\left(Y_{\rm DM}^2 - (Y_{\rm DM}^{\rm eq})^2\right). \tag{4.11}$$

For simplicity, we have omitted the factor $\left(1 + \frac{T}{3h_{\rm eff}}\frac{dh_{\rm eff}}{dT}\right)$ on the right hand side, and we have supposed DM to be symmetric $n_{\rm DM} = n_{\bar{\rm DM}}$.

It is often convenient to extract the temperature dependence from the cross-section by introducing $\langle\sigma\, v_{\rm rel}\rangle = \sigma_0\, x^{-n}$, such that Eq. (4.11) becomes

$$\frac{dY_{\rm DM}}{dx} = -\frac{\lambda}{x^{2+n}}\left(Y_{\rm DM}^2 - (Y_{\rm DM}^{\rm eq})^2\right), \qquad \lambda = M_{\rm pl}\,M_{\rm DM}\,\sigma_0\sqrt{8\pi^2 g_*/45} \tag{4.12}$$

with $g_*^{1/2} \equiv \frac{h_{\rm eff}}{g_{\rm eff}^{1/2}}\left(1 + \frac{1}{3}\frac{T}{h_{\rm eff}}\frac{dh_{\rm eff}}{dT}\right)$. Then, the DM abundance today is given by

$$\Omega_{\rm DM}h^2 = 2_{{\rm DM}\neq \bar{DM}}\frac{h^2\,s_0\,M_{\rm DM}}{3M_{\rm pl}^2 H_0^2}Y_{\rm DM}^{\infty}, \tag{4.13}$$

where s_0 is the entropy today and $H_0/h = 100$ km/s/Mpc. We can either choose the PDG value $s_0 = 2891.2\,{\rm cm}^{-3}$ [34] or the SM prediction $s_0 = 2913\,{\rm cm}^{-3}$ which uses $N_{\rm eff} \simeq 3.045$ [35–37]. The factor 2 stands for DM as being counted as $\Omega_{\rm DM} + \Omega_{\bar{\rm DM}}$. It must be removed if DM is its own anti-particle.

4.1.2 *Freeze-In Versus Freeze-Out*

A simple DM model:
We now propose to compute the DM abundance today in a simple model, where DM is a dirac fermion X which couples to a vector boson V_μ with a coupling constant $\alpha_{\rm D} = g_D^2/4\pi$

$$\mathscr{L} \supset g_D\bar{X}\gamma^\mu X\,V_\mu. \tag{4.14}$$

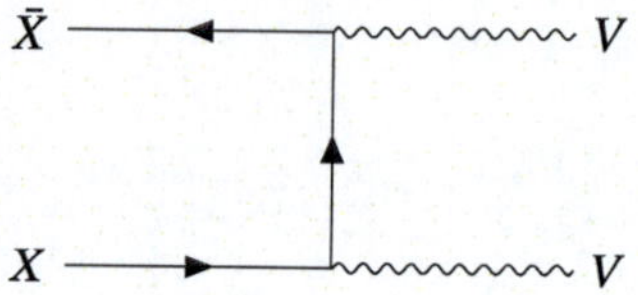

Fig. 4.1 Annihilation of a pair a DM particles X into a pair of hidden vector bosons V. The Feynman diagrams are designed with Tikz-Feynman [38]

We suppose that the vector boson V_μ interacts with the SM through a kinetic coupling with the hypercharge SM boson B_μ

$$\mathscr{L} \supset \varepsilon\, F_{\rm D}\, F_{\rm Y}, \tag{4.15}$$

where ε is the coupling constant, $F_{\rm D\,\mu\nu} = \partial_\mu V_\nu - \partial_\nu V_\mu$ and $F_{\rm Y\,\mu\nu} = \partial_\mu B_\nu - \partial_\nu B_\mu$. We suppose $M_{\rm DM} > m_{\rm V}$ such that X can annihilate with its antiparticle $\bar{X}$ into a pair of V_μ, cf. Fig. 4.1. The corresponding cross-section can easily be computed with FeynCalc [39]

$$\sigma v_{\rm rel} \simeq \begin{cases} \dfrac{\pi\alpha_D^2}{M_{\rm DM}^2}, & S \simeq 4M_{\rm DM}^2, \\ \\ \dfrac{\pi\alpha_D^2}{S}, & S \gg 4M_{\rm DM}^2, \end{cases} \tag{4.16}$$

where $S = -(p_1 + p_2)^2$ is the Mandelstam variable. The thermal average over the thermal distribution of the incoming particles in Eq. (4.2) can be computed with the Gelmini & Gondolo formula [40]

$$\langle \sigma v_{\rm rel} \rangle = \frac{1}{8M_{\rm DM}^4\, T K_2^2\left(\frac{M_{\rm DM}}{T}\right)} \int_{4M_{\rm DM}^2}^{\infty} (S - 4M_{\rm DM}^2)\sqrt{S} K_1\left(\frac{\sqrt{S}}{T}\right) \sigma\, dS, \tag{4.17}$$

where K_α are the hyperbolic Bessel functions of the second kind.[2]

Solutions of the Boltzmann equation (numerical):
The solutions of the Boltzmann equation in Eq. (4.12) are shown in Fig. 4.2, and are compared with the thermal abundance in purple. We can see two different regimes

1. **Freeze-in regime** (blue): The DM-SM interaction strength α_D is so small that DM never reaches thermal equilibrium. Instead, it is constantly produced until the temperature drops below the DM mass, around $x_{\rm FI} \simeq 2$, where then the DM production becomes exponentially suppressed by a Boltzmann factor $\exp(-M_{\rm DM}/T)$. Assuming $M_{\rm DM} = 1$ TeV, the abundance of DM fits the Planck

[2] We can show that it is a relatively good approximation to simply use Eq. (4.16) with $S \simeq 4M_{\rm DM}^2 + 9T^2$, instead of integrating out Eq. (4.17) for each temperature.

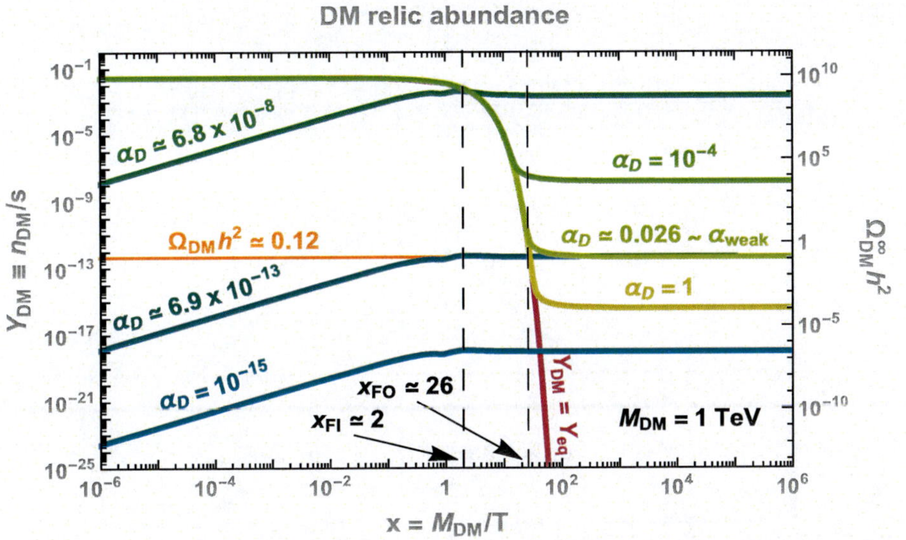

Fig. 4.2 Solutions of the Bolzmann equations for different coupling constant α_D. On the one hand, for small coupling constant, namely for $\alpha_D \lesssim 6.8 \times 10^{-8}$ here in the $U(1)_D$ model (blue shaded lines), the dark matter particle never reaches thermal equilibrium but is instead constantly produced until its production rate becomes smaller than the expansion rate. This is the 'freeze-in' mechanism. On the other hand, for larger coupling constant (gold shaded lines), the dark matter thermalized with the SM. Its abundance today depends on the amount of self-annihilation at the time when it becomes non-relativistic. This is the 'freeze-out' mechanism

data for $\alpha_D \simeq 6.9 \times 10^{-13}$. The corresponding DM candidates are called **Feebly-Interacting Dark Matter (FIMP)** [41, 42].

2. **Freeze-out regime** (yellow): When the DM-SM coupling is larger than $\alpha_D \gtrsim 6.8 \times 10^{-8}$, DM reaches thermal equilibrium. As explained in the previous paragraph, when the temperature becomes smaller than the DM mass, DM production becomes exponentially suppressed. Therefore, DM annihilates massively (drop of the purple line) until its annihilation rate $\Gamma_{\rm ann} \simeq n_{\rm DM}\langle\sigma v_{\rm rel}\rangle$ becomes smaller than the expansion rate of the universe H around $x_{\rm FO} \simeq 30$ (final plateau). For a mass close to the TeV scale, the needed coupling to have the correct relic abundance is close to the weak coupling $\alpha_W \simeq 1/30 \simeq 0.03$. Such DM candidates, supposedly charged under $SU(2)_L$ are called **Weakly-Interacting Dark Matter (WIMP)**. The possible connection between the measured DM abundance and the hierarchy problem which motivates new physics around the TeV scale has been called the **WIMP miracle**. In the literature however, the term WIMP is sometimes enlarged to any **frozen-out thermal DM**: a particle which was once at thermal equilibrium with the SM and whose abundance is set via the freeze-out mechanism. We give more details in Sect. 4.2.

The transition between the FI and the FO regime can be appreciated in Fig. 4.3. The DM abundance grows with the coupling constant in the FI regime, reaches a

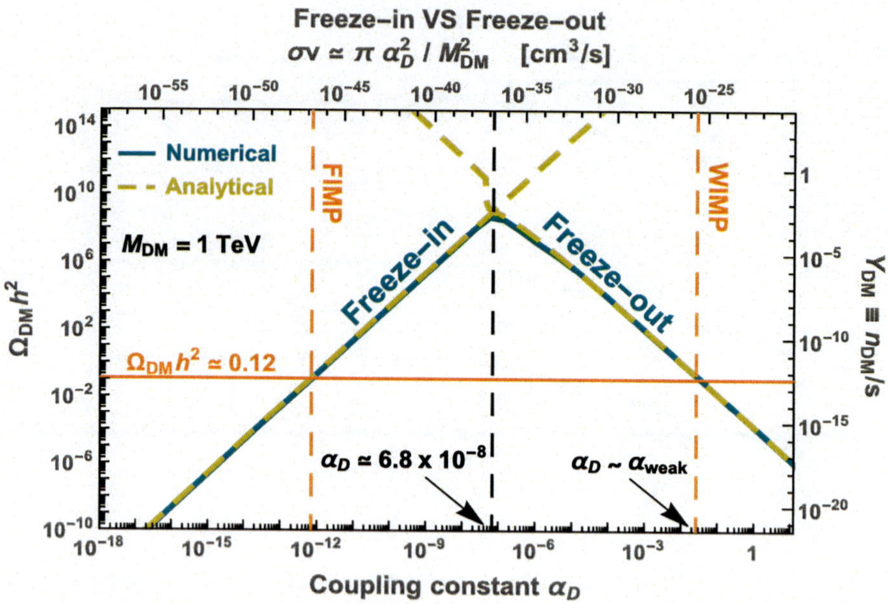

Fig. 4.3 Dark matter abundance as a function of the coupling constant α_D. At small coupling constant, the abundance is set by freeze-in, and the corresponding DM particle is called a **FIMP**. At larger coupling constant, the abundance is set by freeze-out and the corresponding DM particle is called a **WIMP**

maximum and switches to the FO regime where, this time, the abundance decreases with the coupling constant.

Solutions of the Boltzmann equation (analytical):
The DM abundance can also be derived analytically after integrating Eq. (4.12) upon assuming $Y_{\rm DM} \ll Y_{\rm DM}^{\rm eq}$ in the FI regime and upon assuming $Y_{\rm DM} \gg Y_{\rm DM}^{\rm eq}$ in the FO regime

$$Y_{\rm DM} \simeq \begin{cases} \dfrac{\lambda}{x_{\rm FI}^2}\left(Y_{\rm DM}^{\rm eq}(x_{\rm FI})\right)^2, & \text{[freeze-in]}, \\[2ex] \dfrac{(n+1)\,x_{\rm FO}^{n+1}}{\lambda}, & \text{[freeze-out]}, \end{cases} \tag{4.18}$$

where λ is defined in Eq. (4.12). We numerically find the 'freeze-in' temperature $x_{\rm FI} \simeq 2$. The freeze-out temperature $x_{\rm FO}$ which at first-order obeys the equation $n_{\rm DM}\langle\sigma v_{\rm rel}\rangle \simeq H$, is precisely computed semi-analytically in [43]. It is found to be solution of the root equation

$$x_{\rm FO} = \ln\left(0.19\,(n+1)\,\frac{g_{\rm DM}}{\sqrt{g_{\rm eff}}}\,M_{\rm pl}\,M_{\rm DM}\,\sigma_0\right) + \left(n+\frac{1}{2}\right)\ln x_{\rm FO}, \tag{4.19}$$

where g_{DM} is the effective number of degrees of freedom present in g_{DM}^{eq} (e.g. $n_{DM} = \frac{3}{4} \cdot 4$ for a Dirac fermion). See Lee and Weinberg 1977 [44] for a pioneering work on freeze-out calculation.

4.1.3 Exceptions

We discuss thermal production scenarios which deviate from the vanilla freeze-out and freeze-in studied in the previous section.

Resonant DM:
In 1990, Griest and Seckel [45] studied three exceptions to the freeze-out calculation: resonance, threshold and co-annihilation. Annihilation takes place near a pole of the cross-section $S = m_V^2$, where m_V is the mediator mass. The depletion of the DM abundance is enhanced and the cross-section at zero-temperature $\langle \sigma v \rangle_0$ is potentially much smaller than 4×10^{-26} cm^3/s, cf. Eq. (4.24). The maximal effect is obtained when the freeze-out temperature is $x_{FO} = (3u - 2)/(2u(1 - u))$ where $u \equiv (2M_{DM}/m_V)^2$ [45]. For $x_f = 25$, we get $u \simeq 0.98$.

Forbidden DM:
The second exception of Griest and Seckel [45] is when annihilation products are heavier than DM $m_V > M_{DM}$. Hence, DM annihilation at zero-temperature is forbidden [46, 47]. During freeze-out, the kinematic threshold is reached by the tail of the thermal distribution $S > 4m_V^2$ or by 3-to-2 scatterings [48], which allows thermal DM to be in the MeV–10 GeV mass range, and can become a target of low-recoil direct detection experiments [49–51].

Co-annihilating DM:
The last exception of Griest and Seckel [45] is when DM has partner(s) whose mass $m_{\rm partner}$ satisfies $m_{\rm partner} - M_{DM} \lesssim T_{FO}$. Hence the freeze-out of DM and of its partner(s) are inter-connected and a system of Boltzmann equations must be solved. The co-annihilating specie(s) can be either parasitic or symbiotic according to whether they impede or improve the efficiency of DM annhilation [31]. A well-known application is the computation of the relic abundance of the lightest supersymmetric particle (LSP) when the next-to-LSPs are close in mass [52].

Asymmetric DM:
Some mechanism violates the DM quantum number and generates a non vanishing DM asymmetry $n_{DM} \gg n_{\overline{DM}}$ in the very early universe such that the DM comoving abundance is later set by n_{DM}/s. It is interesting to imagine that the same mechanism generating the baryon number, see Sect. 3.5.2 in Chap. 3, could also generate the DM number in such a way that $n_B = n_{DM}$. Hence the observed relation $\Omega_{DM} = 5\Omega_b$ would imply that $M_{DM} \simeq 5$ GeV. In this work, we do not discuss further the possibility that DM is asymmetric and we instead refer the interested reader to existing reviews [15, 53, 54].

Self-annihilating DM:
If DM self-annihilates with itself according to [55]

$$\underbrace{\chi\cdots\chi}_{n} \to \chi\chi, \qquad n \geq 3, \tag{4.20}$$

then DM freezes-out through a cannibalistic phase during which the comoving number of particle N decreases but the comoving entropy $S \propto a^3 T^{1/2} e^{-M_{\rm DM}/T}$ remains constant. The exothermic cannibal scatterings make the temperature decrease much slower than in the standard case

$$\frac{T}{M_{\rm DM}} \simeq \frac{1}{3\log a}. \tag{4.21}$$

$n \geq 3$ scatterings are generally in competition with $n = 2$ scatterings [56, 57]. 3-to-2 scattering arises naturally for Strongly-Interacting Massive Particle (SIMP) in pion EFT from the Wess-Zumino-Witten term [58]. The fact that the confining scale required to reproduce the correct DM abundance is close to the QCD scale has been called the SIMP miracle [59]

$$\begin{cases} \Gamma_{3\to 2} = n_{\rm DM}^2 \langle\sigma v^2\rangle_{3\to 2}\Big|_{T=T_{\rm FO}} = H(T_{\rm FO}), \\ \langle\sigma v^2\rangle_{3\to 2} \sim \alpha_{\rm s}^3/M_{\rm DM}^5 \end{cases} \implies \quad M_{\rm DM} \sim M_{\rm pl}^{1/3} T_{\rm eq}^{2/3} \alpha_{\rm s} \sim 100 \text{ MeV}. \tag{4.22}$$

In the original SIMP model [58], it is assumed that DM and SM are still at thermal equilibrium when DM self-annihilation freezes-out. Instead if DM and SM thermally decouple earlier, then the DM mass prediction becomes dependent on the DM-SM coupling ε instead of the self-coupling $\alpha_{\rm s}$. Such scenario has been called Elastically Decoupling Relic (ELDER) [60, 61] and its realization in $U(1)_{\rm D}$ model kinetically coupled to $U(1)_Y$, cf. Appendix 5.A in Chap. 5, has been called Kinetically Decoupling Relic (KINDER) [62, 63].

Semi-annihilating DM:
DM particles ψ_i annihilate through $\psi_i\psi_j \to \psi_k\phi$ where ϕ is unstable [64]. It has a specific indirect detection signature [65], see Sect. 4.3.4, and can evade the unitarity bound on the mass of DM [66–68].

Homeopathic DM:
The standard lore is that the temperature in the universe evolves adiabatically such that the comoving entropy $S = sa^3$ is conserved. In presence of a mechanism increasing the entropy of the universe by $S \to DS$, the DM relic abundance gets diluted by $1/D$, cf. Chap. 5

$$S \to DS \qquad \implies \qquad \Omega_{\rm DM} \to \frac{\Omega_{\rm DM}}{D}. \tag{4.23}$$

McDonald was the first in 1989 to propose the idea to dilute DM with entropy injection using the decay of a massive particle [69]. This has motivated numerous studies [70–81]. Entropy can also be injected by the reheating phase following a supercooled first-order phase transition (1stOPT) [82–84].

Bubble-wall-interacting DM:
Interactions between the thermal plasma and bubble walls during a 1stOPT can drastically change the DM abundance calculation. E.g. gluon string formation followed by fragmentation [85, 86], see Chap. 7, DM filtering [87, 88], DM squeezing [89, 90], DM production by wall-wall collisions [91] or plasma-wall scattering [92].

4.2 The WIMP Paradigm

Assuming that the universe evolves adiabatically, the freeze-out abundance of a symmetric particle is given by Eqs. (4.13) and (4.18)

$$\frac{\Omega_{\rm DM}^{\rm FO}h^2}{0.1186} \simeq \frac{3.8\times 10^{-9}\ {\rm GeV}^{-2}}{\langle \sigma v_{\rm rel}\rangle} \simeq \frac{4.4\times 10^{-26}\ {\rm cm}^3/{\rm s}}{\langle \sigma v_{\rm rel}\rangle}, \tag{4.24}$$

where we have set $M_{\rm DM} = 1$ TeV and $n = 0$. We have included the factor 2 in Eq. (4.13) due to DM being non-self-conjugate.

4.2.1 *Motivations*

We here discuss the two main reasons why the 'frozen-out thermal dark matter', has been the subject of such extensive studies for the last decades.

(1) **A generic prediction**: A first remarkable feature of the freeze-out mechanism is that the final DM abundance in Eq. (4.24) does not depend on the DM mass.[3] Hence, the prediction for the DM abundance is very generic, and only depends on the value of the annihilation cross-section which must be of order 10^{-26} cm^3/s. More precise values depending on the number of fermionic degrees of freedom and on the velocity-dependence of the cross-section are given in Table 4.1.
(2) **The WIMP miracle**: If we assume the generic expression $\langle \sigma v_{\rm rel}\rangle \simeq \frac{\pi\alpha^2}{M_{\rm DM}^2}$ for the DM anihilation cross-section, then we recover the DM value 10^{-26} cm^3/s with $M_{\rm DM} \sim$ TeV and $\alpha \sim \alpha_W \simeq 1/30$, see Figs. 4.2 and 4.3, which are exactly the values hinted by the 'Naturalness' or 'Hierarchy' problem, see Sect. 2.3.1 in Chap. 2. The common point of most of the solutions to the Naturalness problem

[3] A slight dependence of the DM abundance on the DM mass arises from the change in effective number of relativistic degrees of freedom, $\Omega_{\rm DM} \propto x_{\rm FO}^{n+1} g_*^{-1/2}(M_{\rm DM}/x_{\rm FO})$, cf. Eq. (4.18), and from the logarithmical dependence of $x_{\rm FO}$ on $M_{\rm DM}$, cf. Eq. (4.19). See e.g. [93] for a plot showing this dependence.

is the prediction of new physics at the TeV scale, e.g. susy partners in Supersymmetry, resonances and top partners for Composite Higgs, Kaluza-Klein modes for models with extra-dimensions. This often comes along with the prediction of a Dark Matter candidate if one of the new particles is protected against decay with a symmetry, e.g. R-parity in SUSY or Z_3-symmetry in warped GUT scenario [94, 95], which are both needed for proton stability, or the KK parity in models with universal extra-dimensions [96, 97], or the T-parity in little Higgs models [98]. The connection between the measured DM abundance and solutions to the Hierarchy problem is often called the **WIMP miracle**.

4.2.2 *The WIMP Abundance*

We now study how the abundance of thermal DM varies with respect to its mass $M_{\rm DM}$, for a fixed coupling. We assume that the Dark Matter freezes-out after self-annihilation through weak interactions with the generic cross-section [99]

$$\sigma v_{\rm rel} \simeq \kappa\, \pi\, \alpha_{\rm w}^2 \frac{M_{\rm DM}^2}{(S - m_{Z'}^2)^2 + m_{Z'}^2 \Gamma_{Z'}^2}, \tag{4.25}$$

with $\alpha_{\rm w} \equiv g_{\rm w}^2/4\pi \simeq 1/30$ and where κ is a model-dependent parameter[4] which we allow to vary within the range $10^{-4} \lesssim \kappa \lesssim 10^4$. The vector boson Z' can be either the SM boson[5] Z_0 or an hidden vector boson. For simplicity, we assume its decay width to be $\Gamma_{Z'} \simeq m_{Z'}$. The model in Sect. 4.1.2 whose cross-section is given in Eq. (4.16) is encompassed by Eq. (4.25) in the limit $M_{\rm DM} \gtrsim m_{Z'}$.

The DM abundance computed from Eq. (4.13) in the freeze-out regime is displayed in Fig. 4.4 for a wide range of masses. We can see three different behaviors.

1. $M_{\rm DM} \gtrsim T_{\rm FO}$ and $M_{\rm DM} \gtrsim m_{Z'}$: this is the case already studied earlier in Sect. 4.1.2. The annihilation cross-section is $\sigma v_{\rm rel} = \pi \alpha_D^2 / M_{\rm DM}^2$ such that the DM abundance, see Eq. (4.24), is $\Omega_{\rm DM} \propto M_{\rm DM}^2$. This is the **vanilla WIMP scenario** where $M_{\rm DM} \sim$ TeV for electroweak-like coupling. The mass can not be larger than ~ 100 TeV due to unitarity, see Sect. 4.3.4.
2. $M_{\rm DM} \gtrsim T_{\rm FO}$ and $M_{\rm DM} \lesssim m_{Z'}$: the annihilation cross-section goes like $\sigma v_{\rm rel} \propto M_{\rm DM}^2$ such that the abundance, set by the FO mechanism in Eq. (4.24), decreases like $\Omega_{\rm DM} \propto M_{\rm DM}^{-2}$.
3. $M_{\rm DM} \lesssim T_{\rm FO}$: this is the **Warm Dark Matter scenario**, which we discuss in the next section.

[4] In the case where DM would be neutrinos (excluded by various bounds), we have $\sigma v_{\rm rel} \simeq \frac{c_2}{2\pi} G_{\rm F}^2 M_{\rm DM}^2$, with $c_2 \simeq 5$ [43] and $G_{\rm F} \equiv \sqrt{2} g_{\rm w}^2 / 8 m_{\rm W}^2 \simeq 1.166 \times 10^{-5}\,{\rm GeV}^{-2}$, which corresponds to Eq. (4.25) with $\kappa \simeq 4$.

[5] The case where DM annihilates directly into the SM gauge boson Z_0 is well constrained by direct detection and collider bounds [23, 100]. Note however the possibility to revive this scenario using DM dilution via entropy injection [79].

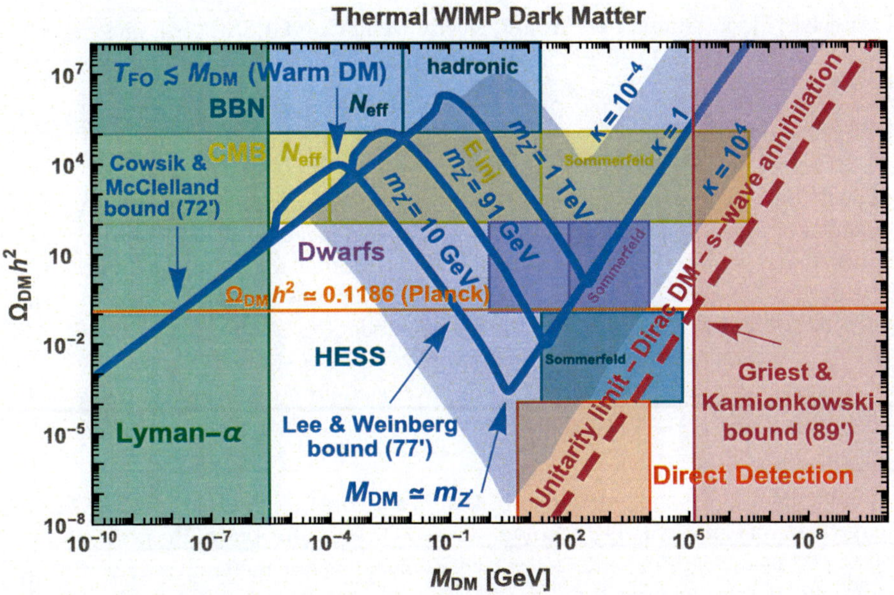

Fig. 4.4 Abundance of WIMP thermal Dark Matter with generic annihilation cross-section defined in Eq. (4.25). Thermal Warm DM is contrained by Lyman-α forests [101–103]. Thermal Cold DM is constrained by BBN [104–111], CMB [112–114], indirect detection experiments, among gamma-ray from Dwarfs by Fermi-Lat [115], gamma-ray from Galactic Center by HESS [116], or neutrino by ANTARES [117–119], and direct detection constraints from XENON1T [120, 121]. The BBN bounds from photo-dissociation due to hadronic injection stops around the deuterium threshold 2.22 MeV. The CMB bounds due to energy injection stops at the WDM limit, when the DM abundance becomes independent of the annihilation cross-section below $T_{FO} \lesssim M_{DM}$ (here we considered the case $m_{Z'} \simeq 10$ GeV). Lower masses are constrained by the N_{eff} bound. The regions labeled "Sommerfeld" include the Sommerfeld enhancement factor of the annihilation cross-section, cf. Sect. 4.3. For all the constraints, we have assumed $\Omega_{DM}h^2 \simeq 0.12$. Masses larger than $M_{DM} \gtrsim$ 100 TeV are constrained by unitarity, see Sect. 4.3.4. The requirement of not overclosing the universe implies all sorts of bounds. The Cowsik-McClelland upper bound on hot DM, $M_{DM} \lesssim 15$ eV [122]. The Lee-Weinberg lower bound on cold DM with weak-scale mediator, $M_{DM} \gtrsim 1$ GeV [44, 123–125]. The Griest-Kamionkowski upper bound due to unitarity, $M_{DM} \lesssim 100$ GeV, see Sect. 4.3.4

4.2.3 Minimal WIMP Under Pressure

Indirect detection:

Indirect detection experiments search for the products of the self-annihilation or decay of dark matter particles in region of high dark matter density, e.g. the milky-way galaxy or its satellites, the dwarf galaxies. We refer to Appendix 5.C of Chap. 5 for a computation of the cosmic-ray flux at Earth. DM could be detected indirectly through an excess of cosmic rays. A major difficulty inherent in such searches is that various astrophysical sources can mimic the signal expected from DM. In Fig. 4.4, we show DM constraints resulting from measurements of gamma rays fluxes from Dwarfs by

Fermi-lat [115], gamma rays from Galactic Center by HESS [116], and neutrinos by ANTARES [117–119]. For reviews on indirect detection, we call attention to [126–136].

Direct detection:
Direct detection experiments aim to observe low-energy recoils (typically a few keVs) of nuclei induced by interactions with particles of dark matter. The non-detection of would-be scintillation light or would-be phonons induced by such recoils implies an upper bound on the DM-proton cross-section. In order to limit interferences from cosmic rays, such experiments are conducted deep underground where the background is minimized. In Fig. 4.4, we show constraints on DM from XENON1T [120, 121]. For reviews on direct detection, we refer to [137–143]. In recent years, a lot of experiments with lower energy recoils have been proposed which allow to search for sub-GeV DM. We refer the interested reader to the dedicated reviews [49–51].

Collider:
Another approach is the production of DM in a laboratory. Particle colliders like the Large Hadron Collider (LHC) could produced DM in collisions of the proton beams. Since DM have negligible interactions with normal visible matter, it may be detected indirectly as missing energy and momentum that escape the detectors. For reviews on DM searches at colliders, we draw attention to [144–150],

BBN and CMB:
An alternative approach to the detection of dark matter particles in nature is to look for deviation from the ΛCDM expectation in BBN and CMB observables. At large DM mass, the DM annihilation cross-section is constrained from the amount of energy injection at the BBN [104–111] or CMB epoch [112–114, 151, 152]. At smaller DM masses or in presence of light mediators, the dark sector could contribute to the effective number of degrees of freedom in the early Universe, potentially detectable in BBN or CMB observations [153–161]. We show the BBN and CMB constraints on thermal DM in Fig. 4.4. In the next Sect. 4.2.4, we discuss other cosmological probes of DM, at later times, in large scale structure observables.

This thesis:
Under the various contraints, it seems that minimal WIMP models are under pressure. The parameter space can however be re-opened using secluded models with an heavy/feebly interacting mediator [73, 77, 153, 162–168]. In this work, we study the possibility to relax the DM overclosure bound above $\sim$100 TeV, by diluting the DM abundance using entropy injection. We explore two directions.

1. Entropy injection due to the decay of the DM mediator, see Chap. 5. The universe is reheated after a matter era.
2. Entropy injection due to a supercooled 1st order cosmological phase transition, see Chap. 7. The universe is reheated after a vacuum-dominated era.

Such scenarios involve non standard cosmologies and therefore can be probed using the associated imprint on the GW spectrum from Cosmic Strings if such GW are observed with future GW detectors, see Chaps. 8 and 9.

4.2.4 Warm Dark Matter

In the previous section we have discussed impact of DM in BBN and CMB observables. DM could also lead to deviation from ΛCDM much later, during structure formation. We will discuss the possibility that DM is warm (WDM), meaning that it was relativistic during a substantial amount of time in the early universe and structure formation is damped at small scales. The lowest scale at which structures can form is limited by the comoving distance over which DM has moved during cosmologicall history, the so-called free-streaming length $\lambda_{\rm FS}$.

Free-streaming length:
On general grounds, after being produced the velocity of DM evolves according to

$$v(a) = \begin{cases} 1, & a < a_{\rm NR} \\ \left(\frac{a_{\rm NR}}{a}\right)^{\frac{1}{2}}, & a_{\rm NR} < a < a_{\rm KD}, \\ \frac{a_{\rm KD}}{a}, & a_{\rm KD} < a, \end{cases} \tag{4.26}$$

where $a_{\rm NR}$ and $a_{\rm KD}$ are the scale factors when DM becomes non-relativistic and decouples kinetically from the thermal bath, respectively. As a result of their thermal velocity, the WDM particles slow down the growth of structures at a scale smaller than their **free-streaming length**, $\lambda_{\rm FS}$, defined by [43]

$$\lambda_{\rm FS}(t) = \int_0^t \frac{v(t')}{a(t')} dt'. \tag{4.27}$$

We are interested in its value just after matter-radiation equality at $t_{\rm eq}$ when structures start to form

$$\lambda_{\rm FS}(t_{\rm eq}) = \int_0^{t_{\rm NR}} \frac{dt}{a(t)} + \int_{t_{\rm KD}}^{t_{\rm NR}} \frac{a_{\rm NR}^{\frac{1}{2}}}{a^{\frac{3}{2}}(t)} dt + \int_{t_{\rm KD}}^{t_{\rm eq}} \frac{a_{\rm KD}}{a^2(t)} dt \tag{4.28}$$

$$\simeq 2\frac{t_{\rm NR}}{a_{\rm NR}} \left[1 + 2\left(\frac{t_{\rm KD}}{t_{\rm NR}}\right)^{\frac{1}{4}} \left(1 - \left(\frac{t_{\rm NR}}{t_{\rm KD}}\right)^{\frac{1}{4}} + \frac{1}{4} \ln \frac{t_{\rm eq}}{t_{\rm KD}}\right)\right], \tag{4.29}$$

$$\simeq 2\frac{t_{\rm eq}}{a_{\rm eq}} \frac{T_{\rm eq}}{T_{\rm NR}} \left[1 + 2\left(\frac{T_{\rm NR}}{T_{\rm KD}}\right)^{\frac{1}{2}} \left(1 - \left(\frac{T_{\rm KD}}{T_{\rm NR}}\right)^{\frac{1}{2}} + \frac{1}{2} \ln \frac{T_{\rm KD}}{T_{\rm eq}}\right)\right], \tag{4.30}$$

where we used $a(t) = (t/t_{\rm NR})^{\frac{1}{2}} a_{\rm NR}$. We now consider two possible scenarios of warm dark matter: relativistic freeze-out and delayed kinetic decoupling.

Relativistic freeze-out:
When $M_{\rm DM} \lesssim T_{\rm FO}$, the DM freezes-out while being relativistic, cf. left part of Fig. 4.4, similar to SM neutrinos which decouple from the $e^{\pm}$, γ bath around

$T_{\text{dec},\,\nu} \sim 1$ MeV. In that case the abundance grows linearly with the DM mass, cf. Eq. (3.43) in Chap. 3

$$\Omega_{\text{DM}} h^2 \simeq \frac{M_{\text{DM}}}{94\ \text{eV}} \frac{11}{4} \left(\frac{T_{\text{DM}}}{T_\gamma} \right)^3 . \tag{4.31}$$

The ratio of temperatures accounts for any possible asymmetric entropy injection in one sector and not the other. We suppose that kinetic decoupling takes place during the relativistic phase

$$T_{\text{KD}} > T_{\text{NR}}, \tag{4.32}$$

such that there is no extra heating due to interaction with the thermal bath in the non-relativistic phase, cf. second line in Eq. (4.26). This implies that the DM temperature can evolve differently from the SM temperature due to non-adiabatic processes in one of the two sectors, see e.g. neutrinos heating in Sect. 3.2.2 of Chap. 3. Hence we denote the DM temperature with a superscript 'DM'. Fixing $T_{\text{KD}}^{\text{DM}} = T_{\text{NR}}^{\text{DM}}$ in Eq. (4.30) leads to

$$\lambda_{\text{FS}}(t_{\text{eq}}) \simeq 2 \frac{t_{\text{eq}}}{a_{\text{eq}}} \frac{T_{\text{eq}}^{\text{DM}}}{T_{\text{NR}}^{\text{DM}}} \left[1 + \ln \frac{T_{\text{NR}}^{\text{DM}}}{T_{\text{eq}}^{\text{DM}}} \right] . \tag{4.33}$$

If the DM is thermally produced then the average DM momentum follows

$$\langle p \rangle = 3.15\, T_{\text{DM}} \tag{4.34}$$

and DM becomes non-relativistic at the temperature[6]

$$T_{\text{NR}}^{\text{DM}} \simeq M_{\text{DM}}/3. \tag{4.35}$$

Upon plugging $a_{\text{eq}}^{-1} = 3402(26)$, $t_{\text{eq}} = 51.1(8)$ kyrs and $T_{\text{eq}} \simeq 0.8$ eV [34] into Eq. (4.33), we obtain

$$\lambda_{\text{FS}}(t_{\text{eq}}) \simeq 0.69\ \text{Mpc} \left(\frac{3\ \text{keV}}{M_{\text{DM}}} \right) \left(\frac{T_{\text{eq}}^{\text{DM}}}{T_{\text{eq}}} \right) \left[1 + 0.12 \ln \left(\frac{M_{\text{DM}}}{3\ \text{keV}} \right) \left(\frac{T_{\text{eq}}}{T_{\text{eq}}^{\text{DM}}} \right) \right] . \tag{4.36}$$

We can use Eq. (4.31) to relate the DM temperature $T_{\text{eq}}^{\text{DM}}$ to the photon temperature T_{eq}

$$T_{\text{eq}}^{\text{DM}} = \left(\frac{4}{11} \frac{94\ \text{eV}}{M_{\text{DM}}} \Omega_{\text{DM}} h^2 \right)^{1/3} T_{\text{eq}}, \tag{4.37}$$

[6] WDM bounds from structure-formation can also be applied to non-thermal DM candidate. For those cases, the thermal relation $\langle p \rangle \big|_{\text{thermal}} = 3.15\, T_{\text{DM}}$ and the temperature Eq. (4.35) when DM becomes non-relativistic must be replaced accordingly.

so that we finally obtain[7]

$$\lambda_{\rm FS}(t_{\rm eq}) \simeq 0.098\ {\rm Mpc}\left(\frac{\Omega_{\rm DM}h^2}{0.12}\right)^{1/3}\left(\frac{3\ {\rm keV}}{M_{\rm DM}}\right)^{4/3}. \tag{4.38}$$

As a result, the measurement of the **matter power spectrum** using Lyman-α forests provides a lower bound on the mass of thermal WDM [101–103]

$$\lambda_{\rm FS}(t_{\rm eq}) \lesssim 0.1\ {\rm Mpc} \qquad \rightarrow \qquad M_{\rm WDM}\Big|_{\rm ly-\alpha} \gtrsim 3\ {\rm keV}. \tag{4.39}$$

Potential candidates for WDM are sterile neutrinos, see [169–171] for reviews. Note that the previous bounds only applies for thermal DM, namely when the relativistic momentum obeys $\langle p \rangle = 3.15\ T$. Otherwise, they must be derived case by case, e.g. resonant production of sterile neutrino from active-sterile neutrino mixing [172, 173], hidden sector hotter than the SM [174], WDM production from primordial black hole evaporation [175, 176], or FIMP DM [177].

Delayed kinetic decoupling:
In presence of large interactions with the thermal bath, after being produced the temperature of DM continues to follow the temperature of the plasma until the momentum transfer rate Γ between DM and the thermal bath

$$\Gamma = n_{\rm rad}\langle \sigma_{\rm rad-DM} v_{\rm rel}\rangle \frac{T}{M_{\rm DM}}, \tag{4.40}$$

falls below the Hubble rate. This defines the **kinetic decoupling** temperature $T_{\rm KD}$. We suppose that kinetic decoupling is so delayed that it takes place deep inside the non-relativistic phase

$$T_{\rm KD} \ll T_{\rm NR}, \tag{4.41}$$

so that Eq. (4.30) becomes

$$\lambda_{\rm FS}(t_{\rm eq}) \simeq 4\frac{t_{\rm eq}}{a_{\rm eq}}\frac{T_{\rm eq}}{T_{\rm NR}}\left(\frac{T_{\rm NR}}{T_{\rm KD}}\right)^{\frac{1}{2}} \ln\frac{T_{\rm KD}}{T_{\rm eq}}, \qquad T_{\rm NR} \simeq M_{\rm DM}/3. \tag{4.42}$$

Plugging $a_{\rm eq}^{-1} = 3402(26)$, $t_{\rm eq} = 51.1(8)$ kyrs and $T_{\rm eq} \simeq 0.8$ eV [34], we obtain

$$\lambda_{\rm FS}(t_{\rm eq}) \simeq 2.5\ {\rm kpc}\left(\frac{\rm keV}{T_{\rm KD}}\right)^{1/2}\left(\frac{\rm MeV}{M_{\rm DM}}\right)^{1/2} \ln\frac{T_{\rm KD}}{T_{\rm eq}}. \tag{4.43}$$

Suppression of structure formation at the kpc scale is too small for being probed by Lyman-α observations. However it translates into a suppression of the number of

[7] The full expression which keeps track of the log is $\lambda_{\rm FS}(t_{\rm eq}) \simeq 0.098$ Mpc $\left(\frac{\Omega_{\rm DM}h^2}{0.12}\right)^{1/3}\left(\frac{3\ {\rm keV}}{M_{\rm DM}}\right)^{4/3}\left[1 + 0.097\ln\left(\frac{0.12}{\Omega_{\rm DM}}\right)^{1/3}\left(\frac{M_{\rm DM}}{3\ {\rm keV}}\right)^{4/3}\right]$.

halos with mass smaller than [178–185]

$$M_{\rm cut} \simeq \frac{4\pi}{3}\rho_{\rm M,0}\lambda^3_{\rm FS}(t_{\rm eq}) \tag{4.44}$$

$$\simeq 3 \times 10^3\, M_\odot \left(\frac{T_{\rm KD}}{\rm keV}\right)\left(\frac{M_{\rm DM}}{\rm MeV}\right), \tag{4.45}$$

where $\rho_{\rm M}$ is the cosmological matter density today. Using Press-Schechter formalism, Ref. [185] found a value for $M_{\rm cut}$ which is larger than Eq. (4.45) by 4 orders of magnitude.

We now evaluate the kinetic decoupling temperature in two different scenarios. A first possibility is that DM X remains coupled to dark radiation V_μ through Thomson scattering $XV \to XV$ with cross-section $\sigma_{\rm T} = 8\pi\alpha_{\rm D}^2/(3M_{\rm DM}^2)$, e.g. [167]. Injecting in Eq. (4.40) gives

$$T_{\rm kd} \simeq 1\ {\rm MeV}\left(\frac{0.03}{\alpha_{\rm D}}\right)\left(\frac{M_{\rm DM}}{1\ {\rm TeV}}\right)^{3/2}\left(\frac{g_*}{100}\right)^{1/4}. \tag{4.46}$$

A second possibility is an interaction between DM and SM radiation. We consider the existence of a kinetic mixing ε between $U(1)_{\rm D}$ and $U(1)_{\rm Y}$. Then electrons e with electric charge q acquire a $U(1)_{\rm D}$ charge εq, cf. Appendix 5.A in Chap. 5. Dark matter particles X can be be kept in kinetic equilibrium through $e\,X \to e\,X$ scattering with t-channel dark photon V_μ exchange. At temperature $m_{\rm V} \gtrsim T \gtrsim m_e$, the cross-section reads [186]

$$\langle \sigma_{\rm el} v_{\rm rel}\rangle \sim \frac{\varepsilon^2 \alpha_{\rm EM}\alpha_{\rm D} T^2}{m_{\rm V}^4}, \tag{4.47}$$

and the resulting kinetic decoupling temperature is

$$T_{\rm kd} \simeq {\rm Max}\left[m_e,\ 1\ {\rm MeV}\left(\frac{10^{-3}}{\varepsilon}\right)^{1/2}\left(\frac{0.03}{\alpha_{\rm D}}\right)^{1/4}\left(\frac{m_{\rm V}}{30\ {\rm MeV}}\right)\left(\frac{M_{\rm DM}}{1\ {\rm TeV}}\right)^{1/4}\left(\frac{g_*}{10.75}\right)^{1/8}\right]. \tag{4.48}$$

Below $T < m_e$, the electron number density is Boltzmann suppressed and kinetic coupling can not be maintained. Kinetic decoupling temperature lower than a keV has been considered using interaction of DM with various dark fermion[8] [185, 187, 190–200]. It has been shown that the suppression of the number of halos lighter than Eq. (4.45) can solve the **missing satellite, cusp-core** and **too-big-too-fail** problems [185, 187, 190–200]. We refer to Sect. 3.5.4 in Chap. 3 for a review of small-scale problems.

Finally, let's mention that a tight coupling between DM and a dark radiation component leads to acoustic oscillations and diffusion damping for DM, analogous

[8] DM interactions with SM neutrinos, proposed in [187] are severely constrained [188, 189].

to the baryon-photon fluid, modifying the linear matter power spectrum. We refer to [196, 201–205] for dedicated studies. We also call attention to [206, 207] in which the fundamental parameters of DM at the Lagrangian level are mapped into physical effective parameters that shape the linear matter power spectrum.

4.3 Heavy WIMP

Assuming a typical DM annihilation cross-section of the form $\sigma_{\rm ann} v_{\rm rel} = \pi\alpha^2/M_{\rm DM}^2$, we can see that the condition of satisfying the correct DM relic abundance in Eq. (4.24) leads to

$$\alpha \simeq 0.3 \left(\frac{M_{\rm DM}}{10\ \text{TeV}}\right). \tag{4.49}$$

Hence, the larger the DM mass, the larger the coupling constant, which leads to two difficulties.

1. The validity of perturbation theory breaks down. Non-perturbative effects have to be taken into account: the Sommerfeld enhancement of the annihilation cross-section and the formation of bound-states.
2. The unitarity of the theory is violated when the probability, $P_{\rm ann} \propto \alpha^2$, for a given pair of DM particles to annihilate is larger than 1.

4.3.1 *Breakdown of Perturbation Theory*

An infinite number of exchanged mediator:
In order to simplify what follows in the present paragraph, we replace the charged fermion DM X by a charged scalar X and the vector boson V_μ by a scalar boson ϕ, and the interaction in Eq. (4.14) becomes

$$\mathscr{L} \supset g\,\phi\, X^\dagger X. \tag{4.50}$$

We now want to study what happens when the coupling constant $\alpha = g^2/(4\pi)^2$ becomes large, which corresponds to the mass of the corresponding thermal Dark Matter candidate being close to the unitarity bound.

Using Feynman rules, we can estimate the amplitude of the double boson exchange relative to the amplitude of the single boson exchange, see Fig. 4.5

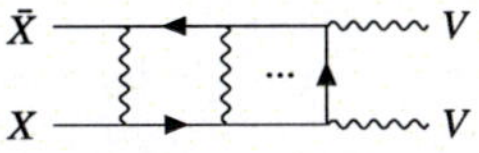

Fig. 4.5 Infinite ladder of mediator exchanges, which become relevant in the long range/small velocity/large coupling limit

$$\frac{\alpha\, m_X^2}{(p_1-p_3)^2-m_V^2} + \int \frac{dk^4}{(2\pi)^4}\frac{\alpha\, m_X^2}{k^2-m_V^2}\frac{\alpha\, m_X^2}{(p_3-p_1+k)^2-m_V^2}\frac{1}{(p_1-k)^2-m_X^2}\frac{1}{(p_2+k)^2-m_X^2}. \tag{4.51}$$

In the non-relativistic limit, the typical exchanged momentum and energy between the two incoming particles are of the order of the Bohr momentum $|\vec{k}| \sim \mu\alpha$ and of the would-be bound-state energy $k_0 \sim \mu\alpha^2/2$ [208, 209]. We have introduced the reduced mass $\mu = m_X/2$. Also, the external momenta are of order $p_\mu = \left(m_X + \mu v_{\rm rel}^2/2,\ \pm\mu\vec{v}_{\rm rel}\right)$, with $\mu v_{\rm rel}^2/2 \ll \mu\vec{v}_{\rm rel}$, where we have introduced the relative velocity $v_{\rm rel}$. Hence, we can write

$$\frac{\alpha\, m_X^2}{(\mu\alpha)^2+m_V^2} + 2\pi^2\frac{(\mu\alpha^2/2)}{2\pi}\frac{(\mu\alpha)^3}{(2\pi)^3}\frac{\alpha\, m_X^2}{(\mu\alpha)^2+m_V^2}\frac{\alpha\, m_X^2}{(\mu\alpha)^2+m_V^2}\frac{1}{(\mu v_{\rm rel})^2}\frac{1}{(\mu v_{\rm rel})^2}. \tag{4.52}$$

which in the limit $\mu\alpha \gg m_V$, leads to

$$\frac{1}{\alpha} + \frac{1}{\alpha}\frac{1}{8\pi^2}\left(\frac{\alpha}{v_{\rm rel}}\right)^4. \tag{4.53}$$

We can see that in the **long range/small velocity/large coupling** limit, $\mu\alpha \gg m_V$ and $\alpha \gg v_{\rm rel}$, the perturbation theory breaks down and we must account for all the ladder diagrams.

The relevance of the infinite ladder of mediator exchanges can be understood as bound states (BS) being present in the theory. Non-perturbative effects arise when those bound states start to play a role in the dynamics. Hence, we must account for the presence of non-perturbative effects when the following condition are satisfied.

BS existence:
The BS are present in the theory when the range of the Dark force $1/m_V$, where m_V is the mediator mass, is larger than the size of the would-be bound state system $(\mu\alpha)^{-1}$, μ being the reduced mass $M_{\rm DM}/2$ of the 2-body system

$$\text{BSE:} \qquad \frac{\alpha M_{\rm DM}}{2m_V} \geq 0.84. \tag{4.54}$$

The numerical factor 0.84 is found after numerically solving the Schrödinger equation in the presence of a Yukawa potential [210].

BS formation 1:
Those bound states can form when the kinetic energy of the system in its center of mass $\mu v_{\rm rel}^2/2$ is less than the ground state binding energy $\mu\alpha^2/2$

$$\text{BSF 1:} \qquad \alpha \geq v_{\rm rel}. \tag{4.55}$$

Otherwise the system remains unbound and the two particles just pass through each other. An equivalent formulation of Eq. (4.55) is to say that the incoming wave packets can form BS only if their spatial extension, the De Broglie wavelength $(\mu v_{\rm rel})^{-1}$, is larger than the typical bound state size $(\mu\alpha)^{-1}$.

BS formation 2:
In order to relax to the ground state of negative energy $-\mu\alpha^2/2$, the system with initial kinetic energy $+\mu v_{\rm rel}^2/2$, must emit of vector boson of energy $\mu v_{\rm rel}^2/2 + \mu\alpha^2/2 \lesssim \mu\alpha^2$. This is impossible if the mediator is too massive, hence the second BS formation condition

$$\text{BSF 2:} \qquad m_V \leq M_{\rm DM}(v_{\rm rel}^2 + \alpha^2)/4 \qquad \rightarrow \qquad \frac{M_{\rm DM}\alpha^2}{4m_V} \gtrsim 1. \tag{4.56}$$

Coulomb approximation:
We can neglect the mass of the mediator when the range of the Dark force $1/m_V$ is larger than the De Broglie wavelength $(\mu v_{\rm rel})^{-1}$. This is the Coulomb approximation

$$\text{Coulomb:} \qquad \frac{M_{\rm DM} v_{\rm rel}}{2m_V} \geq 1. \tag{4.57}$$

4.3.2 Sommerfeld Enhancement

The enhancement of the annihilation cross section at low velocities due to the presence of long-range interactions was first proposed by Sommerfeld in 1931 [211]. Sommerfeld enhancement became the center of the attention in the DM community [212–226] in 2008 after the data release from PAMELA satellite [227, 228] showing an overabundant spectrum of cosmic ray positrons with energies between 10 GeV and a few TeV. The positron excess was later confirmed by Fermi-LAT [229] and AMS-02 [230]. Today, the DM explanation of the positron excess seems ruled out by CMB distortion [151, 152, 231], measurements of gamma-ray [232–236] and neutrino fluxes [237–239], and the asphericity of DM halo [240, 241]. Instead, an astrophysical origin seems more likely, e.g. pulsars [242–244] or supernova remnants [245, 246].

On the good side, the enthusiasm for the PAMELA excess has motivated important theoretical works on the theory of Sommerfeld enhancement [219, 247–251]. This has allowed to push computations of the thermal relic abundance [252–254] and of the cosmic-ray flux [77, 126, 167, 255–257] to the multi-TeV DM mass region, and to more precisely determine the parameter space of neutralino DM [258–265].

Warm-up with classical gravity:
Before giving more details about the nature of quantum interactions in the long range/small velocity/large coupling limit, we first introduce its analog in classical

gravity, as proposed in [219]. Consider a point particle in free fall nearby a star of radius R. Neglecting gravity, only particles with impact parameter b smaller than R can hit the star. Hence, the cross-section for the particle to crash on the star is just the geometrical cross-section $\sigma_0 = \pi R^2$. However, after including the **long range** Newton's force, particles can crash on the star for larger impact parameter. Then, the cross-section is $\pi b_{\max}^2$, where $b_{\max}$ is the largest impact parameter leading to a crash. The **conservation of angular momentum** allows to write

$$m\, v\, b_{\max} = m\, v(R)\, R \quad \rightarrow \quad v(R) = \frac{b_{\max}}{R} v, \tag{4.58}$$

where v is the velocity at infinity. Together with **conservation of energy**, Eq. (4.58) gives

$$\frac{1}{2} m\, v^2 = \frac{1}{2} m\, v(R)^2 - \frac{G\, M\, m}{R}, \tag{4.59}$$

$$v^2 = v^2 \frac{b_{\max}^2}{R^2} - 2\frac{G\, M}{R}, \tag{4.60}$$

$$1 = \frac{b_{\max}^2}{R^2} - \left(\frac{v_{\text{esc}}}{v}\right)^2, \tag{4.61}$$

where $v_{\text{esc}}^2 = 2G_{\text{N}}M/R$ is the escape velocity from the surface of the star. Hence, we get

$$\sigma = \pi^2\, b_{\max} = \sigma_0 \left(1 + \frac{v_{\text{esc}}^2}{v^2}\right). \tag{4.62}$$

We conclude that a long-range Newton's interaction **enhances** the cross-section σv by a factor growing as $1/v$. This is a classical counterpart of the **Sommerfeld enhancement** which we discuss in the next section.

Quantum Mechanics:
We now discuss the effects of the presence of a long range interaction, in the small velocity/large coupling limit, in Quantum Mechanics.[9] Suppose that a non-relativistic particle is moving along the z direction with the wavefunction

$$\psi_k^{(0)}(\vec{r}) = e^{ikz}, \tag{4.63}$$

and can be annihilated and converted into an other state due to a short-range interaction at the origin: $H_{\text{ann}} = U_{\text{ann}}\delta(\vec{r})$ [219]. It is easy to convince oneself that the rate of the process will be proportional to the density of presence at the origin $|\psi_k^{(0)}(\vec{r})|^2 = 1$. If we now add a long-range interaction $V(\vec{r})$, the wavefunction of the particle will be distorted and this will change the annihilation rate. More precisely, the wavefunction $\psi_k(\vec{r})$ obeys the Schrodinger equation

[9] 'Small velocity' implies that we can safely work in the non-relativistic limit.

$$\left[-\frac{1}{2M}\nabla^2 + V(\vec{r}) + U_{\text{ann}}\delta(\vec{r})\right]\psi_k = \varepsilon_k\, \psi_k. \tag{4.64}$$

Since the annihilation by the short range potential H_{ann} takes place locally, in $r = 0$, the effective effect of the long-range force is to modify the value of the wave-function at the origin. This changes the cross-section

$$\sigma = \sigma_0\, S_{\text{ann}}, \tag{4.65}$$

by a factor called the **Sommerfeld factor**

$$S_{\text{ann}} = \frac{|\psi_k(0)|^2}{|\psi_k^{(0)}(0)|^2}. \tag{4.66}$$

σ_0 is the perturbative cross-section, computed when taking only into account the short-range potential.

Hulthen potential:
The Sommerfeld enhancement with a Yukawa potential has no analytical solution and must be computed numerically, see Appendix 4.B. However, an analytical solution exists when replacing the Yukawa potential by the Hulthén potential

$$V_H = -\frac{\alpha\, m_*\, e^{-m_* r}}{(1 - e^{-m_* r})}, \qquad \text{where} \qquad m_* = \frac{\pi^2}{6} m_V. \tag{4.67}$$

We report here the resulting Sommerfeld enhancement factor for arbitrary l-wave process [249] (see also [247])

$$S_H^{(l)} = \left|\frac{\Gamma(a^-)\Gamma(a^+)}{\Gamma(1+l+2i\omega))}\frac{1}{l!}\right|^2, \tag{4.68}$$

with $a^\pm \equiv 1 + l + i\omega(1 \pm \sqrt{1 - x/\omega})$, $x \equiv 2\alpha/v_{\text{rel}}$ and $\omega \equiv M_{\text{DM}}v_{\text{rel}}/(2m_*)$. Explicitly, for s-wave annihilation one finds [250]

$$S_H^{(l=0)} = \frac{2\pi\alpha}{v_{\text{rel}}}\frac{\sinh(\pi M_{\text{DM}}v_{\text{rel}}/m_*)}{\cosh(\pi M_{\text{DM}}v_{\text{rel}}/m_*) - \cosh\pi\sqrt{M_{\text{DM}}^2 v_{\text{rel}}^2/m_*^2 - 4M_{\text{DM}}\alpha/m_*}}. \tag{4.69}$$

Coulomb potential:
In the Coulomb limit, $M_{\text{DM}}v_{\text{rel}}/2m_V > 1$, the potential in Eq. (4.67) reduces to the Coulomb potential, and the associated Sommerfeld enhancement factor in Eq. (4.68) becomes

$$S^{\mathrm{C}}_{\mathrm{ann}} = \frac{2\pi\alpha}{v_{\mathrm{rel}}} \frac{1}{1 - e^{2\pi\alpha/v_{\mathrm{rel}}}} P_l \quad \text{with} \quad P_l = \begin{cases} 1 & \text{if } l = 0, \\ \prod_{k=1}^{l} \left[1 + \frac{1}{k^2}\frac{\alpha^2}{v_{\mathrm{rel}}^2}\right] & \text{if } l > 0. \end{cases} \tag{4.70}$$

4.3.3 Bound-State-Formation

The presence of a long-range force also allows for the formation of bound-states. Significant theoretical works on bound state formation (BSF) were accomplished by Petraki et al. [209, 210], which were then applied to study various DM models with a particular focus on Higgs portal [266–272]. Impact of thermal theoretic effects on BSF have been studied by Biondini et al. [273–279] and Binder et al. [280–284], see also [285, 286]. BSF opens additional annihilation channels which deplete further the relic-abundance during freeze-out [287] and add features to the cosmic-ray flux at Earth [288–291]. The maximal mass of thermal DM compatible with the unitarity limit was recomputed more precisely accounting for those non-perturbative effects in [77, 287, 292, 293]. The theory of BSF has been applied to the study of various topics such as colored dark sectors [78, 267, 294–301], asymmetric DM [292, 302, 303], or dark nucleosynthesis [304].

The capture into the ground-state:
When the conditions Eqs. (4.55) and (4.56) are realized the incoming particles are slow enough to get trapped in a bound state and the mediator is light enough to be emitted.

We will consider only capture into the ground state, with quantum numbers $\{n\ell m\} = \{100\}$. Capture to other bound states is possible but is always subdominant either with respect to capture to the ground state or to the DM annihilation [209, 210]. For BSF via emission of a vector boson, the bound-state angular momentum ℓ differs by one unit with respect to that of the scattering state. Therefore, the capture into the $\{100\}$ state is a p-wave process. This is important for the velocity dependence of σ_{BSF} at small DM velocities, as we discuss below. The BSF cross-sections can be expressed as

$$\sigma_{\mathrm{BSF}} v_{\mathrm{rel}} = \frac{\pi \alpha_D^2}{M_{\mathrm{DM}}^2} S_{\mathrm{BSF}} \Pi_{\mathrm{BSF}}, \tag{4.71}$$

where Π_{BSF} includes both the phase-space suppression due to $m_{\mathrm{V}} > 0$, and the enhancement due to the third polarization of the vector boson, $\Pi_{\mathrm{BSF}} = s_{\mathrm{ps}}^{1/2}(3 - s_{\mathrm{ps}})/2$, with $s_{\mathrm{ps}} = 1 - 16 m_{\mathrm{V}}^2 / [M_{\mathrm{DM}}(v_{\mathrm{rel}}^2 + \alpha_D^2)]^2$. S_{BSF} can be computed analytically only in the Coulomb regime [209, 210], which is specified by the condition $M_{\mathrm{DM}} v_{\mathrm{rel}}/2 \lesssim m_{\mathrm{V}}$, cf. (4.57)

$$S^{\mathrm{C}}_{\mathrm{BSF}} = \frac{2\pi/\varepsilon_v}{1 - e^{-2\pi/\varepsilon_v}} \frac{2^9}{3} \frac{e^{-4\arctan(\varepsilon_v)/\varepsilon_v}}{(1 + \varepsilon_v^2)^2}, \tag{4.72}$$

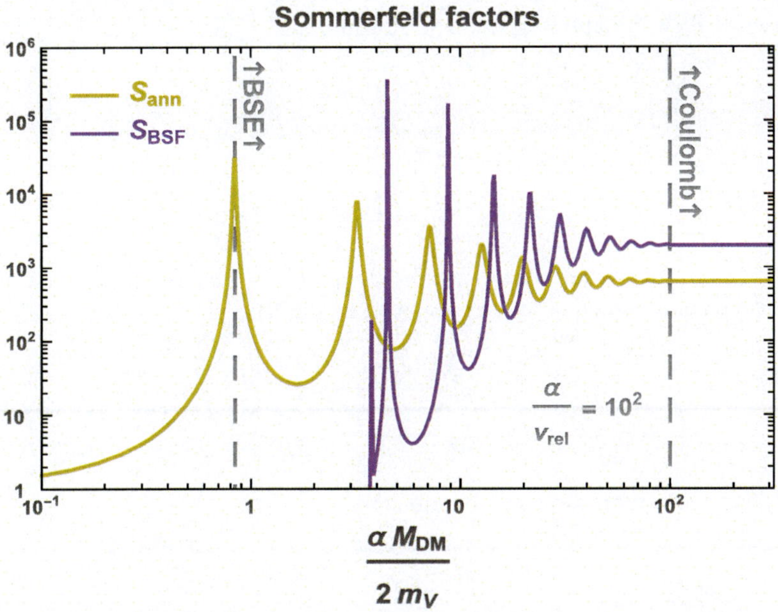

Fig. 4.6 Sommerfeld enhancement factor (gold), cf. Sect. 4.3.2 and Bound-State-Formation factor (purple), cf. Sect. 4.3.3. We assume a Dirac fermion interacting with a vector mediator and only include the capture in the ground state. With dashed vertical lines, we highlight the BSE and Coulomb regime, defined in Eqs. (4.54) and (4.57)

where $\varepsilon_v = v_{rel}/\alpha$. The BSF factor in the Yukawa regime is plotted in Fig. 4.6.[10]

The role of the DM velocity:
In the light mediator / large coupling limit, $2m_V/M_{DM} \lesssim v_{rel} \lesssim \alpha_D$, both S^C_{ann} and S^C_{BSF} scale as $\sim 1/v_{rel}$, and their ratio is $S^C_{BSF}/S^C_{ann} \simeq 3.13$, so that BSF dominates over the direct DM annihilation, cf. Fig. 4.6.

However, away from the Coulomb limit $v_{rel} \lesssim 2m_V/M_{DM}$, S_{ann} and S_{BSF} depend also on $\alpha_D M_{DM}/m_V$, and display a resonant structure along this parameter, as mentioned above [210]. In this regime, and away from resonances, the $1/v_{rel}$ behavior in the Coulomb regime switches to $v^{2\ell}_{rel}$, so that S_{ann} saturates to a constant value while S_{BSF} becomes insignificant, cf. Fig. 4.6.

Close to resonances, the cross sections instead grow with decreasing velocity as $1/v^2_{rel}$, i.e. faster than in the Coulomb limit. This scaling can eventually cause an unphysical violation of the partial-wave unitarity bound on the total inelastic cross section, $\sigma^{(\ell)}_{inel} v_{rel} < 4\pi(2\ell+1)/(v_{rel} M^2_{DM})$, cf. Sect. 4.3.4. The unitarity limit is also seemingly violated for very large α_D. To ensure that our computations do not violate unitarity, we impose the unitarity bound for $\ell = 0$ and $\ell = 1$, as a hard cutoff on the annihilation and BSF cross-sections respectively.

[10] I thank Kallia Petraki for providing the tables for computing the BSF factor. See [209, 210] for the derivation.

The role of BSF in depleting the DM abundance:
The formation of bound states (BSF),

$$\chi + \overline{\chi} \begin{array}{l} \xrightarrow{\sigma_{\rm BSF}/4} (\chi\overline{\chi})^{\uparrow\downarrow}_{\rm BS} + \gamma_{\rm D}, \\ \xrightarrow[3\,\sigma_{\rm BSF}/4]{} (\chi\overline{\chi})^{\uparrow\uparrow}_{\rm BS} + \gamma_{\rm D}, \end{array} \tag{4.73}$$

is either followed by their decay into vector bosons $\gamma_{\rm D}$ (we can neglect the capture into excited states which is suppressed by $1/n^2$ [287]),

$$\begin{aligned} (\chi\overline{\chi})^{\uparrow\downarrow}_{\rm BS} &\xrightarrow{\Gamma_{\uparrow\downarrow}} \gamma_{\rm D}\gamma_{\rm D}, \\ (\chi\overline{\chi})^{\uparrow\uparrow}_{\rm BS} &\xrightarrow{\Gamma_{\uparrow\uparrow}} \gamma_{\rm D}\gamma_{\rm D}\gamma_{\rm D}, \end{aligned} \tag{4.74}$$

or by their photo-ionization due to ambient $\gamma_{\rm D}$,

$$(\chi\overline{\chi})_{\rm BS} + \gamma_{\rm D} \xrightarrow{\sigma_{\rm ion}} \chi + \overline{\chi}. \tag{4.75}$$

BSF counts as a second process participating in the DM depletion only if bound states decay before they get ionized, $\Gamma_{\rm dec} \gtrsim \Gamma_{\rm ion}$. On one side, since the capture into the ground state dominates, the corresponding decay rates are [305]

$$\Gamma_{\uparrow\downarrow} = \alpha^5\mu \quad \text{and} \quad \Gamma_{\uparrow\uparrow} = c_\alpha \alpha^5 \mu, \tag{4.76}$$

with $\mu \equiv m_\chi/2$ and $c_\alpha = 0.12\alpha$. On the other side, the photo-ionization rate can be evaluated using the Milne relation

$$\frac{\sigma_{\rm ion}}{\sigma_{\rm BSF}} = \frac{\mu^2 v_{\rm rel}^2}{2\omega^2}, \tag{4.77}$$

with $\omega = \Delta + \mu v_{\rm rel}^2/2$ and $\Delta = \alpha^2\mu/2$. Indeed, the thermally averaged photo-ionization rate can be expressed as

$$\Gamma_{\rm ion} = \frac{2\cdot 4\pi}{(2\pi)^3} \int_0^\infty \frac{d\omega\, \omega^2}{e^{\omega/T} - 1} \sigma_{\rm ion} = \frac{\mu}{\pi^2} \int_0^\infty \frac{d\omega\, \omega}{e^{\Delta/T} e^{\omega/T} - 1} \sigma_{\rm BSF}. \tag{4.78}$$

One can show [287] that the BS decay rate $\Gamma_{\rm dec}$ is faster than the photo-ionization rate $\Gamma_{\rm ion}$ when the temperature is lower than the BS binding energy $\Delta = \alpha^2\mu/2$,

$$\Gamma_{\rm dec} \gtrsim \Gamma_{\rm ion} \quad \rightarrow \quad T \lesssim \alpha^2\mu/2. \tag{4.79}$$

Upon plugging the freeze-out temperature, $T \simeq M_{\rm DM}/30$, we conclude that BSF participates in the DM depletion if

$$\text{BSF depletes DM if:} \qquad \alpha \gtrsim 0.3. \tag{4.80}$$

Hence, for large coupling $\alpha \gtrsim 0.3$, in the coulomb limit where $S^{\rm C}_{\rm BSF}/S^{\rm C}_{\rm ann} \simeq 3.13$, we expect BSF to additionally deplete the DM abundance by a factor 3 related to the annihilation channel alone. Note however situations where DM depletion through BSF can be more efficient, e.g. thermal field theory corrections [280–284], co-annihilation [285], non-thermal DM [286].

4.3.4 The Unitary Bound

The unitarity of the S-matrix $S^\dagger S$ is just the matrix form of the conservation of the occupation probability of a quantum state $\sum_f P_{i\to f}$. We can split the channels into elastic and inelastic scatterings $P_{\rm elast} + P_{\rm inelast} = 1$. Thanks to the orthonormality properties of the spherical harmonics, the conservation of the probability is even satisfied for each partial wave J, $P^{(J)}_{\rm elast} + P^{(J)}_{\rm inelast} = 1$.[11] Next, the inelastic cross-section of a 2-to-2 process can be decomposed into partial waves with each coefficient being a function of $P^{(J)}_{\rm inelast}$

$$\sigma_{\rm ine} = \sum \sigma^{(J)}_{\rm ine} \quad \text{with} \quad \sigma^{(J)}_{\rm ine} = \frac{\pi(2J+1)}{p_i^2} P^{(J)}_{\rm inelast}. \tag{4.81}$$

This is how in 1990, Griest and Kamionkowski converted the unitarity requirement to an upper-bound on the annihilation cross-section [306] (see Appendix 4.A for a re-derivation)

$$P^{(J)}_{\rm inelast} \lesssim 1 \qquad \to \qquad \sigma^{(J)}_{\rm ine} \lesssim \frac{\pi(2J+1)}{p_i^2}. \tag{4.82}$$

In the early universe, the momentum of the particle i can be written as

$$p_i^2 = E_i^2 - M_{\rm DM}^2 = (\gamma_i^2 - 1) M_{\rm DM}^2 = \frac{M_{\rm DM}^2 v_i^2}{(1-v_i^2)} \approx M_{\rm DM}^2 v_i^2, \tag{4.83}$$

Then the relative velocity $v_{\rm rel}$ is easily related to the individual velocity v_i in the center of mass

$$v_{\rm rel}^2 = (\vec{v}_1 - \vec{v}_2)^2 = 4 v_i^2, \tag{4.84}$$

hence leading to

[11] Partial waves are defined as eigenstates of the square of the angular momentum $\mathbf{J}^2$.

Table 4.1 Precise values of WIMP annihilation cross-section which reproduce the correct DM abundance and the corresponding upper bound on the mass derived from unitarity. We did not include the effect of delayed annihilation and bound states formation which are sub-dominant [287]. Note that the similarity of $M_{\rm DM}^{\rm uni}$ for Majorana and $n = 0$ with $M_{\rm DM}^{\rm uni}$ for Dirac and $n = -1/2$ comes from the almost exact cancellation between the Dirac-to-Majorana factor 2 and the velocity dependence of the non-perturbative (Sommerfeld-enhanced) cross-section $\langle \sigma v_{\rm rel} \rangle \propto 1/v_{\rm rel}$

	Annihilation cross-section which reproduces the correct DM abundance $\langle \sigma v_{\rm rel} \rangle_{\rm FO}$	Maximal DM mass compatible with unitarity $M_{\rm DM}^{\rm uni}$ (TeV)
Majorana and $n = 0$ (s-wave)	2.4×10^{-26} cm^3/s	138
Majorana and $n = -1/2$ (s-wave + Sommerfeld)	1.1×10^{-26} cm^3/s 1	197
Dirac and $n = 0$ (s-wave)	4.9×10^{-26} cm^3/s	96
Dirac and $n = -1/2$ (s-wave + Sommerfeld)	2.4×10^{-26} cm^3/s	137

$$\sigma_{\rm ine}^{(L)} v_{\rm rel} \le \frac{4\pi(2J+1)}{M_{\rm DM}^2 v_{\rm rel}}. \tag{4.85}$$

This leads to an upper bound on the mass of thermal DM. Indeed, assuming that DM freezes-out by annihilating through the partial wave J, we can write

$$\langle \sigma v_{\rm rel} \rangle_{\rm FO} \le \langle \sigma v_{\rm rel} \rangle_{\rm uni}^{\rm max} \quad \to \quad M_{\rm DM}^2 \le \frac{4\pi(2J+1)}{\langle \sigma v_{\rm rel} \rangle_{\rm FO}} \langle \frac{1}{v_{\rm rel}} \rangle, \tag{4.86}$$

where the required annihilation cross-section in order to obtain the correct DM relic abundance, $\langle \sigma v_{\rm rel} \rangle_{\rm FO}$, can be computed from Eqs. (4.13) and (4.18). From using $\langle 1/v_{\rm rel} \rangle = \sqrt{x_{\rm FO}/\pi}$ [306], we obtain

$$M_{\rm DM} \lesssim \begin{cases} 138\sqrt{2J+1}\ \text{TeV}\ f(n), & \text{Majorana}, \\ 96\sqrt{2J+1}\ \text{TeV}\ f(n), & \text{Dirac}, \end{cases} \tag{4.87}$$

where $f(n)$ is a weak function of n, where n encodes the velocity dependence of the cross-section $\sigma = \sigma_0 x^{-n}$, normalized at $f(n=0) = 1$ (e.g. we find $f(n = -1/2) \simeq \sqrt{2}$). We give some values of the maximal thermal DM mass allowed by unitarity, for different velocity dependence n, in Table 4.1.

Appendices

4.A Unitary Bound on Cross-Sections

Clarification of the computation of the unitary bound on $2 \to 2$ cross-section in [306] using results from [307].

4.A.1 Partial-Wave Expansion of the Cross-Section

We consider the scattering $1+2 \to 3+4$. The transition probability amplitude between the two asymptotic states $\langle i|S|f\rangle = \langle p_1, \lambda_1; p_2, \lambda_2|S|p_3, \lambda_3; p_4, \lambda_4\rangle$ can be decomposed as

$$S_{if} = \delta_{if} + i(2\pi)^4\delta^{(4)}(P_i - P_f)T_{if} \tag{4.88}$$

where the matrix element T_{if} contains the non-trivial part. The cross-section reads

$$d\sigma(ab \to cd) = \frac{1}{2E_1 2E_2 v_{\rm rel}} d\,\text{Lips}(\text{s}; \text{P}_3, \text{P}_4) \tag{4.89}$$

where

$$2E_1 2E_2\, v_{\rm rel} = \sqrt{(p_1 \cdot p_2)^2 - m_1 m_2} \equiv 2\lambda^{1/2}(s, m_1^2, m_2^2) \tag{4.90}$$

In the rest frame of particule a, we have $v_{\rm rel} = p_2/E_2 = v_2$ while in the center of mass frame we have

$$v_{\rm rel} = \frac{p\sqrt{s}}{E_1 E_2}. \tag{4.91}$$

We notice that the center of mass initial p_i and final p_f momenta can be expressed as

$$p_i = (4s)^{-1/2}\lambda^{1/2}(s, m_1^2, m_2^2), \quad p_f = (4s)^{-1/2}\lambda^{1/2}(s, m_3^2, m_4^2) \tag{4.92}$$

The Lorentz-invariant phase space is

$$d\,\text{Lips}(\text{s}, \text{P}_3, \text{P}_4) = \frac{d^3 p_3}{(2\pi)^3}\frac{1}{2E_3}\frac{d^3 p_4}{(2\pi)^3}\frac{1}{2E_4}(2\pi)^4\,\delta(P_1 + P_2 - P_3 - P_4) \tag{4.93}$$

$$= \frac{1}{16\pi^2}\frac{1}{E_3 E_4} p^2 dp\, d\Omega\, \delta(s^{1/2} - E_1 - E_2). \tag{4.94}$$

Now using

$$E = (m_3^2 + p^2)^{1/2} + (m_4^2 + p^2)^{1/2} \tag{4.95}$$

$$dE = (E_3^{-1} + E_4^{-1})\, p\, dp = E_3^{-1} E_4^{-1} E\, p\, dp, \tag{4.96}$$

the LIPS becomes

$$d\,\mathrm{Lips}(\mathrm{s}, \mathrm{P}_1, \mathrm{P}_2) = \frac{1}{16\pi^2}\frac{\mathrm{p}}{\mathrm{E}}\delta(\mathrm{s}^{1/2} - \mathrm{E})\mathrm{dEd}\Omega \tag{4.97}$$

and

$$\boxed{d\,\mathrm{Lips}(\mathrm{s}, \mathrm{P}_1, \mathrm{P}_2) = \frac{1}{16\pi^2}\frac{\mathrm{p}}{\sqrt{\mathrm{s}}}\mathrm{d}\Omega} \tag{4.98}$$

Then the differential cross section reads

$$d\sigma(12 \to 34) = \frac{p_f}{4p_i s}\sum_f \left|\frac{T_{if}}{4\pi}\right|^2 d\Omega \tag{4.99}$$

$$= \frac{\pi p_f}{2p_i s}\sum_f \left|\frac{T_{if}}{4\pi}\right|^2 d\cos\theta \tag{4.100}$$

Now, we expand $T_{if}(\theta)$ in terms of Legendre polynomials of $\cos\theta$:

$$T_{if}(\theta) = 8\pi s^{1/2}\sum_{L=0}^{\infty}(2L+1)P_L(\cos\theta)T_{if,L}(s). \tag{4.101}$$

Using the orthogonality relation

$$\frac{1}{2}\int dx\, P_{L'}(x)P_L(x)(2L+1) = \delta_{LL'} \tag{4.102}$$

we obtain the cross section in terms of partial waves $\sigma = \sum \sigma_L$

$$\sigma_L = 4\pi(2L+1)\sum_f \frac{p_f}{p_i}\left|T_{if,L}\right|^2, \tag{4.103}$$

which averaged over polarizations becomes

$$\boxed{\sigma_L = \frac{4\pi(2L+1)}{(2s_1+1)(2s_2+1)}\sum_\lambda\sum_f \frac{p_f}{p_i}\left|T_{if,L}\right|^2.} \tag{4.104}$$

4.A.2 *Unitarity of the Partial-Wave Expansion*

The unitarity means that the Hamiltionian should be Hermitian $H^\dagger = H$. Since the S-matrix is $S = e^{-iHt}$, it implies

$$S^\dagger S = 1 \tag{4.105}$$

and yields to

$$
\begin{aligned}
&- i(2\pi)^4\delta^{(4)}(P_i - P_f)(T_{if} - T^*_{fi}) \\
&= \sum_n \left(\prod_{k=1}^{n} \int \frac{d^3 p_k}{(2\pi)^3} \frac{1}{2E_k} \right) T_{ik}T^*_{fk}(2\pi)^4\delta^{(4)}(P_i - P_k)(2\pi)^4\delta^{(4)}(P_f - P_k),
\end{aligned} \tag{4.106}
$$

where we used Eq. (4.88). We get the generalized optical theorem

$$
\boxed{-i(T_{if} - T^*_{fi}) = \sum_n \left(\prod_{k=1}^{n} \int \frac{d^3 p_k}{(2\pi)^3} \frac{1}{2E_k} \right) T_{ik}T^*_{fk}(2\pi)^4\delta^{(4)}(P_i - P_k)} \tag{4.107}
$$

which holds order by order. Remark: taking $i = f = A$, we obtain

$$
\mathrm{Im\ T(A \rightarrow A) = m_1 \sum_X \Gamma(A \rightarrow X) = m_1 \Gamma_{tot}} \tag{4.108}
$$

if A is a 1-particle state and

$$
\mathrm{Im\ T(A \rightarrow A) = 2E_{cm}\, p_{cm} \sum_X \sigma(A \rightarrow X)} \tag{4.109}
$$

if A is a 2-particles state. In the case of the scattering $12 \rightarrow 34$, by using Eq. (4.98) we get

$$
-i(T_{if}(\Omega) - T^*_{fi}(\Omega)) = \sum_k \int d\ \mathrm{Lips(s; k) T_{ik} T^*_{fk}} \tag{4.110}
$$

$$
= \sum_k \frac{1}{16\pi^2} \frac{p_k}{\sqrt{s}} \int d\Omega' T_{ik}(\Omega') T^*_{fk}(\Omega'') \tag{4.111}
$$

where Ω is the solid angle between p_1 and p_3, Ω' is the solid angle between p_1 and p_k and Ω'' is the solid angle between p_3 and p_k. See Fig. 4.7 for a visual representation.

Now using the expansion in Legendre polynomials in Eq. (4.101), the following property of Legendre polynomials

$$
\int d\phi' P_L(\cos\theta'') = 2\pi P_L(\cos\theta) P_L(\cos\theta') \tag{4.112}
$$

as well as the orthogonality relation in Eq. (4.102), we can write

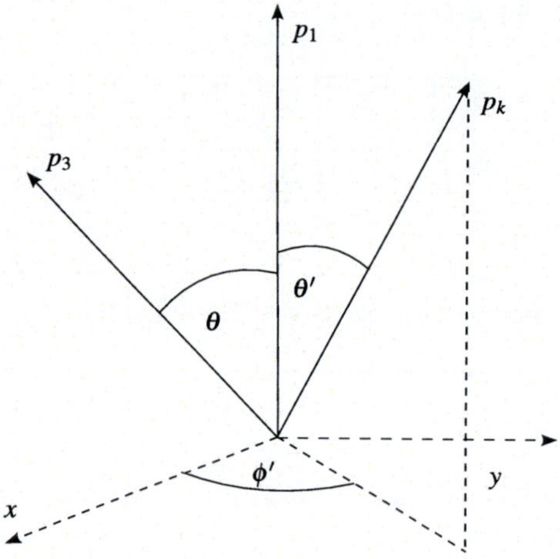

Fig. 4.7 Solid angles Ω, Ω' and Ω'' which appear in the unitarity equation Eq. (4.111)

$$-8\pi s^{1/2}\sum_L (2L+1)P_L(x)i(T_{if,L}-T^*_{fi,L}) \tag{4.113}$$

$$=\sum_k p_k\, 4s^{1/2}\int dx' \sum_{L'}(2L'+1)P_{L'}(x')T_{ik,L'}\sum_L (2L+1)2\pi P_L(x)P_L(x')T^*_{fk,L} \tag{4.114}$$

$$=16\pi s^{1/2}\sum_k p_k \sum_{L'}(2L'+1)P_{L'}(x)T_{ik,L'}T^*_{fk,L'}, \tag{4.115}$$

with $x=\cos\theta$ and $x'=\cos\theta'$. Finally, we get

$$T_{if,L}-T^*_{fi,L}=2i\sum_k p_k\, T_{ik,L}T^*_{fk,L}. \tag{4.116}$$

Using matrix notation, this can be rewritten as

$$\boxed{T_L - T_L^\dagger = 2iT_L\tilde{p}T_L^\dagger} \tag{4.117}$$

with $\tilde{p}=\mathrm{diag}(\ldots p_k\ldots)$ where p_k is the 3-momentum of the intermediate state k. Defining $S_L=1+2i\tilde{p}^{1/2}T_L\tilde{p}^{1/2}$, we see that the partial-wave unitarity can also be written $S_LS_L^\dagger=1$ or

$$\left|S_{el,L}\right|+\sum_f\left|S_{i\neq f,L}\right|^2=1, \tag{4.118}$$

where $S_{el,L}$ stands for the elastic channel, $i=f$. If we define $S_{el,L}=\eta_L e^{2i\delta_L}$, where δ_L is a real phase shift and η_L is an inelaticity factor, $0\le\eta_L\le 1$. Then $\left|S_{el,L}\right|^2=\eta_L^2$, and $\sum_f\left|S_{i\neq f,L}\right|^2=1-\eta_L^2$. Finally, using $T_{el,L}=(S_{el,L}-1)/2ip$ and $T_{f\neq i,L}=$

$S_{f\neq i,L}/2i(p_i p_f)^{1/2}$ and the standard formula for the unpolarized cross section in terms of partial waves $\sigma = \sum \sigma_L$, where σ_L is given in Eq. (4.104), we find

$$\sigma_{r,L} = 4\pi \frac{2L+1}{(2s_1+1)(2s_2+1)} \sum_{\lambda} \sum_{f\neq i} \frac{p_f}{p_i} \left|T_{if,L}\right|^2 \tag{4.119}$$

$$= \frac{\pi(2L+1)(1-\eta_L)}{p_i^2}. \tag{4.120}$$

Here $\sigma_{r,L}$ is the *reaction* cross section, that is, the total cross section minus the elastic piece. It has a maximum when $\eta_L = 0$, so we conclude that

$$\sigma_L(a+b \to c+d) \leq \pi(2L+1)/p_i^2. \tag{4.121}$$

In the early Universe,

$$p_i^2 = E^2 - m_X^2 = \frac{m_X^2 v_{\rm rel}^2}{4(1-v_{\rm rel}^2/4)} \approx \frac{m_X^2 v_{\rm rel}^2}{4}, \tag{4.122}$$

where we used $v_i = p_i/E$ and $v_{\rm rel} = 2v_i$.
So $\sigma_L v_{\rm rel} \leq (\sigma_L)_{\rm max}\, v_{\rm rel}$ where

$$\boxed{(\sigma_L)_{\rm max} v_{\rm rel} \approx \frac{4\pi(2L+1)}{m_X^2 v_{\rm rel}}} \tag{4.123}$$

4.B Computation of the Sommerfeld Factor

4.B.1 The Schrödinger Equation

The non-relativistic potential between the DM wave-functions can be computed as the Fourier transform

$$V(r) = -\frac{1}{4i M_{\rm DM}^2} \int \frac{d\vec{k}^3}{(2\pi)^3} \mathscr{W}(\vec{k})\, e^{-i\vec{k}\vec{r}}, \tag{4.124}$$

of the four-point function amplitude $\mathscr{W}(\vec{k})$ with one-boson mediator exchange [209, 269]. We are interested in the situations where the computation in Eq. (4.124) leads to the static Yukawa potential (long range force)

$$V(\mathbf{r}) = -\alpha \frac{e^{-m_V r}}{r}. \tag{4.125}$$

The DM wave-functions $\psi_k(r)$ are solutions of the Schrödinger equation, which upon neglecting the short range potential reads

$$\left[-\frac{1}{2\mu}\nabla^2 - \alpha\frac{e^{-m_V r}}{r}\right]\psi_k = \frac{k^2}{2\mu}\psi_k, \tag{4.126}$$

where μ is the reduced mass of the $X\overline{X}$ system $\mu = m_X/2$.
We expand $\psi_k(r)$ on Legendre polynomials

$$\psi_k(r) = \frac{1}{k}\sum_i i^l(2l+1)e^{i\delta_l}P_l(\cos\theta)R_{kl}(r), \tag{4.127}$$

$$R_{kl}(r) = \frac{\chi_{kl}(r)}{r}, \tag{4.128}$$

and we introduce dimensionless variables

$$x = \kappa r, \tag{4.129}$$

$$\kappa = \mu\alpha = (\text{Bohr radius})^{-1}, \tag{4.130}$$

$$\xi = \kappa/m_V = \frac{\alpha m_X}{2m_V} = (\text{Force range / BS size}), \tag{4.131}$$

$$\zeta = \kappa/k = \alpha/v_{\text{rel}} = (\text{De Broglie } \lambda \text{ / BS size}), \tag{4.132}$$

such that the Schrödinger equation becomes

$$\chi''(x) + \left[-\frac{l(l+1)}{x^2} + \frac{1}{\zeta^2} + \frac{2e^{-x/\xi}}{x}\right]\chi(x) = 0. \tag{4.133}$$

4.B.2 *Coulomb Potential*

The solution for a massless mediator $\xi = \infty$ is found to be (see Sect. 36 of [308], see also [209, 210])

$$R_{kl}(r) = \frac{C_k}{(2l+1)!}(2kr)^l e^{ikr}F(i/k + l + 1, 2l + 2, -2ikr), \tag{4.134}$$

with the confluent hypergeometric function being such that

$$F(\alpha, \gamma, 0) = 1. \tag{4.135}$$

If we normalize at infinity with the condition

$$\psi_k(r) \to \frac{\sin(kr+\delta)}{r}, \tag{4.136}$$

the normalization constant C_k is found to be

$$C_k = \sqrt{\frac{\pi}{2}} \frac{2\sqrt{k}}{\sqrt{e^{2\pi/k}-1}} \prod_{s=1}^{l} \sqrt{s^2 + \frac{1}{k^2}}. \tag{4.137}$$

Since $R_{kl}(r)$ vanishes for $l \neq 0$, the Sommerfeld enhancement factor, defined in Eq. 4.66, is just

$$S_{\text{ann}} = \left| \frac{R_{k,l=0}(0)}{k} \right|^2, \tag{4.138}$$

which can be expressed as

$$S_{\text{ann}}(\zeta) = \frac{2\pi\zeta}{1 - e^{-2\pi\zeta}}. \tag{4.139}$$

4.B.3 Yukawa Potential

The case with finite vector boson mass does not have any analytical solution. We need to numerically integrate the Schrödinger equation Eq. (4.133) with the boundary conditions

$$\chi(x) \underset{x\to 0}{\longrightarrow} 0, \tag{4.140}$$

$$\chi(x) \underset{x\to \infty}{\longrightarrow} \sin(x/\zeta + \delta). \tag{4.141}$$

The second boundary condition in Eq. (4.141) is difficult to impose numerically. Instead, it is preferable to rely on an different formulation of the Sommerfeld enhancement factor, but equivalent to Eq. (4.66) [219]

$$\tilde{S}_{\text{ann}} = \frac{|\tilde{\chi}(\infty)|^2}{|\tilde{\chi}(0)|^2}, \tag{4.142}$$

where $\tilde{\chi}(x)$ is solution of Eq. (4.133) with boundary conditions

$$\begin{aligned} \tilde{\chi}(x) &\underset{x\to 0}{\longrightarrow} 1, \\ \tilde{\chi}'(x) &\underset{x\to \infty}{\longrightarrow} \frac{i}{\zeta}\tilde{\chi}(x). \end{aligned} \tag{4.143}$$

The latter boundary conditions in Eq. (4.143) are easier to implement numerically than the former in Eqs. (4.140) and (4.141).

Proof:
The equivalence between

$$S_{\text{ann}} = |\psi(0)|^2 = \left|\frac{1}{k}\frac{\chi(0)}{dr}\right|^2, \qquad \text{and} \qquad \tilde{S}_{\text{ann}} = \frac{|\tilde{\chi}(\infty)|^2}{|\tilde{\chi}(0)|^2}, \tag{4.144}$$

can be demonstrated by using the conservation of the Wronskian

$$W(r) = \chi_1'(r)\chi_2(r) - \chi_1(r)\chi_2'(r), \tag{4.145}$$

where χ_1 and χ_2 are solutions of the Schrödinger equation in Eq. (4.133) with boundary conditions

$$\begin{aligned} &\chi_1(x) \underset{x\to 0}{\longrightarrow} 0, && \chi_2(x) \underset{x\to 0}{\longrightarrow} A, \\ &\chi_1(x) \underset{x\to\infty}{\longrightarrow} \sin(x/\zeta + \delta), && \chi_2(x) \underset{x\to\infty}{\longrightarrow} \cos(x/\zeta + \delta). \end{aligned} \tag{4.146}$$

First, notice that $\tilde{\chi}$ introduced Eq. (4.143) can be written as $\tilde{\chi}(x) = \frac{1}{A}(\chi_2(x) + i\chi_1(x))$, such that

$$\tilde{S}_{\text{ann}} = \frac{1}{|A|^2}. \tag{4.147}$$

Second, notice that χ, introduced in Eqs. (4.140) and (4.141), is simply χ_1, such that we can compute S_{ann} from the conservation of the Wronskian

$$W(0) = \chi_1'(0)A = W(\infty) = -k \quad \Rightarrow \quad S_{\text{ann}} = \frac{1}{|A|^2}. \tag{4.148}$$

Hence, the two definitions in Eq. (4.144) are equivalent.

References

1. Jungman G, Kamionkowski M, Griest K (1996) Supersymmetric dark matter. Phys Rep 267:195–373. ArXiv:hep-ph/9506380
2. Bergström L (2000) Nonbaryonic dark matter: observational evidence and detection methods. Rep Prog Phys 63:793. ArXiv:hep-ph/0002126
3. Bergstrom L (2009) Dark matter candidates. New J Phys 11:105006. ArXiv:0903.4849
4. Bergstrom L (2012) Dark matter evidence, particle physics candidates and detection methods. Annalen Phys 524:479–496 ArXiv:1205.4882
5. Olive KA (2003) TASI lectures on dark matter. In: Theoretical advanced study institute in elementary particle physics (TASI 2002): particle physics and cosmology: the quest for physics beyond the standard model(s), pp 797–851, 1. ArXiv:astro-ph/0301505

6. Bertone G, Hooper D, Silk J (2005) Particle dark matter: evidence, candidates and constraints. Phys Rep 405:279–390 ArXiv:hep-ph/0404175
7. Murayama H (2007) Physics beyond the standard model and dark matter. In: Les houches summer school—session 86: particle physics and cosmology: the fabric of spacetime, p 4. ArXiv:0704.2276
8. Steffen FD (2009) Dark matter candidates—axions, neutralinos, gravitinos, and axinos. Eur Phys J C 59:557–588. ArXiv:0811.3347
9. Hooper D (2010) Particle dark matter. In: Theoretical advanced study institute in elementary particle physics: the dawn of the LHC Era, pp 709–764. ArXiv:0901.4090
10. Feng JL (2010) Dark matter candidates from particle physics and methods of detection. Ann Rev Astron Astrophys 48:495–545. ArXiv:1003.0904
11. Roos M, Dark matter: the evidence from astronomy, astrophysics and cosmology. ArXiv:1001.0316
12. Bertone G (2010) The moment of truth for WIMP Dark Matter. Nature 468:389–393 ArXiv:1011.3532
13. Schnee RW (2011) Introduction to dark matter experiments. In: Theoretical advanced study institute in elementary particle physics: physics of the large and the small, pp 775–829. ArXiv:1101.5205
14. Peter AHG, Dark matter: a brief review. ArXiv:1201.3942
15. Zurek KM (2014) Asymmetric dark matter: theories, signatures, and constraints. Phys Rep 537:91–121. ArXiv:1308.0338
16. Profumo S (2013) Astrophysical probes of dark matter. In: Theoretical advanced study institute in elementary particle physics: searching for new physics at small and large scales, pp 143–189. ArXiv:1301.0952I
17. Lisanti M (2017) Lectures on dark matter physics. In: Theoretical advanced study institute in elementary particle physics: new frontiers in fields and strings, pp 399–446. ArXiv:1603.03797
18. Alexander J et al (2016) Dark sectors 2016 workshop: community report, p 8 1608.08632
19. Bauer M, Plehn T (2019) Yet another introduction to dark matter: the particle physics approach, vol 959 of lecture notes in physics. Springer, Berlin. https://doi.org/10.1007/978-3-030-16234-4
20. Bertone G, Hooper D (2018) History of dark matter. Rev Mod Phys 90:045002. ArXiv:1605.04909
21. Freese K (2017) Status of dark matter in the Universe. Int J Mod Phys 1:325–355. ArXiv:1701.01840
22. Roszkowski L, Sessolo EM, Trojanowski S (2018) WIMP dark matter candidates and searches—current status and future prospects. Rep Prog Phys 81:066201. ArXiv:1707.06277
23. Arcadi G, Dutra M, Ghosh P, Lindner M, Mambrini Y, Pierre M et al, The waning of the wimp? a review of models, searches, and constraints. ArXiv:1703.07364
24. Battaglieri M et al (2017) US cosmic visions: new ideas in dark matter 2017: community report. In: U.S. cosmic visions: new ideas in dark matter, p 7. ArXiv:1707.04591
25. Bertone G, Tait TM (2018) A new era in the search for dark matter. Nature 562:51–56. ArXiv:1810.01668
26. Blanco C, Escudero M, Hooper D, Witte SJ (2019) Zmediated WIMPs: dead, dying, or soon to be detected? JCAP 11:024. ArXiv:1907.05893
27. Silk J et al (2010) Particle dark matter: observations, models and searches. Cambridge University Press, Cambridge. https://doi.org/10.1017/CBO9780511770739
28. Einasto J (2013) Dark matter and cosmic web story. World Scientific. https://doi.org/10.1142/8928
29. Majumdar D (2014) Dark matter: an introduction. Taylor & Francis Group, Boca Raton, p 8
30. Vavilova IB, Bolotin YL, Boyarsky AM, Danevich FA, Kobychev VV, Tretyak VI et al (2015) Dark matter: observational manifestation and experimental searches, vol 3. Dark energy and dark matter in the Universe. Akademperiodyka, Kyiv, Ukraine

31. Profumo S (2017) An introduction to particle dark matter. World Scientific. https://doi.org/10.1142/q0001
32. Essig R, Feng J, Zurek K (eds) (2019) Proceedings, of a simons symposium: illuminating dark matter. Kruen, Germany, Springer. https://doi.org/10.1007/978-3-030-31593-1
33. Mambrini Y (2021) Particles in the dark universe: a student's guide to particle physics and cosmology. Springer International Publishing
34. Particle Data Group collaboration, Tanabashi M et al (2018) Review of particle physics. Phys Rev D98:030001
35. Mangano G, Miele G, Pastor S, Pinto T, Pisanti O, Serpico PD (2005) Relic neutrino decoupling including flavor oscillations. Nucl Phys B729:221–234. ArXiv:hep-ph/0506164
36. de Salas PF, Pastor S (2016) Relic neutrino decoupling with flavour oscillations revisited. JCAP 1607:051. ArXiv:1606.06986
37. Escudero Abenza M, Precision early universe thermodynamics made simple: N_{eff} and neutrino decoupling in the standard model and beyond. ArXiv:2001.04466
38. Ellis J (2017) TikZ-Feynman: Feynman diagrams with TikZ. Comput Phys Commun 210:103–123. ArXiv:1601.05437
39. Shtabovenko V, Mertig R, Orellana F, FeynCalc 9.3: New features and improvements. arxiv:2001.04407
40. Gondolo P, Gelmini G (1991) Cosmic abundances of stable particles: improved analysis. Nucl Phys B 360:145–179
41. Hall LJ, Jedamzik K, March-Russell J, West SM (2010) Freeze-in production of FIMP dark matter. JHEP 03:080. ArXiv:0911.1120
42. Bernal N, Heikinheimo M, Tenkanen T, Tuominen K, Vaskonen V (2017) The dawn of FIMP dark matter: a review of models and constraints. Int J Mod Phys A 32:1730023. ArXiv:1706.07442
43. Kolb EW, Turner MS (1990) The early universe. Front Phys 69:1–547
44. Lee BW, Weinberg S (1977) Cosmological lower bound on heavy neutrino masses. Phys Rev Lett 39:165–168
45. Griest K, Seckel D (1991) Three exceptions in the calculation of relic abundances. Phys Rev D 43:3191–3203
46. D'Agnolo RT, Ruderman JT (2015) Light dark matter from forbidden channels. Phys Rev Lett 115:061301. ArXiv:1505.07107
47. D'Agnolo RT, Liu D, Ruderman JT, Wang P-J (2021) Forbidden dark matter annihilations into Standard Model particles. JHEP 06:103. ArXiv:2012.11766
48. Cline JM, Liu H, Slatyer T, Xue W (2017) Enabling forbidden dark matter. Phys Rev D 96:083521. ArXiv:1702.07716
49. Dolan MJ, Kahlhoefer F, McCabe C, Schmidt-Hoberg K (2015) A taste of dark matter: Flavour constraints on pseudoscalar mediators. JHEP 03:171. ArXiv:1412.5174
50. Krnjaic G (2016) Probing light thermal dark-matter with a higgs portal mediator. Phys Rev D 94:073009. ArXiv:1512.04119
51. Knapen S, Lin T, Zurek KM (2017) Light dark matter: models and constraints. Phys Rev D 96:115021. ArXiv:1709.07882
52. Edsjo J, Gondolo P (1997) Neutralino relic density including coannihilations. Phys Rev D 56:1879–1894. ArXiv:hep-ph/9704361
53. Kaplan DE, Luty MA, Zurek KM (2009) Asymmetric dark matter. Phys Rev D 79:115016. ArXiv:0901.4117
54. Petraki K, Volkas RR (2013) Review of asymmetric dark matter. Int J Mod Phys A 28:1330028. ArXiv:1305.4939
55. Carlson ED, Machacek ME, Hall LJ (1992) Self-interacting dark matter. Astrophys J 398:43–52
56. Pappadopulo D, Ruderman JT, Trevisan G (2016) Dark matter freeze-out in a nonrelativistic sector. Phys Rev D 94:035005. ArXiv:1602.04219
57. Farina M, Pappadopulo D, Ruderman JT, Trevisan G (2016) Phases of cannibal dark matter. JHEP 12:039. ArXiv:1607.03108

58. Hochberg Y, Kuflik E, Murayama H, Volansky T, Wacker JG (2015) Model for Thermal relic dark matter of strongly interacting massive particles. Phys Rev Lett 115:021301. ArXiv:1411.3727
59. Hochberg Y, Kuflik E, Volansky T, Wacker JG (2014) Mechanism for thermal relic dark matter of strongly interacting massive particles. Phys Rev Lett 113:171301. ArXiv:1402.5143
60. Kuflik E, Perelstein M, Lorier NR-L, Tsai Y-D (2016) Elastically decoupling dark matter. Phys Rev Lett 116:221302. ArXiv:1512.04545
61. Kuflik E, Perelstein M, Lorier NR-L, Tsai Y-D (2017) Phenomenology of ELDER dark matter. JHEP 08:078. ArXiv:1706.05381
62. Fitzpatrick PJ, Liu H, Slatyer TR, Tsai Y-D, New pathways to the relic abundance of vector-portal dark matter. ArXiv:2011.01240
63. Fitzpatrick PJ, Liu H, Slatyer TR, Tsai Y-D, New thermal relic targets for inelastic vector-portal dark matter. ArXiv:2105.05255
64. D'Eramo F, Thaler J (2010) Semi-annihilation of dark matter. JHEP 06:109. ArXiv:1003.5912
65. D'Eramo F, McCullough M, Thaler J (2013) Multiple gamma lines from semi-annihilation. JCAP 04:030. ArXiv:1210.7817
66. Kim H, Kuflik E (2019) Superheavy thermal dark matter. Phys Rev Lett 123:191801. ArXiv:1906.00981
67. Kramer ED, Kuflik E, Levi N, Outmezguine NJ, Ruderman JT (2021) Heavy thermal dark matter from a new collision mechanism. Phys Rev Lett 126:081802. ArXiv:2003.04900
68. Bian L, Liu X, Xie K-P, Probing superheavy dark matter with gravitational waves. arxiv:2107.13112
69. McDonald J (1991) WIMP densities in decaying particle dominated cosmology. Phys Rev D 43:1063–1068
70. Giudice GF, Kolb EW, Riotto A (2001) Largest temperature of the radiation era and its cosmological implications. Phys Rev D 64:023508. ArXiv:hep-ph/0005123
71. Patwardhan AV, Fuller GM, Kishimoto CT, Kusenko A (2015) Diluted equilibrium sterile neutrino dark matter. Phys Rev D92:103509. ArXiv:1507.01977
72. Berlin A, Hooper D, Krnjaic G (2016) PeV-scale dark matter as a thermal relic of a decoupled sector. Phys Lett B760:106–111. ArXiv:1602.08490
73. Berlin A, Hooper D, Krnjaic G (2016) Thermal dark matter from a highly decoupled sector. Phys Rev D94:095019. ArXiv:1609.02555
74. Bramante J, Unwin J (2017) Superheavy thermal dark matter and primordial asymmetries. JHEP 02:119. ArXiv:1701.05859
75. Hamdan S, Unwin J (2018) Dark matter freeze-out during matter domination. Mod Phys Lett A33:1850181. ArXiv:1710.03758
76. Allahverdi R, Osiński JK (2019) Nonthermal dark matter from modified early matter domination. Phys Rev D 99:083517. ArXiv:1812.10522
77. Cirelli M, Gouttenoire Y, Petraki K, Sala F (2019) Homeopathic dark matter, or how diluted heavy substances produce high energy cosmic rays. JCAP 02:014. ArXiv:1811.03608
78. Contino R, Mitridate A, Podo A, Redi M (2019) Gluequark dark matter. JHEP 02:187. ArXiv:1811.06975
79. Chanda P, Hamdan S, Unwin J (2020) Reviving Z and higgs mediated dark matter models in matter dominated freeze-out. JCAP 01:034. ArXiv:1911.02616
80. Contino R, Podo A, Revello F (2021) Composite dark matter from strongly-interacting chiral dynamics. JHEP 02:091. ArXiv:2008.10607
81. Asadi P, Kramer ED, Kuflik E, Slatyer TR, Smirnov J, Glueballs in a thermal squeezeout model. ArXiv:2203.15813
82. Konstandin T, Servant G (2011) Cosmological consequences of nearly conformal dynamics at the TeV scale. JCAP 12:009. ArXiv:1104.4791
83. Hambye T, Strumia A, Teresi D (2018) Super-cool dark matter. JHEP 08:188. ArXiv:1805.01473
84. Baldes I, Garcia-Cely C (2019) Strong gravitational radiation from a simple dark matter model. JHEP 05:190. ArXiv:1809.01198

85. Baldes I, Gouttenoire Y, Sala F (2021) String fragmentation in supercooled confinement and implications for dark matter. JHEP 04:278. ArXiv:2007.08440
86. Baldes I, Gouttenoire Y, Sala F, Servant G, Supercool composite dark matter beyond 100 TeV. arxiv:2110.13926
87. Baker MJ, Kopp J, Long AJ, Filtered dark matter at a first order phase transition. arxiv:1912.02830
88. Chway D, Jung TH, Shin CS (2020) Dark matter filtering-out effect during a first-order phase transition. Phys Rev D 101:095019. ArXiv:1912.04238
89. Asadi P, Kramer ED, Kuflik E, Ridgway GW, Slatyer TR, Smirnov J, Thermal squeezeout of dark matter. ArXiv:2103.09827
90. Asadi P, Kramer ED, Kuflik E, Ridgway GW, Slatyer TR, Smirnov J, Accidentally asymmetric dark matter. ArXiv:2103.09822
91. Falkowski A, No JM (2013) Non-thermal dark matter production from the electroweak phase transition: multi-TeV WIMPs and 'Baby-Zillas'. JHEP 02:034. ArXiv:1211.5615
92. Azatov A, Vanvlasselaer M, Yin W, Dark matter production from relativistic bubble walls. arxiv:2101.05721
93. Steigman G, Dasgupta B, Beacom JF (2012) Precise relic WIMP abundance and its impact on searches for dark matter annihilation. Phys Rev D86:023506. ArXiv:1204.3622
94. Agashe K, Servant G (2004) Warped unification, proton stability and dark matter. Phys Rev Lett 93:231805. ArXiv:hep-ph/0403143
95. Agashe K, Servant G (2005) Baryon number in warped GUTs: model building and (dark matter related) phenomenology. JCAP 02:002. ArXiv:hep-ph/0411254
96. Servant G, Tait TM (2003) Is the lightest Kaluza-Klein particle a viable dark matter candidate? Nucl Phys B 650:391–419. ArXiv:hep-ph/0206071
97. Hooper D, Profumo S (2007) Dark matter and collider phenomenology of universal extra dimensions. Phys Rep 453:29–115. ArXiv:hep-ph/0701197
98. Birkedal A, Noble A, Perelstein M, Spray A (2006) Little higgs dark matter. Phys Rev D 74:035002. ArXiv:hep-ph/0603077
99. Baltz EA (2004) Dark matter candidates. eConf C040802:L002. ArXiv:astro-ph/0412170
100. Escudero M, Berlin A, Hooper D, Lin M-X (2016) Toward (Finally!) ruling out Z and higgs mediated dark matter models. JCAP 12:029. ArXiv:1609.09079
101. Viel M, Lesgourgues J, Haehnelt MG, Matarrese S, Riotto A (2005) Constraining warm dark matter candidates including sterile neutrinos and light gravitinos with WMAP and the Lyman-alpha forest. Phys Rev D 71:063534. ArXiv:astro-ph/0501562
102. Baur J, Palanque-Delabrouille N, Yàche C, Magneville C, Viel M (2016) Lyman-alpha forests cool warm dark matter. JCAP 08:012. ArXiv:1512.01981
103. Iř sič V et al (2017) New constraints on the free-streaming of warm dark matter from intermediate and small scale Lyman-α forest data. Phys Rev D 96:023522. ArXiv:1702.01764
104. Jedamzik K (2004) Neutralinos and big bang nucleosynthesis. Phys Rev D 70:083510. ArXiv:astro-ph/0405583
105. Jedamzik K, Pospelov M (2009) Big bang nucleosynthesis and particle dark matter. New J Phys 11:105028. ArXiv:0906.2087
106. Hisano J, Kawasaki M, Kohri K, Moroi T, Nakayama K, Sekiguchi T (2011) Cosmological constraints on dark matter models with velocity-dependent annihilation cross section. Phys Rev D 83:123511. ArXiv:1102.4658
107. Kawasaki M, Kohri K, Moroi T, Takaesu Y (2015) Revisiting big-bang nucleosynthesis constraints on dark-matter annihilation. Phys Lett B 751:246–250. ArXiv:1509.03665
108. Hufnagel M, Schmidt-Hoberg K, Wild S (2018) BBN constraints on MeV-scale dark sectors. Part I. Sterile decays. JCAP 02:044. ArXiv:1712.03972
109. Hufnagel M, Schmidt-Hoberg K, Wild S, BBN constraints on MeV-scale dark sectors. Part II. Electromagnetic decays. ArXiv:1808.09324
110. Depta PF, Hufnagel M, Schmidt-Hoberg K, Wild S (2019) BBN constraints on the annihilation of MeV-scale dark matter. JCAP. 04:029. ArXiv:1901.06944

111. Depta PF, Hufnagel M, Schmidt-Hoberg K, Updated BBN constraints on electromagnetic decays of MeV-scale particles. ArXiv:2011.06519
112. Slatyer TR (2016) Indirect dark matter signatures in the cosmic dark ages. I. Generalizing the bound on s-wave dark matter annihilation from Planck results. Phys Rev D93:023527. ArXiv:1506.03811
113. Slatyer TR (2016) Indirect dark matter signatures in the cosmic dark ages II. Ionization, heating and photon production from arbitrary energy injections. Phys Rev D93:023521. ArXiv:1506.03812
114. Elor G, Rodd NL, Slatyer TR, Xue W (2016) Model-independent indirect detection constraints on hidden sector dark matter. JCAP 1606:024. ArXiv:1511.08787
115. Fermi-LAT collaboration, Ackermann M et al (2015) Searching for dark matter annihilation from milky way dwarf spheroidal galaxies with six years of fermi large area telescope data. Phys Rev Lett 115:231301. ArXiv:1503.02641
116. H.E.S.S. collaboration, Abdallah H et al (2016) Search for dark matter annihilations towards the inner Galactic halo from 10 years of observations with H.E.S.S. Phys Rev Lett 117:111301. ArXiv:1607.08142
117. Albert A et al (2017) Results from the search for dark matter in the Milky Way with 9 years of data of the ANTARES neutrino telescope. Phys Lett B769:249–254. ArXiv:1612.04595
118. ANTARES collaboration, Adrián-Martínez S et al (2016) A search for secluded dark matter in the sun with the ANTARES neutrino telescope. JCAP 05:016. ArXiv:1602.07000
119. ANTARES collaboration, Albert A et al, Search for secluded dark matter towards the Galactic Centre with the ANTARES neutrino telescope. ArXiv:2203.06029
120. XENON collaboration, Aprile E et al (2018) Dark matter search results from a one ton-year exposure of XENON1T. Phys Rev Lett 121:111302. ArXiv:1805.12562
121. XENON collaboration, Aprile E et al (2019) Light dark matter search with ionization signals in XENON1T. Phys Rev Lett 123:251801. ArXiv:1907.11485
122. Cowsik R, McClelland J (1972) An upper limit on the neutrino rest mass. Phys Rev Lett 29:669–670
123. Hut P (1977) Limits on masses and number of neutral weakly interacting particles. Phys Lett B 69:85
124. Sato K, Kobayashi M (1977) Cosmological constraints on the mass and the number of heavy lepton neutrinos. Prog Theor 58:1775
125. Vysotsky M, Dolgov A, Zeldovich Y (1977) Cosmological restriction on neutral lepton masses. JETP Lett 26:188–190
126. Cirelli M, Corcella G, Hektor A, Hutsi G, Kadastik M, Panci P et al (2011) PPPC 4 DM ID: a poor particle physicist cookbook for dark matter indirect detection. JCAP 03:051. ArXiv:1012.4515
127. Bringmann T, Weniger C (2012) Gamma ray signals from dark matter: concepts, status and prospects. Phys Dark Univ 1:194–217. ArXiv:1208.5481
128. Cirelli M (2012) Indirect searches for dark matter: a status review. Pramana 79:1021–1043. ArXiv:1202.1454
129. Conrad J (2014) Indirect detection of WIMP dark matter: a compact review. In: Interplay between particle and astroparticle physics, p 11. ArXiv:1411.1925
130. Gaskins JM (2016) A review of indirect searches for particle dark matter. Contemp Phys 57:496–525. ArXiv:1604.00014
131. Hooper D (2019) TASI lectures on indirect searches for dark matter. PoS TASI2018:010. ArXiv:8120.2029
132. Slatyer TR (2018) Indirect detection of dark matter. In: Theoretical advanced study institute in elementary particle physics: anticipating the next discoveries in particle physics, pp 297–353. ArXiv:1710.05137
133. Slatyer TR (2021) Les Houches lectures on indirect detection of dark matter. In: Les Houches summer school on dark matter, p 9. ArXiv:2109.02696
134. Strigari LE (2018) Dark matter in dwarf spheroidal galaxies and indirect detection: a review. Rep Prog Phys 81:056901. ArXiv:1805.05883

135. Ando S et al (2022) Snowmass2021 cosmic frontier: synergies between dark matter searches and multiwavelength/multimessenger astrophysics. In: 2022 snowmass summer study, p 3. ArXiv:2203.06781
136. Boddy KK et al (2022) Astrophysical and cosmological probes of dark matter. In: 2022 snowmass summer study, p 3. ArXiv:2203.06380
137. Baudis L (2012) Direct dark matter detection: the next decade. Phys Dark Univ 1:94–108. ArXiv:1211.7222
138. Saab T (2013) An introduction to dark matter direct detection searches and techniques. In: Theoretical advanced study institute in elementary particle physics: the dark secrets of the terascale, pp 711–738. ArXiv:1203.2566
139. Cushman P et al (2013) Working group report: WIMP dark matter direct detection. In: Proceedings, 2013 community summer study on the future of U.S. particle physics: snowmass on the Mississippi (CSS2013). Minneapolis, MN, USA, p 10. ArXiv:1310.8327
140. Liu J, Chen X, Ji X (2017) Current status of direct dark matter detection experiments. Nat Phys 13:212–216. ArXiv:1709.00688
141. Lin T (2019) Dark matter models and direct detection. PoS 333:009. ArXiv:1904.07915
142. Schumann M (2019) Direct detection of WIMP dark matter: concepts and status. J Phys G 46:103003. ArXiv:1903.03026
143. Billard J et al, Direct detection of dark matter—APPEC committee report. ArXiv:2104.07634
144. Kane G, Watson S (2008) Dark matter and LHC: what is the connection? Mod Phys Lett A 23:2103–2123. ArXiv:0807.2244
145. Kahlhoefer F (2017) Review of LHC dark matter searches. Int J Mod Phys A 32:1730006. ArXiv:1702.02430
146. Buchmueller O, Doglioni C, Wang LT (2017) Search for dark matter at colliders. Nat Phys 13:217–223. ArXiv:1912.12739
147. Morgante E (2018) Simplified dark matter models. Adv High Energy Phys 2018:5012043. ArXiv:1804.01245
148. Boveia A, Doglioni C (2018) Dark matter searches at colliders. Ann Rev Nucl Part Sci 68:429–459. ArXiv:1810.12238
149. Ilten P, Soreq Y, Williams M, Xue W (2018) Serendipity in dark photon searches. JHEP 06:004. ArXiv:1801.04847
150. Criado JC, Djouadi A, Perez-Victoria M, Santiago J (2021) A complete effective field theory for dark matter. JHEP 07:081. ArXiv:2104.14443
151. Zavala J, Vogelsberger M, White SDM (2010) Relic density and CMB constraints on dark matter annihilation with Sommerfeld enhancement. Phys Rev D 81:083502. ArXiv:0910.5221
152. Hannestad S, Tram T (2011) Sommerfeld enhancement of DM Annihilation: resonance structure, freeze-out and CMB spectral bound. JCAP 01:016. ArXiv:1008.1511
153. Ackerman L, Buckley MR, Carroll SM, Kamionkowski M (2009) Dark matter and dark radiation. Phys Rev D 79:023519. ArXiv:0810.5126
154. Baumann D, Green D, Meyers J, Wallisch B (2016) Phases of new physics in the CMB. JCAP 01:007. ArXiv:1508.06342
155. Chacko Z, Cui Y, Hong S, Okui T (2015) Hidden dark matter sector, dark radiation, and the CMB. Phys Rev D 92:055033. ArXiv:1505.04192
156. Chacko Z, Cui Y, Hong S, Okui T, Tsai Y (2016) Partially acoustic dark matter, interacting dark radiation, and large scale structure. JHEP 12:108. ArXiv:1609.03569
157. CMB-S4 collaboration, Abazajian KN et al, CMB-S4 science book, 1st edn. arxiv:1610.02743
158. Ko P, Tang Y (2017) Residual non-abelian dark matter and dark radiation. Phys Lett B 768:12–17. ArXiv:1609.02307
159. Ko P, Tang Y (2016) Light dark photon and fermionic dark radiation for the Hubble constant and the structure formation. Phys Lett B 762:462–466. ArXiv:1608.01083
160. Banerjee A, Jain B, Dalal N, Shelton J (2018) Tests of neutrino and dark radiation models from galaxy and CMB surveys. JCAP 01:022. ArXiv:1612.07126
161. Brust C, Cui Y, Sigurdson K (2017) Cosmological constraints on interacting light particles. JCAP 08:020. ArXiv:1703.10732

162. Pospelov M, Ritz A, Voloshin MB (2008) Secluded WIMP dark matter. Phys Lett B 662:53–61. ArXiv:0711.4866
163. Pospelov M, Ritz A (2009) Astrophysical signatures of secluded dark matter. Phys Lett B 671:391–397. ArXiv:0810.1502
164. Hambye T (2009) Hidden vector dark matter. JHEP 01:028. ArXiv:0811.0172
165. Arkani-Hamed N, Finkbeiner DP, Slatyer TR, Weiner N (2009) A theory of dark matter. Phys Rev D 79:015014. ArXiv:0810.0713
166. Chu X, Hambye T, Tytgat MHG (2012) The four basic ways of creating dark matter through a portal. JCAP 05:034. ArXiv:1112.0493
167. Cirelli M, Panci P, Petraki K, Sala F, Taoso M (2017) Dark matter's secret liaisons: phenomenology of a dark U(1) sector with bound states. JCAP 05:036. ArXiv:1612.07295
168. Hambye T, Tytgat MHG, Vandecasteele J, Vanderheyden L (2019) Dark matter from dark photons: a taxonomy of dark matter production. Phys Rev D 100:095018. ArXiv:1908.09864
169. Abazajian K et al, Light sterile neutrinos: a white paper. ArXiv:1204.5379
170. Drewes M et al (2017) A white paper on keV sterile neutrino dark matter. JCAP 01:025. ArXiv:1602.04816
171. Böser S, Buck C, Giunti C, Lesgourgues J, Ludhova L, Mertens S et al (2020) Status of light sterile neutrino searches. Prog Part Nucl Phys 111:103736. ArXiv:1906.01739
172. Shi X-D, Fuller GM (1999) A new dark matter candidate: nonthermal sterile neutrinos. Phys Rev Lett 82:2832–2835. ArXiv:astro-ph/9810076
173. Laine M, Shaposhnikov M (2008) Sterile neutrino dark matter as a consequence of nuMSM-induced lepton asymmetry. JCAP 06:031. ArXiv:0804.4543
174. Hambye T, Lucca M, Vanderheyden L (2020) Dark matter as a heavy thermal hot relic. Phys Lett B 807:135553. ArXiv:2003.04936
175. Baldes I, Decant Q, Hooper DC, Lopez-Honorez L, Non-cold dark matter from primordial black hole evaporation. ArXiv:2004.14773
176. Auffinger J, Masina I, Orlando G, Bounds on warm dark matter from Schwarzschild primordial black holes. ArXiv:2012.09867
177. D'Eramo F, Lenoci A, Lower mass bounds on FIMPs. ArXiv:2012.01446
178. Chen X-I, Kamionkowski M, Zhang X-M (2001) Kinetic decoupling of neutralino dark matter. Phys Rev D 64:021302. ArXiv:astro-ph/0103452
179. Hofmann S, Schwarz DJ, Stoecker H (2001) amping scales of neutralino cold dark matter. Phys Rev D 64:083507. ArXiv:astro-ph/0104173
180. Loeb A, Zaldarriaga M (2005) The small-scale power spectrum of cold dark matter. Phys Rev D 71:103520. ArXiv:astro-ph/0504112
181. Green AM, Hofmann S, Schwarz DJ (2005) The first wimpy halos. JCAP 08:003. ArXiv:astro-ph/0503387
182. Profumo S, Sigurdson K, Kamionkowski M (2006) What mass are the smallest protohalos? Phys Rev Lett 97:031301. ArXiv:astro-ph/0603373
183. Bertschinger E (2006) The effects of cold dark matter decoupling and pair annihilation on cosmological perturbations. Phys Rev D 74:063509. ArXiv:astro-ph/0607319
184. Bringmann T (2009) Particle models and the small-scale structure of dark matter. New J Phys 11:105027. ArXiv:0903.0189
185. Hooper D, Kaplinghat M, Strigari LE, Zurek KM (2007) MeV dark matter and small scale structure. Phys Rev D 76:103515. ArXiv:0704.2558
186. Feng JL, Kaplinghat M, Tu H, Yu H-B (2009) Hidden charged dark matter. JCAP 07:004. ArXiv:0905.3039
187. van den Aarssen LG, Bringmann T, Pfrommer C (2012) Is dark matter with long-range interactions a solution to all small-scale problems of Λ CDM cosmology? Phys Rev Lett 109:231301. ArXiv:1205.5809
188. Ahlgren B, Ohlsson T, Zhou S (2013) Comment on Is dark matter with long-range interactions a solution to all small-scale problems of Λ cold dark matter cosmology? Phys Rev Lett 111:199001. ArXiv:1309.0991

189. Laha R, Dasgupta B, Beacom JF (2014) Constraints on new neutrino interactions via light abelian vector bosons. Phys Rev D 89:093025. ArXiv:1304.3460
190. Dasgupta B, Kopp J (2014) Cosmologically safe eV-scale sterile neutrinos and improved dark matter structure. Phys Rev Lett 112:031803. ArXiv:1310.6337
191. Bringmann T, Hasenkamp J, Kersten J (2014) Tight bonds between sterile neutrinos and dark matter. JCAP 07:042. ArXiv:1312.4947
192. Bertoni B, Ipek S, McKeen D, Nelson AE (2015) Constraints and consequences of reducing small scale structure via large dark matter-neutrino interactions. JHEP 04:170. ArXiv:1412.3113
193. Ko P, Tang Y (2014) $\nu\Lambda$ MDM: a model for sterile neutrino and dark matter reconciles cosmological and neutrino oscillation data after BICEP2. Phys Lett B 739:62–67. ArXiv:1404.0236
194. Cherry JF, Friedland A, Shoemaker IM, Neutrino portal dark matter: from dwarf galaxies to icecube. ArXiv:1411.1071
195. Chu X, Dasgupta B (2014) Dark radiation alleviates problems with dark matter halos. Phys Rev Lett 113:161301. ArXiv:1404.6127
196. Buckley MR, Zavala J, Cyr-Racine F-Y, Sigurdson K, Vogelsberger M (2014) Scattering, Damping, and acoustic oscillations: simulating the structure of dark matter halos with relativistic force carriers. Phys Rev D 90:043524. ArXiv:1405.2075
197. Binder T, Covi L, Kamada A, Murayama H, Takahashi T, Yoshida N (2016) Matter power spectrum in hidden neutrino interacting dark matter models: a closer look at the collision term. JCAP 11:043. ArXiv:1602.07624
198. Bringmann T, Ihle HT, Kersten J, Walia P (2016) Suppressing structure formation at dwarf galaxy scales and below: late kinetic decoupling as a compelling alternative to warm dark matter. Phys Rev D 94:103529. ArXiv:1603.04884
199. Tang Y (2016) Interacting scalar radiation and dark matter in cosmology. Phys Lett B 757:387–392. ArXiv:1603.00165
200. Agrawal P, Cyr-Racine F-Y, Randall L, Scholtz J (2017) Dark catalysis. JCAP 08:021. ArXiv:1702.05482
201. Boehm C, Fayet P, Schaeffer R (2001) Constraining dark matter candidates from structure formation. Phys Lett B 518:8–14. ArXiv:astro-ph/0012504
202. Boehm C, Riazuelo A, Hansen SH, Schaeffer R (2002) Interacting dark matter disguised as warm dark matter. Phys Rev D 66:083505. ArXiv:astro-ph/0112522
203. Boehm C, Schaeffer R (2005) Constraints on dark matter interactions from structure formation: Damping lengths. Astron Astrophys 438:419–442. ArXiv:astro-ph/0410591
204. Cyr-Racine F-Y, Sigurdson K (2013) Cosmology of atomic dark matter. Phys Rev D 87:103515. ArXiv:1209.5752
205. Cyr-Racine F-Y, de Putter R, Raccanelli A, Sigurdson K (2014) Constraints on large-scale dark acoustic oscillations from cosmology. Phys Rev D 89:063517. ArXiv:1310.3278
206. Cyr-Racine F-Y, Sigurdson K, Zavala J, Bringmann T, Vogelsberger M, Pfrommer C (2016) ETHOS—an effective theory of structure formation: From dark particle physics to the matter distribution of the Universe. Phys Rev D 93:123527. ArXiv:1512.05344
207. Vogelsberger M, Zavala J, Cyr-Racine F-Y, Pfrommer C, Bringmann T, Sigurdson K (2016) ETHOS—an effective theory of structure formation: dark matter physics as a possible explanation of the small-scale CDM problems. Mon Not R Astron Soc 460:1399–1416. ArXiv:1512.05349
208. Hoyer P (2014) Bound states—from QED to QCD, p 2. ArXiv:1402.5005
209. Petraki K, Postma M, Wiechers M (2015) Dark-matter bound states from Feynman diagrams. JHEP 06:128. ArXiv:1505.00109
210. Petraki K, Postma M, de Vries J (2017) Radiative bound-state-formation cross-sections for dark matter interacting via a Yukawa potential. JHEP 04:077. ArXiv:1611.01394
211. Sommerfeld A (1931) Über die Beugung und Bremsung der Elektronen. Annalen Phys 403:257–330
212. Hisano J, Matsumoto S, Nojiri MM, Saito O (2005) Non-perturbative effect on dark matter annihilation and gamma ray signature from galactic center. Phys Rev D 71:063528. ArXiv:hep-ph/0412403

213. Cirelli M, Strumia A, Tamburini M (2007) Cosmology and astrophysics of minimal dark matter. Nucl Phys B 787:152–175. ArXiv:0706.4071
214. Cirelli M, Strumia A (2008) Minimal dark matter predictions and the PAMELA positron excess. PoS IDM2008:089. ArXiv:0808.3867
215. Cirelli M, Kadastik M, Raidal M, Strumia A (2009) Model-independent implications of the e+-, anti-proton cosmic ray spectra on properties of Dark Matter. Nucl Phys B 813:1–21. ArXiv:0809.2409
216. Cholis I, Dobler G, Finkbeiner DP, Goodenough L, Weiner N (2009) The case for a 700+ GeV WIMP: cosmic ray spectra from ATIC and PAMELA. Phys Rev D 80:123518. ArXiv:0811.3641
217. Cholis I, Goodenough L, Weiner N (2009) High energy positrons and the WMAP haze from exciting dark matter. Phys Rev D 79:123505. ArXiv:0802.2922
218. Kamionkowski M, Profumo S (2008) Early annihilation and diffuse backgrounds in models of weakly interacting massive particles in which the cross section for pair annihilation is enhanced by 1/v. Phys Rev Lett 101:261301. ArXiv:0810.3233
219. Arkani-Hamed N, Finkbeiner DP, Slatyer TR, Weiner N (2009) A theory of dark matter. Phys Rev D79:015014. ArXiv:0810.0713
220. Donato F, Maurin D, Brun P, Delahaye T, Salati P (2009) Constraints on WIMP dark matter from the high energy PAMELA $\bar{p}/p$ data. Phys Rev Lett 102:071301. ArXiv:0810.5292
221. Lattanzi M, Silk JI (2009) Can the WIMP annihilation boost factor be boosted by the Sommerfeld enhancement? Phys Rev D 79:083523. ArXiv:0812.0360
222. March-Russell JD, West SM (2009) WIMPonium and boost factors for indirect dark matter detection. Phys Lett B 676:133–139. ArXiv:0812.0559
223. Bedaque PF, Buchoff MI, Mishra RK (2009) Sommerfeld enhancement from Goldstone pseudo-scalar exchange. JHEP 11:046. ArXiv:0907.0235
224. Finkbeiner DP, Goodenough L, Slatyer TR, Vogelsberger M, Weiner N (2011) Consistent scenarios for cosmic-ray excesses from sommerfeld-enhanced dark matter annihilation. JCAP 05:002. ArXiv:1011.3082
225. Slatyer TR, Toro N, Weiner N (2012) Sommerfeld-enhanced annihilation in dark matter substructure: consequences for constraints on cosmic-ray excesses. Phys Rev D 86:083534. ArXiv:1107.3546
226. Liu Z-P, Wu Y-L, Zhou Y-F (2013) Sommerfeld enhancements with vector, scalar and pseudoscalar force-carriers. Phys Rev D 88:096008. ArXiv:1305.5438
227. PAMELA collaboration, Adriani O et al (2009) An anomalous positron abundance in cosmic rays with energies 1.5-100 GeV. Nature 458:607–609. ArXiv:0810.4995
228. Chang J et al (2008) An excess of cosmic ray electrons at energies of 300–800 GeV. Nature 456:362–365
229. Fermi-LAT collaboration, Ackermann M et al (2012) Measurement of separate cosmic-ray electron and positron spectra with the Fermi Large Area Telescope. Phys Rev Lett 108:011103. ArXiv:1109.0521
230. AMS collaboration, Aguilar M et al (2013) First result from the alpha magnetic spectrometer on the international space station: precision measurement of the positron fraction in primary cosmic rays of 0.5–350 GeV. Phys Rev Lett 110:141102
231. Slatyer TR, Padmanabhan N, Finkbeiner DP (2009) CMB constraints on WIMP annihilation: energy absorption during the recombination epoch. Phys Rev D 80:043526. ArXiv:0906.1197
232. Profumo S, Jeltema TE (2009) Extragalactic inverse compton light from dark matter annihilation and the pamela positron excess. JCAP 07:020. ArXiv:0906.0001
233. Belikov AV, Hooper D (2010) The contribution of inverse compton scattering to the diffuse extragalactic gamma-ray background from annihilating dark matter. Phys Rev D 81:043505. ArXiv:0906.2251
234. Kawasaki M, Kohri K, Nakayama K (2009) Diffuse gamma-ray background and cosmic-ray positrons from annihilating dark matter. Phys Rev D 80:023517. ArXiv:0904.3626
235. Cirelli M, Panci P, Serpico PD (2010) Diffuse gamma ray constraints on annihilating or decaying Dark Matter after Fermi. Nucl Phys B 840:284–303. ArXiv:0912.0663

236. Abazajian KN, Harding JP (2012) Constraints on WIMP and sommerfeld-enhanced dark matter annihilation from HESS observations of the galactic center. JCAP 01:041. ArXiv:1110.6151
237. Spolyar D, Buckley MR, Freese K, Hooper D, Murayama H, High energy neutrinos as a test of leptophilic dark matter. ArXiv:0905.4764
238. Buckley MR, Freese K, Hooper D, Spolyar D, Murayama H (2010) High-energy neutrino signatures of dark matter decaying into leptons. Phys Rev D 81:016006. ArXiv:0907.2385
239. Hisano J, Nakayama K, Yang MJS (2009) Upward muon signals at neutrino detectors as a probe of dark matter properties. Phys Lett B 678:101–106. ArXiv:0905.2075
240. Feng JL, Kaplinghat M, Yu H-B (2010) Halo shape and relic density exclusions of sommerfeld-enhanced dark matter explanations of cosmic ray excesses. Phys Rev Lett 104:151301. ArXiv:0911.0422
241. Buckley MR, Fox PJ (2010) Dark matter self-interactions and light force carriers. Phys Rev D 81:083522. ArXiv:0911.3898
242. Hooper D, Blasi P, Serpico PD (2009) Pulsars as the sources of high energy cosmic ray positrons. JCAP 01:025. ArXiv:0810.1527
243. Yuksel H, Kistler MD, Stanev T (2009) TeV gamma rays from geminga and the origin of the gev positron excess. Phys Rev Lett 103:051101. ArXiv:0810.2784
244. Profumo S (2011) Dissecting cosmic-ray electron-positron data with Occam's Razor: the role of known Pulsars. Central Eur J Phys 10:1–31. ArXiv:0812.4457
245. Blasi P (2009) The origin of the positron excess in cosmic rays. Phys Rev Lett 103:051104. ArXiv:0903.2794
246. Blasi P, Serpico PD (2009) High-energy antiprotons from old supernova remnants. Phys Rev Lett 103:081103. ArXiv:0904.0871
247. Iengo R (2009) Sommerfeld enhancement: general results from field theory diagrams. JHEP 05:024. ArXiv:0902.0688
248. Iengo R, Sommerfeld enhancement for a Yukawa potential. ArXiv:0903.0317
249. Cassel S (2010) Sommerfeld factor for arbitrary partial wave processes. J Phys G 37:105009. ArXiv:0903.5307
250. Slatyer TR (2010) The Sommerfeld enhancement for dark matter with an excited state. JCAP 02:028. ArXiv:0910.5713
251. Blum K, Sato R, Slatyer TR (2016) Self-consistent calculation of the sommerfeld enhancement. JCAP 06:021. ArXiv:1603.01383
252. Dent JB, Dutta S, Scherrer RJ (2010) Thermal relic abundances of particles with velocity-dependent interactions. Phys Lett B 687:275–279. ArXiv:0909.4128
253. Feng JL, Kaplinghat M, Yu H-B (2010) Sommerfeld enhancements for thermal relic dark matter. Phys Rev D 82:083525. ArXiv:1005.4678
254. Hryczuk A, Iengo R, Ullio P (2011) Relic densities including Sommerfeld enhancements in the MSSM. JHEP 03:069. ArXiv:1010.2172
255. Pieri L, Lattanzi M, Silk J (2009) Constraining the Sommerfeld enhancement with Cherenkov telescope observations of dwarf galaxies. Mon Not R Astron Soc 399:2033. ArXiv:0902.4330
256. Das A, Dasgupta B (2017) Selection rule for enhanced dark matter annihilation. Phys Rev Lett 118:251101. ArXiv:1611.04606
257. Boddy KK, Kumar J, Strigari LE, Wang M-Y (2017) Sommerfeld-Enhanced J-factors for dwarf spheroidal galaxies. Phys Rev D 95:123008. ArXiv:1702.00408
258. Hryczuk A, Iengo R (2012) The one-loop and Sommerfeld electroweak corrections to the Wino dark matter annihilation. JHEP 01:163. ArXiv:1111.2916
259. Hryczuk A (2011) The Sommerfeld enhancement for scalar particles and application to sfermion co-annihilation regions. Phys Lett B 699:271–275. ArXiv:1102.4295
260. Beneke M, Hellmann C, Ruiz-Femenia P (2015) Non-relativistic pair annihilation of nearly mass degenerate neutralinos and charginos III. Computation of the Sommerfeld enhancements. JHEP 05:115. ArXiv:1411.6924
261. Beneke M, Hellmann C, Ruiz-Femenia P (2015) Heavy neutralino relic abundance with Sommerfeld enhancements—a study of pMSSM scenarios. JHEP 03:162. ArXiv:1411.6930

262. Beneke M, Szafron R, Urban K (2020) Wino potential and Sommerfeld effect at NLO. Phys Lett B 800:135112. ArXiv:1909.04584
263. Beneke M, Szafron R, Urban K (2021) Sommerfeld-corrected relic abundance of wino dark matter with NLO electroweak potentials. JHEP 02:020. ArXiv:2009.00640
264. Rinchiuso L, Rodd NL, Moult I, Moulin E, Baumgart M, Cohen T et al (2018) Hunting for heavy winos in the galactic center. Phys Rev D 98:123014. ArXiv:1808.04388
265. Baumgart M, Cohen T, Moulin E, Moult I, Rinchiuso L, Rodd NL et al (2019) Precision photon spectra for wino annihilation. JHEP 01:036. ArXiv:1808.08956
266. Harz J, Petraki K (2018) Higgs enhancement for the dark matter relic density. Phys Rev D 97:075041. ArXiv:1711.03552
267. Harz J, Petraki K (2018) Radiative bound-state formation in unbroken perturbative non-Abelian theories and implications for dark matter. JHEP 07:096. ArXiv:1805.01200
268. Harz J, Petraki K (2019) Higgs-mediated bound states in dark-matter models. JHEP 04:130. ArXiv:1901.10030
269. Oncala R, Petraki K (2019) Dark matter bound states via emission of scalar mediators. JHEP 01:070. ArXiv:1808.04854
270. Oncala R, Petraki K (2020) Dark matter bound state formation via emission of a charged scalar. JHEP 02:036. ArXiv:1911.02605
271. Oncala R. Petraki K (2021) Bound states of WIMP dark matter in Higgs-portal models. Part I. Cross-sections and transition rates. JHEP 06:124. ArXiv:2101.08666
272. Oncala R, Petraki K (2021) Bound states of WIMP dark matter in Higgs-portal models. Part II. Thermal decoupling. JHEP 08:069. ArXiv:2101.08667
273. Kim S, Laine M (2017) On thermal corrections to near-threshold annihilation. J Cosmol Astropart Phys 01:013. https://doi.org/10.1088/1475-7516/2017/01/013. arXiv:1609.00474
274. Biondini S, Laine M (2017) Re-derived overclosure bound for the inert doublet model. J High Energy Phys 08:047. https://doi.org/10.1007/JHEP08(2017)047. arXiv:1706.01894
275. Biondini S, Laine M (2018) Thermal dark matter co-annihilating with a strongly interacting scalar. J High Energy Phys 04:072. https://doi.org/10.1007/JHEP04(2018)072. arXiv:1801.05821
276. Biondini S, Kim S, Laine M (2019) Non-relativistic susceptibility and a dark matter application. J Cosmol Astropart Phys 10:078. https://doi.org/10.1088/1475-7516/2019/10/078. arXiv:1908.07541
277. Biondini S, Vogl S (2019) Coloured coannihilations: Dark matter phenomenology meets non-relativistic EFTs. J High Energy Phys 02:116. https://doi.org/10.1007/JHEP02(2019)016. arXiv:1811.02581
278. Biondini S, Shtabovenko V (2021) Non-relativistic and potential non-relativistic effective field theories for scalar mediators. J High Energy Phys 08:114. https://doi.org/10.1007/JHEP08(2021)114. arXiv:2106.06472
279. Biondini S, Shtabovenko V (2022) Bound-state formation, dissociation and decays of darkonium with potential non-relativistic Yukawa theory for scalar and pseudoscalar mediators. J High Energy Phys 03:172. https://doi.org/10.1007/JHEP03(2022)172. arXiv:2112.10145
280. Binder T, Covi L, Mukaida K (2018) Dark matter sommerfeld-enhanced annihilation and bound-state decay at finite temperature. Phys Rev D 98:115023. ArXiv:1808.06472
281. Binder T, Mukaida K, Petraki K (2020) Rapid bound-state formation of dark matter in the Early Universe. Phys Rev Lett 124:161102. ArXiv:1910.11288
282. Binder T, Blobel B, Harz J, Mukaida K (2020) Dark matter bound-state formation at higher order: a non-equilibrium quantum field theory approach. JHEP 09:086. ArXiv:2002.07145
283. Binder T, Filimonova A, Petraki K, White G, Saha equilibrium for metastable bound states and dark matter freeze-out. ArXiv:2112.00042
284. Binder T, Mukaida K, Scheihing-Hitschfeld B, Yao X (2022) Non-abelian electric field correlator at NLO for dark matter relic abundance and quarkonium transport. JHEP 01:137. ArXiv:2107.03945
285. Garny M, Heisig J (2022) Bound-state effects on dark matter coannihilation: pushing the boundaries of conversion-driven freeze-out. Phys Rev D 105:055004. ArXiv:2112.01499

286. Bollig J, Vogl S, Impact of bound states on non-thermal dark matter production. ArXiv:2112.01491
287. von Harling B, Petraki K (2014) Bound-state formation for thermal relic dark matter and unitarity. JCAP 12:033. ArXiv:1407.7874
288. Pearce L, Petraki K, Kusenko A (2015) Signals from dark atom formation in halos. Phys Rev D 91:083532. ArXiv:1502.01755
289. Baldes I, Calore F, Petraki K, Poireau V, Rodd NL (2020) Indirect searches for dark matter bound state formation and level transitions. SciPost Phys 9:068. ArXiv:2007.13787
290. Geller M, Telem O (2021) Self-destructing atomic dark matter. Phys Rev D 104:035010. ArXiv:2001.11514
291. Eiger J, Geller M (2021) Detecting dark photons from atomic rearrangement in the galaxy. JHEP 04:016. ArXiv:2010.11205
292. Baldes I, Petraki K (2017) Asymmetric thermal-relic dark matter: Sommerfeld-enhanced freeze-out, annihilation signals and unitarity bounds. JCAP 09:028. ArXiv:1703.00478
293. Smirnov J, Beacom JF (2019) TeV-scale thermal WIMPs: unitarity and its consequences. Phys Rev D 100:043029. ArXiv:1904.11503
294. El Hedri S, Kaminska A, de Vries M (2017) A sommerfeld toolbox for colored dark sectors. Eur Phys J C 77:622. ArXiv:1612.02825
295. El Hedri S, Kaminska A, de Vries M, Zurita J (2017) Simplified phenomenology for colored dark sectors. JHEP 04:118. ArXiv:1703.00452
296. Mitridate A, Redi M, Smirnov J, Strumia A (2017) Cosmological implications of dark matter bound states. JCAP 05:006. ArXiv:1702.01141
297. Mitridate A, Redi M, Smirnov J, Strumia A (2017) Dark matter as a weakly coupled dark baryon. JHEP 10:210. ArXiv:1707.05380
298. De Luca V, Mitridate A, Redi M, Smirnov J, Strumia A (2018) Colored dark matter. Phys Rev D 97:115024. ArXiv:1801.01135
299. Gross C, Mitridate A, Redi M, Smirnov J, Strumia A (2019) Cosmological abundance of colored relics. Phys Rev D 99:016024. ArXiv:1811.08418
300. Redi M, Tesi A (2019) Cosmological production of dark nuclei. JHEP 04:108. ArXiv:1812.08784
301. Becker M, Copello E, Harz J, Mohan KA, Sengupta D, Impact of sommerfeld effect and bound state formation in simplified t-channel dark matter models. ArXiv:2203.04326
302. Baldes I, Cirelli M, Panci P, Petraki K, Sala F, Taoso M (2018) Asymmetric dark matter: residual annihilations and self-interactions. SciPost Phys 4:041. ArXiv:1712.07489
303. Mahbubani R, Redi M, Tesi A (2020) Indirect detection of composite asymmetric dark matter. Phys Rev D 101:103037. ArXiv:1908.00538
304. Mahbubani R, Redi M, Tesi A (2021) Dark nucleosynthesis: cross-sections and astrophysical signals. JCAP 02:039. ArXiv:2007.07231
305. Stroscio M (1975) Positronium: review of the theory. Phys Rep 22:215–277
306. Griest K, Kamionkowski M (1990) Unitarity limits on the mass and radius of dark matter particles. Phys Rev Lett 64:615
307. Pilkuhn HM (2005) Relativistic quantum mechanics. Theoretical and mathematical physics. Springer, Berlin, Germany. https://doi.org/10.1007/3-540-28522-9
308. Landau LD, Lifshitz EM (2013) Quantum mechanics: non-relativistic theory, vol 3. Elsevier

Chapter 5
Homeopathic Dark Matter

5.1 Introduction

The much anticipated physics beyond the Standard Model (BSM) has so far not been discovered at experiments, that have then pushed the scale of many related theories beyond the TeV range. How can we gain an experimental access to such energy scales? High energy cosmic rays currently constitute a privileged access to that realm, as opposed to the more subtle one offered by various precision measurements. Many telescopes are indeed now observing such cosmic rays (HESS II, HAWC, VERITAS, MAGIC, TAIGA, ANTARES, ICECUBE, BAIKAL, AMS, CALET, DAMPE), and many are planned for the near future (CTA, LHAASO, KM3NET, HERD, ISS-CREAM). Using this data optimally to advance our knowledge of fundamental physics is therefore of pressing importance.

Dark matter (DM) is arguably the strongest evidence for BSM physics that can be potentially probed by these observations. New physics (NP) sectors at scales beyond 10–100 TeV, addressing other open questions in our understanding of Nature, have long been known to provide DM candidates in that same mass range, see for example [2, 3]. Moreover, the empty-handed searches for weakly interacting massive particles (WIMPs) motivate exploring new regimes of WIMP models, including DM masses larger than about 10 TeV, see e.g. [4, 5]. But masses much larger than this are challenged, in the wide class of DM models where thermal freeze-out sets the relic abundance, by the so-called unitarity bound [6]: the DM annihilation cross section is bounded from above by the unitarity of the S matrix, and this translates into an upper bound on the DM mass of $O(100)$ TeV, see e.g. [7, 8].

However, the injection of entropy in the SM bath after DM freeze-out can dilute its abundance, and thus open the possibility of obtaining thermal relic DM beyond the unitarity bound. Such a dilution of relics takes place in well motivated NP models, see e.g. [9–11]. Its consequences for DM have been studied since a long time [12], also in relation with heavy DM [13] (in the rest of this chapter, 'heavy DM' will be used as a synonym of DM with a mass at and beyond 10–100 TeV). Recent

Y. Gouttenoire, *Beyond the Standard Model Cocktail*, Springer Theses,
https://doi.org/10.1007/978-3-031-11862-3_5

years have seen a growing interest toward the possibility to evade the Griest and Kamionkowski unitarity bound [6] on the mass of thermal DM. For example, DM dilution after a matter era [1, 13–23], or having a dark sector being much cooler than SM [24, 25], DM becoming heavy only after freezing-out [26], DM annihilating with one spectator field [27, 28] or with many of them [29], DM forming an extended object which undergoes a second annihilation stage [30–32]. Mechanisms involving phase transitions include the possibility of a short inflationary stage associated with gluon-string-mediated quark-wall interactions [33, 34], perturbative DM mass generation [?] [35, 36], DM filtered [37, 38] or squeezed-out [39, 40] by non-relativistic bubble wall motion, DM produced by elastic bubble-bubble collisions [41] or perturbative plasma interactions with relativistic walls [42]. We refer the reader to [43] for a review on heavy DM.

These DM scenarios offer therefore an important physics goal for high energy cosmic ray telescopes, which could test BSM physics at an energy scale where no other experiment has direct access. However, in the standard WIMP-inspired DM scenarios, this is complicated by the treatment of electroweak (EW) radiation at large DM masses. The SM spectra from DM annihilations are indeed governed by an expansion in $\sim \alpha_w/\pi \, \log^2(M_{\rm DM}/m_W)$, that should be resummed when this number is order one (analogoulsy to what is done for QCD radiation), which happens at $M_{\rm DM} \approx 100$ TeV. At present this resummation constitutes a technical challenge, and limits the reliability of the interpretation of high energy cosmic ray data in terms of DM.[1]

As pointed out recently in [15, 16, 48], a class of DM models that allows to evade the unitarity limit consists of DM annihilating into mediators belonging to a dark sector that themselves decay into SM particles. If their lifetime is long enough and they are sufficiently heavy, the mediators may temporarily dominate the energy density of the Universe, so that, when they eventually decay, they inject significant entropy in the SM bath and thus dilute any relic, including DM. A frozen-out overabundant DM can then be diluted to the observed density. This allows for a smaller cross section at freeze-out time, and therefore a larger DM mass, relaxing the unitarity limit.

We observe that the same class of models circumvents also the challenge of reliably computing the SM spectra from heavy DM annihilations. Indeed, the energy scale relevant for computing the SM spectra is now the mass of the mediator m_V, rather than $M_{\rm DM}$. For mediators lighter enough than 100 TeV, $\alpha_w/\pi \, \log^2(m_V/m_W) \ll 1$, so one can use the spectra with EW corrections computed as in [44, 46] (see [49] for a more recent study) and then boost them from the mediator rest frame to the DM one.

From the two points of view we just discussed, these *secluded DM* models appear to constitute an ideal target for telescopes observing high energy cosmic rays, and would allow them to reliably test for the first time annihilating DM beyond 100 TeV. However, to our knowledge, only two experimental analyses of secluded DM models

[1] For example, two of the tools often used for this purpose, PPPC4DMID [44] and Pythia [45], respectively include EW radiation only at first-order [46] and lack radiation processes among gauge bosons like $W^* \to W\gamma$ [47], and thus cannot be completely trusted for heavy DM.

have been performed, which are specific to neutrinos from DM in the Sun and reach DM masses up to 10 TeV [50, 51]. Are current and planned telescopes sensitive to DM models of this kind? In this study we show this to be the case, thus enriching the physics case of these experiments, and opening a new window on BSM theories at the high energy frontier.

The discussion is organized as follows. In Sect. 5.2 we present a quantitative model-independent study of dilution in secluded DM models, and of the limits from Big-Bang Nucleosynthesis (BBN). For concreteness, we then consider a specific model of DM charged under a dark $U(1)$, which we discuss in Sect. 5.3. We study its signals in Sect. 5.4, where we analyse the constraints from observations of CMB, 21cm, neutrinos, gamma rays, antiprotons, electrons and positrons. In Sect. 5.5 we summarise our results and indicate possible future directions.

5.2 Relaxing the Unitarity Bound By Injecting Entropy

We consider the dark photon, as the mediator of the DM which is charged under $U(1)_D$. The dark photon is produced from the annihilation of dark matter at freeze-out. When long-lived and heavy, the dark photon dominates the energy density of the universe before it decays into SM radiation. During the decay, non-relativistic degrees of freedom held in the dark photon are converted into relativistic degrees of freedom held in the radiation, hence creating entropy.

5.2.1 *The Start of the Matter Era*

The temperature $T_{\rm dom}$ at which the heavy relic dominates the energy density of the universe must satisfy $\rho_{\rm rad} \simeq \rho_V$ (already introduced in Eq. (9.2))

$$\frac{\pi^2}{30} g_{\rm SM} T_{\rm dom}^4 \simeq m_V f_V \frac{2\pi^2}{45} g_{\rm SM} T_{\rm dom}^3 \quad \rightarrow \quad T_{\rm dom} \simeq \frac{4}{3} f_V \, m_V. \tag{5.1}$$

f_V is the ratio of the mediator number density over the SM entropy density before decay

$$f_V \equiv \frac{n_V^{\rm before}}{s_{\rm SM}^{\rm before}} = \frac{45\,\zeta(3)}{2\pi^4} \frac{g_D^{\rm before}}{g_{\rm SM}^{\rm before}} r_{\rm before}^3 = \frac{45\,\zeta(3)}{2\pi^4} \frac{\tilde{g}_D}{\tilde{g}_{\rm SM}} \tilde{r}^3 \simeq 0.0169 \left(\frac{\tilde{g}_D}{6.5}\right) \tilde{r}^3, \tag{5.2}$$

Our discussion here will encompass both possibilities, and only assume that the dark sector and the SM were *not* in kinetic equilibrium when the SM plasma had a temperature $T_{\rm SM} < \widetilde{T}_{\rm SM}$.[2] The subsequent evolution of T_D and $T_{\rm SM}$ is controlled by the conservation of the entropy density of the two sectors separately; their ratio reads

[2] $\widetilde{T}_{\rm SM} \gtrsim$ few MeV could be justified by the requirement to not ruin BBN, depending on the specific model. It is not our purpose here to elaborate on this observation.

$$r \equiv \frac{T_D}{T_{SM}} = \left(\frac{g_{SM}}{g_D}\frac{\tilde{g}_D}{\tilde{g}_{SM}}\right)^{\frac{1}{3}} \tilde{r}\,, \tag{5.3}$$

where g_{SM} and g_D are the numbers of relativistic degrees of freedom in the SM and in the dark sector, and the $\sim$ on top of a symbol denotes that the corresponding quantity is evaluated at $T_{SM} = \widetilde{T}_{SM}$. $\tilde{r} = 1$ corresponds to the case where the SM and the dark plasma were in thermal equilibrium for $T_{SM} > \widetilde{T}_{SM}$.

5.2.2 *The End of the Matter Era*

In order to preserve the light element abundance, the mediator must decay before BBN starts. This places an upper limit on the possible dilution. The study in [52] shows that the lifetime of an abundant particle, decaying 30% or more hadronically, must be smaller than 0.03 s. In order to incorporate the modification of the cosmology due to the mediator dominating the energy density of the universe and injecting entropy, we rewrite the upper bound on the lifetime, $\tau_V < 0.03$ s, as a lower bound on the SM temperature at the time of decay, $T_{SM}^{decay} > 5$ MeV. Therefore, we shall now describe the evolution of the SM temperature in our non-standard cosmology.

The evolution of the SM temperature $T_{SM}(z)$ in the presence of the cold relic is driven by a combination of redshift and entropy injection

$$T_{SM}(z) = \frac{T_{SM}^{before}}{z}\left(\frac{g_{SM}(T_{SM}^{before})}{g_{SM}\left(T_{SM}(z)\right)}\right)^{1/3}\left(\frac{S_{SM}(x)}{S_{SM}^{before}}\right)^{1/3}, \tag{5.4}$$

where $z(x) \equiv a(x)/a^{before}$, $x = t/\tau_V$ and [53]

$$\frac{S_{SM}(x)}{S_{SM}^{before}} = \left(1 + \frac{4}{3}\frac{f_V m_V}{(g_{SM}^{before})^{1/3}\, T_{SM}^{before}}\int_0^x d\tilde{x}\, g_{SM}^{1/3}(\tilde{x})\, z(\tilde{x})\, e^{-\tilde{x}}\right)^{3/4}. \tag{5.5}$$

Integrating the Friedmann equation assuming that the cold relic dominates the universe energy density, gives

$$z(x) \simeq \left(\frac{3}{2}\frac{x}{x_{before}}\right)^{2/3}, \tag{5.6}$$

where $x_{before} = \Gamma_V\sqrt{3M_{Pl}^2/\rho_V^{before}}$ and $\rho_V^{before} = m_V f_V (2\pi^2/45) g_{SM}^{before} T_{SM,before}^3$ is the energy density in the mediators before decay. For $x \leq 1$, for simplicity we use $e^{-\tilde{x}} \simeq 1$ in Eq. (5.5). Using Eqs. (5.6) and (5.4) then becomes

$$T_{\rm SM}(x) \simeq 1.25\sqrt{M_{\rm Pl}\Gamma_V}\,\frac{(g_{\rm SM}^{\rm dec})^{1/12}}{g_{\rm SM}^{1/3}}\frac{x_{\rm inj}^{5/12}}{x^{2/3}}\left(1+\left(\frac{x}{x_{\rm inj}}\right)^{5/3}\right)^{1/4}, \tag{5.7}$$

where

$$x_{\rm inj} \simeq 1.43\left(\frac{M_{\rm Pl}^2}{g_{\rm SM}^{\rm dec}\, f_V^4\, m_V^4\, \tau_V^2}\right)^{1/5} \simeq \left(\frac{T_{\rm dec}^{\rm after}}{T_{\rm dom}}\right)^{4/5}, \qquad T_{\rm inj} \simeq T_{\rm dec}^{\rm after}\left(\frac{T_{\rm dom}}{T_{\rm dec}^{\rm after}}\right)^{1/5}. \tag{5.8}$$

$x_{\rm inj}$ corresponds to the time where entropy injection becomes significant, which we define as $S_{\rm SM}(x_{\rm inj})/S_{\rm SM}^{\rm before} = 2^{3/4}$, see Eq. (5.5). Note that Eq. (5.7) implies $T_{\rm SM} \propto x^{-2/3}$ for $x \ll x_{\rm inj}$, as expected for a matter-dominated Universe, and $T_{\rm SM} \propto x^{-1/4}$ for $x \gg x_{\rm inj}$, as expected for a matter-dominated Universe during a period of entropy dilution (in which $T \propto a^{-3/8}$). We emphasise that, in deriving Eq. (5.7), we have assumed a matter dominated universe. That is an excellent approximation for $x \ll 1$. However, around $x \simeq 1$, the radiation energy content of the universe is non-negligible, and therefore our result for the temperature at time of decay, $T_{\rm SM}^{\rm decay}$, is a priori less accurate. We also determine the temperature evolution by solving the Friedmann equations (with decay) numerically, and show our results in Fig. 5.1. Our analytical determination of $x_{\rm inj}$ and $T_{\rm SM}^{\rm inj}$ reproduces very well the point where the numerical solution changes slope, as also shown in Fig. 5.1. Setting $x = 1$ in Eq. (5.7) and assuming a significant entropy injection $x_{\rm inj} \ll 1$, we get the well known temperature below which standard cosmology is recovered [53]

$$T_{\rm SM}^{\rm decay} \approx 1.25\frac{(g_{\rm SM}^{\rm dec})^{1/12}}{g_{\rm SM}^{1/3}}\sqrt{M_{\rm Pl}\Gamma_V}. \tag{5.9}$$

We first note that $T_{\rm SM}^{\rm decay}$ reproduces fairly well its numerical determination, see Fig. 5.1. $T_{\rm SM}^{\rm decay}$ is independent of the mass of the relic (except via the dependence of Γ_V on m_V), its number of degrees of freedom or its initial temperature. Furthermore, under the very good approximation where $g_{\rm SM}^{\rm dec} = g_{\rm SM}$, $T_{\rm SM}^{\rm decay}$ is almost equal to the temperature of a radiation-dominated universe with age τ_V (set by $H(T_{\rm SM}) = 1/2\tau_V$), $T_{\rm rad}(\tau_V) \simeq 1.23\sqrt{M_{\rm Pl}\Gamma_V}/g_{\rm SM}^{1/4}$.[3] The net result is that if a cold relic has dominated the energy density of the universe and has subsequently injected entropy during its decay, then the temperature of the universe at decay time, $T_{\rm SM}^{\rm decay}$, corresponds roughly to that in the standard cosmology (namely without early matter domination nor entropy injection) at time $t = \tau_V$. The only difference with respect to the standard case is that the universe has expanded more: this is the dilution effect.

[3] This corroborates the validity of the approximation, common in the literature, where it is assumed that the cold relic decays instantaneously when $H = \Gamma_V$.

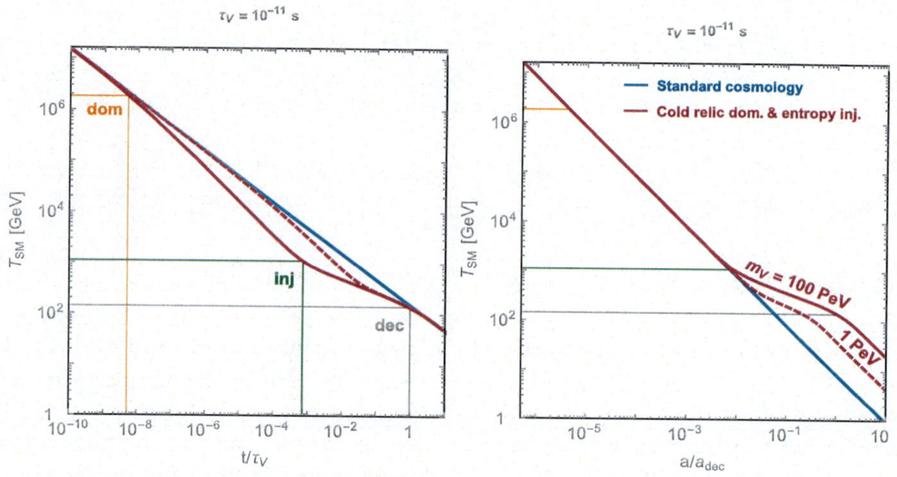

Fig. 5.1 Evolution of the SM temperature with time (left) and with scale factor (right), for $m_V = 100$ PeV (continuous purple) and 1 PeV (dashed purple), computed after integrating the full Friedmann equation. The evolution of the SM temperature in the absence of dilution is shown as a continuous blue line for comparison. Just after the cold relic dominates the energy density of the universe ('dom'), the temperature decreases with time as $T \propto t^{-2/3}$, i.e. faster than in a radiation-dominated Universe for which $T \propto t^{-1/2}$. However, when entropy injection starts to become significant ('inj'), the temperature decreases more slowly $T \propto t^{-1/4}$. The faster redshift during the period of matter domination is counterbalanced by the slower redshift during the period of entropy injection such that the temperature of the SM at the time of decay ('dec') is independent of the mass of the relic and is the same as in standard cosmology. Concerning the scale factor, the temperature starts to depart from the standard evolution, $T \propto a^{-3/8}$ instead of $T \propto a^{-1}$, only after the entropy injection has started. The three moments described above, 'dom' 'dec' and 'inj', are indicated for the case $m_V = 100$ PeV with thin lines, obtained analytically as described in the text

5.2.3 *Dilution by Entropy Injection*

The dilution factor:
We first review the dilution via entropy injection. When a cold relic (in our case, the mediator) decays into SM particles, its energy density is transferred to the SM bath. The non-relativistic degrees of freedom of the mediator are converted into relativistic degrees of freedom of the SM. This heat production increases the total entropy of the Universe and dilutes the frozen-out comoving DM number density n_X/s_{tot}. We shall use the superscripts 'before' and 'after' to denote the times before V decays but after DM freezes-out ($t \ll \tau_V$), and after V decays ($t \gtrsim \tau_V$), respectively. Then the comoving DM density today n_X^0/s_{tot}^0 is related to the comoving DM density at freeze-out $n_X^{\text{FO}}/s_{\text{tot}}^{\text{FO}}$ through[4]

[4] As is conventional, the uppercase letters S and N refer to the comoving entropy and number densities whereas the lowercase s and n refers to the entropy and number densities, such that $S = sa^3$ and $N = na^3$ where a is the scale factor.

$$\frac{n_X^0}{s_{\rm tot}^0} = \frac{n_X^0}{s_{\rm SM}^0} = \frac{N_X^{\rm after}}{S_{\rm SM}^{\rm after}} = \frac{1}{D_{\rm SM}}\frac{N_X^{\rm before}}{S_{\rm SM}^{\rm before}} = \frac{1}{D_{\rm SM}}\frac{n_X^{\rm FO}}{s_{\rm SM}^{\rm FO}} = \frac{1}{D}\frac{n_X^{\rm FO}}{s_{\rm tot}^{\rm FO}}, \tag{5.10}$$

where we used $N_X^{\rm after} = N_X^{\rm before}$ and defined the dilution factors

$$D_{\rm SM} \equiv S_{\rm SM}^{\rm after}/S_{\rm SM}^{\rm before} \quad \text{and} \quad D \equiv S_{\rm SM}^{\rm after}/S_{\rm tot}^{\rm before}, \tag{5.11}$$

which are of course related via

$$D = \frac{D_{\rm SM}}{1 + (\tilde{g}_D/\tilde{g}_{\rm SM})\tilde{r}^3}. \tag{5.12}$$

$D = 1$ corresponds to no dilution of the DM density.

The instantaneous approximation:
If we assume that the decay occurs instantaneously when $H \simeq \Gamma_V$, then we can neglect the universe expansion and the energy density is conserved through the decay. Therefore, we can deduce the temperature of the universe just before the decay $T_{\rm dec}^{\rm before}$

$$\rho_{\rm dec}^{\rm before} = \rho_{\rm dec}^{\rm after} \quad \rightarrow \quad m_V\, f_{\rm V}\, \frac{2\pi^2}{45} g_{\rm SM}\left(T_{\rm dec}^{\rm before}\right)^3 = \frac{\pi^2}{30} g_{\rm SM}\left(T_{\rm dec}^{\rm after}\right)^4, \tag{5.13}$$

$$\rightarrow \quad \frac{4}{3}\, m_V\, f_{\rm V} = \frac{\left(T_{\rm dec}^{\rm after}\right)^4}{\left(T_{\rm dec}^{\rm before}\right)^3}, \tag{5.14}$$

$$\rightarrow \quad \left(T_{\rm dec}^{\rm before}\right)^3 = \frac{\left(T_{\rm dec}^{\rm after}\right)^4}{T_{\rm dom}}. \tag{5.15}$$

We deduce the dilution factor

$$D_{\rm SM} \equiv \frac{S^{\rm after}}{S^{\rm before}} \simeq \left(\frac{T_{\rm dec}^{\rm after}}{T_{\rm dec}^{\rm before}}\right)^3 \simeq \frac{T_{\rm dom}}{T_{\rm dec}^{\rm after}} \simeq \frac{4}{3}\, f_{\rm V} \left(\frac{\pi^2 g_{\rm SM}}{90}\right)^{1/4} \frac{m_V}{\sqrt{\Gamma_V m_{\rm pl}}}. \tag{5.16}$$

where S is the total entropy $S = s\, a^3$ and s is the comoving entropy.

The Friedmann equation with decay term:
A more precise computation of the dilution factor involves the second law of thermodynamics together with the resolution of the Friedmann equation in the presence of decaying matter. We refer the reader to [53] for the derivation and we just report here the expression for the dilution factor ($M_{\rm Pl} \simeq 2.4 \times 10^{18}$ GeV, $\Gamma_V = 1/\tau_V$)

$$D_{\rm SM} = \left[1 + 0.77\, (g_{\rm SM}^{\rm dec})^{1/3}\, f_V^{4/3} \left(\frac{m_V^2}{\Gamma_V\, M_{\rm Pl}}\right)^{\frac{2}{3}}\right]^{\frac{3}{4}}, \tag{5.17}$$

which assumes that the mediator dominates the energy density of the Universe when it starts to decay. By comparing Eqs. (5.10) and (5.17), we conclude that the error done by using the instantaneous approximation is only 7 %.

The effect of the dark sector temperature on the dilution:
From Eq. (5.17) we see that for a given mass and lifetime of the mediator, the dependence of the dilution factor on the thermodynamics of the dark sector—its degrees of freedom and its temperature—is dominantly encoded in f_V. We neglect the dependence on $\tilde{r}$ introduced by $g_{\rm SM}^{\rm dec}$, because it is much milder than that arising from f_V, and because here we do not aim at an extremely precise study but rather at a simple analytical understanding. We may then rewrite Eq. (5.17) as

$$D_{\rm SM} \simeq \left[1 + \left(\bar{D}_{\rm SM}^{4/3} - 1\right) \tilde{r}^4\right]^{3/4}, \tag{5.18}$$

where $\bar{D}_{\rm SM}$ encapsulates all the dependence on the model parameters $m_{\rm V}$, Γ_V, $\tilde{g}_D$ and $\tilde{g}_{\rm SM}$,

$$\bar{D}_{\rm SM} \equiv \left[1 + \left(0.23\, (g_{\rm SM}^{\rm dec})^{1/4}\, \frac{\tilde{g}_D}{\tilde{g}_{\rm SM}} \frac{m_{\rm V}}{\sqrt{\Gamma_V M_{\rm Pl}}}\right)^{4/3}\right]^{3/4}, \tag{5.19}$$

where $\bar{D}_{\rm SM} = D_{\rm SM}^{\tilde{r}=1}$. For large values of $\bar{D}_{\rm SM}$ and/or the temperature ratio $\tilde{r}$, Eq. (5.18) implies the simple scaling $D_{\rm SM} \propto \tilde{r}^3$, with D consequently saturating to a constant value, $D \simeq (\tilde{g}_{\rm SM}/\tilde{g}_D)\bar{D}_{\rm SM} \simeq 0.23(g_{\rm SM}^{\rm dec})^{1/4} m_{\rm V}/\sqrt{\Gamma_V M_{\rm Pl}}$ [cf. Eq. (5.12)].

Cross-section:
Ultimately, we are interested in the couplings needed to reproduce the correct relic abundance, since they determine the DM signals. Let $\langle\sigma v_{\rm rel}\rangle^{\rm FO}$ be the thermally averaged effective annihilation cross-section times relative velocity around the time of freeze-out, which is of course determined by these couplings.[5] Large dilution implies that a smaller $\langle\sigma v_{\rm rel}\rangle^{\rm FO}$ is needed. From Eq. (5.18), it is evident that the dilution is larger the warmer the dark sector is, since more energy is stored into the mediators and transferred to the SM once they decay. However, $r_{\rm FO} \neq 1$ also affects $\langle\sigma v_{\rm rel}\rangle^{\rm FO}$ directly, i.e. even in the absence of any significant dilution. We shall now incorporate this effect.

In the presence of a dark plasma, the Hubble parameter is $H/H_{\rm SM} = \sqrt{1 + (g_{\rm D}/g_{\rm SM})\, r^4}$, where $H_{\rm SM}$ is the Hubble parameter in the absence of the dark sector. The comoving DM energy density today is

[5] We use the specification "FO" because if $\sigma v_{\rm rel}$ depends on $v_{\rm rel}$, as is the case with Sommerfeld enhanced processes that are relevant for heavy DM, then its value during freeze-out is of course different than at later times, when the DM indirect detection signals are generated. Moreover, we use the term "effective annihilation cross-section" because the DM density may be depleted via processes other than direct annihilation—in particular, the formation and decay of unstable bound states—as we shall see in Sect. 5.3.

$$\frac{M_{\rm DM} n_X^0}{s_{\rm SM}^0} = \frac{M_{\rm DM}}{D_{\rm SM}} \frac{n_X^{\rm FO}}{s_{\rm SM}^{\rm FO}} \simeq \frac{M_{\rm DM}}{D_{\rm SM}\, s_{\rm SM}^{\rm FO}} \frac{H^{\rm FO}}{\langle \sigma v_{\rm rel}\rangle^{\rm FO}} = \frac{M_{\rm DM} H_{\rm SM}^{\rm FO}}{s_{\rm SM}^{\rm FO}} \frac{\sqrt{1+(g_D^{\rm FO}/g_{\rm SM}^{\rm FO})\, r_{\rm FO}^4}}{D_{\rm SM}} \frac{1}{\langle \sigma v_{\rm rel}\rangle^{\rm FO}}, \tag{5.20}$$

where in the first step we used Eq. (5.10), and in the second step we used the approximate freeze-out condition $n_X^{\rm FO}\langle \sigma v_{\rm rel}\rangle^{\rm FO} \simeq H^{\rm FO}$. The combination $M_{\rm DM} H_{\rm SM}^{\rm FO}/s_{\rm SM}^{\rm FO}$ is proportional to $M_{\rm DM}/T_{\rm SM} = r_{\rm FO} x^{\rm FO}$, where $x^{\rm FO} \equiv M_{\rm DM}/T_D^{\rm FO}$ (not to be confused with $x = t/\tau_V$ defined earlier) itself depends on $r_{\rm FO}$. From the above, we deduce that

$$\frac{\langle \sigma v_{\rm rel}\rangle^{\rm FO}}{\langle \sigma v_{\rm rel}\rangle^{\rm FO}_{\rm SM}} \simeq \frac{x^{\rm FO}}{x^{\rm FO}_{\rm SM}} \frac{r_{\rm FO}\sqrt{1+(g_D^{\rm FO}/g_{\rm SM}^{\rm FO}) r_{\rm FO}^4}}{D_{\rm SM}}, \tag{5.21}$$

where $x^{\rm FO}_{\rm SM}$ and $\langle \sigma v_{\rm rel}\rangle^{\rm FO}_{\rm SM}$ denote respectively the time parameter $x^{\rm FO}$ and the cross-section needed to establish the observed DM density if DM were annihilating directly into SM particles (in which case we identify T_D with $T_{\rm SM}$ in the definition of $x^{\rm FO}$). $D_{\rm SM}$ can be computed from Eqs. (5.18) and (5.19), and the dark-to-SM temperature ratio at freeze-out $r_{\rm FO}$ is related to $\tilde{r}$ via Eq. (5.3).

Let us now discuss the features of Eq. (5.21). For small $\tilde{r}$, the dilution factor is close to unity and Eq. (5.21) implies $\langle \sigma v_{\rm rel}\rangle_{\rm FO} \propto r_{\rm FO}$. This is expected since for smaller $\tilde{r}$ the DM freeze-out is happening earlier, thus $T_{\rm SM}^{\rm FO}$ is larger. This implies that

- the dilution due to the subsequent Hubble expansion is larger, so that the cross-section must be reduced by $r_{\rm FO}^3$ in order to compensate for the deficit in the DM abundance,
- the Hubble expansion rate during freeze-out is faster, so that the cross-section must be enhanced by $r_{\rm FO}^{-2}$ in order to keep $T_D^{\rm FO}$ and then the DM abundance fixed.

As a result $\langle \sigma v_{\rm rel}\rangle_{\rm FO} \propto r_{\rm FO}$. For larger $\tilde{r}$, the dilution factor becomes more significant, thus suppressing $\langle \sigma v_{\rm rel}\rangle_{\rm FO}$. However, for sufficiently large $\tilde{r}$, the dark sector energy density dominates the Universe and the Hubble expansion rate becomes independent of $r_{\rm FO}$, resulting in the suppression of $\langle \sigma v_{\rm rel}\rangle_{\rm FO}$ to saturate to the value

$$\left[\frac{\langle \sigma v_{\rm rel}\rangle^{\rm FO}}{\langle \sigma v_{\rm rel}\rangle^{\rm FO}_{\rm SM}}\right]_{\rm min} \simeq \frac{x^{\rm FO}}{x^{\rm FO}_{\rm SM}} \frac{1}{\bar{D}_{\rm SM}} \frac{\tilde{g}_D/\tilde{g}_{\rm SM}}{\sqrt{g_D^{\rm FO}/g_{\rm SM}^{\rm FO}}} \simeq 4.35\, \frac{x^{\rm FO}}{x^{\rm FO}_{\rm SM}} \frac{\sqrt{\Gamma_V M_{\rm Pl}}}{m_V} \frac{(g_{\rm SM}^{\rm FO}/g_D^{\rm FO})^{1/2}}{(g_{\rm SM}^{\rm dec})^{1/4}}. \tag{5.22}$$

The interplay between $\tilde{r}$ and $\bar{D}_{\rm SM}$ in determining $\langle \sigma v_{\rm rel}\rangle^{\rm FO}$ is depicted in Fig. 5.2 (left panel).

5.2.4 *Impact on Unitary Bound*

Sketch (neglecting r):
In this paragraph, we first drop the dark-to-SM temperarure ratio r. From comparing Eq. (5.21) with the thermally averaged unitarity cross-section, $\langle \sigma v_{\rm rel}\rangle^{\rm FO} =$

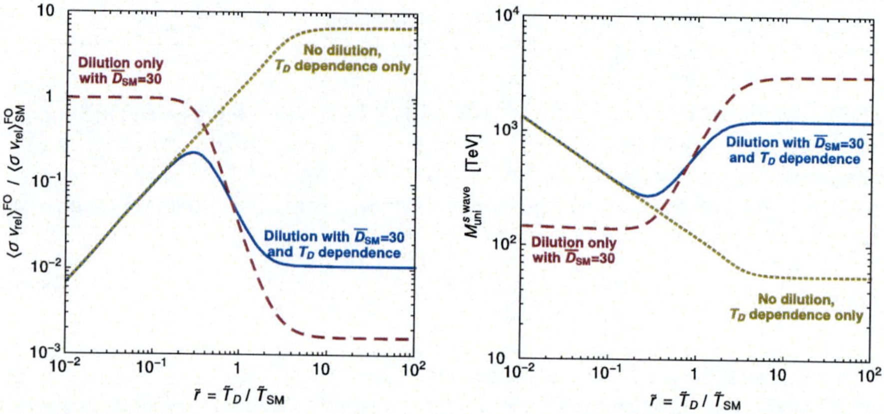

Fig. 5.2 Dependence of the effective annihilation cross-section at freeze-out required to attain the observed DM density (left) and of the s-wave unitarity limit on the mass of thermal relic DM (right), on the dark-to-SM temperature ratio and on the dilution due to the decay of the dark mediators. $\langle\sigma v_{\rm rel}\rangle^{\rm FO}$ is normalized to the corresponding value for DM of the same mass annihilating directly into SM particles with no dilution occurring, $\langle\sigma v_{\rm rel}\rangle^{\rm FO}_{\rm SM}$. The continuous blue lines display the combined effect of the dilution and the temperature ratio. To ease its understanding we also show the effect without dilution, namely $D = 1$ (dotted yellow), and of the dilution only, namely we retain the dependence on $\tilde{r}$ of $\langle\sigma v_{\rm rel}\rangle^{\rm FO}_{\rm SM}$ (dashed purple). Note that higher partial waves may contribute significantly to the depletion of DM in the early Universe, thereby raising the unitarity bound on the DM mass. For definiteness, we have used $\tilde{g}_{\rm SM} = 106.75$, $\tilde{g}_D = 6.5$, $g_D^{\rm FO} = 3$ (cf. Sect. 5.3) and for the left panel we assumed $x^{\rm FO}_{\rm SM} = 29$

$\langle\sigma_{\rm uni} v_{\rm rel}\rangle = 4(2J+1)\sqrt{\pi x^{\rm FO}}/M^2_{\rm uni}$, cf. Sect. 4.3.4, we deduce that the upper bound on the DM mass from unitarity is increased by a factor $\sqrt{D}$

$$M_{\rm DM} \lesssim 137 \text{ TeV } \sqrt{\rm D} \tag{5.23}$$

with D in Eq. (5.10) and where 137 TeV is the standard unitarity bound on Sommerfeld-enhanced Dirac DM, annihilating via s-wave, cf. Table 4.1.

Plugging $m_V = 100$ PeV, $\Gamma_V = (0.1 \text{ s})^{-1}$, $g_{\rm SM} = 106.75$ and $f_V = 10^{-2}$ (cf. Eq. (9.9)), we get the maximal dilution factor compatible with BBN

$$D \simeq 6.2 \times 10^8 \left(\frac{f_V}{10^{-2}}\right)\left(\frac{m_V}{100 \text{ PeV}}\right)\left(\frac{\tau_V}{0.1 \text{ s}}\right)^{1/2}. \tag{5.24}$$

Therefore, for the purpose of setting BBN constraints, we impose $\tau_V < 0.03$ s for all values of the mass m_V, i.e. even for those leading to matter domination and entropy injection. Therefore, the maximal DM mass compatible with both unitarity and BBN is

$$M^{\rm max}_{\rm DM} \simeq 3.5 \text{ EeV} \left(\frac{f_V}{10^{-2}}\right)^{1/2}\left(\frac{m_V}{100 \text{ PeV}}\right)^{1/2}\left(\frac{\tau_V}{0.1 \text{ s}}\right)^{1/4}. \tag{5.25}$$

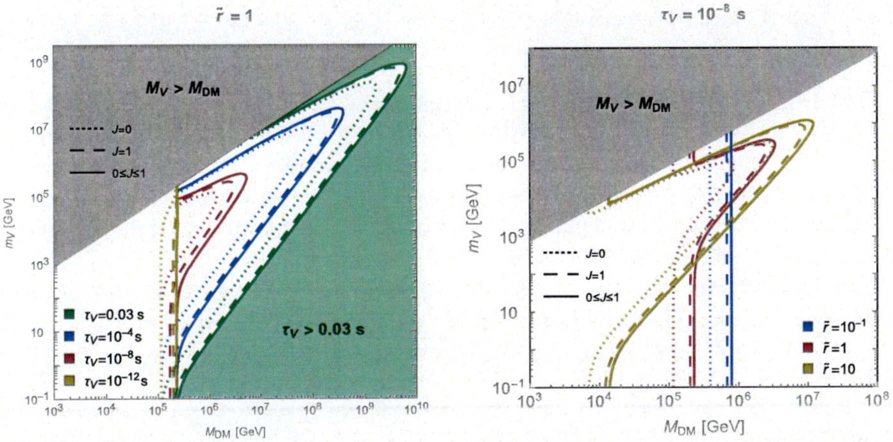

Fig. 5.3 The dependence of the upper bound on the mass of thermal relic DM implied by unitarity, $M_{\rm DM} \leqslant M_{\rm uni}$ (coloured lines), on the mediator mass m_V and lifetime τ_V, and the dark-to-SM temperature ratio $\tilde{r}$. For definiteness, we have used $\tilde{g}_{\rm SM} = 106.75$, $\tilde{g}_D = 6.5$ and $g_D^{\rm FO} = 3$ (cf. Sect. 5.3). *Left:* The dark plasma and the SM are assumed to have the same temperature at early times, $\tilde{r} = 1$. Large m_V and/or τ_V imply that the cosmological energy density carried by the mediators before their decay is large, leading to significant dilution of the frozen-out DM abundance upon their decay and raising $M_{\rm uni}$. However, a lifetime longer than 0.03 s is in conflict with BBN (green-shaded area). Also, the dilution factor gets exponentially suppressed with m_V when m_V is close to $M_{\rm DM}$ (footnote 4), thus bringing $M_{\rm uni}$ back to its value in the standard scenario (for the appropriate $\tilde{r}$). *Right:* For fixed τ_V and small m_V the dilution is negligible; then, larger $\tilde{r}$ implies less efficient depletion of DM (due to less Universe expansion from DM freeze-out, which happens later, till today) and more stringent $M_{\rm uni}$. However, for larger m_V, the dilution due to the mediator decay is significant and increases with $\tilde{r}$, thereby raising $M_{\rm uni}$. Note that for sufficiently large $\tilde{r}$, $M_{\rm uni}$ becomes independent of $\tilde{r}$ (see text for discussion)

The maximal DM mass in Eq. (5.25) can be visualized in Fig. 5.3.

Dependence on the dark-to-SM temperature ratio r:
In the previous paragraph, we have been sloppy about the dependence on the dark-to-SM temperature ratio r. If we now re-introduce this dependence, we obtain

$$M_{\rm uni} \simeq M_{\rm uni,0}^{\rm SM} \left(\frac{x_{\rm SM}^{\rm FO}}{x^{\rm FO}}\right)_{\rm uni}^{1/4} \left[\frac{D_{\rm SM}}{r_{\rm FO}\sqrt{1+(g_D^{\rm FO}/g_{\rm SM}^{\rm FO})r_{\rm FO}^4}}\right]^{1/2} \times \begin{cases} \sqrt{2J+1}, & \text{solely } J \\ J_{\max}+1, & 0 \leqslant J \leqslant J_{\max} \end{cases}, \tag{5.26}$$

where $M_{\rm uni,0}^{\rm SM} \simeq 135$ TeV is the s-wave unitarity limit in the absence of any dilution and if DM is annihilating directly into the SM plasma. The parameters $x^{\rm FO}$ and $x_{\rm SM}^{\rm FO}$ at the unitarity limit, appearing in Eq. (5.26), can be determined by the standard formula,[6] using the unitarity cross-section; we find $x_{\rm SM}^{\rm FO} \simeq 31 + (1/2)\ln[(2J+1)/g_{\rm SM}^{\rm FO}]$ and $x^{\rm FO} - x_{\rm SM}^{\rm FO} = -\ln(M_{\rm uni}/M_{\rm uni}^{\rm SM}) + \ln(r_{\rm FO}^2/\sqrt{1+(g_D^{\rm FO}/g_{\rm SM}^{\rm FO})r_{\rm FO}^4})$.

We recall that $r_{\rm FO}$ and $D_{\rm SM}$ can be computed using Eqs. (5.3), (5.18) and (5.19). This result is consistent with previous results in the limit of no dilution and $\tilde{r} = 1$ [7, 8]. Evidently, $M_{\rm uni}$ depends on the thermodynamics of the dark sector, its temperature and its degrees of freedom; in the presence of dilution, it also depends on the properties of the mediator, its lifetime and mass. In Figs. 5.2 (right panel) and 5.3, we illustrate these relations.

To conclude this section, we find it interesting to report the maximal mass that DM could in principle reach in such models, $M_{\rm uni}^{\rm max}$, being agnostic on the way this limit would be realised. We obtain $M_{\rm uni}^{\rm max}$ from Eq. (5.26) upon maximising $D_{\rm SM}$ (we recall that EeV $= 10^3$ PeV)

$$M_{\rm uni}^{\rm max} \simeq 1.1\,{\rm EeV}\left(g_D^{\rm FO}\frac{m_V}{100\,{\rm PeV}}\left(\frac{\tau_V}{0.03\,{\rm s}}\right)^{\frac{1}{2}}\right)^{\frac{1}{2}}\frac{r_{\rm FO}}{(1+0.009\,g_D^{\rm FO}\,r_{\rm FO}^4)^{\frac{1}{4}}} \times \begin{cases} \sqrt{2J+1}, & \text{solely } J \\ J_{\rm max}+1, & 0 \leqslant J \leqslant J_{\rm max} \end{cases}, \tag{5.27}$$

where for simplicity of the exposition we have omitted the weak log dependence.[7] In the dark $U(1)$ model, with $\tilde{g}_D = 6.5$, $\tilde{r} = 1$, $g_D^{\rm FO} = 3$, $s + p$ wave, this translates to

$$M_{\rm uni}^{\rm max} \simeq 2.4\,{\rm EeV}\left(\frac{m_V}{100\,{\rm PeV}}\right)^{1/2}\left(\frac{\tau_V}{0.03\,{\rm s}}\right)^{1/4}, \tag{5.29}$$

where we have normalized m_V to a value which guarantees that freeze-out happens when V is relativistic (otherwise Eq. (5.29) does not hold).

5.3 The Dark U(1) Model as a Case of Study

5.3.1 The Lagrangian

We consider Dirac fermionic DM X, charged under a dark gauge group $U(1)_D$, under which all the SM particles are neutral. We assume that the dark sector communicates

---

[6] For $\langle\sigma v_{\rm rel}\rangle = \sigma_0 x^{1/2}$, the freeze-out temperature can be estimated by (see e.g. [54], and Ref. [8] for generalization to dark sector freeze-out)

$$x^{\rm FO} \simeq \ln\left[0.095\frac{g_{\rm DM}}{\sqrt{g_{\rm SM}^{\rm FO}}}\frac{r_{\rm FO}^2}{\sqrt{1+(g_D^{\rm FO}/g_{\rm SM}^{\rm FO})r_{\rm FO}^4}}M_{\rm Pl}M_{\rm DM}\sigma_0\right],$$

where $g_{\rm DM}$ are the DM degrees of freedom and $M_{\rm Pl}$ is the reduced Planck mass. .

[7] For completeness, it amounts to multiplying Eq. (5.27) by

$$\left[1 - 0.025\log\left(r_{\rm FO}^{-2}\,g_D{}^{\rm FO}\frac{m_V}{100\,{\rm PeV}}\left(\frac{\tau_V}{0.03\,{\rm s}}\right)^{1/2}\left(1+0.009\,g_D{}^{\rm FO}\,r_{\rm FO}^4\right)^{1/2}\right).\right]^{-1/4} \tag{5.28}$$

with the SM via kinetic mixing between $U(1)_D$ and the hypercharge group $U(1)_Y$. The Lagrangian then reads

$$\mathscr{L} = -\frac{1}{4} F_{D\mu\nu} F_D^{\mu\nu} - \frac{\varepsilon}{2\,c_w} F_{D\mu\nu} F_Y^{\mu\nu} + \bar{X}(i\not{D} - M_{\rm DM})X, \tag{5.30}$$

where $D_\mu = \partial_\mu + i g_D V'_\mu$ is the covariant derivative, with V'_μ being the dark gauge field. (We reserve the symbol V^μ for the state arising after the diagonalisation of the kinetic terms, as described below.) $F_{D,Y}$ are the dark and hypercharge field strength tensors, and c_w is the cosine of the weak angle. We define the dark fine structure constant $\alpha_D \equiv g_D^2/(4\pi)$.

We further assume that the dark photon V^μ obtains a mass $m_{\rm V}$, either via the Stückelberg or the Higgs mechanisms. We shall remain agnostic about which of the two is realized, and only note that in case of the Higgs mechanism, the extra scalar can be decoupled (and thus made irrelevant for the phenomenology discussed in this study) by choosing its charge to be much smaller than one. Dark photons described by Eq. (5.30) and with a Stückelberg mass arise, for example, in string theory constructions, where the predicted ranges for ε and $m_{\rm V}$ include those of interest in our study, see e.g. [55, 56].

5.3.2 *Dark Photon*

The dark photon interacts with SM particles via its kinetic mixing ε with the hypercharge gauge boson. Upon diagonalisation of the kinetic and mass terms[8], its couplings may be written as

$$\mathscr{L} \supset \varepsilon\, e \left(\frac{1}{1-\left(\frac{m_{\rm V}}{m_Z}\right)^2} J^\mu_{\rm em} - \frac{1}{c_w^2} \frac{\left(\frac{m_{\rm V}}{m_Z}\right)^2}{1-\left(\frac{m_{\rm V}}{m_Z}\right)^2} J^\mu_Y + O(\varepsilon^2) \right) V_\mu \tag{5.31}$$

where $J^\mu_{\rm em}$ and J^μ_Y are the usual electromagnetic and hypercharge currents, and where we have denoted by V the mass eigenstate dominated by the gauge eigenstate V'. Equation (5.31) makes transparent both the ε-suppression of the V couplings to SM particles, and the physical limits where V couples as the photon ($m_{\rm V} \ll m_Z$) or as the hypercharge gauge boson ($m_{\rm V} \gg m_Z$). The expansion of Eq. (5.31) in ε is not valid for $m_{\rm V} \simeq m_Z$, where V couples as the Z boson. In our study we use the full expressions valid also in that limit.

In this study we will be interested in dark photon masses larger than 100 MeV, so that tree-level decays to at least a pair of SM particles are always open. We compute the decay widths of the dark photon to all possible two-body final states, and present

[8] This has an effect also on the mass and couplings of the Z boson, that is phenomenologically irrelevant for the small values of ε that we will be interested in, see e.g. [57] for the related constraints.

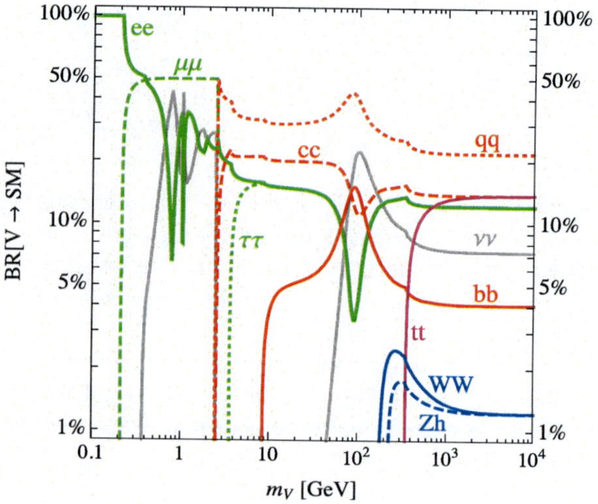

Fig. 5.4 Branching ratios of the dark photon. This figure updates Fig. 4 of [48]

them in Appendix 5.B. For $m_V > \Lambda_{QCD}$, the hadronic decay channels are open, but for $m_V <$ few GeV they cannot be described perturbatively. In that interval, which for definiteness we take to be 350 MeV $< m_V <$ 2.5 GeV, we determine the total decay width of V from measurements of $e^+e^- \to$ hadrons at colliders (see e.g. [57, 58]). Following [48], we assume this width to consist half of μ and half of ν pairs, since these constitute the dominant final states from hadronic decays. We show the resulting branching ratios in Fig. 5.4 and refer to Appendix 5.B for more details.

Throughout this work, we will be interested in the region $M_{DM} > m_V$. The phenomenology of DM is then driven by its annihilations into dark photons, whose tree-level cross section scales as α_D^2. The tree-level cross section for DM annihilation into SM pairs scales as $\varepsilon^2 \alpha_D \alpha_{SM}$, where α_{SM} generically stands for an SM EW coupling, and is therefore negligible for the values of ε we will be interested in.

5.3.3 DM Relic Abundance and Dilution

For each values of the masses M_{DM} and m_V, we compute[9] the value of the dark fine structure constant α_D which leads to the correct value of the DM relic abundance today. We use the procedure in [7] which takes into account both the Sommerfeld enhancement of the direct annihilation and BSF. The novelties of the present work with respect to the most recent analogous computations [7, 48] are the following:

- We take into account the full Yukawa potential for computing the Sommerfeld enhancement and BSF, as opposed to the Coulomb limit only.[10] This is important

[9] We use the latest value of the DM relic abundance today measured by Planck [59], $\Omega_{DM}h^2 = 0.1186 \pm 0.0020$ and the value of the effective number of neutrinos computed in [60], $N_{eff} = 3.046$.

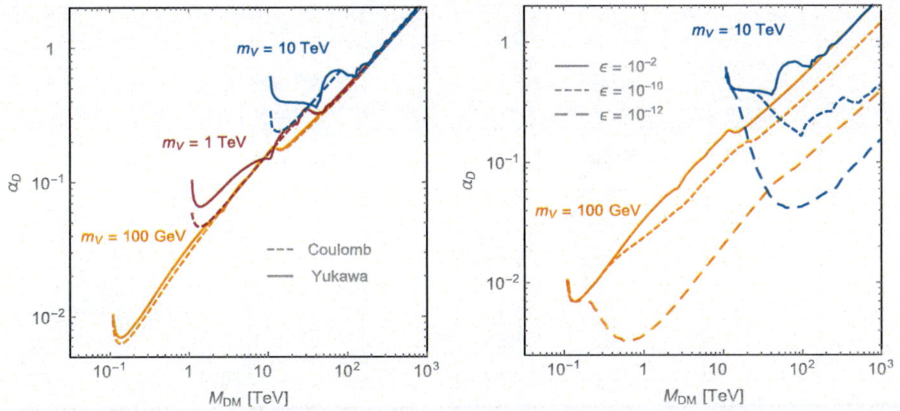

Fig. 5.5 Values of α_D that yield the observed DM abundance as a function of the DM mass, for some reference values of the dark photon mass. Left: our results including the full Yukawa potential (continuous) are compared with the ones in the Coulomb approximation (dashed). Right: comparison of cases with different kinetic mixing ε and therefore with different dilution. The dilution is exponentially suppressed when m_V is close to M_{DM} (see Appendix A of [1])

at large values of the dark photon mass, as seen in the left panel of Fig. 5.5 (we remind that the cross section relevant for the signals scales as α_D^3 away from resonances).

- We study the impact of the entropy dilution due to the dark photon decay, whose effect is shown in the right panel of Fig. 5.5 for some reference values of the kinetic mixing ε (note $\Gamma_V \propto \varepsilon^2$ and see Eq. (5.17) for the dilution).

Equipped with the dark coupling constant that gives the observed DM relic density, we next compute DM annihilation and BSF rates relevant for the cosmic ray signals.

5.3.4 DM Signals

We take both the direct DM annihilation and BSF into account in the computation of the DM relic abundance (and of the unitarity bound), and in the estimation of the DM signals during the cosmic microwave background (CMB), 21 cm, at reionization, in the Milky Way (MW) and in dwarf spheroidal galaxies (dSph).

MW and dSph:
For the MW and dSph we compute S_{ann} and S_{BSF} numerically as in [61]. We then assume a Maxwellian DM velocity distribution $f(\mathbf{v}) = N\,\theta(v - v_{\text{esc}})\,e^{-v^2/v_0^2}$ for each of the interacting particles, where N is fixed by imposing $\int d^3v\, f(\mathbf{v}) = 1$. The velocity $\mathbf{v}$ and the distribution $f(\mathbf{v})$ should not be confused with v_{rel} and the

[10] The Sommerfeld factors with the full Yukawa potential have been computed by Kallia Petraki.

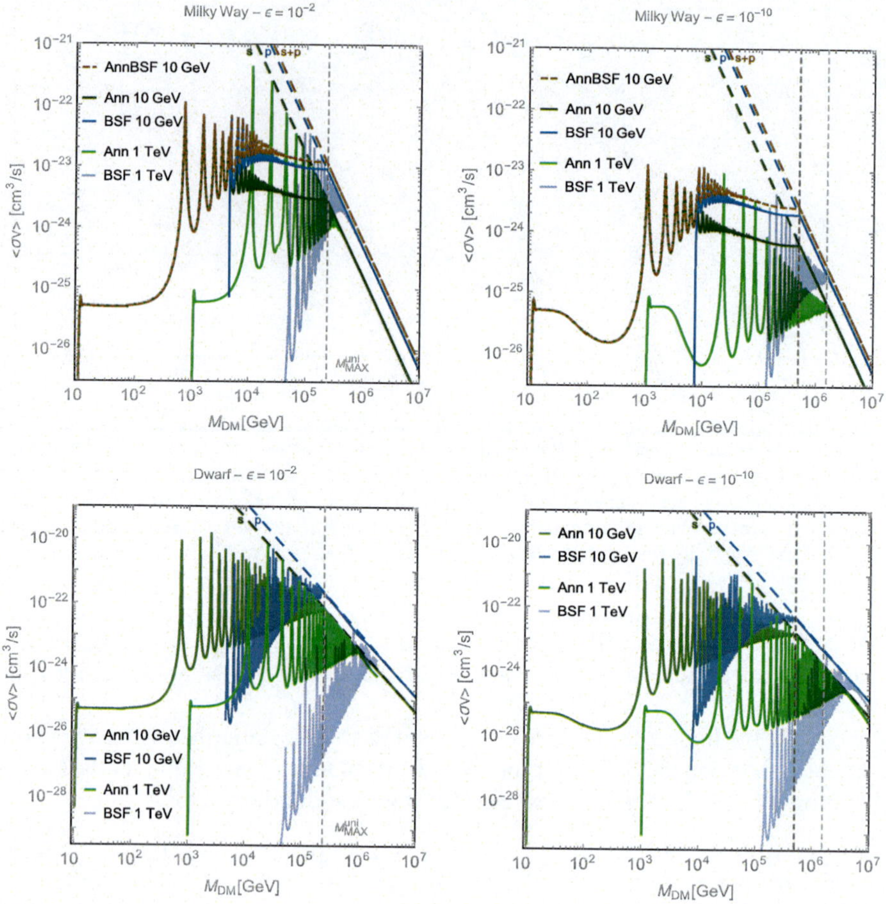

Fig. 5.6 Cross-section averaged over the velocity distribution of the DM in the Milky Way and in Dwarfs versus the dark photon mass. The dashed straight lines are the maximum cross-section values allowed by unitarity for s, p and $s + p$ partial waves. The first column is without entropy dilution whereas in the second column, entropy dilution is present with kinetic mixing $\varepsilon = 10^{-10}$. For m_V close to $M_{\rm DM}$, the dilution is exponentially suppressed (see Appendix A of [1])

distribution over $v_{\rm rel}$; the latter is derived from $\int f(v_1) f(v_2)\, d^3v_1 d^3v_2$ after carrying out the integration over the mean velocity $(\mathbf{v_1} + \mathbf{v_2})/2$. We choose $v_{\rm esc} = 533$ km/s and $v_0 = 220$ km/s for the MW [62], and $v_{\rm esc} = 15$ km/s and $v_0 = 10$ km/s for dSph's [63, 64].

The resulting cross sections $\langle \sigma_{\rm ann} v_{\rm rel} \rangle$ and $\langle \sigma_{\rm BSF} v_{\rm rel} \rangle$ are shown in Fig. 5.6 as a function of $M_{\rm DM}$, for some reference values of m_V. We have fixed α_D to the value that reproduces the correct DM relic abundance, and which depends on $M_{\rm DM}$ and very mildly on m_V (see Sect. 5.3.3). Figure 5.6 displays the features discussed in the previous subsection. Going from smaller to larger $M_{\rm DM}$, the total cross section first coincides with the tree-level one, then the Sommerfeld enhancement of $\sigma_{\rm ann}v_{\rm rel}$ becomes relevant, and for even larger $M_{\rm DM}$ also BSF becomes important. Both $\sigma_{\rm ann}v_{\rm rel}$ and $\sigma_{\rm BSF}v_{\rm rel}$ display resonances, and reach the Coulomb limit at a larger value of $M_{\rm DM}$, which is different between the MW and dSph because the two systems are characterised by different DM velocities. The change of slope around $M_{\rm DM} \simeq 200$ TeV reflects our prescription for not violating the unitarity limit.

CMB and 21 cm:
The evolution of the DM temperature implies that, for the values of the parameters of interest for this study, at redshifts relevant for CMB constraints ($z \approx 600$, see e.g. [65]) one has $v_{\rm rel} \ll 10^{-10}$ [48]. Therefore

(i) the capture into the ground state is negligible because of the $v_{\rm rel}^2$ suppression, and
(ii) the Sommerfeld enhancement of the annihilation processes is well within a saturated regime, so we do not need to compute the velocity average.

At even smaller redshifts, inhomogeneities grow and DM structures start to form, so that the DM velocity acquires a component that depends on the gravitational potential of such structures, that could be larger than the value of $v_{\rm rel}$ relevant for CMB. However, at redshifts relevant for the 21 cm signal ($z \approx 17$, see e.g. [66]), the largest DM halos formed are expected to have masses several orders of magnitude smaller than MW-size galaxies, thus inducing low DM velocities. In addition, the dominant part of the signal from DM annihilations comes from the smallest halos, see e.g. [67, 68]. Therefore, the DM velocities relevant for the 21 cm limits are not expected to be so large that they invalidate (i) and (ii).

The small values of the velocity also imply that the numerical calculation of $S_{\rm ann}$ is impractical. Therefore, to derive the CMB and 21cm constraints in Sect. 5.4.2, we work with the Hulthén potential $V_H = -\alpha_D m_* e^{-m_* r}/(1 - e^{-m_* r})$, that allows for an analytic solution $S_{\rm ann}^H$ (see e.g. [69, 70]). $S_{\rm ann}^H$ approximates well $S_{\rm ann}$ off-resonance, it correctly reproduces the fact that the resonances become denser at larger values of $\alpha_D M_{\rm DM}/m_{\rm V}$, and we choose $m_* = 1.68\, m_{\rm V}$ such that the position of the first resonance coincides with the position of the first resonance in the standard Yukawa case. For both the CMB and 21cm constraints we work with $v = 10^{-11}$.[11]

5.4 Phenomenology

In this Section[12] we present the constraints on the parameter space of the model imposed by the different signals that we consider. For the ease of the reader, we

[11] We have explicitly verified that the regions excluded by CMB do not change for smaller v values, and that those excluded by 21cm observations and allowed by unitarity do not change up to $v \simeq 10^{-8}$ (see Sect. 5.4.2).

[12] Section 5.4 has been mostly written by Filippo Sala, apart from Sects. 5.4.3 and 5.4.3 which have been mostly written by Marco Cirelli.

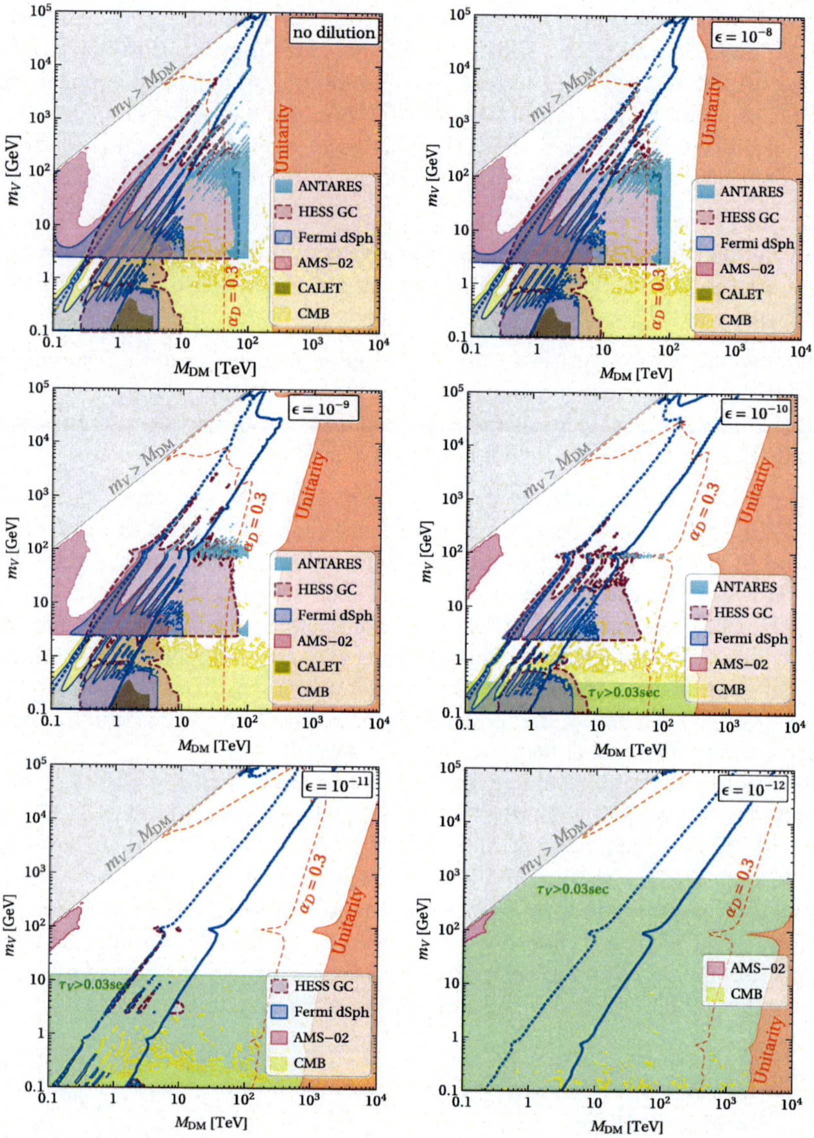

Fig. 5.7 Indirect detection limits. Shaded: excluded by ANTARES (cyan), AMS (magenta), CALET (ocra), PLANCK (light yellow), FERMI (blue), within HESS reach (magenta), violates s+p wave unitarity bound (red), disfavoured by BBN (green), DM annihilation to dark photons kinematically forbidden (gray). On the right of the lines: bound states exist (blue dotted), bound states can form (blue), $\alpha_D > 0.3$ (red dashed, as a rough indication of where higher order in α_D are expected to become important). From top-left to bottom-right: no DM dilution, and then decreasing values of ε corresponding to increasing DM dilution. Note the wider extension in M_{DM} and m_V in the bottom right panel

first present a summary of the results in Fig. 5.7, where each panel corresponds to decreasing values of ε, i.e. increasing dilution ($\Gamma_V \propto \varepsilon^2$ and see Eq. (5.17) for the dilution). A few comments are in order.

- The various indirect detection probes are complementary and collectively cover very large portions of the parameter space. It is remarkable that some of them (neutrinos, γ-ray and CMB searches) are sensitive to DM with a mass beyond $O(10)$ TeV. Multimessenger astronomy is therefore a concrete possibility also for the quest of heavy DM, adding motivation to indirect detection in this mass range.
- The fact that the ANTARES and HESS regions end abruptly at $M_{\rm DM} \simeq 100$ TeV is only a consequence of the largest masses considered in the papers of the experimental collaborations. Our study provides a strong motivation for these experiments to extend their current searches to heavier DM masses, as this would likely test unexplored regions of parameter space. In addition, we point out that our analysis is partly a reinterpretation of the studies performed by the collaborations themselves and as such suffers from the fact that the SM spectra from cascade decays predicted in secluded models differ from the ones used by these collaborations. The above difficulties would be circumvented if public access to the data was possible.
- As expected, the dilution weakens the sensitivities of the various indirect detection probes, since the cross sections needed to obtain the correct DM abundance are smaller than in the standard case. It is interesting to note, however, that for increasing dilution the parameter space accessible to indirect detection shrinks because of BBN constraints. This gives a well-defined window of parameter space to aim at, with current and future telescopes.[13] From the model-building point of view, this shows explicitly that the upper limit on dilution sets a lower limit on the communication with the SM of secluded DM models.

Collider and direct detection searches have instead no power in testing these models, because of the large DM mass and of the small communication with the SM, parameterized by ε. We now move to illustrate in more details the phenomenology of this model.

5.4.1 Constraints on the Kinetic Mixing

For $\varepsilon \lesssim 10^{-9}$ and $m_{\rm V} \gtrsim$ GeV, the $\varepsilon - m_{\rm V}$ parameter space is constrained by the requirement that decays of the dark photon do not spoil BBN. We use here the results of the study [52], that derived BBN constraints on neutral particles decaying at early times. It found that, for particles decaying more than 30% hadronically (like the dark photon in the mass range of our interest), all constraints evaporate for lifetimes of the particle shorter than 0.03 s. Lifetimes as large as 10^2 s become allowed as soon as the abundance of the dark photon, at the time of its decay, drops

[13] Another challenge to test large $M_{\rm DM}$ with gamma ray telescopes is posed by the non-transparency of the Milky Way to photons energies of hundreds of TeV, see e.g. [71] for a recent related study.

below $O(0.1)$ times the critical density. As presented in Sect. 5.2.3, the region we are interested in is that where dark photons dominate the energy density of the Universe just before decaying, to realise sufficient entropy injection. Therefore, we conservatively label as 'disfavoured by BBN' the regions where $\tau_V > 0.03$ s. Note that for mediator masses below the GeV range, where hadronic decay modes close, the BBN constraints weaken significantly [72]. A more detailed analysis, e.g. with a proper mass dependence of the constraints, goes beyond the purpose of this work.

For smaller m_V and/or larger ε, the existence of dark photons is severely constrained also by DM Direct Detection, observations of the supernova SN1987A, beam dump experiments, and neutrino limits from the Sun. As that region is not the focus of interest of this work, we do not discuss these constraints here, and refer the interested reader to the discussion in [48] for the first three, and to [50, 51] for constraints from DM annihilation in the Sun.

5.4.2 DM Constraints from the Early Universe

CMB

DM annihilations at the time of CMB inject energy in the SM bath and could therefore alter the observed CMB spectrum, resulting in stringent limits [73]. Such limits are driven by the ionizing power of the SM final state i, which can be encoded in efficiency factors $f^i_{\rm eff}(M_{\rm DM})$ that we take from [74, 75]. Since these depend mostly on the total amount of energy injected, such limits do not depend on the number of steps between the DM annihilation and the final SM products [76]. Following also our discussion in Sect. 5.3.4, we then place limits as follows

$$\sum_{i=e\bar{e},u\bar{u},\ldots} \langle\sigma^H_{\rm ann} v\rangle\, {\rm BR}(V \to i)\, f^i_{\rm eff} < 8.2 \times 10^{-26}\,\frac{\rm cm^3}{\rm sec}\,\frac{M_{\rm DM}}{100\ {\rm GeV}}, \tag{5.32}$$

where on the left-hand side the dependence on $M_{\rm DM}$ and m_V is implicit. The resulting exclusion is displayed in light yellow in Fig. 5.7 for various possibilities of DM dilution.

21 cm

ΛCDM predicts an absorption signal in the radio band at 21 cm, associated with the relative occupation number of the singlet and triplet hyperfine states of the hydrogen atom at $z \simeq 17$, see e.g. [66]. The EDGES collaboration has recently reported [77] the observation of such a signal, although with an amplitude larger than expected in ΛCDM. Because extra injection of energy in the SM bath would make that amplitude smaller than expected, this observation allows to place limits on DM annihilations and/or decays, that turn out to be competitive with CMB ones [78–82]. Here we do not aim at explaining why the amplitude is observed to be larger than predicted, but rather we simply impose a limit on the DM annihilation following Ref. [78], as

$$\sum_{i=e\bar{e},u\bar{u},\ldots} \langle \sigma_{\rm ann}^{H} v \rangle \, {\rm BR}(V \to i) \, f_{\rm eff}^{i} < 6.3 \times 10^{-26} \, \frac{\rm cm^3}{\rm sec} \, \frac{M_{\rm DM}}{100 \, {\rm GeV}} \,, \tag{5.33}$$

where we have taken into account that DM in our model is not self-conjugate. Among the limits presented in [78], we have chosen the more conservative one, derived by imposing that DM does not reduce the standard amplitude by more than a factor of 4 ('$T_{21} \gtrsim 50$ mK'), and using the more conservative boost due to substructures ('Boost 1').

It is important to stress that the 21 cm limit in Eq. (5.33) is derived assuming an immediate absorption of the energy injected from the SM products of DM annihilations (so that, again, the number of steps in the cascade does not matter). This is however not the case especially when they are very energetic, thus for large DM masses. The effects of delayed energy deposition have been taken into account e.g. in [78, 80], however the resulting limits have not been reported for DM masses beyond 10 TeV. Up to those masses and choosing as an example the $b\bar{b}$ final state, the limits derived as in Eq. (5.33) are stronger than those that take into account delayed energy deposition by a factor of a few ($M_{\rm DM} \simeq 10$ GeV) to a few tens ($M_{\rm DM} \simeq 10$ TeV) [78]. We expect a difference in the same ballpark also in our model, because of the extra softening of the spectra from the extra-steps in the cascade, and because the dominant SM final states (see Fig. 5.4) result in an electron and photon spectra of energies much lower than $M_{\rm DM}$, in qualitative agreement with $b\bar{b}$. Such differences in the limits (between Eq. (5.33) and those that include delayed energy depositions) are of the same order of the uncertainties induced by assumptions like the boost model and the precise value of the bound to apply, see e.g. [78, 80]. In Fig. 8 of [1], we show that the 21 cm-constraints only slightly improves the CMB ones. Waiting for further experimental confirmations of the signal, and for a precise assessment of the 21 cm limit to be settled, we refrain from putting the 21 cm limits on the same footing of the CMB ones, and we show only the latter ones in the summary in Fig. 5.7.

5.4.3 DM Constraints from the Local Universe

Dark Matter interactions (annihilation, formation and decay of bound states) in the local Universe may induce signals at space- and ground-based telescopes, whose observations therefore put limits on the parameter space of our model. Here we study constraints from telescopes observing neutrinos (Sect. 5.4.3), gamma rays (5.4.3), antiprotons (5.4.3) and electrons and positrons (Sect. 5.4.3).

In our model, DM interactions result into 2 (from annihilations and from decays of $B_{\uparrow\downarrow}$) or 3 (from decays of $B_{\uparrow\uparrow}$) energetic dark photons, themselves decaying to SM pairs according to the branching ratios in Fig. 5.4. The resulting 'one-step' SM energy spectra are therefore moved to lower energies compared to the standard case of direct DM annihilation into SM pairs. This constitutes an important input in deriving the constraints discussed in this section, and will be discussed case by case.

The additional and much less energetic V, that is emitted to form the bound state, will be discussed separately in Sect. 5.4.4.

Neutrinos

The ANTARES collaboration derived constraints from the non-observation of excesses in muon neutrino fluxes coming from the Milky Way halo [83]. Those limits were cast assuming direct DM annihilations into various SM pairs, and therefore cannot be directly applied to our case. We circumvent this limitation following [84], that observed that the ANTARES limits are driven by the most energetic part of the neutrino spectra from DM. Therefore they can roughly be applied to one-step signals that result in quarks (up to bottom), because their zero- and one-step neutrino spectra are very similar for neutrino energies close to $M_{\rm DM}$. They can also be applied to one-step signals that result in neutrinos, as is the case in our model for $m_{\rm V}$ larger than a few tens of GeV, with a further caveat that we now discuss. The resulting 'one-step' neutrino spectrum is spread to energies lower than $M_{\rm DM}$, while direct DM annihilation into neutrino pairs would induce a spike at $E_\nu \simeq M_{\rm DM}$. However, the ANTARES finite energy resolution of O(50%) causes a sizeable fraction of 'one-step' neutrinos to fall in the same energy range of those originating from a spike. To summarize, we exclude a point in our parameter space if it does not respect the limit

$$\langle\sigma_{\rm tot} v_{\rm rel}\rangle {\rm BR}(V \to \bar{f} f) < 2\,C\,\langle\sigma_{{\rm DM\,DM}\to \bar{f} f}\, v_{\rm rel}\rangle^{\rm ANTARES}_{\rm limit} \tag{5.34}$$

for at least one of the two final states i) $\bar{f} f = \bar{b}b + \bar{c}c + \cdots + \bar{u}u$, that we compare with the $\bar{b}b$ limit from ANTARES because of the very similar spectrum, and ii) $\bar{f} f = \bar{\nu}_\mu \nu_\mu$, where we include the effect of oscillations (ANTARES presents limits for DM annihilation into $\bar{\nu}_\mu \nu_\mu$, so that the effective flux of muon neutrinos at the telescope is reduced to roughly 40% of the initial one). The factor of 2 in Eq. (5.34) accounts for the fact that, in our model, DM is not self-conjugate. We also choose $C = 1$ for quark final states and $C = 2$ for neutrino ones, to very roughly account for the neutrinos that will be properly recognized to have $E_\nu < M_{\rm DM}$.[14]

The resulting limits are shown in cyan in Fig. 5.7 for various values of the kinetic mixing ε (and therefore of DM dilution). We find also an excluded region for $m_{\rm V} <$ 2.5 GeV, but we do not show it as it is an artifact of the way we modeled dark photon decays in that mass region, see [84] for more details. Our limits can be considered conservative in the sense that one should have summed over the neutrino fluxes from all final states (this would be possible if ANTARES data were public), and aggressive in the sense that ANTARES assumes a DM density profile which is rather peaked towards the Galactic Center. We do not show the analogous ICECUBE limits [85] because they are not provided for $M_{\rm DM} > 1$ TeV, and are weaker than those of ANTARES. The latter aspect could be caused by the fact that ICECUBE does not perform as good as ANTARES for neutrinos coming from the Galactic Center, which drive the exclusions (see also [86] for a derivation of DM limits from IceCube data).

[14] This last prescription for neutrinos, together with the precise values of the dark photon branching ratios, are the only aspects in which this analysis departs from the one in [84].

Gamma rays

We consider first constraints from the FERMI satellite observations of several dwarf spheroidal galaxies (dSphs) [87], which are the strongest ones on our model among those derived from FERMI data (see [84]). We use the results of [88], which derived limits on secluded DM models starting from public FERMI data from dSphs observations, and presented them for different assumptions on the decay products of the mediator and of its mass. In particular, we exclude a point in our parameter space if it does not satisfy

$$\langle\sigma_{\rm tot} v_{\rm rel}\rangle {\rm BR}(V \to \bar{f} f) < 2\, \langle\sigma_{{\rm DM\,DM}\to f\bar{f}}\, v_{\rm rel}\rangle^{\gamma}_{\rm limit} \tag{5.35}$$

for at least one of the final states $\bar{f} f = \bar{b}b, \bar{q}q, \tau^+\tau^-, \ldots$, and where again the factor of 2 accounts for the fact that, in our model, DM is not self-conjugate. The resulting exclusions are shown in dark blue in Fig. 5.7, for various values of ε. We find that these exclusions do not significantly depend on the assumption on $m_{\rm V}$, among the two presented in [88]. As a cross-check we note that the excluded regions are in very good agreement with those derived in [48], for the same model in the case of no dilution, and which made direct use of FERMI data.

Moving to higher energy gamma rays, we now consider observations of gamma rays from the GC with the HESS telescope [89]. We use the results in [88] that, analogously to what done for FERMI dSphs, derived HESS limits on secluded DM models. An important caveat is that they are derived from the public data used in the HESS analysis [90] (corresponding to 112 h of observation time), but presented assuming a rescaling of the 112 h limits with the observation time of 254 h on which [89] is based (because these extra data are not public). Therefore we only treat these limit as an indication of the sensitivity, to this model, that could be achieved by the HESS telescope. In this spirit, we extend them up to $M_{\rm DM} = 70$ TeV, because this is the maximal DM mass reached in the HESS analysis [89]. We derive the sensitivities as in Eq. (5.35), where now of course $\langle\sigma_{{\rm DM\,DM}\to f\bar{f}}\, v_{\rm rel}\rangle^{\gamma}_{\rm limit}$ are the exclusions given in [88], and show them in purple in Fig. 5.7. We stress that the HESS sensitivity, from observations of the GC, would completely evaporate if the DM density profile has a core of size larger than 500 pc, and that it would be substantially weakened for cores down to a few tens of pc [91] (both possibilities are currently allowed both by simulations [92–94] and by data [95–97]).

Other ground-based telescopes have observed high energy gamma rays and cast bounds on annihilations of heavy DM, e.g. VERITAS [98], MAGIC [99] and HAWC [100, 101]. We do not show the resulting limits here because they are derived assuming no steps in the DM annihilations, and because they are anyway weaker than the HESS sensitivity that we show.

Antiprotons

We now turn to the constraints imposed by galactic antiproton observations. The method follows closely the previous analyses[15] in [48, 84, 105, 106]. For completeness, however, we briefly review it here.

Consistently with the previous studies, we use the AMS measurement [107] of the antiproton to proton ratio over the range from 1 to 450 GeV. The data are rather well explained by astrophysical antiproton sources, at least within the uncertainties related to $\bar{p}$ production in the interstellar medium, to their propagation process in the galactic environment and to the modulation of their flux due to solar activity[16]. As for the DM contribution, we adopt an Einasto profile for the distribution in the MW galactic halo and we set the local dark matter density to $\rho_{sun} = 0.42$ GeV/cm^3 [115]. We choose a 'MAX' propagation scheme, as it is the one favored by the fit of the astrophysical background. Upon imposing that the DM contribution does not worsens the astrophysics-only fit by more than $\Delta\chi^2 = 9$, we obtain conservative DM bounds.

The excluded regions are highlighted in magenta in Fig. 5.7. They consist of an area below $M_{\rm DM} \lesssim 1$ TeV, where the DM $\bar{p}$ contribution falls into the relatively low energy and high precision portion of the AMS data, and a series of funnels at $M_{\rm DM} \gtrsim 1$ TeV, corresponding to the resonances of the annihilation and BSF cross-sections. The excluded regions terminate for $m_{\rm V} \lesssim 2.5$ GeV, consistently with the fact that the BRs of a light V into protons and antiprotons are inhibited (see the discussion in Sect. 5.3.2 and Appendix 5.B).

Electrons and positrons

We derive constraints also from the measurements of the 'all-electron flux' (i.e.: $e^+ + e^-$) by the CALET experiment [116] onboard the International Space Station, which cover the range 10 GeV – 3 TeV. Measurements of the same observable have also been produced by the DAMPE satellite [117], by AMS [118] and by HESS [119]. The CALET data agree well with AMS but the former extend to higher energies: since our focus is on very heavy DM, we use CALET. The data by HESS, which are an update of former measurements [120, 121] with extended energy range, have so far been presented only at conferences. They display large systematic uncertainties and we checked that they are currently non competitive (in terms of constraining power), so that we do not use them here. Finally, the data by DAMPE are somewhat higher in normalization then those by CALET and AMS, in the range 50–1000 GeV, and have a harder spectrum. As this might be due to energy miscalibration in DAMPE, we use CALET only.

The astrophysical fluxes of $e^+ + e^-$ are poorly modeled at the high energies under consideration. We therefore decide to proceed, as customary in these case, with two methods: a conservative one and a more aggressive one. In the conservative approach, we assume no astrophysical background, we compute the DM $e^+ + e^-$ and we impose that it does not exceed any CALET data point by more than 2σ. This yields the excluded regions in ocra in Fig. 5.7, which sit in the ballpark of

[15] See also [102–104].

[16] Possible claimed excesses at low energies [108–113] are now shown to have a limited statistical significance [114].

$M_{\rm DM} \sim$ few TeV and $m_{\rm V} \lesssim 500$ MeV. They shrink and disappear as soon as the dilution parameter $\varepsilon \lesssim 10^{-10}$. In the aggressive approach, we include an astrophysical $e^+ + e^-$ background flux modeled with a power-law fitted on the CALET data over the whole energy range. We find $E^3\, \Phi^{e^+ + e^-}_{\rm astro} = 3.4 \times 10^{-2} E^{-0.18}$ GeV^2/(cm^2 s sr), with $\chi^2 = 16.4$ for 38 d.o.f., in reasonably good agreement with the analogous power law fit quoted in [116]. Then, in analogy with the procedure followed for antiprotons, we impose that the DM contribution does not worsens the modeled-astrophysics-only fit by more than $\Delta\chi^2 = 9$ and we obtain the regions drawn in yellow in Fig. 10 of [1]. Not surprisingly, these regions are much more extended than those derived in the conservative approach. We stress, however, that these regions should be interpreted more as a *reach* of the $e^+ + e^-$ measurements *under the assumption* that the astrophysical flux is well modeled by the *empirical* fit described above. In order to be conservative, we do not show them in the summary of Fig. 5.7.

On galactic subhalos and on DM limits from clusters of galaxies

In all the limits presented in this Sect. 5.4.3, we have not included a possible 'boost' of the DM signals from the existence of DM clumps. We summarize their possible impact here.

In principle, the presence of DM subhalos is indeed expected to enhance the annihilation flux [122–125]. However, in the central regions of the Milky Way subhalos are typically tidally disrupted by both the GC and the Galactic disk, so that such a boost turns out to be negligible for signals, e.g. in gamma rays, originating from the inner kiloparsecs (see [126] for a recent assessment). Boost factors up to a few tens of percent are possible for signals coming from dwarf spheroidal galaxies, even though, for conservative assumptions on the mass index, they become again negligible [127]. Boosts of signals coming from our neighborhood in the MW, relevant for DM searches in charged cosmic rays, are found not to exceed a few tens of percent for conservative assumptions [126, 128–130].

On the basis of the above, we find it justified to neglect the impact of DM subhalos as far as limits are concerned. On the other hand, we underline that the interplay of a possible DM signal in charged cosmic rays with gamma observations could provide precious indications on DM subhalo existence and properties.

DM subhalos might have a more dramatic impact on DM signals from galaxy clusters [126, 127], leading to an interesting interplay with other signals from heavy annihilating DM [131]. However, DM limits from galaxy clusters are either negligible with respect to all those considered here [132], or at most comparable for aggressive assumptions on the DM subhalo properties [133]. Therefore, while a possible ID signal from a cluster could in principle be ascribed to annihilating DM, we do not discuss these limits further here.

We finally comment that additional DM subhalos could arise in cosmological histories with a matter-dominated phase after DM (chemical and thermal) decoupling, see [134–136] for quantitative studies of this possibility. The investigation of the formation and survival of these subhalos in our model, while interesting, goes beyond the purpose of this study. In the spirit of being conservative on this matter we neglect it in our ID constraints.

5.4.4 On the Dark Photon Emitted During the Formation of Bound States

So far we have not discussed the signals from the low-energy dark photon emitted during BSF, which could potentially strengthen the indirect detection constraints. This V carries away the binding energy $E_B = \alpha_D^2 M_{\rm DM}/4$, where we neglect the kinetic energy $v_{\rm rel}^2 M_{\rm DM}/4$ because $\alpha_D \gtrsim v_{\rm rel}$ when BSF is important. It thus produces a spectrum of SM final states analogous to the one produced by a DM of mass $E_B/2$ annihilating directly into SM pairs (the same SM pairs that V decays into). However, DM with such low mass would have number density larger by $M_{\rm DM}/(E_B/2) = 8/\alpha_D^2$.

We may then recast the existing upper bounds, $\overline{\sigma v}_{\rm rel}$, on (non-self-conjugate) DM annihilating into SM pairs as follows

$$\langle \sigma_{\rm BSF} v_{\rm rel} \rangle (M_{\rm DM}, m_{\rm V}) < \left(\frac{8}{\alpha_D^2}\right)^2 \times \overline{\sigma v}_{\rm rel}(M = \alpha_D^2 M_{\rm DM}/8)\,. \tag{5.36}$$

Even for large values of the dark fine structure constant, the prefactor above is rather large ($64/\alpha_D^4 \gtrsim 10^3$ for $\alpha_D \lesssim 0.5$), and the existing indirect detection bounds (see e.g. [73, 87] for some strongest ones) do not yield any further constraints on our parameter space – except perhaps on top of the resonances. We shall therefore not consider this low-energy dark photon further, but instead we refer to [137] for a dedicated study.

Note that the CMB and 21 cm constraints scale proportionally to the DM mass [cf. Eqs. (5.32) and (5.33)], which partly mitigates the loosening of the constraints seen in (5.36). However, as noted earlier, BSF via vector emission is a p-wave process, and becomes entirely negligible for the very low $v_{\rm rel}$ during CMB.

5.5 Summary and Outlook

High energy telescopes constitute our unique direct access to energies much above a TeV, therefore they are in a privileged position to test BSM physics at those scales. Annihilating heavy Dark Matter is a particularly motivated example of such BSM physics. The existing and upcoming telescope data at unprecedented energies, together with the growing theoretical interest in DM with masses beyond 10–100 TeV, make it particularly timely to explore the associated phenomenology. In this study we performed a step in this direction.

We pointed out that models where DM pairs annihilate mostly in 'dark' mediators, themselves decaying into SM particles, allow to naturally evade the unitarity limit on the DM mass and, at the same time, to reliably compute cosmic ray signals. They therefore overcome the two main obstacles to the use of data of high energy telescopes to test annihilating heavy Dark Matter. Encouraged by this observation, we then studied in detail several aspects of the cosmological history and of the indirect detection signals of these models, which we summarise as follows:

1. We computed the dilution of relics induced by (i) the entropy injection from decays of the mediators after DM freeze-out, and (ii) the possibility that the SM and dark sector were not in thermal equilibrium at early times. We then determined how (i) and (ii) affect (a) the thermal freeze-out cross section and, in turn, the maximal DM mass compatible with unitarity, see Figs. 5.2 and 5.3; (b) the maximal dilution compatible with BBN constraints, see Sect. 5.2.4. Our study is the first, to our knowledge, to systematically explore the combination of these effects, therefore we reported ready-to-use analytical results (on dilution, freeze-out cross section, BBN limits, maximal DM mass allowed by unitarity) throughout Sect. 5.2.3.
2. As a case study we then considered a model of fermion DM charged under a spontaneously broken dark $U(1)$, where the role of the mediator is played by the associated massive vector. We computed the values of the dark gauge coupling that results in the correct DM abundance, over a wide mass range and for different values of the kinetic mixing and therefore of dilution (see Fig. 5.5 right). Our freeze-out computation improves over previous ones in that it includes long-range interactions beyond the Coulomb approximation, which is particularly relevant for heavy DM and mediator masses (see Fig. 5.5 left). The resulting cross sections relevant for DM indirect detection are shown in Fig. 5.6.
3. We finally studied the indirect detection phenomenology of the Dark $U(1)$ model, both with and without dilution, see Fig. 5.7 for a summary. We found the encouraging result that multimessenger searches for heavy DM are possible, offering different lines of attack to test these models. Our findings also give reasons to think that some telescopes (especially neutrino and gamma ray) have an unexplored potential to test annihilating DM with a mass of $O(100)$ TeV. This motivates an experimental effort in this direction and, importantly, the grant of public access to experimental data in some form. The latter would also overcome the problem, in interpreting experimental results, that some DM models (e.g. secluded ones) result in SM spectra different from the benchmarks used by the collaborations.

Our results open the doors to many possible future investigation. Concerning high energy cosmic rays, it would be interesting to determine the sensitivity of future experiments like CTA [138], LHAASO, TAIGA, KM3NET, HERD, ISS- CREAM on secluded DM models, and to explore whether annihilations of heavy secluded DM could explain the highest energy neutrinos observed by ICECUBE. It would also be intriguing to study the interplay of indirect detection signals with other effects of entropy injections, e.g. on the baryon asymmetry. On a more theory side, two exciting directions of exploration regard the connection, of the small portal of these models with the SM and/or of the heavy DM mass, with the solution of other problems of the SM. We plan to return to some of these aspects in future work.

Appendices

5.A $U(1)_D$ Coupled to Hypercharge $U(1)_Y$

5.A.1 *The Lagrangian*

We consider the Dark matter charged under $U(1)_D$ and interacting through a Dark mediator V_μ, the Dark photon, with mass m_V. The Dark photon is kinematically coupled to the $U(1)_Y$ gauge boson B_μ. If the temperature is below 100 GeV, then the electroweak symmetry is broken

$$L = \bar{\psi}_D i \not{D} \psi_D - m_{DM} \bar{\psi}_D \psi_D \tag{5.37}$$

$$- \frac{1}{4} F_{D\mu\nu} F_D^{\mu\nu} - \frac{1}{2} m_V^2 V_\mu V^\mu \tag{5.38}$$

$$- \frac{1}{4} W^3_{\mu\nu} W^{3\mu\nu} - \frac{1}{4} B_{\mu\nu} B^{\mu\nu} - \frac{1}{2} m_Z^2 Z_\mu Z^\mu \tag{5.39}$$

$$+ \frac{\varepsilon}{c_w} B_{\mu\nu} F_D^{\mu\nu} \tag{5.40}$$

where

$$F_D^{\mu\nu} = \partial^\mu V^\nu - \partial^\nu V^\mu, \qquad W^{3\mu\nu} = \partial^\mu W^{3\nu} - \partial^\nu W^{3\mu}, \tag{5.41}$$

$$B^{\mu\nu} = \partial^\mu B^\nu - \partial^\nu B^\mu, \qquad Z_\mu = c_w W_{3\mu} - s_w B_\mu, \tag{5.42}$$

with $D_\mu = \partial_\mu + i g_D V_\mu, c_w = \cos\theta_w, s_w = \sin\theta_w$, and θ_w the weak mixing angle. W^3_μ is the $SU(2)_L$ gauge boson associated with the generator T_3. The coupling constant ε is supposed to be small: $\varepsilon \ll 1$. For later use, we defined the Dark fine structure constant $\alpha_D = g_D^2/4\pi$.

5.A.2 *Gauge Eigenstates Versus Mass Eigenstates*

Definitions:
The gauge eigenstates are the vector fields which transform under the adjoint representation of their gauge group. In our case they are V_μ, W^3_μ and B_μ or V_μ, Z_μ and A_μ modulo a rotation. If we define

$$G_\mu = \begin{pmatrix} V_\mu \\ B_\mu \\ W^3_\mu \end{pmatrix}, \qquad H_\mu = \begin{pmatrix} V_\mu \\ Z_\mu \\ A_\mu \end{pmatrix} \tag{5.43}$$

then the orthogonal matrix R which rotates H_μ into $G_\mu = R \cdot H_\mu$, is equal to

$$R = \begin{pmatrix} 1 & 0 & 0 \\ 0 & -s_W & c_W \\ 0 & c_W & s_W \end{pmatrix}. \tag{5.44}$$

Now, let's rewrite the kinetic terms as

$$-\frac{1}{4}B^{\mu\nu}B_{\mu\nu} = -\frac{1}{2}\left(\partial^\mu B^\nu \partial_\mu B_\nu - \partial^\mu B^\nu \partial_\nu B_\mu\right) = B_\mu k^{\mu\nu} B_\nu \tag{5.45}$$

with $k^{\mu\nu} = \frac{\eta^{\mu\nu}\partial^2 - \partial^\mu\partial^\nu}{2}$.

We use it to write the Lagrangian as

$$L = \bar{\psi}_D i \not{D} \psi_D + G^t_\mu K k^{\mu\nu} G_\nu - \frac{1}{2} H^t_\mu M \eta^{\mu\nu} H_\nu \tag{5.46}$$

with

$$K = \begin{pmatrix} 1 & \frac{\varepsilon}{c_W} & 0 \\ \frac{\varepsilon}{c_W} & 1 & 0 \\ 0 & 0 & 1 \end{pmatrix} \tag{5.47}$$

and

$$M = \begin{pmatrix} m_V^2 & 0 & 0 \\ 0 & m_Z^2 & 0 \\ 0 & 0 & 0 \end{pmatrix}. \tag{5.48}$$

We define the mass eigenstates as the fields with a **kinetic** matrix K equal to the **identity** matrix and with a **mass** matrix M which is **diagonal**.

Let's explain how we compute them.

1. First, we find new fields $\tilde{G}_\mu$, linear combinations of G_μ, with **canonical** kinetic terms, i.e. the kinetic matrix is the **identity** matrix I. We are looking for the transformation matrix $P_G \in M(3, \mathbb{R})$ defined as $G_\mu = P_G \tilde{G}_\mu$ and which fulfills

$$P^t_G K P_G = I. \tag{5.49}$$

The Lagrangian becomes

$$L = \bar{\psi}_D i \not{D} \psi_D + \tilde{G}^t_\mu k^{\mu\nu} \tilde{G}_\nu - \frac{1}{2} H^t_\mu M\, \eta^{\mu\nu} H_\nu. \tag{5.50}$$

Introducing $\tilde{H}_\mu$ defined by $\tilde{G}_\mu = R\tilde{H}_\mu$ with R defined in Eq. (5.44) allows us to write

$$L = \bar{\psi}_D i \not{D} \psi_D + \tilde{H}_\mu{}^t k^{\mu\nu} \tilde{H}_\nu - \frac{1}{2} \tilde{H}_\mu{}^t P^t_H M P_H\, \eta^{\mu\nu} \tilde{H}_\nu \tag{5.51}$$

with

$$P_H = R^{-1} P_G R. \tag{5.52}$$

2. Second, we find new fields $\tilde{\tilde{H}}_\mu$, linear combinations of $\tilde{H}_\mu$, with still a **canonical** kinetic matrix but with a **diagonal** mass matrix. These are the mass eigenstates. One needs to find the transformation matrix $Q \in M(3, \mathbb{R})$ defined by $\tilde{H}_\mu = Q\tilde{\tilde{H}}_\mu$ and which verifies

$$\begin{aligned} Q^{-1}Q = 1 \quad &\rightarrow \quad Q \in O(3), \\ Q^t P_H^t M P_H Q = D. \end{aligned} \tag{5.53}$$

with D a diagonal matrix.
Indeed, the lagrangian becomes

$$L = \bar{\psi}_D i \not{D} \psi_D + \tilde{\tilde{H}}_\mu Q^{-1} Q k^{\mu\nu} \tilde{\tilde{H}}_\mu - \frac{1}{2} \tilde{\tilde{H}}_\mu Q^t P_H^t M P_H Q \, \eta^{\mu\nu} \tilde{\tilde{H}}_\mu, \tag{5.54}$$

$$= \bar{\psi}_D i \not{D} \psi_D + \tilde{\tilde{H}}_\mu \left(k^{\mu\nu} - \frac{1}{2} D \, \eta^{\mu\nu} \right) \tilde{\tilde{H}}_\mu. \tag{5.55}$$

Computation of the mass eigenstates:
Let's now apply the recipe. First, we need to solve Eq. (5.49) in order to find $P_G \in M(3, \mathbb{R})$ that we choose to parametrize as

$$P_G = \begin{pmatrix} R_1 \cos\theta_1 & R_1 \sin\theta_1 & 0 \\ R_2 \cos\theta_2 & R_2 \sin\theta_2 & 0 \\ 0 & 0 & 1 \end{pmatrix}. \tag{5.56}$$

since only the restriction to $(V, B)_\mu$ is not diagonal.
Equation (5.49) yields to

$$R_1^2 \cos^2\theta_1 + R_2^2 \cos^2\theta_2 + 2\frac{\varepsilon}{c_w} R_1 R_2 \cos\theta_1 \cos\theta_2 = 1, \tag{5.57}$$

$$R_1^2 \sin^2\theta_1 + R_2^2 \sin^2\theta_2 + 2\frac{\varepsilon}{c_w} R_1 R_2 \sin\theta_1 \sin\theta_2 = 1, \tag{5.58}$$

$$R_1^2 \sin 2\theta_1 + R_2^2 \sin 2\theta_2 + 2\frac{\varepsilon}{c_w} R_1 R_2 \sin(\theta_1 + \theta_2) = 0. \tag{5.59}$$

We can check that $\theta_1 = 0$ and $R_1 = R_2 = R$ are allowed by the last equations which simplify to

$$R^2 \left(\cos^2\theta_2 + 1 + 2\frac{\varepsilon}{c_w} \cos\theta_2 \right) = 1, \tag{5.60}$$

$$R^2 \sin^2\theta_2 = 1, \tag{5.61}$$

$$\cos\theta_2 + \frac{\varepsilon}{c_w} = 0. \tag{5.62}$$

We finally get

$$P_G = \begin{pmatrix} \frac{1}{\sqrt{1-\left(\frac{\varepsilon}{c_W}\right)^2}} & 0 & 0 \\ -\frac{1}{\sqrt{1-\left(\frac{\varepsilon}{c_W}\right)^2}}\frac{\varepsilon}{c_W} & 1 & 0 \\ 0 & 0 & 1 \end{pmatrix}, \tag{5.63}$$

from which and from Eq. (5.52) we compute P_H

$$P_H = \begin{pmatrix} \frac{1}{\sqrt{1-\left(\frac{\varepsilon}{c_W}\right)^2}} & 0 & 0 \\ \frac{t_W\varepsilon}{\sqrt{1-\left(\frac{\varepsilon}{c_W}\right)^2}} & 1 & 0 \\ -\frac{\varepsilon}{\sqrt{1-\left(\frac{\varepsilon}{c_W}\right)^2}} & 0 & 1 \end{pmatrix}. \tag{5.64}$$

One deduces the new mass matrix

$$P_H^t M P_H = \begin{pmatrix} \frac{m_V^2+t_W^2\varepsilon^2 m_Z^2}{1-\left(\frac{\varepsilon}{c_W}\right)^2} & \frac{t_W\varepsilon}{\sqrt{1-\left(\frac{\varepsilon}{c_W}\right)^2}}m_Z^2 & 0 \\ \frac{t_W\varepsilon}{\sqrt{1-\left(\frac{\varepsilon}{c_W}\right)^2}}m_Z^2 & m_Z^2 & 0 \\ 0 & 0 & 0 \end{pmatrix}. \tag{5.65}$$

The last step is to diagonalize it.

We find the diagonal mass matrix

$$D = \begin{pmatrix} m_Z^2\delta^2\left(1+\varepsilon^2\frac{1-\left(\frac{\delta}{c_W}\right)^2}{1-\delta^2}\right) & 0 & 0 \\ 0 & m_Z^2\left(1+\varepsilon^2\frac{t_W^2}{1-\delta^2}\right) & 0 \\ 0 & 0 & 0 \end{pmatrix} + O(\varepsilon^4), \tag{5.66}$$

and the matrix with the eigenvectors as its columns at leading orders in ε

$$Q = \begin{pmatrix} -1 & \frac{t_W\varepsilon}{1-\delta^2} & 0 \\ \frac{t_W\varepsilon}{1-\delta^2} & 1 & 0 \\ 0 & 0 & 1 \end{pmatrix} + O(\varepsilon^2) \tag{5.67}$$

with $\delta \equiv m_V/m_Z$.

When $\varepsilon \to 0$, the matrix D becomes

$$\lim_{\varepsilon\to 0} D = \begin{pmatrix} m_V^2 & 0 & 0 \\ 0 & m_Z^2 & 0 \\ 0 & 0 & 0 \end{pmatrix} = M. \tag{5.68}$$

From this, we can identifies the first, second and third coordinate of $\tilde{\tilde{H}}_\mu$ to be the new Dark photon $\tilde{\tilde{V}}_\mu$, the new $U(1)_Y$ boson $\tilde{\tilde{Z}}_\mu$ and the new photon $\tilde{\tilde{A}}_\mu$ and we write

$$\tilde{\tilde{H}}_\mu = \begin{pmatrix} \tilde{\tilde{V}}_\mu \\ \tilde{\tilde{Z}}_\mu \\ \tilde{\tilde{A}}_\mu \end{pmatrix}. \tag{5.69}$$

We finally deduce the transformation between the gauge eigenstates H_μ and the mass eigenstates $\tilde{\tilde{H}}_\mu$ at leading orders in ε

$$H_\mu = P_H Q \tilde{\tilde{H}}_\mu \quad \rightarrow \quad \begin{pmatrix} V_\mu \\ Z_\mu \\ A_\mu \end{pmatrix} = \begin{pmatrix} -1 & \frac{1}{1-\delta^2} t_w \varepsilon & 0 \\ \frac{\delta^2}{1-\delta^2} t_w \varepsilon & 1 & 0 \\ \varepsilon & 0 & 1 \end{pmatrix} \begin{pmatrix} \tilde{\tilde{V}}_\mu \\ \tilde{\tilde{Z}}_\mu \\ \tilde{\tilde{A}}_\mu \end{pmatrix}. \tag{5.70}$$

Not visible in 5.70 is the emergence of an additional $O(2)$ symmetry in the massless limit $m_V \rightarrow 0$, which allows to decouple the dark photon from the SM current [139].

5.A.3 *Dark Photon Interaction with SM*

Now let's look at how the interaction terms between gauge fields and SM matter fields are modified.

$$L_{int} = e J^\mu_{em} A_\mu + \frac{g}{c_w} J^\mu_z Z_\mu \tag{5.71}$$

$$= e J^\mu_{em} \left(\varepsilon \tilde{\tilde{V}}_\mu + \tilde{\tilde{A}}_\mu \right) + \frac{g}{c_w} J^\mu_z \left(\frac{\delta^2}{1-\delta^2} t_w \varepsilon \tilde{\tilde{V}}_\mu + \tilde{\tilde{Z}}_\mu \right) + O(\varepsilon^2) \tag{5.72}$$

$$= e J^\mu_{em} \tilde{\tilde{A}}_\mu + \frac{g}{c_w} J^\mu_z \tilde{\tilde{Z}}_\mu + \varepsilon \left(e J^\mu_{em} + g \frac{s_w}{c_w^2} \frac{\delta^2}{1-\delta^2} J^\mu_z \right) \tilde{\tilde{V}}_\mu + O(\varepsilon^2) \tag{5.73}$$

We can express J_Z^μ as $J_Z^\mu = c_w^2 J_{em} - J_Y$ and rewrite the interaction terms between the Dark photon $\tilde{\tilde{V}}_\mu$ and the SM matter fields as

$$\varepsilon\left[\left(e + g s_w \frac{\delta^2}{1-\delta^2}\right) J_{em}^\mu - g \frac{s_w}{c_w^2} \frac{\delta^2}{1-\delta^2} J_Y\right] \tilde{\tilde{V}}_\mu + O(\varepsilon^2). \tag{5.74}$$

Using $e = \frac{gg'}{\sqrt{g^2+g'^2}}$, $c_w = \frac{g}{\sqrt{g^2+g'^2}}$ and $s_w = \frac{g'}{\sqrt{g^2+g'^2}}$, we finally get

$$\varepsilon\left[e \frac{1}{1-\delta^2} J_{em}^\mu - \frac{e}{c_w^2} \frac{\delta^2}{1-\delta^2} J_Y\right] \tilde{\tilde{V}}_\mu + O(\varepsilon^2). \tag{5.75}$$

This can be rewritten as

$$\sum_f g_f \tilde{\tilde{V}}_\mu \bar{f} \gamma^\mu f, \qquad g_f = \varepsilon e\left[Q_f \frac{1}{1-\delta^2} - \frac{Y_f}{c_w^2} \frac{\delta^2}{1-\delta^2}\right] + O(\varepsilon^2). \tag{5.76}$$

The mixing term between $U(1)_Y$ and $U(1)_D$ in the Lagrangian in Eq. (5.40) couples the Dark photon to SM fermions.

5.B Dark Photon Decay Widths

We have computed the decay width of the dark photon V into pairs of SM fermions, at tree-level, in the small ε limit. The final expression reads

$$\Gamma(V \to \bar{f} f) = \frac{N_f}{24\pi} m_V \sqrt{1 - 4\delta_f^{-2}} \left[g_{f_L}^2 + g_{f_R}^2 - \delta_f^{-2}(g_{f_L}^2 + g_{f_R}^2 - 6 g_{f_L} g_{f_R})\right] \tag{5.77}$$

with

$$g_f = \varepsilon e\left[Q_f \frac{1}{1-\delta_Z^2} - \frac{Y_f}{c_w^2} \frac{\delta_Z^2}{1-\delta_Z^2}\right] + O(\varepsilon^2), \tag{5.78}$$

$$\delta_f = m_V/m_f, \quad \delta_Z = m_V/m_Z, \tag{5.79}$$

$$e = \frac{g_2 g_y}{\sqrt{g_2^2 + g_y^2}}, \quad c_w = \frac{g_2}{\sqrt{g_2^2 + g_y^2}}. \tag{5.80}$$

N_f stands for the color number and is equal to 3 for quarks and 1 for leptons. Q_f, Y_f, g_2 and g_y are respectively the electric charge of the fermion f in unit of e, the hypercharge of f, the gauge coupling constant of $SU(2)_L$ and that of $U(1)_Y$. We have also computed the decay widths of the dark photon, at tree-level, into Zh

$$\Gamma_{V\to Zh} = \varepsilon^2 \left(\frac{g_y}{g_2}\right)^2 \frac{g_y^2+g_2^2}{192\pi} m_V \left(\frac{\delta_Z^2}{1-\delta_Z^2}\right)^2 \lambda_{Zh,V}^{3/2} \left(1+12\lambda_{Zh,V}^{-1}\delta_Z^{-2}\right) \quad (5.81)$$

where the 2-body phase space function $\lambda_{Zh,V}$ is defined as

$$\lambda_{Zh,V} = \left(1-\delta_Z^{-2}-\delta_h^{-2}\right)^2 - 4\left(\delta_Z\delta_h\right)^{-2} \quad (5.82)$$

with $\delta_h = m_V/m_h$, and into WW

$$\Gamma_{V\to WW} = \varepsilon^2 \left(\frac{g_y}{g_2}\right)^2 \frac{g_y^2+g_2^2}{192\pi} m_V \left(\frac{\delta_Z^2}{1-\delta_Z^2}\right)^2 \left(1-4\delta_W^{-2}\right)^{3/2} \left(1+20\delta_W^{-2}+12\delta_W^{-4}\right). \quad (5.83)$$

with $\delta_W = m_V/m_W$. We have verified that these expressions agree both with the Goldstone equivalence theorem [140]

$$\lim_{m_{V_D}\gg m_Z, m_h} \Gamma_{V_D\to Zh} = \Gamma_{V_D\to\pi^3 h} \quad (5.84)$$

$$\lim_{m_{V_D}\gg m_W} \Gamma_{V_D\to W^- W^+} = \Gamma_{V_D\to\pi^-\pi^+}. \quad (5.85)$$

and with the existing literature, e.g. [141, 142].

In our calculations, we use the expressions above in the region $m_V < 350$ MeV and $m_V > 2.5$ GeV. For 350 MeV$< m_V < 2.5$ GeV we use the perturbative tree-level expression in Eq. (5.77) for the width into leptons, and for the width into hadrons we instead use measurements of $R(s) = \sigma(e^+e^- \to \text{hadrons})/\sigma(e^+e^- \to \mu^+\mu^-)$, that we extract from [57], and where $\sqrt{s}$ is the energy of the collision in the center of mass frame. These scattering processes are dominated by a virtual photon exchange in s-channel, therefore the quantum numbers of the final state coincide with those of the dark photon, making their use justified for our purpose. Therefore we write

$$\Gamma(V \to \text{hadrons}) = R(s = m_V^2) \times \Gamma(V \to \mu^+\mu^-), \quad (5.86)$$

where we take $\Gamma(V \to \mu^+\mu^-)$ from Eq. (5.77). The hadronic decay chains in $e^+e^- \to$ hadrons end dominantly in charged pions [58, 143], that have BR($\pi^\pm \to \mu\nu_\mu$) $> 99.9\%$. Therefore, following [48], we assume for simplicity that the V final states, for 350 GeV $< m_V < 2.5$ GeV, consist 50 % of $\mu\bar{\mu}$ and 50 % of $\nu\bar{\nu}$.

5.C Gamma Ray from DM Annihilation

Let's compute the contribution to the detected gamma spectrum due to the Dark Matter annihilation in the Milky way. The number of DM pairs in the volume d^3r at position $\vec{r}$ in the MW is

$$d^3r\frac{1}{2}\left(\frac{\rho_{DM}(\vec{r})}{M_{\rm DM}}\right)^2 \tag{5.87}$$

where $\rho_{DM}(\vec{r})$ is the DM energy density at $\vec{r}$. We have assumed self-conjugate DM. In the opposit case where particles can only interact with anti-particles, the $1/2$ must be replaced by $1/4$.

If we suppose that a pair of DM annihilates into a pair of SM particles f as shown in Fig. 5.8 with a cross-section $\langle\sigma v\rangle_f$, then multiplying the last expression by $\langle\sigma v\rangle_f$ gives the number of DM annihilations into f per second

$$d^3r\frac{1}{2}\left(\frac{\rho_{DM}(\vec{r})}{M_{\rm DM}}\right)^2\sum_f\langle\sigma v\rangle_f. \tag{5.88}$$

Now multiplying by the number of photons produced per second, in energy bin ΔE centered on E, per DM annihilation into SM particles f,

$$\Delta E\frac{dN_{f\to\gamma}}{dE}, \tag{5.89}$$

we get the number of photons produced in volume d^3r, at position $\vec{r}$, in energy bin ΔE centered on E

$$d^3r\frac{1}{2}\left(\frac{\rho_{DM}(\vec{r})}{M_{\rm DM}}\right)^2\sum_f\langle\sigma v\rangle_f\Delta E\frac{dN_{f\to\gamma}}{dE}. \tag{5.90}$$

However, as described in Fig. 5.9 a telescope with a surface $S_{\rm tel}$, receiving signals within a solid angle $\Delta\Omega$ and within an energy bin ΔE detects only a fraction

$$\frac{\Delta\Omega}{4\pi}\frac{S_{\rm tel}}{4\pi r^2} \tag{5.91}$$

of the total number of photons produced. So this telescope detects a number of photons per second which is

$$S_{tel}\Delta\Omega\Delta E\frac{d\phi_\gamma}{d\Omega dE}=\int d^3r\frac{1}{2}\left(\frac{\rho_{DM}(\vec{r})}{M_{\rm DM}}\right)^2\sum_f\langle\sigma v\rangle_f\Delta E\frac{dN_{f\to\gamma}}{dE}\frac{\Delta\Omega}{4\pi}\frac{S_{\rm tel}}{4\pi r^2}. \tag{5.92}$$

where ϕ_γ is the integrated γ flux in cm^2/s.

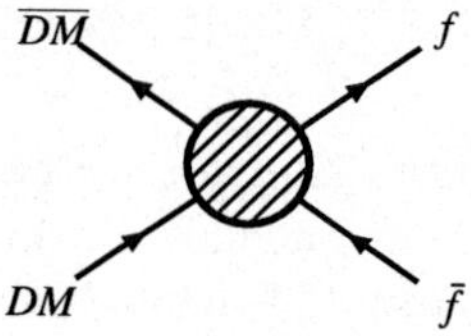

Fig. 5.8 DM annihilation into SM

Finally the detected gamma flux per solid angle unit and per energy unit, assuming self-conjugate DM, is

$$\frac{d\phi_\gamma}{d\Omega dE} = \frac{1}{2}\frac{r_\odot}{4\pi}\left(\frac{\rho_\odot}{M_{\rm DM}}\right)^2 J_{\rm ann}\sum_f \langle\sigma v\rangle_f \frac{dN_{f\to\gamma}}{dE} \tag{5.93}$$

where

$$J_{\rm ann} = \int \frac{ds}{r_\odot}\left(\frac{\rho_{DM}(\vec{r})}{\rho_\odot}\right)^2 \tag{5.94}$$

is the so-called $J_{\rm ann}$ factor for annihilation, $r_\odot = 8.33$ kpc is the Earth distance to the GC and $\rho_\odot = 0.3\ \text{GeV/cm}^3$ is the supposed DM energy density at the Earth position.

A similar treatment for DM decay leads to

$$\frac{d\phi_\gamma}{d\Omega dE} = \frac{r_\odot}{4\pi}\left(\frac{\rho_\odot}{M_{\rm DM}}\right) J_{\rm dec}\sum_f \Gamma_f \frac{dN_{f\to\gamma}}{dE} \tag{5.95}$$

where

$$J_{\rm dec} = \int \frac{ds}{r_\odot}\left(\frac{\rho_{DM}(\vec{r})}{\rho_\odot}\right). \tag{5.96}$$

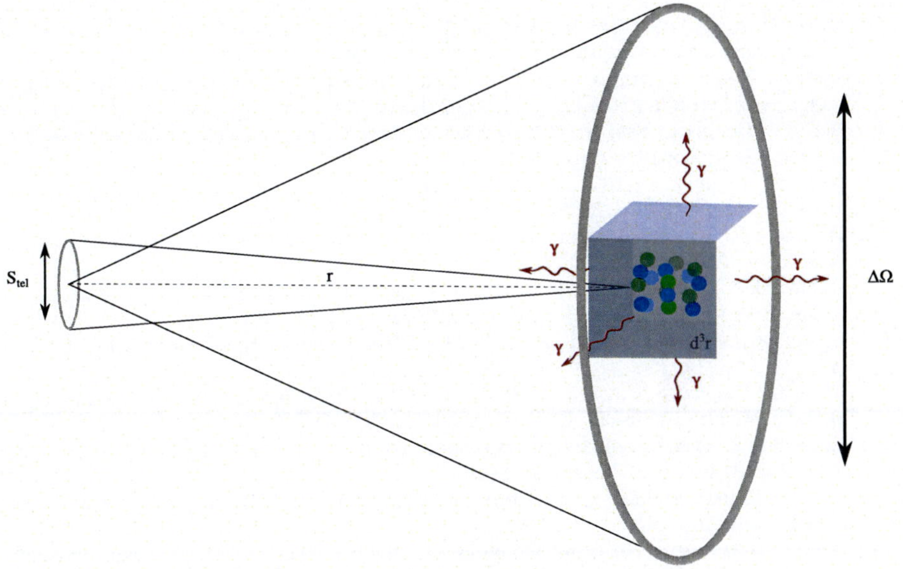

Fig. 5.9 The detected γ-ray flux is the fraction $\frac{\Delta\Omega}{4\pi}\frac{S_{\rm tel}}{4\pi r^2}$ of the total emission

References

1. Cirelli M, Gouttenoire Y, Petraki K, Sala F (2019) Homeopathic dark matter, or how diluted heavy substances produce high energy cosmic rays. JCAP 02:014. arxiv:1811.03608
2. Dimopoulos S, Giudice GF, Pomarol A (1996) Dark matter in theories of gauge mediated supersymmetry breaking. Phys Lett B 389:37–42. arxiv:hep-ph/9607225
3. Antipin O, Redi M, Strumia A (2015) Dynamical generation of the weak and Dark Matter scales from strong interactions. JHEP 01:157. arxiv:1410.1817
4. Antipin O, Redi M, Strumia A, Vigiani E (2015) Accidental composite dark matter. JHEP 07:039. arxiv:1503.08749
5. Del Nobile E, Nardecchia M, Panci P (2016) Millicharge or decay: a critical take on minimal dark matter. JCAP 1604:048. arxiv:1512.05353
6. Griest K, Kamionkowski M (1990) Unitarity limits on the mass and radius of dark matter particles. Phys Rev Lett 64:615
7. von Harling B, Petraki K (2014) Bound-state formation for thermal relic dark matter and unitarity. JCAP 12:033. arxiv:1407.7874
8. Baldes I, Petraki K (2017) Asymmetric thermal-relic dark matter: Sommerfeld-enhanced freeze-out, annihilation signals and unitarity bounds. JCAP 09:028. arxiv:1703.00478
9. de Carlos B, Casas JA, Quevedo F, Roulet E (1993) Model independent properties and cosmological implications of the dilaton and moduli sectors of 4-d strings. Phys Lett B 318:447–456. arxiv:hep-ph/9308325
10. Banks T, Kaplan DB, Nelson AE (1994) Cosmological implications of dynamical supersymmetry breaking. Phys Rev D 49:779–787. arxiv:hep-ph/9308292
11. Moroi T, Randall L (2000) Wino cold dark matter from anomaly mediated SUSY breaking. Nucl Phys B 570:455–472. arxiv:hep-ph/9906527
12. McDonald J (1991) WIMP densities in decaying particle dominated cosmology. Phys Rev D 43:1063–1068
13. Giudice GF, Kolb EW, Riotto A (2001) Largest temperature of the radiation era and its cosmological implications. Phys Rev D 64:023508. arxiv:hep-ph/0005123

14. Patwardhan AV, Fuller GM, Kishimoto CT, Kusenko A (2015) Diluted equilibrium sterile neutrino dark matter. Phys Rev D92:103509. arxiv:1507.01977
15. Berlin A, Hooper D, Krnjaic G (2016) PeV-scale dark matter as a thermal relic of a decoupled sector. Phys Lett B760:106–111. arxiv:1602.08490
16. Berlin A, Hooper D, Krnjaic G (2016) Thermal dark matter from a highly decoupled sector. Phys Rev D94:095019. arxiv:1609.02555
17. Bramante J, Unwin J (2017) Superheavy thermal dark matter and primordial asymmetries. JHEP 02:119. arxiv:1701.05859
18. Hamdan S, Unwin J (2018) Dark matter freeze-out during matter domination. Mod Phys Lett A33:1850181. arxiv:1710.03758
19. Allahverdi R, Osiński JK (2019) Nonthermal dark matter from modified early matter domination. Phys Rev D 99:083517. arxiv:1812.10522
20. Contino R, Mitridate A, Podo A, Redi M (2019) Gluequark dark matter. JHEP 02:187. arxiv:1811.06975
21. Chanda P, Hamdan S, Unwin J (2020) Reviving Z and higgs mediated dark matter models in matter dominated freeze-out. JCAP 01:034. arxiv:1911.02616
22. Contino R, Podo A, Revello F (2021) Composite dark matter from strongly-interacting chiral dynamics. JHEP 02:091. arxiv:2008.10607
23. Asadi P, Kramer ED, Kuflik E, Slatyer TR, Smirnov J, Glueballs in a thermal squeezeout model. arxiv:2203.15813
24. Heurtier L, Huang F (2019) Inflaton portal to a highly decoupled EeV dark matter particle. Phys Rev D 100:043507. arxiv:1905.05191
25. Hambye T, Lucca M, Vanderheyden L (2020) Dark matter as a heavy thermal hot relic. Phys Lett B 807:135553. arxiv:2003.04936
26. Davoudiasl H, Mohlabeng G (2020) Getting a THUMP from a WIMP. JHEP 04:177. arxiv:1912.05572
27. Kramer ED, Kuflik E, Levi N, Outmezguine NJ, Ruderman JT (2021) Heavy thermal dark matter from a new collision mechanism. Phys Rev Lett 126:081802. arxiv:2003.04900
28. Bian L, Liu X, Xie K-P, Probing superheavy dark matter with gravitational waves. arxiv:2107.13112
29. Kim H, Kuflik E (2019) Superheavy thermal dark matter. Phys Rev Lett 123:191801. arxiv:1906.00981
30. Harigaya K, Ibe M, Kaneta K, Nakano W, Suzuki M (2016) Thermal relic dark matter beyond the unitarity limit. JHEP 08:151. arxiv:1606.00159
31. Mitridate A, Redi M, Smirnov J, Strumia A (2017) Dark matter as a weakly coupled dark baryon. JHEP 10:210. arxiv:1707.05380
32. Geller M, Iwamoto S, Lee G, Shadmi Y, Telem O, Dark quarkonium formation in the early universe. arxiv:1802.07720
33. Baldes I, Gouttenoire Y, Sala F (2021) String fragmentation in supercooled confinement and implications for dark matter. JHEP 04:278. arxiv:2007.08440
34. Baldes I, Gouttenoire Y, Sala F, Servant G, Supercool composite dark matter beyond 100 TeV. arxiv:2110.13926
35. Hambye T, Strumia A, Teresi D (2018) Super-cool dark matter. JHEP 08:188. arxiv:1805.01473
36. Baldes I, Garcia-Cely C (2019) Strong gravitational radiation from a simple dark matter model. JHEP 05:190. arxiv:1809.01198
37. Baker MJ, Kopp J, Long AJ, Filtered dark matter at a first order phase transition. arxiv:1912.02830
38. Chway D, Jung TH, Shin CS (2020) Dark matter filtering-out effect during a first-order phase transition. Phys Rev D 101:095019. arxiv:1912.04238
39. Asadi P, Kramer ED, Kuflik E, Ridgway GW, Slatyer TR, Smirnov J, Thermal squeezeout of dark matter. arxiv:2103.09827
40. Asadi P, Kramer ED, Kuflik E, Ridgway GW, Slatyer TR, Smirnov J, Accidentally asymmetric dark matter. arxiv:2103.09822

41. Falkowski A, No JM (2013) Non-thermal dark matter production from the electroweak phase transition: multi-TeV WIMPs and "Baby-Zillas". JHEP 02:034. arxiv:1211.5615
42. Azatov A, Vanvlasselaer M, Yin W, Dark matter production from relativistic bubble walls. arxiv:2101.05721
43. Carney D et al, Snowmass2021 cosmic frontier white paper: ultraheavy particle dark matter. arxiv:2203.06508
44. Cirelli M, Corcella G, Hektor A, Hutsi G, Kadastik M, Panci P et al (2011) PPPC 4 DM ID: a poor particle physicist cookbook for dark matter indirect detection. JCAP 03:051. arxiv:1012.4515
45. Sjöstrand T, Ask S, Christiansen JR, Corke R, Desai N, Ilten P et al (2015) An Introduction to PYTHIA 8.2. Comput Phys Commun 191:159–177. arxiv:1410.3012
46. Ciafaloni P, Comelli D, Riotto A, Sala F, Strumia A, Urbano A (2011) Weak corrections are relevant for dark matter indirect detection. JCAP 1103:019. arxiv:1009.0224
47. Christiansen JR, Sjöstrand T (2014) Weak gauge boson radiation in parton showers. JHEP 04:115. arxiv:1401.5238
48. Cirelli M, Panci P, Petraki K, Sala F, Taoso M (2017) Dark matter's secret liaisons: phenomenology of a dark U(1) sector with bound states. JCAP 05:036. arxiv:1612.07295
49. Bauer CW, Rodd NL, Webber BR, Dark matter spectra from the electroweak to the planck scale. arxiv:2007.15001
50. ANTARES collaboration, Adrián-Martínez S et al (2016) A search for Secluded Dark Matter in the Sun with the ANTARES neutrino telescope. JCAP 1605:016. arxiv:1602.07000
51. Ardid M, Felis I, Herrero A, Martínez-Mora JA (2017) Constraining secluded dark matter models with the public data from the 79-string IceCube search for dark matter in the Sun. JCAP 1704:010. arxiv:1701.08863
52. Jedamzik K (2006) Big bang nucleosynthesis constraints on hadronically and electromagnetically decaying relic neutral particles. Phys Rev D 74:103509. arxiv:hep-ph/0604251
53. Scherrer RJ, Turner MS (1985) Decaying particles do not heat up the universe. Phys Rev D 31:681
54. Kolb EW, Turner MS (1990) The early universe. Front Phys 69:1–547
55. Jaeckel J, Ringwald A (2010) The low-energy frontier of particle physics. Ann Rev Nucl Part Sci 60:405–437. arxiv:1002.0329
56. Essig R et al (2013) Working group report: new light weakly coupled particles. In: Community summer study 2013: snowmass on the Mississippi, p 10. arxiv:1311.0029
57. Curtin D, Essig R, Gori S, Shelton J (2015) Illuminating dark photons with high-energy colliders. JHEP 02:157. arxiv:1412.0018
58. Buschmann M, Kopp J, Liu J, Machado PAN (2015) Lepton jets from radiating dark matter. JHEP 07:045. arxiv:1505.07459
59. Particle Data Group collaboration, Patrignani C et al (2016) Review of particle physics. Chin Phys C40:100001
60. Mangano G, Miele G, Pastor S, Pinto T, Pisanti O, Serpico PD (2005) Relic neutrino decoupling including flavor oscillations. Nucl Phys B 729:221–234. arxiv:hep-ph/0506164
61. Petraki K, Postma M, de Vries J (2017) Radiative bound-state-formation cross-sections for dark matter interacting via a Yukawa potential. JHEP 04:077. arxiv:1611.01394
62. Piffl T et al (2014) The RAVE survey: the Galactic escape speed and the mass of the Milky Way. Astron Astrophys 562:A91. arxiv:1309.4293
63. McConnachie AW (2012) The observed properties of dwarf galaxies in and around the local group. Astron J 144:4. arxiv:1204.1562
64. Burkert A (2015) The structure and dark halo core properties of dwarf spheroidal galaxies. Astrophys J 808:158. arxiv:1501.06604
65. Finkbeiner DP, Galli S, Lin T, Slatyer TR (2012) Searching for dark matter in the CMB: a compact parameterization of energy injection from new physics. Phys Rev D 85:043522. arxiv:1109.6322
66. Furlanetto SR, The 21-cm Line as a probe of reionization. arxiv:1511.01131

67. Lopez-Honorez L, Mena O, Moliné A, Palomares-Ruiz S, Vincent AC (2016) The 21 cm signal and the interplay between dark matter annihilations and astrophysical processes. JCAP 1608: 004. arxiv:1603.06795
68. Liu H, Slatyer TR, Zavala J (2016) Contributions to cosmic reionization from dark matter annihilation and decay. Phys Rev D94:063507. arxiv:1604.02457
69. Cassel S (2010) Sommerfeld factor for arbitrary partial wave processes. J Phys G 37:105009. arxiv:0903.5307
70. Slatyer TR (2010) The Sommerfeld enhancement for dark matter with an excited state. JCAP 02:028. arxiv:0910.5713
71. Blanco C, Harding JP, Hooper D, Novel gamma-ray signatures of PeV-scale dark matter. arxiv:1712.02805
72. Hufnagel M, Schmidt-Hoberg K, Wild S, BBN constraints on MeV-scale dark sectors. Part II. Electromagnetic decays. arxiv:1808.09324
73. Planck collaboration, Ade PAR et al (2016) Planck 2015 results. XIII. Cosmological parameters. Astron Astrophys 594:A13. arxiv:1502.01589
74. Slatyer TR (2016) Indirect dark matter signatures in the cosmic dark ages. I. Generalizing the bound on s-wave dark matter annihilation from Planck results. Phys Rev D93:023527. arxiv:1506.03811
75. Slatyer TR (2016) Indirect dark matter signatures in the cosmic dark ages II. Ionization, heating and photon production from arbitrary energy injections. Phys Rev D93:023521. arxiv:1506.03812
76. Elor G, Rodd NL, Slatyer TR, Xue W (2016) Model-independent indirect detection constraints on hidden sector dark matter. JCAP 1606:024. arxiv:1511.08787
77. Bowman JD, Rogers AEE, Monsalve RA, Mozdzen TJ, Mahesh N (2018) An absorption profile centred at 78 megahertz in the sky-averaged spectrum. Nature 555:67–70
78. D'Amico G, Panci P, Strumia A, Bounds on dark matter annihilations from 21 cm data. arxiv:1803.03629
79. H. Liu and T. R. Slatyer, *Too Hot, Too Cold or Just Right? Implications of a 21-cm Signal for Dark Matter Annihilation and Decay*, arxiv:1803.09739
80. Cheung K, Kuo J-L, Ng K-W, Tsai Y-LS, The impact of EDGES 21-cm data on dark matter interactions. arxiv:1803.09398
81. Clark S, Dutta B, Gao Y, Ma Y-Z, Strigari LE, 21cm limits on decaying dark matter and primordial black holes. arxiv:1803.09390
82. Mitridate A, Podo A, Bounds on dark matter decay from 21 cm line. arxiv:1803.11169
83. Albert A et al (2017) Results from the search for dark matter in the Milky Way with 9 years of data of the ANTARES neutrino telescope. Phys Lett B769:249–254. arxiv:1612.04595
84. Baldes I, Cirelli M, Panci P, Petraki K, Sala F, Taoso M (2018) Asymmetric dark matter: residual annihilations and self-interactions. SciPost Phys 4:041. arxiv:1712.07489
85. IceCube collaboration, Aartsen MG et al (2017) Search for neutrinos from dark matter self-annihilations in the center of the milky Way with 3 years of IceCube/DeepCore. Eur Phys J C77:627. arxiv:1705.08103
86. El Aisati C, Garcia-Cely C, Hambye T, Vanderheyden L (2017) Prospects for discovering a neutrino line induced by dark matter annihilation. JCAP 1710:021. arxiv:1706.06600
87. Fermi-LAT collaboration, Ackermann M et al (2015) Searching for dark matter annihilation from Milky Way dwarf spheroidal galaxies with six years of fermi large area telescope data. Phys Rev Lett 115:231301. arxiv:1503.02641
88. Profumo S, Queiroz FS, Silk J, Siqueira C (2018) Searching for secluded dark matter with H.E.S.S., Fermi-LAT, and planck. JCAP 1803:010. arxiv:1711.03133
89. H.E.S.S. collaboration, Abdallah H et al (2016) Search for dark matter annihilations towards the inner Galactic halo from 10 years of observations with H.E.S.S. Phys Rev Lett 117:111301. arxiv:1607.08142
90. H.E.S.S. collaboration, Abramowski A et al (2011) Search for a dark matter annihilation signal from the galactic center halo with H.E.S.S. Phys Rev Lett 106:161301. arxiv:1103.3266

91. H.E.S.S. collaboration, Abramowski A et al (2015) Constraints on an annihilation signal from a core of constant dark matter density around the Milky Way center with H.E.S.S. Phys Rev Lett 114:081301. arxiv:1502.03244
92. Di Cintio A, Brook CB, Macciò AV, Stinson GS, Knebe A, Dutton AA et al (2014) The dependence of dark matter profiles on the stellar-to-halo mass ratio: a prediction for cusps versus cores. Mon Not R Astron Soc 437:415–423. arxiv:1306.0898
93. Marinacci F, Pakmor R, Springel V (2014) The formation of disc galaxies in high resolution moving-mesh cosmological simulations. Mon Not R Astron Soc 437:1750–1775. arxiv:1305.5360
94. Tollet E et al (2016) NIHAO—IV: core creation and destruction in dark matter density profiles across cosmic time. Mon Not R Astron Soc 456:3542–3552. arxiv:1507.03590
95. Nesti F, Salucci P (2013) The dark matter halo of the Milky Way, AD 2013. JCAP 1307:016. arxiv:1304.5127
96. Cole DR, Binney J (2017) A centrally heated dark halo for our Galaxy. Mon Not R Astron Soc 465:798–810. arxiv:1610.07818
97. Wegg C, Gerhard O, Portail M (2016) MOA-II Galactic microlensing constraints: the inner Milky Way has a low dark matter fraction and a near maximal disc. Mon Not R. Astron Soc 463:557–570. arxiv:1607.06462
98. VERITAS collaboration, Archambault S et al (2017) Dark matter constraints from a joint analysis of dwarf spheroidal galaxy observations with VERITAS. Phys Rev D95:082001. arxiv:1703.04937
99. MAGIC collaboration, Ahnen ML et al (2018) Indirect dark matter searches in the dwarf satellite galaxy Ursa Major II with the MAGIC Telescopes. JCAP 1803:009. arxiv:1712.03095
100. HAWC collaboration, Albert A et al, Dark matter limits from dwarf spheroidal galaxies with the HAWC gamma-ray observatory. arxiv:1706.01277
101. Abeysekara AU et al, A Search for Dark Matter in the Galactic Halo with HAWC.arxiv:1710.10288
102. Boudaud M, Génolini Y, Derome L, Lavalle J, Maurin D, Salati P et al (2020) AMS-02 antiprotons' consistency with a secondary astrophysical origin. Phys Rev Res 2:023022. arxiv:1906.07119
103. Génolini Y, Boudaud M, Cirelli M, Derome L, Lavalle J, Maurin D et al (2021) New minimal, median, and maximal propagation models for dark matter searches with Galactic cosmic rays. Phys Rev D 104:083005. arxiv:2103.04108
104. Calore F, Cirelli M, Derome L, Genolini Y, Maurin D, Salati P et al, AMS-02 antiprotons and dark matter: trimmed hints and robust bounds. arxiv:2202.03076
105. Boudaud M, Cirelli M, Giesen G, Salati P (2015) A fussy revisitation of antiprotons as a tool for dark matter searches. JCAP 1505:013. arxiv:1412.5696
106. Giesen G, Boudaud M, Génolini Y, Poulin V, Cirelli M, Salati P et al (2015) AMS-02 antiprotons, at last! Secondary astrophysical component and immediate implications for dark matter. JCAP 1509:023. arxiv:1504.04276
107. AMS collaboration, Aguilar M et al (2016) Antiproton flux, antiproton-to-proton flux ratio, and properties of elementary particle fluxes in primary cosmic rays measured with the alpha magnetic spectrometer on the international space station. Phys Rev Lett 117:091103
108. Hooper D, Linden T, Mertsch P (2015) What does the PAMELA antiproton spectrum tell us about dark matter? JCAP 1503:021. arxiv:1410.1527
109. Cuoco A, Krer M, Korsmeier M (2017) Novel dark matter constraints from antiprotons in light of AMS-02. Phys Rev Lett 118:191102. arxiv:1610.03071
110. Cui M-Y, Yuan Q, Tsai Y-LS, Fan Y-Z (2017) Possible dark matter annihilation signal in the AMS-02 antiproton data. Phys Rev Lett 118:191101. arxiv:1610.03840
111. Huang X-J, Wei C-C, Wu Y-L, Zhang W-H, Zhou Y-F (2017) Antiprotons from dark matter annihilation through light mediators and a possible excess in AMS-02 $\bar{p}/p$ data. Phys Rev D95:063021. arxiv:1611.01983
112. Feng J, Zhang H-H (2018) Dark matter search in space: combined analysis of cosmic ray antiproton-to-proton flux ratio and positron flux measured by AMS-02. Astrophys J 858:116. arxiv:1701.02263

113. Arcadi G, Queiroz FS, Siqueira C (2017) The semi-hooperon: gamma-ray and anti-proton excesses in the galactic center. Phys Lett B775:196–205. arxiv:1706.02336
114. Reinert A, Winkler MW (2018) A precision search for WIMPs with charged cosmic rays. JCAP 1801:055. arxiv:1712.00002
115. Pato M, Iocco F, Bertone G (2015) Dynamical constraints on the dark matter distribution in the Milky Way. JCAP 1512:001. arxiv:1504.06324
116. CALET collaboration, Adriani O et al (2017) Energy spectrum of cosmic-ray electron and positron from 10 GeV to 3 TeV observed with the calorimetric electron telescope on the international space station. Phys Rev Lett 119:181101. arxiv:1712.01711
117. DAMPE collaboration, Ambrosi G et al (2017) Direct detection of a break in the teraelectron-volt cosmic-ray spectrum of electrons and positrons. Nature 552:63–66. arxiv:1711.10981
118. AMS collaboration, Aguilar M et al (2014) Precision measurement of the $(e^+ + e)$ flux in primary cosmic rays from 0.5 GeV to 1 TeV with the alpha magnetic spectrometer on the international space station. Phys Rev Lett 113:221102
119. Kerszberg D (2017) The cosmic-ray electron spectrum measured with H.E.S.S. In: ICRC conference. Busan (South Korea)
120. H.E.S.S. collaboration, Aharonian F et al (2008) The energy spectrum of cosmic-ray electrons at TeV energies. Phys Rev Lett 101:261104. arxiv:0811.3894
121. H.E.S.S. collaboration, Aharonian F et al (2009) Probing the ATIC peak in the cosmic-ray electron spectrum with H.E.S.S. Astron Astrophys 508:561. arxiv:0905.0105
122. Silk J, Stebbins A (1993) Clumpy cold dark matter. Astrophys J 411:439–449
123. Bergstrom L, Edsjo J, Gondolo P, Ullio P (1999) Clumpy neutralino dark matter. Phys Rev D 59:043506. arxiv:astro-ph/9806072
124. Ullio P, Bergstrom L, Edsjo J, Lacey CG (2002) Cosmological dark matter annihilations into gamma-rays - a closer look. Phys Rev D 66:123502. arxiv:astro-ph/0207125
125. Berezinsky V, Dokuchaev V, Eroshenko Y (2003) Small—scale clumps in the galactic halo and dark matter annihilation. Phys Rev D 68:103003. arxiv:astro-ph/0301551
126. Stref M, Lavalle J (2017) Modeling dark matter subhalos in a constrained galaxy: global mass and boosted annihilation profiles. Phys Rev D95:063003. arxiv:1610.02233
127. Moliné A, Sánchez-Conde MA, Palomares-Ruiz S, Prada F (2017) Characterization of subhalo structural properties and implications for dark matter annihilation signals. Mon Not R Astron Soc 466:4974–4990. arxiv:1603.04057
128. Lavalle J, Pochon J, Salati P, Taillet R (2007) Clumpiness of dark matter and positron annihilation signal: computing the odds of the galactic lottery. Astron Astrophys 462:827–848. arxiv:astro-ph/0603796
129. Lavalle J, Yuan Q, Maurin D, Bi XJ (2008) Full calculation of clumpiness boost factors for antimatter cosmic rays in the light of Lambda-CDM N-body simulation results. Abandoning hope in clumpiness enhancement?. Astron Astrophys 479:427–452. arxiv:0709.3634
130. Pieri L, Lavalle J, Bertone G, Branchini E (2011) Implications of high-resolution simulations on indirect dark matter searches. Phys Rev D 83:023518. arxiv:0908.0195
131. Murase K, Beacom JF (2012) Constraining very heavy dark matter using diffuse backgrounds of neutrinos and cascaded gamma rays. JCAP 1210:043. arxiv:1206.2595
132. H.E.S.S. collaboration, Abramowski A et al (2012) Search for dark matter annihilation signals from the fornax galaxy cluster with H.E.S.S. Astrophys J 750:123. arxiv:1202.5494
133. Fermi-LAT collaboration, Ackermann M et al (2015) Search for extended gamma-ray emission from the Virgo galaxy cluster with Fermi-LAT. Astrophys J 812:159. arxiv:1510.00004
134. Gelmini GB, Gondolo P (2008) Ultra-cold WIMPs: relics of non-standard pre-BBN cosmologies. JCAP 0810:002. arxiv:0803.2349
135. Erickcek AL, Sigurdson K (2011) Reheating effects in the matter power spectrum and implications for substructure. Phys Rev D 84:083503. arxiv:1106.0536
136. Erickcek AL (2015) The dark matter annihilation boost from low-temperature reheating. Phys Rev D92:103505. arxiv:1504.03335
137. Baldes I, Calore F, Petraki K, Poireau V, Rodd NL (2020) Indirect searches for dark matter bound state formation and level transitions. SciPost Phys 9:068. arxiv:2007.13787

138. Siqueira C (2019) Secluded dark matter in light of the cherenkov telescope array (CTA). Phys Lett B 797:134840. arxiv:1901.11055
139. Holdom B (1986) Two U(1)'s and Epsilon Charge Shifts. Phys Lett B 166:196–198. https://doi.org/10.1016/0370-2693(86)91377-8
140. Chanowitz MS, Gaillard MK (1985) The TeV physics of strongly interacting W's and Z's. Nucl Phys B 261:379–431
141. Bai Y, Berger J, Osborne J, Stefanek BA (2016) Phenomenology of strongly coupled chiral gauge theories. JHEP 11:153. arxiv:1605.07183
142. Ekstedt A, Enberg R, Ingelman G, Löfgren J, Mandal T (2016) Constraining minimal anomaly free U(1) extensions of the standard model. JHEP 11:071. arxiv:1605.04855
143. Whalley MR (2003) A compilation of data on hadronic total cross-sections in e+ e- interactions. J Phys G 29:A1–A133

Chapter 6
First-Order Cosmological Phase Transition

As first observed by Linde and his supervisor in 1972 [1], at high temperature in the Early universe, the Higgs field had a thermal mass and the electroweak (EW) symmetry was effective (we say "restored"). The electromagnetism and the weak interaction constituted a single electroweak interaction, and electroweak and Higgs bosons were continuously created and absorbed in the plasma. This raises the question on how the Higgs field acquired the non-zero vacuum expectation value $\langle H^\dagger H \rangle \neq 0$ responsible for the masses of the fermions and electroweak bosons which we observe today ? (Sect. 2.2.3 in Chap. 2) The nature of the electroweak phase transition (EWPT) was first investigated using lattice simulations at the end of the 90s [2–4]. The authors showed that in the SM the transition is a cross-over, meaning that the order parameter $\langle H^\dagger H \rangle$ varies smoothly with the temperature,[1] if $m_{\rm Higgs} \gtrsim 72$ GeV. In contrast, the scenario where $\langle H^\dagger H \rangle$ would manifest a jump is called a first-order phase transition (1stOPT). Also, the chiral phase transition of QCD is known to be a cross-over in the SM, except if, at the time of the phase transtion, the baryon chemical potential is large [7] or if there are more than $N_f \gtrsim 3$ massless quarks, as shown by Pisarski and Wilczek in 1983 [8].

1stOPT have been the subject of a lot of interest because they can explain the baryon asymmetry (Sect. 3.5.2), they can generate GW which are potentially detectable by future interferometers like LISA, see [9–11] for pionnering works and [12–14] for reviews, they can generate magnetic fields [15–17], topological defects [18–20], primordial black holes [21–33], they can set the abundance of dark matter [34–49], and offer a new access to the scale of supersymmetry breaking [50]. We refer to [51, 52] for other reviews on 1stOPT.

[1] Technically speaking, there is no gauge-invariant order parameter which can distinguish the symmetric high temperature phase and the low temperature Higgs phase [2, 5, 6]. Except along the 1stOPT line in the phase diagram, there is no breaking or restoration of the gauge symmetry across the phase transition. For instance, the gauge boson masses themselves receive finite-temperature corrections and varies smoothly with T.

Y. Gouttenoire, *Beyond the Standard Model Cocktail*, Springer Theses,
https://doi.org/10.1007/978-3-031-11862-3_6

It is rather straightforward to make the EWPT first-order by adding new physics in the scalar sector. Specific realizations include: $|H|^6$ operator [53–61], extra singlet models [60, 62–67], two-Higgs double models [68, 69], models with spontaneous breaking of conformality due to 1-loop quantum correction [40, 61, 70–76], models with warped extra-dimensions [34, 77–85], or strong dynamics [49, 86–91], the last two being related by holography. Generically, we can say that models with polynomial potentials typically lead to weakly first-order phase transitions while models with nearly-conformal potentials can naturally induce strongly first-order phase transitions with large amount of supercooling. We mention [14, 92] for reviews on SM extensions leading to strong 1stOPT.

The present chapter is a review on 1stOPT and not my research. I believe it might be useful for students and researchers desiring learning about this topic. In Sect. 6.1, we show how to compute the effective potential in a given model, and how to compute the tunneling temperature. In Sect. 6.2, we present different methods to compute the velocity of the bubble wall. In Sect. 6.3, we present the GW spectrum resulting from bubble collisions, sound-waves and turbulence. Finally, in Sect. 6.4 we focus on the class of models giving the largest GW signal: models with nearly-conformal potential, which are responsible for supercooled 1stOPT.

All the figures in the chapter are mine (Figs. 6.1, 6.2, 6.3, 6.4, 6.6, 6.7, 6.8, 6.9, 6.10, 6.11, 6.12, 6.13 and 6.14).

6.1 Bubble Nucleation

6.1.1 Effective Potential at Finite Temperature

Tree level:
We start with the following classical effective potential for the SM Higgs

$$V(H) = m^2\,|H|^2 + \lambda\,|H|^4 + \Lambda^{-2}\,|H|^6\,, \tag{6.1}$$

with $H^T = (\chi_1 + i\chi_2,\ \varphi + i\chi_3)/\sqrt{2}$. It is the SM Lagrangian, defined in Sect. 2.2.3 of Chap. 2, augmented with the non-renormalizable operator $|H|^6$ [53–61]. The energy scale Λ sets the range of validity of the model. After expanding the Higgs around its VEV $\varphi = \phi + h$, we obtain

$$V_{\text{tree}}(\phi) = \frac{m^2}{2}\phi^2 + \frac{\lambda}{4}\phi^4 + \frac{1}{8}\frac{\phi^6}{\Lambda^2}. \tag{6.2}$$

One-loop finite-temperature effective potential:
The quantum corrections at one-loop were first computed in 1973 by Coleman and E. J. Weinberg at zero temperature [93] and a few months later by Dolan and Jackiw at finite-temperature [94]. We call attention to the reviews [55, 95] and textbooks [96, 97] for more details. The 1-loop corrections read

$$V_{\text{eff}}(\phi, T) = V_{\text{tree}}(\phi) + \Delta V_1(\phi, T), \tag{6.3}$$

with

$$\Delta V_1(\phi, T) = \sum_{i=\{h,\chi,W,Z,t\}} \frac{n_i T}{2} \sum_{n=-\infty}^{+\infty} \int \frac{d^3\vec{k}}{(2\pi)^3} \log\left[\vec{k}^2 + \omega_n^2 + m_i^2(\phi)\right], \tag{6.4}$$

where $k_E = (\omega_n, \vec{k})$ is the euclidean loop 4-momentum, ω_n are the Matsubara frequencies in the imaginary time formalism, $\omega_n = 2n\pi T$ for bosons (periodic on the euclidean time circle) and $\omega_n = (2n+1)\pi T$ for fermions (anti-periodic on the euclidean time circle). We assumed the Landau gauge ($\xi = \infty$ in R_ξ gauge [2]) where the ghosts decouple and the Goldstones are included.[3] The number of degrees of freedom is then $n_{\{h,\chi,W,Z,t\}} = \{1, 3, 6, 3, -12\}$. The masses $m_i^2(\phi) = \partial^2 V(\phi)/\partial \chi_i^2$ read

$$m_h^2(\phi) = m^2 + 3\lambda\phi^2 + \frac{15}{4}\frac{\phi^4}{\Lambda^2}, \tag{6.6}$$

$$m_\chi^2(\phi) = m^2 + \lambda\phi^2 + \frac{3}{4}\frac{\phi^4}{\Lambda^2}, \tag{6.7}$$

$$m_W^2(\phi) = \frac{g_2^2}{4}\phi^2, \quad m_Z^2(\phi) = \frac{g_1^2 + g_2^2}{4}\phi^2, \quad m_t^2(\phi) = \frac{y_t^2}{2}\phi^2. \tag{6.8}$$

We can decompose the one-loop correction as a sum of a zero-temperature term and a T-dependent one which vanishes when $T \to 0$

---

[2] The gauge-independence of the energy value at the extrema of the effective potential is guaranteed by the so-called Nielsen identity (1975 [98, 99])

$$\frac{\partial V_{\text{eff}}(\phi, \xi)}{\partial \xi} = -C(\phi, \xi)\frac{\partial V_{\text{eff}}(\phi, \xi)}{\partial \phi}. \tag{6.5}$$

It says that the gauge dependence of the effective action is equivalent to a nonlocal field redefinition. As a consequence, the position of the minima are gauge-dependent, e.g. [100, 101], but the nucleation rate as well as the energy profile of the nucleated bubbles are gauge-independent, e.g. [102–104]. I thank Thomas Konstandin for related discussions.

[3] Note that away from the zero-temperature minimum, longitudinal modes of gauge bosons and Goldstones are independent of each other.

$$\Delta V_1^0(\phi) = \sum_{i=\{h,\chi,W,Z,t\}} \frac{n_i}{2} \int \frac{d^4k_E}{(2\pi)^4} \log\left[k_E^2 + m_i^2(\phi)\right], \tag{6.9}$$

$$\Delta V_1^T(\phi, T) = \sum_{i=\{h,\chi,W,Z,t\}} \frac{n_i\, T^4}{2\pi^2} \int_0^\infty dk\, k^2 \log\left[1 \mp \exp -\sqrt{k^2 + m_i^2(\phi)/T^2}\right]. \tag{6.10}$$

Zero-temperature corrections:
Upon imposing a hard cut-off $k_E < \Lambda_{\rm UV}$ and integrating over k_E, Eq. (6.9) gives the so-called Coleman-Weinberg potential

$$\Delta V_1^0(\phi) = \sum_{i=h,\chi,W,Z,t} n_i \frac{m_i^4(\phi)}{64\pi^2}\left[\log\frac{m_i^2(\phi)}{\Lambda_{\rm UV}^2} + \Lambda_{\rm UV}^2 m_i^2(\phi)\right], \tag{6.11}$$

The UV cut-off $\Lambda_{\rm UV}$ disappears after renormalization. For doing this, we introduce counterterms

$$V_{\rm ct} = \delta V_0 + \frac{\delta m^2}{2}\phi^2 + \frac{\delta\lambda}{4}\phi^4, \tag{6.12}$$

and we impose the minimum of the potential to correspond to the Higgs VEV $v = 246$ GeV, and the second derivative to the Higgs mass $m_{\rm Higgs} = 125$ GeV

$$\left.\frac{d(\Delta V_1^0 + V_{ct})}{d\phi}\right|_{\phi=v} = 0, \tag{6.13}$$

$$\left.\frac{d^2(\Delta V_1^0 + V_{ct})}{d\phi^2}\right|_{\phi=v} = m_{\rm Higgs}^2. \tag{6.14}$$

We obtain

$$\Delta V_1^0(\phi) = \sum_{i=\{h,\chi,W,Z,t\}} n_i \frac{1}{64\pi^2}\left[m_i^4(\phi)\left(\log\frac{m_i^2(\phi)}{m_i^2(v)} - \frac{3}{2}\right) + 2m_i^2(v)m_i^2(\phi)\right]. \tag{6.15}$$

Finite temperature corrections:
The finite-temperature contributions in Eq. (6.10) can be re-expressed in term of the $J_{b,f}$ functions

$$\Delta V_1^T(\phi, T) = \sum_{i=\{h,\chi,W,Z,t\}} \frac{n_i\, T^4}{2\pi^2} \int_0^\infty dk\, k^2 \log\left[1 \mp \exp -\sqrt{k^2 + m_i^2(\phi)/T^2}\right] \tag{6.16}$$

$$\equiv \sum_{i=\text{bosons}} \frac{n_i T^4}{2\pi^2} J_b\left(\frac{m_i^2(\phi)}{T^2}\right) + \sum_{i=\text{fermions}} \frac{n_i T^4}{2\pi^2} J_f\left(\frac{m_i^2(\phi)}{T^2}\right). \tag{6.17}$$

It is sometimes convenient to Taylor-expand $J_{b,f}$ in the high-temperature limit

$$J_b(x) \underset{x\to 0}{=} -\frac{\pi^4}{45} + \frac{\pi^2}{12}x - \frac{\pi}{6}x^{3/2} - \frac{x^2}{32}\log\frac{x}{a_b} + \mathcal{O}\left(x^3\log\frac{x^{3/2}}{\text{cst.}}\right), \tag{6.18}$$

$$J_f(x) \underset{x\to 0}{=} \frac{7\pi^4}{360} - \frac{\pi^2}{24}x - \frac{x^2}{32}\log\frac{x}{a_f} + \mathcal{O}\left(x^3\log\frac{x^{3/2}}{\text{cst.}}\right), \tag{6.19}$$

with $\log a_b \simeq 5.4076$ and $\log a_f \simeq 2.6350$. In order to speed-up the numerical computation, it may be useful to use the fitting formula [105][4]

$$J_{B/F}(x) = t_{B(F)} f^{\text{HT}}_{B(F)}(x) + (1 - t_{B(F)}) f^{\text{LT}}(x) \tag{6.20}$$

with

$$J_B^{\text{HT}}(x) = -\frac{\pi^4}{45} + \frac{\pi^2}{12}x - \frac{\pi}{6}x^{3/2} - \frac{1}{32}x^2\left[\log x - \log a_b\right], \tag{6.21}$$

$$J_F^{\text{HT}}(x) = -\frac{7\pi^4}{360} + \frac{\pi^2}{24}x + \frac{1}{32}x^2\left[\log x - \log a_f\right], \tag{6.22}$$

$$J^{\text{LT}}(x) = -\sqrt{\frac{\pi}{2}}x^{3/4}e^{-\sqrt{x}}\left(1 + \frac{15}{8}x^{-1/2} + \frac{105}{128}x^{-1}\right), \tag{6.23}$$

$$t_B(x) = e^{-(x/6.3)^4}, \tag{6.24}$$

$$t_F(x) = e^{-(x/3.25)^4}. \tag{6.25}$$

Daisy corrections:

Perturbative expansion at finite-temperature suffers from infrared divergences coming from the so-called daisy diagrams. At order λ^N, they can be described as N independent daughter loops (or petals) attached to 1 mother loop. Their contributions to the effective potential read [106]

$$\Delta V^N_{\text{daisy}} = -\frac{1}{2}T\sum_n \frac{d^3\vec{k}}{(2\pi)^3}\left[-\frac{\Pi_1}{\omega_n^2 + \vec{k}^2 + m^2}\right]^N, \tag{6.26}$$

where Π_1 the 1-loop self-energy of petals and $\vec{k}$ is the 3-momentum running in the mother loop. We can see that for massless particles $m = 0$ and zero Matsubara modes $n = 0$, each individual daisy diagram with $N \gtrsim 2$ is IR-divergent. In order to cure those divergences, we must resum all of them, which gives

[4] Note that the Eq. (3.9) of [105] has a typo.

$$\Delta V_{\text{daisy}} = \sum_{N=1}^{\infty} V_{\text{daisy}}^{N} = \frac{1}{2} T \sum_{n} \frac{d^3\vec{k}}{(2\pi)^3} \ln\left[1 + \frac{\Pi_1}{\omega_n^2 + \vec{k}^2 + m^2}\right]. \tag{6.27}$$

By comparing the last equation (6.27) with Eq. (6.4), we see that the effect of resumming all the daisy diagrams is a temperature-dependent shift of the square mass by

$$m_b^2(\phi) \to m_b^2(\phi) + \Pi_b(T), \tag{6.28}$$

where $\Pi_b(T)$ is the self-energy of the bosonic field b in the IR limit, $\omega = \vec{p} = 0$. Since fermions have no zero Matsubara modes, they are free from IR-divergences. In the SM + $|H|^6$ theory, these shifts, also known as Debye masses, read [55, 106]

$$\Pi_{h,\chi_i}(T) = \frac{T^2}{4v^2}\left(m_h^2 + 2m_W^2 + m_Z^2 + 2m_t^2\right) - \frac{3}{4} T^2 \frac{v^2}{\Lambda^2}, \tag{6.29}$$

$$\Pi_W(T) = \frac{22}{3} \frac{m_W^2}{v^2} T^2, \tag{6.30}$$

while the shifted masses $\Pi_{Z/\gamma}(T)$ of Z and γ are eigenvalues of the following mass matrix

$$\begin{pmatrix} \frac{1}{4} g_2^2 \phi^2 + \frac{11}{6} g_2^2 T^2 & -\frac{1}{4} g_1^2 g_2^2 \phi^2 \\ -\frac{1}{4} g_1^2 g_2^2 \phi^2 & \frac{1}{4} g_1^2 \phi^2 + \frac{11}{6} g_1^2 T^2 \end{pmatrix}. \tag{6.31}$$

Note that the daisy resummation being based on a high-T expansion, it should not be relevant at large m_i/T. So, in order to prevent the non-physical contributions from gauge boson at large m_i/T, it may be useful to introduce the cut-off [74]

$$.\Pi_i(T) \to \Pi_i(T)\, C(m_i/T), \qquad \text{with} \quad C(x) = x^2/2\, K_2(x), \tag{6.32}$$

$K_2(x)$ being the modified Bessel function of the second kind.

Upon performing the integration in Eq. (6.27) while keeping only the dominant contribution from the zero Matsubara mode, and upon summing over the longitudinal modes of all the vector bosons in the theory, we obtain

$$\Delta V_{\text{daisy}} = \sum_{i=\{h,\chi,W,Z,\gamma\}} \frac{\bar{n}_i T}{12\pi}\left[m_i^3(\phi) - \left(m_i^2(\phi) + \Pi_i(T)\right)^{3/2}\right], \tag{6.33}$$

with $\bar{n}_{\{h,\chi,W,Z,\gamma\}} = \{1, 3, 2, 1, 1\}$. The non-integer power 3/2 gives evidence of the breakdown of perturbation theory.

Total:
We conclude that the total effective potential, in the Landau gauge, at one-loop and finite-temperature reads

$$\begin{aligned} V_{\text{eff}}(\phi, T) =& \frac{m^2}{2}\phi^2 + \frac{\lambda}{4}\phi^4 + \frac{1}{8}\frac{\phi^6}{\Lambda^2} \\ &+ \sum_{i=\{h,\chi,W,Z,t\}} n_i \frac{1}{64\pi^2}\left[m_i^4(\phi)\left(\log\frac{m_i^2(\phi)}{m_i^2(v)} - \frac{3}{2}\right) + 2m_i^2(v)m_i^2(\phi)\right] \\ &+ \sum_{\text{i=bosons}} \frac{n_i T^4}{2\pi^2} J_b\left(\frac{m_i^2(\phi)}{T^2}\right) + \sum_{\text{i=fermions}} \frac{n_i T^4}{2\pi^2} J_f\left(\frac{m_i^2(\phi)}{T^2}\right) \\ &+ \sum_{i=\{h,\chi,W,Z,\gamma\}} \frac{\bar{n}_i T}{12\pi}\left[m_i^3(\phi) - (m_i^2(\phi) + \Pi_i(T))^{3/2}\right], \end{aligned} \tag{6.34}$$

with $n_{\{h,\chi,W,Z,t\}} = \{1, 3, 6, 3, -12\}$, $\bar{n}_{\{h,\chi,W,Z,\gamma\}} = \{1, 3, 2, 1, 1\}$ and

$$m^2 = \frac{m_H^2}{2} - \frac{3}{4}\frac{v^4}{\Lambda^2}, \tag{6.35}$$

$$\lambda = \frac{1}{2}\frac{m_H^2}{v^2} - \frac{3}{2}\frac{v^2}{\Lambda^2}. \tag{6.36}$$

We show the different contributions to the effective potential in Fig. 6.1.

Note that the high-temperature expansion of the effective potential reads [62]

$$V_{\text{eff}}(\phi, T) \overset{T\gg v}{\longrightarrow} D(T^2 - T_0^2)\phi^2 - ET\phi^3 + \frac{\lambda(T)}{4}\phi^4 + \frac{1}{8}\frac{\phi^6}{\Lambda^2}, \tag{6.37}$$

with

$$D = \frac{2m_W^2 + m_Z^2 + 2m_t^2}{8v^2}, \tag{6.38}$$

$$E = \frac{2m_W^3 + m_Z^3}{4\pi v^3}, \tag{6.39}$$

$$T_0^2 = \frac{m_H^2 - 8Bv^3}{4D}, \tag{6.40}$$

$$B = \frac{3}{64\pi^2 v^4}(2m_W^4 + m_Z^4 - 4m_t^4), \tag{6.41}$$

$$\lambda(T) = \lambda - \frac{3}{16\pi^2 v^4}\left(2m_W^4 \log\frac{m_W^2}{A_B T^2} + m_Z^4 \log\frac{m_Z^2}{A_B T^2} - 4m_t^4 \log\frac{m_t^2}{A_F T^2}\right) \tag{6.42}$$

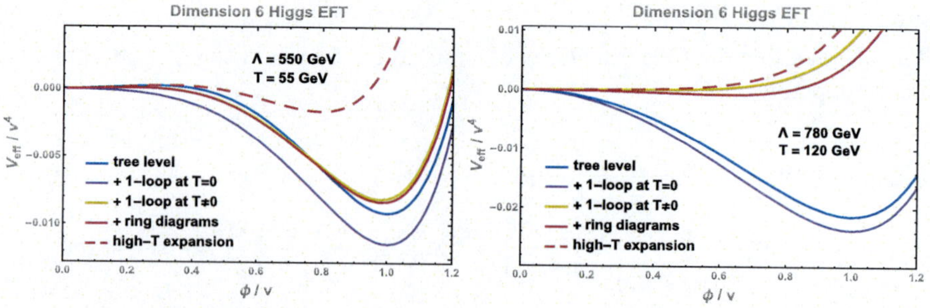

Fig. 6.1 Effective potential in the Landau gauge at one-loop and finite-temperature of the SM + dimension 6 EFT, for two different choices of temperature and cut-off

where $v = 246$ GeV, $\log A_B = \log a_b - 3/2$ and $\log A_F = \log a_f - 3/2$. As shown in Fig. 6.1, the high-temperature expansion of the effective potential is only a good approximation at small ϕ/T.

6.1.2 *Tunneling Rate*

As shown in Fig. 6.1, in the presence of finite-temperature corrections, the effective potential can manifest a barrier separating the false vacuum (in $\phi_{\rm FV} \approx 0$) and the true vacuum (in $\phi_{\rm TV} \approx v$). Below the critical temperature T_c, the potential energy in the true minimum becomes lower than the potential energy in the false vacum, which therefore becomes **metastable**. The decay rate of the false vacuum was first computed by Coleman and Callan in 1977 at zero temperature [107, 108] and extended to finite-temperature by Linde in 1980 [109, 110].[5] We refer the reader to Coleman's lecture for more details [112]. Based on those results, the decay rate can be expressed as

$$\Gamma(T) \simeq \max\left[T^4\left(\frac{S_3/T}{2\pi}\right)^{3/2}\exp\left(-S_3/T\right),\quad R_0^{-4}\left(\frac{S_4}{2\pi}\right)^2\exp\left(-S_4\right)\right], \quad (6.43)$$

where S_3 and S_4 are the 3- and 4-dimensional Euclidean action of the $O(3)$- and $O(4)$-symmetric tunneling solutions. The former is **thermally-induced** while the later is purely **quantum**. R_0 is the bubble radius. The tunneling rate via $O(3)$ bounce dominates for bubble radius larger than the imaginary time periodicity $R \gtrsim 1/T$, while the tunneling rate for $O(4)$ bounce dominates in the opposite case [109, 110].

[5] Gravitational corrections were computed for the first time in 1980 by Coleman and De Luccia [111].

The S_3 and S_4 bounce actions read

$$\frac{S_3}{T} = 4\pi \int dr\, r^2 \left[\frac{1}{2}\phi'(r)^2 + V(\phi(r))\right], \tag{6.44}$$

$$S_4 = 2\pi^2 \int dr\, r^3 \left[\frac{1}{2}\phi'(r)^2 + V(\phi(r))\right], \tag{6.45}$$

where $V(\phi)$ is the effective potential discussed in Sect. 6.1.1, and $\phi(r)$ is the field configuration which interpolates between the two asymptotic vacuum. Extremization of the action leads to the **Euclidean** equation of motion

$$\phi''(r) + \frac{d-1}{r}\phi'(r) = \frac{dV}{d\phi}, \tag{6.46}$$

with **boundary conditions**

$$\phi'(0) = 0, \qquad \text{and} \qquad \lim_{r\to\infty} \phi(r) = 0. \tag{6.47}$$

$O(3)$ bounces have $d = 3$ dimensions and are described by the coordinate $r = \vec{x}$, while $O(4)$ bounces have $d = 4$ dimensions and are described by the coordinate $r = \sqrt{\vec{x}^2 + t^2}$.

It is worth mentionning the existence of a **stochastic approach** of quantum tunneling, introduced by Linde et al. in 1991 [113, 114]. According to this picture, one treats the field ϕ in the false vacuum state as a **random gaussian variable**, which can move over the barrier to the true vacuum after developing large quantum fluctuations [113–117]. This approach does not rely on classically forbidden tunneling path as in Eq. (6.46), but instead it relies on a **real-time description** which has the advantage to be extendable to time-dependent scenarios.

Numerical solutions:
The bounce solution $\phi(r)$ is equivalent to the trajectory of a classical point particle in an inverted potential $-V(\phi)$. Finding the bounce solution $\phi(r)$ can be realized numerically by guessing an initial field value $\phi(0)$ between the zero of the potential ϕ_{zero}, defined by $V(\phi_{\text{zero}}) = V(\phi_{\text{FV}})$, and the true vacuum ϕ_{TV}. If the field trajectory $\phi(r)$ **overshoots** the false vacuum ϕ_{FV}, we must decrease $\phi(0)$, while if it **undershoots** the false vacuum, we must increase $\phi(0)$.

Once the numerical bounce solution is obtained, we can check a posteriori that it corresponds to a stationary point of the action. To do so, let us first define a one-parameter family of functions

$$\phi_\lambda(x) = \bar{\phi}(x/\lambda) \tag{6.48}$$

where $\bar{\phi}(x)$ is the bounce solution. The corresponding $O(d)$ bounce action reads

$$S_d(\lambda) = \lambda^{d-2} \int dx^d \frac{1}{2}(\partial_\mu \bar{\phi})^2 + \lambda^d \int dx^d V(\bar{\phi}) \tag{6.49}$$

$$= \lambda^{d-2} K + \lambda^d U \tag{6.50}$$

Since $\bar{\phi}$ is solution of the equation of motion, $S_d(\lambda)$ must be stationary for $\lambda = 1$, which implies [112]

$$K = -\frac{d}{d-2} U. \tag{6.51}$$

The last equality allows to quantify the degree of error of the numerical solution. For instance, we can define the **error parameter** ϵ

$$\epsilon = \frac{(d-2)K + d\,U}{(d-2)K - d\,U}. \tag{6.52}$$

Over the last decade, various numerical packages have been released to the community: CosmoTransitions [118], AnyBubble [119], BubbleProfiler [120], SimpleBounce [121], FindBounce [122, 123], as well as semi-analytical approaches [124–126].

Semi-analytical solutions:
For a polynomial potential of the type

$$V(\phi) = a\phi^2 - b\phi^3 + \frac{\lambda}{4}\phi^4, \tag{6.53}$$

semi-analytical expressions for the O_3- and O_4-bounce actions exist (Adams 1993 [127], see also [128–130] for more recent use)

$$S_3 = \frac{8\pi b}{\lambda^{3/2}} \frac{8\sqrt{\delta}}{81(2-\delta)^2}(\beta_1\delta + \beta_2\delta^2 + \beta_3\delta^3), \tag{6.54}$$

$$S_4 = \frac{4\pi^2}{3\lambda} \frac{1}{(2-\delta)^3}(\alpha_1\delta + \alpha_2\delta^2 + \alpha_3\delta^3), \tag{6.55}$$

with

$$\delta \equiv \frac{2\lambda a}{b^2}, \tag{6.56}$$

$$\beta_1 = 8.2938, \quad \beta_2 = -5.5330, \quad \beta_3 = 0.8180, \tag{6.57}$$

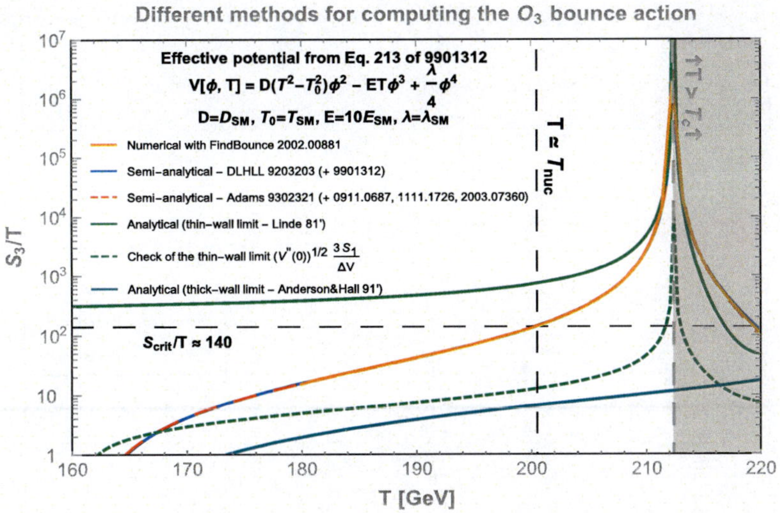

Fig. 6.2 We consider the high-temperature expansion of the effective potential of the SM EFT in Eq. (6.37), with $\Lambda \to 0$ and $E \to 10E$. The semi-analytical solutions for the O_3-bounce action, in Eqs. (6.54) and (6.59), are very good approximations of the solution from the numerical package FindBounce [122, 123]. On the other side, the thin-wall, cf. Eq. (6.68), and thick-wall limit, cf. Eq. (6.73), are very bad approximation for the potential which we consider. More precisely, the thin-wall limit ceases to be valid when the quantity in Eq. 3.7 of [110] (dashed green line) is smaller than $\sim 10^2$

$$\alpha_1 = 13.832, \quad \alpha_2 = -10.819, \quad \alpha_3 = 2.0765. \tag{6.58}$$

Another semi-analytical expression for S_3 which we can be found in the litterature is the one that DLHLL derived in 1992 [131] (see also [95])

$$S_3 = \frac{13.72\, a^{3/2}}{b^2} f(\delta/2), \tag{6.59}$$

with

$$f(x) = 1 + \frac{x}{4}\left[1 + \frac{2.4}{1-x} + \frac{0.26}{(1-x)^2}\right]. \tag{6.60}$$

In Fig. 6.2 we show that the two semi-analytical expressions in Eq. (6.54) and Eq. (6.59) fit very well the solution of the package FindBounce [122, 123]. As a toy model of the effective potential we consider the high temperature expansion of the SM EFT in Eq. (6.37), with $\Lambda \to \infty$, where we modified $E \to 10E$ in order to increase the barrier strength and to make the transition strongly first order.

In the next section, we provide analytical approximations of the bounce action in the so-called thin-wall and thick-wall limit.

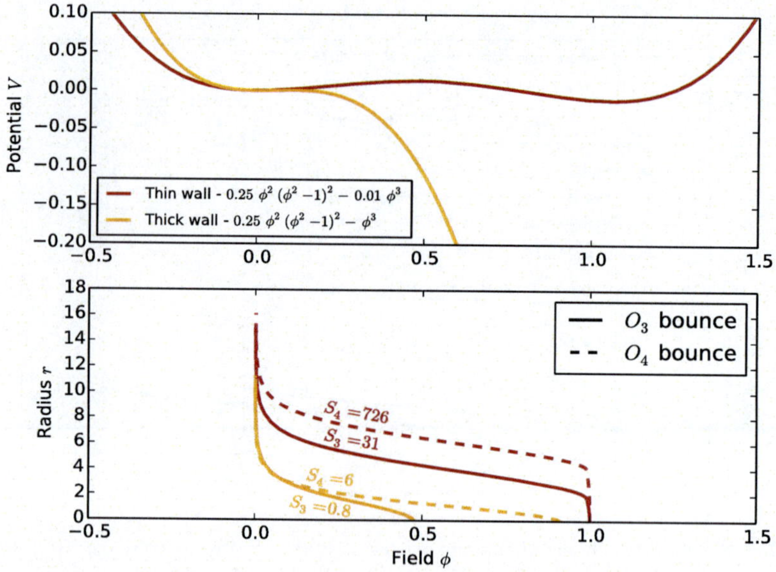

Fig. 6.3 **Top:** Thin-walled potential (Large barrier compared to the potential energy difference between the minima) versus thick-walled potential (small barrier compared to the potential energy difference between the two minima). The true minimum of the thick-walled potential does not appear on the plot. **Bottom:** Corresponding O_3- and O_4- symmetrical bounce profile $\phi(r)$, solution of the Euclidean equation of motion in Eq. (6.46). Computed with CosmoTransition [118]

6.1.3 *Thin-Wall and Thick-Wall Limits*

Thin-wall limit:

The first class of potential where the bounce action can be estimated analytically are potentials where the extrema are **nearly-degenerate** and separated by a **large barrier**. The wall thickness is as thin as possible in order to minimize the region of space sitting on top of the potential barrier. This is the **thin-wall** limit (red line in Fig. 6.3). In order for the trajectory to be able to reach the false vacuum, the initial field value $\phi(0)$ has to start close to the true vacuum $\phi_{\rm TV}$ where the potential is very flat. Hence, the particle sits there for a long time before evolving significantly. Hence, we can neglect the damping term in the equation of motion in Eq. (6.46)

$$\frac{d^2\phi}{dr^2} = V'(\phi) \tag{6.61}$$

which, after multiplying both sides by $d\phi/dr$ and using the first boundary condition in Eq. (6.47), leads to

$$\frac{d\phi}{dr} = \sqrt{2(V(\phi) - V(\phi_{\rm TV}))} \tag{6.62}$$

Therefore, the $O(4)$-symmetric bounce action can be estimated as

$$S_4 = 2\pi^2 \int dr\, r^3 \left[\frac{1}{2}\left(\frac{d\phi}{dr}\right)^2 + V(\phi)\right] \tag{6.63}$$

$$= \underbrace{-2\pi^2 \frac{R^4}{4} \Delta V}_{r \ll R} + \underbrace{2\pi^2 R^3 S_1}_{r \sim R} \tag{6.64}$$

where $\Delta V > 0$ is the potential energy difference and where S_1 is the action of a one-dimensional bounce

$$S_1 = \int_0^\infty dr \left[\frac{1}{2}\left(\frac{d\phi}{dr}\right)^2 + V(\phi)\right] = \int_{\phi_{\rm TV}}^{\phi_{\rm FV}} d\phi \sqrt{2(V(\phi) - V(\phi_{\rm TV}))}. \tag{6.65}$$

Physically, S_1 corresponds to the energy of the bubble wall per unit of area (the **wall tension**). The **critical radius** R_c of the bubble at nucleation is the one which minimizes the action

$$\frac{\delta S_4}{\delta R} = 0 \quad \rightarrow \quad -2\pi^2\, R_c^3\, \Delta V + 6\pi^2\, R_c^2\, S_1 = 0. \tag{6.66}$$

By injecting the solution R_c into Eq. (6.64), we obtain [107]

$$R_c = \frac{3 S_1}{\Delta V} \qquad \text{and} \qquad S_4 = \frac{27\pi^2\, S_1^4}{2\, \Delta V^3}. \tag{6.67}$$

It is very straightforward to generalize this result to the $O(3)$-symmetric bounce action in Eq. (6.44) and we find [110]

$$R_c = \frac{2 S_1}{\Delta V} \qquad \text{and} \qquad S_3 = \frac{16\pi\, S_1^3}{3\, \Delta V^2}. \tag{6.68}$$

Whether the tunneling completes via $O(3)$ or $O(4)$ bounce depends on the ratio

$$\frac{S_4}{S_3/T} = \frac{81\pi}{32} \frac{S_1}{\Delta V} T \sim 3\, R_c\, T. \tag{6.69}$$

Thick-wall limit:
When the **potential barrier is small** compared to the potential energy difference, the wall thickness is as large as possible in order to minimize the gradient energy. This is the **thick-wall** limit (yellow line in Fig. 6.3). The $O(4)$-symmetric bounce action reads

$$S_4 = \pi^2\, R^3\, \delta R \left(\frac{\delta\phi}{\delta R}\right)^2 + \frac{\pi^2}{2} R^4\, \bar{V} \tag{6.70}$$

where $\delta\phi$ is the field excursion while δR is the wall thickness. In the thick-wall limit, we can set $\delta R = R$. The quantity $\bar{V} < 0$ is the potential energy averaged in the bubble volume. We approximate it as $\bar{V} = V(\phi_*) - V(\phi_{\rm FV})$ where ϕ_* is the initial field value, at the center of the bubble.[6] Additionally, we assume (without lack of generality) that the false vacuum energy and the corresponding field value (at asymptotic distance) are zero, $V(\phi_{\rm FV}) = 0$ and $\phi_{\rm FV} = 0$, such that we can write $\bar{V} = V(\phi_*)$ and $\delta\phi = \phi_*$. Hence, we obtain

$$S_4 = \pi^2 R^2\, \delta R\, \phi_*^2 + \frac{\pi^2}{2} R^4\, V(\phi_*), \tag{6.71}$$

where ϕ_* is understood as the value which minimizes the bounce action. Upon solving for the **critical bubble radius** R_c, solution of $\delta S_4/\delta R = 0$, and reinjecting into Eq. (6.71), we obtain [62, 78, 79]

$$R_c^2 = \frac{\phi_*^2}{-V(\phi_*)} \quad \text{and} \quad S_4 = \frac{\pi^2}{2}\frac{\phi_*^4}{-V(\phi_*)}. \tag{6.72}$$

Repeating straightforwardly the same steps for the $O(3)$-symmetrical bounce leads to

$$R_c^2 = \frac{\phi_*^2}{-2\,V(\phi_*)} \quad \text{and} \quad S_3 = \frac{4\pi}{3}\frac{\phi_*^3}{\sqrt{-2\,V(\phi_*)}}. \tag{6.73}$$

Whether the tunneling completes via $O(3)$ or $O(4)$ bounce depends on the ratio

$$\frac{S_4}{S_3/T} = \frac{3\pi}{4}\frac{\phi_*}{\sqrt{-2\,V(\phi_*)}}T \sim 2\,R_c\,T. \tag{6.74}$$

In Fig. 6.4, we compare the nucleation temperature $T_{\rm nuc}$, discussed in the next section in Sect. 6.1.4, in the dim 6 Higgs EFT model, computed using the thick-wall formula for S_3 to the one computed using a public numerical package,[7] as well as to the one which can be found in the literature. The agreement is rather acceptable.

6.1.4 *Temperature at Which the Phase Transition Completes*

Nucleation temperature:
At first approximation, we can consider the phase transition to complete when the number of bubbles nucleated per Hubble volume and per Hubble time is of order 1

$$\Gamma(T_{\rm nuc}) \simeq H(T_{\rm nuc})^4. \tag{6.75}$$

[6] In the thick-wall limit, the initial field value ϕ_* is generally close to the zero of the potential $V(\phi_{\rm zero}) = V(\phi_{\rm FV})$.

[7] I thank Victor Guada and Jorinde Van de Vis for useful discussions regarding the computation of the bounce action with existing numerical packages.

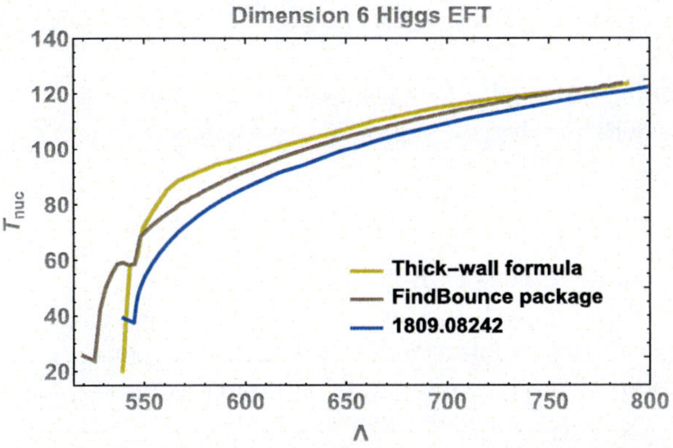

Fig. 6.4 Nucleation temperature, assuming O_3 tunneling, in the dimension 6 Higgs EFT model, which effective potential $V_{\rm eff}$ is given by Eq. (6.34). We compare the results obtained with the (easy-to-use) FindBounce package [122, 123], the results obtained from the thick-wall formula in Eq. (6.73) and the results from [60]

Percolation temperature:
In this paragraph we give a more thorough analysis of the temperature for which regions of false vacuum are entirely converted into true vacuum following the early work of [132–134, 134, 135] and the more recent of work of [60]. Starting from $\Gamma(t')\,a(t')^3$ which is the number of bubbles formed per unit of time and per unit of comoving volume and from $\frac{4\pi}{3}\left(\int_{t'}^{t}\frac{d\tilde{t}}{a(\tilde{t})}\right)^3$ which is the comoving volume of a bubble which has been nucleated at t' and has expanded at the speed of light until t, we deduce the volume fraction converted into the true vacuum at time t, including **multiple counting** when bubbles overlap [8]

$$I(t) = \int_{t_c}^{t} dt'\,\Gamma(t')\,a(t')^3\,\frac{4\pi}{3}\left(\int_{t'}^{t}\frac{d\tilde{t}}{a(\tilde{t})}\right)^3. \tag{6.76}$$

We can also interpret $I(t)$ as the average number of bubbles inside which, a given point is contained. It can be larger than 1 because it includes overlapping. However, multiple counting can been subtracted by **exponentiation**.[9] Hence, we define

[8] Note that we have neglected the initial size of the bubble just after it nucleates. Including it would give $I(t) = \int_{t_c}^{t} dt'\,\Gamma(t')\,\frac{4\pi}{3}\left(\int_{t'}^{t}\frac{a(t')}{a(\tilde{t})}\,d\tilde{t} + R(t')\right)^3 = \frac{4\pi}{3}\int_{T_{\rm nuc}}^{T_c}\frac{dT'}{T'}\,\frac{\Gamma(T')}{T'^3\,H(T')}\left(\int_{T_{\rm nuc}}^{T'}\frac{d\tilde{T}}{H(\tilde{T})} + R(T')\,T'\right)^3$ where R' is the bubble radius at nucleation. The correction is of order $R(T')\,H' \sim \frac{\rm TeV}{M_{\rm Pl}}$ and is completely negligible.

[9] The probability $P(t)$ that a given point remains in the false vacuum is the $N \to \infty$ limit of the product $\prod_{n=1}^{N}\left(1 - dt\,\frac{dI(t_n)}{dt}\right)$ where $dt = \frac{t-t_c}{N}$ and $1 - dt\,\frac{dI(t_n))}{dt}$ is the probability of remaining in the false vacuum between $t_n = t_c + (t-t_c)\,\frac{n}{N}$ and t_{n+1}.

$$P(t) \equiv e^{-I(t)} \tag{6.77}$$

which is the probability that a given point remains in the false vacuum at time t. From this, we can define the percolation to occur when

$$P(t) \lesssim 1/e \rightarrow I(t) \gtrsim 1. \tag{6.78}$$

Do bubbles percolate during an inflating period?
During inflation it can happen that the probability $1 - P(t)$ that a given point has been converted to the true vacuum tends to 1, whereas the physical volume remaining in the false vacuum (here given per unit of comoving volume)

$$\mathcal{V}_{\text{false}}(t) \equiv a(t)^3 \, P(t), \tag{6.79}$$

goes to infinity [136]. In order to avoid this to occur,[10] we add to Eq. (6.78) a second condition which is that the decrease of the probability to remain in the false vacuum $P(t)$ drops **faster** than the increase of the false vacuum volume due to inflation [135]

$$\frac{d\mathcal{V}_{\text{false}}(t)}{dt} \lesssim 0. \tag{6.80}$$

Upon using the adiabatic relations $H\,dt = -dT/T$ and $d(T\,a) = 0$, we can re-express the criteria in Eq. (6.78) and in Eq. (6.80) solely in terms of the temperature. Hence, the **percolation temperature** T_{per} is obtained from

$$\frac{4\pi}{3} \int_{T_{\text{per}}}^{T_c} \frac{dT'}{T'} \frac{\Gamma(T')}{T'^3 \, H(T')} \left(\int_{T_{\text{per}}}^{T'} \frac{d\tilde{T}}{H(\tilde{T})} \right)^3 \simeq 1, \tag{6.81}$$

after checking that

$$3 + T_{\text{per}} \frac{dI(T)}{dT} \bigg|_{T_{\text{per}}} < 0. \tag{6.82}$$

Note however that the distinction between nucleation temperature and percolation temperature is relevant only for tuned polynomial potential in the supercool regime [60]. Particularly, the distinction is not relevant for supercooled phase transitions which are naturally generated by non-polynomial potential, see Refs. [49, 74] and Sect. 6.4.

---

[10] In the study [137], the authors explore the alternative possibility that a 1stOPT occuring during inflation completes after the end of inflation. They show that a continuous nucleation of bubbles during inflation would generate a large spectrum of primordiale scalar fluctuation which could source observable GW at second order in perturbations.

6.2 Bubble Propagation

The vacuum energy difference between the broken phase and the symmetric phase acts as a negative pressure which accelerates the bubble wall. If coupled to the scalar field driving the PT, the particles entering the broken phase acquire a mass term proportional to the scalar VEV. Hence, the scalar field gradient along the wall profile induces a force acting on the particles in the plasma. The terminal velocity of the bubble wall results from the equilibrium between the accelerating vacuum pressure and the retarding friction pressure. In this section we present different formalism which can be found in the literature for computing the bubble wall velocity. In Sect. 6.2.1, we derive the equilibrium condition between the accelerating pressure and the friction pressure from the conservation of the energy-momentum tensor of the system plasma+scalar field. In Sect. 6.2.2, we derive the retarding pressure at local-thermal-equilibrium in the field-theoretic approach. In Sect. 6.2.3, we present a field-theoretic computation of the friction pressure in the presence of out-of-equilibrium conditions. In Sect. 6.2.4, we compute the friction pressure in the limit where collisions are neglected. We discuss the important NLO correction in gauge coupling constant in Sect. 6.2.5. Finally, we deduce the speed of the bubble wall in Sect. 6.2.6. Part of the results of this section are relevant for the work on string fragmentation in supercooled confinement in Chap. 7, where the new effects which we point out, string fragmentation and deep-inelastic-scattering depends on the value of the bubble wall Lorentz factor γ_{wp}.

6.2.1 *Equation of Motion for the Scalar Field*

The dynamics of the scalar field during the bubble propagation can be derived from the conservation of the energy-momentum tensor

$$\partial^\mu T^{\text{total}}_{\mu\nu} = \partial^\mu \left(\sum_i T^{\text{plasma}}_{i,\,\mu\nu} + T^{\phi}_{\mu\nu} \right) = 0, \tag{6.83}$$

where i runs over the species in the plasma. The covariant formulation of the energy-momentum tensor of the plasma in terms of the one-particle state occupation number $f_i(\vec{p},\, x)$ reads

$$T^{\text{plasma}}_{i,\,\mu\nu} = \int \frac{d^3p}{(2\pi)^3} \frac{p_\mu p_\nu}{E_i} f_i(\vec{p},\, x). \tag{6.84}$$

$f_i(\vec{p},\, x)$ is solution of the Boltzmann equation (Liouville equation + collision)

$$\frac{df}{dt} = \text{coll} \quad \rightarrow \quad \partial_t f + \dot{\vec{x}} \cdot \partial_{\vec{x}} + \dot{\vec{p}} \cdot \partial_{\vec{p}} = \text{coll}, \tag{6.85}$$

which follows from the time conservation of the number of particles $N = \int d^3r\, d^3p\, f(\vec{r}, \vec{p}, t)$ up to the collision term "coll". Now, supposing that the wall velocity is constant $v_w = \text{cst}$ (the wall has reached equilibrium), we have

$$\partial_t f = v_w\, \partial_z f \quad (= 0 \text{ in the wall frame}), \tag{6.86}$$

$$\dot{z} = \frac{p_z}{E}, \tag{6.87}$$

$$\dot{p}_z = -\frac{\partial_z m(z)^2}{2E} \quad (\text{the problem is time-invariant: } \dot{E} = 0). \tag{6.88}$$

we can rewrite Eq. (6.85) as (e.g. [138])

$$\left(\frac{p_z}{E}\partial_z - \frac{\partial_z m^2}{2E}\partial_{p_z}\right) f(p_z, z) = \text{coll.} \tag{6.89}$$

We deduce the divergence of Eq. (6.84)

$$\partial^\mu T^{\text{plasma}}_{i,\,\mu\nu} + \text{coll}_\text{i} = \int \frac{d^3p}{(2\pi)^3} p_\nu \left(\frac{1}{E} p^\mu \partial_\mu f_i(\vec{p}, x) - \frac{f_i(\vec{p}, x)}{2E^3}\partial_\mu m^2\right) \tag{6.90}$$

$$= \frac{1}{2}\partial_\mu m^2 \int \frac{d^3p}{(2\pi)^3} p_\nu \left(-\frac{1}{E}\partial_{p_\mu} f_i(\vec{p}, x) - \partial_{p_\mu}\left(\frac{1}{E}\right) f_i(\vec{p}, x)\right) \tag{6.91}$$

$$= \frac{1}{2}\partial_\nu m^2 \int \frac{d^3p}{(2\pi)^3}\frac{1}{E}\, f_i(\vec{p}, x). \tag{6.92}$$

where in the second line we have used the Boltzmann equation in Eq. (6.89) whereas in the last line we have performed an integration by part. In addition, the energy-momentum tensor of the classical field background is

$$T^\phi_{\mu\nu} = \partial_\mu\phi\partial_\nu\phi - g_{\mu\nu}\left(\frac{1}{2}\partial_\rho\phi\partial^\rho\phi - V_0(\phi)\right), \tag{6.93}$$

where V_0 is the effective potential at zero temperature. Its divergence reads

$$\partial^\mu T^\phi_{\mu\nu} = \partial_\nu\phi\left(\Box\phi + \frac{dV_0}{d\phi}\right). \tag{6.94}$$

Upon plugging Eqs. (6.94) and (6.92) into Eq. (6.83), we obtain the equation of motion for the scalar field at constant bubble wall velocity

$$\Box\phi + \frac{dV_0}{d\phi} + \sum_i \frac{dm_i^2}{d\phi}\int \frac{d^3p}{(2\pi)^3}\frac{1}{2E}\, f_i(\vec{p}, x) = 0. \tag{6.95}$$

Note that the collision terms are absent since they cancel out when they are summed over all the species [139]. The last term can be expressed as

$$\frac{dm^2}{d\phi}\int\frac{d^3p}{(2\pi)^3}\frac{1}{2E}f(\vec{p},x) = -\frac{1}{3}\frac{dm^2}{d\phi}\int\frac{d^3p}{(2\pi)^3}\frac{p}{2}\cdot\frac{d}{dp}\frac{f_i(\vec{p},x)}{E} \tag{6.96}$$

$$= -\frac{1}{3}\frac{dm^2}{d\phi}\int\frac{d^3p}{(2\pi)^3}\frac{p}{2}\cdot\frac{dE}{dp}\cdot\frac{dm^2}{dE}\cdot\frac{d\phi}{dm^2}\cdot\frac{d}{d\phi}\frac{f_i(\vec{p},x)}{2E} \tag{6.97}$$

$$= -\frac{dP_{\text{plasma}}}{d\phi}, \tag{6.98}$$

where P_{plasma} is the thermal pressure

$$P_{\text{plasma}} = \int\frac{d^3p}{(2\pi)^3}\frac{p^2}{3E}\sum_i f_i(\vec{p},x), \tag{6.99}$$

In the presence of collisions, $f_i(\vec{p},x)$ can be highly out-of-equilibrium. We now express Eq. (6.100) in spherical coordinates (we use the $SO(3)$ symmetry to reduce the 3 Cartesian coordinates to the radial r coordinate [11])

$$\frac{\partial^2\phi}{\partial t^2} - \frac{1}{r^2}\frac{\partial}{\partial\phi}\left[r^2\frac{\partial\phi}{\partial r}\right] + \frac{\partial V}{\partial\phi} = \frac{dP_{\text{plasma}}}{d\phi}. \tag{6.101}$$

Going in the wall frame, we can drop the time derivative in the previous Eq. (6.101). Then, upon multiplying by $d\phi/dz$ and integrating over z across the wall, we find the equilibrium condition

$$\Delta V_{\text{vac}} = \Delta P_{\text{plasma}}, \qquad \text{(equilibrium condition)} \tag{6.102}$$

where ΔV_{vac} and ΔP_{plasma} are respectively the difference of effective potential at zero-temperature (vacuum energy difference) and the difference between the pressure from plasma effects between the two phases. In order to write Eq. (6.102), we used that the wall profile flattens at infinity $\int\phi''\phi'\,dz = \left[\frac{\phi'^2}{2}\right]_{-\infty}^{+\infty}$ and we have neglected the Laplace pressure ΔP_{Lapace} due to the surface tension of the bubble [141] which goes to zero at large radius/time

[11] Note that we could also use the $SO(3,1)$ symmetry of Eq. (6.100) which allows to trade r and t for the time-like light-cone coordinate $s = \sqrt{t^2 - r^2}$ [140]

$$\frac{\partial^2\phi}{\partial s^2} + \frac{3}{s}\frac{\partial\phi}{\partial s} + \frac{\partial V}{\partial\phi} = \frac{dP_{\text{plasma}}}{d\phi}. \tag{6.100}$$

In this coordinate, the trajectory of the bubble wall, $r = v_w\,t$, satisfies $s = t/\gamma_w$.

$$\Delta P_{\text{Laplace}}(t) = \int_{\text{across the wall}} dz \, \frac{2}{r}\left(\frac{d\phi}{dr}\right)^2 \underset{r\to\infty}{\longrightarrow} 0. \tag{6.103}$$

We can generally expect the retarding pressure from plasma effects P_{plasma} to be an increasing function of the wall velocity v_w such that the wall accelerates under the vacuum pressure ΔV until it compensates with P_{plasma}.

6.2.2 *Friction Pressure at Local Thermal Equilibrium*

Energy-momentum conservation in the stationary regime:
In this section, we derive the friction pressure at local thermal equilibrium. Under the assumption that the bubble wall reaches a constant velocity, the energy-momentum tensor in the bubble frame is time-invariant and its conservation law in the bubble frame reads [142]

$$\Delta\left\langle T^p_{z0}\right\rangle + \Delta\left\langle T^b_{z0}\right\rangle = 0, \tag{6.104}$$

$$\Delta\left\langle T^p_{zz}\right\rangle + \Delta\left\langle T^b_{zz}\right\rangle = 0. \tag{6.105}$$

where the subscripts p and b denote the plasma and bubble components, respectively. We can write the stress-energy tensor of the plasma in the plasma frame as

$$\left\langle (T^p_{00})_{\text{plasma frame}}\right\rangle = \rho_p, \qquad \left\langle (T^p_{zz})_{\text{plasma frame}}\right\rangle = P_p, \tag{6.106}$$

such that after a Lorentz boost to the bubble frame, it becomes

$$\left\langle T^p_{zz}\right\rangle = (\gamma^2 - 1)(\rho_p + P_p) + P_p, \qquad \left\langle T^p_{z0}\right\rangle = \beta\gamma^2(\rho_p + P_p), \tag{6.107}$$

where $\gamma = 1/\sqrt{1-\beta^2}$. For the bubble components, we obtain

$$\left\langle T^b_{zz}\right\rangle \equiv P_b, \qquad \left\langle T^b_{z0}\right\rangle = 0, \tag{6.108}$$

where we have set $\rho_b + P_b = 0$ since a scalar condensate has no entropy. Eqs. (6.104) and (6.105) integrated along the z direction become

$$\Delta(\beta\gamma^2\, T\, S) = 0, \tag{6.109}$$

$$\Delta((\gamma^2 - 1)\, T\, S) + \Delta P_p + \Delta P_b = 0, \tag{6.110}$$

where $S = (\rho_p + P_p)/T$ is the entropy of the plasma.

First energy-momentum conservation equation:
The first equation implies that whenever $\Delta(TS) \neq 0$, the velocity of the plasma can not be uniform along the fluid profile. The temperature and velocity of fluid just outside the bubble T_+, β_+ and just inside the bubble T_-, β_- must be determined by solving the two equations in Eqs. (6.109) and (6.110). We refer to [143, 144] for such a treatment. Instead, here we follow [142] and make the approximation

$$\Delta T \simeq 0. \tag{6.111}$$

This is the **local-thermal-equilibrium** (l.t.e.) condition.

Second energy-momentum conservation equation:
Equation (6.110) implies that the pressure difference in the plasma frame at l.t.e. reads [142][12]

$$\Delta P \equiv \Delta(P_p + p_b) = -(\gamma^2 - 1)\, T\, \Delta S \tag{6.112}$$

Hence, at local thermal equilibrium, the pressure difference is given by the entropy difference between the two phases ΔS (in the plasma frame).

Entropy in quantum field theory:
The entropy is related to the component of the energy-momentum tensor which for a free scalar field, reads

$$T_{\mu\nu} = \partial_\mu\phi\partial_\nu\phi + g_{\mu\nu}\mathscr{L}_\phi \tag{6.113}$$

with

$$\mathscr{L}_\phi = -\frac{1}{2}g_{\mu\nu}\partial_\mu\phi\partial_\nu\phi - \frac{1}{2}m^2\phi^2. \tag{6.114}$$

The thermal average, after renormalization, reads

$$\langle T_{\mu\nu}\rangle_{\text{ren}} = -\eta_{\mu\nu}\frac{m^4}{64\pi^2}\left[\ln\left(\frac{m^2}{4\pi\mu^2}\right) + \gamma_{\text{E}} - \frac{3}{2}\right] + \frac{\eta_\mu\nu}{6\pi^2\beta^5 m}\left[\partial_z J_{\text{B}}(b,\, z)\right]_{z=\beta m} \tag{6.115}$$

$$+\delta^0_\mu\delta^0_\nu\left(\frac{2}{3\pi^2\beta^5\, m}\left[\partial_z J_{\text{b}}(6,\, z)\right]_{z=\beta m} + \frac{m}{2\pi^2\beta^3}\left[\partial_z J_{\text{b}}(4,\, z)\right]_{z=\beta m}\right), \tag{6.116}$$

with

$$J_{\text{b/f}}(n,\, z) \equiv \int_0^\infty dx\, x^{n-2}\ln\left(1 \mp e^{-\sqrt{x^2+z^2}}\right). \tag{6.117}$$

[12] I thank Bogumiła Świeżewska for a valuable correspondence.

We deduce the change in entropy between the massless phase to the massive phase

$$T\Delta S = \frac{2\pi^2}{45\beta^4} - \frac{2}{3\pi^2\beta^5\, m}\left[\partial_z J_b(6,\, z)\right]_{z=\beta m} - \frac{m}{2\pi^2\beta^3}\left[\partial_z J_b(4,\, z)\right]_{z=\beta m}. \quad (6.118)$$

Results:
Since ΔS is independent of γ, this implies

$$\Delta P \propto \gamma^2. \quad (6.119)$$

Hence a bubble can not run-away to an arbitrary large γ value at local thermal equilibrium. However, we generally expect the local thermal equilibrium condition to be quickly violated as soon as $\Gamma \lesssim$ few, due to the mass of particles varying faster than the collision time scale as the wall passes by. In the following sections, we discuss different possibilities to account for the deviation from local thermal equilibrium. In [142], the temperature and the fluid velocity is assumed to be constant across the wall. Even though this contradicts energy-momentum conservation [145], it can be a good approximation in the limit where particles of the plasma interact strongly with each others [146]. In [143, 144], the results of [142] and [145] are merged to compute the wall velocity at local equilibrium with varying temperature and fluid velocity across the wall.

Bubble wall velocity:
The pressure difference defined from Eq. (6.112) accounting for both scalar field and plasma contribution coincides with the effective potential at finite temperature $V_{\text{eff}}(\phi,\, T)$ defined in Eq. (6.34). Hence the bubble velocity can be computed from

$$\Delta V_{\text{eff}}(T) = -(\gamma^2 - 1)\, T\, \Delta S. \quad (6.120)$$

The friction pressure and bubble velocity will be shown in Figs. 6.6 and 6.7.

6.2.3 *Friction Pressure Close to Local Thermal Equilibrium*

A thermal field theoretic approach:
At thermal equilibrium, the thermal pressure P_{plasma} is directly related to the finite-temperature corrections ΔV_{plasma} to the effective potential [97]

$$\Delta V_{\text{plasma}} = -P_{\text{plasma}} \quad (6.121)$$

Therefore, splitting the particle distribution functions into an equilibrium part plus some deviation δf_i yields

$$\Box\phi + \frac{dV(\phi, T)}{d\phi} + K(\phi) = 0, \tag{6.122}$$

where $V(\phi, T)$ is the **effective potential at finite temperature**, introduced in Sect. 6.1.1, and gives the **force** driving the wall. The last term encodes the deviation from local thermal equilibrium

$$K(\phi) \equiv \sum_i \frac{dm_i^2}{d\phi} \int \frac{d^3p}{(2\pi)^3} \frac{1}{2E} \delta f_i(\vec{p}, x), \tag{6.123}$$

and is responsible for **friction.** Upon considering the Ansatz

$$\phi(z) = \frac{\phi_0}{2}\left[1 + \tanh\left(\frac{z}{L_w}\right)\right] \tag{6.124}$$

for the wall profile in the wall frame (ϕ_0 being the scalar VEV) and taking the moments

$$\int_{-\infty}^{\infty} dz \left[\text{l.h.s of Eq. (6.122)}\right] \times \phi' = 0, \tag{6.125}$$

$$\int_{-\infty}^{\infty} dz \left[\text{l.h.s of Eq. (6.122)}\right] \times (2\phi - \phi_0)\phi' = 0, \tag{6.126}$$

we obtain

$$\frac{\Delta V_T}{T^4} = F, \tag{6.127}$$

$$-\frac{2}{15(T\,L_w)^2}\left(\frac{\phi_0}{T}\right)^3 + \frac{W}{T^5} = G, \tag{6.128}$$

where ΔV_T denotes the difference in finite-temperature effective potential between the two phases and W is the integral

$$W \equiv \int_0^{\phi_0} \frac{dV(\phi, T)}{d\phi} (2\phi - \phi_0)\, d\phi. \tag{6.129}$$

ΔV_T measures the **pressure difference** while W measures the **pressure gradient**.

The functions F and G measure the deviation from local thermal equilibrium and can be computed in close-to-equilibrium **thermal field theory** [138, 139, 147–152]. The resolution of the Eqs. (6.127) and (6.128) allows to determine the **wall velocity** v_w and the **wall thickness** L_w.[13]

An hydrodynamic approach:
We can also choose to describe the friction term in Eq. (6.123) in a fluid approach, for instance, starting from the ansatz [145, 153–158]

$$K(\phi) = \eta\, u^\mu\, \partial_\mu \phi, \tag{6.130}$$

where η is a model-dependent parameter encoding the friction efficiency ($[\eta] =$ energy). For instance in the SM, we expect $\eta \approx 3\,T$ [145]. The fluid velocity v, the temperature T and the scalar profile ϕ obey to the following combined set of equations

$$\partial_z^2 \phi - \frac{\partial V(\phi,\, T)}{\partial \phi} + \eta\, u^\mu\, \partial_\mu \phi = 0, \tag{6.131}$$

$$\partial_z \left[\omega \gamma^2 v\right] = 0, \tag{6.132}$$

$$\partial_z \left[\frac{1}{2}(\partial_z \phi)^2 + \omega \gamma^2 v^2 + p\right] = 0, \tag{6.133}$$

where $\omega = \rho + p$ is the fluid enthapy. The last two equations results from the conservation of the components T_{z0} and T_{00} of the energy-momentum tensor. See Ref. [145] for a determination of the wall velocity in this approach.

6.2.4 *Friction Pressure in the Ballistic Approximation*

Collisionless regime:
For fast moving wall, we can expect large departure from local thermal equilibrium to take place, such that close-to-equilibrium thermal field techniques or fluid approaches, as mentioned in the previous section, break down. This is particularly true when the time scale, $L_w/\gamma_w\, v_w$, over which the background change, is much shorter than the collision time scale, $1/\Gamma_{\rm coll} \sim 1/\alpha^2\, T$, where T is the plasma temperature and α is the typical coupling constant responsible for the interactions

$$\frac{L_w}{\gamma_w\, v_w} \ll \frac{1}{\alpha^2\, T}, \qquad \text{(ballistic regime).} \tag{6.134}$$

[13] Since plasma effects enters the scalar field equation of motion in Eq. (6.122), it was expected that the wall thickness would depend on the plasma properties and wall velocity.

v_w and γ_w denotes the wall velocity and Lorentz factor with respect to the bubble center. In that regime, we can neglect the collision term "coll″ in the Boltzmann equation in Eq. (6.89) which becomes

$$\left(\frac{p_z}{E}\partial_z - \frac{\partial_z m^2}{2E}\partial_{p_z}\right) f(p_z, z) = 0, \tag{6.135}$$

and we can consider the particles as evolving independently from each others, this is the **ballistic regime**.

Solution of the Liouville equation:
The solution of the Liouville equation in Eq. (6.135) is any function of the form

$$f(\vec{r}, \vec{p}, t) = f(p_z^2 + m(z)^2, \vec{p}_\perp). \tag{6.136}$$

Its exact form is determined using the **boundary conditions** at $z \to \pm\infty$, where local thermal equilibrium is well-satisfied. Indeed, far from the wall, the collision term "coll" is large such that local thermal equilibrium is satisfied and the occupation number is

$$f(z \to \pm\infty) = \frac{1}{e^{\beta p_\mu u^\mu} - 1} = \frac{1}{e^{\beta\gamma(E - v p_z)} \mp 1}, \tag{6.137}$$

with a minus/plus sign for bosons/fermions. The second term is Lorentz-invariant whereas the third one is expressed in the wall frame, with $v \equiv v_w$ and $\gamma \equiv \gamma_w$.

From Eqs. (6.136) and (6.137), we obtain

$$f = \begin{cases} \dfrac{1}{e^{\beta\gamma\left(E - v\sqrt{p_z^2 + m(z)^2}\right)} \mp 1} & \text{if coming from the symmetric phase,} \\[2ex] \dfrac{1}{e^{\beta\gamma\left(E + v\sqrt{p_z^2 + m(z)^2 - m_0^2}\right)} \mp 1} & \text{if coming from the broken phase,} \end{cases} \tag{6.138}$$

where m_0 is the mass of the particle deep inside the broken phase (inside the bubble).

The **physical intuition** behind the final solutions is that $E = \sqrt{p_z^2 + p_\perp^2 + m(z)^2}$ and $p_\perp$ are conserved across the bubble wall, since time- and $\perp$- translation are symmetries of the problem. This implies that $\sqrt{p_z^2 + m(z)^2}$ is conserved. Note that p_z is not conserved since invariance under z-translation is broken (spontaneously) by the wall. Conservation of the number of particles passing through the wall implies that f must depend on $\sqrt{p_z^2 + m(z)^2}$ and not on p_z.

Three contributions to the pressure:
In the ballistic approach, the **local** pressure due to plasma effects, which enters the bubble wall equilibrium condition in Eq. (6.102), is

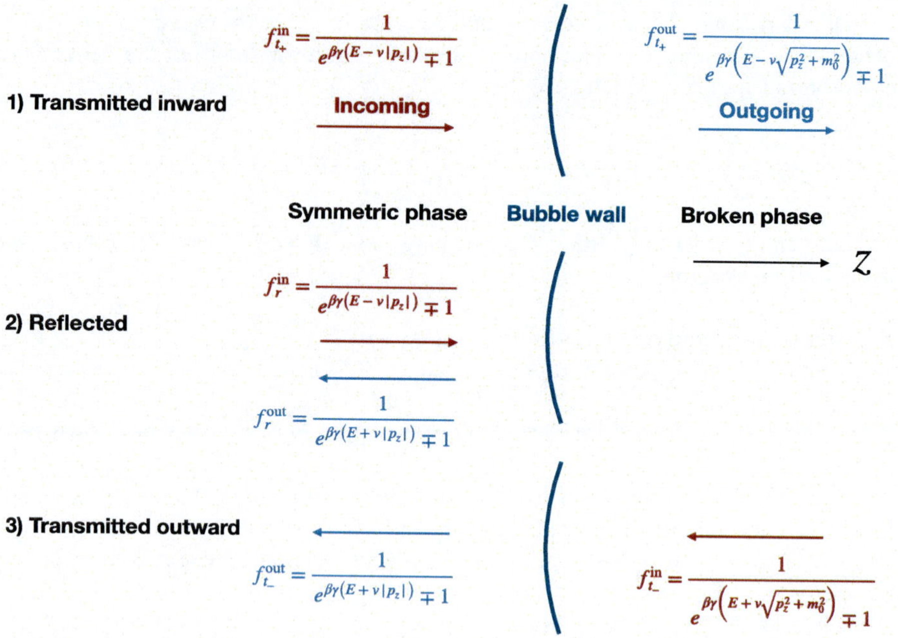

Fig. 6.5 Three contributions to the ballistic pressure. The occupancy functions f are solutions of the Liouville equation is Eq. (6.136) with the boundary condition in Eq. (6.137). We distinguish incoming and outgoing particles, with respect to the bubble wall, with red and blue colors respectively

$$P_{\text{plasma}} = \int \frac{d^3p}{(2\pi)^3} \frac{p_z^2}{E} f(\vec{r}, \vec{p}, t), \tag{6.139}$$

and it receives **three contributions**, as displayed in Fig. 6.5 and detailed below [131, 138, 142, 159].

1. First, there is the contribution due to the particles going from the **symmetric** to the **broken** phase, which reads

$$\begin{aligned} \Delta P_{t+} = & \int_{p_z > m} \frac{d^3p}{(2\pi)^3} \frac{1}{e^{\beta\gamma(\sqrt{p_r^2 + p_z^2} - vp_z)} \mp 1} \left(\frac{p_z^2}{\sqrt{p_r^2 + p_z^2}} \right) \\ & - \int_{p_z > 0} \frac{d^3p}{(2\pi)^3} \frac{1}{e^{\beta\gamma(\sqrt{p_r^2 + p_z^2 + m^2} - v\sqrt{p_z^2 + m^2})} \mp 1} \left(\frac{p_z^2}{\sqrt{p_r^2 + p_z^2 + m^2}} \right). \end{aligned} \tag{6.140}$$

This is the difference between the local pressure in the deep symmetric phase minus the pressure in the deep broken phase, due to particles going in the $+z$ direction. We have assumed the z-axis pointing in the broken phase direction. We have denoted m as the mass deep inside the broken phase ($m \equiv m_0$).

2. Second, we have the contribution coming from the particles getting **reflected**, which reads

$$\Delta P_r = \int_{-m<p_z<m} \frac{d^3p}{(2\pi)^3} \frac{1}{e^{\beta\gamma(\sqrt{p_r^2+p_z^2}-vp_z)} \mp 1} \left(\frac{p_z^2}{\sqrt{p_r^2+p_z^2}} \right). \tag{6.141}$$

We have defined $\Delta P_r = P_{\text{sym}}^{|p_z|<m} - P_{\text{turn}}^{p_z=0}$ where $P_{\text{sym}}^{|p_z|<m}$ is the pressure in the deep symmetric phase resulting from the particles whose momentum is insufficient to climb up to the potential energy of the wall, $|p_z| < m$, considered either before ($p_z > 0$) or after reflection ($p_z < 0$). $P_{\text{turn}}^{p_z=0}$ is the pressure at the turning-point position and is identically zero since the phase space of $p_z = 0$ is zero.

3. Finally, there is the contribution due to the particles passing from the **broken** to the **symmetric** phase

$$\Delta P_{t-} = \int_{p_z<0} \frac{d^3p}{(2\pi)^3} \frac{1}{e^{\beta\gamma(\sqrt{p_r^2+p_z^2+m^2}+v\sqrt{p_z^2})} \mp 1} \left(\frac{p_z^2}{\sqrt{p_r^2+p_z^2+m^2}} \right)$$
$$- \int_{p_z<-m} \frac{d^3p}{(2\pi)^3} \frac{1}{e^{\beta\gamma(\sqrt{p_r^2+p_z^2}+v\sqrt{p_z^2-m^2})} \mp 1} \left(\frac{p_z^2}{\sqrt{p_r^2+p_z^2}} \right). \tag{6.142}$$

This is the difference between the local pressure in the deep broken phase minus the pressure in the deep symmetric phase, due to particles going in the $-z$ direction. We can show that this last contribution is **suppressed** like $\Delta P_{t-} \propto e^{-\beta\gamma m}$ or stronger [142], so most of the time it is negligible.

Upon changing variable $p_r\, dp_r = E\, dE$, integrating over E between p_z and ∞, and performing a shift $p_z^2 \to p_z^2 - m^2$, we obtain

$$\Delta P_{t+} = \mp \frac{1}{4\pi^2} \frac{m^3}{\beta\gamma M_-^3} \left[\int_{M_-}^{\infty} dx\, x^2 \log\left[1 \mp e^{-x}\right] - \int_0^{\infty} dx\, x^2 \log\left[1 \mp e^{-\sqrt{x^2+M_-^2}}\right] \right], \tag{6.143}$$

$$\Delta P_r = \mp \frac{1}{4\pi^2} \frac{m^3}{\beta\gamma} \left[\frac{1}{M_-^3} \int_0^{M_-} dx\, x^2 \log\left[1 \mp e^{-x}\right] + \frac{1}{M_+^3} \int_0^{M_+} dx\, x^2 \log\left[1 \mp e^{-x}\right] \right], \tag{6.144}$$

$$\Delta P_{t-} = \mp \frac{1}{4\pi^2} \frac{1}{\beta^4\gamma^4} \int_0^{\infty} dx\, x \left[\sqrt{x^2 + (\beta\gamma m)^2} - x \right] \log\left[1 \mp e^{-\sqrt{x^2+(\beta\gamma m)^2}-v x}\right], \tag{6.145}$$

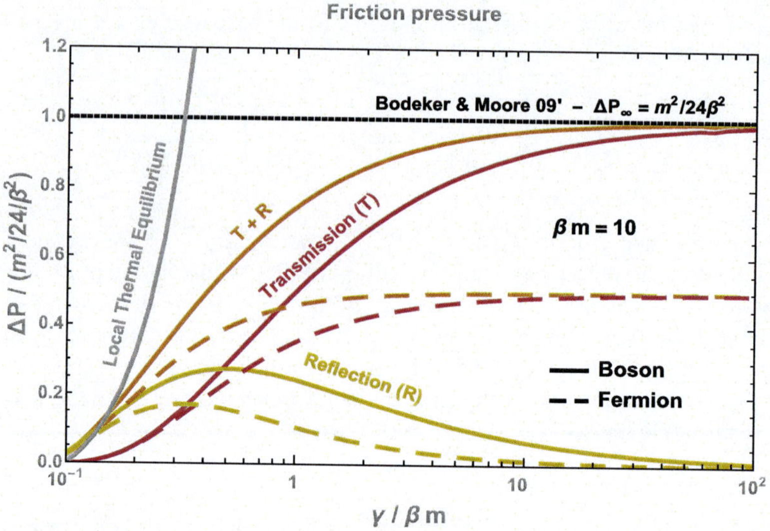

Fig. 6.6 Retarding friction pressure on the bubble wall versus the Lorentz factor of the wall. The pressure computed using local thermal equilibrium (gray lines), cf. Eq. (6.112), increases as $\Delta P_{\text{lte}} \propto \gamma^2$, implying that the bubble wall would never run away. However, for $\gamma \gtrsim 1$, local thermal equilibrium is violated. Instead, for relativistic wall velocities the ballistic approximation (brown lines) is more trustable. We decompose the later into the contribution from particles entering the bubble (purple lines, Eq. (6.143) and case 1 in Fig. 6.5) and the contribution from the particles reflected off the wall (yellow lines, Eq. (6.144) and case 2 in Fig. 6.5), the particles being either bosons (solid) or fermions (dashed), both having a unique degree of freedom. The third contribution to the ballistic pressure, due to particles going from the broken phase to the symmetric phase (Eq. (6.145) and case 3 in Fig. 6.5), is suppressed like $e^{-\beta\gamma m}$ and is not represented. We can see that the ballistic pressure saturates to the Bodeker and Moore value from 2009 [160] at large γ (the asymptotic pressure receives an extra factor 1/2 for fermions)

where $M_- = m\,\beta\,\gamma\,(1-v)$ and $M_+ = m\,\beta\,\gamma\,(1+v)$. We compare the different contributions to the ballistic pressure in Fig. 6.6. As expected the dominant contribution comes from particles entering the bubbles ($t+$) and the contribution coming from particles going out of the bubble ($t-$) is negligible (not shown on the plot). The contribution from the **reflected** particles is **relevant** for $\gamma \lesssim \beta\,m$.

Total ballistic pressure:
Upon combining the three contributions,[14] we obtain the **total pressure** in the ballistic approximation [142]

[14] In order to get Eqs. (6.148) and (6.149) we have used the relation

$$\int_0^M dx\, x^2 \log(1-e^{-x}) = -\frac{\pi^4}{45} + M^2\,\mathrm{Li}_2(e^{-M}) + 2M\,\mathrm{Li}_3(e^{-M}) + 2\,\mathrm{Li}_4(e^{-M}), \qquad (6.146)$$

and

$$\Delta P^b_{\text{ballistic}} = \frac{\pi^2}{90\beta^4}\left[4(\gamma^2-1)+1\right] - \frac{1}{4\pi^2\beta^4\gamma^4}\Bigg[-\frac{J_b(\beta\gamma(1-v)m)}{(1-v)^3}$$
$$+\frac{\left[M_+^2\,\mathrm{Li}_2(e^{-M_+})+2M_+\,\mathrm{Li}_3(e^{-M_+})+2\,\mathrm{Li}_4(e^{-M_+})\right]}{(1+v)^3}$$
$$+\int_0^\infty dx\, x\left[\sqrt{x^2+(\beta\gamma m)^2}-x\right]\ln\left[1-e^{-\sqrt{x^2+(\beta\gamma m)^2}-v\,x}\right]\Bigg], \tag{6.148}$$

for **bosons** and

$$\Delta P^f_{\text{ballistic}} = \frac{7}{8}\frac{\pi^2}{90\beta^4}\left[4(\gamma^2-1)+1\right] + \frac{1}{4\pi^2\beta^4\gamma^4}\Bigg[-\frac{J_f(\beta\gamma(1-v)m)}{(1-v)^3}$$
$$+\frac{\left[M_+^2\,\mathrm{Li}_2(-e^{-M_+})+2M_+\,\mathrm{Li}_3(-e^{-M_+})+2\mathrm{Li}_4(-e^{-M_+})\right]}{(1+v)^3}$$
$$+\int_0^\infty dx\, x\left[\sqrt{x^2+(\beta\gamma m)^2}-x\right]\ln\left[1+e^{-\sqrt{x^2+(\beta\gamma m)^2}-v\,x}\right]\Bigg], \tag{6.149}$$

for **fermions.**[15] The functions J_b and J_f where already defined in Eq. (6.17) and $\mathrm{Li}_s(z) = \sum_{n=1}^{\infty}(z^n/n^s)$ are the polylogarithm functions.

Ultra-relativistic limit:
In contrast to the l.t.e pressure presented in Sect. 6.2.2, the ballistic pressure saturates at large γ. In this regime, the only relevant contribution to the ballistic pressure arises from particles entering the bubble, cf. Eq. (6.143), which in the large γ limit, reads

$$\Delta P_{t+} = \mp\frac{1}{4\pi^2}\frac{m^3}{\beta\gamma M_-^3}\int_{M_-}^{\infty} dx\, x\left(x-\sqrt{x^2-M_-^2}\right)\log\left[1\mp e^{-x}\right] \tag{6.150}$$

$$\underset{\gamma\to\infty}{\longrightarrow}\ \frac{1}{24}\frac{m^2}{\beta^2}\begin{cases}1 & \text{bosons,}\\ 1/2 & \text{fermions.}\end{cases} \tag{6.151}$$

This is the result found in 2009 by Bodeker and Moore [160].

$$\int_0^M dx\, x^2\log(1+e^{-x}) = \frac{7\pi^4}{360} + M^2\,\mathrm{Li}_2(-e^{-M}) + 2M\,\mathrm{Li}_3(-e^{-M}) + 2\,\mathrm{Li}_4(-e^{-M}). \tag{6.147}$$

[15] Reference [142] only studied the boson case.

Non-relativistic limit:
At small v the ballistic pressure does not vanish, see Fig. 6.6, in contrast to the pressure at local thermal equilibrium in Eq. (6.112), which scales as $P_{\mathrm{lte}} \propto v^2$ at small v. It was expected that the ballistic approximation breaks down at small v where we can no longer neglect particle interactions. A possible prescription for an easy-to-use retarding pressure, which vanishes at small v and saturates at large v is

$$P_{\mathrm{lte+ballistic}}(v) \;=\; \mathrm{Min}\,[\Delta P_{\mathrm{lte}}(v),\;P_{\mathrm{ballistic}}(v)] \tag{6.152}$$

where P_{lte} is the l.t.e pressure defined in Eqs. (6.112) and (6.118), and $P_{\mathrm{ballistic}}$ is the ballistic pressure defined in Eqs. (6.148) and (6.149).

6.2.5 *Friction Pressure at NLO*

Transition splitting:
In the previous section, we have presented the retarding pressure in the ballistic approximation where particle interactions were neglected. In this section we introduce a possible correction which arises in the presence of a finite gauge coupling constant. The correction term, which is called NLO pressure, arises from the possibility for the incoming particle to radiate a soft boson which gets a mass in the broken phase [161]

$$\mathscr{P}_{1\to 2} = \sum_a \nu_a \int \frac{d^3p}{(2\pi)^3} f_a(p)\,\frac{p_z}{p_0} \times \sum_{bc} \int dP_{a\to bc} \times (p_{z,s} - k_{z,h} - q_{z,h}), \tag{6.153}$$

where h, s stands for the 'Higgsed' and the symmetric phases. p and q are the momenta of the incoming particle before and after the splitting while k is the momentum of the radiated boson. We summed over all the species a likely to participate in the process, ν_a being their number of degrees of freedom. The differential splitting probability is given by

$$\int dP_{a\to bc} \equiv \int \frac{d^3k}{(2\pi)^3 2k^0}\frac{d^3q}{(2\pi)^3 2q^0}\,\langle\phi|\mathscr{T}|k,q\rangle\,\langle k,q|\mathscr{T}|\phi\rangle\,, \tag{6.154}$$

with the transition element

$$\langle k,q|\mathscr{T}|p\rangle = \int d^4x\,\langle k,q|\mathscr{H}_{\mathrm{int}}|p\rangle\,, \tag{6.155}$$

$$= (2\pi)^3\,\delta(\vec{p}_\perp - \vec{k}_\perp - \vec{q}_\perp)\,\delta(p_0 - k_0 - q_0)\,\mathscr{M}, \tag{6.156}$$

where

$$\mathcal{M} \equiv \int dz\, \chi_k^*(z)\chi_q^*(z)\chi_k(z)V(z)\chi_p(z). \tag{6.157}$$

We obtain [161]

$$\mathcal{P}_{1\to 2} = \sum_{a,bc} \nu_a \int \frac{d^3p}{(2\pi)^3 2p_0} f_a(p) \frac{d^3k}{(2\pi)^3 2k^0} \frac{d^3q}{(2\pi)^3 2q^0} [1 \pm f_k][1 \pm f_q]\,(p_{z,s} - k_{z,h} - q_{z,h})$$
$$\times (2\pi)^3\, \delta(\vec{p}_\perp - \vec{k}_\perp - \vec{q}_\perp)\, \delta(p_0 - k_0 - q_0)\, |\mathcal{M}|^2. \tag{6.158}$$

Now we assume $p_z \simeq p_0, q_z \simeq q_0 \simeq p_0$ and $k_z \simeq k_0 - \dfrac{m_V^2(z) + k_\perp^2}{2k_0}$, from which we get

$$p_{z,s} - k_{z,h} - q_{z,h} \simeq \frac{m_V^2(z) + k_\perp^2}{2k_0}, \tag{6.159}$$

and

$$\mathcal{P}_{1\to 2} = \sum_{a,\,bc} \nu_a \int \frac{d^3p}{(2\pi)^3(2p_0)^2} f_a(p) \frac{d^2k_\perp}{(2\pi)^2} \frac{dk_0}{(2\pi)2k^0} [1 \pm f_k][1 \pm f_{p-k}] \frac{m_V^2(z) + k_\perp^2}{2k_0}\, |\mathcal{M}|^2. \tag{6.160}$$

WKB approximation:
Next, we make use of the WKB approximation,

$$\chi_k(z) \simeq \exp\left(i \int_0^z k_z(z')dz'\right) \simeq e^{ik_0 z} \exp\left(-\frac{i}{2k_0} \int_0^z (m^2(z') + k_\perp^2)\, dz'\right), \tag{6.161}$$

which allows to write the product of wave functions in terms of a phase-dependent quantity A.

$$\chi_a(z)\chi_b^*(z)\chi_c^*(z) = \exp\left(\frac{i}{2p^0} \int_0^z A(z')\, dz'\right) \tag{6.162}$$

with

$$-A = m_a^2 - \frac{m_b^2 + k_\perp^2}{x} - \frac{m_c^2 + k_\perp^2}{1 - x} \simeq \frac{k_\perp^2 + m_b^2}{x}. \tag{6.163}$$

We have introduced the variable $x \equiv k^0/p^0$ and assumed $x \ll 1$ in the last equality. We can now split the integral over z across the wall in Eq. (6.157) into a contribution from the broken phase and a contribution from the symmetric phase. We assume that the vertices V and WKB phases A on each side of the wall are z-independent and we denote them by (V_h, A_h) and (V_s, A_s), such that we obtain

$$\mathcal{M} \simeq V_s \int_{-\infty}^{0} dz \exp\left[iz\frac{A_s}{2p^0}\right] + V_h \int_{0}^{\infty} dz \exp\left[iz\frac{A_h}{2p^0}\right] = 2ip^0\left(\frac{V_h}{A_h} - \frac{V_s}{A_s}\right). \tag{6.164}$$

Radiation of a soft transverse boson:
It can be shown [161] that the most important process contributing to the pressure at large p_0 is $X(p) \to V_T(k)\ X(q)$ where V_T is a transverse vector boson. The corresponding vertex function is phase-independent, $V_h = V_s$, and equal to

$$|V^2| = 4\, g^2\, C_2[R]\, \frac{1}{x^2}\, k_\perp^2, \tag{6.165}$$

where g is the gauge coupling constant and $C_2[R]$ is the second casimir of the representation R of $SU(N)$

$$C_2[R] = \begin{cases} \frac{N^2-1}{2N} & \text{if R= fundamentale,} \\ N & \text{if R= adjoint.} \end{cases} \tag{6.166}$$

Therefore, Eq. (6.164) becomes

$$|\mathcal{M}|^2 \simeq 16\, g^2\, C_2[R]\, p_0^2\, \frac{m_V^4}{k_\perp^2\,(k_\perp^2 + m_V^2)^2}. \tag{6.167}$$

where we have replaced $m_V \equiv m_b$. The $k_\perp$ integral becomes

$$\int \frac{d^2k_\perp}{(2\pi)^2}\, \frac{1}{k_\perp^2\,(k_\perp^2 + m_V^2)} = \frac{\ln\left(1 + \frac{m_V^2}{k_*^2}\right)}{4\,\pi\, m_V^2}, \tag{6.168}$$

where k_* is the IR cut-off on $k_\perp$.

Final NLO pressure:
Finally, injecting the last two equations into Eq. (6.160) yields

$$\mathcal{P}_{1\to 2} = \sum_{a,bc} \nu_a \int \frac{d^3p}{(2\pi)^3} f_a(p) \frac{dk_0}{(2\pi)k_0^2} [1 \pm f_k][1 \pm f_{p-k}]\, g^2\, C_2[R]\, m_V^2\, \frac{\ln\left(1 + \frac{m_V^2}{k_*^2}\right)}{4\pi}. \tag{6.169}$$

The Pauli blocking or Bose enhancing factor $1 \pm f_{p-k}$ is of order 1, while $1 \pm f_k$ sums to 1 when considering both absorption and emission processes. Hence, the result simplifies to

$$\mathscr{P}_{1\to 2} = \sum_a \nu_a\, b_a\, \frac{\zeta(3)}{\pi^2} T^3\, \gamma_{\rm wp}\, 8\pi\, \alpha\, C_2[R]\, m_V\, \frac{\ln\left(1+\frac{m_V^2}{k_*^2}\right)}{k_*/m_V}, \tag{6.170}$$

where $b_a = 1$ (3/4) for bosons (fermions) and $\alpha \equiv g^2/4\pi$. The Lorentz factor $\gamma_{\rm wp}$ between the the wall and the plasma comes from d^3p. Taking $k_* \sim m_V$, we obtain

$$\gamma \Delta P_{\rm NLO} \equiv \mathscr{P}_{1\to 2} \simeq \sum_a \nu_a\, b_a\, \frac{\zeta(3)}{\pi^2}\, T^3\, \gamma_{\rm wp}\, 4\pi\, \alpha\, C_2[R]\, m_V. \tag{6.171}$$

Beyong NLO:
While this chapter was completed, this paper [162] came out, where the authors aim at improving the B&M17 result [161] in Eq. (6.170) by resumming soft boson emission processes to all orders in perturbation theory and by obtaining a gauge-invariant matrix element. The authors find that the exchanged momentum $\Delta p \sim m_V$ in Eq. (6.159) becomes $\Delta p \sim \gamma T$, which implies that the NLO pressure $\mathscr{P}_{1\to 2} \sim \alpha\, \gamma_{\rm wp}\, m_V\, T^3$ becomes $\mathscr{P}_{1\to 2} \sim \alpha\, \gamma_{\rm wp}^2\, T^4$. As pointed out in [163], the result from [162] does not vanish as it should in the symmetry restoration limit $\langle \phi \rangle \to 0$.

Later in [164], together with Ryusuke Jinno and Filippo Sala, we showed that by breaking energy-momentum conservation, (or by preventing the onshellness of the momenta at the splitting vertex), the presence of the wall boundary prevents the trivial applicability of the Ward identity at the level of the single splitting vertex function. Without any additional ingredient standing for the presence of the wall in the Feynmann rules, gauge invariance is not expected, which is at odd with the finding of [162]. Instead, the B&M17 result is not gauge invariant but assumes physical polarizations. In [164], we show that the average exchanged momentum in which both real and virtual leading-log emissions are resummed is identical to the naive expectation using the perturbative splitting probability as in the original B&M17 paper [161]. Additionally, we compute the IR cut-off k_* which regulates the logarithmic divergence, we point out that a dominant fraction of radiated gauge bosons are too soft to enter the broken phase and discuss their fate, we compute effects from wall finite thickness and we discuss various approximations entering the computation.

6.2.6 *Speed of the Wall*

At LO:
As already discussed at the end of Sect. 6.2.4, a possible prescription for the pressure at LO in gauge coupling constant, which presents the correct behavior at small v (goes to zero) and at large v (saturates) is

$$\Delta P_{\rm LO}(v) = \sum_a \nu_a \, {\rm Min}\left[\Delta P^{b/f}_{\rm lte}(v),\ \Delta P^{b/f}_{\rm ballistic}(v)\right] \tag{6.172}$$

where we summed over all the species in the plasma with ν_a degrees of freedom. $\Delta P^{b/f}_{\rm lte}$ is the l.t.e pressure defined in Eqs. (6.112) and (6.118). $\Delta P^{b/f}_{\rm ballistic}$ is the ballistic pressure given by Eqs. (6.148) and (6.149). We might worry that for intermediate bubble wall velocities (with a range of values might be large), Eq. (6.172) is too simplistic and instead one should use a close-to-equlibrium approach, along the line of Sect. 6.2.3. However, such computations go beyond the scope of this chapter whose goal is not to perform thorough computations but instead to introduce different approaches available in the litterature.

The bubble wall reaches a finite velocity v when the retarding pressure $\Delta P_{\rm LO}(v)$ compensates for the accelerating vacuum pressure $\Delta V_{\rm vac}$, cf. Eq. (6.102). In the top panel of Fig. 6.7, we give the bubble wall velocity with respect to the almost model-independent variables

$$\alpha \equiv \frac{\Delta V_{\rm vac}}{\rho_{\rm rad}}, \qquad \text{and} \qquad \Delta V_{\rm vac}/\phi^4, \tag{6.173}$$

where $\phi = 246$ GeV. In the brown region, we have

$$\Delta V_{\rm vac} > \Delta P_{\rm LO}(v = 1), \tag{6.174}$$

and the bubble is predicted to run-away (to keep accelerating until collisions with neighboring bubbles) at LO in the gauge coupling constant g.

We add the NLO term:
In the bottom panel of Fig. 6.7, we add the pressure resulting from NLO effets (in gauge coupling constant g), $\gamma \Delta P_{\rm NLO} \sim \gamma\, g^2\, f\, T^3_{\rm nuc}$, computed in the previous Sect. 6.2.5. As a result, the bubble wall reaches a terminal velocity given by

$$\gamma^{\rm NLO}_{\rm eq} = \frac{\Delta V - P_{\rm LO}}{P_{\rm NLO}} \sim \frac{\Delta V}{g^2\, f\, T^3_{\rm nuc}} \sim 10^{18} \left(\frac{f}{\rm TeV}\frac{\rm MeV}{T_{\rm nuc}}\right)^3, \tag{6.175}$$

where f is the scalar field vev.

Since the NLO pressure is only known in the large γ limit, cf. Eq. (6.171), we propose to define the LO+NLO pressure with the following trick

$$\Delta P_{\rm LO+NLO}(v) = \Delta P_{\rm LO}(v) \left[1 + \frac{\Delta P_{\rm NLO}(\gamma)}{\Delta P_{\rm LO}(v = 1)}\right], \tag{6.176}$$

where $P_{\rm LO}(v)$ is the LO ballistic pressure defined in Eq. (6.172) and $P_{\rm NLO}(\gamma)$ is NLO pressure in the large γ limit, defined in Eq. (6.171).

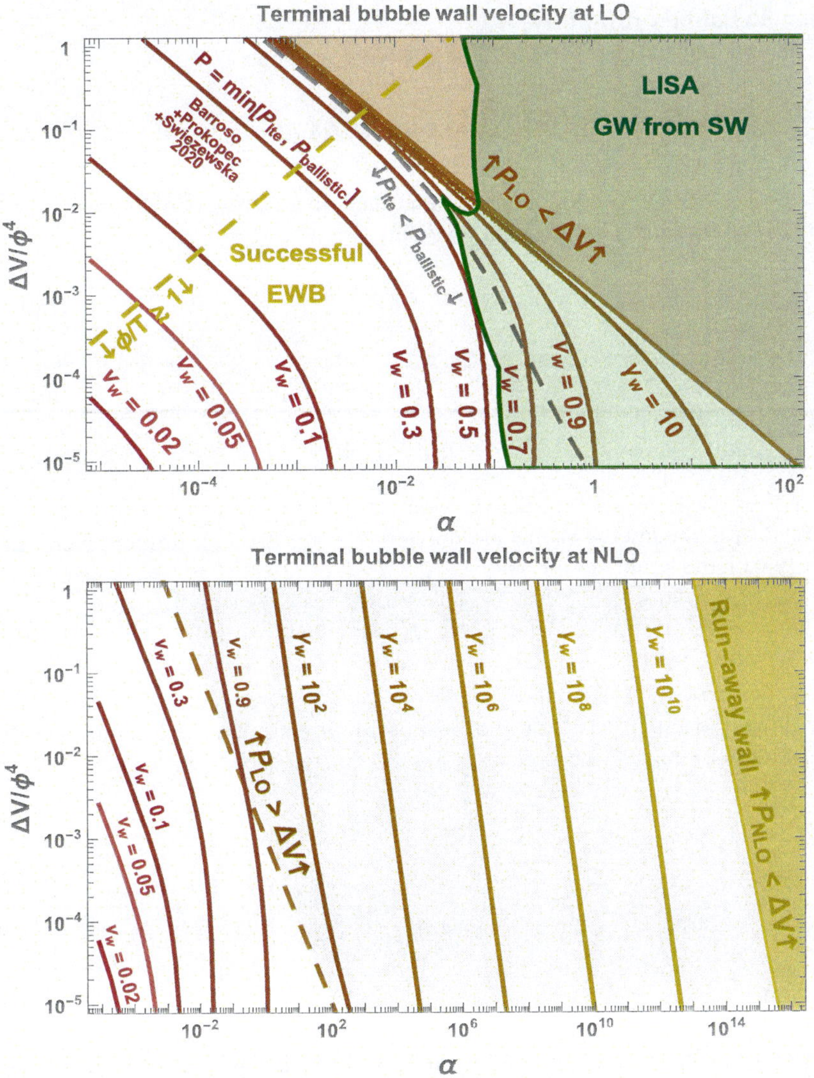

Fig. 6.7 **Top:** Bubble wall velocity at equilibrium $\Delta P_{\rm LO}(v) = \Delta V_{\rm vac}$ with $\Delta P_{\rm LO}$ defined in Eq. (6.172), in a minimal extension of the SM including the contributions from t, $W^{\pm}$, Z and h. The reach of LISA on the GW spectrum from sound-waves (SW), see Sect. 6.3.3 requires large α. We might worry that the corresponding large wall velocity $v_w \gtrsim 0.5$ could jeopardize successful EW baryogenesis (EWBG). However, successful EWBG only requires the fluid velocity at the wall v_+ to be sub-sonic $v_+ \lesssim 0.5$, which can be much smaller than v_w for strong phase transition (large α) [165]. Note also, the claim that EWBG can also work for super-sonic v_+ in [166–169]. **Bottom:** We include the NLO pressure $P_{\rm NLO}(v) = \Delta V_{\rm vac}$ using the trick explained along Eq. (6.176) for $v_w \gtrsim 0.9$. In the yellow region, the vacuum energy is larger than the NLO pressure at the time of bubble collisions, such that bubbles never stop accelerating until they collide, cf. Eq. (6.186)

When do bubbles run-away?
The energy gained upon formation of a bubble of radius R is

$$E_{\text{bubble}} \simeq \frac{4}{3}\pi R^3 \Delta V_{\text{vac}}, \tag{6.177}$$

where ΔV_{vac} is the vacuum energy of the transition. As the bubble grows, the later energy is stored in the bubble wall given by

$$E_{\text{wall}} \simeq 4\pi R^2 \gamma\, \sigma, \tag{6.178}$$

where σ is the surface energy of the wall (wall tension) in the wall frame. After equating Eqs. (6.177) and (6.178), we obtain the wall Lorentz factor at collision in the run-away regime

$$\gamma_{\text{coll}} = \frac{\Delta V_{\text{vac}}\, R_{\text{coll}}}{3\,\sigma}, \tag{6.179}$$

where R_{coll} is the bubble radius at collision. Hence, the wall Lorentz factor grows linearly with the bubble radius (see e.g. the review in [61]). The wall tension decomposes as the product

$$\sigma = L_{\text{w}}\, E_{\text{w}}, \tag{6.180}$$

of L_{w}, the wall thickness, and E_{w}, the averaged energy density in the wall. E_{w} is the kinetic energy density of the field oscillating around the true minima. From conservation of energy, we expect

$$E_{\text{w}} \simeq \Delta V_{\text{vac}}, \tag{6.181}$$

such that

$$\gamma_{\text{coll}} = \frac{R_{\text{coll}}}{3\,L_{\text{w}}}. \tag{6.182}$$

In order for the wall to run-away (i.e. $\dot{\gamma} \neq 0$) at the collision time, we need to lower the friction pressure as much as possible and therefore to lower the nucleation temperature. Hence, situations where run-away walls are expected correspond to phase transitions with a lot of supercooling, meaning that

$$\alpha \gg 1, \qquad \text{(supercooled phase transition)}. \tag{6.183}$$

Typically, supercooled phase transitions are generated by shallow ($T = 0$)-potentials, namely potentials where the curvature f close to the true vacuum is much larger than the curvature close to the false vacuum, $f\,\exp -c/\epsilon$ where $\epsilon \ll 1$ and $c = O(1)$. In that case, the tunneling exit point is very close to the false minimum and the bounce action is only sensitive to the scale $f\,\exp -c/\epsilon$ [34]. Hence, we expect the wall thickness (or equivalently the bubble size at nucleation) to be

$$L_w \simeq T_{\rm nuc}^{-1}. \tag{6.184}$$

Later in Sect. 6.4, we study in detail two classes of potential potentially leading to large amount of supercooling: the Coleman-Weinberg potential and the light-dilaton potential.

Finally, Eq. (6.182) becomes[16]

$$\gamma_{\rm coll} \simeq T_{\rm nuc}\, \beta^{-1}, \tag{6.185}$$

where β is the rate of change of the tunneling rate (inverse of the bubble propagation time).

Upon comparing Eqs. (6.185) with (6.175), we find that collisions occur before the terminal velocity is reached (run-away bubbles) when

$$\gamma_{\rm coll} \lesssim \gamma_{\rm eq} \quad \Longrightarrow \quad T_{\rm nuc} \lesssim T_{\rm eq} \equiv \left(\frac{f^4}{M_{\rm pl}}\frac{\beta}{H}\right)^{1/3} \simeq 10\,{\rm MeV}\left(\frac{f}{\rm TeV}\right)^{4/3}\left(\frac{\beta/H}{10}\right)^{1/3}. \tag{6.186}$$

If $T_{\rm nuc} \lesssim T_{\rm eq}$, then most of the vacuum energy is used for accelerating the bubble walls and the GW spectrum is dominated by the scalar field contribution. If $T_{\rm nuc} \gtrsim T_{\rm eq}$, then the terminal velocity is reached and most of the vacuum energy is converted into thermal and kinetic energy of the plasma via the friction, leading to a GW spectrum dominated by sound waves and turbulence contributions (see e.g. the review in [61]).

Bubble wall velocity from heating and cooling rates:
We have derived the bubble wall velocity from the equilibrium condition between the friction pressure and the expanding pressure, cf. Eq. (6.102). We would like to mention a completely different approach - and as far as we can tell which gives a different result—for deriving the bubble wall velocity, which does not even need the computation of the friction pressure [9, 48]. Instead the wall velocity is found as resulting from a detailed balance between the rates at which the bubble wall heats and cools the surrounding plasma. On the one hand, the plasma around the wall boundary is heated due to the latent heat conversion at a rate

$$\dot{T}_{\rm heat} \sim l\,\Lambda\left(-\frac{dR}{dt}\right)\times\frac{dT}{d\rho} \sim \Lambda\,\frac{dR}{dt}, \tag{6.187}$$

where $-dR/dt$ is the outward-oriented wall velocity, Λ^{-1} is the wall thickness and $l \sim \Lambda^4$ is the latent heat. On the other hand, due to the temperature gradient the same plasma cools down at a rate

[16] Note that the difference between Eq. (6.185) and [14] is due to the choice of the authors of [14] of setting $L_w \sim f$ instead of $L_w \sim T_{\rm nuc}$. For typical potentials leading to large supercooling, the later choice is the correct one, see App. 7.A.

$$\dot{T}_{\rm cooling} \sim -K\nabla^2 T \sim \Lambda^{-1}\frac{T_c - T}{\Lambda^2}, \tag{6.188}$$

where we assume that the transport coefficient K and the gradient length scale are both given by Λ^{-1}. The wall velocity results from the balance between the two thermal rates [9, 48]

$$\dot{T}_{\rm heat} \sim \dot{T}_{\rm cooling} \implies v_w \equiv \dot{R} \simeq \frac{T - T_c}{T_c}, \tag{6.189}$$

where we have assumed $T_c \simeq \Lambda$. This results assume that the thermalization rate due to latent heat conversion is faster than the expansion rate of the bubble v_w/R, and therefore assume small bubble wall velocity $v_w \ll 1$. The evolution of the temperature T is a combination of Hubble cooling and latent heat conversion

$$\dot{T} = -HT + \frac{dT}{dx}\dot{x}, \qquad \text{with} \quad \frac{dT}{dx} = \frac{l}{d\rho(T)/dT} = 10^{-2}T_c\, dx, \tag{6.190}$$

where x is the fraction of converted false vacuum (it starts at 0 and ends at 1). We can write x as

$$x = n_b\frac{4\pi}{3}R^3 = \left(\frac{R}{R_c}\right)^3, \tag{6.191}$$

where n_b is the number density of bubbles, $R(t)$ the mean bubble radius at a given time, and R_c is the mean bubble radius at percolation. If the latent heat is large enough to balance against Hubble cooling (still assuming an instantaneous thermalization rate), then the temperature can be kept constant $\dot{T} = 0$ between nucleation and percolation, so that the bubble wall velocity is set by the expansion rate of the universe

$$\dot{x} \simeq 100H \implies \dot{R} \simeq 30\,HRc\left(\frac{R_c}{R}\right)^2 \simeq 3\times 10^{-4}\left(\frac{HRc}{10^{-5}}\right)\left(\frac{R_c}{R}\right)^2. \tag{6.192}$$

.

6.3 GW Generation

In Sect. 3.4 of Chap. 3, we have introduced Gravitational Waves (GW) as linearized solutions of the Einstein equation. We have discussed the possibility that the universe today is filled with a Stochastic Gravitational-Wave Background (SGWB) from primordial origin, which we have computed in the quadrupole approximation. In the present section, we focus on the case where the SGWB is sourced by bubble dynamics during a cosmological 1stOPT.

6.3.1 *The GW Spectrum for a Generic Source*

The GW from the energy-momentum tensor correlator:
The energy fraction contained in SGWB today is given by its energy fraction at the time of its emission $*$ redshifted until today as a radiation fluid

$$h^2\Omega_{\rm GW}(k) = \frac{h^2}{\rho_c}\left(\frac{a_p}{a_0}\right)^4 \left(\frac{d\rho_{\rm GW}}{d\log k}\right)_* . \tag{6.193}$$

ρ_c is the critical density today and $\rho_{\rm GW}$ is the energy density of GW [170, 171]

$$\rho_{\rm GW} = \frac{\langle \dot{h}_{\rm ij}(\vec{x},\,\eta)\dot{h}_{\rm ij}(\vec{x},\,\eta)\rangle}{32\pi G a^2}. \tag{6.194}$$

η is the conformal time and $h_{\rm ij}(k,\,t)$ is the tensor spatial perturbation of the FLRW metric

$$ds^2 = dt^2 - a(t)^2(\delta_{\rm ij} + h_{\rm ij})d\vec{x}^2 = a(t)^2(d\eta^2 - (\delta_{\rm ij} + h_{\rm ij})d\vec{x}^2) \tag{6.195}$$

in the transverse-traceless (TT) gauge, so that $\partial_{\rm i} h_{\rm ij} = 0$ and $h_{\rm ii} = 0$. We define the Fourier modes as $h_{\rm ij}(\vec{k},\,\eta) = \int \frac{d\vec{k}^3}{(2\pi)^3} h_{\rm ij}(\vec{k},\,\eta)e^{-i\vec{k}.\vec{x}}$ and the characteristic amplitude $h(k,\,\eta)$ from the correlation function of the metric perturbation

$$\left\langle \dot{h}_{\rm ij}(\vec{k},\,\eta)\dot{h}_{\rm ij}(\vec{q},\,\eta)\right\rangle = (2\pi)^3\delta^{(3)}(\vec{k} - \vec{q})\,|\dot{h}(k,\,\eta)|^2. \tag{6.196}$$

from which we deduce

$$\langle \dot{h}_{\rm ij}(\vec{x},\,\eta)\dot{h}_{\rm ij}(\vec{x},\,\eta)\rangle = \int d\ln k\,\frac{k^3}{2\pi^2}\,|\dot{h}(k,\,\eta)|^2 \tag{6.197}$$

and

$$\left(\frac{d\rho_{\rm GW}}{d\log k}\right)_* = \frac{k^3|\dot{h}(k,\,\eta)|^2}{64\pi^3 G a^2} = \frac{k^5|h'(k,\,\eta)|^2}{64\pi^3 G a^2}. \tag{6.198}$$

The dot stands for the derivative with respect to the conformal time η and the prime for the derivative with respect to the dimensionless conformal time $\tilde{\eta} \equiv k\eta$. We define the conformal Hubble factor $\mathcal{H} \equiv \dot{a}/a = H/a$. The linearised Einstein equation in the FLRW background leads to [13]

$$\ddot{h}_{\rm ij}(k,\,\eta) + 2\mathcal{H}\dot{h}_{\rm ij}(k,\,\eta) + k^2 h_{\rm ij}(k,\,\eta) = 16\pi G a^2\,\Pi^{\rm TT}_{\rm ij}(\vec{k},\,\eta)) \tag{6.199}$$

where $\Pi^{\rm TT}_{\rm ij}(\vec{k},\,\eta)$ is the transverse and traceless part of the anisotropic stress. It is easier to compare the different terms of the equation if we use the dimensionless conformal time $\tilde{\eta} = k\eta$

$$h''_{ij}(k,\eta) + 2\frac{\mathscr{H}}{k} h'_{ij}(k,\eta) + h_{ij}(k,\eta) = \frac{16\pi G a^2}{k^2}\,\Pi^{TT}_{ij}(\vec{k},\tilde{\eta}) \tag{6.200}$$

For sub-Hubble modes $k \gg \mathscr{H}$, we can neglect the first derivative and the solution reads

$$h_{ij}(k,\eta) = \frac{16\pi G}{k^2} \int_{\tilde{\eta}_{in}}^{\tilde{\eta}_{fi}} d\tilde{\eta}'\, G(\tilde{\eta}-\tilde{\eta}')\, a^2\, \Pi^{TT}_{ij}(\vec{k},\tilde{\eta}')) \tag{6.201}$$

where $G(\tilde{\eta}-\tilde{\eta}') = \cos(\tilde{\eta}'-\tilde{\eta})$ is the Green function of the operator $\left[\frac{d}{d\tilde{\eta}^2}+1\right]$ with the boundary conditions $G(0) = 0$ and $G'(0) = 0$. We define $\tilde{\Pi}_{ij}(\vec{k},\eta))$ as $\Pi^{TT}_{ij}(\vec{k},\eta)) = \rho_s\, \tilde{\Pi}_{ij}(\vec{k},\eta))$ where ρ_s is the energy density of the source, e.g. due to the gradient of the scalar field or the compressional and turbulent motion of the plasma. Then

$$\left\langle h'_{ij}(\vec{k},\tilde{\eta}) h'_{ij}(\vec{q},\tilde{\eta})\right\rangle = \left(\frac{16\pi G}{k^2}\right)^2 \rho_s^2 \int_{\tilde{\eta}_{in}}^{\tilde{\eta}_{fi}} d\tilde{\eta}_1 \int_{\tilde{\eta}_{in}}^{\tilde{\eta}_{fi}} d\tilde{\eta}_2 \cos(\tilde{\eta}_1-\tilde{\eta})\cos(\tilde{\eta}_2-\tilde{\eta})$$
$$a(\tilde{\eta}_1)^2 a(\tilde{\eta}_2)^2\, \tilde{\Pi}_{ij}(\vec{k},\tilde{\eta}_1)\, \tilde{\Pi}_{ij}(\vec{q},\tilde{\eta}_2) \tag{6.202}$$

which after averaging over time becomes

$$\left\langle h'_{ij}(\vec{k},\tilde{\eta}) h'_{ij}(\vec{q},\tilde{\eta})\right\rangle \simeq \frac{1}{2}\left(\frac{16\pi G a^2}{k^2}\right)^2 \rho_s^2 \int_{\tilde{\eta}_{in}}^{\tilde{\eta}_{fi}} d\tilde{\eta}_1 \int_{\tilde{\eta}_{in}}^{\tilde{\eta}_{fi}} d\tilde{\eta}_2 \cos(\tilde{\eta}_1-\tilde{\eta}_2)$$
$$a(\tilde{\eta}_1)^2 a(\tilde{\eta}_2)^2\, \tilde{\Pi}_{ij}(\vec{k},\tilde{\eta}_1)\, \tilde{\Pi}_{ij}(\vec{q},\tilde{\eta}_2) \tag{6.203}$$

We define the unequal-time correlator of the transverse-traceless energy stress tensor as

$$\left\langle \tilde{\Pi}_{ij}(\vec{k},\tilde{\eta}_1)\tilde{\Pi}_{ij}(\vec{q},\tilde{\eta}_2)\right\rangle = (2\pi)^3 \delta^{(3)}(\vec{k}-\vec{q})\, \tilde{\Pi}(\vec{k},\tilde{\eta}_1,\tilde{\eta}_2) \tag{6.204}$$

such that we can write

$$\left(\frac{d\rho_{GW}}{d\log k}\right)_* = \frac{2G}{\pi a^2}\rho_s^2 k \int_{\tilde{\eta}_{in}}^{\tilde{\eta}_{fi}} d\tilde{\eta}_1 \int_{\tilde{\eta}_{in}}^{\tilde{\eta}_{fi}} d\tilde{\eta}_2 \cos(\tilde{\eta}_1-\tilde{\eta}_2)\, a(\tilde{\eta}_1)^2 a(\tilde{\eta}_2)^2\, \tilde{\Pi}(\vec{k},\tilde{\eta}_1,\tilde{\eta}_2). \tag{6.205}$$

For short-lasting source we can approximate $\tilde{\eta}_{in} \simeq \tilde{\eta}_{fi}$ and we obtain (e.g. [172])

$$\left(\frac{d\rho_{\rm GW}}{d\log k}\right)_* = \frac{2Ga^2}{\pi}\,\rho_s^2\,k\int_{\tilde\eta_{\rm in}}^{\tilde\eta_{\rm fi}} d\tilde\eta_1 \int_{\tilde\eta_{\rm in}}^{\tilde\eta_{\rm fi}} d\tilde\eta_2\,\cos(\tilde\eta_1-\tilde\eta_2)\,\tilde\Pi(\vec k,\tilde\eta_1,\tilde\eta_2). \tag{6.206}$$

After its emission at time $*$, the GW energy density redshifts as radiation, such that its abundance today reads

$$h^2\Omega_{\rm GW}(k) = \frac{h^2}{\rho_c}\left(\frac{a_*}{a_0}\right)^4 a_*^2\,(\rho_{\rm tot}^*)^2\,K^2\,\frac{2G}{\pi}\,k\int_{\tilde\eta_{\rm in}}^{\tilde\eta_{\rm fi}} d\tilde\eta_1 \int_{\tilde\eta_{\rm in}}^{\tilde\eta_{\rm fi}} d\tilde\eta_2\,\cos(\tilde\eta_1-\tilde\eta_2)\,\tilde\Pi(\vec k,\tilde\eta_1,\tilde\eta_2). \tag{6.207}$$

We introduced the fraction of total energy density that gets converted into the energy of the source

$$K \equiv \frac{\rho_s^*}{\rho_{\rm tot}^*}, \tag{6.208}$$

where $\rho_{\rm tot}^*$ is the total energy density just before the phase transition (in the symmetric phase). For convenience, we can define the dimensionless quantity

$$\Delta(k) = \frac{3}{4\pi^2}\,a_*^2\int_{\tilde\eta_{\rm in}}^{\tilde\eta_{\rm fi}} d\tilde\eta_1 \int_{\tilde\eta_{\rm in}}^{\tilde\eta_{\rm fi}} d\tilde\eta_2\,\cos(\tilde\eta_1-\tilde\eta_2)\,\frac{k}{\beta}\left(\beta^3\,\tilde\Pi(\vec k,\tilde\eta_1,\tilde\eta_2)\right). \tag{6.209}$$

where β is the time variation of the nucleation rate

$$\beta \equiv \left.\frac{dS}{dt}\right|_* = -H_*T_*\left.\frac{dS}{dT}\right|_*, \tag{6.210}$$

such that the SGWB abundance today becomes

$$\Omega_{\rm GW}(k)\,h^2 = h^2\frac{\rho_{\rm tot}^*}{\rho_c}\left(\frac{a_*}{a_0}\right)^4\left(\frac{H_*}{\beta}\right)^2 K^2\,\Delta(k). \tag{6.211}$$

with $H_*^2 = 8\pi G\rho_{\rm tot}^*/3$. The GW emission time denoted by $*$ coincides with the bubble percolation time

$$a_* = a_{\rm perc}, \tag{6.212}$$

which can be different from the nucleation temperature, see discussion Sect. 6.1.4, and the end of reheating $a_{\rm reh}$.

The GW propagation from the percolation epoch to today:
Upon assuming an adiabatic evolution from today a_0 up to reheating $a_{\rm reh}$, i.e. $h_{\rm eff}\, T^3\, a^3 = \text{cst}$, and possibly assuming the domination of the universe by a scalar field redshifting as $\rho \propto a^{-n}$ between the end of reheating $a_{\rm reh}$ and the time of bubble nucleation a_*, we obtain

$$\Omega_{\rm GW}(k)\, h^2 = F_{\rm gw,\,0}\, h^2 \left(\frac{H_*}{\beta}\right)^2 K^2\, \Delta(k), \tag{6.213}$$

with

$$\begin{aligned} F_{\rm gw,\,0}\, h^2 &\equiv \Omega\, h^2 \left(\frac{h_{\rm eff,\,0}}{h_{\rm eff,\,reh}}\right)^{4/3} \frac{g_{\rm eff,\,reh}}{g_{\gamma,\,0}} \left[\frac{\rho_{\rm reh}}{\rho_* \,(1+\alpha)}\right]^{\frac{4-n}{n}}, \\ &\simeq 1.657 \times 10^{-5} \left(\frac{100}{g_{\rm eff,\,reh}}\right)^{1/3} \left[\frac{\rho_{\rm reh}}{\rho_* \,(1+\alpha)}\right]^{\frac{4-n}{n}}, \end{aligned} \tag{6.214}$$

where $\rho_{\rm reh}/\rho_* = g_{\rm eff,\,reh}/g_{\rm eff,\,*}\; T_{\rm reh}^4/T_*^4$. The last factor $[\cdots]^{\frac{4-n}{n}}$ accounts for the extra-redshift[17] of the universe as $\rho \propto a^{-n}$ between bubble nucleation and reheating, and only differs from unity if the lifetime of the scalar field driving the PT is longer than a Hubble time. For a scalar field oscillating in a potential $V \propto \phi^{2p}$, the equation of state averaged over the oscillations is $\bar{\omega} = (p-1)/(p+1)$ [173], so the resdhift parameter is $n = 3(1+\bar{\omega}) = 6p/(1+p)$. Note that for instantaneous reheating, $a_{\rm reh} = a_*$, the reheating temperature reads

$$T_{\rm reh} = \left(\frac{g_{\rm eff*}}{g_{\rm reh}}\right)^{1/4} (1+\alpha)^{1/4}\, T_*. \tag{6.215}$$

We made used of $g_{\rm eff,\,p} = h_{\rm eff,\,p}$, $g_{\gamma,\,0} = 2$, $h_{\rm eff,\,0} = 3.94$ (which assumes $N_{\rm eff} \simeq 3.045$ [174–176]) and $\Omega\, h^2 \simeq 2.473 \times 10^{-5}$ [177].

From doing similar maths, the GW frequency today reads

$$f_0 = \frac{a_*}{a_0} f_* = 1.65 \times 10^{-5}\,\text{Hz} \left(\frac{T_{\rm reh}}{100\,\text{GeV}}\right) \left(\frac{g_{\rm eff,\,reh}}{100}\right)^{1/6} \frac{f_*}{H_*} \left[\frac{\rho_* \,(1+\alpha)}{\rho_{\rm reh}}\right]^{\frac{n-2}{2n}}, \tag{6.216}$$

---

[17] Equation (6.214) follows from $F_{\rm gw,\,0} = (a_{\rm reh}/a_0)^4 (a_*/a_{\rm reh})^4 (\rho_{\rm reh}/\rho_c)(\rho^*_{\rm tot}/\rho_{\rm reh})$ with $\rho^*_{\rm tot}/\rho_{\rm reh} = (a_{\rm reh}/a_*)^n$.

6.3.2 Contribution from the Scalar Field

The envelope approximation:
Bubble collisions resulting from first-order phase transition are known for producing GW for 4 decades (Witten 1984 [9, 10]). The first computations of the GW spectrum have been carried out in the early 90s [11, 178, 179], with the simulation of the collision of two bubbles on a lattice and the analytical computation of the GW spectrum. Along these lines, the authors introduced the **envelope approximation** where GW are sourced by the anisotropic stress-tensor of the scalar field gradient localised on infinitely thin shells in which the collided portions have been removed. Since then, other analytical studies [172, 180–183] have refined the results found earlier. In the envelope approximation, the stochastic GW background (SGWB) sourced by the gradient energy of the scalar field reads (LISA paper 2015 [12], derived by Huber+ in 2008 [181])

$$\Omega_{\text{Œ}}h^2 = F_{\text{gw}}\,h^2 \left(\frac{H_*}{\beta}\right)^2 \left(\frac{\kappa_\phi \alpha}{1+\alpha}\right)^2 \left(\frac{0.11 v_{\text{w}}^3}{0.42 + v_{\text{w}}^2}\right) S_\phi(f_*), \tag{6.217}$$

where $S_\phi(f_*)$ describes the spectral shape

$$S_\phi(f_*) = \frac{3.8(f_*/f_\phi)^{2.8}}{1 + 2.8(f_*/f_\phi)^{3.8}}, \tag{6.218}$$

$$\frac{f_\phi}{H_*} = \left(\frac{0.62}{1.8 - 0.1 v_w + v_w^2}\right)\left(\frac{\beta}{H_*}\right). \tag{6.219}$$

The prefactor $F_{\text{gw}}\,h^2$ is given in Eq. (6.214). f_* denotes the frequency at emission and is related to the frequency today f_0 through Eq. (6.216). The parameters α, β and κ_ϕ are defined by

$$\alpha \equiv \frac{\Delta V}{\rho^*_{\text{tot}}}, \tag{6.220}$$

$$\kappa_\phi \equiv \frac{\dot{\phi}^2/2 + (\nabla\phi)^2/2}{\Delta\theta}, \tag{6.221}$$

$$\beta \equiv \left.\frac{dS}{dt}\right|_* = -H_* T_* \left.\frac{dS}{dT}\right|_*, \tag{6.222}$$

where ρ^*_{tot} is the total energy density in the symmetric phase (sometimes noted e_+) and where ΔV is the zero-temperature energy difference between the false vacuum and the true vacuum.

The IR slope is f^3 (fixed by causality [184–187]) while the UV slope is f^{-1}. The peak frequency is given by $f_* \simeq \beta/5$. In purple lines of Fig. 6.8, we compare the formula of Huber+ 2008 for $v_w = 1$ to more recent studies, Jinno+ 2016 [172] and Konstandin 2017 [183], and show that there are only little differences.

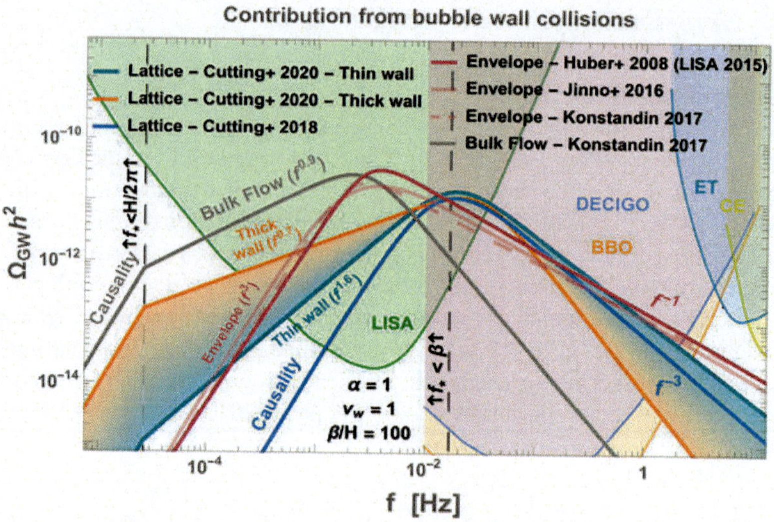

Fig. 6.8 Comparison of the different expression for the SGWB sourced by bubble collisions (scalar field contribution), available in the litterature. The envelope approximation predicts a f^3 and f^{-1} scaling in the IR and the UV respectively, c.f Huber-Konstandin 2008 [181] (version in the LISA paper 2015 [12]), Jinno-Takimoto 2016 [172] and Konstandin 2017 [183]. First deviations from the envelope approximation are found by Lattice simulations run in vacuum by Cutting+ in 2018 [188]: a $f^{-1.5}$ at higher frequency and a larger peak frequency. Later in 2020, new deviations from the envelope approximation are found by Cutting+ [189]: a dependence on the wall profile (thick versus thin wall) and an enhancement in the IR (see footnote 18) due to the shells of transverse-traceless stress-energy tensor still propagating after the collision (especially for thick walls). This confirms the predictions from the analytical bulk flow model (Jinno-Takimoto 2017 [190] and Konstandin 2017 [183]). Note that whether the IR slope at $f \lesssim \beta$ is enhanced or not, at much lower frequency, $f \lesssim H/2\pi$, the slope must converge to k^3 due to causality [184–187]

Beyond the envelope:
The envelope approximation was first challenged by the lattice simulation of Cutting+ in 2018 [188], which showed that the peak frequency is 5 times higher and the UV slopes is steeper ($f^{1.5}$) (blue line in Fig. 6.8). Later, Cutting+ 2020 [189] showed that the IR and UV slopes depend on the wall profile (orange-to-blue gradient in Fig. 6.8) and that the IR slope is softer[18] due to the long-lasting free propagation of the shells of anisotropic energy-momentum tensor after the collision. This confirms the analytical work of Jinno+ 2017 [190] and Konstandin 2017 [183] (bulk flow model, see gray line of Fig. 6.8). Note that Cutting+ 2020 [189] finds that the IR enhancement is stronger for thick-walled bubbles, since for thin-walled bubbles, after collision the scalar field can be trapped back in the false vacuum. Hence, instead

[18] The dynamical range in Cutting+ 2018 [188] was not large enough to estimate the spectral index of the IR slope. Therefore, the authors simply imposed f^3 in order to respect causality. In Cutting+ 2020 [189], both the kinetic $\dot{\phi}^2/2$ and gradient energy $(\nabla\phi)^2/2$ are considered (while only gradient energy was considered in 2018), which allows to reduce the artificial oscillations in the IR slope and to see the hint of the shallower power law [191]. .

of propagating freely, the shells of energy-momentum tensor dissipates via multiple re-bounces of the walls, see [140, 192–194]. Note that whether the IR slope at $f \lesssim \beta$ is enhanced or not, at much lower frequency, $f \lesssim H/2\pi$, the slope must converge to k^3 due to causality [184–187]. Finally, it has been shown that scalar field oscillations after bubble collisions can also contribute to the GW spectrum [195], however the time scale is set by the scalar mass and the signal is Planck-suppressed $\propto \beta/m_\phi$ [188].

Only relevant for runaway bubbles:
Following the study of Bodeker and Moore in 2017, which shows the presence of a NLO contribution to the retarding pressure scaling linearly with γ, cf. Sect. 6.2.5, in most of the models studied in the litterature, bubbles do not run-away and the scalar field contribution to the GW signal is **irrelevant**. Instead, most of the vacuum energy is converted into **fluid kinetic energy**, see next section. As already discussed along Eq. (6.186), only phase transitions with a lot of **supercooling**, $\alpha \gg 1$, (e.g. $T_{\rm nuc} \lesssim 10$ MeV for EWPT) have run-away bubbles,[19] such that the released vacuum energy is contained in the kinetic+gradient energy of the scalar field, or equivalently in the wall kinetic energy, and not in the fluid motion.

The energy transfer between the vacuum energy and the kinetic energy of the bubble wall, is given by [61]

$$\kappa_\phi = \begin{cases} \frac{\gamma_{\rm eq}}{\gamma_{\rm coll}}\left[1 - \left(\frac{\alpha_\infty}{\alpha}\right)^2\right] \simeq 0, & \gamma_{\rm coll} > \gamma_{\rm eq}, \quad \text{(constant wall velocity at coll.)} \\ 1 - \frac{\alpha_\infty}{\alpha} \simeq 1, & \gamma_{\rm coll} \leq \gamma_{\rm eq}, \quad \text{(run} - \text{away wall at coll.)} \end{cases} \tag{6.223}$$

where $\alpha_\infty \equiv \Delta P_{\rm LO}/\rho_{\rm rad}(T_{\rm nuc})$. We recall that $\gamma_{\rm eq}$ and $\gamma_{\rm coll}$ are the Lorentz factors when bubbles reach finite velocity and when bubbles collide, respectively. They are given by Eqs. (6.175) and (6.185), which we report here

$$\gamma_{\rm eq} \simeq \frac{\Delta V}{g^2\, f\, T_{\rm nuc}^3}, \tag{6.224}$$

$$\gamma_{\rm coll} \simeq T_{\rm nuc}\, \beta^{-1}. \tag{6.225}$$

Constraints:
We show the currents and future constraints on the GW signal generated by bubble collision during a supercooled first-order phase transition in Fig. 6.9. We consider pulsar-timing arrays EPTA [196], (current), NANOGrav [197] (current) and SKA [198] (planned), Space-based interferometer LISA [199] (planned), Earth-based interferometers LIGO O2 [200] (current), LIGO O5 (planned), Einstein Telescope [201, 202] (waiting for approval) and Cosmic Explorer [203] (waiting for approval). We fixed the signal-to-noise ratio SNR = 10 and the observation time to $T = 10$ years, except for LIGO O2 ($T = 268$ days) and LIGO O5 ($T = 1$ year). See Appendix 6.A for details about the computation of the power-law-integrated-sensitivity curves.

[19] Or strong phase transitions without gauge bosons.

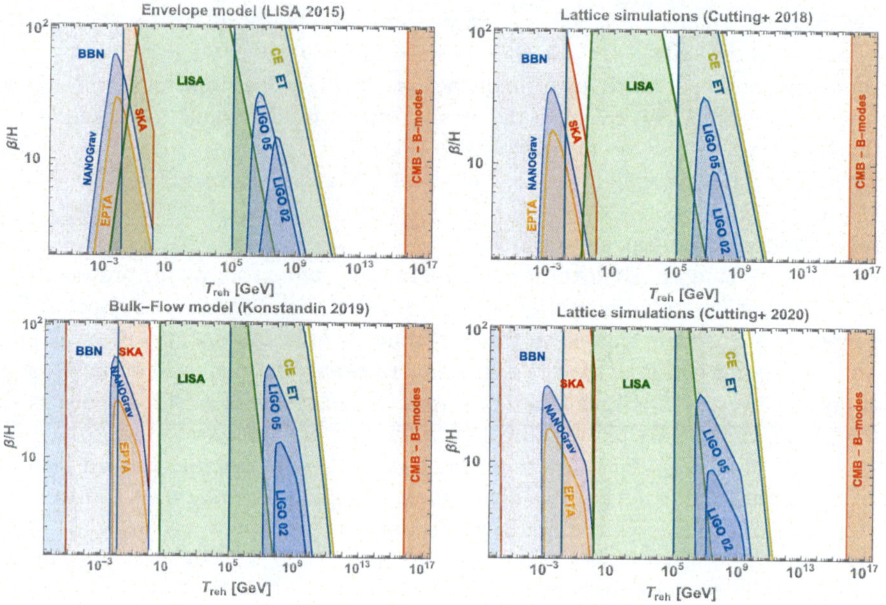

Fig. 6.9 Sensitivity regions of various experiments on vacuum-dominated first-order phase transitions with $\alpha \gg 1$ (Supercooled phase transitions). For such PT, we expect $\beta/H \sim 10$. GW are sourced by the scalar field gradient during bubble collisions, details on the different models/simulations are given in Fig. 6.8 and in text. The power-law integrated sensitivity curves of the different experiments are computed in Appendix 6.A

6.3.3 *Contributions from Sound Waves and Turbulence*

In the presence of an interaction between the plasma and the scalar field driving the phase transition, the expanding bubble wall, if subsonic, generates a shock wave in front of the wall (deflagration fluid profile) and, if supersonic, generates a rarefaction wave behind the wall (detonation fluid profile) [145]. The first studies of the GW spectrum sourced by sound shells goes back to the 90s [11]. Since then, the hydrodymical simulations have been performed [204–208], complemented by analytical investigations of the GW production from sound shell overlap [209, 210]

As sound waves propagate, the flow becomes turbulent and non-linearities forms, triggering energy cascades to smaller scales and leading to another source of GW production [211–221].

The stochastic GW background sourced by sound waves, based on the lattice work of [204, 206, 207] and sourced by turbulence, cf. LISA paper 2019 [14], reads

$$h^2\Omega_{\rm sw} = F_{\rm gw}\, h^2\, 0.049\ (H_*\tau_{\rm sw}) \left(\frac{H_* R_*}{c_s}\right) K_{\rm sw}^2\, S_{\rm sw}(f), \tag{6.226}$$

$$h^2\Omega_{\rm turb} = F_{\rm gw}\, h^2\; 11.9\; (1 - H_*\tau_{\rm sw}) \left(\frac{H_* R_*}{c_s}\right) K_{\rm turb}^{3/2}\, S_{\rm turb}(k). \tag{6.227}$$

with the spectral shapes

$$S_{\rm sw}(f_*) = \left(\frac{f_*}{f_{\rm sw}}\right)^3 \left(\frac{1}{1+\frac{3}{4}(f_*/f_{\rm sw})^2}\right)^{7/2}, \tag{6.228}$$

$$S_{\rm turb}(f_*) = \frac{(f_*/f_{\rm turb})^3}{(1+(f_*/f_{\rm turb}))^{11/3}(1+8\pi f/H_*)}, \tag{6.229}$$

$$\frac{f_{\rm sw}}{H_*} = \frac{1.57}{H_* R_*}, \tag{6.230}$$

$$\frac{f_{\rm turb}}{H_*} = \frac{2.38}{H_* R_*}. \tag{6.231}$$

The prefactor $F_{\rm gw}\, h^2$ is given in Eq. (6.214). f_* denotes the frequency at emission and is related to the frequency today f_0 through Eq. (6.216). The radius at collision is related to the duration of the phase transition β^{-1} though

$$R_* = \frac{(8\pi)^{1/3}}{\beta}\, \mathrm{Max}(v_w,\, c_s). \tag{6.232}$$

$K_{\rm sw}$ and $K_{\rm turb}$ are the efficiency of energy transfer from the total energy density (in the symmetric phase) $\rho^*_{\rm tot}$ to the energy density of sound waves and turbulence.

We discuss $K_{\rm turb}$ in the next paragraph and dedicate the entire next section, Sect. 6.3.4, to $K_{\rm sw}$, mainly based on the work of [145, 222, 223]. The result for $K_{\rm sw}$ is given by Eq. (6.265) for arbitrary speed of sound c_s and more specifically by Eq. (6.278) in the case of the well-studied Bag model in which $c_s = 1/\sqrt{3}$.

$\tau_{\rm sw}$ is the life-time of sound-waves, and will also be defined later in Eq. (6.288) after having introduced the averaged fluid velocity U_f.

Efficiency transfer to turbulence:
For turbulence, lattice simulations at small fluid velocity $U_f \lesssim 0.05$ have found [206, 207]

$$K_{\rm turb} \equiv \epsilon_{\rm turb}\, K_{\rm sw}, \qquad \text{with } \epsilon_{\rm turb} \simeq 5\%. \tag{6.233}$$

The precise value of $\epsilon_{\rm turb}$ as well as its dependence on α remains to be determined. As hinted by recent simulations Cutting+ 19 [208], we expect $\epsilon_{\rm turb}$ to increase with α. See also the recent works [220, 221]. The life-time of sound-waves $\tau_{\rm sw}$ coincides with the time after which non-linearities start to form and turbulence motion takes place. This explains the prescription of [130] of adding the factor $(1 - H_*\tau_{\rm sw})$ in Eq. (6.227).

In the next section, we discuss how to compute the efficiency of the energy transfer to sound waves kinetic energy $K_{\rm sw}$, needed in order to compute the GW signal in Fig. 6.12.

6.3.4 Energy Transfer to Sound-Waves

Sound-waves as a source of GW:
The energy momentum tensor of the fluid at first-order in the fluid gradient $\partial_\mu u_\nu$ expansion (perfect fluid approximation) is

$$T_{\mu\nu} = w\, u_\mu u_\nu - g_{\mu\nu}\, p \tag{6.234}$$

where $u_\mu = (\gamma,\ \gamma\,\vec{v})$ is the fluid 4-velocity and $w = e + p$ is the fluid enthalpy density. e and p are the total energy and pressure densities ($e \equiv \rho_{\rm tot}$). GW are sourced by the transverse-traceless-spatial part of $T_{\mu\nu}$, cf. Sect. 3.4.1, which for one bubble, has only one component along rr, and we get

$$K_{\rm sw} = \frac{\rho_{\rm fl}}{\rho^*_{\rm tot}}, \qquad \rho_{\rm fl} = \left\langle w\,\gamma^2\,v^2 \right\rangle, \tag{6.235}$$

with

$$\left\langle w\,\gamma^2\,v^2 \right\rangle = \frac{3}{\xi_w^3}\int d\xi\,\xi^2\,v^2\,\gamma^2\,w. \tag{6.236}$$

We have assumed that the profiles $v(\xi)$ and $w(\xi)$ are self-similar and only depend on $\xi = r/t$ where r is the bubble radius and t is the nucleation time. $\rho^*_{\rm tot}$ is the total energy density in the symmetric phase far from the wall (sometimes noted $e_n \equiv \rho^*_{\rm tot}$).

The velocity and enthalpy density profile of the sound-shell:
The velocity and enthalpy profile of the fluid can be computed using local conservation of the energy-momentum tensor [145]

$$\partial_\mu T^{\mu\nu} = u^\nu\,\partial_\mu(u^\mu\,w) + u^\mu\,w\,\partial_\mu u^\nu - \partial^\nu p = 0. \tag{6.237}$$

which for a self-similar profile described by $\xi = r/t$ and a constant speed of sound $c_s^2 = dp/de$, becomes

$$2\frac{v}{\xi} = \gamma^2(1 - v\,\xi)\left[\frac{\mu^2}{c_s^2} - 1\right]\partial_\xi v, \tag{6.238}$$

$$\frac{\partial_v w}{w} = 2w(1 + c_s^2)\frac{v}{\xi}\frac{\xi - v}{(\xi - v)^2 - c_s^2(1 - \xi v)^2}, \tag{6.239}$$

where μ is the boosted fluid velocity

$$\mu(\xi, v) = \frac{\xi - v}{1 - \xi v}. \tag{6.240}$$

Additionally due the change of pressure between the two phases, the velocity in the **symmetric** phase v_+ and in the **broken** phase v_- at the wall position $\xi_w \equiv v_w$, in the wall frame, do not coincide, but can be related using the two matching equations [145]

$$\partial_z T^{zz} = 0, \quad \rightarrow \quad \omega_+ v_+^2 \gamma_+^2 + p_+ = \omega_- v_-^2 \gamma_-^2 + p_-, \tag{6.241}$$

$$\partial_z T^{z0} = 0, \quad \rightarrow \quad \omega_+ v_+ \gamma_+^2 = \omega_- v_- \gamma_-^2. \tag{6.242}$$

In the **Bag** model where the energy and pressure densities read [145]

$$e_+ = a_+ T_+^4 + \epsilon, \qquad e_- = a_- T_-^4, \tag{6.243}$$

$$p_+ = a_+ T_+^4/3 - \epsilon, \qquad p_- = a_- T_-^4/3, \tag{6.244}$$

we obtain

$$v_+ = \frac{1}{1+\alpha_+}\left[\left(\frac{v_-}{2} + \frac{1}{6v_-}\right) \pm \sqrt{\left(\frac{v_-}{2} + \frac{1}{6v_-}\right)^2 + \alpha_+^2 + \frac{2}{3}\alpha_+ - \frac{1}{3}}\right], \tag{6.245}$$

with

$$\alpha_+ \equiv \frac{\epsilon}{a_+ T_+^4}. \tag{6.246}$$

Note that the value of α_+ at the wall is different from the input value α_N far ahead from the wall, which we simply denote by α. They are related through

$$\alpha \equiv \alpha_N = \alpha_+ \frac{w(\xi = v_w^+)}{w_N}. \tag{6.247}$$

Depending on the value of v_w, the fluid profile behaves in three different ways: deflagration (shock-front, $v_w < c_s$ and $v_- = v_w$), hybrid (shock-front and rarefaction wave, $c_s < v_w < \xi_J, v_- = c_s$) and detonation (rarefaction wave, $v_w > \xi_J, v_+ = v_w$). The transition between deflagration and hybrid occurs at the speed of sound, which in the Bag model reads

$$c_s^2 = 1/3, \tag{6.248}$$

while the transition between detonation and hybrid occurs at the Jouguet velocity ξ_J

$$\xi_J = \frac{1 + \sqrt{\alpha_+(2 + 3\alpha_+}}{\sqrt{3}(1 + \alpha_+)}, \tag{6.249}$$

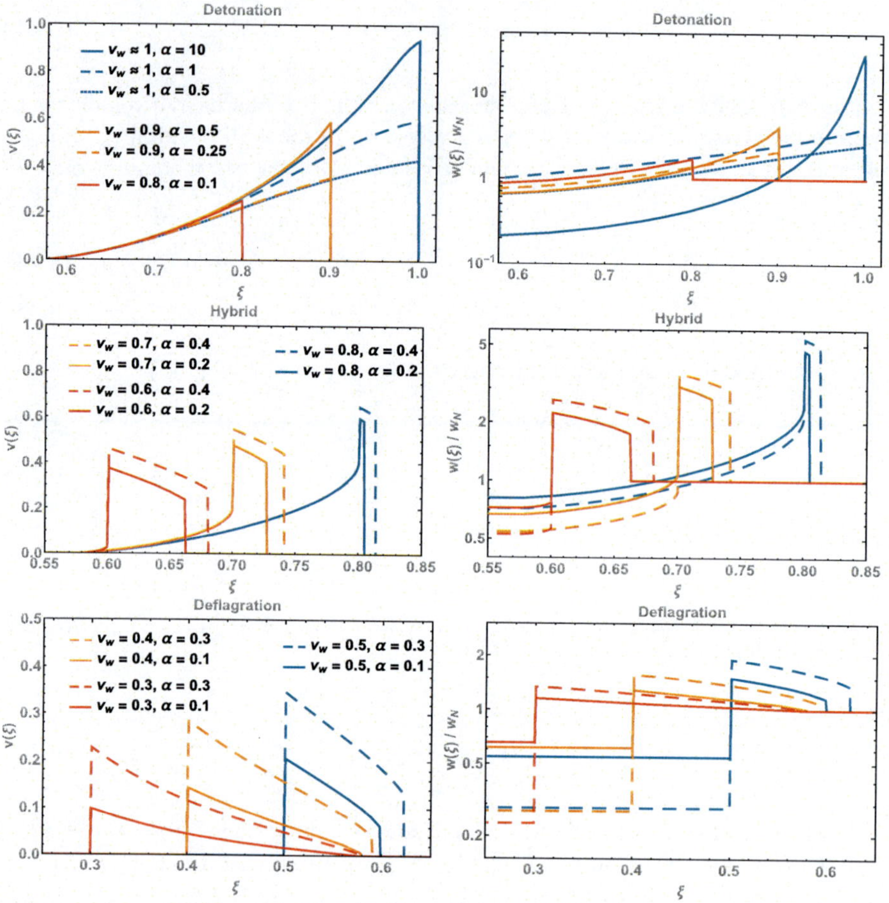

Fig. 6.10 Velocity (in the plasma frame) and enthalpy profile found after integration of the hydrodynamic equations in Eqs. (6.238) and (6.239) with the matching conditions in Eq. (6.245). Detonations, hybrids and deflagrations are defined by the boundary conditions $v_+ = v_w$ (only exists if $v_w > \xi_J$), $v_- = c_s$ (only exists if $\xi_J > v_w > c_s$) and $v_- = v_w$ (only exists if $c_s > v_w$), respectively. See [130, 145] for details about the computation of the hydrodynamic profile

obtained after setting $v_- = c_s$ in Eq. (6.245). We compute some profile solutions[20] for given values of v_w and α and plot them in Fig. 6.10.

[20] I thank Felix Giese and Jorinde Van de Vis for useful discussions regarding the computation of the fluid profile.

The potential energy sourcing the bubble expansion:
The first thermodynamics law at constant volume (take a volume much larger than the bubble) in the presence of a heat transfer $\delta(e_Q V)$ reads

$$\delta(e_Q V) = d(eV) + p dV \quad \xrightarrow{V=\text{cst}} \quad \delta e_Q = de = \frac{dw}{1+c_s^2}, \tag{6.250}$$

where in the last term we have introduced the speed of sound

$$c_s^2 = \frac{dp}{de}. \tag{6.251}$$

Hence, we can identify

$$e_Q = w/(1+c_s^2), \tag{6.252}$$

as the thermal energy density of the fluid. First, observe that the component T_{00} of the energy-momentum tensor can be split into

$$T_{00} = w\gamma^2 - p = w\gamma^2 v^2 + e \tag{6.253}$$

$$= w\gamma^2 v^2 + \frac{w}{1+c_s^2} + \frac{c_s^2}{1+c_s^2}\left(e - \frac{p}{c_s^2}\right). \tag{6.254}$$

Second, note that the integral of T^{00} around a single bubble

$$E = 4\pi \int_0^R dr\, r^2\, T_{00}, \tag{6.255}$$

must be constant for R much larger than the bubble radius. As a result, we get the following energy budget[21]

$$e_{\text{K}} + \Delta e_{\text{Q}} = -\Delta e_{\text{pot}}, \tag{6.256}$$

where

$$\Delta e_{\text{pot}} = \frac{3}{\xi_{\text{max}}^3} \int_0^{\xi_{\text{max}}} d\xi\, \xi^2\, (e_{\text{pot, n}} - e_{\text{pot}}), \tag{6.257}$$

$$e_{\text{K}} = \frac{3}{\xi_{\text{max}}^3} \int_0^{\xi_{\text{max}}} d\xi\, \xi^2\, w\, \gamma^2\, v^2, \tag{6.258}$$

[21] These lines are inspired from Appendix B.2 of [210] where the energy budget is realized in the restricted case $c_s^2 = 1/3$.

$$\Delta e_Q = \frac{3}{\xi_{\rm max}^3} \int_0^{\xi_{\rm max}} d\xi\, \xi^2\, (\frac{w_{\rm n}}{1+c_s^2} - \frac{w}{1+c_s^2}), \tag{6.259}$$

and where $e_{\rm pot}$ is defined thought the pseudotrace $\bar{\theta}$ of the energy-momentum tensor

$$e_{\rm pot} = \frac{c_s^2}{1+c_s^2}\bar{\theta}, \qquad \text{with} \qquad \bar{\theta} \equiv \left(e - \frac{p}{c_s^2}\right). \tag{6.260}$$

The upper integral boundary is $\xi_{\rm max} = \max(v_w,\ \xi_{\rm sh})$, with $\xi_{\rm sh}$ being the outermost position of the velocity profile. Quantities with the subscript n are evaluated in the symmetric phase, far from the wall. Equation (6.256) allows to interpret the **pseudotrace anomaly** $\frac{c_s^2}{1+c_s^2}\bar{\theta}$ as the **potential energy** available for the transformation. As the bubble expands, it gets converted into the fluid kinetic energy density $w\,\gamma^2\,v^2$ and the fluid thermal energy density $w/(1+c_s^2)$.

In the Bag model where, by definition, the energy and pressure densities read Eqs. (6.243) and (6.244), the potential energy which drives the phase transition coincides with the **vacuum energy**

$$\Delta e_{\rm pot} = \epsilon, \qquad \text{(Bag model)} \tag{6.261}$$

However for generic models, the pressure and the energy density are functions of the type $a + b\,T + c\,T^2 + c\,T^3 + d\,T^4 + \cdots$. Therefore, the net distinction between a purely relativistic component and a purely vacuum component, characteristic of the Bag model, is not possible anymore. Generally, the pressure is given by the effective potential at finite-temperature $p = -V_{\rm eff}(\phi,\ T)$ and the energy follows from the identity $e \equiv T\frac{\partial p}{\partial T} - p$, such that the potential energy in Eq. (6.260) driving the bubble expansion reads

$$\Delta e_{\rm pot} \simeq e_{\rm pot,\,s} - e_{\rm pot,\,b}, \tag{6.262}$$

with

$$e_{\rm pot,\,i} = \left[V_{\rm eff} - \frac{T}{1+\frac{1}{c_s^2}} \frac{dV_{\rm eff}}{dT} \right]_{\rm phase\ i}. \tag{6.263}$$

Note that the speed of sound

$$c_s^2 = \frac{dp/dT}{de/dT}, \tag{6.264}$$

depends on the phase (symmetric s or broken b). Idem for the temperature T_s or T_b.

Efficiency parameter for arbitrary c_s

We have seen in Eq. (6.256) that the pseudotrace $\bar{\theta}$ pops up naturally when computing the energy budget of the bubble expansion for arbitrary sound of speed c_s.

Additionally, the authors of [222, 223] have shown that, for arbitrary c_s, the pseudotrace[22] is the good quantity which appears in the matching equations at the wall position $T^+_{\mu\nu} = T^-_{\mu\nu}$, in Eqs. (6.241) and (6.242), after performing a Taylor expansion in $T_- = T_+ + \delta T$.

For a **detonation** profile,[23] this motivates the re-writing of the energy transfer parameter $K_{\rm sw}$ in Eq. (6.235), as

$$K_{\rm sw} = \frac{\rho_{\rm fl}}{e_+} = \Gamma\, \alpha_{\bar{\theta}}\, \kappa_{\bar{\theta}}, \qquad \text{(arbitrary } c_s \text{ model)} \tag{6.265}$$

with

$$\text{pseudotrace:} \qquad \bar{\theta} \equiv e - p/c_s^2, \tag{6.266}$$

$$\text{pseudotrace difference:} \qquad D\bar{\theta} = \bar{\theta}_+(T_+) - \bar{\theta}_-(T_+), \tag{6.267}$$

$$\text{speed of sound:} \qquad c_s^2 \equiv \left.\frac{dp_-/dT}{de_-/dT}\right|_{T_+}, \tag{6.268}$$

$$\text{adiabaticity parameter:} \qquad \Gamma = w_+/e_+, \tag{6.269}$$

$$\text{strength parameter:} \qquad \alpha_{\bar{\theta}} = \frac{D\bar{\theta}}{4w_+}, \tag{6.270}$$

$$\text{efficiency parameter:} \qquad \kappa_{\bar{\theta}}(c_s,\, \alpha_{\bar{\theta}},\, v_w) = \frac{4\rho_{\rm fl}}{D\bar{\theta}}, \tag{6.271}$$

$$\text{fluid kinetic energy density:} \qquad \rho_{\rm fl} = \langle w\, \gamma^2\, v^2 \rangle. \tag{6.272}$$

e_+, w_+ and T_+ are the energy, enthalpy density and temperature in the symmetric phase at the wall position, which for the detonation case, coincide with the quantities in the symmetric phase, far from the wall, e_n, w_n, T_n. This allows to cast all the model-dependence within the three quantities c_s, $\alpha_{\bar{\theta}}$, ξ_w such that the function $\kappa_{\bar{\theta}}(c_s,\, \alpha_{\bar{\theta}},\, v_w)$ is model-independent and provided in Appendix A of [222] in the detonation case.

In the Bag model, we have

$$c_s^2 = 1/3, \tag{6.273}$$

$$\Gamma = \frac{4}{3}\frac{1}{1+\alpha_\epsilon}, \tag{6.274}$$

$$D\bar{\theta} = 4\epsilon, \tag{6.275}$$

---

[22] Before the pseudo-trace $\bar{\theta}$, the simple trace (or trace anomaly) $\theta \equiv e - 3p$ was first advocated by [204, 206, 207].

[23] For the deflagration case, we refer to [223].

$$\alpha_{\bar{\theta}} = \frac{3}{4}\alpha_{\epsilon}, \tag{6.276}$$

$$\kappa_{\bar{\theta}} = \kappa_{\epsilon}, \tag{6.277}$$

or equivalently

$$K_{\text{sw}} = \frac{\rho_{\text{fl}}}{e_n} = \frac{\alpha_{\epsilon}\kappa_{\epsilon}}{1+\alpha_{\epsilon}}, \quad \text{(Bag model)}, \tag{6.278}$$

where[24]

$$\alpha_{\epsilon} \equiv \frac{\epsilon}{a_+ T_+^4}, \tag{6.279}$$

$$\kappa_{\epsilon}(\alpha, v_w) \equiv \frac{\rho_{\text{fl}}}{\epsilon}, \tag{6.280}$$

are the usual α and κ parameters [145], which however are only relevant for the (restricted) Bag model. The function $\kappa(\alpha, v_w)$ is provided in Appendix A of [145] for detonation, deflagration and hybrid profiles.

For general models, the pressure is given by the effective potential at finite-temperature $p = -V_{\text{eff}}(\phi, T)$ and the energy follows from the identities $e \equiv T\frac{\partial p}{\partial T} - p$, such that the parameters of the **arbitary** c_s **model**, c_s, $\alpha_{\bar{\theta}}$ and Γ, reduce to

$$c_s^2 = \left[\frac{dV_{\text{eff}}/dT}{Td^2V_{\text{eff}}/dT^2}\right]_{\phi=\phi_{\text{true}},\, T_+}, \tag{6.281}$$

$$\alpha_{\bar{\theta}} = \frac{1}{4}\left[1-(1+\frac{1}{c_s^2})\frac{DV_{\text{eff}}}{T\partial DV_{\text{eff}}/\partial T}\right]_{T_+}, \tag{6.282}$$

$$\Gamma = \left[\frac{TdV_{\text{eff}}//dT}{TdV_{\text{eff}}/dT - V_{\text{eff}}}\right]_{Œ=0,\, \text{T}_+}, \tag{6.283}$$

$$DV_{\text{eff}} = V_{\text{eff}}(\phi = 0, T_+) - V_{\text{eff}}(\phi = \phi_{\text{true}}, T_+). \tag{6.284}$$

Average fluid velocity

Another averaged quantity which we can compute once we know the profile $v(\xi)$ and $w(\xi)$ is the enthalpy-averaged root-mean-square velocity of the fluid

$$\bar{\gamma}_f^2\,\bar{U}_f^2 \equiv \frac{1}{\bar{w}V}\int_V T_{ii}\,d^3x = \frac{\langle w\,\gamma^2\,v^2\rangle}{\bar{w}} \tag{6.285}$$

The average fluid velocity in Eq. (6.285) can be related to the efficiency parameter $\kappa_{\bar{\theta}}$ in Eq. (6.271) and κ_{ϵ} in Eq. (6.280)

---

[24] In the bag model, which is defined by the energy e and pressure p density being given by a radiation component $\propto T^4$ plus a vacuum component $\propto T^0$, we have $\alpha_{\epsilon} \equiv \frac{\Delta\theta}{e_+}$ where $\theta = \frac{1}{4}(e-3p)$ is the trace anomaly, e_+ energy density in the symmetry phase, and Δ being the phase difference [145]. From using the thermodynamic identity $e = T\frac{\partial p}{\partial T} - p$, we obtain $\theta = -p + \frac{T}{4}\frac{\partial p}{\partial T}$.

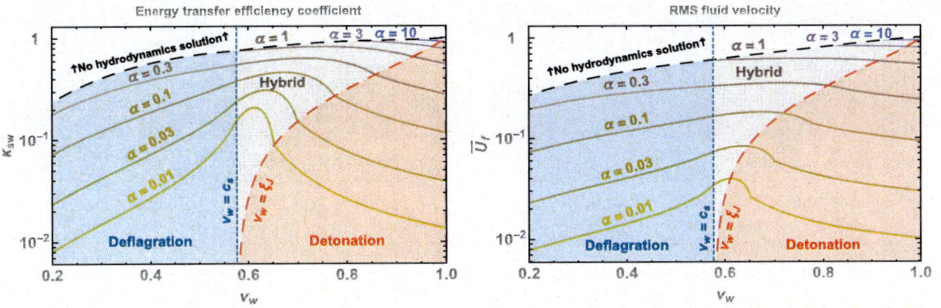

Fig. 6.11 **Left**: Efficiency coefficient κ in the Bag model, defined in Eq. (6.280), of the energy transfer from vacuum energy to fluid bulk motion. The function is given in Appendix A of [145]. **Right**: Enthalpy-averaged root-mean-square fluid velocity in the bubble-center frame, defined in Eq. (6.285), in the case of the Bag model

$$\bar{\gamma}_f^2\,\bar{U}_f^2 = \begin{cases} \kappa_{\bar{\theta}}\alpha_{\bar{\theta}} & \text{(Arbitrary } c_s \text{ model)}, \\ \frac{3}{4}\kappa_\epsilon\alpha_\epsilon & \text{(Bag model: } c_s^2 = 1/3), \end{cases} \tag{6.286}$$

or equivalently[25]

$$\bar{U}_f^2 = \begin{cases} \dfrac{\kappa_{\bar{\theta}}\alpha_{\bar{\theta}}}{1+\kappa_{\bar{\theta}}\alpha_{\bar{\theta}}} & \text{(Arbitrary } c_s \text{ model)}, \\ \dfrac{3\kappa_\epsilon\alpha_\epsilon/4}{1+3\kappa_\epsilon\alpha_\epsilon/4} & \text{(Bag model: } c_s^2 = 1/3), \end{cases} \tag{6.287}$$

We plot the functions κ_ϵ and $\bar{U}_f$ in the case of the Bag model in Fig. 6.11.

Life-time of sound-waves and onset of turbulence:
The averaged fluid velocity is an important quantity since it allows to determine the lifetime $\tau_{\rm sw}$ of the sound-waves, after which shocks form and non-linearities develop [207, 219]

$$\tau_{\rm sw} \sim L_f/\bar{U}_f, \tag{6.288}$$

where L_f is the characteristic length of the fluid flow. It is approximately given by the mean bubble separation at percolation R_*, such that [130]

$$H\tau_{\rm sw} \sim \frac{H\,R_*}{\bar{U}_f} \simeq \frac{(8\pi)^{1/3}\text{Max}(v_w,\, c_s)}{\bar{U}_f}\left(\frac{H}{\beta}\right) \tag{6.289}$$

[25] Note the difference with the litterature [14, 130, 204, 206, 207] where the authors do the approximations $\bar{\gamma}_f^2\,\bar{U}_f^2 \to \bar{U}_f^2$ and $\frac{\bar{w}}{\bar{e}} = \frac{4}{3(1+\alpha_\epsilon)} \to \frac{4}{3}$, which implies $\bar{U}_f^2 = \dfrac{\kappa_{\bar{\theta}}\alpha_{\bar{\theta}}}{1+\kappa_{\bar{\theta}}\alpha_{\bar{\theta}}} \to \frac{3}{4}\dfrac{\kappa_\epsilon\alpha_\epsilon}{1+\alpha_\epsilon}$.

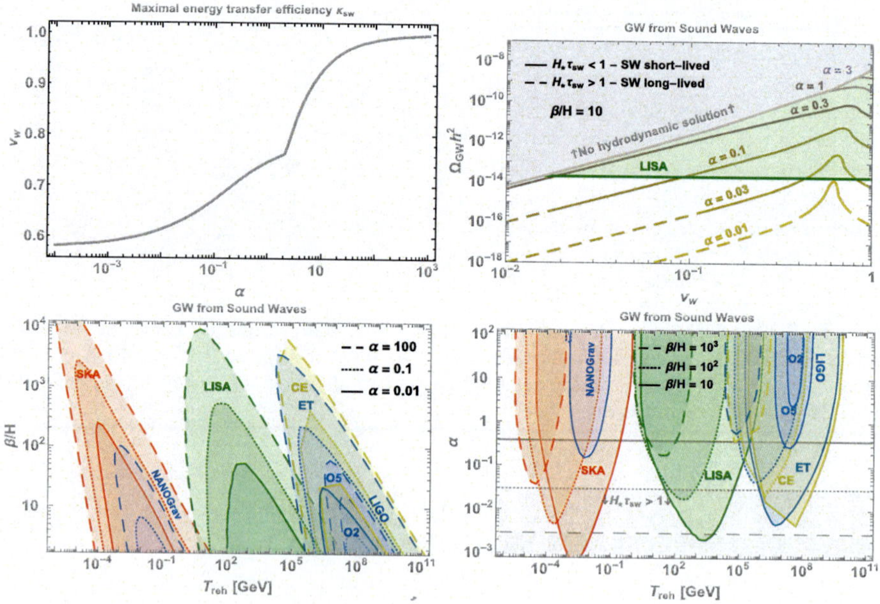

Fig. 6.12 Reach of current and future experiments on thermal 1stOPT. GW are sourced by the acoustic waves, cf. formula of LISA paper 2019 in Eq. (6.226). **Top left:** Wall velocity which maximizes the energy transfer efficiency, see left panel of Fig. 6.11. **Top right:** Dependence of the peak amplitude of the GW signal on the wall velocity. In the gray region, corresponding to $\alpha > \frac{1}{3}(1 - v_w)^{-13/10}$, the fluid profile has no hydrodynamic solution [145]. **Bottom:** The wall velocity v_w is chosen in order to maximize the GW signal. We consider the Bag model, such that $c_s = 1/\sqrt{3}$, $\alpha = \alpha_\epsilon$ and $\kappa_{\rm sw} = \kappa_\epsilon$, cf. definitions around Eq. (6.265). The contribution from turbulence is not included

Constraints on GW from sound-waves:
We show the current and future constraints on the GW signal generated by acoustic waves during a thermal first-order phase transition in Fig. 6.12. We consider pulsar-timing arrays NANOGrav [197] (current) and SKA [198] (planned), Space-based interferometer LISA [199] (planned), Earth-based interferometers LIGO O2 [200] (current), LIGO O5 (planned), Einstein Telescope [201, 202] (waiting for approval) and Cosmic Explorer [203] (waiting for approval). We fixed the signal-to-noise ratio SNR = 10 and the observation time to $T = 10$ years, except for LIGO O2 ($T =$ 268 days) and LIGO O5 ($T = 1$ year). See Appendix 6.A for details about the computation of the power-law-integrated-sensitivity curves.

Note with bottom-right panel of Fig. 6.12 that except for the poorly motivated case where $\beta/H \sim 10$ and $\alpha \ll 1$, the life-time of sound-waves corresponding to the reach of experiments is never longer than a Hubble time $H_* \tau_{\rm sw} < 1$. This is in contrast to what was initially claimed [12, 213]. See the extensive discussion in [130].

6.4 Supercooling from a Nearly-Conformal Sector

In the previous sections of the chapter, we have discussed the rate of bubble nucleation, the speed of the bubble wall and the GW generation due to bubble collisions. We know study in details two classes of potential leading to a supercooled first-order phase transition: the Coleman-Weinberg potential in the weakly-coupled scenario and the light-dilaton potential in the strongly-coupled scenario. For both cases, we compute the bounce action, the nucleation temperature and the GW parameters α and β. The results for the light-dilaton potential are relevant for Chap. 7 about string fragmentation in supercooled confinement and follow-ups [49].

6.4.1 *Weakly-coupled Scenario: The Coleman-Weinberg Potential*

The lagrangian:
We extend the SM with a new gauge group $SU(2)_X$ whose generators define three dark gauge bosons X. We also add a scalar ϕ transforming as a doublet of $SU(2)_X$. We suppose that the tree level Lagrangian is **scale-invariant**, so that the bare mass terms are forbidden and the only interaction terms between ϕ and the SM Higgs H which we can write are [40, 74]

$$\mathscr{L} \supset -V = -\lambda_{\rm h} H^4 - \lambda_{\rm \sigma} \phi^4 - \lambda_{\rm h\sigma} \phi^2 H^2. \tag{6.290}$$

with

$$H = \exp\left(i\sigma^i G^i/v\right) \begin{pmatrix} 0 \\ h/\sqrt{2} \end{pmatrix}, \qquad \phi = \exp\left(i\sigma^i G^i_X/v\right) \begin{pmatrix} 0 \\ \sigma/\sqrt{2} \end{pmatrix}, \qquad i \in 1, 2, 3. \tag{6.291}$$

G^i and G^i_X are the Golsdstone bosons whereas h and σ are the massive states after taking the vevs $\langle h \rangle = v$ and $\langle \sigma \rangle = f$.

Radiatively-induced symmetry breaking:
In the absence of bare mass term, spontenous symmetry breaking of $SU(2)_X$ can occur due to radiative effects [224]. Indeed, the integration of the β function

$$\beta_{\lambda_\sigma} \equiv \frac{d\lambda_\sigma}{d\ln\mu} \simeq \frac{1}{(4\pi)^2} \frac{9g_x^4}{8}, \tag{6.292}$$

gives the one-loop Coleman-Weinberg potential [40, 74]

$$V_{1-\rm loop}(\sigma) \simeq \beta_{\lambda_\sigma} \frac{\sigma^4}{4} \left(\ln\frac{\sigma}{f} - \frac{1}{4}\right), \tag{6.293}$$

which has a minimum at $\langle\sigma\rangle = f$.
As a consequence

1. The $SU(2)_X$ gauge bosons receive the mass[26]

$$M_X = g_X\, f/2 \tag{6.294}$$

where g_X is the $SU(2)_X$ coupling constant.
2. The higgs gets the vev

$$v = f\sqrt{\frac{-\lambda_{\rm h\sigma}}{2\lambda_{\rm h}}}. \tag{6.295}$$

The mass matrix of the $\sigma - h$ system is

$$\frac{\partial V}{\partial h\partial\sigma}|_{(v,\,\omega)} = \begin{pmatrix} 2\lambda_{\rm h}v^2 & \lambda_{\rm h\sigma}vf \\ \lambda_{\rm h\sigma}vf & \beta_{\lambda_\sigma}f^2 - \lambda_{\rm h\sigma}^2\frac{f^2}{4\lambda_{\rm h}} \end{pmatrix}. \tag{6.296}$$

In the limit of small $h - \sigma$ mixing $\lambda_{\rm h\sigma} \ll 1$, the mass eigenvalues reduce to

$$m_h^2 \simeq 2\lambda_{\rm h}v^2, \tag{6.297}$$

$$m_\sigma^2 \simeq \beta_{\lambda_\sigma}\, f^2. \tag{6.298}$$

Finite-temperature contributions:
At finite temperature the potential for σ receives 1-loop thermal corrections from the 3 gauge bosons + the Daisy contribution from the 3 longitudinal modes, cf. Sect. 6.1.1

$$V_T(\sigma,\ T) = V_{\rm 1-loop}^T + V_{\rm Daisy} = \frac{9\,T^4}{2\pi^2}J_B\left(\frac{M_X^2}{T^2}\right) + \frac{T}{4\pi}\left[M_X^3 - \left(M_X^2 + \Pi_X\right)^{3/2}\right]. \tag{6.299}$$

where the dark gauge boson thermal mass

$$\Pi_X = \frac{5}{6}g_X^2T^2\ C(M_X/T), \tag{6.300}$$

is cutted-off with $C(x) = x^2/2\ K_2(x)$ to prevent non-physical contributions from gauge boson at large M_X/T [74].

At high temperature, the scalars get positive thermal masses such that the minimum of the potential is in $\langle\sigma\rangle = \langle h\rangle = 0$. When the radiation energy density $\pi^2 g_* T^4/30$ gets smaller than the vacuum energy

[26] The Dark gauge bosons transform as a triplet under a custodial $SO(3)$ symmetry so their masses are degenerate and DM has 9°C of freedom.

$$V_\Lambda \simeq \frac{9M_X^4}{8(4\pi)^2}, \tag{6.301}$$

a period of inflation starts, this is the **supercooling** period. It happens when the temperature reaches the critical temperature

$$T_{\text{inf}} \simeq \frac{M_X}{8.4}\left(\frac{116}{g_*}\right)^{1/4}. \tag{6.302}$$

Bounce action:
Close to the false vacuum the Coleman-Weinberg potential is very flat such that the thermal barrier in Eq. (6.299) is well approximated by its high-temperature expansion, cf. Eq. (6.18)

$$V_T(\sigma,\ T) \overset{T\gg\sigma}{\longrightarrow} \frac{m_{\text{eff}}(T)^2}{2} - \frac{\lambda_{\text{eff}}}{4}\sigma^4, \tag{6.303}$$

with

$$m_{\text{eff}}(T)^2 = \frac{3}{16}\, g_X^2\, T^2, \tag{6.304}$$

$$\lambda_{\text{eff}} = -\beta_{\lambda_{\sigma}}\left(\ln\frac{\sigma}{f} - \frac{1}{4}\right), \qquad \beta_{\lambda_{\sigma}} \simeq \frac{1}{(4\pi)^2}\frac{9g_x^4}{8}. \tag{6.305}$$

We now use the thick-wall formula introduced in Sect. 6.1.3 in order to evaluate the O_3- and O_4- bounce actions. Upon injecting Eqs. (6.303) into (6.73), we obtain

$$\frac{S_3}{T} = \frac{16\pi}{3\,\lambda_{\text{eff}}}\frac{m_{\text{eff}}}{T}, \qquad \text{and} \qquad \phi_* = 2\frac{m_{\text{eff}}}{\sqrt{-\lambda_{\text{eff}}}}. \tag{6.306}$$

We recall that ϕ_* is the field value after tunneling. In terms of the physical parameters, the O_3-bounce action reads

$$\frac{S_3}{T} \simeq \frac{A}{\log\left(\frac{M}{T}\right)} \quad \text{with} \quad A = \frac{1018}{g_X^3} \quad \text{and} \quad M = 5.65\, f. \tag{6.307}$$

We now repeat the same steps in the O_4 case, so that we inject Eq. (6.303) into Eq. (6.72). However, by doing that we get

$$S_4 = 0, \tag{6.308}$$

which means that in the presence of the thermal barrier, there is no O_4 bounce. However, if we neglect the thermal barrier and set $m_{\text{eff}} = 0$ in Eq. (6.303), then the thick-wall formula in Eq. (6.72) gives

$$S_4 \simeq \frac{2\pi^2}{\lambda_{\text{eff}}}. \tag{6.309}$$

Note that due to the scale-invariance of $V = -\lambda_{\text{eff}}\sigma^4/4$, no exit point ϕ^* of the tunneling is singled out in Eq. (6.72) which leads to a degenerate family of instantons with arbitrary ϕ^* [225]. They are the so-called Fubini bounces and their tunneling actions can be computed exactly [226, 227]

$$S_4 = \frac{8\pi^2}{3\lambda_{\text{eff}}}, \tag{6.310}$$

which confirms the good convenience of the thick-wall formula. We conclude that tunneling occurs via O_3 when

$$\frac{S_3}{T} \lesssim S_4 \quad \rightarrow \quad m_{\text{eff}} \lesssim T \quad \rightarrow \quad g_X \lesssim 2.7. \tag{6.311}$$

In the top left panel of Fig. 6.13, we compare the O_3- and O_4- bounce actions computed with the thick-wall formula and we confirm the previous estimate in Eq. (6.311).

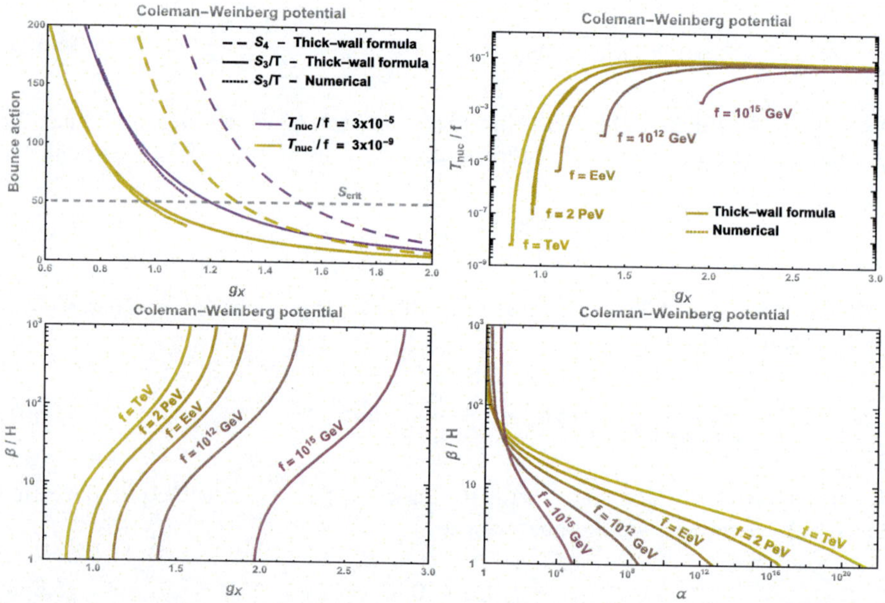

Fig. 6.13 **Top left:** We compare the O_3- and O_4- bounce action computed with the thick-wall formula, cf. Eqs. (6.307) and (6.309), to the O_3-bounce action computed with our own shooting method. **Top right:** Nucleation temperature computed with the thick-wall formula for S_3/T. The end points correspond to the last possible nucleation before eternal inflation, cf. Eq. (6.319). **Bottom:** GW parameters α and β

We also compare the analytical O_3-bounce action to the one computed numerically. For this, we implemented our own undershoot-overshoot algorithm.[27] Note that in order to improve the convergence of the code, it can be useful to modify the potential as follows

$$\tilde{V}(\sigma) = V(\sigma) - V(0), \tag{6.312}$$

$$\tilde{V}(\sigma) = 0 \qquad \sigma < 0, \tag{6.313}$$

$$\tilde{V}(\sigma) = V(\sigma) - V(0) - \sigma^4 + f^4 \qquad \sigma > f. \tag{6.314}$$

We numerically find that the size of the bounce is of order $R \sim 10/T$.

Nucleation temperature:
The nucleation happens when the tunneling rate in Eq. (6.43) becomes comparable to the Hubble expansion rate per Hubble volume

$$\frac{S_3(T_{\text{nuc}})}{T_{\text{nuc}}} \simeq 4 \ln \frac{T_{\text{nuc}}}{H(T_{\text{nuc}})} + \frac{3}{2} \ln \frac{S_3/T}{2\pi} \tag{6.315}$$

with

$$H^2(T) = H_\Lambda^2 + H_{\text{rad}}^2 = \frac{1}{(4\pi)^2} \frac{9 M_X^4}{24 M_{\text{pl}}^2} + \frac{\pi^2 g_* T^4}{90 M_{\text{pl}}^2}, \quad g_* = 116.75. \tag{6.316}$$

From injecting Eq. (6.307) into Eq. (6.315), we obtain

$$T_{\text{nuc}} = \sqrt{H_\Lambda\, M} \left(\frac{2\pi}{S_3/T_{\text{nuc}}} \right)^{3/16} \exp\left(\frac{1}{2} \sqrt{-A + \left(\ln \frac{M}{H_\Lambda} + \frac{3}{8} \ln \frac{S_3/T_{\text{nuc}}}{2\pi} \right)^2} \right). \tag{6.317}$$

Neglecting the $S_3/2\pi$ terms, we conclude that there is no nucleation solution when [90]

$$A \gtrsim \ln \frac{M}{H_\Lambda}, \tag{6.318}$$

and that the minimal nucleation temperature is

$$T_{\text{nuc}}^{\text{min}} \simeq \sqrt{H_\Lambda\, M} \simeq 0.1 \left(\frac{f}{M_{\text{pl}}} \right)^{1/2} f. \tag{6.319}$$

In top right panel of Fig. 6.13, we show the nucleation temperature computed with the thick-wall formula.

[27] I thank Iason Baldes, Sebastian Bruggisser, Benedict Von Harling and Oleksii Matsedonskyi for useful discussions related to the numerical computations of the bounce action.

Catalysis by QCD effects:
The period of super-cooling ends when the phase transition occurs and the vacuum energy becomes zero. If the tunneling rate is large enough, the phase transition happens by bubble nucleation, otherwise it can be trigged by the QCD phase transition, see the pioneering paper [228] and more recent works [40, 73].[28] Indeed when quarks condensates, the Yukawa coupling induces a linear term for the Higgs potential and the latter gets a temperature dependent vev [228]. Then $\langle h \rangle$ induces a negative mass term for the dark scalar $m_\sigma^2 = \lambda_{h\sigma} \langle h \rangle^2 /2$ and σ starts rolling down as soon as the Higgs-mediated quarks condensate contribution becomes comparable to the thermal mass $m_\sigma^2(T) = \frac{3}{16} g_X^2 T^2$. If $-\lambda_{h\sigma}$ is large enough, it happens immediately at the QCD phase transition when $T_{\rm QCD} = 85\,\text{MeV}$[29] otherwise it happens at [40]

$$T_{\rm end}^{\rm QCD} \approx \frac{0.1\,\langle h \rangle}{M_X/\text{TeV}} \tag{6.320}$$

Then the temperature at the end of super-cooling is

$$T_{\rm end} = \text{Max}\left(T_{\rm nuc},\ \text{Min}\left(T_{\rm end}^{\rm QCD},\ T_{\rm QCD}\right)\right). \tag{6.321}$$

Catalysis by gravity effects:
When the bubble size at nucleation $R_0 \sim 10/T$ is of the same order as the Hubble horizon H^{-1}[111, 231], the bounce action receives important correction in $R_0 H$. Intuitively, when $T \lesssim H$, the quantum fluctuations induced by the de Sitter temperature $T_{\rm dS} = H/2\pi$ becomes larger than the thermal barrier induced by the plasma temperature T such that we expect the tunneling to be triggered by gravity effects when

$$T \sim H. \tag{6.322}$$

See [232] for a related study.

Catalysis by primordial black holes:
The idea that the bounce action of 1stOPT could be enhanced by the presence of a black hole (BH) is credited to Hiscock in 1987 [233]. More recently, the idea has been the object of numerous studies [234–245]. The optimum BH mass $M_{\rm PBH}$ is a compromise between the enhancement of the bounce action (which decreases with $M_{\rm PBH}$) and the BH lifetime though evaporation (which increases with $M_{\rm PBH}$ and which should not be much smaller than the false vacuum lifetime). For instance, the maximal effect on the metastability of the electroweak vacuum is reached for BH masses around $M_{\rm PBH} \sim 10^4 - 10^8\ M_{\rm pl} \sim 1\ \text{g} - 1\ \text{kg}$, depending on the Higgs

[28] For studies of the QCD effects on a 1stOPT arising from a strongly-coupled sector, see the pioneering paper [229] in which the gluon condensate induces a new term in the dilaton potential, and more recent works [88, 230]. We discuss catalysis by QCD effects in confining PT at the end of Sect. 6.4.2.

[29] This is the temperature at which the QCD phase transition occurs when the quarks are massless.

potential parameters [237], while the nucleation rate of electroweak-like 1stOPT could be boosted in the presence of BH lighter than $M_{\rm PBH} \lesssim 10^{15}\ M_{\rm pl} \sim 10^{9}$ g [245]. The impact of seed BH on the nucleation rate of nearly-conformal potential remains to be investigated.

Catalysis by topological defects:
In 1981, Steinhardt showed that cosmological 1stOPT could be seeded by monopoles [246, 247]. If the monopole core is made of the false vacuum and that the later is sufficiently metastable, then the monopole is unstable under fluctuations of the core radius, causing it to expand. This effect called **monopole dissociation** has been explored further in [248–251]. Other works have studied the possibility to catalyze cosmological phase transitions with cosmic strings [252–256] and domain walls [257].

Reheating:
The reheating temperature can be shown to be

$$T_{\rm RH} = T_{\rm inf}\ \text{Min}\left(1,\ \frac{\Gamma}{H}\right)^{1/2} \tag{6.323}$$

After the period of super-cooling, the dark scalar field oscillates around its minima and reheates the Universe by transfering its energy into SM. However the dark scalar being coupled to the Higgs, a part of the energy will go into oscillation of the Higgs around its minima. The equation of motion of the scalar fields are

$$\ddot{\sigma} + (3H + \Gamma_\sigma)\dot{\sigma} = -\frac{\partial V}{\partial \sigma} \tag{6.324}$$

$$\ddot{h} + (3H + \Gamma_h)\dot{h} = -\frac{\partial V}{\partial h} \tag{6.325}$$

Which at first-order around the minima becomes

$$\ddot{h} + (3H + \Gamma_h)\dot{h} = m_h^2(h - v) + \lambda_{h\sigma}(\sigma - \sigma_0) \tag{6.326}$$

$$\ddot{\sigma} + (3H + \Gamma_\sigma)\dot{\sigma} = m_\sigma^2(\sigma - \sigma_0) + \lambda_{h\sigma}(h - v) \tag{6.327}$$

with $\lambda_{h\sigma} = \partial^2 V/\partial h \partial\sigma = -m_h^2(v/f)^2$. In the limit $\Gamma_\sigma,\ H \ll \Gamma_h \ll M_\sigma$, one gets the frequency of the normal modes $\omega \simeq m_h \pm i(3H + \Gamma_h)/2$ and $f \simeq m_\sigma \pm i(3H + \Gamma)/2$ with [40].

$$\Gamma = \Gamma_h \sin^2\theta + \Gamma_\sigma \cos^2\theta \tag{6.328}$$

where $\theta = -v/f$.

GW signal:
In order to estimate the GW signal, we first need to compute the parameters α and β, cf. Eqs. (6.220) and (6.222)

$$\alpha = \frac{\Delta V}{\rho_{\rm rad}(T_{\rm nuc})} \simeq 2\times 10^{-4}\frac{100}{g_*}\left(\frac{M_X}{T_{\rm nuc}}\right)^4, \qquad (6.329)$$

$$\beta/H \simeq -4 + T\frac{d(S_3/T)}{dT}\Big|_{T_{\rm nuc}} = -4 + \frac{S_3/T}{\log\frac{M}{T}}\Big|_{T_{\rm nuc}}. \qquad (6.330)$$

Note the non-polynomial relation $\beta/H \propto 1/\log\alpha$ characteristic of nearly-conformal potential. In contrast, for polynomial potentials in the thin-wall limit, we expect $\beta/H \propto \alpha^{-2}$. We show α and β in Fig. 6.13.

6.4.2 *Strongly-Coupled Scenario: The Light-Dilaton Potential*

The dilaton potential:
In this section we suppose that spontaneous symmetry breaking arises from the condensation of a nearly-conformal sector, the broken symmetry being scale invariance. If the source of explicit breaking of scale invariance is small, the spontaneous breaking of scale invariance generates a pseudo Nambu-Goldstone boson, the dilaton which we parametrize as [258]

$$\chi(x) = f e^{\frac{\sigma(x)}{f}}, \qquad (6.331)$$

where $\sigma(x)$ transforms non-linearly under the scale transformation $\sigma(x) \to \sigma\lambda x + \log\lambda$. We are interested in the case where the dilaton χ is the lightest state of the nearly-conformal sector such that we can integrate out heavier degrees of freedom and describe the confined phase with the dilaton only [49, 87]. We suppose that the Higgs VEV is small compared to the one of the dilaton so that we neglect the field excursion in the Higgs direction. The only operator invariant under dilatations $x_\mu \to x_\mu/\lambda$, $\chi \to \lambda\chi$ is χ^4 and the dilaton has vanishing vev. Spontaneous breaking of scale invariance is only possible after adding an explicit source of scale invariance breaking [259–264]. So we add a slightly relevant operator

$$\epsilon\,\mathcal{O}_\epsilon, \qquad (6.332)$$

with scaling dimension $d = 4 + \gamma_\epsilon \lesssim 4$ and the potential for the dilaton becomes

$$V_\chi^{T=0}(\chi) = c_\chi g_\chi^2 \chi^4 - \epsilon(\chi)\chi^4. \qquad (6.333)$$

where g_χ is the typical coupling constant of the strong sector and c_χ is an order 1 constant. The source of explicit breaking of scale invariance in Eq. (6.332) leads to the running of the coupling constant ϵ

$$\frac{\partial \epsilon}{\partial \log \mu} \simeq \gamma_\epsilon \epsilon \tag{6.334}$$

where μ is the renormalization scale. This yields a non-zero vev f for the dilaton χ

$$\left.\frac{dV}{d\chi}\right|_{\chi=f} = 0 \quad \rightarrow \quad \epsilon(\chi) = \frac{c_\chi g_\chi^2}{1+\gamma_\epsilon/4}\left(\frac{\chi}{f}\right)^{\gamma_\epsilon}, \tag{6.335}$$

and also implies a none-vanishing mass for the dilaton m_χ whose smallness is tied to the smallness of the anomalous dimension γ_ϵ

$$\left.\frac{d^2V}{d\sigma^2}\right|_{\sigma=0} = m_\chi^2 \quad \rightarrow \quad \gamma_\epsilon \simeq -\frac{1}{4c_\chi}\frac{m_\chi^2}{g_\chi^2 f^2}. \tag{6.336}$$

The dilaton potential reduces to [49, 87]

$$V_\chi^{\mathrm{T=0}}(\chi) = c_\chi\, g_\chi^2\, \chi^4 \left(1 - \frac{1}{1+\gamma_\epsilon/4}\left(\frac{\chi}{f}\right)^{\gamma_\epsilon}\right), \tag{6.337}$$

with

$$\gamma_\epsilon \simeq -\frac{1}{4}\frac{y_\chi^2}{c_\chi\, g_\chi^2}, \tag{6.338}$$

and $y_\chi \equiv m_\chi/f$. From Eq. (6.338), the conformal limit $|\gamma_\epsilon| \ll 1$ coincides with the light dilaton limit $m_\chi/f \ll 1$. Later in Eq. (6.352) we will deduce from holography that $g_\chi = 4\pi/N$. Hence, at fixed $|\gamma_\epsilon|$ the light dilaton limit coincides with the large N limit.

Note that in the conformal limit $|\gamma_\epsilon| \ll 1$, the dilaton potential at zero-temperature reduces to the Coleman-Weinberg potential in Eq. (6.293)

$$V_\chi^{\mathrm{T=0}}(\chi) \overset{\gamma_\epsilon \ll 1}{=} -\gamma_\epsilon\, c_\chi\, g_\chi^2\, \chi^4 \log\left(\frac{\chi}{f}\right) \simeq \frac{m_\chi^2}{f^2}\frac{\chi^4}{4}\log\left(\frac{\chi}{f}\right). \tag{6.339}$$

However, the different nature of the thermal corrections between the weakly-coupled and the strongly-coupled scenario - presence of the thermal mass in T^2 at the tunneling point in the former case and absence of the finite-temperature contributions at the tunneling point in the latter case - make the bounce action behaving differently, see below.

Non-canonical kinetic terms:
Note that dual theories in 5D motivate the presence of non-canonical kinetic terms, cf. Eq. (6.348)

$$\mathscr{L}_{\rm K} = \frac{\kappa^2}{2}\partial_\mu\chi\partial^\mu\chi. \tag{6.340}$$

After rescaling

$$\frac{\kappa^2}{2}\partial_\mu\chi\partial^\mu\chi \xrightarrow{\chi\to\chi/\kappa} \frac{1}{2}\partial_\mu\chi\partial^\mu\chi, \tag{6.341}$$

the dilaton potential becomes

$$V_\chi^{\rm T=0}(\chi) = \frac{c_\chi\, g_\chi^2}{\kappa^4}\chi^4\left(1 - \frac{1}{1+\gamma_\epsilon/4}\left(\frac{\chi}{\kappa\, f}\right)^{\gamma_\epsilon}\right), \tag{6.342}$$

with

$$\gamma_\epsilon \simeq -\frac{\kappa^2}{4}\frac{y_\chi^2}{c_\chi\, g_\chi^2}, \tag{6.343}$$

and $y_\chi \equiv m_\chi/f$. In the rest of the chapter, we keep $\kappa = 1$, and we refer to [49] for the study of $\kappa \neq 1$.

Holography—the confined phase:
The strongly-coupled 4D theory explicited above can be mapped to a weakly-coupled 5D theory with AdS_5 metric (see the AdS-CFT correspondence pioneering articles [265, 266] and some reviews about holography [267–269])

$$ds^2 = e^{-2ky}\eta_{\mu\nu}dx^\mu dx^\nu + dy^2, \tag{6.344}$$

$$= \frac{L^2}{z^2}\left(\eta_{\mu\nu}dx^\mu dx^\nu + dz^2\right), \tag{6.345}$$

where $k = 1/L$ is the AdS curvature and $z = L\, e^{ky}$ is the coordinate along the fifth dimension. The AdS/CFT correspondence relate the AdS curvature in Planck unit to the rank of the 4D gauge group

$$(M_5 L)^3 = \frac{N^2}{16\pi^2}. \tag{6.346}$$

We consider the Randall-Sundrum scenario [270, 271], which is a slice of AdS_5 bounded by two branes, a UV-brane located at $y = 0$ and a IR-brane located at $y_{\rm IR}$. The use of the RS scenario for solving the hierarchy problem, is discussed in Sect. 2.3.1. The position of the IR brane along the 5th dimension $y_{\rm IR}$ defines the radion field

$$\mu = \frac{1}{z_{\rm IR}} = \frac{e^{-k\, y_{\rm IR}}}{L}, \tag{6.347}$$

whose kinetic term is

$$\mathscr{L}_K = 12(M_5 L)^3 (\partial\mu)^2 = 12\frac{N^2}{16\pi^2}(\partial\mu)^2. \tag{6.348}$$

Upon adding a field ϕ_{GW} of mass m in the 5D bulk, the radion field receives a stabilizing potential (Goldberger-Wise mechanism [272])

$$V_{GW}(\mu) = v_{IR}^2\,\mu^4\left[(4+2\gamma_\epsilon)\left(1-\frac{v_{UV}}{v_{IR}}\left(\frac{\mu}{\mu_0}\right)^{\gamma_\epsilon}\right)^2+\delta\right], \tag{6.349}$$

with

$$\gamma_\epsilon = \sqrt{4+m^2/k^2}-2 \simeq m^2/k^2, \tag{6.350}$$

and where μ_0 is the UV scale, δ is a dimensionless parameter, v_{UV} and v_{IR} are the vevs in Planck unit of the Goldberger-Wise field ϕ_{GW} on the Planck and IR branes.

The quantity γ_ϵ in Eq. (6.350), related to the mass m of the light 5D field, is the holographic dual of the anomalous dimension $\gamma_\epsilon = d-4$ of the slightly relevant 4D scalar operator in Eq. (6.332). The 5D radion potential in Eq. (6.349) is the holographic dual of the 4D light-dilaton potential in Eq. (6.337). Note however that the 5D and 4D potentials do not exactly coincide since in addition to the term $(\chi/f)^{\gamma_\epsilon}$, the 5D potential contains a term $(\chi/f)^{2\gamma_\epsilon}$, which can generate a barrier at zero-temperature.

Upon matching the non-canonical kinetic terms in 5D and 4D

$$\mathscr{L}_{K,\,5D} = 12\frac{N^2}{16\pi^2}(\partial\mu)^2 \quad \text{and} \quad \mathscr{L}_{K,\,4D} = \frac{\kappa^2}{2}\frac{1}{g_\chi^2}\partial_\mu\chi\partial^\mu\chi, \tag{6.351}$$

we obtain

$$\kappa = \sqrt{24} \quad \text{and} \quad g_\chi = \frac{4\pi}{N}. \tag{6.352}$$

The interaction coupling constant scales like $g_\chi \propto 1/N$, which corresponds to the glueball scaling at large N [273]. Finally, the AdS/CFT dictionary also identifies the 5D Kaluza-Klein resonances to the bound-states of the 4D theory (e.g. [268]) whose masses are

$$m_{kk} \simeq (n+\frac{1}{4})\,\pi\mu, \qquad n = 1,\,2,\,\cdots \tag{6.353}$$

and which after rescaling of the kinetic terms $\frac{\kappa^2}{g_\chi^2}\frac{1}{2}(\partial\mu)^2 \to \frac{1}{2}(\partial\chi)^2$ become

$$m_{kk} \simeq (n+\frac{1}{4})\,\frac{\pi\,g_\chi}{\kappa}\chi, \tag{6.354}$$

in conformity with the rescaling of the dilaton mass in Eq. (6.343).

Holography—the deconfined phase:
In the 5D perspective, at high enough temperature the IR brane is replaced by a black-hole horizon at position z_h. The corresponding metric called AdS_5-Schwarzschild is solution of the Einstein equation with compact Euclidean time [77]

$$ds^2 = -\frac{L^2}{z^2}\left(-f(z)dt^2 + dx_i^2 + \frac{1}{f(z)}dz^2\right), \qquad f(z) = 1 - \left(\frac{z}{z_h}\right)^4. \tag{6.355}$$

and its free energy density is [77]

$$F_{\rm BH}(T_{\rm BH}) = 2\pi^4 (M_5 L)^3 \left(3T_{\rm BH}^4 - 4T\, T_{\rm BH}^3\right) = \frac{\pi^2}{8} N^2 \left(3T_{\rm BH}^4 - 4T\, T_{\rm BH}^3\right), \tag{6.356}$$

where $T_{\rm BH} = 1/\pi z_h$ is the black-hole temperature. The free energy density of the deconfined phase is minimized when $T = T_{\rm BH}$ due to the existence of a conical singularity when $T \neq T_{\rm BH}$. Note that the deconfined phase and the confined phase coincide when both the black hole horizon and the IR-brane are sent to infinity, which corresponds to $T_{\rm BH} = 0 = \mu$. Hence, the total free energy density can be constructing by analytically extending the zero-temperature potential in Eq. (6.349) to negative value of the field μ using Eq. (6.356) with $\mu = -T_{\rm BH}$ [77]. Using holography, the finite-temperature corrections in the 4D theory can be incorporated by extending the zero-temperature potential in Eq. (6.337) to negative value of the field χ using Eq. (6.356) with $\chi = -T_{\rm BH}$.

We refer to [77] for the first study of the phase transition in Randall-Sundrum scenarios and [34, 78–85, 88, 274] for later works on holographic phase transitions. For studies of the dual 4D confinement, we call attention to [86–91].

The deconfined phase in 4D:
We now describe another approach for modelling the free energy between the deconfined phase and the confine phase, borrowed from [78, 87], which relies on the usual thermal corrections in 4D QFT. In the 4D perspective, the effective potential receives finite-temperature corrections generated by the particles in the plasma.

1. When $T < g_\chi \chi$, the finite-temperature effects are the standard SM thermal corrections, which we neglect.
2. When $T > g_\chi \chi$, we expect the existence of heavier bound states as well as eventually their deconfined constituents and the EFT which only describes the SM, the Higgs and the dilaton, breaks done.

We expect the free energy density of the CFT phase to scale like

$$F_{\rm CFT}(\chi = 0) \simeq -b\, N^2\, T^4. \tag{6.357}$$

Instead of looking for a finer description of the strong sector which would be very model dependent, we assume that we can approximate the deconfined phase by a $\mathcal{N} = 4\, SU(N)$ large N super-YM dual to an AdS-Schwarzschild space-time where in that particular case, we have,[30,31] cf. Eq. (6.356)

$$b = \frac{\pi^2}{8}. \tag{6.358}$$

Without a precise UV description, we are unable to determine the potential for $0 < \chi < T/g_\chi$, so we assume that the dilaton still exists, that its zero-temperature potential still holds, and that it receives thermal corrections, coming from the N CFT bosons [78, 87]

$$\Delta V_T^{1-\text{loop}}(\chi) = \sum_{\text{CFT bosons}} \frac{nT^4}{2\pi^2} J_B\left(\frac{m_{\text{CFT}}^2}{T^2}\right) \tag{6.359}$$

with mass

$$m_{\text{CFT}} = g_\chi \chi, \qquad \text{with} \;\; g_\chi = 4\pi/N. \tag{6.360}$$

In order to recover the free energy Eq. (6.357) in the deconfined phase, we set

$$\sum_{\text{CFT bosons}} n = \frac{45\,N^2}{4}. \tag{6.361}$$

Finally the total potential for the dilaton is

$$V_{\text{tot}}(\chi) = V_\chi(\chi) + \Delta V_T^{1-\text{loop}}(\chi) \tag{6.362}$$

where $V_\chi(\chi)$ and $\Delta V_T^{1-\text{loop}}(\chi)$ are given by Eqs. (6.342) and (6.359). The zero-temperature energy difference between the false vacuum and the true vacuum reads

$$\Delta V \simeq \frac{m_\chi^2 f^2}{16}. \tag{6.363}$$

According to Eq. (6.18), the high-temperature expansion of Eq. (6.362) reads

$$V_{\text{tot}}(\chi) \simeq -\frac{\pi^2}{8} N^2\, T^4 + 15\pi^2 T^2 \chi^2 + \frac{m_\chi^2}{f^2}\frac{\chi^4}{4} \log\left(\frac{\chi}{f}\right) + \cdots, \tag{6.364}$$

[30] Note that the free-energy of an interacting gluons gas is lower than the free-energy of a none-interacting gluons gas, $F = -\frac{\pi^2}{90} 2N^2 T^4$, by a factor 5, in conformity with the intuition that switching on the strong interaction would stabilize the phase.

[31] We neglect the number of degrees of freedom g_{SM} of the SM and of the techni-quarks g_q compared to the effective number of the techni-gluons $\frac{45}{4}N^2$, namely, $-\frac{\pi^2}{90}(g_{\text{SM}} + g_q + \frac{45}{4}N^2)T^4 \simeq -\frac{\pi^2}{8}N^2 T^4$.

where we have assumed the conformal limit $|\gamma_\epsilon| \ll 1$ in Eq. (6.339) and used Eq. (6.360). From Eq. (6.364), we can see that in the large N limit, the large number of degrees of freedom in the deconfined phase leads to a large negative free energy at $\chi = 0$. However, the $1/N$ scaling of the strong coupling g_χ implies a smaller contribution to the thermal mass $m_{\rm CFT}$ in Eq. (6.360), which then becomes negligible. This is an important difference with respect to the weakly-coupled Coleman-Weinberg scenario in Eq. (6.303) for which the thermal mass plays a significant role in the bounce action. While the bounce action of the weakly-coupled scenario was dominated by O_3, we will now see that in the large N strongly-coupled scenario,[32] the bounce action can be dominated by O_4.

Bounce action:
We numerically checked that the bounce action in the small T and small $|\gamma_\epsilon|$ limit, is insensitive to the shape of the free energy between the thermal CFT phase and the confined phase. This is due to the thermal corrections being absent at the tunneling point which makes the bounce action only depend on the depth of the thermal CFT free energy and not on its exact shape.[33] In other world, the bounce action can simply be determined from the boundary conditions [81, 89, 90]

$$\chi'(0) = 0, \qquad \chi'(\chi \to 0) = \sqrt{b}\, T^2. \tag{6.365}$$

However, we will not use these boundary conditions but instead we will rely on the thick-wall formula.[34] For O_3-bounce, Eq. (6.73) with $V(\phi_{\rm FV}) = -bN^2T^4$ and Eq. (6.339) reads[35]

$$S_3 = c_* \, \underset{\chi_*}{\rm Min} \frac{4\pi}{3} \frac{\chi_*^3}{\sqrt{-2\left(b\, N^2\, T^4 - \lambda_*\, \chi_*^4\right)}}, \tag{6.366}$$

with

$$\lambda_* = |\gamma_\epsilon|\; c_\chi\, g_\chi^2 \log \frac{f}{\chi_*}, \tag{6.367}$$

which gives

$$S_3/T = c_* \, \frac{2\pi}{3^{1/4}} \frac{(b\, N^2)^{1/4}}{\lambda_*^{3/4}}, \tag{6.368}$$

---

[32] We recall that due to Eq. (6.338), the large N limit coincides with the light-dilaton limit $m_\chi/f \ll 1$ and with the conformal limit $|\gamma_\epsilon| \ll 1$.

[33] See also [89, 90] where the authors show that the contribution to the bounce action coming from the free energy in the thermal CFT phase is negligible.

[34] For small $|\gamma_\epsilon| \ll 1$, the barrier is very small compared to the potential energy difference such that the bounce can be estimated with the thick-wall formula.

[35] We assume that the kinetic terms are canonically normalized ($\kappa = 1$).

$$\chi_* = \left(\frac{3\,b\,N^2}{\lambda_*}\right)^{1/4} T. \tag{6.369}$$

c_* is a coefficient which we add and which we fit on the numerical solution. We set

$$c_* \simeq 2, \tag{6.370}$$

for both the O_3- and O_4- bounce actions. The thick-wall formula for the O_4-bounce in Eq. (6.73), reads

$$S_4 = c_*\,\underset{\chi_*}{\text{Min}} \frac{\pi^2}{2} \frac{\chi_*^4}{-\left(b\,N^2\,T^4 - \lambda_*\,\chi_*^4\right)}, \tag{6.371}$$

which gives

$$S_4 = c_*\,\frac{\pi^2}{2}\frac{1}{\lambda_*}, \tag{6.372}$$

$$\chi_* = \frac{T}{T_c}\,f. \tag{6.373}$$

where T_c is the critical temperature, defined when the two minima of the free energy are equal

$$b\,N^2\,T_c^4 = \frac{m_\chi^2\,f^2}{16}, \quad \rightarrow \quad T_c = \left(\frac{m_\chi^2\,f^2}{16\,b\,N^2}\right)^{1/4} = \left(\frac{|\gamma_\epsilon|\,c_\chi\,g_\chi^2}{4\,b\,N^2}\right)^{1/4} f \tag{6.374}$$

We conclude that

$$\frac{S_4}{S_3/T} = \frac{T}{b\,N^2\,\lambda_*} = \frac{f^2}{m_\chi^2}\frac{4}{b\,N^2}\frac{1}{\log\frac{T_c}{T}}. \tag{6.375}$$

Hence, at low temperature we expect the O_4 bounce to overtake over the O_3 bounce, see top left panel of Fig. 6.14. We refer to Appendix B of [49] for more details on the computation of the bounce action in the light-dilaton model.

Nucleation temperature:
The nucleation happens when the tunneling rate in Eq. (6.43) becomes comparable to the Hubble expansion rate per Hubble volume. For O_4-dominance, the nucleation temperature is solution of

$$S_4(T_{\text{nuc}}) \simeq 4\ln\frac{R_0^{-1}}{H(T_{\text{nuc}})} + \frac{1}{2}\ln\frac{S_4}{2\pi} \tag{6.376}$$

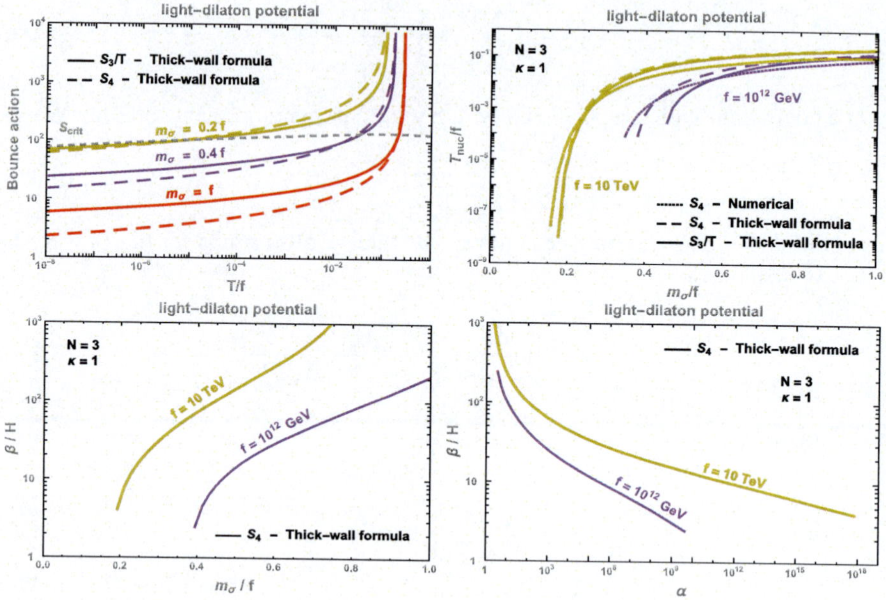

Fig. 6.14 **Top left:** We compare the O_3- and O_4- bounce action computed with the thick-wall formula, cf. Eqs. (6.368) and (6.372), to the O_4-bounce action computed with our own shooting method. **Top right:** Nucleation temperature computed with the thick-wall formula for S_4. The end points correspond to the last possible nucleation before eternal inflation, cf. Eq. (6.381). **Bottom:** GW parameters α and β

where R_0 is the bubble radius which according to the thick-wall formula in Eq. (6.72), reads

$$R_0 = \left(\frac{f/m_\chi}{\sqrt{b}\, N \log \frac{T_c}{T}} \right)^{1/2} T^{-1} \sim T^{-1}. \tag{6.377}$$

The Hubble parameter is, cf. Eqs. (6.361) and (6.363)

$$H^2(T) = H_\Lambda^2 + H_{\rm rad}^2 = \frac{m_\chi^2 f^2/16}{3M_{\rm pl}^2} + \frac{\pi^2 g_* T^4}{90 M_{\rm pl}^2}, \quad g_* = 106.75 + \frac{45\, N^2}{4}. \tag{6.378}$$

From rewriting the S_4 bounce action in Eq. (6.372) as

$$S_4 = \frac{A}{\log T_c/T} \qquad \text{with} \quad A = c_* \, \frac{\pi^2}{2} \frac{1}{|\gamma_\epsilon| \; c_\chi \, g_\chi^2} = c_* \, 2\pi^2 \frac{f^2}{m_\chi^2}. \tag{6.379}$$

From injecting it into Eq. (6.376), we obtain

$$T_{\text{nuc}} = \sqrt{H_\Lambda\, T_c}\, \left(\frac{2\pi}{S_4}\right)^{1/4} \exp\left(\frac{1}{2}\sqrt{-A + \left(\ln\frac{T_c}{H_\Lambda} + \frac{1}{2}\ln\frac{S_4}{2\pi}\right)^2}\right). \tag{6.380}$$

Neglecting the $S_4/2\pi$ terms, we conclude that there is no nucleation solution when

$$A \gtrsim \ln\frac{T_c}{H_\Lambda}, \tag{6.381}$$

and that the minimal nucleation temperature is

$$T_{\text{nuc}}^{\text{min}} \simeq \sqrt{H_\Lambda\, T_c} \simeq 0.1\left(\frac{m_\chi}{f}\right)^{3/4}\left(\frac{f}{M_{\text{pl}}}\right)^{1/2} f. \tag{6.382}$$

In top right panel of Fig. 6.14, we show the nucleation temperature computed with the thick-wall formula. Note that from inverting Eq. (6.379), we can write

$$\left(\frac{T_{\text{nuc}}}{T_c}\right)^4 \simeq \text{Exp}\left(-c_*\,\frac{2\,\pi^2}{S_{\text{crit}}}\,\frac{1}{\kappa^2 y_\chi^2}\right), \tag{6.383}$$

where $y_\chi \equiv m_\chi/f$. The advantage of Eq. (6.383) is to make explicit that the nucleation temperature is exponentially suppressed for small dilaton mass m_χ, or equivalently for small anomalous coupling $|\gamma_\epsilon|$. Note also that from using Eq. (6.343) and $g_\chi = 4\pi/N$, we can rewrite Eq. (6.383) as

$$\left(\frac{T_{\text{nuc}}}{T_c}\right)^4 \simeq \text{Exp}\left(-c_*\,\frac{N^2}{32\,c_\chi\,|\gamma_\epsilon|\,S_{\text{crit}}}\right), \tag{6.384}$$

and we recover that at fixed γ_ϵ, the nucleation temperature scales like $T_{\text{nuc}}/f \propto e^{-N^2}$ [88].

Catalysis by QCD effects:

With Eq. (6.381), we have concluded that there is no nucleation solution for too small dilaton mass m_χ, such that the universe gets stuck in the false vacuum and inflates eternally. We know discuss the possibility, first pointed out by [85], that the QCD PT triggers the light-dilaton or holographic PT. In 4D, it comes from the possibility for the gluon condensate to generate a potential for the dilaton. If the CFT couples to the $SU(3)_c$ of the SM, we expect it to modify the QCD beta function [85, 88]

$$\frac{dg_3}{d\ln\mu} = \beta_3 = \frac{g_3^2}{16\pi^2}\left(-\frac{11}{3}N_c + \frac{2}{3}N_f + \alpha N\right). \tag{6.385}$$

Hence, in the presence of the deconfined CFT phase, we expect the QCD scale Λ_{QCD} to be modified as follows [85, 88]

$$\Lambda_{QCD}^{dec} = \Lambda_{QCD} \left(\frac{\Lambda_{QCD}}{f}\right)^{\frac{n}{1-n}}, \quad \text{with} \quad n = \frac{\beta_3(\mu < f) - \beta_3(\mu > f)}{\beta_3(\mu < f)}, \tag{6.386}$$

where $\beta_3(\mu < f)$ and $\beta_3(\mu > f)$ are the QCD beta functions without and with the αN correction, respectively. Note that for $0 < n < 1$, we have $\Lambda_{QCD}^{dec} < \Lambda_{QCD}$. Additionally, when the QCD phase transition takes place, we expect it to generate a potential for the dilaton with a negative sign

$$V_{QCD}(\chi) \simeq -\Lambda_{QCD}(\chi)^4 \quad \text{with} \quad \Lambda_{QCD} \sim \chi^n, \tag{6.387}$$

Note that this provides an elegant solution to the graceful exit problem of eternal inflation. See [230, 274] for other applications of the same mechanism.

Appendices

6.A Sensitivity Curves of GW Detectors

6.A.1 The Signal-to-Noise Ratio

The total output of a detector is given by the GW signal plus the noise, $h(t) + n(t)$ where the level of noise $n(t)$ is measured by its noise spectral density $S_n(f)$ [13].

$$\langle \tilde{n}^*(f)\tilde{n}(f') \rangle \equiv \delta(f - f')\, S_n(f). \tag{6.388}$$

We define the detector sensitivity $\Omega_{sens}(f)$ as the magnitude of the SGWB energy density which would mimick the noise spectral density $S_n(f)$

$$\Omega_{sens}(f) = \frac{2\pi^2}{3H_0^2} f^3\, S_n(f). \tag{6.389}$$

The capability of an interferometer to detect a SGWB of energy density $\Omega_{GW}(f)$ after an observation time T is measured by the signal-to-noise ratio (SNR) [275]

$$\text{SNR} = \sqrt{T \int_{f_{min}}^{f_{max}} df \left[\frac{\Omega_{GW}(f)}{\Omega_{sens}(f)}\right]^2}. \tag{6.390}$$

6.A.2 *The Power-Law Integrated Sensitivity Curve*

Assuming a power law spectrum

$$\Omega_{\rm GW}(f) = \Omega_\beta \left(\frac{f}{f_{\rm ref}}\right)^\beta , \tag{6.391}$$

with spectral index β, amplitude Ω_β and reference frequency $f_{\rm ref}$, we deduce from Eq. (6.390) the amplitude Ω_β needed to reach a given SNR after a given observation time T

$$\Omega_\beta = \frac{\rm SNR}{\sqrt{T}} \left(\int_{f_{\rm min}}^{f_{\rm max}} df \left[\frac{h^2}{h^2\Omega_{\rm sens}(f)} \left(\frac{f}{f_{\rm ref}}\right)^\beta \right]^2 \right)^{-1/2} , \tag{6.392}$$

which upon re-injecting into Eq. (6.391) gives

$$h^2\Omega_{\rm GW}(f) = f^\beta \frac{\rm SNR}{\sqrt{T}} \left(\int_{f_{\rm min}}^{f_{\rm max}} df \left[\frac{f^\beta}{h^2\Omega_{\rm sens}(f)} \right]^2 \right)^{-1/2} . \tag{6.393}$$

For a given pair (SNR, T), one obtains a series in β of power-law integrated curves. One defines the power-law integrated sensitivity curve $\Omega_{PI}(f)$ as the envelope of those functions [276]

$$\Omega_{PI}(f) \equiv \max_\beta \left[f^\beta \frac{\rm SNR}{\sqrt{T}} \left(\int_{f_{\rm min}}^{f_{\rm max}} df \left[\frac{f^\beta}{h^2\Omega_{\rm sens}(f)} \right]^2 \right)^{-1/2} \right] . \tag{6.394}$$

Any SGWB signal $\Omega_{\rm GW}(f)$ which lies above $\Omega_{PI}(f)$ would gives a signal to noise ratio > SNR after an observation time T.

6.A.3 *Results*

For the purpose of our study, we computed the power-law integrated sensitivity curve $\Omega_{PI}(f)$, starting from the noise spectral density in [201] for ET, [203] for CE and [277] for BBO/DECIGO. For pulsar timing arrays EPTA, NANOGrav and SKA, we directly took the sensitivity curves from [198]. The signal-to-noise ratio can be improved by using cross-correlation between multiple detectors, e.g. LIGO-Hanford, LIGO-Livingston, VIRGO but also KAGRA which may join the network at the end of run O3, which began on the 1st of April 2019, or LIGO-India which may be operational for run O5 [278]. We computed the SNR for LIGO from the expression [276]

$$\mathrm{SNR} = \left[2T \int_{f_{\min}}^{f_{\max}} df \frac{\Gamma^2(f) S_{\mathrm{h}}^2(f)}{S_n^1(f) S_n^2(f)} \right]^{1/2}, \tag{6.395}$$

where S_n^1 and S_n^2 are the noise spectral densities of the detectors in Hanford and in Livingston for the runs O2, O4 or O5 and $\Gamma(f)$ is the overlap function between the two LIGO detectors which we took from [279]. The GW power spectral density $S_h(f)$ is related to the GW energy density through

$$S_h(f) = \frac{3H_0^2}{2\pi^2} \frac{\Omega_{\mathrm{GW}}(f)}{f^3}. \tag{6.396}$$

We fixed the signal-to-noise ratio SNR $= 10$ and the observational time $T = 268$ days for LIGO O2, 1 year for LIGO O4 and O5, and 10 years for other sensitivity curves.

As this study was completed, Ref. [280] appeared, where the sensitivity curves may differ from us by a factor of order 1.

References

1. Kirzhnits D, Linde AD (1972) Macroscopic consequences of the weinberg model. Phys Lett B 42:471–474. https://doi.org/10.1016/0370-2693(72)90109-8
2. Kajantie K, Laine M, Rummukainen K, Shaposhnikov ME (1996) Is there a hot electroweak phase transition at m(H) larger or equal to m(W)? Phys Rev Lett 77:2887–2890. arXiv: hep-ph/9605288, https://doi.org/10.1103/PhysRevLett.77.2887
3. Rummukainen K, Tsypin M, Kajantie K, Laine M, Shaposhnikov ME (1998) The Universality class of the electroweak theory. Nucl Phys B 532:283–314. arXiv: hep-lat/9805013, https://doi.org/10.1016/S0550-3213(98)00494-5
4. Csikor F, Fodor Z, Heitger J (1999) Endpoint of the hot electroweak phase transition. Phys Rev Lett 82:21–24. arXiv: hep-ph/9809291, https://doi.org/10.1103/PhysRevLett.82.21
5. Banks T, Rabinovici E (1979) Finite temperature behavior of the lattice abelian higgs model. Nucl Phys B 160:349–379. https://doi.org/10.1016/0550-3213(79)90064-6
6. Fradkin EH, Shenker SH (1979) Phase diagrams of lattice gauge theories with higgs fields. Phys Rev D 19:3682–3697. https://doi.org/10.1103/PhysRevD.19.3682
7. Stephanov M (2006) QCD phase diagram: an Overview. PoS LAT2006:024. arXiv: hep-lat/0701002, https://doi.org/10.22323/1.032.0024
8. Pisarski RD, Wilczek F (1984) Remarks on the chiral phase transition in chromodynamics. Phys Rev D29:338–341. https://doi.org/10.1103/PhysRevD.29.338
9. Witten E (1984) Cosmic separation of phases. Phys Rev D 30:272–285. https://doi.org/10.1103/PhysRevD.30.272
10. Hogan CJ (1986) Gravitational radiation from cosmological phase transitions. Mon Not R Astron Soc 218:629–636

11. Kamionkowski M, Kosowsky A, Turner MS (1994) Gravitational radiation from first order phase transitions. Phys Rev D49:2837–2851. arXiv: astro-ph/9310044, https://doi.org/10.1103/PhysRevD.49.2837
12. Caprini C et al (2016) Science with the space-based interferometer eLISA. II: Gravitational waves from cosmological phase transitions. JCAP 04:001. arXiv: 1512.06239, https://doi.org/10.1088/1475-7516/2016/04/001
13. Caprini C, Figueroa DG (2018) Cosmological backgrounds of gravitational waves. Class Quant Grav 35:163001. arXiv: 1801.04268, https://doi.org/10.1088/1361-6382/aac608
14. Caprini C et al (2020) Detecting gravitational waves from cosmological phase transitions with LISA: an update. JCAP 03:024. arXiv: 1910.13125, https://doi.org/10.1088/1475-7516/2020/03/024
15. Vachaspati T (1991) Magnetic fields from cosmological phase transitions. Phys Lett B 265:258–261. https://doi.org/10.1016/0370-2693(91)90051-Q
16. Zhang Y, Vachaspati T, Ferrer F (2019) Magnetic field production at a first-order electroweak phase transition. Phys Rev D 100:083006. arXiv: 1902.02751, https://doi.org/10.1103/PhysRevD.100.083006
17. Ellis J, Fairbairn M, Lewicki M, Vaskonen V, Wickens A (2019) Intergalactic magnetic fields from first-order phase transitions. JCAP 09:019. arXiv: 1907.04315, https://doi.org/10.1088/1475-7516/2019/09/019
18. Kibble TWB (1976) Topology of cosmic domains and strings. J Phys A9:1387–1398. https://doi.org/10.1088/0305-4470/9/8/029
19. Kibble T, Vilenkin A (1995) Phase equilibration in bubble collisions. Phys Rev D 52:679–688. arXiv: hep-ph/9501266, https://doi.org/10.1103/PhysRevD.52.679
20. Borrill J, Kibble T, Vachaspati T, Vilenkin A (1995) Defect production in slow first order phase transitions. Phys Rev D 52:1934–1943. arXiv: hep-ph/9503223, https://doi.org/10.1103/PhysRevD.52.1934
21. Kodama H, Sasaki M, Sato K (1979) Abundance of primordial holes produced by cosmological first order phase. Trans Prog Theor Phys 68. https://doi.org/10.1143/PTP.68.1979
22. Hawking SW, Moss IG, Stewart JM (1982) Bubble collisions in the very early universe. Phys Rev D 26:2681. https://doi.org/10.1103/PhysRevD.26.2681
23. Jung TH, Okui T, Primordial black holes from bubble collisions during a first-order phase transition. arXiv:2110.04271
24. Garriga J, Vilenkin A, Zhang J (2016) Black holes and the multiverse. JCAP 02:064. arXiv: 1512.01819, https://doi.org/10.1088/1475-7516/2016/02/064
25. Deng H, Garriga J, Vilenkin A (2017) Primordial black hole and wormhole formation by domain walls. JCAP 04:050. arXiv: 1612.03753, https://doi.org/10.1088/1475-7516/2017/04/050
26. Deng H, Vilenkin A (2017) Primordial black hole formation by vacuum bubbles. JCAP 12:044. ArXiv:1710.02865
27. Maeso DN, Marzola L, Raidal M, Vaskonen V, Veermäe H, Primordial black holes from spectator field bubbles. arXiv:2112.01505
28. Gross C, Landini G, Strumia A, Teresi D (2021) Dark matter as dark dwarfs and other macroscopic objects: multiverse relics? JHEP 09:033. arXiv: 2105.02840, https://doi.org/10.1007/JHEP09(2021)033
29. Baker MJ, Breitbach M, Kopp J, Mittnacht L, Primordial Black Holes from First-Order Cosmological Phase Transitions. arXiv:2105.07481
30. Baker MJ, Breitbach M, Kopp J, Mittnacht L, Detailed calculation of primordial black hole formation during first-order cosmological phase transitions. arXiv:2110.00005
31. Kawana K, Xie K-P (2022) Primordial black holes from a cosmic phase transition: the collapse of fermi-balls. Phys. Lett. B 824:136791. arXiv: 2106.00111, https://doi.org/10.1016/j.physletb.2021.136791
32. Liu J, Bian L, Cai R-G, Guo Z-K, Wang S-J, Primordial black hole production during first-order phase transitions. arXiv:2106.05637

33. Hashino K, Kanemura S, Takahashi T, Primordial black holes as a probe of strongly first-order electroweak phase transition. arXiv:2111.13099
34. Konstandin T, Servant G (2011) Cosmological consequences of nearly conformal dynamics at the TeV scale. JCAP 12:009. arXiv: 1104.4791, https://doi.org/10.1088/1475-7516/2011/12/009
35. Falkowski A, No JM (2013) Non-thermal dark matter production from the electroweak phase transition: multi-TeV WIMPs and "Baby-Zillas". JHEP 02:034. arXiv: 1211.5615, https://doi.org/10.1007/JHEP02(2013)034
36. Hambye T, Strumia A (2013) Dynamical generation of the weak and Dark Matter scale. Phys Rev D 88:055022. arXiv: 1306.2329, https://doi.org/10.1103/PhysRevD.88.055022
37. Baker MJ, Kopp J (2017) Dark matter decay between phase transitions at the weak scale. Phys Rev Lett 119:061801. arXiv: 1608.07578, https://doi.org/10.1103/PhysRevLett.119.061801
38. Baker MJ, Breitbach M, Kopp J, Mittnacht L (2018) Dynamic freeze-in: impact of thermal masses and cosmological phase transitions on dark matter production. JHEP 03:114. arXiv: 1712.03962, https://doi.org/10.1007/JHEP03(2018)114
39. Baker MJ, Mittnacht L (2019) Variations on the Vev Flip-Flop: instantaneous freeze-out and decaying dark matter. JHEP 05:070. arXiv: 1811.03101, https://doi.org/10.1007/JHEP05(2019)070
40. Hambye T, Strumia A, Teresi D (2018) Super-cool dark matter. JHEP 08:188. arXiv: 1805.01473, https://doi.org/10.1007/JHEP08(2018)188
41. Bai Y, Long AJ, Lu S (2019) Dark quark nuggets. Phys Rev D 99:055047. arXiv: 1810.04360, https://doi.org/10.1103/PhysRevD.99.055047
42. Heurtier L, Partouche H (2020) Spontaneous freeze out of dark matter from an early thermal phase transition. Phys Rev D 101:043527. arXiv: 1912.02828, https://doi.org/10.1103/PhysRevD.101.043527
43. Baker MJ, Kopp J, Long AJ, Filtered Dark Matter at a First Order Phase Transition. arXiv:1912.02830
44. Azatov A, Vanvlasselaer M, Yin W, Dark Matter production from relativistic bubble walls. arXiv:2101.05721
45. Baldes I, Gouttenoire Y, Sala F (2021) String fragmentation in supercooled confinement and implications for dark matter. JHEP 04:278. arXiv: 2007.08440, https://doi.org/10.1007/JHEP04(2021)278
46. Hong J-P, Jung S, Xie K-P (2020) Fermi-ball dark matter from a first-order phase transition. Phys Rev D 102:075028. arXiv: 2008.04430, https://doi.org/10.1103/PhysRevD.102.075028
47. Asadi P, Kramer ED, Kuflik E, Ridgway GW, Slatyer TR, Smirnov J, Accidentally asymmetric dark matter. arXiv:2103.09822
48. Asadi P, Kramer ED, Kuflik E, Ridgway GW, Slatyer TR, Smirnov J, Thermal squeezeout of dark matter. arXiv:2103.09827
49. Baldes I, Gouttenoire Y, Sala F, Servant G, Supercool composite dark matter beyond 100 TeV. arXiv:2110.13926
50. Craig N, Levi N, Mariotti A, Redigolo D (2020) Ripples in spacetime from broken supersymmetry. JHEP 21:184. arXiv: 2011.13949, https://doi.org/10.1007/JHEP02(2021)184
51. Mazumdar A, White G (2019) Review of cosmic phase transitions: their significance and experimental signatures. Rep Prog Phys 82:076901. arXiv: 1811.01948, https://doi.org/10.1088/1361-6633/ab1f55
52. Hindmarsh, MB, Lüben M, Lumma J, Pauly M, Phase transitions in the early universe. arXiv:2008.09136
53. Grojean C, Servant G, Wells JD (2005) First-order electroweak phase transition in the standard model with a low cutoff. Phys Rev D 71:036001. arXiv: hep-ph/0407019, https://doi.org/10.1103/PhysRevD.71.036001
54. Bodeker D, Fromme L, Huber SJ, Seniuch M (2005) The Baryon asymmetry in the standard model with a low cut-off. JHEP 02:026. arXiv: hep-ph/0412366, https://doi.org/10.1088/1126-6708/2005/02/026

55. Delaunay C, Grojean C, Wells JD (2008) Dynamics of non-renormalizable electroweak symmetry breaking. JHEP 04:029. arXiv: 0711.2511, https://doi.org/10.1088/1126-6708/2008/04/029
56. Huang FP, Gu P-H, Yin P-F, Yu Z-H, Zhang X (2016) Testing the electroweak phase transition and electroweak baryogenesis at the LHC and a circular electron-positron collider. Phys Rev D 93:103515. arXiv: 1511.03969, https://doi.org/10.1103/PhysRevD.93.103515
57. Huang FP, Wan Y, Wang D-G, Cai Y-F, Zhang X (2016) Hearing the echoes of electroweak baryogenesis with gravitational wave detectors. Phys Rev D 94:041702. arXiv: 1601.01640, https://doi.org/10.1103/PhysRevD.94.041702
58. de Vries J, Postma M, van de Vis J, White G (2018) Electroweak baryogenesis and the standard model effective field theory. JHEP 01:089. arXiv: 1710.04061, https://doi.org/10.1007/JHEP01(2018)089
59. Chala M, Krause C, Nardini G (2018) Signals of the electroweak phase transition at colliders and gravitational wave observatories. JHEP 07:062. arXiv: 1802.02168, https://doi.org/10.1007/JHEP07(2018)062
60. Ellis J, Lewicki M, No JM (2019) On the maximal strength of a first-order electroweak phase transition and its gravitational wave signal. JCAP 04:003. arXiv: 1809.08242, https://doi.org/10.1088/1475-7516/2019/04/003
61. Ellis J, Lewicki M, No JM, Vaskonen V (2019) Gravitational wave energy budget in strongly supercooled phase transitions. JCAP 06:024. arXiv: 1903.09642, https://doi.org/10.1088/1475-7516/2019/06/024
62. Anderson GW, Hall LJ (1992) The electroweak phase transition and baryogenesis. Phys Rev D45:2685–2698. https://doi.org/10.1103/PhysRevD.45.2685
63. Choi J, Volkas R (1993) Real higgs singlet and the electroweak phase transition in the standard model. Phys Lett B 317:385–391. arXiv: hep-ph/9308234, https://doi.org/10.1016/0370-2693(93)91013-D
64. Espinosa J, Quiros M (1993) The electroweak phase transition with a singlet. Phys Lett B 305:98–105. arXiv: hep-ph/9301285, https://doi.org/10.1016/0370-2693(93)91111-Y
65. Profumo, S, Ramsey-Musolf MJ, Shaughnessy G (2007) Singlet Higgs phenomenology and the electroweak phase transition. JHEP 08:010. arXiv: 0705.2425, https://doi.org/10.1088/1126-6708/2007/08/010
66. Espinosa JR, Konstandin T, Riva F (2012) Strong electroweak phase transitions in the standard model with a singlet. Nucl Phys B 854:592–630. arXiv: 1107.5441, https://doi.org/10.1016/j.nuclphysb.2011.09.010
67. Beniwal A, Lewicki M, White M, Williams AG (2019) Gravitational waves and electroweak baryogenesis in a global study of the extended scalar singlet model. JHEP 02:183. arXiv: 1810.02380, https://doi.org/10.1007/JHEP02(2019)183
68. Branco G, Ferreira P, Lavoura L, Rebelo M, Sher M, Silva JP (2012) Theory and phenomenology of two-Higgs-doublet models. Phys Rep 516:1–102. arXiv: 1106.0034, https://doi.org/10.1016/j.physrep.2012.02.002
69. Basler P, Krause M, Muhlleitner M, Wittbrodt J, Wlotzka A (2017) Strong first order electroweak phase transition in the CP-conserving 2HDM revisited. JHEP 02:121. arXiv: 1612.04086, https://doi.org/10.1007/JHEP02(2017)121
70. Meissner KA, Nicolai H (2007) Conformal symmetry and the standard model. Phys Lett B 648:312–317. arXiv: hep-th/0612165, https://doi.org/10.1016/j.physletb.2007.03.023
71. Iso S, Okada N, Orikasa Y (2009) Classically conformal B^- L extended standard model. Phys Lett B 676:81–87. arXiv: 0902.4050, https://doi.org/10.1016/j.physletb.2009.04.046
72. Iso S, Okada N, rikasa Y (2009) The minimal B-L model naturally realized at TeV scale. Phys Rev D 80:115007. arXiv: 0909.0128, https://doi.org/10.1103/PhysRevD.80.115007
73. Iso S, Serpico PD, Shimada K (2017) QCD-electroweak first-order phase transition in a supercooled universe. Phys Rev Lett 119:141301. arXiv: 1704.04955, https://doi.org/10.1103/PhysRevLett.119.141301
74. Baldes I, Garcia-Cely C (2019) Strong gravitational radiation from a simple dark matter model. JHEP 05:190. arXiv: 1809.01198, https://doi.org/10.1007/JHEP05(2019)190

75. Marzo C, Marzola L, Vaskonen V (2019) Phase transition and vacuum stability in the classically conformal B–L model. Eur Phys J C 79:601. arXiv: 1811.11169, https://doi.org/10.1140/epjc/s10052-019-7076-x
76. Brdar V, Helmboldt AJ, Lindner M (2019) Strong supercooling as a consequence of renormalization group consistency. JHEP 12:158. arXiv: 1910.13460, https://doi.org/10.1007/JHEP12(2019)158
77. Creminelli P, Nicolis A, Rattazzi R (2002) Holography and the electroweak phase transition. JHEP 03:051. arXiv: hep-th/0107141, https://doi.org/10.1088/1126-6708/2002/03/051
78. Randall L, Servant G (2007) Gravitational waves from warped spacetime. JHEP 05:054. arXiv: hep-ph/0607158, https://doi.org/10.1088/1126-6708/2007/05/054
79. Nardini G, Quiros M, Wulzer A (2007) A confining strong first-order electroweak phase transition. JHEP 09:077. arXiv: 0706.3388, https://doi.org/10.1088/1126-6708/2007/09/077
80. Hassanain B, March-Russell J, Schvellinger M (2007) Warped deformed throats have faster (Electroweak) phase transitions. JHEP 10:089. arXiv: 0708.2060, https://doi.org/10.1088/1126-6708/2007/10/089
81. Konstandin T, Nardini G, Quiros M (2010) Gravitational backreaction effects on the holographic phase transition. Phys Rev D 82:083513. arXiv: 1007.1468, https://doi.org/10.1103/PhysRevD.82.083513
82. Bunk D, Hubisz J, Jain B (2018) A perturbative RS I cosmological phase transition. Eur Phys J C 78:78. arXiv: 1705.00001, https://doi.org/10.1140/epjc/s10052-018-5529-2
83. Dillon BM, El-Menoufi BK, Huber SJ, Manuel JP (2018) Rapid holographic phase transition with brane-localized curvature. Phys Rev D 98:086005. arXiv: 1708.02953, https://doi.org/10.1103/PhysRevD.98.086005
84. Megías E, Nardini G, Quirós M (2018) Cosmological phase transitions in warped space: gravitational waves and collider signatures. JHEP 09:095. arXiv: 1806.04877, https://doi.org/10.1007/JHEP09(2018)095
85. von Harling B, Servant G (2018) QCD-induced electroweak phase transition. JHEP 01:159. arXiv: 1711.11554, https://doi.org/10.1007/JHEP01(2018)159
86. Bruggisser S, Von Harling B, Matsedonskyi O, Servant G (2018) Baryon asymmetry from a composite higgs boson. Phys Rev Lett 121:131801. arXiv: 1803.08546, https://doi.org/10.1103/PhysRevLett.121.131801
87. Bruggisser S, Von Harling B, Matsedonskyi O, Servant G (2018) Electroweak phase transition and baryogenesis in composite higgs models. JHEP 12:099. ArXiv:1804.07314
88. Baratella P, Pomarol A, Rompineve F (2019) The supercooled universe. JHEP 03:100. arXiv: 1812.06996, https://doi.org/10.1007/JHEP03(2019)100
89. Agashe K, Du P, Ekhterachian M, Kumar S, Sundrum R, Cosmological phase transition of spontaneous confinement. arXiv:1910.06238
90. Delle Rose L, Panico G, Redi M, Tesi A (2020) Gravitational waves from supercool axions. JHEP 04:025. arXiv: 1912.06139, https://doi.org/10.1007/JHEP04(2020)025
91. Von Harling B, Pomarol A, Pujolas O, Rompineve F, Peccei-quinn phase transition at LIGO. arXiv:1912.07587
92. Schmitz K, LISA sensitivity to gravitational waves from sound waves. arXiv:2005.10789
93. Coleman SR, Weinberg EJ (1973) Radiative corrections as the origin of spontaneous symmetry breaking. Phys Rev D7:1888–1910. https://doi.org/10.1103/PhysRevD.7.1888
94. Dolan L, Jackiw R (1974) Symmetry behavior at finite temperature. Phys Rev D 9:3320–3341. https://doi.org/10.1103/PhysRevD.9.3320
95. Quiros M (1999) Finite temperature field theory and phase transitions. In: ICTP summer school in high-energy physics and cosmology, pp 187–259, 1. arXiv:hep-ph/9901312
96. Kapusta J, Gale C (2011) Finite-temperature field theory: Principles and applications. Cambridge monographs on mathematical physics. Cambridge University Press. https://doi.org/10.1017/CBO9780511535130
97. Bellac ML (2011) Thermal field theory. Cambridge monographs on mathematical physics. Cambridge University Press, p 3. https://doi.org/10.1017/CBO9780511721700

98. Nielsen N (1975) On the gauge dependence of spontaneous symmetry breaking in gauge theories. Nucl Phys B 101:173–188. https://doi.org/10.1016/0550-3213(75)90301-6
99. Fukuda R, Kugo T (1976) Gauge invariance in the effective action and potential. Phys Rev D 13:3469. https://doi.org/10.1103/PhysRevD.13.3469
100. Di Luzio L, Mihaila L (2014) On the gauge dependence of the standard model vacuum instability scale. JHEP 06:079. arXiv: 1404.7450, https://doi.org/10.1007/JHEP06(2014)079
101. Andreassen A, Frost W, Schwartz MD (2014) Consistent use of the standard model effective potential. Phys Rev Lett 113:241801. arXiv: 1408.0292, https://doi.org/10.1103/PhysRevLett.113.241801
102. Metaxas D, Weinberg EJ (1996) Gauge independence of the bubble nucleation rate in theories with radiative symmetry breaking. Phys Rev D 53:836–843. arXiv: hep-ph/9507381, https://doi.org/10.1103/PhysRevD.53.836
103. Plascencia AD, Tamarit C (2016) Convexity, gauge-dependence and tunneling rates. JHEP 10:099. arXiv: 1510.07613, https://doi.org/10.1007/JHEP10(2016)099
104. Espinosa JR, Garny M, Konstandin T, Riotto A (2017) Gauge-independent scales related to the standard model vacuum instability. Phys Rev D 95:056004. arXiv: 1608.06765, https://doi.org/10.1103/PhysRevD.95.056004
105. Li T, Zhou Y-F (2014) Strongly first order phase transition in the singlet fermionic dark matter model after LUX. JHEP 07:006. arXiv: 1402.3087, https://doi.org/10.1007/JHEP07(2014)006
106. Carrington ME (1992) The effective potential at finite temperature in the standard model. Phys Rev D45:2933–2944. https://doi.org/10.1103/PhysRevD.45.2933
107. Coleman SR (1977) The fate of the false vacuum. 1. Semiclassical theory. Phys Rev D 15:2929–2936. https://doi.org/10.1103/PhysRevD.16.1248
108. Callan J, Curtis G, Coleman SR (1977) The fate of the false vacuum. 2. First quantum corrections. Phys Rev D 16:1762–1768. https://doi.org/10.1103/PhysRevD.16.1762
109. Linde AD (1981) Fate of the false vacuum at finite temperature: theory and applications. Phys Lett B 100:37–40. https://doi.org/10.1016/0370-2693(81)90281-1
110. Linde AD (1983) Decay of the false vacuum at finite temperature. Nucl Phys B216:421. https://doi.org/10.1016/0550-3213(83)90293-6
111. Coleman SR, De Luccia F (1980) Gravitational effects on and of vacuum decay. Phys Rev D 21:3305. https://doi.org/10.1103/PhysRevD.21.3305
112. Coleman S (1985) Aspects of symmetry. Cambridge University Press, Cambridge, U.K. https://doi.org/10.1017/CBO9780511565045
113. Ellis JR, Linde AD, Sher M (1990) Vacuum stability, wormholes, cosmic rays and the cosmological bounds on m(t) and m(H). Phys Lett B 252:203–211. https://doi.org/10.1016/0370-2693(90)90862-Z
114. Linde AD (1992) Hard art of the universe creation (stochastic approach to tunneling and baby universe formation). Nucl Phys B 372:421–442. arXiv: hep-th/9110037, https://doi.org/10.1016/0550-3213(92)90326-7
115. Braden J, Johnson MC, Peiris HV, Pontzen A, Weinfurtner S (2019) New semiclassical picture of vacuum decay. Phys Rev Lett 123:031601. arXiv: 1806.06069, https://doi.org/10.1103/PhysRevLett.123.031601
116. Hertzberg MP, Yamada M (2019) Vacuum decay in real time and imaginary time formalisms. Phys Rev D 100:016011. arXiv: 1904.08565, https://doi.org/10.1103/PhysRevD.100.016011
117. Blanco-Pillado JJ, Deng H, Vilenkin A (2019) Flyover vacuum decay. JCAP 12:001. arXiv: 1906.09657, https://doi.org/10.1088/1475-7516/2019/12/001
118. Wainwright CL (2012) CosmoTransitions: computing cosmological phase transition temperatures and bubble profiles with multiple fields. Comput Phys Commun 183:2006–2013. arXiv: 1109.4189, https://doi.org/10.1016/j.cpc.2012.04.004
119. Masoumi A, Olum KD, Shlaer B (2017) Efficient numerical solution to vacuum decay with many fields. JCAP 01:051. arXiv: 1610.06594, https://doi.org/10.1088/1475-7516/2017/01/051

120. Athron P, Balázs C, Bardsley M, Fowlie A, Harries D, White G (2019) BubbleProfiler: finding the field profile and action for cosmological phase transitions. Comput Phys Commun 244:448–468. arXiv: 1901.03714, https://doi.org/10.1016/j.cpc.2019.05.017
121. Sato R, SimpleBounce : a simple package for the false vacuum decay. arXiv:1908.10868
122. Guada V, Maiezza A, Nemevšek M (2019) Multifield Polygonal Bounces. Phys Rev D 99:056020. arXiv: 1803.02227, https://doi.org/10.1103/PhysRevD.99.056020
123. Guada V, Nemevšek M, Pintar MZ, FindBounce: package for multi-field bounce actions. arXiv:2002.00881
124. Espinosa JR (2018) A fresh look at the calculation of tunneling actions. JCAP 1807:036. arXiv: 1805.03680, https://doi.org/10.1088/1475-7516/2018/07/036
125. Espinosa J (2019) Fresh look at the calculation of tunneling actions including gravitational effects. Phys Rev D 100:104007. arXiv: 1808.00420, https://doi.org/10.1103/PhysRevD.100.104007
126. Espinosa J, Konstandin T (2019) A fresh look at the calculation of tunneling actions in multi-field potentials. JCAP 01:051. arXiv: 1811.09185, https://doi.org/10.1088/1475-7516/2019/01/051
127. Adams FC (1993) General solutions for tunneling of scalar fields with quartic potentials. Phys Rev D 48:2800–2805. arXiv: hep-ph/9302321, https://doi.org/10.1103/PhysRevD.48.2800
128. Kehayias J, Profumo S (2010) Semi-analytic calculation of the gravitational wave signal from the electroweak phase transition for general quartic scalar effective potentials. JCAP 03:003. arXiv: 0911.0687, https://doi.org/10.1088/1475-7516/2010/03/003
129. Caprini C, No JM (2012) Supersonic electroweak baryogenesis: achieving baryogenesis for fast bubble walls. JCAP 01:031. arXiv: 1111.1726, https://doi.org/10.1088/1475-7516/2012/01/031
130. Ellis J, Lewicki M, No JM (2020) Gravitational waves from first-order cosmological phase transitions: lifetime of the sound wave source. JCAP 07:050. arXiv: 2003.07360, https://doi.org/10.1088/1475-7516/2020/07/050
131. Dine M, Leigh RG, Huet PY, Linde AD, Linde DA (1992) Towards the theory of the electroweak phase transition. Phys Rev D 46:550–571. arXiv: hep-ph/9203203, https://doi.org/10.1103/PhysRevD.46.550
132. Guth AH, Tye SHH (1980) Phase transitions and magnetic monopole production in the very early universe. Phys Rev Lett 44:631. https://doi.org/10.1103/PhysRevLett.44.631, https://doi.org/10.1103/PhysRevLett.44.963.2
133. Guth AH, Weinberg EJ (1981) Cosmological consequences of a first order phase transition in the SU(5) grand unified model. Phys Rev D 23:876. https://doi.org/10.1103/PhysRevD.23.876
134. Enqvist K, Ignatius J, Kajantie K, Rummukainen K (1992) Nucleation and bubble growth in a first order cosmological electroweak phase transition. Phys Rev D45:3415–3428. https://doi.org/10.1103/PhysRevD.45.3415
135. Turner MS, Weinberg EJ, Widrow LM (1992) Bubble nucleation in first order inflation and other cosmological phase transitions. Phys Rev D46:2384–2403. https://doi.org/10.1103/PhysRevD.46.2384
136. Guth AH, Weinberg EJ (1983) Could the universe have recovered from a slow first order phase transition?. Nucl Phys B 212:321–364. https://doi.org/10.1016/0550-3213(83)90307-3
137. Barir J, Geller M, Sun C, Volansky T, Gravitational waves from incomplete inflationary phase transitions. arXiv:2203.00693
138. Moore GD, Prokopec T (1995) How fast can the wall move? a Study of the electroweak phase transition dynamics. Phys Rev D 52:7182–7204. arXiv: hep-ph/9506475, https://doi.org/10.1103/PhysRevD.52.7182
139. Konstandin T, Nardini G, Rues I (2014) From Boltzmann equations to steady wall velocities. JCAP 09:028. arXiv: 1407.3132, https://doi.org/10.1088/1475-7516/2014/09/028
140. Jinno R, Konstandin T, Takimoto M, Relativistic bubble collisions—a closer look. arXiv:1906.02588

141. De Gennes P-G, Brochard-Wyart F, Quéré D (2013) Capillarity and wetting phenomena: drops, bubbles, pearls, waves. Springer Science & Business Media
142. Barroso Mancha M, Prokopec T, Swiezewska B, Field theoretic derivation of bubble wall force. arXiv:2005.10875
143. Balaji S, Spannowsky M, Tamarit C, Cosmological bubble friction in local equilibrium. arXiv:2010.08013
144. Ai W-Y, Garbrecht B, Tamarit C, Bubble wall velocities in local equilibrium. arXiv:2109.13710
145. Espinosa JR, Konstandin T, No JM, Servant G (2010) Energy budget of cosmological first-order phase transitions. JCAP 1006:028. arXiv: 1004.4187, https://doi.org/10.1088/1475-7516/2010/06/028
146. Lewicki M, Vaskonen V, Veermäe H, Bubble dynamics in fluids with N-body simulations. arXiv:2205.05667
147. Moore GD (2000) Electroweak bubble wall friction: analytic results. JHEP 03:006. arXiv: hep-ph/0001274, https://doi.org/10.1088/1126-6708/2000/03/006
148. John P, Schmidt MG (2001) Do stops slow down electroweak bubble walls?. Nucl Phys B 598:291–305. arXiv: hep-ph/0002050, https://doi.org/10.1016/S0550-3213(00)00768-9
149. Huber SJ, Sopena M, An efficient approach to electroweak bubble velocities. arXiv:1302.1044
150. Dorsch GC, Huber SJ, Konstandin T (2018) Bubble wall velocities in the standard model and beyond. JCAP 12:034. arXiv: 1809.04907, https://doi.org/10.1088/1475-7516/2018/12/034
151. Friedlander A, Banta I, Cline JM, Tucker-Smith D, Wall speed and shape in singlet-assisted strong electroweak phase transitions. arXiv:2009.14295
152. Dorsch GC, Huber SJ, Konstandin T, A sonic boom in bubble wall friction. arXiv:2112.12548
153. Ignatius J, Kajantie K, Kurki-Suonio H, Laine M (1994) The growth of bubbles in cosmological phase transitions. Phys Rev D 49:3854–3868. arXiv: astro-ph/9309059, https://doi.org/10.1103/PhysRevD.49.3854
154. Megevand A, Sanchez AD (2009) Detonations and deflagrations in cosmological phase transitions. Nucl Phys B 820:7–74. arXiv: 0904.1753, https://doi.org/10.1016/j.nuclphysb.2009.05.007
155. Megevand A, Sanchez AD (2010) Velocity of electroweak bubble walls. Nucl Phys B 825:151–176. arXiv: 0908.3663, https://doi.org/10.1016/j.nuclphysb.2009.09.019
156. Leitao L, Megevand A (2016) Hydrodynamics of ultra-relativistic bubble walls. Nucl Phys B 905:45–72. arXiv: 1510.07747, https://doi.org/10.1016/j.nuclphysb.2016.02.009
157. L. Leitao and A. Megevand, *Gravitational waves from a very strong electroweak phase transition*, https://doi.org/10.1088/1475-7516/2016/05/037*JCAP* **05** (2016) 037, arXiv: 1512.08962
158. Mégevand A, Ramírez S (2018) Bubble nucleation and growth in slow cosmological phase transitions. Nucl Phys B 928:38–71. arXiv: 1710.06279, https://doi.org/10.1016/j.nuclphysb.2018.01.012
159. Liu B-H, McLerran LD, Turok N (1992) Bubble nucleation and growth at a baryon number producing electroweak phase transition. Phys Rev D 46:2668–2688. https://doi.org/10.1103/PhysRevD.46.2668
160. Bodeker D, Moore GD (2009) Can electroweak bubble walls run away?. JCAP 0905:009. arXiv: 0903.4099, https://doi.org/10.1088/1475-7516/2009/05/009
161. Bodeker D, Moore GD (2017) Electroweak bubble wall speed limit. JCAP 1705:025. arXiv: 1703.08215, https://doi.org/10.1088/1475-7516/2017/05/025
162. Höche S, Kozaczuk J, Long AJ, Turner J, Wang Y, Towards an all-orders calculation of the electroweak bubble wall velocity. arXiv:2007.10343
163. Vanvlasselaer AAM, Bubble wall velocity: heavy physics effects. arXiv:2010.02590
164. Gouttenoire Y, Jinno R, Sala F (2022) Friction pressure on relativistic bubble walls. JHEP 05:004. arXiv: 2112.07686, https://doi.org/10.1007/JHEP05(2022)004
165. No JM (2011) Large gravitational wave background signals in electroweak baryogenesis scenarios. Phys Rev D 84:124025. arXiv: 1103.2159, https://doi.org/10.1103/PhysRevD.84.124025

166. Cline JM, Kainulainen K (2020) Electroweak baryogenesis at high bubble wall velocities. Phys Rev D 101:063525. arXiv: 2001.00568, https://doi.org/10.1103/PhysRevD.101.063525
167. Laurent B, Cline JM, Fluid equations for fast-moving electroweak bubble walls. arXiv:2007.10935
168. Dorsch GC, Huber SJ, Konstandin T, On the wall velocity dependence of electroweak baryogenesis. arXiv:2106.06547
169. De Curtis S, Rose LD, Guiggiani A, Muyor AG, Panico G, Bubble wall dynamics at the electroweak phase transition. arXiv:2201.08220
170. Misner CW, Thorne KS, Wheeler JA (1973) Gravitation. W. H. Freeman, San Francisco
171. Maggiore M (2007) Gravitational waves. Vol. 1: theory and experiments. Oxford Master Series in Physics. Oxford University Press
172. Jinno R, Takimoto M (2017) Gravitational waves from bubble collisions: an analytic derivation. Phys Rev D 95:024009. arXiv: 1605.01403, https://doi.org/10.1103/PhysRevD.95.024009
173. Turner MS (1983) Coherent scalar field oscillations in an expanding universe. Phys Rev D28:1243. https://doi.org/10.1103/PhysRevD.28.1243
174. Mangano G, Miele G, Pastor S, Pinto T, Pisanti O, Serpico PD (2005) Relic neutrino decoupling including flavor oscillations. Nucl Phys B729:221–234. arXiv: hep-ph/0506164, https://doi.org/10.1016/j.nuclphysb.2005.09.041
175. de Salas PF, Pastor S (2016) Relic neutrino decoupling with flavour oscillations revisited. JCAP 1607:051. arXiv: 1606.06986, https://doi.org/10.1088/1475-7516/2016/07/051
176. Escudero Abenza M, Precision early universe thermodynamics made simple: N_{eff} and neutrino decoupling in the standard model and beyond. arXiv:2001.04466
177. Particle Data Group collaboration, Tanabashi M et al (2018) Review of particle physics. Phys. Rev. D98:030001. https://doi.org/10.1103/PhysRevD.98.030001
178. Kosowsky A, Turner MS, Watkins R (1992) Gravitational radiation from colliding vacuum bubbles. Phys Rev D45:4514–4535. https://doi.org/10.1103/PhysRevD.45.4514
179. Kosowsky A, Turner MS (1993) Gravitational radiation from colliding vacuum bubbles: envelope approximation to many bubble collisions. Phys Rev D 47:4372–4391. arXiv: astro-ph/9211004, https://doi.org/10.1103/PhysRevD.47.4372
180. Caprini C, Durrer R, Servant G (2008) Gravitational wave generation from bubble collisions in first-order phase transitions: an analytic approach. Phys Rev D77:124015. arXiv: 0711.2593, https://doi.org/10.1103/PhysRevD.77.124015
181. Huber SJ, Konstandin T (2008) Gravitational wave production by collisions: more bubbles. JCAP 09:022. arXiv: 0806.1828, https://doi.org/10.1088/1475-7516/2008/09/022
182. Weir DJ (2016) Revisiting the envelope approximation: gravitational waves from bubble collisions. Phys Rev D93:124037. arXiv: 1604.08429, https://doi.org/10.1103/PhysRevD.93.124037
183. Konstandin T (2018) Gravitational radiation from a bulk flow model. JCAP 03:047. arXiv: 1712.06869, https://doi.org/10.1088/1475-7516/2018/03/047
184. Durrer R, Caprini C (2003) Primordial magnetic fields and causality. JCAP 11:010. arXiv: astro-ph/0305059, https://doi.org/10.1088/1475-7516/2003/11/010
185. Caprini C, Durrer R, Konstandin T, Servant G (2009) General properties of the gravitational wave spectrum from phase transitions. Phys Rev D 79:083519. arXiv: 0901.1661, https://doi.org/10.1103/PhysRevD.79.083519
186. Cai R-G, Pi S, Sasaki M (2020) Universal infrared scaling of gravitational wave background spectra. Phys Rev D 102:083528. arXiv: 1909.13728, https://doi.org/10.1103/PhysRevD.102.083528
187. Hook A, Marques-Tavares G, Racco D, Causal gravitational waves as a probe of free streaming particles and the expansion of the Universe. arXiv:2010.03568
188. Cutting D, Hindmarsh M, Weir DJ (2018) Gravitational waves from vacuum first-order phase transitions: from the envelope to the lattice. Phys Rev D97:123513. arXiv: 1802.05712, https://doi.org/10.1103/PhysRevD.97.123513

189. Cutting D, Escartin EG, Hindmarsh M, Weir DJ, Gravitational waves from vacuum first order phase transitions II: from thin to thick walls. arXiv:2005.13537
190. Jinno R, Takimoto M (2019) Gravitational waves from bubble dynamics: Beyond the Envelope. JCAP 1901:060. arXiv: 1707.03111, https://doi.org/10.1088/1475-7516/2019/01/060
191. Cutting D, Private communication
192. Konstandin T, Servant G (2011) Natural cold baryogenesis from strongly interacting electroweak symmetry breaking. JCAP 1107:024. arXiv: 1104.4793, https://doi.org/10.1088/1475-7516/2011/07/024
193. Lewicki M, Vaskonen V (2020) On bubble collisions in strongly supercooled phase transitions. Phys Dark Univ 30:100672. arXiv: 1912.00997, https://doi.org/10.1016/j.dark.2020.100672
194. Gould O, Sukuvaara S, Weir D (2021) Vacuum bubble collisions: from microphysics to gravitational waves. Phys Rev D 104:075039. arXiv: 2107.05657, https://doi.org/10.1103/PhysRevD.104.075039
195. Child HL, Giblin Jr. JT (2012) Gravitational radiation from first-order phase transitions. JCAP 1210:001. arXiv: 1207.6408, https://doi.org/10.1088/1475-7516/2012/10/001
196. Lentati L et al (2015) European pulsar timing array limits on an isotropic stochastic gravitational-wave background. Mon Not R Astron Soc 453:2576–2598. arXiv: 1504.03692, https://doi.org/10.1093/mnras/stv1538
197. NANOGRAV collaboration, Arzoumanian Z et al (2018) The NANOGrav 11-year data set: pulsar-timing constraints on the stochastic gravitational-wave background. Astrophys J 859:47. arXiv: 1801.02617, https://doi.org/10.3847/1538-4357/aabd3b
198. Breitbach M, Kopp J, Madge E, Opferkuch T, Schwaller P (2019) Dark, cold, and noisy: constraining secluded hidden sectors with gravitational waves. JCAP 1907:007. arXiv: 1811.11175, https://doi.org/10.1088/1475-7516/2019/07/007
199. LISA collaboration, Audley H et al, Laser interferometer space antenna. arXiv:1702.00786
200. LIGO Scientific, VIRGO collaboration, Aasi J et al (2015) Characterization of the LIGO detectors during their sixth science run. Class Quant Grav 32:115012. arXiv: 1410.7764, https://doi.org/10.1088/0264-9381/32/11/115012
201. Hild S et al (2011) Sensitivity studies for third-generation gravitational wave observatories. Class Quant Grav 28:094013. arXiv: 1012.0908, https://doi.org/10.1088/0264-9381/28/9/094013
202. Punturo M et al (2010) The Einstein telescope: a third-generation gravitational wave observatory. Class Quant Grav 27:194002. https://doi.org/10.1088/0264-9381/27/19/194002
203. LIGO Scientific collaboration, Abbott BP et al (2017) Exploring the sensitivity of next generation gravitational wave detectors. Class Quant Grav 34:044001. arXiv: 1607.08697, https://doi.org/10.1088/1361-6382/aa51f4
204. Hindmarsh M, Huber SJ, Rummukainen K, Weir DJ (2014) Gravitational waves from the sound of a first order phase transition. Phys Rev Lett 112:041301. arXiv: 1304.2433, https://doi.org/10.1103/PhysRevLett.112.041301
205. Giblin JT, Mertens JB (2014) Gravitional radiation from first-order phase transitions in the presence of a fluid. Phys Rev D 90:023532
206. Hindmarsh M, Huber SJ, Rummukainen K, Weir DJ (2015) Numerical simulations of acoustically generated gravitational waves at a first order phase transition. Phys Rev D92:123009. arXiv: 1504.03291, https://doi.org/10.1103/PhysRevD.92.123009
207. Hindmarsh M, Huber SJ, Rummukainen K, Weir DJ (2017) Shape of the acoustic gravitational wave power spectrum from a first order phase transition. Phys Rev D96:103520. arXiv: 1704.05871, https://doi.org/10.1103/PhysRevD.96.103520
208. Cutting D, Hindmarsh M, Weir DJ, Vorticity, kinetic energy, and suppressed gravitational wave production in strong first order phase transitions. arXiv:1906.00480
209. Hindmarsh M (2018) Sound shell model for acoustic gravitational wave production at a first-order phase transition in the early Universe. Phys Rev Lett 120:071301. arXiv: 1608.04735, https://doi.org/10.1103/PhysRevLett.120.071301
210. Hindmarsh M, Hijazi M (2019) Gravitational waves from first order cosmological phase transitions in the sound shell model. JCAP 12:062. arXiv: 1909.10040, https://doi.org/10.1088/1475-7516/2019/12/062

211. Kosowsky A, Mack A, Kahniashvili T (2002) Gravitational radiation from cosmological turbulence. Phys Rev D66:024030. arXiv: astro-ph/0111483, https://doi.org/10.1103/PhysRevD.66.024030
212. Caprini C, Durrer R (2006) Gravitational waves from stochastic relativistic sources: Primordial turbulence and magnetic fields. Phys Rev D74:063521. arXiv: astro-ph/0603476, https://doi.org/10.1103/PhysRevD.74.063521
213. Caprini C, Durrer R, Servant G (2009) The stochastic gravitational wave background from turbulence and magnetic fields generated by a first-order phase transition. JCAP 0912:024. arXiv: 0909.0622, https://doi.org/10.1088/1475-7516/2009/12/024
214. Gogoberidze G, Kahniashvili T, Kosowsky A (2007) The spectrum of gravitational radiation from primordial turbulence. Phys Rev D 76:083002. arXiv: 0705.1733, https://doi.org/10.1103/PhysRevD.76.083002
215. Kahniashvili T, Kosowsky A, Gogoberidze G, Maravin Y (2008) Detectability of gravitational waves from phase transitions. Phys Rev D78:043003. arXiv: 0806.0293, https://doi.org/10.1103/PhysRevD.78.043003
216. Kahniashvili T, Campanelli L, Gogoberidze G, Maravin Y, Ratra B (2008) Gravitational radiation from primordial helical inverse cascade MHD turbulence. Phys Rev D78:123006. arXiv: 0809.1899, https://doi.org/10.1103/PhysRevD.78.123006, https://doi.org/10.1103/PhysRevD.79.109901
217. Kahniashvili T, Kisslinger L, Stevens T (2010) Gravitational radiation generated by magnetic fields in cosmological phase transitions. Phys Rev D81:023004. arXiv: 0905.0643, https://doi.org/10.1103/PhysRevD.81.023004
218. Kalaydzhyan T, Shuryak E (2015) Gravity waves generated by sounds from big bang phase transitions. Phys Rev D91:083502. arXiv: 1412.5147, https://doi.org/10.1103/PhysRevD.91.083502
219. Pen U-L, Turok N (2016) Shocks in the early universe. Phys Rev Lett 117:131301. arXiv: 1510.02985, https://doi.org/10.1103/PhysRevLett.117.131301
220. Niksa P, Schlederer M, Sigl G (2018) Gravitational waves produced by compressible MHD turbulence from cosmological phase transitions. Class Quant Grav 35:144001. arXiv: 1803.02271, https://doi.org/10.1088/1361-6382/aac89c
221. A. Roper Pol, S. Mandal, A. Brandenburg, T. Kahniashvili and A. Kosowsky, *Numerical Simulations of Gravitational Waves from Early-Universe Turbulence*, arXiv:1903.08585
222. Giese F, Konstandin T, van de Vis J, Model-independent energy budget of cosmological first-order phase transitions. arXiv:2004.06995
223. Giese F, Konstandin T, Schmitz K, Van De Vis J (2021) Model-independent energy budget for LISA. JCAP 01:072. arXiv: 2010.09744, https://doi.org/10.1088/1475-7516/2021/01/072
224. Gildener E, Weinberg S (1976) Symmetry breaking and scalar Bosons. Phys Rev D13:3333. https://doi.org/10.1103/PhysRevD.13.3333
225. Espinosa J (2019) Tunneling without bounce. Phys Rev D 100:105002. arXiv: 1908.01730, https://doi.org/10.1103/PhysRevD.100.105002
226. Fubini S (1976) A new approach to conformal invariant field theories. Nuovo Cim A 34:521. https://doi.org/10.1007/BF02785664
227. Lipatov L (1977) Divergence of the perturbation theory series and the quasiclassical theory. Sov Phys JETP 45:216–223
228. Witten E (1981) Cosmological Consequences of a light higgs Boson. Nucl Phys B177:477–488. https://doi.org/10.1016/0550-3213(81)90182-6
229. von Harling B, Petraki K (2014) Bound-state formation for thermal relic dark matter and unitarity. JCAP 12:033. arXiv: 1407.7874, https://doi.org/10.1088/1475-7516/2014/12/033
230. Bloch IM, Csáki C, Geller M, Volansky T, Crunching away the cosmological constant problem: dynamical selection of a small Λ. arXiv:1912.08840
231. Joti A, Katsis A, Loupas D, Salvio A, Strumia A, Tetradis N et al (2017) (Higgs) vacuum decay during inflation. JHEP 07:058. arXiv: 1706.00792, https://doi.org/10.1007/JHEP07(2017)058

232. Lewicki M, Pujolàs O, Vaskonen V, Escape from supercooling with or without bubbles: gravitational wave signatures. arXiv:2106.09706
233. Hiscock W (1987) Can black holes nucleate vacuum phase transitions?. Phys Rev D 35:1161–1170. https://doi.org/10.1103/PhysRevD.35.1161
234. Gregory R, Moss IG, Withers B (2014) Black holes as bubble nucleation sites. JHEP 03:081. arXiv: 1401.0017, https://doi.org/10.1007/JHEP03(2014)081
235. Burda P, Gregory R, Moss I (2015) Gravity and the stability of the Higgs vacuum. Phys Rev Lett 115:071303. arXiv: 1501.04937, https://doi.org/10.1103/PhysRevLett.115.071303
236. Burda P, Gregory R, Moss I (2015) Vacuum metastability with black holes. JHEP 08:114. arXiv: 1503.07331, https://doi.org/10.1007/JHEP08(2015)114
237. Burda P, Gregory R, Moss I (2016) The fate of the Higgs vacuum. JHEP 06:025. arXiv: 1601.02152, https://doi.org/10.1007/JHEP06(2016)025
238. Rajantie A, Stopyra S (2017) Standard model vacuum decay with gravity. Phys Rev D 95:025008. arXiv: 1606.00849, https://doi.org/10.1103/PhysRevD.95.025008
239. Canko D, Gialamas I, Jelic-Cizmek G, Riotto A, Tetradis N (2018) On the catalysis of the electroweak vacuum decay by black holes at high temperature. Eur Phys J C 78:328. arXiv: 1706.01364, https://doi.org/10.1140/epjc/s10052-018-5808-y
240. Gorbunov D, Levkov D, Panin A (2017) Fatal youth of the Universe: black hole threat for the electroweak vacuum during preheating. JCAP 10:016. arXiv: 1704.05399, https://doi.org/10.1088/1475-7516/2017/10/016
241. Mukaida K, Yamada M (2017) False vacuum decay catalyzed by black holes. Phys Rev D 96:103514. arXiv: 1706.04523, https://doi.org/10.1103/PhysRevD.96.103514
242. Kohri K, Matsui H (2018) Electroweak vacuum collapse induced by vacuum fluctuations of the higgs field around evaporating black holes. Phys Rev D 98:123509. arXiv: 1708.02138, https://doi.org/10.1103/PhysRevD.98.123509
243. Oshita N, Yamada M, Yamaguchi M (2019) Compact objects as the catalysts for vacuum decays. Phys Lett B 791:149–155. arXiv: 1808.01382, https://doi.org/10.1016/j.physletb.2019.02.032
244. Dai D-C, Gregory R, Stojkovic D (2020) Connecting the higgs potential and primordial black holes. Phys Rev D 101:125012. arXiv: 1909.00773, https://doi.org/10.1103/PhysRevD.101.125012
245. El-Menoufi BK, Huber SJ, Manuel J, Black holes seeding cosmological phase transitions. arXiv:2006.16275
246. Steinhardt PJ (1981) Monopole and vortex dissociation and decay of the false vacuum. Nucl Phys B 190:583–616. https://doi.org/10.1016/0550-3213(81)90449-1
247. Steinhardt PJ (1981) Monopole dissociation in the early universe. Phys Rev D 24:842. https://doi.org/10.1103/PhysRevD.24.842
248. Hosotani Y (1983) Impurities in the early universe. Phys Rev D 27:789. https://doi.org/10.1103/PhysRevD.27.789
249. Kumar B, Yajnik U (2010) Graceful exit via monopoles in a theory with O'Raifeartaigh type supersymmetry breaking. Nucl Phys B 831:162–177. arXiv: 0908.3949, https://doi.org/10.1016/j.nuclphysb.2010.01.011
250. Kumar B, Paranjape MB, Yajnik UA (2010) Fate of the false monopoles: induced vacuum decay. Phys Rev D 82:025022. arXiv: 1006.0693, https://doi.org/10.1103/PhysRevD.82.025022
251. Agrawal P, Nee M, The Boring Monopole. arXiv:2202.11102
252. Yajnik UA (1986) Phase transition induced by cosmic strings. Phys Rev D 34:1237–1240. https://doi.org/10.1103/PhysRevD.34.1237
253. Yajnik UA, Padmanabhan T (1987) Analytical approach to string induced phase transitiOn. Phys Rev D 35:3100. https://doi.org/10.1103/PhysRevD.35.3100
254. Dasgupta I (1997) Vacuum tunneling by cosmic strings. Nucl Phys B 506:421–435. arXiv: hep-th/9702041, https://doi.org/10.1016/S0550-3213(97)00546-4
255. Kumar B, Yajnik UA (2009) On stability of false vacuum in supersymmetric theories with cosmic strings. Phys Rev D 79:065001. arXiv: 0807.3254, https://doi.org/10.1103/PhysRevD.79.065001

256. Lee B-H, Lee W, MacKenzie R, Paranjape MB, Yajnik UA, Yeom D-H (2013) Battle of the bulge: decay of the thin, false cosmic string. Phys Rev D 88:105008. arXiv: 1310.3005, https://doi.org/10.1103/PhysRevD.88.105008
257. Blasi S, Mariotti A, Domain walls seeding the electroweak phase transition. arXiv:2203.16450
258. Goldberger WD, Grinstein B, Skiba W (2008) Distinguishing the Higgs boson from the dilaton at the Large Hadron Collider. Phys Rev Lett 100:111802. arXiv: 0708.1463, https://doi.org/10.1103/PhysRevLett.100.111802
259. Contino R, Pomarol A, Rattazzi R
260. Chacko Z, Mishra RK (2013) Effective theory of a light dilaton. Phys Rev D87:115006. arXiv: 1209.3022, https://doi.org/10.1103/PhysRevD.87.115006
261. Bellazzini B, Csaki C, Hubisz J, Serra J, Terning J (2014) A naturally light dilaton and a small cosmological constant. Eur Phys J. C74:2790. arXiv: 1305.3919, https://doi.org/10.1140/epjc/s10052-014-2790-x
262. Coradeschi F, Lodone P, Pappadopulo D, Rattazzi R, Vitale L (2013) A naturally light dilaton. JHEP 11:057. arXiv: 1306.4601, https://doi.org/10.1007/JHEP11(2013)057
263. Megias E, Pujolas O (2014) Naturally light dilatons from nearly marginal deformations. JHEP 08:081. arXiv:1401.4998, https://doi.org/10.1007/JHEP08(2014)081
264. Pomarol A, Pujolas O, Salas L, Holographic conformal transition and light scalars. arXiv:1905.02653
265. Maldacena JM (1999) The large N limit of superconformal field theories and supergravity. Int J Theor Phys 38:1113–1133. arXiv: hep-th/9711200, https://doi.org/10.1023/A:1026654312961
266. Witten E (1998) Anti-de Sitter space and holography. Adv Theor Math Phys 2:253–291. arXiv: hep-th/9802150, https://doi.org/10.4310/ATMP.1998.v2.n2.a2
267. Rattazzi R, Zaffaroni A (2001) Comments on the holographic picture of the Randall-Sundrum model. JHEP 04:021. arXiv: hep-th/0012248, https://doi.org/10.1088/1126-6708/2001/04/021
268. Rattazzi R (2003) Cargese lectures on extra-dimensions. In: Particle physics and cosmology: the interface. Proceedings, NATO Advanced Study Institute, School, Cargese, France, pp 461–517. arXiv:hep-ph/0607055
269. Gherghetta T (2011) A holographic view of beyond the standard model physics. In: Physics of the large and the small, TASI 09, proceedings of the theoretical advanced study institute in elementary particle physics. Boulder, Colorado, USA, pp 165–232. arXiv:1008.2570. https://doi.org/10.1142/9789814327183_0004DOI
270. Randall L, Sundrum R (1999) A large mass hierarchy from a small extra dimension. Phys Rev Lett 83:3370–3373. arXiv: hep-ph/9905221, https://doi.org/10.1103/PhysRevLett.83.3370
271. Randall L, Sundrum R (1999) An alternative to compactification. Phys Rev Lett 83:4690–4693. arXiv: hep-th/9906064, https://doi.org/10.1103/PhysRevLett.83.4690
272. Goldberger WD, Wise MB (1999) Modulus stabilization with bulk fields. Phys Rev Lett 83:4922–4925. arXiv: hep-ph/9907447, https://doi.org/10.1103/PhysRevLett.83.4922
273. Witten E (1979) Baryons in the 1/n Expansion. Nucl Phys B 160:57–115. DOI:https://doi.org/10.1016/0550-3213(79)90232-3
274. Fujikura K, Nakai Y, Yamada M (2020) A more attractive scheme for radion stabilization and supercooled phase transition. JHEP 02:111. arXiv: 1910.07546, https://doi.org/10.1007/JHEP02(2020)111
275. Maggiore M (2000) Gravitational wave experiments and early universe cosmology. Phys Rep 331:283–367. arXiv: gr-qc/9909001, https://doi.org/10.1016/S0370-1573(99)00102-7
276. Thrane E, Romano JD (2013) Sensitivity curves for searches for gravitational-wave backgrounds. Phys Rev D 88:124032. arXiv: 1310.5300, https://doi.org/10.1103/PhysRevD.88.124032
277. Yagi K, Seto N (2011) Detector configuration of DECIGO/BBO and identification of cosmological neutron-star binaries. Phys Rev D83:044011. arxiv:1101.3940, https://doi.org/10.1103/PhysRevD.95.109901, https://doi.org/10.1103/PhysRevD.83.044011

278. KAGRA, LIGO Scientific, VIRGO collaboration, Abbott BP et al (2018) Prospects for observing and localizing gravitational-wave transients with advanced LIGO, advanced virgo and KAGRA. Living Rev Rel 21:3. arXiv: 1304.0670, https://doi.org/10.1007/s41114-018-0012-9, https://doi.org/10.1007/lrr-2016-1
279. LIGO Scientific, VIRGO collaboration, Abadie J et al (2012) Upper limits on a stochastic gravitational-wave background using LIGO and Virgo interferometers at 600–1000 Hz. Phys Rev D85:122001. arXiv:1112.5004, https://doi.org/10.1103/PhysRevD.85.122001
280. Mingarelli CMF, Taylor SR, Sathyaprakash BS, Farr WM, Understanding $\Omega_{\rm gw}(f)$ in Gravitational Wave Experiments. arXiv:1911.09745

Chapter 7
String Fragmentation in Supercooled Confinement and Implications for Dark Matter

7.1 Introduction

The possible existence of new confining sectors is motivated by most major failures of our understanding of Nature at a fundamental level. First, the stability of particle Dark Matter can be elegantly achieved as an accident if it is a composite state of a new strongly-coupled sector, similarly to proton stability in QCD, see e.g. [2]. The hierarchy problem of the Fermi scale is solved via dimensional transmutation by new confining gauge theories, whose currently most appealing incarnation is that of composite Higgs models [3, 4]. Analogous composite pictures can UV-complete [5–7] twin-Higgs scenarios [8], and so ameliorate also the little hierarchy problem. A rationale to understand the SM hierarchies of masses and CKM mixing angles is provided by partial compositeness of the SM fermions [9]. Finally, new confining sectors play crucial roles in addressing the strong CP problem [10, 11], the baryon asymmetry [12, 13], etc.

Given their ubiquity, it makes sense to look for predictions of confining sectors that do not depend on the specific way they address a given SM issue. Cosmology naturally offers such a playground, in association with the confinement phase transition (PT) in the early universe. The low-density QCD phase transition would for example be strongly first-order if the strange or more quarks had smaller masses [14], with associated signals in gravitational waves [15, 16]. New confining sectors could also well feature a similar PT. In addition, the confinement transition could be supercooled, a property that for example arises naturally in 5-dimensional (5D) duals of 4D confining theories [17–19].

Generically, supercooling denotes a PT in which bubble percolation occurs significantly below the critical temperature. Here we are interested in the case where a cosmological PT becomes sufficiently delayed so that the radiation energy density becomes subdominant to the vacuum energy. The universe then experiences a stage of inflation until the PT completes [20]. This implies a dilution of any pre-existing relic, such as dark matter (DM), the baryon or other asymmetries, topological defects, and gravitational waves, see e.g. [21–23].

Y. Gouttenoire, *Beyond the Standard Model Cocktail*, Springer Theses,
https://doi.org/10.1007/978-3-031-11862-3_7

In this paper we point out an effect that, to our knowledge, had been so far missed: when the fundamental quanta of the strong sector enter the expanding bubbles of the confined phase, their relevant distance can be much larger than the inverse of the confinement scale, thus realising a situation whose closest known analogues are perhaps QCD jets in particle colliders or cosmic ray showers. We anticipate that our attempt to model this phenomenon implies an additional production mechanism of any composite resonance—string fragmentation followed by deep inelastic scattering—which introduces a mismatch between the dilution of composite and other relics. This opens new model building and phenomenological avenues, which we begin exploring here in a model independent manner for the case of composite DM. The application of our findings to a specific model, namely composite dark matter with dilaton mediated interactions, will appear elsewhere [24].

7.2 Synopsis

Due to the numerous effects which will be discussed in the following sections, it is perhaps useful for the reader that we summarise the overall picture in a few paragraphs. We begin in the deconfined phase in which the techniquanta TC of the new strong sector (which we will call quarks and gluons) are in thermal equilibrium. Their number density normalised to entropy takes a familiar form

$$Y_{\rm TC}^{\rm eq} = \frac{45\,\zeta(3)\,g_{\rm TC}}{2\pi^4 g_s}, \tag{7.1}$$

where $g_{\rm TC}$ (g_s) are the degrees of freedom of the quarks and gluons (entropic bath) respectively. Next a period of supercooling occurs, in which the universe finds itself in a late period of thermal inflation, which is terminated by bubble nucleation. As is known from previous studies, such a phase will dilute the number density of primordial particles. The dilution factor is given by

$$D^{\rm SC} = \left(\frac{T_{\rm nuc}}{T_{\rm start}}\right)^3 \frac{T_{\rm RH}}{T_{\rm start}}, \tag{7.2}$$

where $T_{\rm nuc}$ is the nucleation temperature, $T_{\rm start} \propto f$ is the temperature at which the thermal inflation started, $T_{\rm RH}$ is the temperature after reheating, and f is the energy scale of confinement. We assume reheating to occur within one Hubble time, so that $T_{\rm RH} \propto f$. The supercooled number density of quarks and gluons then becomes

$$Y_{\rm TC}^{\rm SC} = D^{\rm SC}\, Y_{\rm TC}^{\rm eq} \propto \left(\frac{T_{\rm nuc}}{f}\right)^3. \tag{7.3}$$

For completeness, the details entering Eq. (7.3) will be rederived in Sect. 7.3.

When the fundamental techniquanta are swept into the expanding bubbles, they experience a confining force. Because $f \gg T_{\rm nuc}$ in the supercooled transition, the distance between them is large compared to the size of the composite states ψ (which we will equivalently call 'hadrons'). The field lines attached to a quark or gluon then find it energetically more convenient to form a flux tube oriented towards the bubble wall, rather than directly to the closest neighbouring techniquantum, which is in general much further than the wall (see Fig. 7.2). The string or flux tube connecting the quark or the gluon and the wall then fragments, producing a number of hadrons inside the wall. Additionally, because of charge conservation, techniquanta must be ejected outside the wall to compensate (see Fig. 7.3). The process is conceptually analogous to the production of a pair of QCD partons at colliders, and we model it as such. The details are explained in Sect. 7.4. The result is an increase of the yield of composite particles, compared to the naive estimate following directly from Eq. (7.3), by a string fragmentation factor $K^{\rm string}$,

$$Y_{\psi}^{\rm SC+string} = K^{\rm string} D^{\rm SC}\, Y_{\rm TC}^{\rm eq} \propto \left(\frac{T_{\rm nuc}}{f}\right)^3 \times \text{logs}\left(\frac{\gamma_{\rm wp} T_{\rm nuc}}{f}\right), \tag{7.4}$$

where $\gamma_{\rm wp} > f/T_{\rm nuc} \gg 1$ is the Lorentz factor of the bubble wall at the time the quarks enter.

The Lorentz factor is estimated in Sect. 7.5. In Sect. 7.6 we show that our picture can be relevant already for $T_{\rm nuc}/T_{\rm start} \lesssim 1$. The quarks ejected from the bubbles are treated in detail in Sect. 7.7. We find they enter neighbouring bubbles and confine there into hadrons. Acting as a cosmological catapult, string fragmentation at the wall boundary gives a large boost factor to the newly formed hadrons, such that their momenta in the plasma frame can be $\gg f$.

The composite states and their decay products can next undergo scatterings with other particles they encounter, e.g. with particles of the preheated 'soup' after the bubbles collide. Since the associated center-of-mass energy can be much larger than f, the resulting deep inelastic scatterings (DIS) increase the number of hadrons. We explore this in detail in Sect. 7.8. The resulting effect on the yield can be encapsulated in a factor $K^{\rm DIS}$, and reads

$$Y_{\psi}^{\rm SC+string+DIS} = K^{\rm DIS} D^{\rm SC}\, Y_{\rm TC}^{\rm eq} \propto \left(\frac{T_{\rm nuc}}{f}\right)^3 \gamma_{\rm wp} \xrightarrow{\text{if runaway}} \left(\frac{T_{\rm nuc}}{f}\right)^4 \frac{M_{\rm Pl}}{m_*}, \tag{7.5}$$

where $M_{\rm Pl}$ is the Planck mass and $m_* = g_* f$ is the mass scale of hadrons. The last proportionality holds in the regime of runaway bubble walls, relevant for composite DM.

Finally the late-time abundance of the long-lived and stable hadrons, if any, evolves depending on their inelastic cross section in the thermal bath $\langle \sigma v_{\rm rel} \rangle$, and on $Y_{\psi}^{\rm SC+string+DIS}$ as an initial condition at $T_{\rm RH}$. We compute it in Sect. 7.9 by solving the associated Boltzmann equations.

By combining all the above effects we arrive at an estimate of the final relic abundance of the composite states. Our findings impact their abundance by several orders of magnitude, as can be seen in Fig. 7.9 for the concrete case where the relic is identified with DM. The formalism leading to this estimate can readily be adapted for other purposes. For example, if ψ instead decays out-of-equilibrium, it could source the baryon asymmetry. The estimate of $Y_\psi^{\text{SC+string+DIS}}$ would then act as the first necessary step for the determination of the baryonic yield.

7.3 Supercooling Before Confinement

7.3.1 Strongly Coupled CFT

Although striving to remain as model independent as possible in our discussion, we shall be making a minimal assumption that the confined phase of the strongly coupled theory can be described as an EFT with a light scalar χ, e.g. a dilaton. The scalar VEV, $\langle\chi\rangle$, then parametrizes the local value of the strong scale. It can be thought of as a scalar condensate of the strong sector, such as a glueball- or pion-like state. The scalar VEV at the minimum of its zero-temperature potential is identified with $\langle\chi\rangle = f$, where f is the confinement energy scale, while $\langle\chi\rangle = 0$ at large enough temperatures. In order to have strong supercooling, we require the approximate (e.g. conformal) symmetry to be close to unbroken, thus justifying the lightness of the associated pseudo-Nambu-Goldstone boson (e.g. the dilaton [25]). That supercooling occurs with a light dilaton is known from a number of previous studies [17–19], see [26–34] for studies in a confining sector and [35–44] for studies of holographic dual 5D warped extra dimension models.

7.3.2 Thermal History

The vacuum energy before the phase transition is given by

$$\Lambda_{\text{vac}}^4 \equiv c_{\text{vac}} f^4, \tag{7.6}$$

with some model dependent $c_{\text{vac}} \sim \mathcal{O}(0.01)$ constant. The radiation density is given by

$$\rho_{\text{rad}} = \frac{g_R \pi^2}{30} T^4, \tag{7.7}$$

where g_R counts the effective degrees of freedom of the radiation bath. We define $g_R \equiv g_{Ri}$ (g_{Rf}) in the deconfined (confined) phase. Now consider the case of strong supercooling. The universe will enter a vacuum-dominated phase at a temperature

$$T_{\rm start} = \left(\frac{30\, c_{\rm vac}}{g_{Ri}\, \pi^2} \right)^{1/4} f, \tag{7.8}$$

provided the phase transition has not yet taken place beforehand. The vacuum domination signals a period of late-time inflation. The phase transition takes place at the nucleation temperature, $T_{\rm nuc}$, when the bubble nucleation rate becomes comparable to the Hubble factor. Following the phase transition, the dilaton undergoes oscillations and decay, reheating the universe to a temperature

$$T_{\rm RH} = \left(\frac{g_{Ri}}{g_{Rf}} \right)^{1/4} T_{\rm start}, \tag{7.9}$$

At this point the universe is again radiation dominated. We have assumed the decay to occur much faster than the expansion rate of the universe such that we can neglect a matter-dominated phase [21].

7.3.3 *Dilution of the Degrees of Freedom*

Now consider some fundamental techniquanta of the strong sector, e.g. techniquarks or technigluons (for simplicity we always refer to them as quarks and gluons). Prior to the phase transition the number density of techniquanta follows a thermal distribution for massless particles

$$n_{\rm TC}^{\rm eq} = g_{\rm TC} \frac{\zeta(3)}{\pi^2} T^3, \tag{7.10}$$

where $g_{\rm TC}$ denotes the degrees of freedom of the quanta under consideration. The entropy density is given by

$$s = \frac{2\pi^2 g_s}{45} T^3, \tag{7.11}$$

where g_s are the total entropic degrees of freedom.[1] The number density normalized to entropy before the phase transition,

$$Y_{\rm TC}^{\rm eq} = \frac{45 \zeta(3)\, g_{\rm TC}}{2\pi^4 g_s}, \tag{7.12}$$

remains constant up to the point when the phase transition takes place. The entropy density then increases during reheating giving

$$Y_{\rm TC}^{\rm SC} = D^{\rm SC}\, Y_{\rm TC}^{\rm eq}, \tag{7.13}$$

[1] In a picture with N_f flavours of quarks in fundamental representations of an $SU(N)$ confining gauge group, one has $g_q = 2N_f N$, $g_g = 2(N^2 - 1)$, $g_{\rm TC} = g_g + 3g_q/4$, $g_s = g_g + 7g_q/8$.

when we find ourselves back in the radiation-dominated phase. The dilution factor from the additional expansion during the vacuum-dominated phase can be derived by finding the increase in entropy between $T_{\rm nuc}$ and $T_{\rm RH}$. It reads

$$D^{\rm SC} \equiv \left(\frac{T_{\rm nuc}}{T_{\rm start}}\right)^3 \left(\frac{T_{\rm RH}}{T_{\rm start}}\right) \simeq \frac{g_{Ri}}{c_{\rm vac}^{3/4}\, g_{Rf}^{1/4}} \left(\frac{T_{\rm nuc}}{f}\right)^3 . \tag{7.14}$$

If the quarks and gluons were non-interacting following the phase transition, the yield today would be given by the above formula. (In the presence of interactions the above would be taken as an initial condition at $T_{\rm RH}$ for the Boltzmann equations describing the effects of number changing interactions between reheating and today.) The picture would then be analogous to that studied, in a theory without confinement, in [21]. The picture is completely changed, however, for supercooled confining phase transitions, which we elucidate next.

7.4 Confinement and String Fragmentation

7.4.1 Where Does Confinement Happen?

Bubble wall profile:
y described by the Klein-Gordon equation [45]

$$\frac{d^2\chi}{ds^2} + \frac{3}{s}\frac{d\chi}{ds} + \frac{dV}{d\chi} = 0, \tag{7.15}$$

where $s^2 = t^2 - r^2$ is the light-cone coordinate and V is the scalar potential. A sketch of a typical bubble profile for close-to-conformal potentials is shown in Fig. 7.1. The key point here is that the wall thickness is

$$L_{\rm w} \lesssim \frac{1}{T_{\rm nuc}}, \tag{7.16}$$

as shown by numerical computations and analytical estimates, see Appendix 7.A for a calculation in an explicit example.

Confinement time scale:
The techniquanta (quarks and gluons) constitute a plasma with temperature of order $T_{\rm nuc}$ before entering the bubble. Once they enter the bubbles, they could in principle either confine in a region close to the bubble wall where $\langle\chi\rangle \ll f$, or approach as free particles the region where χ has reached its zero-temperature expectation value $\langle\chi\rangle = f$. To determine this, let us define a 'confinement rate' and a 'confinement length' as

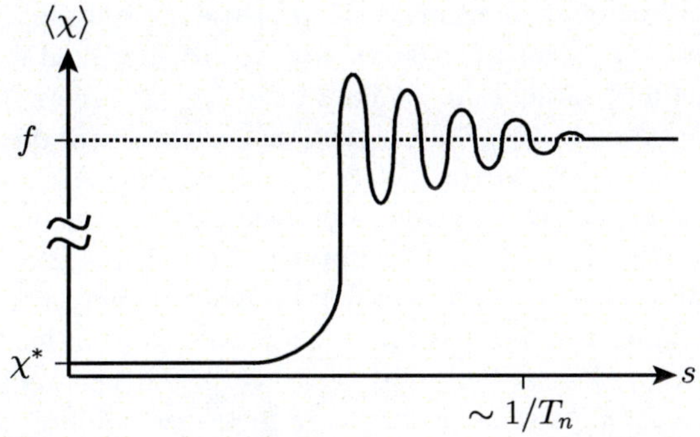

Fig. 7.1 A typical wall profile found in close-to-conformal potentials. After nucleating by tunneling to the exit point, $\chi^* \ll f$, the field rolls down and undergoes damped oscillations around the minimum of its potential. The typical wall thickness is $L_w \sim 1/T_{\rm nuc}$

$$\Gamma_{\rm conf} = L_{\rm conf}^{-1} = n_{\rm TC}\, v_{\rm TC}\, \sigma_{\rm conf}, \tag{7.17}$$

where $n_{\rm TC}$ and $v_{\rm TC}$ are, respectively, the number density and the relative Møller velocity of the techniquanta $v_{\rm TC} \equiv \left[|\mathbf{v}_1 - \mathbf{v}_2|^2 - |\mathbf{v}_1 \times \mathbf{v}_2|^2\right]^{1/2}$ [46], and $\sigma_{\rm conf}$ is a 'confining cross section'. We want to compare $L_{\rm conf}$ with the length of the bubble wall, defined as the distance over which χ varies from its value at the exit point, $\langle\chi\rangle = \chi^* \ll f$, to $\langle\chi\rangle = f$. Of course we need to perform this comparison in the same Lorentz frame, so we emphasise our definition of L_w as the bubble-wall length in the bubble-wall frame, and $L_{\rm p} = L_{\rm w}/\gamma_{\rm wp}$ as the bubble-wall length in the frame of the center of the bubble, which coincides with the plasma frame, and where $\gamma_{\rm wp}$ is the boost factor between the two frames. Let us now move to the confinement timescale of Eq. (7.17). Since we expect confinement to happen 'as soon as possible', we assume the related cross section to be close to the unitarity limit [47],

$$\sigma_{\rm conf} \sim \frac{4\pi}{T_{\rm nuc}^2}\,. \tag{7.18}$$

Since $n_{\rm TC}^2 v_{\rm TC}$ is Lorentz invariant [46, 48], one then has that $n_{\rm TC}\, v_{\rm TC}$ transforms under boosts as $n_{\rm TC}^{-1}$. The boost to apply in this case is $\gamma_{\rm wp}$, because by definition the string forms after confinement, so we can treat the plasma frame as the center-of-mass frame of the techniquanta. Combining this with the Lorentz invariance of the cross section, we obtain

$$\Gamma_{{\rm conf},w} = n_{\rm TC,\,w}\, v_{\rm TC,\,w}\, \sigma_{\rm conf} = \frac{n_{\rm TC,\,p}\, v_{\rm TC,\,p}}{\gamma_{\rm wp}}\, \sigma_{\rm conf} \sim \frac{4\pi\, T_{\rm nuc}}{\gamma_{\rm wp}}, \tag{7.19}$$

where in the last equality we have used that the average relative speed and density of the techniquanta in the plasma frame satisfy, respectively, $v_{{\rm TC},\,p} \simeq 1$ and $n_{{\rm TC},\,p} \sim T_{\rm nuc}^3$, because they are relativistic. This in turn implies

$$L_{\rm conf,\,w} \sim \frac{\gamma_{\rm wp}}{4\pi} L_{\rm w}. \tag{7.20}$$

Confinement takes place deep inside the bubble:
For the regimes of supercooling we are interested in, the phase transition is of detonation type and the Lorentz factor $\gamma_{\rm wp}$ is orders of magnitude larger than unity. Therefore, $L_{\text{conf},\,w} \gg L_{\rm w}$ such that confinement does not happen in the outermost bubble region where $\langle\chi\rangle \ll f$. This conclusion is solid in the sense that it would be strengthened by using a confinement cross section smaller than what assumed in Eq. (7.18), which is at the upper end of what is allowed by unitarity. The end effect of the above discussion, is that for practical purposes, we can consider the wall profile to be a step-like function between the deconfined phase, $\langle\chi\rangle = 0$, and confined phase, $\langle\chi\rangle = f$. Furthermore, as we shall discuss below, the quarks will not confine directly in pairs but rather form fluxtubes pointing toward the bubble wall as they penetrate the $\langle\chi\rangle = f$ region of the bubble.

The ballistic approximation is valid:
Equation (7.20), together with the large wall-Lorentz-factors encountered in this study, implies that we can safely neglect the interactions between neighbouring techniquanta during the time when they cross the bubble wall. This is the so-called ballistic regime, see e.g. [49], which will be useful for deriving the friction pressure in Sect. 7.5.

7.4.2 *Fluxtubes Attach to the Wall Following Supercooling*

A hierarchy of scale:
Upon entering the region $\langle\chi\rangle = f$ of expanding bubbles, the techniquanta experience a confinement potential much stronger than in the region close to the wall. This can be easily understood by taking the long-distance potential of the Cornell form [50–59]

$$E_{\rm TC} = c_{\rm TC}\, f^2\, d_c, \tag{7.21}$$

where d_c is the techniquanta seperation in their ‘center of interaction frame’ (or equivalently ‘string center of mass frame’),[2] and $c_{\rm TC}$ is an adimensional constant,[3] $c_{q\bar{q}} \simeq 10$ in QCD [59]. A crucial point regarding the string energy in this context, besides the fact it grows proportionally to χ^2, is that the inter-quanta distance is large compared to the natural confinement scale, i.e. $d_c \gg f^{-1}$, due to the supercooling. Indeed the distance between quanta outside the wall, in the plasma and wall frames

---

[2] Lattice simulations find that the QCD potential at $d_c \gtrsim$ fm saturates to a constant, a behavior which is interpreted in terms of pair creation of quarks from the vacuum, see e.g. the recent [60]. Therefore this realises an outcome that, for our purposes, coincides with having $E_{\rm TC} \propto d_c$ to larger distances. Lattice simulations with quarks only as external sources [61], so without sea quarks (‘quenched’), find that the linear regime of the QCD Cornell potential extends up to the maximal distances probed, namely $d_c \simeq 3$ fm in the results reported in [61].

[3] $c_{\rm TC}$ does not hide any ‘coupling dimension’, indeed in units where $\hbar \neq 1$, $[f] = (\text{energy}/\text{distance})^{\frac{1}{2}}$.

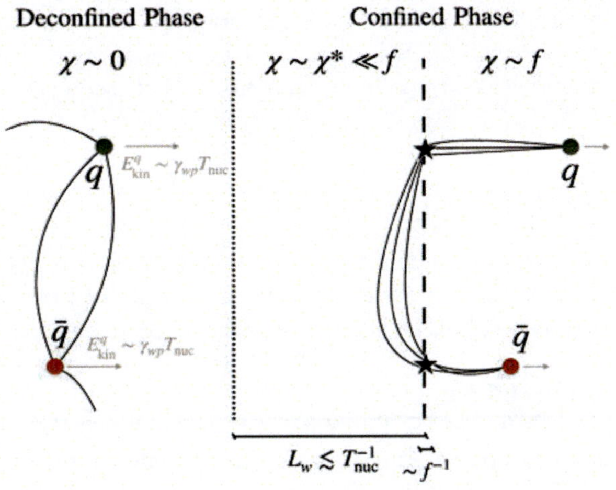

Fig. 7.2 Quarks entering the bubble as seen in the frame of the bubble wall, together with the associated field lines and quantities defined in the text. The rest energy of the string is minimized if the fluxtubes in the region $\chi = f$ point to the bubble wall, rather than if they point to the closest color charge

respectively, scales as $d_p \sim T_{\rm nuc}^{-1}$ and $d_w \sim \gamma_{\rm wp}^{-1/3} T_{\rm nuc}^{-1}$. Since $\gamma_{\rm wp} \ll (f/T_{\rm nuc})^3$ (see Sect. 7.5) and $d_c \geq d_w$ (because $d_c = d_p$ outside the wall, and because the quarks and gluons cannot be accelerated upon entering so d_w is Lorentz contracted with respect to d_p), one ends up with $d_c \gg f^{-1}$. What happens then to the techniquanta and to the fields connecting them?

Flux tubes minimize their energy:
In a picture without hierarchy of scales, the fields would compress in fluxtubes connecting different charges, 'isolated' in pairs or groups to form color-singlets. Here, we argue that the fluxtubes have another option, which is energetically preferable: that of orienting themselves towards the direction of minimal energy, i.e. as perpendicular as possible to the bubble-wall,[4] and to keep a 'looser' connection in the outer region where $\chi \ll f$. Indeed, a straight-line connection between techniquanta would result in a much longer portion of fluxtubes in the region $\langle\chi\rangle = f$, with respect to our picture of fluxtubes perpendicular to the wall. Via Eq. (7.21), this would in turn imply a much higher cost in energy, disfavoring that option. We stress that, in our picture, the fluxtubes are still connecting techniquanta in such a way to form an overall color singlet, just these fluxtubes minimise their length in the region $\chi = f$, and partly live in a region $\chi \simeq \chi^* \ll f$. This picture is visualised in Fig. 7.2. Note the nearest neighbour quark from the plasma may also be located outside the bubble.

Condensed matter analogy:
An interesting analogue to the picture above is the vortex string of magnetic flux in the Landau-Ginzburg model of superconductivity. To match onto confinement dynamics a dual superconductor is pictured, in which the external colour-electric field—rather than the magnetic field—is expelled by the Meissner effect [62]. Here

[4] We wish to express our gratitude to Benedict von Harling, Oleksii Matsedonskyi, and Philip Soerensen, for discussions which lead us to develop the picture we employ in this paper.

the bubble of confining phase corresponds to the superconductor from which the colour-electric field is expelled. Quarks entering the bubble then map onto magnetic monopoles being fired into a regular superconductor.

7.4.3 String Energy and Boost Factors

To possibly be quantitative on the implications of the picture we just outlined, we first need to determine the string energy and the Lorentz boosts among the frames of the plasma, wall and center-of-mass of the string.

String end-points

Let us define as TC_i the quark or gluon that constitutes an endpoint, inside the bubble, of a fluxtube pointing towards the wall, and ★ the end-point of the fluxtube on the wall. The energy of the incoming techniquantum in the wall frame is $E_{i,\text{w}} = 3\gamma_{\text{wp}} T_{\text{nuc}}$, where for simplicity we have averaged over their angle with respect to the wall. We assume ★ to be at rest or almost, and to carry some $\mathcal{O}(1)$ fraction of the inertia of the string. Hence the respective four-momenta are

$$p_{i,\text{w}} = \begin{pmatrix} 3\,\gamma_{\text{wp}} T_{\text{nuc}} \\ \sqrt{9\,\gamma_{\text{wp}}^2 T_{\text{nuc}}^2 - m_i^2} \end{pmatrix}, \quad p_{\bigstar,w} = \begin{pmatrix} m_f \\ \varepsilon f \end{pmatrix}, \qquad \varepsilon \ll 1, \qquad m_i \simeq m_f = qf, \ \ q \leq \frac{1}{2}. \tag{7.22}$$

String center-of-mass:

Then we define the center-of-mass of the string as the one of TC_i and ★, and find

$$E_{\text{CM}} = |p_{\bigstar,w} + p_{i,\text{w}}| \simeq \sqrt{3\,\gamma_{\text{wp}}\, T_{\text{nuc}}\, f}\,, \tag{7.23}$$

where the second expression is valid up to relative orders $(\gamma_{\text{wp}} f/T_{\text{nuc}})^{-1} \ll 1$. By employing a Lorentz boost between the wall and center-of-mass frames, and imposing $p_{i,c} = -p_{\bigstar,c}$, we find

$$\gamma_{\text{wc}} \simeq \sqrt{3\,\gamma_{\text{wp}} \frac{T_{\text{nuc}}}{f}}\,. \tag{7.24}$$

On the right-hand side of the equations above we have omitted a factor of $\sqrt{2(q-\varepsilon)}$, in (7.23), and of $1/\sqrt{2(q-\varepsilon)}$, in (7.24), because for simplicity we take these to be ≈ 1 from now on (as per the benchmark $q = 1/2$, $\varepsilon = 0$). Finally we determine the boost between the center-of-mass frame of the string and the plasma frame as

$$\gamma_{\text{cp}} \simeq \frac{\gamma_{\text{wp}}}{2\gamma_{\text{wc}}} = \frac{1}{2}\sqrt{\frac{\gamma_{\text{wp}}}{3}\frac{f}{T_{\text{nuc}}}}, \tag{7.25}$$

which is valid up to a relative order $(\gamma_{\text{wp}} f/T_{\text{nuc}})^{-1} \ll 1$ (Fig. 7.3).

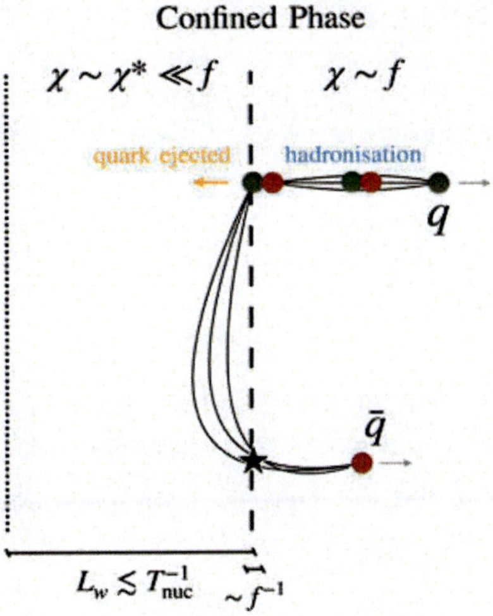

Fig. 7.3 The string inside the wall breaks, producing hadrons (Sect. 7.4.4), and a quark is ejected from the wall (Sect. 7.4.6)

7.4.4 Hadrons from String Fragmentation: Multiplicity and Energy

The fluxtubes connecting a quark or gluon to the wall will fragment and form hadrons, singlet under the new confining gauge group. We would now like to determine:

- The number of hadrons formed per fluxtube.
- The momenta of said hadrons.

Collider analogy:
We start by noticing that the process of formation of a fluxtube, in our picture, is analogous to two color charges in an overall-singlet state, TC_i and ★, moving apart with a certain energy $E_{\rm CM}$, where $E_{\rm CM} = \sqrt{3\,\gamma_{\rm wp}\,T_{\rm nuc}\,f}$ in the modelling of Sect. 7.4.3. This physical process appears entirely analogous to what would happen in a collider that produces a pair of techniquanta of the new confining force, starting from an initial singlet state. In light of this observation, we then decide to model the process by analogy with a very well-studied process observed in Nature, that of QCD-quark pair production at electron-positron colliders, where the analogy lies also in the fact that the initial state electron-positron pairs is in a color singlet state. Needless to say, a BSM confining sector needs not behave as QCD in terms of number and momenta of hadrons produced per scattering, see e.g. [63]. However, QCD constitutes a well studied and tested theory, so that we find it reasonable to use it as our benchmark. Moreover, we anticipate from Sect. 7.8 that our final result for the cosmological abundance of hadrons, in the assumption of efficient-enough interactions between them and the SM, will only depend on the initial available energy $E_{\rm CM}$. This suggests that, within that assumption, our final findings hold for confining sectors that distribute this energy over a number of hadrons different from QCD.

Numerical simulations:
We use Pythia v8.2 [66] interfaced to MadGraph v2.7.0 [65] to simulate the process $e^-e^+ \to q\bar{q}$ for different center-of-mass energies, and MadAnalysis v1.8.34 [64] to extract from these simulations both the total number of hadrons produced per scattering and their energy distribution. We thus recover known QCD results and display them in Fig. 7.4. We translate them to our picture by replacing the units of a GeV$\simeq 4\pi f_\pi$ used by Pythia, with the generic mass of a composite state $m_* = g_* f$, where $1 \le g_* \le 4\pi$ is some strong effective coupling. These results can be summarised as follows:

- The number of hadrons produced per fluxtube grows logarithmically in $E_{\rm CM}$.
- The distribution of hadron energies is such that its root square mean coincides, to a percent level accuracy, with the average energy per hadron

$$\bar{E}_{\rm hadr} = \frac{E_{\rm CM}}{\langle N_{\rm hadr}\rangle}. \tag{7.26}$$

This will support, in Sect. 7.8.3, our simplifying assumption that all hadrons produced by the string fragmentation carry an energy of order $\bar{E}_{\rm hadr}$.

Results from the literature:
The multiplicity of QCD hadrons from various scattering processes has been the object of experimental and theoretical investigation, since the late 1960s [67]. We now leverage such studies both to check the results of our simulation and to obtain

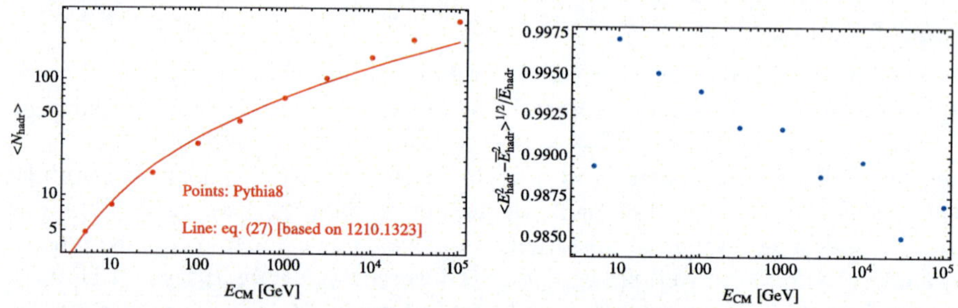

Fig. 7.4 Left: average hadron multiplicity per single QCD scattering $e^-e^+ \to q\bar{q}$. Right: Root square mean of the hadron energy per single QCD scattering $e^-e^+ \to q\bar{q}$, where $\bar{E}_{\rm hadr} = E_{\rm CM}/\langle N_{\rm hadr}\rangle$ is the average hadron energy per scattering. Dots in both plots are extracted via MadAnalysis v1.8.34 [64] by simulations with MadGraph v2.7.0 [65] plus Pythia v8.2 [66], the line in the left-hand plot displays Eq. (7.28). Results are expressed as a function of the center-of-mass energy of the scattering in GeV, to export them to our cosmological picture we simply substitute GeV $\simeq 4\pi f_\pi$ by $m_* = g_* f$, and use $E_{\rm CM} = \sqrt{3\,\gamma_{\rm wp}\, T_{\rm nuc}\, f}$

analytical control over them. Collider studies have typically focused on the multiplicity of charged QCD resonances per scattering, $\langle n_{\rm ch}\rangle$. In particular, works such as [68, 69] have carried out the exercise of collecting the most significant measurements of $\langle n_{\rm ch}\rangle$ and 'filling' the missing phase space—not covered by detectors—with the output of MC programs, thus obtaining a full-phase-space quantity. We take as our starting point the result provided in [69] from pp collisions, which reads

$$\langle n_{\rm ch}\rangle(E_{\rm CM}) = a + b\log\frac{E_{\rm CM}}{m_*} + c\log^2\frac{E_{\rm CM}}{m_*} + d\log^3\frac{E_{\rm CM}}{m_*}, \tag{7.27}$$

with $(a, b, c, d) = (0.95, 0.37, 0.43, 0.04)$. Here, as already explained, we substituted the normalisation of a GeV with $m_* = g_* f$.

Our modelling:
To obtain the total number of hadrons from e^+e^- collisions we proceed as follows. First, most hadrons coming out from hard scatterings consist in the lightest ones, i.e. the pions. Second, the total number of pions produced is very well approximated by $3\langle n_{\rm ch}\rangle/2$, because of isospin conservation. By the first argument, this coincides with very good approximation to the total number of hadrons produced. Third, the multiplicity of composite states from e^+e^- collisions has been found to roughly match the one from pp collisions, upon increasing the e^+e^- energy by a factor of 2, see e.g. Sect. 2.2 in [70].[5] We then model the total number of composite states produced, per string fragmentation, as

$$N_\psi^{\rm string}(E_{\rm CM}) \simeq \frac{3}{2}\langle n_{\rm ch}\rangle(2E_{\rm CM})\exp(-3m_*/E_{\rm CM}) + 1\,, \tag{7.28}$$

where we have multiplied by an exponential and added one to smoothen $N_\psi^{\rm string}(E_{\rm CM})$ to 1 as $E_{\rm CM} \to m_*$, because this physical regime was not taken into account in [69]. In the left-hand panel of Fig. 7.4 one sees that Eq. (7.28) reproduces the results of our Pythia simulation for $E_{\rm CM}$ smaller than a few TeV rather well. This was to be expected since Eq. (7.27) was determined in [69] from fits to data up to that energy. It is not the purpose of this paper to improve on this fit, as stated above, we simply use the above results as a check of our Pythia simulation.

[5] This is qualitatively understood by the fact that, in purely leptonic initial states, there is more energy available to produced hadrons, while in the case with protons in the initial state much energy is carried over by the initial hadron remnant.

7.4.5 *Enhancement of Number Density from String Fragmentation*

Production of composite states:
Prior to (p)reheating, we then have a yield of composite states given by the yield of strings, which can be estimated from Eq. (7.13), multiplied by the number of composite states per string

$$K^{\rm string} = \begin{cases} \frac{\frac{3}{4} g_q N_\psi^{\rm string}(E_{\rm CM}) + g_g (N_\psi^{\rm string}(E_{\rm CM}) - 1)}{g_{\rm TC}} & \text{heavy composite state,} \\ N_\psi^{\rm string}(E_{\rm CM}) & \text{light composite state}, \end{cases} \tag{7.29}$$

where $N_\psi^{\rm string}(E_{\rm CM})$ is given by Eq. (7.28) and $E_{\rm CM} = \sqrt{3\,\gamma_{\rm wp}\,T_{\rm nuc}\,f}$ in Eq. (7.23). We have distinguished the cases where the composite state of interest is heavier or lighter than the glueballs (e.g. the analogous of a proton or a pion in QCD). In the former case, the -1 we added to the factor multiplying g_g accounts for the fact that, if the final composite states produced by string fragmentation do not undergo other additional interactions, then glueballs decay to the light composite states and do not contribute to the final yield of any heavy composite state of quarks. The yield of composite states ψ then reads

$$Y_\psi^{\rm SC+string} = Y_{\rm TC}^{\rm eq}\, D^{\rm SC}\, K^{\rm string} \propto \left(\frac{T_{\rm nuc}}{f}\right)^3 \times \log s\left(\frac{\gamma_{\rm wp} T_{\rm nuc}}{f}\right). \tag{7.30}$$

The appearance of $Y_{\rm TC}^{\rm eq}$ in Eq. (7.30) accounts for string formation from both quarks and gluons. Hence, not only is the number of ψ's enhanced by the string fragmentation, relative to the case with no confinement, but also by the possibility of gluons to form strings. $K^{\rm string}$ and $Y_\psi^{\rm SC+string}$ are plotted in Fig. 7.8.

Hadrons are highly boosted in the plasma frame:
The hadrons formed after string fragmentation schematically consist of two equally abundant groups. Hadrons in the first group, which for later convenience we call 'Population A', move towards the bubble wall with an average energy

$$E_{\rm A,p} \simeq 2\gamma_{\rm cp}\frac{E_{\rm CM}}{N_\psi^{\rm string}(E_{\rm CM})} \simeq \frac{\gamma_{\rm wp} f}{N_\psi^{\rm string}(E_{\rm CM})}, \tag{7.31}$$

where we have boosted the energy per hadron of Eq. (7.26) to the plasma frame with the $\gamma_{\rm cp}$ of Eq. (7.25), and also used Eqs. (7.23) and (7.29). We conclude from Eq. (7.31) that the newly formed hadrons have large momenta in the plasma frame. The formation of a gluon string between the incoming techniquanta and the wall acts as a cosmological catapult which propels the string fragments in the direction the wall is moving. Hadrons in the second group move, in the wall frame, towards the bubble wall center, and their energy in the plasma frame is negligible compared to (7.31).

Note that if only one hadron is produced on average per every string, then it would roughly be at rest in the center-of-mass frame of the string, with an energy (mass) of order $E_{\rm CM}$. In the plasma frame, its energy would then read $E_{\rm p} \simeq \gamma_{\rm cp} E_{\rm CM} \simeq \gamma_{\rm wp} f/2$. As we will see in Sect. 7.8, the impact of this hadron on the final yield would then be captured by our expressions.

Following this first stage of string fragmentation, the composite states, and/or their decay products, can undergo further interactions with remnant particles of the bath, preheated or reheated plasma, and among themselves. Such interactions may change the ultimate yield of the relic composite states. Before taking these additional effects into account in Sect. 7.8, in the next sections we complete the modelling we proposed above, by describing the behaviour of the ejected quarks and deriving the Lorentz factor of the wall, $\gamma_{\rm wp}$.

7.4.6 Ejected Quarks and Gluons and Their Energy Budget

So far we dealt with what happens inside the bubble wall. The process we described apparently does not conserve color charge: we started with a physical quark or gluon with a net color charge entering the bubble, and we ended up with a system of hadrons which is color neutral. Where has the color charge gone?

The necessity of ejecting a quark or gluon:
To understand this, it is convenient to recall the physical modelling behind the process of string fragmentation that converts the initial fluxtube into hadrons, see e.g. the original Lund paper [71]. When the fluxtube length, in its center-of-mass frame, becomes of order f^{-1}, the string breaks at several points via the nucleation of quark-antiquark pairs from the vacuum. Now consider, in our cosmological picture, the quark-antiquark pair nucleated closest to the bubble wall. One of the two—say the antiquark—forms a hadron inside the wall. The only thing that can happen to the quark is for it to be ejected from the wall, because of the lack of charge partners inside the wall. This process, somehow reminiscent of black hole evaporation, thus allows for charge to be conserved. The momentum of the ejected quark, in the wall frame, has to be some order-one fraction of the confinement scale f, because that is the only energy scale in the process. For definiteness, in the following we will take this fraction to be a half. This picture is visualized in Fig. 7.3, and it is analogous if TC_i is a gluon instead of a quark.

Energy of the ejected quark or gluon:
One then has one ejected quark (at least) or gluon per fluxtube, thus per quark or gluon that initially entered. Therefore, the number of techniquanta outside the bubble wall does not diminish upon expansion of the bubble. This population of ejected techniquanta is energetically as important as that of hadrons inside the bubble. Indeed the energy of an ejected quark or gluon (or quark pair), in the plasma frame, reads

$$E_{\rm ej,p} \simeq \gamma_{\rm wp} f. \tag{7.32}$$

This is of the same order as the total energy in the hadrons from the fragmentation of a single string,

$$E_{\rm A,p}^{\rm tot} = \frac{N_\psi^{\rm string}(E_{\rm CM})}{2} E_{\rm A,p} \simeq \gamma_{\rm wp} \frac{f}{2}, \tag{7.33}$$

obtained by multiplying $E_{\rm A,p}$ of Eq. (7.31) times half of the total number of hadrons produced per string (i.e. we included only the energetic ones). The population of ejected techniquanta cannot therefore be neglected in the description of the following evolution of this cosmological system.

7.5 Bubble Wall Velocities

The wall boost in the plasma frame, $\gamma_{\rm wp}$, affects many key properties of our scenario, from the ejection of techniquanta to the number and energy of the hadrons produced by string fragmentation. It is the purpose of this section to study the possible values it can take over the PT.

Final results:
As bubbles are nucleated and start to expand, $\gamma_{\rm wp}$ starts growing as well. If nothing slows down the bubble-wall acceleration, then $\gamma_{\rm wp}$ keeps growing until its value at the time of bubble-wall collision, $\gamma_{\rm wp}^{\rm runaway}$. Sources of friction that could prevent this runaway regime are given by the equivalent, in this scenario, of the so-called leading order (LO) and next-to-leading order (NLO) contributions of [72, 73] respectively. We find it convenient to report right away our final result for the maximal possible value of $\gamma_{\rm wp}$,

$$\gamma_{\rm wp}^{\rm max} \simeq {\rm Min}\left[1.7 \frac{10}{\beta/H}\left(\frac{0.01}{c_{\rm vac}}\right)^{\frac{1}{2}} \frac{T_{\rm nuc}}{f}\frac{M_{\rm Pl}}{f},\ 1.0 \times 10^{-3} \frac{c_{\rm vac}}{0.01}\frac{80}{g_{\rm TC}}\left(\frac{f}{T_{\rm nuc}}\right)^3\right], \tag{7.34}$$

where the first entry is associated to $\gamma_{\rm wp}^{\rm runaway}$, and the second to the boost as limited by the LO pressure, $\gamma_{\rm wp}^{\rm LO}$. $\gamma_{\rm wp}^{\rm LO}$ is always smaller than $\gamma_{\rm wp}^{\rm NLO}$ in the parameter space of our interest, so that $\gamma_{\rm wp}^{\rm NLO}$ does not enter Eq. (7.34). We learn that in the regime of very strong supercooling and/or of very large confinement scale f, which will be the most relevant one for the DM abundance, bubble walls run away. The behaviour of $\gamma_{\rm wp}$ is illustrated in Fig. 7.5.

The impact on GW:
The behaviour of $\gamma_{\rm wp}$ also has important consequences for the gravitational wave signal from the phase transition [74, 75]. If $\gamma_{\rm wp}^{\rm max} = \gamma_{\rm wp}^{\rm runaway}$ then the vacuum energy is converted into kinetic energy of the bubble walls [76]. The gravitational wave (GW) spectrum sourced by scalar field gradient is traditionally computed in the

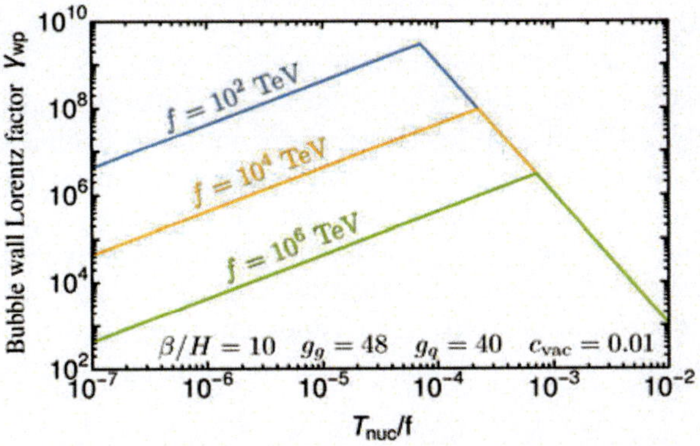

Fig. 7.5 The Lorentz factor of the wall at bubble percolation for various values of f and amounts of supercooling, assuming $\beta/H = 10$. For extreme supercooling (on the left side of the plot) γ_{wp} is in the runaway regime. In this regime, larger f or smaller T_{nuc} leads to a smaller distance over which the bubble can accelerate. The former because of the smaller Hubble horizon and the latter due to the larger bubble size at nucleation. Therefore γ_{wp} decreases for more supercooling

envelope approximation [77–79]. However, the latest lattice results [80, 81] suggest an enhancement of the GW spectrum at low frequency due to the free propagation of remnants of bubble walls after the collision, the IR slope $\propto k^3$ becoming close to $\propto k^1$. This confirms the predictions from the analytical bulk flow model [82, 83]. Note that the IR-enhancement is stronger for thick-walled bubbles [80], which is the case relevant for nearly-conformal potential leading to strong supercooling, and thus for the PT considered here. (Instead, for thin-walled bubbles, after collision the scalar field can be trapped back in the false vacuum [12, 45, 84]. Instead of propagating freely, the shells of energy-momentum tensor remain close to the collision point and dissipate via multiple bounces of the walls.) Irrespectively of whether the IR slope at $f \lesssim \beta$ is $\propto k^3$ or $\propto k^1$, at much lower frequency, $f \lesssim H$, the slope must converge to k^3 due to causality [85–87]. Oscillations of the condensate following the PT can provide an additional source of GW [88]. However, instead of β^{-1} the time scale is set by the inverse scalar mass $\sim f^{-1}$ and the signal is Planck-suppressed $\propto \beta/f$ [89].

If instead, $\gamma_{wp}^{max} = \gamma_{wp}^{NLO}$, the vacuum energy is converted into thermal and kinetic energy of the particles in the plasma already prior to the bubble wall collision. The contribution from sound waves or turbulence [74, 75], however, in supercooled transitions is not yet clearly understood. Indeed, current hydrodynamical simulations, which aim to capture the contribution of the bulk motion of the plasma to the gravitational wave signal, do not yet extend into the regime in which the energy density in radiation is subdominant to the vacuum [90]. And analytical studies of shock-waves in the relativistic limit have just started [91]. In any case, we expect supercooled transitions to provide promising avenue for detection in future GW observatories. We now proceed to a detailed derivation of Eq. (7.34).

Linear growth:
The energy gained upon formation of a bubble of radius R is $E_{\rm bubble} = \frac{4}{3}\pi R^3 \Delta V_{\rm vac}$, where $\Delta V_{\rm vac}$ is the difference between the vacuum energy density outside and inside the bubble. The energy lost upon formation of a bubble of radius R is $E_{\rm wall} \simeq 4\pi R^2 \gamma_{\rm wp}\sigma_w$, where σ_w is the surface energy density of the wall (surface tension) in the wall frame. If a bubble nucleates and expands, its energy $E_{\rm bubble}$ is transferred to the wall energy $E_{\rm wall}$. As soon as a nucleated bubble contains the region $\chi \simeq f$, neither $\Delta V_{\rm vac}$ nor σ_w change upon bubble expansion. Indeed both are a function of the bubble wall profile, which does not change in that regime (also see Fig. 7.1). We thus recover the well-known property that $\gamma_{\rm wp}$ grows linearly in R,

$$\gamma_{\rm wp} = \frac{R}{R_0} \sim T_{\rm nuc}\, R\,, \tag{7.35}$$

where R_0 is a normalisation of the order of the minimal radius needed for a bubble to nucleate, and where in the second relation we have used $R_0 \gtrsim L_w \sim T_{\rm nuc}^{-1}$ because we assumed the nucleated bubble to contain the region $\chi \simeq f$. A more precise treatment can be found, e.g. in the recent [76], which confirms the parametric dependence of Eq. (7.35).

At collision time:
In a runaway regime, i.e. for small enough retarding pressure on the bubble walls, $\gamma_{\rm wp}$ at collisions then reads

$$\gamma_{\rm wp}^{\rm runaway} \sim T_{\rm nuc}\,\beta^{-1} \simeq 1.7\,\frac{10}{\beta/H}\left(\frac{0.01}{c_{\rm vac}}\right)^{\frac{1}{2}}\frac{T_{\rm nuc}}{f}\frac{M_{\rm Pl}}{f}, \tag{7.36}$$

where β^{-1} is the average radius of bubbles at collision, $H \simeq \Lambda_{\rm vac}^2/(\sqrt{3}M_{\rm Pl})$, and the value $\beta/H \simeq 10$ is a benchmark typical of supercooled phase transitions [18, 24, 28, 36, 37, 42, 92], which we employ from now on.

The bubbles swallow most of the volume of the universe, and thus most techniquanta, when their radius is of the order of their average radius at collision β^{-1}. Therefore, in the regime of runaway bubble walls, the relevant $\gamma_{\rm wp}$ for all the physical processes of our interest (hadron formation from string fragmentation, quark ejection, etc.) will be some order one fraction of $\gamma_{\rm wp}^{\rm runaway}$. For simplicity, in the runaway regime we will then employ the simplifying relation $\gamma_{\rm wp} = \gamma_{\rm wp}^{\rm runaway}$. This will not only be a good-enough approximation for our purposes, but it will also allow to clearly grasp the parametric dependence of our novel findings. Moreover, a more precise treatment, to be consistent, would need to be accompanied by a more precise solution for $\gamma_{\rm wp}$ than that of Eq. (7.35), i.e. we would need to specify the potential driving the supercooled PT and solve for $\gamma_{\rm wp}$. As the purpose of this paper is to point out effects which are independent of details of the specific potential, we leave a more precise treatment to future work.

7.5.1 LO Pressure

Origin:
By LO pressure we mean the pressure from the partial conversion—of the quark's momenta before entering the bubbles—into hadron masses [72], plus that from the ejection of quarks. We use the subscript LO in reference to [72, 73], because this pressure is of the form $\mathscr{P}_{\mathrm{LO}} \sim \Delta m^2 T^2$, where Δm is the rest energy of the flux tube between the incoming techni-quanta and the wall. However, in contrast to [72, 73], here the pressure arises from non-perturbative effects.

Momentum transfer:
The momentum exchanged with the wall, upon hadronization of a single entering quark plus the associated quark ejection, reads in the wall frame

$$\Delta p_{\mathrm{LO}} = E_{\mathrm{in}} - \sqrt{E_{\mathrm{in}}^2 - \Delta m_{\mathrm{in}}^2} + E_{\mathrm{ej}} \simeq f \,, \tag{7.37}$$

where $E_{\mathrm{in}} \simeq 3\,\gamma_{\mathrm{wp}} T_{\mathrm{nuc}}$ is the energy of the incoming quark, Δm_{in}^2 is the fraction of that energy that is converted into 'inertia' of the string, and $E_{\mathrm{ej}} \simeq f/2$ is the energy of the ejected quark or gluon. In the second equality, we have used $\Delta m_{\mathrm{in}}^2 \simeq E_{\mathrm{CM}}^2 \simeq 3\,\gamma_{\mathrm{wp}}\, T_{\mathrm{nuc}}\, f$ from Eq. (7.23) and $\gamma_{\mathrm{wp}} \gg f/T_{\mathrm{nuc}}$. Note that Δp_{LO} is independent of p_{in}.

Pressure:
In light of Sect. 7.4.1, we can safely consider a collision-less approach and neglect the interactions between neighboring quarks. The associated pressure is given by

$$\mathscr{P}_{\mathrm{LO}} = \sum_a g_a \int \frac{d^3 p_{\mathrm{in}}}{(2\pi)^3} \frac{1}{e^{|p_{\mathrm{in}}|/T_{\mathrm{nuc}}} \pm 1} \Delta p_{\mathrm{LO}}, \tag{7.38}$$

where g_a is the number of internal degrees of freedom of a given species a of the techniquanta. Upon using Eq. (7.37), we get

$$\mathscr{P}_{\mathrm{LO}} \simeq \frac{\zeta(3)}{\pi^2}\, g_{\mathrm{TC}}\, \gamma_{\mathrm{wp}}\, T_{\mathrm{nuc}}^3 f \,, \tag{7.39}$$

where we remind that $g_{\mathrm{TC}} = g_g + \frac{3 g_q}{4}$. This result can be understood intuitively from $\mathscr{P}_{\mathrm{LO}} \sim n_{\mathrm{TC},w} \Delta p_{\mathrm{LO}}$, where γ_{wp} enters through $n_{\mathrm{TC},w}$ [72]. Note that, in the absence of ejected particles, the pressure would have been a half of our result in Eq. (7.39).

Terminal velocity:
The resulting upper limit on γ_{wp} is obtained by imposing that the LO pressure equals that of the internal pressure from the difference in vacuum energies,

$$\mathcal{P}_{\text{expand}} = c_{\text{vac}} f^4 , \tag{7.40}$$

and reads

$$\gamma_{\text{wp}}^{\text{LO}} = c_{\text{vac}} \frac{\pi^2}{\zeta(3)} \frac{1}{g_{\text{TC}}} \left(\frac{f}{T_{\text{nuc}}} \right)^3 . \tag{7.41}$$

We finally remark that $\mathcal{P}_{\text{LO}}$ grows linearly in γ_{wp}, unlike in 'standard' PTs where it is independent of the boost. The reason lies in the fact that the effective mass Δm_{in}^2 grows with γ_{wp}, whereas in 'standard' PTs it is constant in γ_{wp}. Our results then imply that, in confining phase transitions, the LO pressure is in principle enough to ensure the bubble walls do not runaway asymptotically. This is to be contrasted with non-confining PTs, where the asymptotic runaway is only prevented by the NLO pressure.[6]

7.5.2 *NLO Pressure*

Origin:
The NLO pressure comes from the techniquanta radiating a soft gluon [73] which itself forms a string attached to the wall in the broken phase.

Result:
We derive it in detail in Sect. 6.2.5. We find

$$\mathcal{P}_{\text{NLO}} \simeq \left(g_g C_2[g] + \frac{3}{4} g_q C_2[q] \right) \frac{8\zeta(3)}{\pi} \frac{g_{\text{conf}}^2}{4\pi} \varepsilon_{\text{ps}} \frac{\log\left(1 + \frac{m_g^2}{k_*^2}\right)}{k_*/m_g} \gamma_{\text{wp}} T_{\text{nuc}}^3 m_g , \tag{7.42}$$

where $C_2[g, q]$ are the second Casimirs of the representations of gluons and quarks under the confining group (if $SU(N)$, $C_2[g] = N$, $C_2[q] = (N^2 - 1)/2N$), g_{conf} is the gauge coupling of the confining group, $\varepsilon_{\text{ps}} \leq 1$ encodes the suppression from phase-space saturation of the emitted soft quanta g, important for large coupling g_{conf}, m_g is an effective mass of the soft radiated gluons responsible for this pressure, and k_* the IR cut-off on the momentum radiated in the direction parallel to the wall.

Vector boson mass:
As we model the masses of our techniquanta as the inertia that their fluxtube would gain inside the bubble, these masses increase with increasing momentum of the techniquanta, in the wall frame. The NLO pressure is caused by emission of gluons 'soft' with respect to the incoming quanta. Their would-be mass m_g upon entering the wall cannot, therefore, be as large as that of the incoming quanta that emit them,

[6] In our scenario, bubble walls can still run away until collision for some values of the parameters, and we anticipate they will. Unlike in non-confining PTs, the scaling of our LO pressure with γ_{wp} implies they could not runaway indefinitely if there were no collisions.

$\Delta m_{\text{in}} \simeq \sqrt{3\,\gamma_{\text{wp}}\,T_{\text{nuc}}\,f}$. At the same time, the effective gluon mass should at least allow for the formation of one hadron inside the wall, therefore we assume it to be of the order of the confinement scale, $m_g \sim f$. The fact that m_g does not grow with γ_{wp} while Δm_{in} does, is the reason why unlike in non-confining phase transitions, we find here that $\mathscr{P}_{\text{NLO}}$ and $\mathscr{P}_{\text{LO}}$ have the same scaling in γ_{wp} and in the amount of supercooling.

NLO pressure is sub-leading:
By making the standard [73] choice $k_* \simeq m_g$, and assuming $\varepsilon_{\text{ps}} g_{\text{conf}}^2 < 1$, we then find that $\mathscr{P}_{\text{NLO}} \ll \mathscr{P}_{\text{LO}}$ in the entire parameter space of our interest. Thus, for simplicity, we do not report the NLO limit on $\gamma_{\text{wp}}^{\text{max}}$ in Eq. (7.34).

Recently, Ref. [93] performed a resummation of the log-enhanced radiation that leads to the scaling $\mathscr{P}_{\text{NLO}} \propto g_{\text{conf}}^2 \gamma_{\text{wp}}^2 T_{\text{nuc}}^4$. By using the analogue of that result for confining theories, we find that $\mathscr{P}_{\text{NLO}}$ dominates over $\mathscr{P}_{\text{LO}}$ in some region of parameter space, and therefore that the values of the parameters for which bubble walls run away slightly change. Still, even by using that resummed result, we find that the region relevant for DM phenomenology corresponds to the region where bubble walls run away, so that the difference between the results of [73, 93] does not impact the DM abundance. As observed in [94], the pressure as determined in [93] does not tend to zero when the order parameter of the transition goes to zero, casting a shadow on that result. As pointed out in [95], this might result from the violation of energy-momentum conservation due to the presence of the wall not being correctly accounted. Therefore, both for this issue as well as for the limited impact on the DM abundance that we will discuss later, we content ourselves with a treatment analogous to [73] in our paper.

Summary and runaway condition:
At small supercooling (i.e. not too small T_{nuc}/f) the bubble wall velocity reaches an equilibrium value set by the LO pressure. At larger supercooling bubble walls collide before reaching their terminal LO velocity, and γ_{wp} is set by the runaway value Eq. (7.36). By comparing Eq. (7.41) with Eq. (7.36), we find that bubble walls run away for

$$\frac{T_{\text{nuc}}}{f} \lesssim 1.2 \times 10^{-4} \left(\frac{80}{g_{\text{TC}}} \frac{\beta/H}{10} \frac{f}{\text{PeV}}\right)^{\frac{1}{4}} \left(\frac{c_{\text{vac}}}{0.01}\right)^{\frac{3}{8}} \tag{7.43}$$

The bubble wall Lorentz factor is plotted in Fig. 7.5 against the amount of supercooling.

7.5.3 Ping-Pong Regime

Condition to enter:
For even a single hadron to form inside the bubble, one needs $E_{\rm CM} \geq m_\pi$, where π is the lightest hadron of the new confining sector (e.g. a pseudo-goldstone boson). Via Eq. (7.23), this implies

$$\gamma_{\rm wp} \gtrsim \gamma_{\rm wp}^{\rm enter} = \frac{m_\pi^2}{3\,T_{\rm nuc}\,f}\,. \tag{7.44}$$

Contribution to the pressure:
For $\gamma_{\rm wp} \lesssim \gamma_{\rm wp}^{\rm enter}$, which holds at least in the initial stages of the bubble expansion, the quarks and gluons are reflected and induce a pressure

$$\mathcal{P}_{\rm refl} \sim n_{{\rm TC},w} \times \Delta p_{{\rm TC},w} \sim T_{\rm nuc}^3 \gamma_{\rm wp} \times \gamma_{\rm wp} T_{\rm nuc} \sim \gamma_{\rm wp}^2 T_{\rm nuc}^4\,. \tag{7.45}$$

This is to be compared with Eq. (7.40), $\mathcal{P}_{\rm expand} = c_{\rm vac} f^4$, which implies the bubble wall could in principle be limited by this pressure to $\gamma_{\rm wp} \sim (f/T_{\rm nuc})^2$. Nevertheless, as $(f/T_{\rm nuc})^2 \gg \gamma_{\rm wp}^{\rm enter}$, this pressure ceases to exist at an earlier stage of the expansion, namely once $\gamma_{\rm wp} = \gamma_{\rm wp}^{\rm enter}$. Hence the maximum Lorentz factor remains encapsulated by Eq. (7.34).

Ping-pong regime:
In some extreme regions of parameter space, however, one could have $\gamma_{\rm wp}^{\rm max} < \gamma_{\rm wp}^{\rm enter}$, so that all techniquanta in the plasma are reflected at least once before entering a bubble. We leave a treatment of this 'ping-pong' regime to future work.

7.6 Amount of Supercooling Needed for Our Picture to Be Relevant

Intuition about the limit of no supercooling:
In the limit of no supercooling, one does not expect the fluxtubes to attach to the bubble wall, but rather to connect the closest charges that form a singlet and induce their confinement. In other words, in the limit of no supercooling one expects the picture of confinement to be the one of 'standard phase transitions'. By continuity, there should exist a value of $T_{\rm nuc}$, smaller than f, such that the our picture ceases to be valid, and one instead recovers the more familiar confinement among closest color charges. We now wish to determine it. In order to do so, we note that the absence of ejected techniquanta is a necessary condition for the above to hold, therefore we now phrase the problem in terms of absence of ejected techniquanta.

Rate of detachment of ★:
We propose and analyse some effects that could lead to fluxtubes detaching from the bubble walls without ejecting particles. To take place, these effects need to happen before the end-point of the fluxtube on the wall, ★, ceases to exist, i.e. when the string breaking inside the bubble has already taken place and a quark is ejected. So we start by computing the rate $\Gamma_{\text{det}\bigstar}$ of detachment of ★, the point where the fluxtubes is attached to the wall, from the wall itself. To estimate it, we again borrow the modelling of the classic paper on string fragmentation [71].

The distances between the several points of breaking of a given string (that connects in our case TC_i and ★) are space-like. In the frame of each point of breaking, that breaking is itself the first to happen, a time of order N/f after the string formation (we adopt the scaling for strong sector gauge groups $SU(N)$ [96, 97]). This time therefore also applies to the outermost breaking point in our picture, i.e. that closest to the wall, whose frame approximately coincides with the wall frame. We remind the reader that the outermost breaking is the one that nucleates the quark or gluon that is eventually ejected. The rate we need can therefore be estimated as the inverse of the nucleation time of the outermost pair,

$$\Gamma_{\text{det}\bigstar,\text{w}} = \tau^{-1}_{\text{det}\bigstar,\text{w}} \simeq f/N\,. \tag{7.46}$$

We now enumerate and model effects that could lead to fluxtubes detaching from the bubble walls without ejecting techniquanta, and compare their time scales with Eq. (7.46).

1. **Flux lines overlap** The faster a bubble-wall, the denser and thus the closer together in the wall frame are the quarks and gluons entering it. Eventually, they could get closer than the typical transverse size of a fluxtube $d_{\text{tr}} \simeq f^{-1}$ [98]. When that happens, the fluxtubes between different color charges have a non-negligible overlap. We expect that in this situation it will not be clearly preferable energetically for these strings to attach directly to the wall. Thus there would be no ejected techniquanta. This situation is of course realised also in the case of small supercooling f/T_{nuc}, in addition to and independently of the case of fast bubble-walls.
   We then obtain a rate of 'string breaking by fluxtube overlap', Γ_{overlap}, as follows. We define an effective associated cross section as the area of a circle on the wall, centered on any ★ and with radius d_{tr},

$$A_{\text{overlap}} = \pi d_{\text{tr}}^2 \simeq \pi f^{-2}\,. \tag{7.47}$$

   The associated rate then reads

$$\Gamma_{\text{overlap}} = A_{\text{overlap}}\, v\, n_{\text{TC,w}} \simeq \frac{\gamma_{\text{wp}}\zeta(3) g_{\text{TC}}}{\pi}\frac{T_{\text{nuc}}^3}{f^2}\,, \tag{7.48}$$

   where $n_{\text{TC,w}} = \gamma_{\text{wp}} n_{\text{TC,p}}$ is the density of techniquanta in the wall frame, $g_{\text{TC}} = g_g + 3g_q/4$, and we have used that they are relativistic $v = 1$. The condition of no ejected techniquanta then reads

$$\Gamma_{\text{overlap}} > \Gamma_{\text{det}\bigstar,\text{w}} \Rightarrow \gamma_{\text{wp}} \gtrsim \frac{2.6}{g_{\text{TC}} N} \left(\frac{f}{T_{\text{nuc}}} \right)^3 . \tag{7.49}$$

2. The entire fluxtube connecting real color charges, so including its portion in the region $\chi \simeq \chi^* \ll f$ (see Fig. 7.2), could enter the region $\chi = f$ before its portions in the region $\chi = f$ break and form hadrons, and eject particles. We see two ways this could happen.

   2.1 **Attractive interaction between neighboring flux lines**
   The points $\bigstar$ are not static, because they move by the force exerted by the part of the string which is outside the wall, in the layer where $\langle \chi \rangle \simeq \chi^*$. Defining $y_\bigstar$ as the transverse distance, on the wall, between two $\bigstar$ points connected by a fluxtube, one has

$$\frac{d^2 y_\bigstar}{dt^2} = \frac{F}{m_\bigstar} \sim -\frac{dE_{q\bar{q}}/dy}{f} \simeq -c_{q\bar{q}} \frac{\chi^{*2}}{f} \sim -c_{q\bar{q}} \frac{T_{\text{nuc}}^2}{f} , \tag{7.50}$$

   where, consistently with our previous treatments, we have assigned to $\bigstar$ an inertia $m_\bigstar \sim f$. If $y_\bigstar$ goes to zero in a time shorter than the breaking time $\tau_{\text{det}\bigstar,\text{w}} \sim N f^{-1}$, then the two fluxtubes connect and become fully contained in the region $\chi = f$ before they break and form hadrons, and thus there are no ejected techniquanta. To determine this condition, we assume initially static points $\bigstar$, and thus we only need the initial distance between them $y_\bigstar(t = 0) \simeq (\gamma_{\text{wp}} n_{\text{TC,p}})^{-1/3}$. We then obtain

$$y_\bigstar(t = \tau_{\text{det}\bigstar,\text{w}}) \simeq (\gamma_{\text{wp}} n_{\text{TC,p}})^{-1/3} - c_{q\bar{q}} \frac{T_{\text{nuc}}^2}{f} \frac{\tau_{\text{det}\bigstar,\text{w}}^2}{2} . \tag{7.51}$$

   The resulting condition for no ejected quarks reads

$$y_\bigstar(t = \tau_{\text{det}\bigstar,\text{w}}) < 0 \Rightarrow \gamma_{\text{wp}} \gtrsim \frac{6.6 \times 10^{-2}}{g_{\text{TC}} N^6} \left(\frac{10}{c_{q\bar{q}}} \right)^3 \left(\frac{f}{T_{\text{nuc}}} \right)^9 . \tag{7.52}$$

   2.2 **Limit of no distortion of the flux lines**
   When the string portion in the region $\chi \simeq \chi^* \ll f$ has a small enough length $d_\bigstar$, the possibility that it is pulled inside the region $\chi = f$ could be energetically more convenient than the one of our picture, where it stays outside and instead energy goes in increasing the length of the strings that are perpendicular to the wall. The energy price, for the string portion in the region $\chi \simeq \chi^* \ll f$ to enter the region $\chi = f$, reads in the wall frame

$$\Delta E_{\text{pull-in,w}} \simeq c_{q\bar{q}} (f^2 - \chi^{*2}) d_\bigstar \simeq c_{q\bar{q}} f^2 d_\bigstar , \tag{7.53}$$

where we stress that the length of the string portion $d_\bigstar$ is transverse to the bubble-wall velocity and therefore is not Lorentz contracted in the process of being pulled into the bubble. In the wall frame, it reads $d_\bigstar \simeq (\gamma_{\rm wp} n_{\rm TC,p})^{-1/3}$ The transition between $\chi \simeq \chi^* \ll f$ and $\chi = f$ is exponentially fast in the proper coordinate s (see Appendix 7.A), and happens over an interval (a distance, in the wall frame) $L_{\rm f} \sim f^{-1}$. The energy price of Eq. (7.53) should therefore be compared with the one to stretch two strings, inside the wall, by an amount $L_{\rm f}$:

$$\Delta E_{\rm stretch,w} \simeq 2c_{q\bar{q}} f^2 L_{\rm f}/\gamma_{\rm wc} \sim 2c_{q\bar{q}} f \left(\frac{f}{3\,\gamma_{\rm wp} T_{\rm nuc}}\right)^{1/2}, \tag{7.54}$$

where we have used that the string length in the expression for $E_{q\bar{q}}$, Eq. (7.21), has to be evaluated in the string center-of-mass frame, and that $\gamma_{\rm wc} \simeq \sqrt{3\,\gamma_{\rm wp} T_{\rm nuc}/f}$ from Eq. (7.24). Therefore, it is energetically more convenient to pull the fluxtube inside the region $\chi \simeq f$, and so to have no ejected quarks, if

$$\Delta E_{\rm pull\text{-}in} < \Delta E_{\rm stretch} \Rightarrow \gamma_{\rm wp} \lesssim 0.035\, g_{\rm TC}^2 \left(f\, L_f\right)^6 \left(\frac{T_{\rm nuc}}{f}\right)^3. \tag{7.55}$$

Contrary to the previous two possibilities to have no ejected quarks, Eqs. (7.49) and (7.52), the possibility in Eq. (7.55) imposes an upper limit on $\gamma_{\rm wp}$. We anticipate that, in the regimes of supercooling interesting for our work $T_{\rm nuc}/f \ll 1$, Eq. (7.55) cannot be satisfied consistently with $\gamma_{\rm wp} > 1$, so that it is not relevant for our work.

Summary of required supercooling

In the regime where $T_{\rm nuc} \gtrsim f$, we expect that neither ejection of techniquanta nor string fragmentation should take place, and that the standard picture of quarks and gluons confining with their neighbors should be recovered (which we dub the 'standard phase transition'). More precisely, if any of Eqs. (7.49), (7.52) and (7.55) hold, we depart from our picture in at least one regard. By demanding none of these inequalities hold, we expect the new effects of our study, namely flux line attached to the wall, string fragmentation, quark ejection and deep inelastic scattering, to take place. In the non-runaway regime, we require

$$c_{\rm vac} \lesssim \frac{0.32}{N} \quad \text{and} \tag{7.56}$$

$$\frac{T_{\rm nuc}}{f} \lesssim {\rm Min}\left[0.19\left(\frac{5}{N}\right)\left(\frac{0.01}{c_{\rm vac}}\right)^{1/6}\left(\frac{10}{c_{q\bar{q}}}\right)^{1/2},\ \frac{0.12}{f L_f}\left(\frac{c_{\rm vac}}{0.01}\right)^{1/6}\left(\frac{90}{g_{\rm TC}}\right)^{1/2}\right],$$

for our picture to hold. In the runaway regime, we instead require

$$\frac{T_{\rm nuc}}{f} \lesssim$$
$$\mathrm{Min}\Bigg[6.1\times 10^{-5}\left(\frac{\beta/H}{10}\right)^{1/4}\left(\frac{c_{\rm vac}}{0.01}\right)^{1/8}\left(\frac{90}{g_{\rm TC}}\right)^{1/4}\left(\frac{5}{N}\right)^{1/4}\left(\frac{f}{10~\mathrm{TeV}}\right)^{1/4},$$
$$6.4\times 10^{-3}\left(\frac{\beta/H}{10}\right)^{1/10}\left(\frac{c_{\rm vac}}{0.01}\right)^{1/20}\left(\frac{10}{c_{q\bar q}}\right)^{3/10}\left(\frac{90}{g_{\rm TC}}\right)^{1/10}\left(\frac{5}{N}\right)^{3/5}\left(\frac{f}{10~\mathrm{TeV}}\right)^{1/10},$$
$$\frac{1.2\times 10^{6}}{(fL_f)^3}\left(\frac{10}{\beta/H}\right)^{1/2}\left(\frac{0.01}{c_{\rm vac}}\right)^{1/4}\left(\frac{90}{g_{\rm TC}}\right)\left(\frac{10~\mathrm{TeV}}{f}\right)^{1/2}\Bigg]. \tag{7.57}$$

for our picture to hold. Here we have used $\gamma_{\rm wp}$ in Eq. (7.34). The conditions are visually summarised in Fig. 7.6.

In light of this figure, we conclude that some new effects pointed out in our study are also relevant in confining phase transitions where $T_{\rm nuc} \sim T_{\rm start} \sim T_c$ (see Appendix 7.A for the definition of the critical temperature T_c), e.g. [99–105], provided $c_{\rm vac}$ is small enough. A possible impact on the QCD phase transition, e.g. [106–115], remains to be investigated.

Averaged quantities only:
We conclude this section by also stressing that all the conditions above refer to averaged quantities, and therefore do not take into account the leaks from tails of distributions. These leaks could for example imply that there are a few strings that hadronise without ejecting particles, even if all conditions Eqs. (7.49), (7.52) and (7.55) are violated. As these strings constitute a small minority of the total ones, these effects

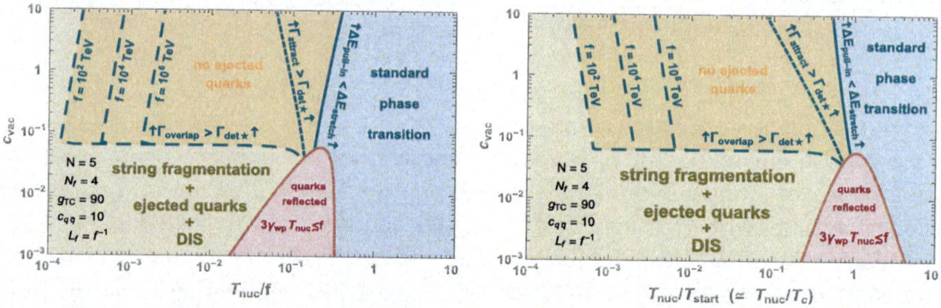

Fig. 7.6 **Blue Region**: The incoming techniquanta confine with their neighbours as in the standard picture of phase transitions that are not supercooled. **Olive Region**: All three inequalities, (7.49), (7.52), (7.55), are violated and the new effects pointed out in this study, i.e. string fragmentation, ejection of techniquanta and deep inelastic scattering, should be taken into account. **Orange Region**: At least one but not all of the inequalities above hold, therefore there are no ejected techniquanta. The dynamics taking place in this region remains to be investigated. **Purple Region**: Quarks are too weakly energetic to enter the bubbles, see. Sect. 7.5.3. **Left of solid line**: Eq. (7.55) is violated and it is energetically favourable for the flux lines to be distorted. **Left of dotted line**: Eq. (7.52) is violated we can neglect the attractive interactions between neighboring flux lines. **Left of dashed lines**: Eq. (7.49) is violated and we can neglect the overlap of neighbouring flux lines. The two plots only differ through their horizontal axis, see Sect. 7.3.2 for the definitions of $c_{\rm vac}$ and $T_{\rm start}$, and Appendix 7.A for that of T_c. To avoid the unphysical values $\gamma_{\rm wp} < 1$, we have added 1 to Eq. (7.34)

have a negligible impact on the phenomenology we discuss. They could however be important in studying other situations of supercooled confinement. Though certainly interesting, the exploration of these effects goes beyond the scope of this paper.

7.7 Ejected Quarks and Gluons

7.7.1 Density of Ejected Techniquanta

In the wall frame, since we have one ejected quark or gluon per each incoming one, we find

$$n_{\rm ej,w} = n_{\rm TC,w}(r_{\rm ej}) = \gamma_{\rm wp}(r_{\rm ej}) n_{\rm TC,p}\,, \tag{7.58}$$

where $n_{\rm TC,p} = g_{\rm TC}\zeta(3)T_{\rm nuc}^3/\pi^2$ is the density of the diluted bath in the plasma frame. The density of ejected techniquanta then depends on the time passed since bubble wall nucleation, or equivalently on the bubble radius at the time of ejection $r_{\rm ej}$, via $\gamma_{\rm wp}(r)$ (see Sect. 7.5). In the plasma frame, and at a given distance D from the center of the bubble, we then have[7]

$$n_{\rm ej,p}(D) = 2\gamma_{\rm wp}^2(r_{\rm ej})\left(\frac{r_{\rm ej}}{D}\right)^2 n_{\rm TC,p}\,, \tag{7.59}$$

where we have included the surface dilution from the expansion between the radius at which a given quark has been ejected, $r_{\rm ej}$, and the radius D where we are evaluating $n_{\rm ej,p}$.

Radial dependence:
It is convenient to express $n_{\rm ej,p}$ as a function of the radial distance x from the bubble wall in the plasma frame, where for definiteness $x = 0$ denotes the position of the wall and $x = L_{\rm ej,p}$ the position of the techniquanta ejected first (which constitute the outermost layer). In order to do so, we determine the relation between the position x of a quark and the radius $r_{\rm ej}(x)$ when it has been ejected. We assume that the bare mass of the quarks is small enough such that they move at the speed of light, like the gluons. The wall at $x = 0$, instead, moves at a speed $v_{\rm wall} \simeq 1 - 1/(2\gamma_{\rm wp}^2)$ (we have used the relativistic limit $\gamma_{\rm wp} \gg 1$), dependent on its radius. The coordinate x of a given layer of ejected particles can then be found by integrating the difference between the world line of an ejected particle and that of the wall,

$$x = \int_{t_{\rm ej}}^{t_D} dt(1 - v_{\rm wall}) \simeq \int_{t_{\rm ej}}^{t_D} \frac{1}{2\gamma_{\rm wp}^2(t)} \simeq \frac{1}{2T_{\rm nuc}^2}\left(\frac{1}{t_{\rm ej}} - \frac{1}{t_D}\right), \tag{7.60}$$

[7] The factor 2 arises when we boost the quark current $(\gamma_{\rm wp}\, n_{\rm TC,p}, \gamma_{\rm wp}\, \beta\, n_{\rm TC,p})$, with $\beta = e_r$, from the wall to the plasma frame.

where we defined t_D and $t_{\rm ej}$ as the times when the bubble radius is respectively D and $r_{\rm ej}$, and we used $\gamma_{\rm wp}(t) \simeq T_{\rm nuc} t$, cf. Eq (7.35), valid up to relative orders $1/\gamma_{\rm wp}^2 \ll 1$. It is convenient to rewrite Eq. (7.60) as

$$r_{\rm ej}(x) \simeq \frac{D}{1 + 2\,T_{\rm nuc}^2\,D\,x}\,. \tag{7.61}$$

We finally obtain

$$n_{\rm ej,p}(x) = \frac{2\,\gamma_{\rm wp}^2(x)}{\left(1 + 2\,T_{\rm nuc}^2\,D\,x\right)^2} n_{\rm TC,p} \simeq \frac{2\,T_{\rm nuc}^2\,D^2}{\left(1 + 2\,T_{\rm nuc}^2\,D\,x\right)^2} n_{\rm TC,p}\,, \tag{7.62}$$

where the last equality is valid as long as the bubbles run away, i.e. as long as Eq. (7.35) $\gamma_{\rm wp} \simeq T_{\rm nuc}\,r$ holds.

Thickness of the layer of ejected techniquanta:
Our result Eq. (7.62) implies that the highest density, of ejected techniquanta, is located in the shell within a distance of the bubble wall

$$L_{\rm ej,p}^{\rm eff} \simeq \frac{1}{2\,T_{\rm nuc}^2\,D}\,. \tag{7.63}$$

The density of ejected quarks $n_{\rm ej,p}(x)$ extends to $x = L_{\rm ej,p}$, i.e. to the outermost ejected layer, that we now show to be much larger than $L_{\rm ej,p}^{\rm eff}$. Indeed, $L_{\rm ej,p}$ can be related to the time $t_{\rm first}$ of ejection of the first techniquanta (corresponding to $\gamma_{\rm wp} \simeq m_\pi/T_{\rm nuc}$, Eq. (7.44)). Using $t_i \gg t_{\rm first}$ and $t_{\rm first} \sim m_\pi/T_{\rm nuc}^2$, we find

$$L_{\rm ej,p} \simeq \frac{1}{t_{\rm first}\,2\,T_{\rm nuc}^2} \sim \frac{1}{f}\,, \tag{7.64}$$

where for simplicity we have assumed $m_\pi \simeq f$ as in QCD. As long as $L_{\rm ej,p} \gg L_{\rm ej,p}^{\rm eff}$, as it holds for our estimate Eq. (7.64), the value of $L_{\rm ej,p}$ does not affect any of the results of this paper.[8] The density profile of Eq. (7.62) is shown in Fig. 7.7.

Sanity check:
As a check of our result Eq. (7.62), we verify that one has one ejected quark or gluon per each one that entered the bubble. Indeed, we compute

[8] One could easily envisage situations in which m_π differs sizeably from f, e.g. because pions are much lighter or because of a possible dependence of the mass of the lightest resonances on the number of colours N. The exploration of if and how this possibility would affect our results (for example the conclusion that $L_{\rm ej,p} \gg L_{\rm ej,p}^{\rm eff}$), while certainly interesting, goes beyond the purposes of this paper.

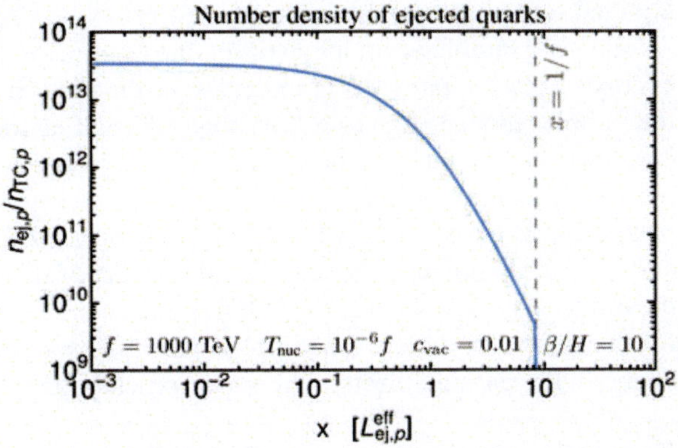

Fig. 7.7 The density of the ejected quarks in front of the bubble wall as a function of the distance x in front of the bubble wall, Eq. (7.62), for an example parameter point. Here we have used the relation $\gamma_{wp} \simeq D T_{nuc}$. The distance to the outermost techniquanta $L_{ej,p} \approx 1/f$ is also shown

$$4\pi D^2 \int_0^{L_{ej,p}} dx\, n_{ej,p}(x) = \frac{4}{3}\pi D^3 n_{TC,p}\,, \tag{7.65}$$

where we have assumed $D \gg f/T_{nuc}^2$, i.e. we have placed ourselves deep in the regime where hadrons can form inside bubbles (see Eq. (7.44)). Equation (7.65) guarantees that the number of ejected techniquanta in the layer of thickness $L_{ej,p}$ is equal to the total number of techniquanta that entered the bubble up to radius D.

Interactions between ejected quarks:
Let us finally comment why, we think, interactions among the ejected techniquanta cannot much alter their density. The density of the particles in the incoming bath does not change out of their own interactions. In the wall frame, both the density and the relative momentum of the ejected techniquanta are of the same order of those of the particles in the incoming bath. Therefore, we analogously expect that the density of the ejected techniquanta would also not change after ejection. Since what will matter for the following treatment is the energy in the ejected techniquanta, rather than how this energy is spread among the various degrees of freedom, we content ourselves with this qualitative understanding and leave a more precise treatment to future work.

7.7.2 Scatterings of Ejected Quarks and Gluons Before Reaching Other Bubbles

Before possibly reaching other expanding bubble-walls and their ejected techniquanta, ejected quarks and gluons could undergo scatterings with particles from the supercooled bath at temperature T_{nuc}, and with techniquanta ejected from other bubbles. In this section we study the effects of these scatterings.

Ejected techniquanta are energetic:
As soon as a bubble occupies an order one fraction of its volume at collision, the total energy in ejected particles is much larger than that in the supercooled bath outside the bubble. Indeed, we have seen that for each quark or gluon in the supercooled bath that enters a bubble, there is at least an ejected one, and that the energy ejected per each incoming particle is much larger than the energy per each particle in the bath, $E_{\text{ej,p}} \simeq \gamma_{\text{wp}} f \gg T_{\text{nuc}}$, Eq. (7.32). Assuming the degrees of freedom in quarks and gluons are not an extremely small fraction of those in the diluted medium, then the diluted medium outside the bubbles does not have enough energy to act as a bath for the ejected particles. This implies that most ejected particles keep most of their energy upon passing through the supercooled bath.

Energy transfer between ejected techniquanta and diluted bath:
By reversing the logic above, the ejected particles can deposit in the supercooled bath an energy much larger than its initial one. Pushing this to the extreme, the ejected techniquanta could make the bath move away from the bubble wall, thus making our treatment so far valid only in the first stages of bubble expansion. In order to assess this, we estimate the rate of transferred energy between ejected techniquanta and particles from the bath outside the bubbles,

$$\Gamma_{\text{ej-bath}} = n_{\text{ej}} \int_{\Delta E_{\min}}^{\Delta E_{\max}} d\Delta E \frac{d\sigma v}{d\Delta E} \Delta E \simeq n_{\text{ej}} \int_{-s}^{t_{\text{IR}}} dt \frac{d\sigma}{dt} \sqrt{-t}\,, \tag{7.66}$$

where n_{ej} is the density of ejected techniquanta, ΔE is the energy transferred per single scattering, and where in the second equality we have taken the limit of relativistic particles $v \simeq 1$ and small energy transfer per single scattering ΔE, so that the Mandaelstam variable t can be expressed as $t \simeq -\Delta E^2$. The quantity $d\sigma/dt$ depends on the specific model under consideration, in particular it depends both on whether the ejected particle is a quark or a gluon, and on the identity of the scatterer in the bath outside the bubbles. For definiteness, we model it as the cross section for fermion-fermion scattering mediated by a light vector with some effective coupling $\sqrt{4\pi\alpha_{\text{eff}}}$,

$$\frac{d\sigma}{dt} = \frac{4\pi\alpha_{\text{eff}}^2}{t^2}\,. \tag{7.67}$$

We then obtain

$$\Gamma_{\text{ej-bath}} = n_{\text{ej}} \frac{8\pi\alpha_{\text{eff}}^2}{\sqrt{-t_{\text{IR}}}}. \tag{7.68}$$

$\Gamma_{\text{ej-bath}}$ is of course not Lorentz invariant, it depends on the frame via the density of ejected techniquanta n_{ej} determined in Sect. 7.7.1.

Impact on diluted bath:
The average energy transferred to a particle in the diluted bath at position D, when this particle goes across the layer of ejected techniquanta (so before it reaches the wall and initiates the processes described in Sect. 7.4), then reads

$$Q_{\text{ej-bath}} \equiv \int_0^{L_{\text{ej,p}}} dx\, \Gamma_{\text{ej-bath,p}}(x)\,, \tag{7.69}$$

where we remind that the spatial coordinate x is the distance between a given layer of ejected techniquanta and the wall at $x = 0$. Upon use of Eqs. (7.68) and (7.62), we can then evaluate the average energy transferred to an incoming particle from the diluted bath, Eq. (7.69), as

$$Q_{\text{ej-bath}} \simeq \frac{8\pi\alpha_{\text{eff}}^2}{3} \frac{D\, n_{\text{TC,p}}}{\sqrt{-t_{\text{IR}}}}\,. \tag{7.70}$$

Note that the product $D\, n_{\text{TC,p}}$ is Lorentz-invariant, so that $Q_{\text{ej-bath}}$ is indeed a Lorentz-invariant quantity. To learn whether particles from the diluted bath are prevented from entering the wall, because of the interaction with the ejected techniquanta, we compare the energy they exchange with them upon passing their layer with their initial energy in the wall frame,[9] $E_{i,w} \simeq 3\,\gamma_{\text{wp}} T_{\text{nuc}}$,

$$\frac{Q_{\text{ej-bath}}}{E_{i,w}} \simeq \frac{8\zeta(3)}{9\pi} \alpha_{\text{eff}}^2\, g_{\text{TC}} \frac{T_{\text{nuc}} D}{\gamma_{\text{wp}}} \frac{T_{\text{nuc}}}{\sqrt{-t_{\text{IR}}}}\,. \tag{7.71}$$

The novel physical picture we described in Sects. 7.4 and 7.7 is valid as long as $Q_{\text{ej-bath}}/E_{i,w} \ll 1$. As seen in Sect. 7.5, γ_{wp} initially grows linearly with the bubble radius, $\gamma_{\text{wp}} \simeq T_{\text{nuc}}\, D$, until the retarding pressure possibly becomes effective. It will turn out in Sect. 7.9 that the runaway regime of linear growth is the one relevant for the phenomenology we will discuss. In that regime, the condition $Q_{\text{ej-bath}}/E_{i,w} \ll 1$ translates into $T_{\text{nuc}}/\sqrt{-t_{\text{IR}}} \ll 1$.

IR cut-off:
The quantity $-t_{\text{IR}}$ is the IR cutoff of the scattering, $-t_{\text{IR}} \equiv m_V^2$, with m_V some effective mass of the mediator responsible for the interactions that exchange momentum. In the absence of mass scales, which is the case for example for the SM photon and for the gluons, the effective mass m_V is equal to the plasma mass of these particles in the thermal bath. If the only bath was the diluted one, one would have $m_{V,\text{therm}}^2 \sim \alpha_{\text{eff}}\, n_{\text{TC,p}}/\langle E_{\text{TC,p}}\rangle \sim T_{\text{nuc}}^2$ (see e.g. [116]). However, the process of our interest here happens in the much denser bath of ejected techniquanta, $n_{\text{ej,p}} \gg n_{\text{TC,p}}$, so that we indeed expect $m_{V,\text{therm}}^2 \gg T_{\text{nuc}}^2$, so that $Q_{\text{ej-bath}}/E_{i,w} \ll 1$ and our picture so far is valid. More precisely, the screening mass for non-equilibrium systems scales

[9] Had we chosen another frame, we would have had to include the wall velocity in the condition.

as [117] ($f(p)$ is the non-equilibrium phase space distribution of the particles in the system)

$$m_{V,\text{therm}}^2 \simeq g_{\text{TC}}\alpha_{\text{eff}} \int \frac{f(p)}{|p|} \sim \frac{n_{\text{ej,p}}}{\langle E_{\text{ej,p}}\rangle} \sim \frac{\gamma_{\text{wp}}}{f/T_{\text{nuc}}} T_{\text{nuc}}^2 \gg T_{\text{nuc}}^2\,, \tag{7.72}$$

where we have used $\langle E_{\text{ej,p}}\rangle \sim \gamma_{\text{wp}} f$ and $n_{\text{ej,p}} \sim (D/L_{\text{ej,p}}) n_{\text{TC,p}} \sim D^2 T_{\text{nuc}}^5 \sim \gamma_{\text{wp}}(D)^2 T_{\text{nuc}}^3$. Equations (7.71) and (7.72) teach us that, in the regions of parameter space where $\gamma_{\text{wp}} \gg f/T_{\text{nuc}}$, the energy received by each particle in the diluted bath, from scatterings with the ejected techniquanta, is much smaller than their energy in the wall frame $E_{i,w} \simeq 3\,\gamma_{\text{wp}} T_{\text{nuc}}$.[10] Since $E_{i,w}$ was the crucial input quantity for our treatment in Sect. 7.4, the picture that emerged there is not affected by these scatterings.

Energy transferred to techniquanta ejected from other bubbles:
Finally, before ejected techniquanta can possibly enter another expanding bubble, they also have to pass through the layer of the techniquanta ejected from that other bubble. To investigate this, one can use the result derived above, Eq. (7.70), with the specification that now D is the maximal radius reached on average by expanding bubbles, because the shells of ejected quarks and gluons meet just before the bubble walls do. We then find that the average energy transferred is much smaller than the energy of an ejected techniquanta in the plasma frame $\simeq \gamma_{\text{wp}} f$,

$$\frac{Q_{\text{ej-ej}}}{\gamma_{\text{wp}} f} \simeq \frac{8\zeta(3)}{3\pi}\alpha_{\text{eff}}^2\, g_{\text{TC}} \frac{T_{\text{nuc}}}{f}\frac{T_{\text{nuc}} D}{\gamma_{\text{wp}}}\frac{T_{\text{nuc}}}{\sqrt{-t_{\text{IR}}}} \ll 1\,. \tag{7.75}$$

Hence, for the purpose of determining the average energy of ejected quarks when they enter another bubble, one can safely ignore the interactions between the two shells.

[10] One could be worried that in the outer shell of size $L_{\text{ej,p}} \sim 1/f$, the thermal mass $m_{V,\text{therm}}$ is much smaller than its value in the densest region in $x \simeq 0$, explicited in Eq. (7.72), such that $\frac{Q_{\text{ej-bath}}}{E_{i,w}}$ becomes larger than 1. We can check that it is not the case by including the x-dependence of $m_{V,\text{therm}}$, Eq. (7.72), in the integral in Eq. (7.69)

$$Q_{\text{ej-bath, p}} = \int_0^{L_{\text{ej,p}}} dx\; n_{\text{ej, p}} \frac{8\pi\alpha_{\text{eff}}^2}{\sqrt{n_{\text{ej,p}}/\langle E_{\text{ej,p}}\rangle}}\,, \tag{7.73}$$

where $n_{\text{ej,p}}$ is defined in Eq. (7.62), $L_{\text{ej,p}} \sim 1/f$ in Eq. (7.64), $D \sim \gamma_{\text{wp}}/T_{\text{nuc}}$ in Eq. (7.35), and $E_{\text{ej,p}} \sim \langle E_{\text{ej,p}}\rangle \sim \gamma_{\text{wp}} f$. We compute the integral in Eq. (7.73) and obtain

$$\frac{Q_{\text{ej-bath}}}{E_{i,w}} \simeq \begin{cases} \frac{8\sqrt{2\zeta(3)}}{9}\alpha_{\text{eff}}^2\sqrt{g_{\text{TC}}\frac{f/T_{\text{nuc}}}{\gamma_{\text{wp}}}} \ll 1, & \text{if } \gamma_{\text{wp}} \gtrsim f/T_{\text{nuc}} \\ \frac{8\sqrt{2\zeta(3)}}{3}\alpha_{\text{eff}}^2\sqrt{g_{\text{TC}}\frac{\gamma_{\text{wp}}}{f/T_{\text{nuc}}}} \ll 1, & \text{if } \gamma_{\text{wp}} \lesssim f/T_{\text{nuc}}, \end{cases} \tag{7.74}$$

This confirms that the energy of incoming particles, $E_{i,w} \simeq 3\gamma_{\text{wp}} T_{\text{nuc}}$, is not affected by the shell of ejected quarks.

7.7.3 Ejected Techniquanta Enter Other Bubbles (and Their Pressure on Them)

Ejected techniquanta are squeezed:
In the plasma frame, all ejected techniquarks are contained within a shell of length given by Eq. (7.64) $L_{\mathrm{ej,p}} \sim 1/f$, and most of them lie within a length given by Eq. (7.63) $L^{\mathrm{eff}}_{\mathrm{ej,p}} \sim 1/(T^2_{\mathrm{nuc}} D) \ll 1/f$. In the frame of the wall of the bubble they are about to enter, these lengths are further shrunk, so that ejected techniquarks are closer to each other than $1/f$ by several orders of magnitude. Therefore we expect no phenomenon of string fragmentation when they enter other bubbles. So each ejected particle, upon entering another bubble, forms a hadron with one or more of its neighbours. This also implies there is no further ejection of other techniquanta. Each of these hadrons carries an energy equal to that of the techniquanta that formed it, of order $\gamma_{\mathrm{wp}} f$ in the plasma frame.

Contribution to the retarding pressure:
This conversion of ejected techniquanta into hadrons results in another source of pressure on the bubble walls, that acts for the relatively short time during which the bubble wall swallows the layer of ejected techniquanta. In the frame of the bubble wall that they are entering, the energy of each ejected quark or gluon reads $E_{\mathrm{ej,w2}} \simeq 2\gamma^2_{\mathrm{wp}} f$. We then proceed analogously to what done in Sect. 7.5.1, and compute

$$\Delta p^{\mathrm{ej}}_{\mathrm{LO}} = E_{\mathrm{ej,w2}} - \sqrt{E^2_{\mathrm{ej,w2}} - \Delta m^2_{\mathrm{in}}} \simeq \frac{f}{4\,\gamma^2_{\mathrm{wp}}}\,, \tag{7.76}$$

$$P^{\mathrm{ej}}_{\mathrm{LO}} \simeq n_{\mathrm{ej,w2}}\,\Delta p^{\mathrm{ej}}_{\mathrm{LO}} \simeq \frac{\zeta(3)}{\pi^2}\, g_{\mathrm{TC}}\,\gamma_{\mathrm{wp}} T^3_{\mathrm{nuc}} f\,, \tag{7.77}$$

where we have used $g_{\mathrm{TC}} = g_g + 3g_q/4$, $\Delta m^2_{\mathrm{in}} \simeq f^2$, $n_{\mathrm{ej,w2}} \simeq 2\gamma_{\mathrm{wp}} n_{\mathrm{ej,p}}$ and, for simplicity, the peak value $n_{\mathrm{ej,p}} \simeq 2\,\gamma^2_{\mathrm{wp}} n_{\mathrm{TC,p}}$ of Eq. (7.62). The population of techniquanta ejected from other bubbles thus exert, on a given bubble wall, a pressure comparable to that exerted by the techniquanta incoming from the bath at LO, cf. Eq. (7.39). Therefore, the pressure from ejected techniquanta does not alter the picture described so far—a fortiori—because it is exerted only just before bubble walls collide and not throughout their entire expansion.

7.7.4 Ejected Techniquanta Heat the Diluted SM Bath

In Sect. 7.7.2 we found that the scatterings between ejected techniquanta and the diluted bath do not quantitatively change the picture of string fragmentation described in Sect. 7.4. These scatterings may however affect the properties of the particles, in the diluted bath, that do not confine. These particles include all the SM ones that are not charged under the new confining group, so that for simplicity we denote them as

'SM'. By a derivation analogous to the one that lead us to Eq. (7.70), we find that the average energy they exchange with the ejected quarks reads

$$Q_{\text{ej-SM}} \simeq \frac{8\pi\alpha_{\text{SM}}^2}{3}\frac{D\,n_{\text{TC,p}}}{\sqrt{-t_{\text{IR}}}} \sim \alpha_{\text{SM}}^2\,g_{\text{TC}}\left(\frac{\gamma_{\text{wp}}}{f/T_{\text{nuc}}}\right)^{\frac{1}{2}} f\,, \tag{7.78}$$

where we have used $T_{\text{nuc}}D \simeq \gamma_{\text{wp}}$ and $-t_{\text{IR}} \sim T_{\text{nuc}}^3\gamma_{\text{wp}}/f$, cf. Eq. (7.72). We have denoted by α_{SM} an effective coupling between SM particles and the techniquanta, which is model-dependent.

Now assume the techniquarks carry SM charges, e.g. as expected in composite Higgs models. Then, in the wall frame, the fractional change of energy is of course similar to that derived in Eq. (7.71) for the incoming techniquanta. However the incoming techniquanta next undergo string fragmentation, and Eq. (7.71) does not affect that energy balance for $\gamma_{\text{wp}} \gg f/T_{\text{nuc}}$. In other words, string fragmentation renders this energy transfer irrelevant for the techniquanta, while the SM particles neutral under the confining group just proceed undisturbed so they keep track of it. In particular, $Q_{\text{ej-SM}}$, is much larger than the latter energy in the plasma frame $\sim T_{\text{nuc}}$, and may even be slightly larger than the confinement scale f.[11]

This need not be the case, however, as the new techniquanta may be very weakly interacting with the SM. As they cannot interact too weakly, otherwise our assumption of instantaneous reheating would not hold, for simplicity we ignore this case in what follows and we assume that some techniquarks carry SM charges.

7.8 Deep Inelastic Scattering in the Early Universe

The physical picture described so far results in a universe that, before (p)reheating from bubble wall collisions, contains three populations of particles.

- **Population A.** Arises from hadronisation following string fragmentation. It consists of $N_\psi^{\text{string}}/2$ hadrons per quark or gluon in the initial bath, each on average with energy

$$E_A \simeq \frac{\gamma_{\text{wp}} f}{N_\psi^{\text{string}}(E_{\text{CM}})}, \tag{7.79}$$

  in the plasma frame, and of roughly the same number of hadrons with much smaller energy. (The latter can be thought as coming from the half of the string closer to the center of the bubble wall.) The physics resulting in this population is described in Sect. 7.4, see Eq. (7.31) for E_A and Eq. (7.28) for $N_\psi^{\text{string}}(E_{\text{CM}} = \sqrt{3\,\gamma_{\text{wp}}\,T_{\text{nuc}}\,f})$.

[11] As already anticipated, in the regime of interest for DM phenomenology we will find that bubble walls run away, so that $\gamma_{\text{wp}}^{\text{max}}$ is (much) smaller than $\sim 10^{-3}(f/T_{\text{nuc}})^3$, see Eq. (7.34). Note also that, for Eq. (7.78) only, $g_{\text{TC}} = 3g_q/4$, i.e. the gluon contribution to heating the SM is negligible because they cannot carry SM charge.

- **Population B.** Comes from the hadronisation of the ejected techniquanta. This population consists in $\sim$ one hadron per quark or gluon in the initial bath, each with energy
$$E_B \simeq \gamma_{\rm wp} f \, . \tag{7.80}$$
So this population carries an energy of the same order of that of population A. Its physics is described in Sect. 7.7, the energy E_B is that of the initial quark or gluon, Eq. (7.32).
- **Population C.** Consists of the particles that do not feel the confinement force, that we denote 'SM' for simplicity, each with a model-dependent energy given by Eq. (7.78), and whose total energy is much smaller than that in populations A and B.

The direction of motion of all these populations points, on average, out of the centers of bubble nucleation.

Hadrons from both populations A and B have large enough energies, in the plasma frame, that showers of the new confining sector are induced when they (or their decay products) scatter with the other particles in the universe and/or among themselves. These deep inelastic scatterings (DIS):

- Increase the number density of composite states.
- Decrease the momentum of each of these states with respect to the initial one $|p_\psi|$.

Hence, such effects need to be taken into account to find the yield of any long-lived hadron.

The evolution of our physical system would require solving Boltzmann equations for the creation and dynamics of populations A, B and C in a universe in which preheating is occurring, and of the interactions of populations A, B and C among themselves and with the preheated particles produced from bubble wall collisions. While certainly interesting, such a refined treatment goes beyond the purpose of this paper. In this Section, we aim rather at a simplified yet physical treatment, in order to obtain an order-of-magnitude prediction for the yield of long-lived hadrons.

7.8.1 Scatterings Before (P)reheating

We begin by considering the interactions among populations A, B and C.

Number densities of scatterers:
Let us define L_X, with $X = A, B, C$, the effective thickness of the shells containing populations A, B, and C respectively. For example, $L_{\rm B,p} = L^{\rm eff}_{\rm ej,p}$ of Eq. (7.63). We know that population A(B) consists on average of $N_\psi^{\rm string}/2$ hadrons (one hadron) per each quark or gluon in the initial diluted bath, and that population C is the initial diluted SM population. By conservation of the number of particles, we then obtain the number densities

$$n_A \simeq \frac{K^{\text{string}}}{2} \times \frac{D}{3L_A} n_{\text{TC}}, \qquad n_B \simeq \frac{D}{3L_B} n_{\text{TC}}, \qquad n_C \simeq \frac{D}{3L_C} n_{\text{SM}}, \tag{7.81}$$

where D is the average radius of a bubble at collision and we have used $L_X \ll D$.

Energy transferred between scatterers:
We now determine the average momentum, transferred to a particle from population X, upon going across a shell of population Y. In order to do so, we use our result Eq. (7.68) for the rate of transferred energy and compute

$$Q_{\text{Y}\to\text{X}} \simeq \Gamma_{\text{Y}\to\text{X}} L_Y \simeq n_{\text{Y}} L_{\text{Y}} \frac{8\pi\alpha^2_{\text{X-Y}}}{\sqrt{-t_{\text{IR}}}} \simeq \frac{8\zeta(3)}{3\pi} \alpha^2_{\text{X-Y}} g_{\text{Y}} \frac{T_{\text{nuc}}}{\sqrt{-t_{\text{IR}}}} \gamma_{\text{wp}} T_{\text{nuc}}\,, \tag{7.82}$$

where $\alpha_{\text{X-Y}}$ is the effective interaction strength of the scatterings of interest, g_{Y} the number of degrees of freedom in density of population Y (where we include a factor of $N_\psi^{\text{string}}/2$ for Y = A), and we have used the relation $T_{\text{nuc}} D = \gamma_{\text{wp}}$ valid in the runaway regime. We conclude that:

- **Populations A and B.** The energies of the hadrons of population A and B in the plasma frame, respectively $\gamma_{\text{wp}} f / N_\psi^{\text{string}}$ and $\gamma_{\text{wp}} f$, are both much larger than the energy they can exchange with any of the other baths among A,B,C, by a factor that scales parametrically as f/T_{nuc} or larger (because for all populations we have $-t_{\text{IR}} \sim n/\langle E\rangle > T^2_{\text{nuc}}$, see the discussion in Sect. 7.7.2). Therefore these elastic scatterings are not effective in reducing the energy of the hadrons of either population A or population B.
- **Population C.** On the contrary, $Q_{\text{A,B}\to\text{C}}$ can be of the same order of the energy of each particle in population C, Eq. (7.78), which therefore are significantly slowed down by these interactions. Importantly for our treatment, this does not alter the fact that population C was energetically subdominant with respect to populations A and B.

No significant DIS between populations A, B and C:
Finally, we determine whether any of the scatterings among particles in populations A,B,C could result in significant hadron production, via deep inelastic scattering. A single scattering event potentially results in a shower of the new confining sector if the exchanged momentum is larger than the confinement scale, $t^2 > f^2$. This condition is allowed by kinematics, because the center-of-mass energy of the scatterings between any of the populations above is much larger than f. A significant amount of DIS happens if the DIS scattering rate $\Gamma^{\text{DIS}}_{\text{Y}\to\text{X}}$ of a particle from population X, upon going across a shell of population Y, is much larger than the inverse of the length of the shell Y. We then compute

$$\Gamma^{\text{DIS}}_{\text{Y}\to\text{X}} L_{\text{Y}} \simeq n_{\text{Y}} \sigma_{\text{X-Y}} v\, L_{\text{Y}} \simeq \frac{4\zeta(3)}{3\pi} \alpha^2_{\text{X-Y}} g_{\text{Y}} \frac{\gamma_{\text{wp}}}{(f/T_{\text{nuc}})^2}\,, \tag{7.83}$$

where again we have used the runaway relation $T_{\rm nuc} D = \gamma_{\rm wp}$ and, for definiteness, we have assumed the scattering cross section has the form of Eq. (7.67). Therefore, no significant DIS happens in the regions where $\gamma_{\rm wp} \ll (f/T_{\rm nuc})^2$. This condition will turn out to be always satisfied in the parameter space of our interest, so we can ignore the DIS among populations A, B and C in what follows.

7.8.2 *Scatterings with the (P)reheated Bath*

By *preheating*, we intend the stage between the time when bubble walls collide and start to produce particles (e.g. from the resulting profile of the condensate), and the *reheating* time when these particles have thermalised into a bath. We now discuss the scatterings of populations A and B with the particles produced at preheating, that we have assumed to be efficient. The contribution of population C to the final yield of hadrons is subdominant with respect to the one of populations A and B because, as seen in Sects. 7.7.4 and 7.8.1, the total energy in population C is much smaller than that in populations A and B.

Energy of the (p)reheated bath:
The preheated particles are produced with energies, in the plasma frame, of the order of the mass of the scalar condensate,[12]

$$\langle E_{\rm prh} \rangle \simeq m_\chi < f \,. \tag{7.84}$$

Their total energy scales as

$$E_{\rm prh}^{\rm tot} \sim f^4\, V\,, \tag{7.85}$$

with V the volume of a large enough region of the universe. For comparison, the total energy in populations A and B scales as

$$E_{\rm A,B}^{\rm tot} \sim \gamma_{\rm wp} f\, T_{\rm nuc}^3\; V\,, \tag{7.86}$$

which is much smaller than $f^4 V$ because $\gamma_{\rm wp} \ll (f/T_{\rm nuc})^3$, Eq. (7.34). So the preheated particles can act as a thermal bath for all the other populations A, B and C, because the energy of A, B, and C is subdominant in the energy budget of the universe.

Inelastic versus elastic scattering:
Scatterings of hadrons (or their decay products) with the preheated bath will, there-

[12] In the picture we have in mind, non-perturbative effects such as Bose enhancement or parametric resonance (see e.g. [118]) are not relevant: the first because the SM particles are interacting, thus they exchange momentum and do not occupy the same phase space cells; the second because the variation of their masses from the dilaton's oscillations is smaller than their mass at the minimum. Note that, unlike what occurs in many inflationary scenarios, we expect only a small hierarchy $T_{\rm RH} \lesssim \langle E_{\rm prh} \rangle$.

fore, eventually slow down and thermalise populations A and B. However, these scatterings can also exchange energies much larger than f, thus inducing deep inelastic scatterings. Indeed their center-of-mass energy squared reads

$$s_{\rm A,B} \simeq 2m_\chi E_{\rm A,B}\,, \tag{7.87}$$

where $E_{\rm A} \simeq \gamma_{\rm wp} f/N_\psi^{\rm string}(E_{\rm CM})$ and $E_{\rm B} \simeq \gamma_{\rm wp} f$. Eq. (7.87) is the result of our simplifying assumption to neglect masses and to average to zero scattering angles with particles in a bath: define $p_E = E(1,\,\hat{E})$, $p_{\rm prh} = m_\chi(1,\,\hat{m})$, then $s = (p_E + p_{\rm prh})^2 \simeq 2E\,m_\chi(1 - \hat{E}\cdot\hat{m}) \simeq 2E\,m_\chi$. We now determine if those center of mass energies are entirely available for particle production via DIS, or if instead they are reduced by several low-momentum-exchange interactions. In order to do so, we evaluate the rate of energy loss of a particle from population A or B, $\Gamma^{\rm loss}_{\rm A,B}$, as the ratio between the rate of energy it exchanges with the preheated bath, that we evaluate analogously to Eq. (7.68), and its initial energy $E_{\rm A,B}$. We then compare this quantity with the rate for a deep inelastic scattering to happen with the full energy available $s^{1/2}_{\rm A,B}$,

$$\frac{\Gamma^{\rm loss}_{\rm A,B}}{\Gamma^{\rm DIS}_{\rm A,B}} \simeq \frac{n_{\rm prh} 8\pi\alpha^2_{\rm eff}/(E_{A,B}\sqrt{-t_{\rm IR}})}{n_{\rm prh} 4\pi\alpha^2_{\rm eff}/s_{A,B}} \simeq \frac{m_\chi}{\sqrt{-t_{\rm IR}}} \sim \frac{1}{\sqrt{c_{\rm vac}}}\frac{m_\chi^2}{f^2}\,. \tag{7.88}$$

In the last equality, we have again used the screening mass for non-equilibrium systems [117]

$$-\,t_{\rm IR} \sim \frac{n_{\rm prh}}{\langle E_{\rm prh}\rangle} \sim c_{\rm vac}\frac{f^4}{m_\chi^2}\,, \tag{7.89}$$

where we have used that by conservation of energy $n_{\rm prh} \sim \rho_{\rm RH}/\langle E_{\rm prh}\rangle$, and where we have expressed the energy density of the reheated bath $\rho_{\rm RH}$ using the results of Sect. 7.3.2.

We conclude that, if

$$\left(\frac{m_\chi}{f}\right)^2 \ll c_{\rm vac}^{1/2}\,, \tag{7.90}$$

the full center-of-mass energies $s_{\rm A,B}$ are available for deep inelastic scattering, i.e. populations A and B do not lose a significant amount of their energy via interactions with the preheated bath. For simplicity, in what follows we assume this model-dependent property to hold.

7.8.3 *Enhancement of Hadron Abundance Via DIS*

The picture: a cascade of DIS:
The number of composite states arising from a hard scattering depends on how the strings fragment, so on the same physics that set the abundance of the composite

states when the techniquanta cross the bubble walls, discussed in Sect. 7.4.4. Each scattering, depending on its center-of-mass energy, produces a number $N_\psi^{\rm string}$ of hadrons ψ, that we model in the same was as in Eq. (7.28). Given the large initial energies $s_{\rm A,B}$, the daughter hadrons typically still have enough energy to themselves induce further deep inelastic scatterings with the particles in the preheated bath, and hence additional hadron production. Analogously, SM particles produced in such DIS typically have large enough energies to also initiate showers of the new confining force with their subsequent scatterings. This process iterates until the average energy of scatterings drops below the confinement scale.

Number of hadrons produced per scattering:
For reasons given in Sect. 7.4.4, together with simplicity, we assume that the available energy $\sqrt{s}$ at each scattering splits equally among all the outcoming particles. We then write the average of this number as

$$N^{\rm DIS}(s) = N_\psi^{\rm string}(\sqrt{s}/2)\,, \tag{7.91}$$

where the factor of 2 in the argument of $N_\psi^{\rm string}$ arises because Eq. (7.28), which defines $N_\psi^{\rm string}$, assumes that $\sqrt{s}$ is the center of mass energy of the scattering of two particles neutral under the new confining force. If a hadron is included among the two scatterers, then QCD studies find that the final number of hadrons can be obtained by just halving the energy in the center of mass frame [70], also see footnote 5.[13]

Energies of produced hadrons:
Explicitly, we assume $E'_{\rm com} = \sqrt{s}/N^{\rm DIS}$, where $E'_{\rm com}$ is the energy of any outgoing particle (SM and/or composite) in the center-of-mass frame of the scattering. To iterate to many scatterings, we write $E'_{\rm com}$ in the plasma frame as $E' = \gamma'\, E'_{\rm com}(1 - \hat{v}' \cdot \hat{v})$, where γ' and $\hat{v}'$ are the associated Lorentz boost and its direction, and $\hat{v}$ is the direction of motion of the outgoing particle in the center-of-mass frame of the scattering. By averaging $\hat{v}' \cdot \hat{v}$ to zero for simplicity, we obtain

$$E' = \gamma'\, E'_{\rm com}\,. \tag{7.92}$$

We then determine γ' by observing that the energy of each particle, in the center-of-mass frame of the scattering, is both $E_{\rm com} = \sqrt{s}/2$ and $E_{\rm com} = \gamma' E_{\rm prh}(1 + \hat{v}' \cdot \hat{E}_{\rm com})$, where $E_{\rm prh}$ is the energy in the plasma frame of the particles in the preheated bath. By averaging $\hat{v}' \cdot \hat{E}_{\rm com}$ to zero for simplicity, we obtain the Lorentz boost

[13] Note that if a hadron instead decays to two SM particles before it scatters, which is model-dependent, then $\sqrt{s}/2$ is again the good argument for the function $N_\psi^{\rm string}$, because then one has two particles each with half the initial energy, but both neutral under the new confining force. In this case, however, Eq. (7.91) becomes $N^{\rm DIS}(s) = 2N_\psi^{\rm string}(\sqrt{s}/2)$. When iterating the treatment to many scatterings, we find that this extra factor of 2 does not impact the final abundance of hadrons, which can be understood by thinking that the same initial energy is spread faster to zero.

$$\gamma' \simeq \frac{\sqrt{s}}{2\langle E_{\rm prh}\rangle}\,. \tag{7.93}$$

Using Eq. (7.87) for s we finally obtain

$$E'_{\rm A,B} \simeq \frac{1}{N^{\rm DIS}} E_{\rm A,B}\,. \tag{7.94}$$

(If we did not average over angles, we would have obtained $E'_{\rm A,B} = (E_{\rm A,B}/N^{\rm DIS})(1-\hat{v}'\cdot\hat{v})(1-\hat{E}\cdot\hat{m})/(1+\hat{v}'\cdot\hat{E}_{\rm com})$). So, after a hard scattering the energy of each outgoing particle in the plasma frame is roughly the initial energy divided by a factor $N^{\rm DIS}$. The subsequent s is then reduced by the same factor, ensuring a convergence of $N^{\rm DIS}(s)$ to unity, via Eq. (7.28), after only a few iterations. This also teaches us that the average energy of the particles, produced this way, quickly decreases to values lower than about m_*.

Number of hadrons produced by a chain of DIS:
Let us now estimate the yield of final hadrons by following the above arguments. Assuming interactions are fast enough, also those following the first one happen with preheated particles of the same average energy $\langle E_{\rm prh}\rangle$. Now define the number of states (both composite and not) N_k produced at the $k^{\rm th}$ interaction. This can be expressed as

$$N_k(s) \simeq N^{\rm DIS}\left(\frac{s}{N_{k-1}\times N_{k-2}\times\cdots\times N_1}\right), \tag{7.95}$$

where we remind the reader that the function $N^{\rm DIS}$ is obtained from Eqs. (7.27) and (7.91). Starting from a single resonance produced from the fragmentation of strings between quanta inside the bubble, after this chain of scattering processes one obtains a total number of resonances given by the product $\prod_k N_k(s)$. We find numerically that this product can be expressed as

$$K^{\rm DIS}_{\rm A,B} \simeq \frac{s_{\rm A,B}}{m_*^2}\,. \tag{7.96}$$

In other words, the iterative process we described converts the initial available energy into the rest mass of hadrons m_*. Since our aim here is not to achieve a more precise treatment, we refrain from refining the assumption that the momenta are distributed evenly among the particles coming out of a scattering process. In the same spirit of building a physically-clear picture without drowning in model-dependent details, we do not cover here the possibility that every scattering produces, in addition to the composite states, a comparable or larger amount of SM particles. (That would result in $N^{\rm DIS} > N_\psi^{\rm string}$ and in a faster degrowth of the available scattering energy to m_* at each step.) In addition to simplicity, this can be justified by observing that, in the limit of large number of degrees of freedom in the dark sector, our assumption that they carry SM charges will make their production dominant with respect to the one of SM particles.

Additional comments:
We conclude our derivations with two comments concerning its validity.

- If the full center-of-mass energies are not available for DIS, i.e. if Eq. (7.90) does not hold, then one could use the same result $K^{\rm DIS}_{\rm A,B}$ of Eq. (7.96), upon substituting $s_{\rm A,B} = 2E_{\rm A,B}m_\chi$ with the largest energy for which $\Gamma^{\rm loss}_{\rm A,B} \ll \Gamma^{\rm DIS}_{\rm A,B}$, that can be derived via Eq. (7.88).
- We have ignored the production of heavy particles from the collisions of bubble walls [37, 119–121]. This is justified as it has been shown that it only occurs when the minima of the potential are nearly degenerate and seperated by a sizable barrier [122, 123], which is not the case for the close-to-conformal potentials we have in mind. Hence we expect only particles lighter than the scalar condensate to be produced during reheating following the wall collision.

7.8.4 DIS Summary

The yield of hadrons, resulting from the processes of deep inelastic scattering described above, receives contributions from:

- **Population A.** That is, the hadrons produced from string fragmentation as described in Sect. 7.4. Their contribution reads

$$Y_{\rm A}^{\rm SC+string+DIS} \simeq \frac{1}{2} K_{\rm A}^{\rm DIS} N_{\psi}^{\rm string}(E_{\rm CM}) D^{\rm SC} Y_{\rm TC}^{\rm eq} \simeq \frac{\gamma_{\rm wp} f m_\chi}{m_*^2} D^{\rm SC} Y_{\rm TC}^{\rm eq}, \tag{7.97}$$

where we have used $K_{\rm A}^{\rm DIS} = s_{\rm A}/m_*^2$, cf. Eq. (7.96), with $s_{\rm A} \simeq 2m_\chi E_{\rm A} \simeq 2m_\chi \gamma_{\rm wp} f / N_\psi^{\rm string}(E_{\rm CM})$, cf. Eqs. (7.87) and (7.79). Note that the above expression captures also the regime where each string fragmentation produces on average one hadron, because the energy of that single hadron is roughly $\gamma_{\rm wp} f/2$, see the related discussion in Sect. 7.4.5.
- **Population B.** That is, the hadrons produced out of the techniquanta ejected from the bubbles, described in Sect. 7.7. Their contribution reads

$$Y_{\rm B}^{\rm SC+string+DIS} \simeq K_{\rm B}^{\rm DIS} D^{\rm SC} Y_{\rm TC}^{\rm eq} \simeq 2\,\frac{\gamma_{\rm wp} f m_\chi}{m_*^2} D^{\rm SC} Y_{\rm TC}^{\rm eq}, \tag{7.98}$$

where we have used $K_{\rm B}^{\rm DIS} = s_{\rm B}/m_*^2$, cf. Eq. (7.96), with $s_{\rm B} \simeq 2m_\chi E_{\rm B} \simeq 2m_\chi \gamma_{\rm wp} f$, cf. Eqs. (7.87) and (7.80).

Thus, the combined contribution to the total hadron yield is given by

$$Y^{\rm SC+string+DIS} \simeq K^{\rm DIS} D^{\rm SC}\, Y_{\rm TC}^{\rm eq} \simeq 3\,\frac{\gamma_{\rm wp}\, f\, m_\chi}{m_*^2} D^{\rm SC}\, Y_{\rm TC}^{\rm eq}, \tag{7.99}$$

where we have defined

$$K^{\rm DIS} = \frac{1}{2} K_{\rm A}^{\rm DIS} N_{\psi}^{\rm string}(E_{\rm CM}) + K_{\rm B}^{\rm DIS} \,. \tag{7.100}$$

Note finally that, in the regime of runaway bubble-walls, one obtains the parametric scaling $Y^{\rm SC+string+DIS} \propto (T_{\rm nuc}/f)^3 \gamma_{\rm wp}$. Which is much larger than the simple supercooling dilution, $\sim (T_{\rm nuc}/f)^3$, in the regions of parameter space where our analysis holds, namely for $\gamma_{\rm wp} > f/T_{\rm nuc}$.

7.9 Supercooled Composite Dark Matter

7.9.1 Initial Condition for Thermal Evolution

Finally, all unstable resonances decay either to SM or to the long-lived or stable hadrons, which we take to form DM. To obtain the yield of any such hadron i at the onset of reheating, one should use the expression

$$Y_i^{\rm SC+string+DIS} = {\rm BR}_i \, K^{\rm DIS} D^{\rm SC} Y_{\rm TC}^{\rm eq} \,, \tag{7.101}$$

where $K^{\rm DIS}$, $D^{\rm SC}$ and $Y_{\rm TC}^{\rm eq}$ are defined, respectively, in Eqs. (7.100), (7.14) and (7.12). $\rm BR_i$ is a pseudo-branching ratio, of the energy available to the confining techniquanta, into ψ_i particles. Estimates of $\rm BR_i$ for the cases where ψ_i is a meson and a baryon are given in Appendix 7.B, which show a broad range of underlying-model dependent values are possible, albeit with a large uncertainty. For example, in a QCD-like theory where ψ_i is a baryon with mass $\sim 4\pi f$ and the pions have mass $\sim f$, one obtains values $\rm BR_i \sim 10^{-6}$, while larger values $\rm BR_i$ are obtained for baryon-pion mass rations closer to one, or if ψ_i is a meson. Hence, we will take $\rm BR_i$ to be a free parameter.

For completeness, the supercooling plus string and supercooling yields read

$$Y_i^{\rm SC+string} = {\rm BR}_i \, K^{\rm string} D^{\rm SC} Y_{\rm TC}^{\rm eq}, \tag{7.102}$$

$$Y_i^{\rm SC} = {\rm BR}_i \, D^{\rm SC} \frac{\frac{3}{4} g_q}{g_{\rm TC}} Y_{\rm TC}^{\rm eq} \,, \tag{7.103}$$

where $K^{\rm string}$ is defined in Eq. (7.29). We have included a factor $\frac{3}{4} g_q / g_{\rm TC}$ in $Y_i^{\rm SC}$ to account for the fact that, in the case of no string fragmentation nor DIS, gluons do not contribute to the final abundance of heavy composite states of quarks. It would be absent if one was interested in light composite states of quarks. The yield of the various contributions is shown in Fig. 7.8.

It will turn out that the measured DM abundance is achieved in the regime of runaway bubble walls. In that regime, the resulting expression for the DM yield has

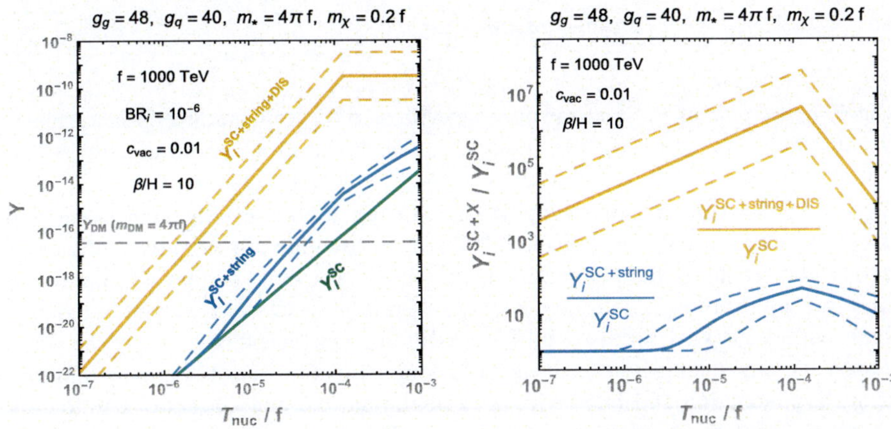

Fig. 7.8 Left: the yields following supercooling Eq. (7.103), string fragmentation Eq. (7.102) and deep inelastic scattering Eq. (7.101). The yield matching the observed relic abundance of DM for $m_{DM} = 4\pi f$ is also shown. The dashed lines show the effect of varying γ_{wp} by an order of magnitude in either direction around Eq. (7.34). This illustrates the sensitivity of the yield to our determination of γ_{wp}. Right: ratios of the same yields. The peak corresponds to the maximum γ_{wp}

a simple parametric form that eventually results in the DM abundance being independent of the DM mass, if it is to match onto observation $Y_{DM} \simeq 0.43\,\text{eV}/M_{DM}$ [124], which we find convenient to report here. By using Eqs. (7.12), (7.14), (7.36), and (7.99), with $g_{Rf} = g_{SM} = 106.75$, we find

$$Y_{i,\text{runaway}}^{\text{SC+string+DIS}} \simeq 0.43\,\frac{\text{eV}}{m_*} \times \frac{\text{BR}_i}{10^{-6}}\,\frac{g_{\text{TC}}}{120}\left(\frac{0.01}{c_{\text{vac}}}\right)^{\frac{5}{4}}\frac{m_\chi/f}{0.2}\,\frac{4\pi}{g_*}\left(\frac{T_{\text{nuc}}/f}{10^{-5.7}}\right)^4. \tag{7.104}$$

7.9.2 *Thermal Contribution*

To complete our discussion, we must still determine the effects on the yield of any DM interactions with the thermal bath after supercooling, DIS, and reheating. The importance of thermal effects following reheating was already pointed out in [21] (therein dubbed the subthermal contribution). Following the phase transition and particle production through DIS, the SM bath and the DM have returned to kinetic equilibrium. The scattering energy is now insufficient to break the resonances, but these may still annihilate into SM particles or be produced in the inverse process. Thus, just after the reheating, the DM abundance evolves according to the well known Boltzmann equation [125]

$$\frac{dY_{\mathrm{DM}}}{dx} = -\sqrt{\frac{8\pi^2 g_{\mathrm{SM}}}{45}} \frac{M_{\mathrm{pl}} M_{\mathrm{DM}} \langle \sigma v_{\mathrm{rel}} \rangle}{x^2} \left(Y_{\mathrm{DM}}^2 - Y_{\mathrm{DM}}^{\mathrm{eq}\,2} \right), \tag{7.105}$$

where we use $x \equiv M_{\mathrm{DM}}/T$ as the time variable, and M_{pl} is the reduced Planck mass. For simplicity we only consider velocity independent cross sections here. As an intitial condition we take the relic abundance at the reheat temperature, $Y_{\mathrm{DM}}(T_{\mathrm{RH}}) = Y_{\mathrm{DM}}^{\mathrm{SC+string+DIS}}$, estimated following string fragmentation and DIS enhancement in Eq. (7.99). For our plots we solve the Boltzmann equation numerically. If the cross section and reheating temperatures are sufficiently large the system will be driven back into equilibrium. The relic density is then largely set by freezeout dymanics, albeit with somewhat different initial conditions. On the other hand, if the cross section and reheat temperatures are small enough, the relic density is set by dilution, string fragmentation and DIS, with only negligible thermal corrections following reheating. Using the dilution mechanism of the PT, of course, we can avoid the usual unitarity constraint on the maximum thermal relic DM mass [47] (see e.g. [126, 127] for recent appraisals).

7.9.3 *Dark Matter Relic Abundance*

We now combine all our results together and determine the amount of supercooling required to match the observed relic abundance $Y_{\mathrm{DM}} \simeq 0.43\,\mathrm{eV}/M_{\mathrm{DM}}$. Examples are shown in Fig. 7.9 for some representative choices of the parameters. From these figures we can draw a number of conclusions.

(i) If we assume $\langle \sigma v_{\mathrm{rel}} \rangle \propto 1/M_{\mathrm{DM}}^2$, thermal effects will necessarily dominate if the DM is light enough. This occurs because T_{RH} cannot realistically be arbitrarily suppressed below f, for sensible choices of g_{Ri} and g_{Rf}. This regime corresponds to the point in which the contours turn vertical in Fig. 7.9. At which value of M_{DM} this occurs depends on the precise choice for $\langle \sigma v_{\mathrm{rel}} \rangle$. For definiteness, in Fig. 7.9 we choose $\langle \sigma v_{\mathrm{rel}} \rangle = 4\pi/M_{\mathrm{DM}}^2$ as typical of baryon scatterings in a strongly coupled sector. Thermal effects can of course be further suppressed if we depart from the efficient reheating assumption made here [21].

(ii) String fragmentation and DIS lead to large corrections to the composite DM relic density, compared to the naive supercooling dilution. This implies a mismatch between the relic abundances of primordial elementary and composite relics alluded to before. Whether the composite or elementary relic would have the greater abundance depends on the details of confinement (for elementary relics $\mathrm{BR}_i = 1$). If the composite relic is say, a light meson which is produced abundantly, the multiplicative DIS process can be highly efficient in populating these states following the PT. This implies we require much more supercooling to match onto the observed DM relic abundance. On the other hand, if the composite relic is some heavy state, perhaps a baryon, it could be produced in a highly suppressed rate both in string fragmentation and DIS. In this latter case,

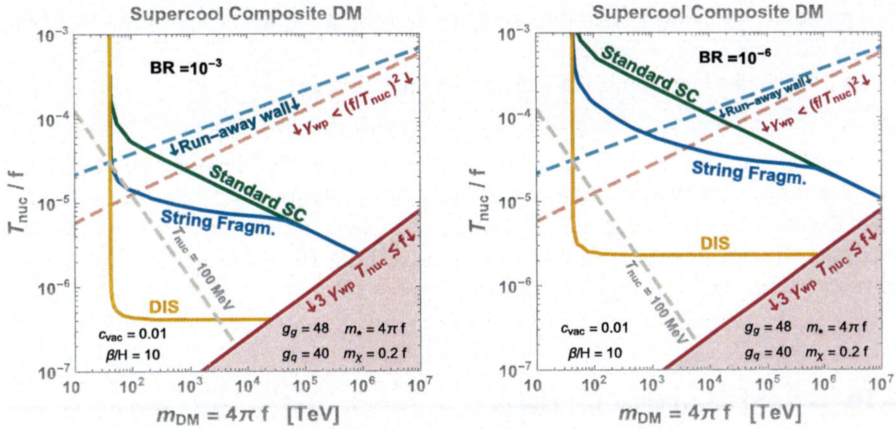

Fig. 7.9 Solid lines: supercooling $T_{\rm nuc}/f$ and DM mass $M_{\rm DM}$ required to obtain the observed DM abundance. The parameters chosen imply a reheating temperature $T_{\rm RH} \simeq 0.13\, f$, see Sect. 7.3.2. All lines include the thermal contribution discussed in Sect. 7.9.2. The line with initial condition $Y_i^{\rm SC+string+DIS}$ of Eq. (7.101) corresponds to the yellow contour. For comparison, we show in green (blue) the contour that one would obtain by skipping directly from the supercooling (supercooling plus string fragmentation) step to the thermal corrections, respectively Eqs. (7.103) and (7.102). All contours converge at some $M_{\rm DM}$ where thermal effects following reheating become dominant, of the order of $M_{\rm DM} \approx 100$ TeV because we fixed $\langle \sigma v_{\rm rel} \rangle = 4\pi/M_{\rm DM}^2$. Below this mass, the relic density is necessarily suppressed compared to the observed DM density, due to efficient DM annihilation after reheating. In the purple region $3\,\gamma_{\rm wp}\,T_{\rm nuc} \lesssim f$ the quarks are reflected by the first wall they encounter, but may enter the bubbles in following stages of their evolution, and the DM abundance lines ignore possible modifications arising from this 'ping-pong' effect. They also ignore that, for values of $\gamma_{\rm wp}$ only slightly larger than $f/T_{\rm nuc}$ and depending on other model-dependent parameters, the energetics of our treatment may be more complicated, see Eqs. (7.71) and (7.72). The dashed gray line delimits the area $T_{\rm nuc} < O(100)$ MeV where the supercooled phase transition could happen because of QCD dynamics. The dashed light blue line indicates the regimes where bubble walls run away, cf. Eq. (7.43). The dashed purple line indicates the regime where $\gamma_{\rm wp} < (f/T_{\rm nuc})^2$, and the fact it lies above the horizontal part of the DIS line confirms that our treatment has been consistent when ignoring the DIS of Eq. (7.83)

the required amount of supercooling to match onto the DM relic density is also reduced. The two cases are illustrated with two different assumptions for the branching ratios in Fig. 7.9.[14]

(iii) In some cases, we find $T_{\rm nuc} \lesssim 100$ MeV, as delineated in Fig. 7.9. Thus QCD effects could assist in completing the PT [21, 38, 129, 130]. On the other hand, if QCD effects help the transition to occur, they can also suppress the eventual gravitational wave signature [92] (simply because the QCD effects increase the

[14] We checked that the DIS line is unaffected if we use the treatment of the NLO pressure of [93], instead of the one of [73] that we have employed in this paper, cf. Sects. 7.5.2 and 6.2.5. On the other hand, the run-away wall dashed line, and hence the string fragmentation line, could be affected by this choice. For simplicity as well as in light of the criticism of [93] appeared in [94, 95, 128], we employed the results of [73] in this paper.

tunneling probability and thus will act to shorten the timescale of the PT). The details will depend on the physics entering the effective potential of the scalar χ and need to be studied in a model dependent way.

Together with the gravitational wave signal from the PT, there may also be model dependent collider, direct, and indirect detection signatures associated with the DM from the strongly coupled sector. We will investigate these further, together with their interplay with the novel string fragmentation and DIS effect, in a concrete realisation of such a confining sector in a companion paper [24].

7.10 Discussion and Outlook

The possible existence of a new confining sector of Nature is motivated by several independent problems of the Standard Model of particle physics and by cosmology. This encourages the identification of predictions of confining sectors, that are independent of the specific problem they solve. One such prediction is the possibility that the finite temperature phase transition in the early universe, between the deconfined and confined phase, is supercooled. This possibility has received a lot of attention in recent years, see e.g. [1, 21, 24, 28–32, 38, 76, 92, 130].

In this paper, we have pointed out and modelled a novel dynamical picture taking place in every supercooled confining phase transition, that (to our knowledge) had been missed in the literature. This novel picture stems from the observation that, when fundamental techniquanta of the confining sector are swept into expanding bubbles of the new confining phase, the distance between them is large with respect to the confinement scale. Therefore the energy of the fluxtubes connecting techniquanta is so large that string breaking produces many hadrons per fluxtube, with large momenta in the plasma (CMB) frame, in a sense analogously to QCD hadrons produced in electron-positron collisions at colliders. These hadrons and their decay products subsequently undergo scatterings with other particles in the universe, with center-of-mass energies much larger than both the confinement scale and the temperature that the universe reaches after reheating. The dynamics just described is partly pictured in Figs. 7.2 and 7.3.

The processes of string fragmentation and 'deep inelastic scatterings in the sky', synthetised above, have a plethora of implications. A key quantity to study them is the pressure on the bubble walls induced by this novel dynamics, which we have determined in Sect. 7.5, see Eq. (7.34) and Fig. 7.5 for the resulting bubble-wall velocities. An interesting aspect of our findings is that the so-called 'leading-order' pressure is proportional to the boost factor of the bubble wall, unlike in the case of non-confining supercooled PTs [72, 73].

We then quantified the values of supercooling below which one recovers the 'standard phase transition', where confinement happens between nearest charges. By relying on the modelling we proposed in Sect. 7.6 we found, interestingly, that the PT does not proceed in the 'standard' way already for minor supercooling, i.e. if bubbles are nucleated and expand just after vacuum energy starts to dominate. Our proposed dynamics should not only be employed in the large supercooling region, but also in the minor supercooling one depending on the value of another model-dependent parameter, see Fig. 7.6. The regimes in between these regions (one being the 'ping-pong' regime of Sect. 7.5.3) will be studied in future work, to not charge this paper with too much content.

Next, we have focussed on the implications of our dynamical picture for the abundance of long-lived or stable particles that are composite states of the new confining sector. They are summarised in the Synopsis, Sect. 7.2, and a quantitatively accurate expression of the final yield of a given composite particle is given in Eq. (7.104), for concreteness in the regime where bubble walls run away. Compared to the simple dilution of relics induced by supercooling of non-confinement transitions, these processes enhance their abundance by parametrically large factors. Therefore they have to be taken into account whenever a property of the universe, e.g. the DM and/or the baryon abundance, depends on the final yield of hadrons. As an example, their dramatic impact on the abundance of supercooled composite DM can be seen in Fig. 7.9.

Concerning DM in particular, this study constitutes a novel production mechanism of DM with mass beyond the unitarity bound [47]. It would be interesting and timely to study its experimental signals, given the new wave of telescopes that is starting to take data of high-energy neutrinos and gamma rays (e.g. KM3NeT, LHAASO, CTA) and given their potential in testing heavy DM, e.g. see [131]. One such study will appear in a forthcoming publication [24].

During the course of carrying out this study we have made a number of simplifications, for the purpose of obtaining a general and clear enough picture of the physics involved. For example, the various populations of particles created by this novel dynamics, such as the ejected techniquanta and the hadrons that follow the bubble walls, could be better described by Boltzmann equations, by the use of simulations etc., rather than with our simple treatment that focused on their average properties.

Finally, this dynamics opens broader and exciting avenues of investigation, that we think deserve exploration. For example, it would be interesting to study its interplay with recent interesting ideas regarding phase transitions [12, 17, 18, 31, 33, 37, 122, 132–136], or its impact on the production of gravitational waves in supercooled confining phase transitions. As for the latter, our study of the bubble wall Lorentz factor in Sect. 7.5 constitutes a necessary first step.

Appendices

7.A Wall Profile of the Expanding Bubbles

7.A.1 *The Light-Dilaton Potential*

Zero-temperature:
In this section we suppose that confinement occurs from the condensation of a nearly-conformal strongly-interacting sector, when an approximate scale-invariance gets spontaneously broken. If the source of explicit breaking is small, the spontaneous breaking of scale invariance generates a pseudo Nambu-Goldstone boson, the dilaton which we parameterize as [137]

$$\chi(x) = f e^{\frac{\sigma(x)}{f}}, \tag{7.106}$$

where f is the confining scale and where $\sigma(x)$ transforms non-linearly $\sigma(x) \to \sigma(\lambda x) + \log \lambda$ under the scale transformation $x \to \lambda x$. Its potential is given by [28]

$$V_\chi^{\mathrm{T=0}}(\chi) = c_\chi\, g_\chi^2\, \chi^4 \left[1 - \frac{1}{1+\gamma_\varepsilon/4}\left(\frac{\chi}{f}\right)^{\gamma_\varepsilon}\right], \tag{7.107}$$

with

$$\gamma_\varepsilon \simeq -\frac{1}{4}\frac{m_\chi^2}{c_\chi\, g_\chi^2\, f^2} < 1, \tag{7.108}$$

where m_χ is the dilaton mass, and c_χ is a constant of order 1, which we fix to $c_\chi = 1$. The dilaton coupling constant g_χ is chosen to reproduce the glueball normalization

$$g_\chi \simeq \frac{4\pi}{N}, \tag{7.109}$$

with N being the rank of the confining gauge group. The validity of the EFT relies on the smallness of the parameter $|\gamma_\varepsilon| \ll 1$ (here taken negative) which controls the size of the explicit breaking of scale invariance, and thus of the dilaton mass.

Note that in the limit where $|\gamma_\varepsilon| \ll 1$, the dilaton potential at zero-temperature reduces to the Coleman-Weinberg potential [138], i.e.

$$V_\chi^{\mathrm{T=0}}(\chi) \overset{|\gamma_\varepsilon|\ll 1}{=} -\gamma_\varepsilon\, c_\chi\, g_\chi^2\, \chi^4 \log\left(\frac{\chi}{f}\right). \tag{7.110}$$

Thermal corrections:
To model thermal effects, we follow [18, 28], and consider the finite-temperature corrections generated by the particles charged under the confining force (the CFT bosons)

$$V_T(\chi, T) = \sum_{i \in \text{CFT } bosons} \frac{nT^4}{2\pi^2} J_B\left(\frac{m_i^2}{T^2}\right), \qquad \text{with} \quad m_i \simeq g_\chi \, \chi. \tag{7.111}$$

The total number of CFT bosons n is fixed to[15]

$$\sum_{\text{CFT bosons}} n = \frac{45\, N^2}{4} \equiv \tilde{g}_g, \tag{7.112}$$

in order to recover the free energy of $\mathcal{N} = 4$ $SU(N)$ large N super-YM dual to an AdS-Schwarzschild space-time [17]

$$V_T(0, T) \simeq -b\, N^2\, T^4, \qquad \text{with} \quad b = \frac{\pi^2}{8}. \tag{7.113}$$

By writing Eq. (7.113), we have neglected the contribution from the fermions present in the plasma. For simplicity, we suppose that the dilaton degree of freedom χ still exists in the deconfined phase, such that the total potential for the dilaton is

$$V_{\text{tot}}(\chi, T) = V_\chi(\chi) + V_T(\chi, T), \tag{7.114}$$

where $V_\chi(\chi)$ and $V_T(\chi, T)$ are given by Eqs. (7.107) and (7.111). We plot the potential in Fig. 7.10. The supercooling stage starts when the energy density becomes vacuum-dominated

$$\frac{\pi^2}{30} g_{\text{Ri}} T_{\text{start}}^4 \simeq c_{\text{vac}} f^4 \quad \Longrightarrow \quad T_{\text{start}} \simeq \left(\frac{30 c_{\text{vac}}}{g_{\text{Ri}} \pi^2}\right)^{1/4} f, \tag{7.115}$$

with $c_{\text{vac}} = \dfrac{m_\sigma^2}{16 f^2}$ and $g_{\text{Ri}} = g_{\text{SM}} + g_{\text{TC}}$ where (see Eq. (7.112))

$$g_{\text{TC}} = g_{\text{q}} + \tilde{g}_g \simeq \frac{45 N^2}{4}. \tag{7.116}$$

7.A.2 *The Wall Profile*

Space-like region: the bounce profile:
We solve the tunneling temperature by solving the equation

$$\Gamma(T_{\text{nuc}}) \simeq H(T_{\text{nuc}})^4. \tag{7.117}$$

[15] The effective number of gluons in the deconfined phase $45N^2/4$ being different from $2(N^2 - 1)$ is a property valid at thermal equilibrium. It results from the peculiar strongly-coupled dynamics of the CFT. However, due to the large wall Lorentz factor, the CFT gas entering the wall can be considered as collisionless, cf. Sect. 7.4.1. This is why in the main text we consider the number of gluons entering the wall as $g_g = 2(N^2 - 1)$.

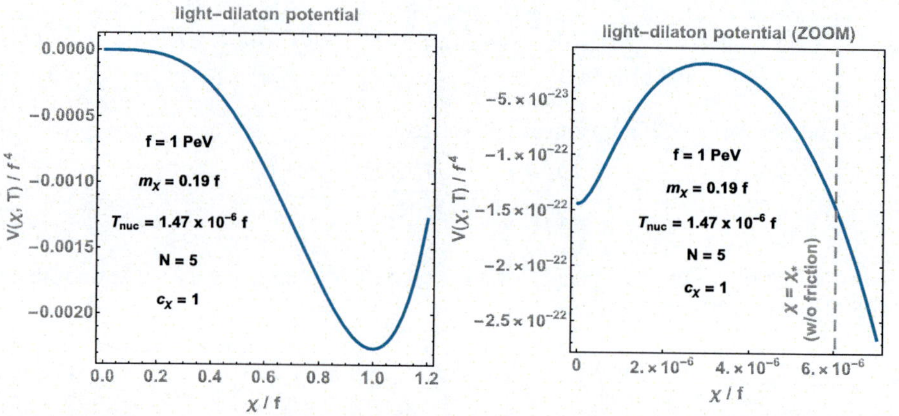

Fig. 7.10 **Left:** light-dilaton potential with temperature corrections in Eq. (7.114). **Right:** Zoom on the thermal barrier. The tunneling point χ_*, in the case where the friction term in the Euclidean equation of motion is neglected, is also shown

with [139, 140]

$$\Gamma(T_{\rm nuc}) = R_0^{-4}\left(\frac{S_4}{2\pi}\right)^2 \exp\left(-S_4\right), \tag{7.118}$$

where $R_0 \sim 1/T_{\rm nuc}$ is the bubble radius at nucleation and S_4 is the O_4-bounce action

$$S_4 = 2\pi^2 \int dr\, r^3 \left[\frac{1}{2}\phi'(r)^2 + V\left(\phi(r)\right)\right], \tag{7.119}$$

which we compute from solving the Euclidean equation of motion ($d = 4$)

$$\phi''(s) + \frac{d-1}{s}\phi'(s) = \frac{dV}{d\phi}, \tag{7.120}$$

with boundary conditions

$$\phi'(0) = 0, \qquad \text{and} \qquad \lim_{r\to\infty} \phi(r) = 0. \tag{7.121}$$

$s = \sqrt{r^2 + t_E^2} = \sqrt{r^2 - t^2}$ is the space-like light-cone coordinate and $t_E = i\,t$ is the Euclidean time.

We plot the bounce profile in the left-hand panel of Fig. 7.11 for given parameters relevant for the study. The value at the center of the bubble—the tunneling point χ_*—can be estimated analytically by energy conservation between $\chi = \chi_*$ and the false vacuum in $\chi = 0$ if we neglect the friction term in the equation of motion in Eq. (7.120),

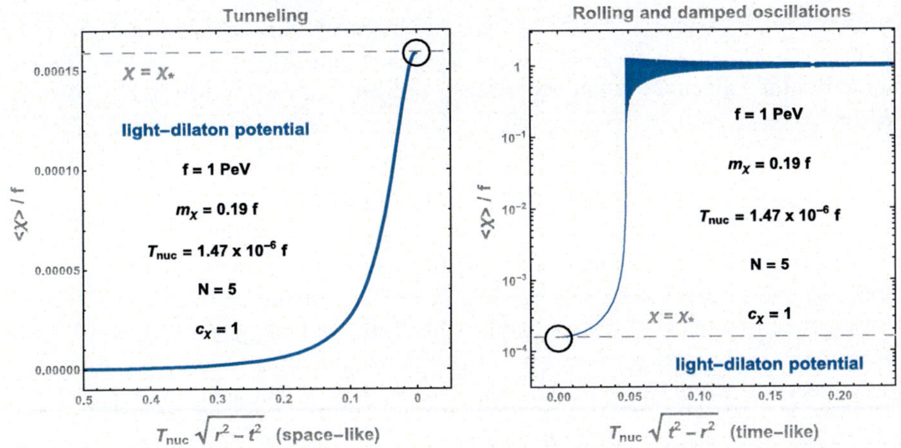

Fig. 7.11 **Left:** Bounce profile at nucleation. It interpolates between the false vacuum $\langle\chi\rangle = 0$ outside the bubble and the release point $\langle\chi\rangle_*$ at the center of the bubble. **Right:** Evolution of the scalar field after tunneling. First, the scalar field rolls along the shallow part of the nearly-conformal potential and then realized oscillations with period $\sim f^{-1}$ with a damping time $\lesssim T_{\rm nuc}^{-1} \gg f^{-1}$. Taking into account the decay of the scalar field would reduce the damping time after the first oscillation. The full bubble wall profile can be obtained after connecting the two figures through the two black circles

$$V_{\rm tot}(\chi_*) \simeq V_{\rm tot}(0), \qquad \rightarrow \qquad \frac{\chi_*}{f} \simeq \frac{1}{\sqrt{2}\log^{1/4}(f/\chi_*)}\frac{T}{T_c}. \tag{7.122}$$

Here (coincidence numeric) T_c is the critical temperature, defined when the two minima of the free energy are equal

$$c_{\rm vac} f^4 + V_T(f,\, T_c) - V_T(0,\, T_c) \equiv 0, \tag{7.123}$$

Note that for confining phase transition with $m_i(f) \gtrsim f$, the quantity $V_T(f,\, T_c)$ in Eq. (7.123) vanishes[16] and T_c is related to the temperature at which supercooling starts $T_{\rm start}$ in Eq. (7.115) through

$$T_c \simeq 3^{1/4}\left(\frac{g_{\rm Ri}}{g_{\rm TC}}\right)^{1/4} T_{\rm start}. \tag{7.125}$$

[16] We recall that the thermal functions in Eq. (7.111) verify the property $\lim_{x\to\infty} J_{B/F}(x) = 0$. From using Eq. (7.115), we observe that

$$4.0\,\frac{m_i(f)}{f}\left(\frac{0.1}{c_{\rm vac}}\right)^{1/4}\left(\frac{g_{\rm Ri}}{80}\right)^{1/4} \gg 1 \qquad \Longrightarrow \qquad m_i(f)/T_{\rm start} \gg 1. \tag{7.124}$$

The connection between T_c and $T_{\rm start}$ in Eq. (7.125) applies for all phase transitions satisfying Eq. (7.124).

$g_{\rm Ri}$ is the total number of relativistic d.o.f in the symmetric phase while $g_{\rm TC}$ only counts those which are involved in the phase transition ($g_{\rm TC} < g_{\rm Ri}$). In the scenario studied in this appendix, upon assuming $g_{\rm Ri} \simeq g_{\rm TC}$ with $g_{\rm TC}$ given in Eq. (7.116), we get

$$T_c = \left(\frac{m_\chi^2 f^2}{16 b N^2} \right)^{1/4} = \left(\frac{|\gamma_\varepsilon|\, c_\chi\, g_\chi^2}{4 b N^2} \right)^{1/4} f. \tag{7.126}$$

The tunneling point χ_* in absence of friction is shown in Fig. 7.10, while the tunneling point from numerically solving the bounce equation is visible in Fig. 7.11. Plugging the numbers chosen for making the plots, we find $\chi_*/f \simeq 6.0 \times 10^{-6}$ for the analytical value and $\chi_*/f \simeq 1.6 \times 10^{-4}$ for the numerical value. This difference was expected since the analytical estimate neglects the friction term in Eq. (7.120).

Time-like region: rolling and damped oscillations:
As soon as the bubble expands, the scalar field starts to roll toward the true vacuum $\langle \chi \rangle = f$ and realize damped oscillations. The field dynamic is captured by the Klein-Gordon equation for an inhomogeneous field

$$\Box \phi - \frac{\partial V}{\partial \phi} = 0. \tag{7.127}$$

We first use the $SO(3)$ symmetry to reduce the 3 Cartesian coordinates to the radial r coordinate

$$\frac{\partial^2 \phi}{\partial r^2} + \frac{2}{r}\frac{\partial \phi}{\partial r} - \frac{\partial^2 \phi}{\partial t^2} - \frac{\partial V}{\partial \phi} = 0. \tag{7.128}$$

We used the Minkowski metric since we can neglect the universe expansion during the time of bubble propagation. We then use the $SO(3,1)$ symmetry which reduces r and t to the time-like light-cone coordinate $s = \sqrt{t^2 - r^2}$ only [45]

$$\frac{\partial^2 \phi}{\partial s^2} + \frac{3}{s}\frac{\partial \phi}{\partial s} + \frac{\partial V}{\partial \phi} = 0. \tag{7.129}$$

Note the opposite sign in front of the potential V between the space-like (or Euclidean) equation of motion in Eq. (7.120) and the time-like (or Minkowskian) equation of motion in Eq. (7.129). Here the damping is purely geometrical, reminiscent of the $SO(3,1)$ symmetry and we do not consider the damping due to the dilaton decay or due to the interaction with the plasma (see e.g. [141]). In right panel of Fig. 7.11, we display the scalar field profile obtained after integration of the time-like equation in Eq. (7.129), using the initial condition $\chi(s = 0) = \chi_*$ given by the bounce solution in Eq. (7.120).

The full bubble wall profile:
The full bubble wall profile is obtained after matching the profile in the space-like region, left panel of Fig. 7.11, with the profile in the time-like region, right panel of Fig. 7.11. One can see that the first confining scale, that the incoming techniquanta are subject to upon entering the wall, is the exit scale

$$\chi_* \gtrsim T_{\text{nuc}}\,. \tag{7.130}$$

Our explicit computation also shows that the length of the section of the bubble wall where $\langle\chi\rangle = \chi_*$, in the wall frame, satisfies

$$L_{\text{w}} \lesssim T_{\text{nuc}}^{-1}\,, \tag{7.131}$$

as we assumed in Eq. (7.16) in the main text. Then, $\langle\chi\rangle$ transits to its zero-temperature value f over a length, in the wall frame, of order f^{-1}.

7.B Example Estimates of the String to DM Branching Ratio

In Sect. 7.4.2, we have discussed that, after supercooling, the quarks enter inside the confined phase, with a typical seperation $\sim T_{\text{nuc}}^{-1}$, much larger than the confining scale f, such that a highly energetic fluxtube forms. We have shown that this string, which is unstable under quark-anti-quark pair nucleation, breaks into K_{string} pieces. The dynamics of strings is then also relevant in the processes of deep inelastic scatterings of Sect. 7.8. In this section, we estimate the branching ratio of a string to a given hadron i, introduced in Eq. (7.101), in two different cases. First, when i is a light meson, in which case we expect the yield of i to be independent of its mass and given by a combinatoric factor implying the number of flavors. Second, when i is a heavy baryon in which case one expects the yield to be Boltzmann suppressed.

7.B.1 Light Meson—Combinatorics

In the limit of large string energy, $E_{\text{CM}} \gg f$, one expects the fragmentation of the string to be democratic with respect to the different bound-states if they are light enough. In that case, the string-to-i branching ratio is given by a combinatoric factor depending on the number of flavors N_f and the number of quark constituents (either 2 for meson and N_{TC} for baryons). In the particular case of a light meson $q_1\bar{q}_2$, one obtains

$$\text{Br}(\text{string} \to i) = \begin{cases} 1/N_f^2, & \text{if } q_1 = q_2, \\ 2/N_f^2, & \text{if } q_1 \neq q_2. \end{cases} \tag{7.132}$$

7.B.2 Heavy Baryon—Boltzmann Suppression

For this example a useful model for us will be the thermal one [142–145], which was able to fit LEP data of particle yields up to a 10% error [146], even with an initial state far from thermal equilibrium. In this model, the yield of heavy mesonic or baryonic resonances is suppressed by a Boltzmann factor [142–145], in which the strong scale plays the usual role of temperature. The yield of heavy resonances can be modelled by

$$\langle N_i \rangle \sim A_i \frac{(2J_i + 1)}{\text{Exp}\,[M_i/B_i]}, \tag{7.133}$$

where M_i and J_i are the mass and spin of the state i respectively. Here A_i is an overall normalisation, which will depend on whether the particle is a pseudoscalar meson, vector meson, or baryon etc. In QCD it was found to differ by $\lesssim 10$ between vector mesons, tensor mesons, and baryons [145]. For these particles B_i was found to be a common factor between the groups, $B_i \equiv B \sim 150$ MeV [145]. Note the pseudoscalar mesons in QCD, however, which are lighter, follow a softer spectrum.

Following the above discussion, we shall construct a toy model for the baryonic particle yield from our string fragmentation. In order to retain some simplicity in our model we will consider all particles to share a common $B_i = m_* = g_* f$. In our toy model we consider $SU(N_c)$ theories, with techniquarks in the fundamental representation, in which baryons will contain N_c quarks. Mesons on the other hand will contain a techni-quark-anti-quark pair independent of N_c. In order to take into account the reduced probability of creating a baryon as opposed to a meson it is therefore suitable to include an additional suppression in the prefactor A_i for baryons [147]

$$p_{\mathcal{B}i} = \begin{cases} \frac{1}{1+2^{N_c-1}/N_c}, & \text{if } i \text{ is a baryon,} \\ 1, & \text{if } i \text{ is a meson.} \end{cases} \tag{7.134}$$

Other than this we take a common $A_i = p_{\mathcal{B}i} A$. Applying energy conservation, we thus find the average number of the composite state i produced per string breaking to be

$$\langle N_i \rangle \simeq \frac{p_{\mathcal{B}i}(2J_i + 1)}{\text{Exp}\,[M_i/m_\pi]} \left(\sum_k \frac{p_{\mathcal{B}j}(2J_k + 1)}{\text{Exp}\,[M_k/m_\pi]} \frac{M_k}{m_\pi} \right)^{-1} \langle N_\psi \rangle \equiv \text{BR}_i\, \langle N_\psi \rangle, \tag{7.135}$$

where the sum runs over all the states in the spectrum, and we remind that π denotes the lightest composite state(s). In this case it is clearly possible to have a highly suppressed BR_i, e.g. $\text{BR}_i \simeq 10^{-6}$ for $m_i = m_* \simeq 4\pi f$, $m_\pi \simeq f$, $N_c = 10$, $N_\pi = 3$.

References

1. Baldes I, Gouttenoire Y, Sala F (2021) String Fragmentation in Supercooled Confinement and Implications for Dark Matter. JHEP 04:278. arxiv:2007.08440, https://doi.org/10.1007/JHEP04(2021)278
2. Antipin O, Redi M, Strumia A, Vigiani E (2015) Accidental composite dark matter. JHEP 07:039. arxiv: 1503.08749, https://doi.org/10.1007/JHEP07(2015)039
3. Contino R (2009) The Higgs as a composite nambu-goldstone Boson. In: Physics of the large and the small, TASI 09, proceedings of the theoretical advanced study institute in elementary particle physics. Boulder, Colorado, USA, pp 235–306. arxiv: 1005.4269, https://doi.org/10.1142/9789814327183_0005
4. Panico G, Wulzer A (2016) The composite nambu-goldstone higgs. Lect Notes Phys 913:1–316. arxiv: 1506.01961, https://doi.org/10.1007/978-3-319-22617-0
5. Geller M, Telem O (2015) Holographic twin higgs model. Phys Rev Lett 114:191801. arxiv: 1411.2974, https://doi.org/10.1103/PhysRevLett.114.191801
6. Barbieri R, Greco D, Rattazzi R, Wulzer A (2015) The composite twin higgs scenario. JHEP 08:161, arxiv: 1501.07803, https://doi.org/10.1007/JHEP08(2015)161
7. Low M, Tesi A, Wang L-T (2015) Twin higgs mechanism and a composite higgs Boson. Phys Rev D91:095012. arxiv: 1501.07890, https://doi.org/10.1103/PhysRevD.91.095012
8. Chacko Z, Goh H-S, Harnik R (2006) The twin higgs: natural electroweak breaking from mirror symmetry. Phys Rev Lett 96:231802. arxiv: hep-ph/0506256, https://doi.org/10.1103/PhysRevLett.96.231802
9. Kaplan DB (1991) Flavor at SSC energies: a new mechanism for dynamically generated fermion masses. Nucl Phys B365:259–278. https://doi.org/10.1016/S0550-3213(05)80021-5
10. Rubakov VA (1997) Grand unification and heavy axion. JETP Lett 65:621–624. arxiv: hep--ph/9703409, https://doi.org/10.1134/1.567390
11. Redi M, Sato R (2016) Composite accidental axions. JHEP 05:104. arxiv: 1602.05427, https://doi.org/10.1007/JHEP05(2016)104
12. Konstandin T, Servant G (2011) Natural cold baryogenesis from strongly interacting electroweak symmetry breaking. JCAP 1107:024. arxiv: 1104.4793, https://doi.org/10.1088/1475-7516/2011/07/024
13. Servant G (2014) Baryogenesis from strong CP violation and the QCD axion. Phys Rev Lett 113:171803. arxiv: 1407.0030, https://doi.org/10.1103/PhysRevLett.113.171803
14. Pisarski RD, Wilczek F (1984) Remarks on the chiral phase transition in chromodynamics. Phys Rev D29:338–341. https://doi.org/10.1103/PhysRevD.29.338
15. Witten E (1984) Cosmic separation of phases. Phys Rev D 30:272–285. https://doi.org/10.1103/PhysRevD.30.272
16. Helmboldt AJ, Kubo J, van der Woude S (2019) Observational prospects for gravitational waves from hidden or dark chiral phase transitions. Phys Rev D100:055025. arxiv: 1904.07891, https://doi.org/10.1103/PhysRevD.100.055025
17. Creminelli P, Nicolis A, Rattazzi R (2002) Holography and the electroweak phase transition. JHEP 03:051. arxiv: hep-th/0107141, https://doi.org/10.1088/1126-6708/2002/03/051
18. Randall L, Servant G (2007) Gravitational waves from warped spacetime. JHEP 05:054. arxiv: hep-ph/0607158, https://doi.org/10.1088/1126-6708/2007/05/054
19. Nardini G, Quiros M, Wulzer A (2007) A confining strong first-order electroweak phase transition. JHEP 09:077. arxiv: 0706.3388, https://doi.org/10.1088/1126-6708/2007/09/077
20. Kolb EW, Wolfram S (1980) Spontaneous symmetry breaking and the expansion rate of the early universe. Astrophys J 239:428. https://doi.org/10.1086/158126
21. Hambye T, Strumia A, Teresi D (2018) Super-cool dark matter. JHEP 08:188. arxiv: 1805.01473, https://doi.org/10.1007/JHEP08(2018)188
22. Barreiro T, Copeland EJ, Lyth DH, Prokopec T (1996) Some aspects of thermal inflation: the Finite temperature potential and topological defects. Phys Rev D54:1379–1392. arxiv: hep-ph/9602263, https://doi.org/10.1103/PhysRevD.54.1379

23. Easther R, Giblin JT, Lim EA, Park W-I, Stewart ED (2008) Thermal inflation and the gravitational wave background. JCAP 0805:013. arxiv: 0801.4197, https://doi.org/10.1088/1475-7516/2008/05/013
24. Baldes I, Gouttenoire Y, Sala F, Servant G, Supercool Composite Dark Matter beyond 100 TeV. arxiv: 2110.13926
25. Bardeen WA, Leung CN, Love ST (1986) The dilaton and chiral symmetry breaking. Phys Rev Lett 56:1230. https://doi.org/10.1103/PhysRevLett.56.1230
26. Paterson AJ (1981) Coleman-Weinberg symmetry breaking in the Chiral SU(N) X SU(N) linear sigma model. Nucl Phys B 190:188–204. https://doi.org/10.1016/0550-3213(81)90489-2
27. Bruggisser S, Von Harling B, Matsedonskyi O, Servant G (2018) Baryon asymmetry from a composite higgs Boson. Phys Rev Lett 121:131801. arxiv: 1803.08546, https://doi.org/10.1103/PhysRevLett.121.131801
28. Bruggisser S, Von Harling B, Matsedonskyi O, Servant G (2018) Electroweak phase transition and baryogenesis in composite higgs models. JHEP 12:099. arxiv: 1804.07314, https://doi.org/10.1007/JHEP12(2018)099
29. Baratella P, Pomarol A, Rompineve F (2019) The supercooled universe. JHEP 03:100. arxiv: 1812.06996, https://doi.org/10.1007/JHEP03(2019)100
30. Agashe K, Du P, Ekhterachian M, Kumar S, Sundrum R, Cosmological phase transition of spontaneous confinement. arxiv: 1910.06238
31. Delle Rose L, Panico G, Redi M, Tesi A (2020) Gravitational waves from supercool axions. JHEP 04:025. arxiv: 1912.06139, https://doi.org/10.1007/JHEP04(2020)025
32. Von Harling B, Pomarol A, Pujolas O, Rompineve F, Peccei-quinn phase transition at LIGO. arxiv: 1912.07587
33. Bloch IM, Csáki C, Geller M, Volansky T, Crunching away the cosmological constant problem: dynamical selection of a small λ. arxiv: 1912.08840
34. Azatov A, Vanvlasselaer M (2020) Phase transitions in perturbative walking dynamics. JHEP 09:085. arxiv: 2003.10265, https://doi.org/10.1007/JHEP09(2020)085
35. Hassanain B, March-Russell J, Schvellinger M (2007) Warped deformed throats have faster(Electroweak) phase transitions. JHEP 10:089. arxiv:0708.2060
36. Konstandin T, Nardini G, Quiros M (2010) Gravitational backreaction effects on the holographic phase transition. Phys Rev D 82:083513. arxiv: 1007.1468, https://doi.org/10.1103/PhysRevD.82.083513
37. Konstandin T, Servant G (2011) Cosmological consequences of nearly conformal dynamics at the TeV scale. JCAP 12:009. arxiv: 1104.4791, https://doi.org/10.1088/1475-7516/2011/12/009
38. von Harling B, Servant G (2018) QCD-induced electroweak phase transition. JHEP 01:159. arxiv: 1711.11554, https://doi.org/10.1007/JHEP01(2018)159
39. Fujikura K, Nakai Y, Yamada M (2020) A more attractive scheme for radion stabilization and supercooled phase transition. JHEP 02:111. arxiv: 1910.07546, https://doi.org/10.1007/JHEP02(2020)111
40. Bunk D, Hubisz J, Jain B (2018) A perturbative RS I cosmological phase transition. Eur Phys J C 78:78. arxiv: 1705.00001, https://doi.org/10.1140/epjc/s10052-018-5529-2
41. Dillon BM, El-Menoufi BK, Huber SJ, Manuel JP (2018) Rapid holographic phase transition with brane-localized curvature. Phys Rev D 98:086005. arxiv: 1708.02953, https://doi.org/10.1103/PhysRevD.98.086005
42. Megías E, Nardini G, Quirós M (2018) Cosmological phase transitions in warped space: gravitational waves and collider signatures. JHEP 09:095. arxiv: 1806.04877, https://doi.org/10.1007/JHEP09(2018)095
43. Megias E, Nardini G, Quiros M (2020) Gravitational imprints from heavy Kaluza-Klein resonances. Phys Rev D 102:055004. arxiv: 2005.04127, https://doi.org/10.1103/PhysRevD.102.055004
44. Agashe K, Du P, Ekhterachian M, Kumar S, Sundrum R (2021) Phase transitions from the fifth dimension. JHEP 02:051. arxiv: 2010.04083, https://doi.org/10.1007/JHEP02(2021)051

45. Jinno R, Konstandin T, Takimoto M, Relativistic bubble collisions—a closer look. arxiv: 1906.02588
46. Møller C (1945) General properties of the characteristic matrix in the theory of elementary particles
47. Griest K, Kamionkowski M (1990) Unitarity limits on the mass and radius of dark matter particles. Phys Rev Lett 64:615. https://doi.org/10.1103/PhysRevLett.64.615
48. Gondolo P, Gelmini G (199) Cosmic abundances of stable particles: improved analysis. Nucl Phys B 360:145–179. https://doi.org/10.1016/0550-3213(91)90438-4
49. Barroso Mancha M, Prokopec T, Swiezewska B, Field theoretic derivation of bubble wall force. arxiv: 2005.10875
50. Eichten E, Gottfried K, Kinoshita T, Kogut JB, Lane KD, Yan T-M (1975) The spectrum of charmonium. Phys Rev Lett 34:369–372. https://doi.org/10.1103/PhysRevLett.34.369, https://doi.org/10.1103/PhysRevLett.36.1276
51. Thorn CB (1934). Asymptotic Freedom in the Infinite Momentum Frame. https://doi.org/10.1103/PhysRevD.20.1934Phys. Rev. D 20 (1979)
52. Greensite J (1989) Monte Carlo evidence for the gluon chain model of QCD string formation. Nucl Phys B315:663–680. https://doi.org/10.1016/0550-3213(89)90007-2
53. Poulis GI (1996) Abelian dominance and adjoint sources in lattice QCD. Phys Rev D54:6974–6985. arxiv: hep-lat/9601013, https://doi.org/10.1103/PhysRevD.54.6974
54. Ko Y-K, Derivation of a quark confinement potential from QCD. arxiv: nucl-th/9901025
55. Greensite J, Thorn CB (2002) Gluon chain model of the confining force. JHEP 02:014. arxiv: hep-ph/0112326, https://doi.org/10.1088/1126-6708/2002/02/014
56. Bardakci K, Thorn CB (2003) A mean field approximation to the world sheet model of planar phi**3 field theory. Nucl Phys B652:196–228. arxiv: hep-th/0206205, https://doi.org/10.1016/S0550-3213(03)00004-X
57. Greensite J, Olejnik S (2003) Coulomb energy, vortices, and confinement. Phys Rev D67:094503. arxiv: hep-lat/0302018, https://doi.org/10.1103/PhysRevD.67.094503
58. Greensite J (2003) The confinement problem in lattice gauge theory. Prog Part Nucl Phys 51:1. arxiv: hep-lat/0301023, https://doi.org/10.1016/S0146-6410(03)90012-3
59. Trawiński AP, Głazek SD, Brodsky SJ, de Téramond GF, Dosch HG (2014) Effective confining potentials for QCD. Phys Rev D 90:074017. arxiv: 1403.5651, https://doi.org/10.1103/PhysRevD.90.074017
60. Bulava J, Hörz B, Knechtli F, Koch V, Moir G, Morningstar C et al (2019) String breaking by light and strange quarks in QCD. Phys Lett B 793:493–498. arxiv: 1902.04006, https://doi.org/10.1016/j.physletb.2019.05.018
61. Bali GS (2001) QCD forces and heavy quark bound states. Phys Rep 343:1–136. arxiv: hep-ph/0001312, https://doi.org/10.1016/S0370-1573(00)00079-X
62. Ripka G (2004) Dual superconductor models of color confinement, vol 639. https://doi.org/10.1007/b94800
63. Strassler MJ, Why unparticle models with mass gaps are examples of hidden valleys. arxiv: 0801.0629
64. Conte E, Fuks B, Serret G (2013) MadAnalysis 5, A user-friendly framework for collider phenomenology. Comput Phys Commun 184:222–256. arxiv: 1206.1599, https://doi.org/10.1016/j.cpc.2012.09.009
65. Alwall J, Frederix R, Frixione S, Hirschi V, Maltoni F, Mattelaer O et al (2014) The automated computation of tree-level and next-to-leading order differential cross sections, and their matching to parton shower simulations. JHEP 07:079. arxiv: 1405.0301, https://doi.org/10.1007/JHEP07(2014)079
66. Sjöstrand T, Ask S, Christiansen JR, Corke R, Desai N, Ilten P et al (2015) An Introduction to PYTHIA 8.2. Comput Phys Commun 191:159–177. arxiv: 1410.3012, https://doi.org/10.1016/j.cpc.2015.01.024
67. Feynman RP (1969) Very high-energy collisions of hadrons. Phys Rev Lett 23:1415–1417. https://doi.org/10.1103/PhysRevLett.23.1415

68. Grosse-Oetringhaus JF, Reygers K (2010) Charged-particle multiplicity in proton-proton collisions. J Phys G37:083001. arxiv: 0912.0023, https://doi.org/10.1088/0954-3899/37/8/083001
69. Kumar A, Singh BK, Srivastava PK, Singh CP (2013) Wounded quarks and multiplicity at relativistic ion colliders. Eur Phys J Plus 128:45. arxiv: 1210.1323, https://doi.org/10.1140/epjp/i2013-13045-9
70. Rosin ML (2006) Energy dependence of the mean charged multiplicity in deep inelastic scattering with ZEUS at HERA. PhD thesis, Wisconsin U., Madison
71. Andersson B, Gustafson G, Ingelman G, Sjostrand T (1983) Parton fragmentation and string dynamics. Phys Rep 97:31–145. https://doi.org/10.1016/0370-1573(83)90080-7
72. Bodeker D, Moore GD (2009) Can electroweak bubble walls run away?. *JCAP* 0905:009. arxiv: 0903.4099, https://doi.org/10.1088/1475-7516/2009/05/009
73. Bodeker D, Moore GD (2017) Electroweak bubble wall speed limit. JCAP 1705:025. arxiv: 1703.08215, https://doi.org/10.1088/1475-7516/2017/05/025
74. Caprini C et al (2016) Science with the space-based interferometer eLISA. II: Gravitational waves from cosmological phase transitions. JCAP 04:001. arxiv: 1512.06239, https://doi.org/10.1088/1475-7516/2016/04/001
75. Caprini C et al (2020) Detecting gravitational waves from cosmological phase transitions with LISA: an update. JCAP 03:024. arxiv: 1910.13125, https://doi.org/10.1088/1475-7516/2020/03/024
76. Ellis J, Lewicki M, No JM, Vaskonen V (2019) Gravitational wave energy budget in strongly supercooled phase transitions. JCAP 06:024. arxiv: 1903.09642, https://doi.org/10.1088/1475-7516/2019/06/024
77. Kosowsky A, Turner MS (1993) Gravitational radiation from colliding vacuum bubbles: envelope approximation to many bubble collisions. Phys Rev D 47:4372–4391. arxiv: astro-ph/9211004, https://doi.org/10.1103/PhysRevD.47.4372
78. Huber SJ, Konstandin T (2008) Gravitational wave production by collisions: more bubbles. JCAP 09:022. arxiv: 0806.1828, https://doi.org/10.1088/1475-7516/2008/09/022
79. Jinno R, Takimoto M (2017) Gravitational waves from bubble collisions: an analytic derivation. Phys Rev D 95:024009. arxiv: 1605.01403, https://doi.org/10.1103/PhysRevD.95.024009
80. Cutting D, Escartin EG, Hindmarsh M, Weir DJ, Gravitational waves from vacuum first order phase transitions II: from thin to thick walls. arxiv: 2005.13537
81. Lewicki M, Vaskonen V, Gravitational wave spectra from strongly supercooled phase transitions. arxiv: 2007.04967
82. Jinno R, Takimoto M (2019) Gravitational waves from bubble dynamics: beyond the Envelope. JCAP 1901:060. arxiv: 1707.03111, https://doi.org/10.1088/1475-7516/2019/01/060
83. Konstandin T (2018) Gravitational radiation from a bulk flow model. JCAP 03:047. arxiv: 1712.06869, https://doi.org/10.1088/1475-7516/2018/03/047
84. Lewicki M, Vaskonen V (2020) On bubble collisions in strongly supercooled phase transitions. Phys Dark Univ 30:100672. arxiv: 1912.00997, https://doi.org/10.1016/j.dark.2020.100672
85. Durrer R, Caprini C (2003) Primordial magnetic fields and causality. JCAP 11:010. arxiv: astro-ph/0305059, https://doi.org/10.1088/1475-7516/2003/11/010
86. Caprini C, Durrer R, Konstandin T, Servant G (2009) General properties of the gravitational wave spectrum from phase transitions. Phys Rev D 79:083519. arxiv: 0901.1661, https://doi.org/10.1103/PhysRevD.79.083519
87. Cai R-G, Pi S, Sasaki M (2020) Universal infrared scaling of gravitational wave background spectra. Phys Rev D 102:083528. arxiv: 1909.13728, https://doi.org/10.1103/PhysRevD.102.083528
88. Child HL, Giblin Jr. JT (2012) Gravitational radiation from first-order phase transitions. JCAP 1210:001. arxiv: 1207.6408, https://doi.org/10.1088/1475-7516/2012/10/001
89. Cutting D, Hindmarsh M, Weir DJ (2018) Gravitational waves from vacuum first-order phase transitions: from the envelope to the lattice. Phys Rev D97:123513. arxiv: 1802.05712, https://doi.org/10.1103/PhysRevD.97.123513

90. Cutting D, Hindmarsh M, Weir DJ, Vorticity, kinetic energy, and suppressed gravitational wave production in strong first order phase transitions. arxiv: 1906.00480
91. Jinno R, Seong H, Takimoto M, Um CM (2019) Gravitational waves from first-order phase transitions: Ultra-supercooled transitions and the fate of relativistic shocks. JCAP 10:033. arxiv: 1905.00899, https://doi.org/10.1088/1475-7516/2019/10/033
92. Baldes I, Garcia-Cely C (2019) Strong gravitational radiation from a simple dark matter model. JHEP 05:190. arxiv: 1809.01198, https://doi.org/10.1007/JHEP05(2019)190
93. Höche S, Kozaczuk J, Long AJ, Turner J, Wang Y (2021) Towards an all-orders calculation of the electroweak bubble wall velocity. JCAP 03:009. arxiv: 2007.10343, https://doi.org/10.1088/1475-7516/2021/03/009
94. A. A. M. Vanvlasselaer, *Bubble wall velocity: heavy physics effects*, arxiv: 2010.02590
95. Gouttenoire Y, Jinno R, Sala F (2022) Friction pressure on relativistic bubble walls. JHEP 05:004. arxiv: 2112.07686, https://doi.org/10.1007/JHEP05(2022)004
96. Casher A, Neuberger H, Nussinov S (1979) Chromoelectric flux tube model of particle production. Phys Rev D 20:179–188. https://doi.org/10.1103/PhysRevD.20.179
97. Armoni A, Shifman M (2003) Remarks on stable and quasistable k strings at large N. Nucl Phys B 671:67–94. arxiv: hep-th/0307020, https://doi.org/10.1016/j.nuclphysb.2003.08.021
98. Bicudo P, Cardoso N, Cardoso M (2019) Pure gauge QCD flux tubes and their widths at finite temperature. Nucl Phys B 940:88–112. arxiv: 1702.03454, https://doi.org/10.1016/j.nuclphysb.2019.01.012
99. Schwaller P (2015) Gravitational waves from a dark phase transition. Phys Rev Lett 115:181101. arxiv: 1504.07263, https://doi.org/10.1103/PhysRevLett.115.181101
100. Aoki M, Goto H, Kubo J (2017) Gravitational waves from hidden QCD phase transition. Phys Rev D 96:075045. arxiv: 1709.07572, https://doi.org/10.1103/PhysRevD.96.075045
101. Aoki M, Kubo J (2020) Gravitational waves from chiral phase transition in a conformally extended standard model. JCAP 04:001. arxiv: 1910.05025, https://doi.org/10.1088/1475-7516/2020/04/001
102. Bigazzi F, Caddeo A, Cotrone AL, Paredes A, Dark holograms and gravitational waves. arxiv: 2011.08757
103. Ares FR, Hindmarsh M, Hoyos C, Jokela N, Gravitational waves from a holographic phase transition. arxiv: 2011.12878
104. Huang W-C, Reichert M, Sannino F, Wang Z-W, Testing the dark confined landscape: from lattice to gravitational waves. arxiv: 2012.11614
105. Halverson J, Long C, Maiti A, Nelson B, Salinas G, Gravitational waves from dark Yang-Mills sectors. arxiv: 2012.04071
106. Buballa M (2005) NJL model analysis of quark matter at large density. Phys Rep 407:205–376. arxiv: hep-ph/0402234, https://doi.org/10.1016/j.physrep.2004.11.004
107. Fukushima K, Hatsuda T (2011) The phase diagram of dense QCD. Rep Prog Phys 74:014001. arxiv: 1005.4814, https://doi.org/10.1088/0034-4885/74/1/014001
108. Fukushima K, Sasaki C (2013) The phase diagram of nuclear and quark matter at high baryon density. Prog Part Nucl Phys 72:99–154. arxiv: 1301.6377 https://doi.org/10.1016/j.ppnp.2013.05.003
109. Herzog CP (2007) A holographic prediction of the deconfinement temperature. Phys Rev Lett 98:091601. arxiv: hep-th/0608151, https://doi.org/10.1103/PhysRevLett.98.091601
110. Schwarz DJ, Stuke M (2009) Lepton asymmetry and the cosmic QCD transition. JCAP 11:025. arxiv: 0906.3434, https://doi.org/10.1088/1475-7516/2009/11/025
111. Schettler S, Boeckel T, Schaffner-Bielich J (2011) Imprints of the QCD phase transition on the spectrum of gravitational waves. Phys Rev D 83:064030. arxiv: 1010.4857, https://doi.org/10.1103/PhysRevD.83.064030
112. Alho T, Järvinen M, Kajantie K, Kiritsis E, Rosen C, Tuominen K (2014) A holographic model for QCD in the Veneziano limit at finite temperature and density. JHEP 04:124. arxiv: 1312.5199, https://doi.org/10.1007/JHEP04(2014)124
113. Ahmadvand M, Bitaghsir Fadafan K (2017) Gravitational waves generated from the cosmological QCD phase transition within AdS/QCD. Phys Lett B 772:747–751. arxiv: 1703.02801, https://doi.org/10.1016/j.physletb.2017.07.039

114. Ahmadvand M, Bitaghsir Fadafan K (2018) The cosmic QCD phase transition with dense matter and its gravitational waves from holography. Phys Lett B 779:1–8. arxiv: 1707.05068, https://doi.org/10.1016/j.physletb.2018.01.066
115. Chen Y, Huang M, Yan Q-S (2018) Gravitation waves from QCD and electroweak phase transitions. JHEP 05:178. arxiv: 1712.03470, https://doi.org/10.1007/JHEP05(2018)178
116. Kapusta J, Gale C (2011) Finite-temperature field theory: principles and applications. Cambridge monographs on mathematical physics. Cambridge University Press. https://doi.org/10.1017/CBO9780511535130
117. Arnold PB, Moore GD, Yaffe LG (2003) Effective kinetic theory for high temperature gauge theories. JHEP 01:030. arxiv: hep-ph/0209353, https://doi.org/10.1088/1126-6708/2003/01/030
118. Kofman L, Linde AD, Starobinsky AA (1997) Towards the theory of reheating after inflation. Phys Rev D56:3258–3295. arxiv: hep-ph/9704452, https://doi.org/10.1103/PhysRevD.56.3258
119. Hawking SW, Moss IG, Stewart JM (1982) Bubble Collisions in the very early universe. Phys Rev D 26:2681. https://doi.org/10.1103/PhysRevD.26.2681
120. Watkins R, Widrow LM (1992) Aspects of reheating in first order inflation. Nucl Phys B374:446–468. https://doi.org/10.1016/0550-3213(92)90362-F
121. Braden J, Bond JR, Mersini-Houghton L (2015) Cosmic bubble and domain wall instabilities I: parametric amplification of linear fluctuations. JCAP 1503:007. arxiv: 1412.5591, https://doi.org/10.1088/1475-7516/2015/03/007
122. Falkowski A, No JM (2013) Non-thermal dark matter production from the electroweak phase transition: multi-TeV WIMPs and 'Baby-Zillas'. JHEP 02:034. arxiv: 1211.5615, https://doi.org/10.1007/JHEP02(2013)034
123. Katz A, Riotto A (2016) Baryogenesis and gravitational waves from runaway bubble collisions. JCAP 1611:011. arxiv: 1608.00583, https://doi.org/10.1088/1475-7516/2016/11/011
124. Planck collaboration, Aghanim N et al, Planck 2018 results. VI. Cosmological parameters. arxiv: 1807.06209
125. Kolb EW, Turner MS (1990) The early universe. Front Phys 69:1–547
126. von Harling B, Petraki K (2014) Bound-state formation for thermal relic dark matter and unitarity. JCAP 12:033. arxiv: 1407.7874, https://doi.org/10.1088/1475-7516/2014/12/033
127. Baldes I, Petraki K (2017) Asymmetric thermal-relic dark matter: Sommerfeld-enhanced freeze-out, annihilation signals and unitarity bounds. JCAP 09:028. arxiv: 1703.00478, https://doi.org/10.1088/1475-7516/2017/09/028
128. Azatov A, Vanvlasselaer M, Yin W, Dark matter production from relativistic bubble walls. arxiv: 2101.05721
129. Witten E (1981) Cosmological consequences of a light higgs Boson. Nucl Phys B177:477–488. https://doi.org/10.1016/0550-3213(81)90182-6
130. Iso S, Serpico PD, Shimada K (2017) QCD-Electroweak first-order phase transition in a supercooled universe. Phys Rev Lett 119:141301. arxiv: 1704.04955, https://doi.org/10.1103/PhysRevLett.119.141301
131. Cirelli M, Gouttenoire Y, Petraki K, Sala F (2019) Homeopathic dark matter, or how diluted heavy substances produce high energy cosmic rays. JCAP 02:014. arxiv: 1811.03608, https://doi.org/10.1088/1475-7516/2019/02/014
132. Ipek S, Tait TM (2019) Early cosmological period of QCD confinement. Phys Rev Lett 122:112001. arxiv: 1811.00559, https://doi.org/10.1103/PhysRevLett.122.112001
133. Bai Y, Long AJ, Lu S (2019) Dark quark nuggets. Phys Rev D 99:055047. arxiv: 1810.04360, https://doi.org/10.1103/PhysRevD.99.055047
134. Baker MJ, Kopp J, Long AJ, Filtered dark matter at a first order phase transition. arxiv: 1912.02830
135. Chway D, Jung TH, Shin CS (2020) Dark matter filtering-out effect during a first-order phase transition. Phys Rev D 101:095019. arxiv: 1912.04238, https://doi.org/10.1103/PhysRevD.101.095019
136. Kitajima N, Takahashi F, Primordial black holes from QCD axion bubbles. arxiv: 2006.13137

137. Goldberger WD, Grinstein B, Skiba W (2008) Distinguishing the Higgs boson from the dilaton at the Large Hadron Collider. Phys Rev Lett 100:111802. arxiv: 0708.1463, https://doi.org/10.1103/PhysRevLett.100.111802
138. Coleman SR, Weinberg EJ (1973) Radiative corrections as the origin of spontaneous symmetry breaking. Phys Rev D7:1888–1910. https://doi.org/10.1103/PhysRevD.7.1888
139. Coleman SR (1977) The fate of the false vacuum. 1. Semiclassical theory. Phys Rev D 15:2929–2936. https://doi.org/10.1103/PhysRevD.16.1248
140. Callan J, Curtis G, Coleman SR (1977) The fate of the false vacuum. 2. First quantum corrections. Phys Rev D 16:1762–1768. https://doi.org/10.1103/PhysRevD.16.1762
141. Dorsch GC, Huber SJ, Konstandin T (2018) Bubble wall velocities in the standard Model and beyond. JCAP 12:034. arxiv: 1809.04907, https://doi.org/10.1088/1475-7516/2018/12/034
142. Chliapnikov PV, Uvarov VA (1995) Striking regularity in meson and baryon production rates in e+ e- annihilations. Phys Lett B345:313–320. https://doi.org/10.1016/0370-2693(94)01613-H
143. Becattini F (1996) A thermodynamical approach to hadron production in e+ e- collisions. Z Phys C69:485–492. https://doi.org/10.1007/BF02907431
144. Pei Y-J (1996) A Simple approach to describe hadron production rates in e+ e- annihilation. Z Phys C72:39–46. https://doi.org/10.1007/s002880050221
145. Chliapnikov PV (1999) Hyperfine splitting in light-flavour hadron production at LEP. Phys Lett B462:341–353. https://doi.org/10.1016/S0370-2693(99)00910-7
146. Andronic A, Beutler F, Braun-Munzinger P, Redlich K, Stachel J (2009) Thermal description of hadron production in e+e- collisions revisited. Phys Lett B675:312–318. arxiv: 0804.4132, https://doi.org/10.1016/j.physletb.2009.04.024
147. Mitridate A, Redi M, Smirnov J, Strumia A (2017) Dark matter as a weakly coupled Dark Baryon. JHEP 10:210. arxiv: 1707.05380, https://doi.org/10.1007/JHEP10(2017)210

Chapter 8
Gravitational Waves from Cosmic Strings

8.1 Introduction

The Standard Model of particle physics needs to be completed to address observational facts such as the matter antimatter asymmetry and the dark matter of the universe, as well as the origin of inflation. These, together with a number of other fundamental theoretical puzzles associated with e.g. the flavour structure of the matter sector and the ultra-violet properties of the Higgs scalar field, motivate extensions of the Standard Model featuring new degrees of freedom and new energy scales. In turn, such new physics can substantially impact the expansion history in the early universe and leads to deviations with respect to the standard cosmological model. Any deviations in the Friedmann equation occurring at temperatures above the MeV remain to date essentially unconstrained.

In the standard cosmological model, primordial inflation is followed by a long period of radiation domination until the more recent transitions to matter and then dark energy domination. Evidence for this picture comes primarily from observations of the Cosmic Microwave Background (CMB) and the successful predictions of Big-Bang Nucleosynthesis (BBN), which on the other hand, do not allow to test cosmic temperatures above $\mathcal{O}(\mathrm{MeV})$.

An exciting prospect for deciphering the pre-BBN universe history and therefore high energy physics unaccessible by particle physics experiments, comes from the possible detection of a stochastic background of gravitational waves (SGWB), originating either from cosmological phase transitions, from cosmic strings or from inflation [2].

Particularly interesting are cosmic strings (CS), which act as a long-lasting source of gravitational waves (GW) from the time of their production, presumably very early on, until today. The resulting frequency spectrum therefore encodes information from the almost entire cosmic history of our universe, and could possibly reveal precious details about the high energy particle physics responsible for a modified universe expansion.

Y. Gouttenoire, *Beyond the Standard Model Cocktail*, Springer Theses,
https://doi.org/10.1007/978-3-031-11862-3_8

There has been a large literature on probes of a non-standard cosmology through the nearly-scale invariant primordial GW spectrum generated during inflation [3–21]. In contrast, little efforts have been invested to use the scale-invariant GW spectrum generated by CS [22–27] while there has been intense activity on working out predictions for the SGWB produced by CS in standard cosmology [28–48].

In this study, we propose to use the detection of a SGWB from local cosmic strings to test the existence of alternative stages of cosmological expansion between the end of inflation and the end of the radiation era. Particularly well-motivated is a stage of early-matter domination era induced by a heavy cold particle dominating the universe and decaying before BBN. Another possibility is a stage of kination triggered by the fast rolling evolution of a scalar field down its potential, e.g. [49, 50] for the pioneering articles. See [51, 52] for a detailed discussion of kination occuring during the standard radiation era. Finally, supercooled confining phase transitions [53–65] can induce some late short stages of inflation inside a radiation era. The latter were motivated at the TeV scale but the properties of the class of scalar potentials naturally inducing a short inflationary era can be applied to any other scale. We will consider these various possibilities and their imprints on the GW spectrum from cosmic strings.

The dominant source of GW emission from a cosmic string network comes from loops which are continuously formed during the network fragmentation. We thus primarily need to compute the loop-production efficiency during the non-standard transition eras. This is crucial for a precise prediction of the turning-point frequency as a signature of the non-standard era. The temperature of the universe at the end of the non-standard era can be deduced from the measurement of these turning point frequencies.

The observational prospects for measuring the SGWB from cosmic string networks at LISA was recently reviewed in [24]. Besides, the effect of particle production on the loop distribution and thus on the SGWB was recently discussed [66, 67] where it was however concluded that the expected cutoff is outside the range of current and planned detectors (see also [68]). Our study integrates these recent developments and goes beyond in several directions:

- We go beyond the so-called scaling regime by computing the time evolution of the string network parameters (long string mean velocity and correlation length) and thus the loop-production efficiency during modifications of the equation of state of the universe. Including these transient effects results in a turning-point frequency smaller by $\mathcal{O}(20)$ compared to the prediction from the scaling regime.[1] As a result, the energy scale of the universe associated with the departure from the standard radiation era that can be probed is correspondingly larger than the one predicted from scaling networks.
- We investigate a large variety of non-standard cosmologies, in particular models where a non-standard era can be rather short inside the radiation era, due for instance to some cold particle temporarily dominating the energy density (short

[1] The turning-point frequency can even be smaller by $\mathcal{O}(400)$ if in a far-future, a precision of the order of 1% can be reached in the measurement of the SGWB, cf. Eq. (8.118).

matter era, see Fig. 8.9) [69] or some very short stage of inflation (for a couple of efolds) due to a high-scale supercooled confinement phase transition, see Fig. 8.13. Such inflationary stages occurring at scales up to 10^{14} GeV could be probed, see Fig. 8.16. Even 1 or 2 e-folds could lead to observable features, see Fig. 8.13. The study of GW signatures of cosmological histories containing a kination era is given in [51, 52].

- We also consider longer low-scale inflation models. For instance, an intermediate inflationary era lasting for $\mathcal{O}(10)$ efolds, the SGWB from cosmic strings completely looses its scale invariant shape and has a peak structure instead, see Fig. 8.14. A TeV scale inflation era can lead to a broad peak either in the LISA or BBO band or even close to the SKA band, depending on the precise value of the string tensions $G\mu$, and the number of efolds N_e.
- We include high-frequency cutoff effects from particle production which can limit observations for small value of the string tension $G\mu \lesssim 10^{-15}$ and high-frequency cutoff from thermal friction, see Fig. 8.4 and top left panel of Fig. 8.6.
- We discuss how to read information about the small-scale structure of CS from the high-frequency tail of the GW spectrum, see Appendix 8.B.6.
- We discuss the comparison between local and global string networks, see Appendix 8.F.

The plan of the chapter is the following. In Sect. 8.2, we recap the key features of CS networks, their cosmological evolution, decay channels and the pulsar timing array constraints on the string tension. Section 8.3 reviews the computation of the SGWB from Nambu-Goto CS. We first discuss the underlying assumptions on the small-scale structure and on the loop distribution and then derive the master formula of the GW frequency spectrum. An important discussion concerns the non-trivial frequency-temperature relation and how it depends on the cosmological scenario. Section 8.4 is devoted to the derivation of the loop production efficiency beyond the scaling regime, taking into account transient effects from the change in the equation of state of the universe. We apply this to predict the SGWB in the standard cosmological model in Sect. 8.5. We then move to discuss non-standard cosmological histories, a short intermediate matter era inside the radiation era in Sect. 8.6, and an intermediate inflationary era in Sect. 8.7. We discuss the specific spectral features in each of these cases and their observability by future instruments. We conclude in Sect. 8.8. Additional details are moved to appendices, such as non-GW constraints on the string tension $G\mu$ in Appendix 8.A, a step-by-step derivation of the GW spectrum as well as the values of its slopes in Appendix 8.B, the formulae of the various turning-point frequencies in Appendix 8.C, the derivation of the equations which govern the evolution of the long-string network in the Velocity-dependent One-Scale (VOS) model in Appendix 8.D, a discussion of the extensions to the original VOS model in Appendix 8.E, the prediction of the GW spectrum from global strings in Appendix 8.F, and the calculation of the integrated power-law sensitivity curves for each experiment in Appendix 6.A.

8.2 Recap on Cosmic Strings

Cosmic strings have been the subject of numerous studies since the pioneering paper [70], see [71–73] for reviews.

8.2.1 *Microscopic Origin of Cosmic Strings*

Fundamental objects:
A cosmic string is an extended object of cosmological size. It can be a fundamental object in superstring theories [74–81]. In superstring theory, strings are the fundamental mandatory new degrees of freedom, which are usually unstable and decay before they could stretch to cosmological scales. However, the so-called F-, D-, and (p, q)-strings can grow large and have cosmological consequences.

Topological defects:
Cosmic strings can also be topological solitons in classical field theories. It is a classical field configuration which arises whenever there is a symmetry breaking $G \to H$ with a non-trivial homotopy group[2] $\pi_1(G/H) \neq \text{Id}$, where 'Id' denotes the identity map, i.e. one can find at least two non-equivalent maps from the circle to the vacuum manifold G/H. For example, any theory with spontaneous breaking of a $U(1)$ symmetry [82, 83] has a string solution, since $\pi_1(U(1)) = \mathbb{Z}$. More complex vacuum manifolds with string solutions can appear in various grand unified theories [84, 85], e.g. $SO(10) \to SO(5) \times \mathbb{Z}_2$. However, we emphasize that there is no string solution in the Standard Model (SM) because the symmetry breaking pattern

$$SU(2)_L \times U(1)_Y \to U(1)_{\text{EM}}, \tag{8.1}$$

has a trivial homotopy group[3]

$$\pi_1(SU(2) \times U(1)/U(1)) \cong \pi_0(U(1)) \cong \text{Id}. \tag{8.2}$$

Thus, the existence of cosmic strings requires physics Beyond the Standard Model (BSM). In this study, we will assume a generic BSM scenario giving rise to cosmic strings without specifying a particular model.

Abelian-Higgs string:
We now present the standard example of field theories with a string-liked solution: the abelian-Higgs (AH) model. It is a field theory with a complex scalar field ϕ

---

[2] The existence of other topological defects, domain walls, monopoles, or textures also relies on the non triviality of an homotopy group, $\pi_n(G/H) \neq \text{Id}$ for $n = 0, 2$ or 3 respectively.

[3] The homotopy group of a coset can be easily computed via the so-called fundamental theorem: $\pi_n(G/H) \cong \pi_{n-1}(H)$ [72].

charged under a $U(1)$ gauge interaction mediated by a gauge field A_μ. Note that the symmetry can also be global in the absence of the gauge boson. The resulting strings solutions corresponding to local and global symmetries are called local and global strings, respectively. The AH lagrangian in $d+1$ space-time dimensions is described by

$$\mathscr{L} = -\frac{1}{4}F_{\mu\nu}^2 + |D_\mu\phi|^2 - \frac{1}{4}\lambda(|\phi|^2 - \eta^2)^2 \qquad \text{for } \mu, \nu = 0, 1, 2, \ldots, d\,, \tag{8.3}$$

where $D_\mu = \partial_\mu - i\,e\,A_\mu$ and where λ and η are parameters of the scalar potential generating the spontaneous symmetry breaking (SSB). η is the vacuum expectation value (VEV) of the scalar field. The energy of the system can be derived as the sum of the energy in the scalar and gauge field

$$E = \int d^dx \left(-\frac{1}{4}F_{\mu\nu}^2 + |D_\mu\phi|^2 + V(\phi)\right). \tag{8.4}$$

It has been shown that a stable field configuration with finite energy density exists, known as the Nielsen-Olesen vortex [86] which are lines where the scalar field sits on the top of its 'mexican hat potential' $V(\phi)$ and approaches its VEV at large distance. When following a closed path around the string, the phase of the complex scalar field returns to its original value after winding around the mexican an integer n number of times. Hence, the angular component of the gradient of the scalar field diverges at large distance

$$\int_\delta^R d^2r \left|\frac{1}{r}\frac{\partial\phi}{d\theta}\right|^2 = \int 2\pi r dr\, n\frac{\eta^2}{r^2} = 2\pi\eta^2\, n \log\frac{R}{\delta} \underset{R\to\infty}{\longrightarrow} \infty \tag{8.5}$$

where $\delta \sim m_\phi^{-1} = (\sqrt{\lambda}\eta)^{-1}$ or $m_A^{-1} = (\sqrt{2}e\eta)^{-1}$ is the string core width. m_ϕ^{-1} and m_A^{-1} are the length scales below which the scalar and gauge field gradients are respectively confined. For local strings, the divergence can be exactly canceled by a judicious gauge choice. However, for global strings, which are a particular realization of the AH model in the absence of gauge field or in the presence of a vanishing gauge coupling $e = 0$, the logarithmic dependence of the energy with the infrared cut-off can not be avoided. Physically, the logarithm divergence is due to the existence of a long-range interaction mediated by the massless Goldstone mode (which is the complex phase of ϕ). Therefore, the energy per unit of length, also known as the 'string tension' reads

$$\mu \approx 2\pi\eta^2 n \times \begin{cases} 1 & \text{for local strings,} \\ \ln\left(\frac{R}{\delta}\right) & \text{for global strings,} \end{cases} \tag{8.6}$$

where the infrared cut-off R can be either the string inter-distance or the causal horizon H^{-1} and where the string core width δ is given by the inverse mass of the radial mode m_ϕ^{-1}. The interaction between strings receives an attractive contribution from the vortex core-overlap and a repulsive contribution (attractive for vortex-anti-vortex interactions) from the circulation of the magnetic lines outside the vertex core. Strings where the attractive interaction dominates are called 'type I' whereas the other ones are called 'type II', in analogy with vortices in superconductors. Type II with winding number $n > 1$ are unstable against decay to $n' < n$. Conversely, type I strings with high winding numbers are stable, but it has been shown in numerical simulations that string interactions could quickly decrease the winding number via 'peeling' [87].

Nambu-Goto string (local):
Masses of the underlying fields determine the size of the core width. However, for local strings this can be rather small compared to the cosmological scale H^{-1} such that an observer is unable to resolve any physics within the core. Hence, it seems reasonable to consider only the effective description of theories in the zero-width limit. For the local abelian-Higgs model, the appropriate effective action is the Nambu-Goto (NG) action [72, 88, 89], which strictly speaking is defined by the area of the world-sheet swept out by the motion of a featureless one-dimensional object,

$$\mathcal{S}_{\rm NG} = -\mu \int dt dl \left(1 - v_\perp^2\right)^{1/2}, \tag{8.7}$$

where $d\tau = (1 - v_\perp)^2 \, dt$ is the proper time of a string segment of length dl. It can also be written in a covariant form

$$\mathcal{S}_{\rm NG} = -\mu \int d^2\zeta \sqrt{-\gamma}, \tag{8.8}$$

where $\gamma \equiv \det \gamma_{ab}$ is the determinant of the metric γ_{ab} induced on the 2D world-sheet, starting from the 4D bulk metric $g_{\mu\nu}$,

$$\gamma_{ab} = g_{\mu\nu} \frac{\partial x^\mu}{\partial \zeta^a} \frac{\partial x^\nu}{\partial \zeta^b}, \tag{8.9}$$

The coordinates $x^\mu(\zeta_a)$, with $\mu = 1, 2, 3, 4$ and $a = 1, 2$, describe the embedding of the 2D world-sheet in the 4D space-time.

Nambu-Goto string (global):
However, global strings whose core extends due to the long-range interaction mediated by the massless Goldstone boson θ, complex phase of the scalar field $\phi = \eta \, e^{i\theta}$, can not be completely described by the NG action. One must account for the interaction between the string and the Golstone field, the so-called Kalb-Ramond action [74, 90]

$$\mathcal{S}_{\mathrm{KR}} = -\mu \int d^2\zeta\sqrt{-\gamma} + \frac{1}{6}\int H^2 d^4x + 2\pi\eta \int B_{\mu\nu} d\sigma^{\mu\nu}, \tag{8.10}$$

where $B_{\mu\nu}$ is the antisymmetric field, defined as the hodge-dual of the Goldstone boson

$$\frac{1}{2}\varepsilon_{\mu\nu\lambda\rho}\partial^{\nu}B^{\lambda\rho} \equiv \eta\, \partial_{\mu}\boldsymbol{\theta}, \tag{8.11}$$

$H^{\mu\nu\lambda} \equiv \partial^{\mu}B^{\nu\lambda} + \partial^{\lambda}B^{\mu\nu} + \partial^{\nu}B^{\lambda\mu}$ is the field strength tensor of $B_{\mu\nu}$ and $d\sigma^{\mu\nu} \equiv \varepsilon^{ab}x^{\mu}_{,a}x^{\nu}_{,b}d^2\zeta$ is the world-sheet area element.

8.2.2 *Cosmic-String Network Formation and Evolution*

Kibble mechanism:
The formation of cosmic strings occurs during a cosmological phase transition associated with spontaneous symmetry breaking, occurring at a temperature, approximately given by the VEV acquired by the scalar field

$$T_{\mathrm{p}} \sim 10^{11}\ \mathrm{GeV}\left(\frac{G\mu}{10^{-15}}\right)^{1/2}. \tag{8.12}$$

CS are randomly distributed and form a network characterized by its correlation length L, which can be defined as

$$L \equiv \sqrt{\mu/\rho_{\infty}}, \tag{8.13}$$

where μ is the string tension, the energy per unit length, and ρ_{∞} is the energy density of long strings. More precisely, long strings form infinite random walks [91] which can be visualized as collections of segments of length L.

Loop chopping:
Each time two segments of a long string cross each other, they inter-commute, with a probability P and form a loop. Loop formation is the main energy-loss mechanism of the long string network. In numerical simulations [92] and analytical modelling [93], the probability of inter-commutation has been found to be $P = 1$ but in some models it can be lower. This is the case of models with extra-dimensions [75, 94], strings with junctions [95] or peeling [87], or the case of highly relativistic strings [96].

Scaling regime:
Just after the network is formed, the strings may interact strongly with the thermal plasma such that their motion is damped. When the damping stops, cosmic strings oscillate and enter the phase of *scaling* evolution. During this phase, the network experiences two competing dynamics:

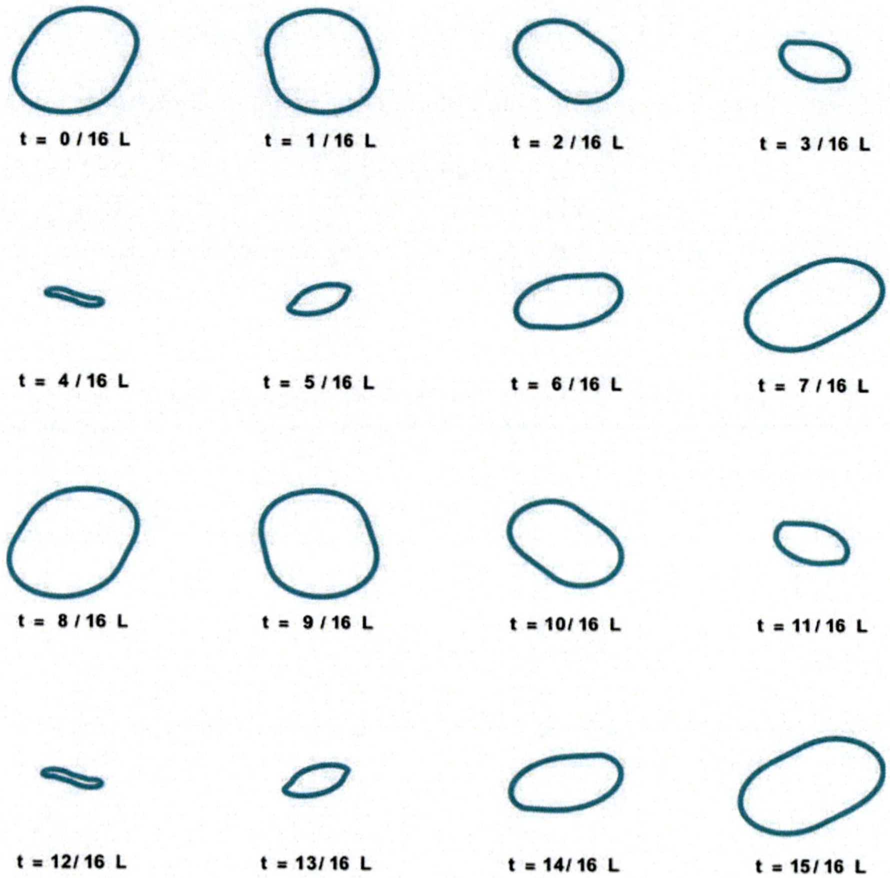

Fig. 8.1 Loop trajectory found after numerically integrating (c++) the Nambu-Goto equation in flat space. The loop initial conditions correspond to the parameters $\phi = 0$ and $\kappa = 0.5$ in Sect. 6.2.4 of [72]. Under the effect of its tension, the loop oscillates: it contracts while converting its potential energy into kinetic energy and vice-versa. We can see that the loop oscillation period is $L/2$. This figure is not included in [1] but instead it has been realized during the redaction of the manuscript

1. Hubble stretching: the correlation length scale stretches due to the cosmic expansion, $L \sim a$.
2. Fragmentation of long strings into loops: a loop is formed after each segment crossing. Right after their formation, loops evolve independently of the network, they oscillate under the effect of the tension, see Fig. 8.1 and start to decay through gravitational radiation and/or particle production.

It is known since a long time ago [97–101], that out of the two competing dynamics, Hubble expansion and loop fragmentation, there is an attractor solution, called the *scaling regime*, where the correlation length scales as the cosmic time,

$$L \sim t. \tag{8.14}$$

Note however that in the case of global-string network, it has been claimed that the scaling property in Eq. (8.14), is logarithmically violated due to the dependence of the string tension on the Hubble horizon [102–109]. Notet that an opposite conclusion has been drawn in [48, 110].

Proof:
We know show that the scaling regime in Eq. (8.14) is an attractor solution, following the approach of [72]. We start by evaluating the loop chopping rate. In a correlation volume L^3, a segment of length L must travel a distance L before encountering another segment. So the collision rate, per unit of volume, is $\frac{v}{L}\cdot\frac{1}{L^3}\sim\frac{1}{L^4}$. At each collision forming a loop, the network looses a loop energy $\mu\, L$. Hence, the master equation for the energy density $\rho_\infty \sim \mu/L^2$ reads

$$\dot{\rho}_\infty = -2\frac{\dot{a}}{a}\rho_\infty - \frac{1}{L^4}\cdot\mu\, L. \tag{8.15}$$

Plugging the ansatz $L = \xi t$ in the above equation leads to

$$\dot{\xi} = \frac{1}{2t}\,(1-\xi)\,. \tag{8.16}$$

Clearly $\xi = 1$ is a fix point. If the density of long string is too high, $\xi < 1$, then one has $\dot{\xi} > 0$ and the loop production becomes more efficient. If the density of long string is too low $\xi > 1$, then one has $\dot{\xi} < 0$ and the loop production becomes less efficient. The exact coefficient $\xi = L/t$ can be computed numerically or analytically, e.g. within the Velocity-dependent One-Scale model presented in Sect. 8.4.

Number of strings:
During the scaling regime, the number of strings per Hubble patch is conserved

$$\frac{\rho_\infty H^{-3}}{\mu L} = \text{constant}. \tag{8.17}$$

Moreover, the energy density of the long-string network, which scales as $\rho_\infty \sim \mu/t^2$, has the same equation-of-state as the main background cosmological fluid $\rho_{\rm bkg} \sim a^{-n}$,

$$\frac{\rho_\infty}{\rho_{\rm bkg}} \sim \frac{a^n}{t^2} \sim \text{constant}, \tag{8.18}$$

where we used $a = t^{2/n}$. Hence, the long-string energy density redshifts as matter during matter domination and as radiation during radiation domination. The scaling regime allows cosmic strings not to dominate the energy density of the universe, unlike other topological defects. The scaling property of a string network has been checked some fifteen years ago in numerical Nambu-Goto simulations [111–114] and more recently with larger simulations [115]. During the scaling regime, the loop production function is scale-free, with a power-law shape, meaning that loops are produced at any size between the Hubble horizon t and the scale $\sim \Gamma G\mu t$, below which the strings have been smoothened by the gravitational backreaction and there is no further segment crossing.

A scale-invariant SGWB:
An essential outcome is the scale-invariance of the Stochastic GW Background generated by loops during the scaling regime [28–44]. We construct the GW spectrum in Sect. 8.3.3 and give more details in Appendix 8.B. Remarkably, the spectrum generated by loops produced during radiation domination is flat, $\propto f^0$, whereas an early matter domination or an early kination-domination era turns the spectral index from f^0 to respectively $f^{-1/3}$ or f^1. As recently pointed out by [116], in the presence of an early matter, the slope f^{-1} predicted by [22, 23], is changed to $f^{-1/3}$ due to the high-k modes. We give more details on the impact of high-k modes on the GW spectrum in the presence of a decreasing slope due to an early matter era, a second period of inflation, particle production, thermal friction or network formation in Appendix 8.B.6. Hence, the detection of the SGWB from CS by LIGO [117], DECIGO, BBO [118], LISA [119, 120], Einstein Telescope [121, 122] or Cosmic Explorer [123] would offer an unique observation window on the equation of state of the Universe at the time when the CS loops responsible for the detected GW are formed.

8.2.3 *Decay Channels of Cosmic Strings*

Cosmic strings can decay in several ways, as we discuss below.

GW radiation from long strings:
Because of their topological nature, straight infinitely-long strings are stable against decay. However, small-scale structures of wiggly long strings can generate gravitational radiation. Intuitively, a highly wiggly string can act as a gas of small loops. The GW emission from long strings can be neglected compared to the GW emission from loops, as loops live much longer than a Hubble time [34, 72]. Indeed, the GW signal emitted by loops is enhanced by the large number of loops (continuously produced). Nambu-Goto numerical simulations have shown that the loop energy density is at least 100 times larger than the long-string energy density [124]. Only for global strings where loops are short-lived due to efficient Goldstone production, the GW emission from long strings can give a major contribution to the SGWB [125–128]. In what follows, we only consider the emission from loops.

GW radiation from loops (local strings):
In contrast to long strings, loops do not contain any topological charge and are free to decay into GW. The GW radiation power is found to be [72]

$$P_{\rm GW} = \Gamma G\mu^2, \tag{8.19}$$

where the total GW emission efficiency Γ is determined from Nambu-Goto simulations, $\Gamma \simeq 50$ [129]. Note that the gravitational power radiated by a loop is independent of its length. This can be understood from the quadrupole formula $P = G/5(Q''')^2$ [30, 130] where the triple time derivative of the quadrupole, $Q''' \propto \text{mass}\,(\text{length})^2/(\text{time})^3 \propto \mu$, is indeed independent of the length. The resulting GW are emitted at frequencies [71, 131]

$$\tilde{f} = \frac{2k}{l}, \qquad k \in \mathbb{Z}^+, \tag{8.20}$$

corresponding to the proper modes k of the loop. The tilde is used to distinguished the frequency emitted at $\tilde{t}$ from the frequency today

$$f = a(\tilde{t})/a(t_0)\ \tilde{f}. \tag{8.21}$$

The frequency dependence of the power spectrum $P_{\rm GW}(k)$ relies on the nature of the loop small-scale structures [132, 133], e.g. kinks or cusps, cf. Fig. 8.2. More precisely, the spectrum of the gravitational power emitted from one loop reads

$$P_{\rm GW}^{(k)} = \Gamma^{(k)} G\mu^2, \qquad \text{with } \Gamma^{(k)} \equiv \frac{\Gamma\, k^{-n}}{\sum_{p=1}^{\infty} p^{-n}}, \quad n = \begin{cases} 4/3 & \text{cusps} \\ 5/3 & \text{kinks} \\ 2 & \text{kink-kink collisions} \end{cases} \tag{8.22}$$

where the spectral index $n = 4/3$ when the small-scale structure is dominated by cusps [30, 38, 134], $n = 5/3$ for kink domination [38], or $n = 2$ for kink-kink collision domination [132, 133]. A discussion on how to read information about the small-scale structure of CS from the GW spectrum, is given in Appendix 8.B.6. In particular, we show that the high-frequency slope of the GW spectrum in the presence of an early matter era, a second period inflation, particle production or network formation, which is expected to be f^{-1} from the fundamental, $k = 1$, GW spectrum alone, is actually given by f^{1-n}. Immediately after a loop gets created, at time t_i with a length $\alpha\, t_i$, its length $l(\tilde{t})$ shrinks through emission of GW with a rate $\Gamma G\mu$

$$l(\tilde{t}) = \alpha t_i - \Gamma G\mu(\tilde{t} - t_i). \tag{8.23}$$

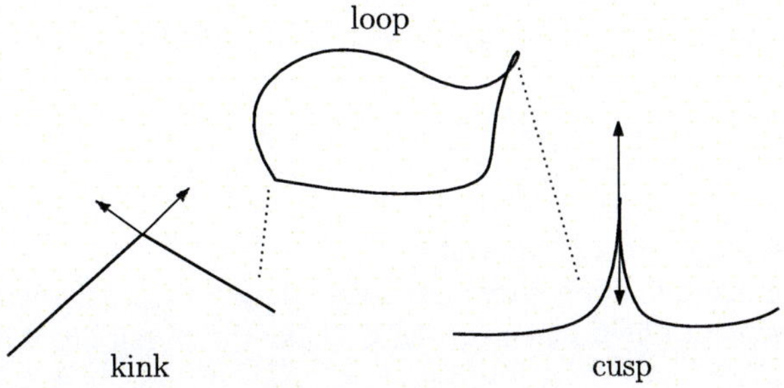

Fig. 8.2 Cartoon showing the geometry of a kink and a cusp which are singular structures formed on loops. The arrows denote the tangent vectors of the string segments

Consequently, the string lifetime due to decay into GW is given by

$$\tau_{\rm GW} = \frac{\alpha\, t_i}{\Gamma G\mu}. \tag{8.24}$$

The superposition of the GW emitted from all the loops formed since the creation of the long-string network generates a Stochastic GW Background. Also, cusp formations can emit high-frequency, short-lasting GW bursts [37, 38, 132, 133, 135]. If the rate of such events is lower than their frequency, they might be subtracted from the SGWB.

Goldstone boson radiation (global strings):
For global strings, the massless Goldstone particle production is the main decay channel. The radiation power has been estimated [72]

$$P_{\rm Gold} = \Gamma_{\rm Gold}\, \eta^2, \tag{8.25}$$

where η is the scalar field VEV and $\Gamma_{\rm Gold} \approx 65$ [27, 90]. We see that the GW emission power in Eq. (8.19) is suppressed by a factor $G\mu$ with respect to the Goldstone emission power in Eq. (8.25). Therefore, for global strings, the loops decay into Goldtone bosons after a few oscillations before having the time to emit much GW [72, 136]. However, as shown in Appendix 8.F, the SGWB from global string is detectable for large values of the string scale, $\eta \gtrsim 10^{14}$ GeV. Other recent studies of GW spectrum from global strings in standard and non-standard cosmology include [26, 27, 105, 137].

A well-motivated example of global string is the axion string coming from the breaking of a $U(1)$ Peccei-Quinn symmetry [90, 138–140]. Its phenomenological interest relies on the possibility to generate the correct Dark Matter abundance from Goldstone production. However, the value of the energy of the emitted Goldstone, needed to compute the axion relic abundance, is still matter of debate [103, 104]. Particularly, the precise dependence of the emitted mean energy with the infrared cut-off H^{-1} is unknown and is not tractable in numerical simulations due to the too large hierarchy between the string core size and the Hubble horizon. Hence, in the case of post-inflationnary Peccei-Quinn breaking, it is currently not understood whether the axion abundance today coming from the decay of the axionic string network supersedes the abundance coming from the vacuum misalignement or not. Note, however [110] which claims that the dependence of the emitted energy with the cut-off is an artefact. Note also the possibility that $1 - 10\%$ of the axion abundance would come from heavy-radial-mode production followed by its decay, and then would be hot [136]. Reference [26] shows the detectability of the GW from the axionic network of QCD axion Dark Matter (DM), after introducing an early-matter era which dilutes the axion DM abundance and increases the corresponding Peccei-Quinn scale η.

Massive particle radiation:
When the string curvature size is larger than the string thickness, one expects the quantum field nature of the CS, like the possibility to radiate massive particles, to

give negligible effects and one may instead consider the CS as an infinitely thin 1-dimensional classical object with tension μ: the Nambu-Goto (NG) string. However, due to the presence of small-scale structures on the strings, regions with curvature comparable to the string core size can develop and the Nambu-Goto approximation breaks down. In that case, massive radiation can be emitted during processes known as cusp annihilation [141] or kink-kink collisions [66]. We discuss massive particle emission in more details in Sect. 8.3.1.

8.2.4 Constraints on the String Tension $G\mu$ from GW Emission

The observational signatures of Nambu-Goto cosmic strings are mainly gravitational. The GW emission can be probed by current and future pulsar timing arrays and GW interferometers, while the static gravitational field around the string can be probed by CMB, 21 cm, and lensing observables, see Appendix 8.A for more details on non-GW probes. The strongest constraints come from pulsar timing array EPTA, $G\mu \lesssim 8 \times 10^{-10}$ [142], and NANOGrav, $G\mu \lesssim 5.3 \times 10^{-11}$ [143]. Comparison with the theoretical predictions from the SGWB from cosmic strings leads to $G\mu \lesssim 2 \times 10^{-11}$ [23, 129] or $G\mu \lesssim 10^{-10}$ [133], even though it can be relaxed to $G\mu \lesssim 5 \times 10^{-7}$ [40], after taking into account uncertainties on the loop size at formation and on the number of emitting modes. Note that it can also be strengthened by decreasing the inter-commutation probability [42, 144, 145].

By using the EPTA sensitivity curve derived in [146], we obtain the upper bound on $G\mu$, one order of magnitude higher, 2×10^{-10}, instead of 2×10^{-11}, cf. Fig. 8.6. This bound becomes $\sim 5 \times 10^{-11}$ by using the NANOGrav sensitivity curve derived in [146]. Another large source of uncertainty is the nature of the GW spectrum generated by a loop, which depends on the assumption on the loop small-scale structure (e.g. the number of cusps, kinks and kink-kink collisions per oscillations) [42, 133]. For instance, the EPTA bound can be strengthened to $G\mu \lesssim 6.7 \times 10^{-14}$ if the loops are very kinky [133]. CS can also emit highly-energetic and short-lasting GW bursts due to cusp formation [37, 38, 132, 133, 135]. From the non-observation of such events with LIGO/VIRGO [147, 148], one can constrain $G\mu \lesssim 4.2 \times 10^{-10}$ with the loop distribution function from [149]. However, the constraints are completely relaxed with the loop distribution function from [124].

8.3 Gravitational Waves from Cosmic Strings

In the main text of this study, we do not consider the case of global strings where the presence of a massless Goldstone in the spectrum implies that particle production is the main energy loss so that GW emission is suppressed [72]. However, we give an overview of the GW spectrum from global strings in Appendix 8.F, which can be detectable for string scales $\eta \gtrsim 10^{14}$ GeV. Other studies of the sensitivity of next generation GW interferometers to GW from global strings are [26, 27, 137].

There has been a long debate in the community whether local cosmic strings mainly loose their energy via GW emission or by particle production. We summarise the arguments and clarify the underlying assumptions below.

8.3.1 Beyond the Nambu-Goto Approximation

Quantum field string simulations:
Quantum field string (Abelian-Higgs) lattice simulations run by Hindmarsh et al. [150–152] have shown that decay into massive radiation is the main energy loss and is sufficient to lead to scaling. Then, loops decay within one Hubble time into scalar and gauge boson radiation before having the time to emit GW. It is suggested that the presence of small-scale structures, kinks and cusps, at the string core size are responsible for the energy loss into particle production. In these regions of large string curvature, the Nambu-Goto approximation, which considers CS as infinitely thin 1-dimensional classical objects, is no longer valid.

However, Abelian-Higgs simulations run by [153–155] have claimed the opposite result, that energy loss into massive radiation is exponentially suppressed when the loop size is large compared to the thickness of the string.

Small-scale structure:
At formation time, loops are not smooth but made of straight segments linked by kinks [156]. Kinks are also created in pairs after each string intercommutation, see [157] or Fig. 2.1 in [158]. The presence of straight segments linked by kinks prevents the formation of cusps. However, backreaction from GW emission smoothens the shapes, hence allowing for the formation of cusps [156] (see Fig. 8.2). Because of the large hierarchy between the gravitational backreaction scale and the cosmological scale H, the effects of the gravitational backreaction on the loop shape are not easily tractable numerically. The effects of backreaction from particle emission are shown in [157]. Nevertheless, it has been proposed since long [159] that the small-scale structures are smoothened below the gravitational backreaction scale $\sim \Gamma\, G\mu\, t$,. Particularly, based on analytical modelling on simple loop models, it has been shown in [160, 161] that due to gravitational backreaction, kinks get rounded off, become closer to cusps and then cusps get weakened. In earlier works, the same authors [162, 163] claimed that whether the smoothening has the time to occur within the loop life time strongly depends on the initial loop shape. In particular, for a four-straight-segment loop, the farther from the square shape, the faster the smoothening, whereas for more general loop shapes, the smoothening may not always occur.

To summarise the last two paragraphs, the efficiency of the energy loss into massive radiation depends on the nature of the small-scale structure, which can be understood as a correction to the Nambu-Goto approximation. The precise nature of the small-scale structure, its connection with the gravitational backreaction scale and the conflict between Nambu-Goto and Abelian Higgs simulations remain to be explained. Moreover, the value of the gravitational backreaction scale itself, see Sect. 8.3.2 is matter of debate. For our study, we follow the proposal of [67] for investigating how the GW spectrum is impacted for two benchmark scenarios: when the small-scale structures are dominated by cusps or when they are dominated by kinks. We give more details in the next paragraph. In Appendix 8.B.6, we show that if the high-frequency slope of the fundamental, $k = 1$, GW spectrum is f^{-1}, as expected in the presence of an early matter era or in the presence of an Heavide cut-off in the loop formation time, then the existence of the high-k modes, turns it to $f^{-1} \rightarrow f^{1-n}$, where n, defined in Eq. (8.22), depends on the small-scale structure. We can therefore read information about the small-scale structure of CS from the high-frequency GW spectrum.

Massive radiation emission:

In the vicinity of a cusp, the topological charge vanishes where the string cores overlap. Hence, the corresponding portions of the string can decay into massive radiation. The length of the overlapping segment has been estimated to be $\sqrt{r\,l}$ [141, 164] where $r \simeq \mu^{-1/2}$ is the string core size and l is the loop length. Hence, the energy radiated per cusp formation is $\mu\sqrt{rl}$, from which we deduce the power emitted from a loop

$$P_{\text{cusp}}^{\text{part}} \simeq N_{\text{c}} \frac{\mu^{3/4}}{l^{1/2}}, \tag{8.26}$$

where N_{c} is the average number of cusps per oscillation, estimated to be $N_{\text{c}} \sim 2$ [156]. Note that the consideration of pseudo-cusps, pieces of string moving at highly relativistic velocities, might also play a role [165, 166].

Even without the presence of cusps, Abelian-Higgs simulations [66] have shown that kink-kink collisions produce particles with a power per loop

$$P_{\text{kink}}^{\text{part}} \simeq N_{\text{kk}} \frac{\varepsilon}{l}, \tag{8.27}$$

where N_{kk} is the average number of kink-kink collisions per oscillation. Values possibly as large as $N_{\text{kk}} \sim O(10^3)$ have been considered in [133] or even as large as 10^6 for the special case of strings with junctions [167], due to kink proliferations [168]. In contrast to the cusp case, the energy radiated per kink-kink collision, ε, is independent of the loop size l and we expect $\varepsilon \sim \mu^{1/2}$.

Upon comparing the power of GW emission in Eq. (8.19) with either Eq. (8.26) or Eq. (8.27), one expects gravitational production to be more efficient than particle production when loops are larger than [67]

$$l \gtrsim l_c \equiv \beta_c \frac{\mu^{-1/2}}{(\Gamma G\mu)^2}, \tag{8.28}$$

for small-scale structures dominated by cusps, and

$$l \gtrsim l_k \equiv \beta_k \frac{\mu^{-1/2}}{\Gamma G\mu}, \tag{8.29}$$

for kink-kink collision domination. β_c and β_k are numbers which depend on the precise refinement. We assume $\beta_c,\ \beta_k \sim O(1)$. Therefore, loops with length smaller than the critical value in Eq. (8.28) or Eq. (8.29) are expected to decay into massive radiation before they have time to emit GW, which means that they should be subtracted when computing the SGWB. Equations (8.28) and (8.29) are crucial to determine the cutoff frequency, as we discuss in Sect. 8.3.4.

The cosmological and astrophysical consequences of the production of massive radiation and the corresponding constraints on CS from different experiments are presented in Sect. 8.A.4

8.3.2 *Assumptions on the Loop Distribution*

The SGWB resulting from the emission by CS loops strongly relies on the distribution of loops. In the present section, we introduce the loop-formation efficiency and discuss the assumptions on the loop-production rate, inspired from Nambu-Goto simulations. The loop-formation efficiency is computed later, in Sect. 8.4.

Loop-formation efficiency:
The SGWB resulting from the emission by CS loops strongly relies on the assumption for the distribution of loops which we now discuss. The equation of motion of a Nambu-Goto string in a expanding universe implies the following evolution equation for the long string energy density, cf. Sect. 8.D

$$\frac{d\rho_\infty}{dt} = -2H(1+\bar{v}^2)\rho_\infty - \left.\frac{d\rho_\infty}{dt}\right|_{\text{loop}}, \tag{8.30}$$

where $\bar{v}$ is the long string mean velocity. The energy loss into loop formation can be expressed as [72]

$$\left.\frac{d\rho_\infty}{dt}\right|_{\text{loop}} \equiv \mu \int_0^\infty l f(l,t)dl \equiv \frac{\mu}{t^3}\tilde{C}_{\text{eff}}, \tag{8.31}$$

with $f(l,t)$ the number of loops created per unit of volume, per unit of time t and per unit of length l and where we introduced the loop-formation efficiency $\tilde{C}_{\text{eff}}$. The loop-formation efficiency $\tilde{C}_{\text{eff}}$ is related to the notation introduced in [22, 23] by

$$\tilde{C}_{\text{eff}} \equiv \sqrt{2}\, C_{\text{eff}}. \tag{8.32}$$

In Sect. 8.4, we compute the loop-formation efficiency C_{eff} as a function of the long string network parameters $\bar{v}$ and L, which themselves are solutions of the Velocity-dependent One-Scale (VOS) equations.

Only loops produced at the horizon size contribute to the SGWB:
As pointed out already a long time ago by [99, 159] and more recently in large Nambu-Goto simulations [124], the most numerous loops are the ones of the size of the gravitational backreaction scale

$$\Gamma G\mu \times t, \tag{8.33}$$

which acts as a cut-off below which, small-scale structures are smoothened and such that smaller loops can not be produced below that scale. However, it has been claimed that only large loops are relevant for GW [37, 124, 169]. In particular, Nambu-Goto numerical simulations realized by Blanco-Pillado et al. [124] have shown that a fraction $\mathcal{F}_\alpha \simeq 10\%$ of the loops are produced with a length equal to a fraction $\alpha \simeq 10\%$ of the horizon size, and with a Lorentz boost factor $\gamma \simeq \sqrt{2}$. The remaining 90% of the energy lost by long strings goes into highly boosted smaller loops whose contributions to the GW spectrum are sub-dominant. Under those assumptions, the number of loops, contributing to the SGWB, produced per unit of time can be computed from the total energy flow into loops in Eq. (8.31)

$$\frac{dn}{dt_i} = \frac{\mathcal{F}_\alpha}{\gamma\, \mu\, \alpha\, t_i} \left.\frac{d\rho_\infty}{dt}\right|_{\text{loop}}, \tag{8.34}$$

with $\mathcal{F}_\alpha = 0.1, \gamma = \sqrt{2}$ and $\alpha = 0.1$. In an appendix of [1], we discuss the possibility to define the loop-size as a fixed fraction of the correlation length L instead of a fixed fraction of the horizon size t. Especially, we show that the impact on the GW spectrum is negligible. The latter can be recast as a function of the loop-formation efficiency $\tilde{C}_{\text{eff}}$ defined in Eq. (8.31)

$$\frac{dn}{dt_i} = \mathcal{F}_\alpha \frac{\tilde{C}_{\text{eff}}(t_i)}{\gamma\, \alpha\, t_i^4}. \tag{8.35}$$

This is equivalent to choosing the following monochromatic horizon-sized loop-formation function

$$f(l,\, t_i) = \frac{\tilde{C}_{\text{eff}}}{\alpha\, t_i^4} \delta(l - \alpha t_i). \tag{8.36}$$

The assumptions leading to Eq. (8.35) are the ones we followed for our study and which are also followed by [22, 23]. Our results strongly depend on these assumptions and would be dramatically impacted if instead we consider the model discussed in the next paragraph.

A second population of smaller loops:
The previous assumption—that the only loops relevant for the GW signal are the loops produced at horizon size—which is inspired from the Nambu-Goto numerical simulations of Blanco-Pillado et al. [124, 170], is in conflict with the results from Ringeval et al. [133, 149, 171]. In the latter works, the loop production function is derived analytically starting from the correlator of tangent vectors on long strings, within the

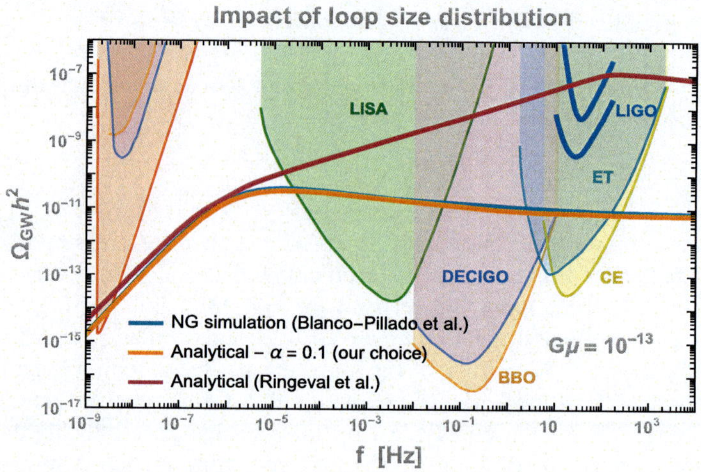

Fig. 8.3 We compare the GW spectrum from CS computed in the NG simulation of Blanco-Pillado et al. [124, 170] to the one computed analytically by Ringeval et al. [133, 149, 171]. In the later case, the cut-off on the loop size is smaller. This acts as a second population of small loops which enhances the GW spectrum at high frequency. The analytical modeling we consider in our work is fitted on the work of Blanco-Pillado et al. and consider a monochromatic loop size distribution peaked on $\alpha = 0.1$. This figure is not included in [1] but instead it has been done during the redaction of the manuscript

Polchinski-Rocha model [172–175]. In the Polchinski-Rocha model, which has been tested in Abelian-Higgs simulations [151], the gravitational back-reaction scale, i.e. the lower cut-off of the loop production function, is computed to be

$$\Upsilon (G\mu)^{1+2\chi} \times t, \tag{8.37}$$

with $\Upsilon \simeq 10$ and $\chi \sim 0.25$. Consequently, the gravitational back-reaction scale in the Polchinski-Rocha model is significantly smaller than the usual gravitational back-reaction scale, commonly assumed to match the gravitational radiation scale, $\Gamma G\mu\, t$. Therefore, the model of Ringeval et al. predicts the existence of a second population of smaller loops which enhances the GW spectrum at high frequency by many orders of magnitude, see Fig. 8.3. However, as raised by [176], the model of Ringeval et al. predicts the amount of long-string energy converted into loops, to be ~200 times larger than the one computed in the numerical simulations of Blanco-Pillado et al. [124]. These discrepancies between Polchinski-Rocha analytical modeling and Nambu-Goto numerical simulations remain to be understood.

8.3.3 *The Gravitational-Wave Spectrum*

For our study, we compute the GW spectrum observed today generated from CS as follows (see Appendix 8.B for a derivation)

$$\Omega_{GW}(f) \equiv \frac{f}{\rho_c}\left|\frac{d\rho_{GW}}{df}\right| = \sum_k \Omega_{GW}^{(k)}(f), \tag{8.38}$$

where

$$\Omega_{GW}^{(k)}(f) = \frac{1}{\rho_c}\cdot\frac{2k}{f}\cdot\frac{\mathcal{F}_\alpha\,\Gamma^{(k)}G\mu^2}{\alpha(\alpha+\Gamma G\mu)}\int_{t_{osc}}^{t_0} d\tilde{t}\,\frac{C_{eff}(t_i)}{t_i^4}\left[\frac{a(\tilde{t})}{a(t_0)}\right]^5\left[\frac{a(t_i)}{a(\tilde{t})}\right]^3\Theta(t_i - t_{osc})\Theta(t_i - \frac{l_*}{\alpha}), \tag{8.39}$$

with

Θ	Heaviside function,
$\mu,\ G,\ \rho_c$	string tension, Newton constant, critical density,
a	scale factor of the universe (we solve the full Friedmann equation for a given energy density content),
k	proper mode number of the loop (effect of high-k modes are discussed in Appendix 8.B.6. For technical reasons, in most of our plots, we restrict to 2×10^4 modes),
Γ	gravitational loop-emission efficiency, ($\Gamma \simeq 50$ [170])
$\Gamma^{(k)}$	Fourier modes of Γ, dependent on the loop small-scale structures, ($\Gamma^{(k)} \propto k^{-4/3}$ for cusps, e.g. [38]) ,
$\mathcal{F}_\alpha$	fraction of loops formed with size $\alpha(\mathcal{F}_\alpha \simeq 0.1)$, cf. Sect. 8.3.2,
C_{eff}	loop-production efficiency, defined in Eq. 8.4.6, (C_{eff} is a function of the long-string mean velocity $\bar{v}$ and correlation length ξ, both computed upon integrating the VOS equations, cf. Sect. 8.4)
α	loop length at formation in unit of the cosmic time, ($\alpha \simeq 0.1$) (we consider a monochromatic, horizon-sized loop-formation function, cf. Sect. 8.3.2),
$\tilde{t}$	the time of GW emission,

f observed frequency today
(related to frequency at emission $\tilde{f}$ through $f\, a(t_0) = \tilde{f}\, a(\tilde{t})$,
related to loop length 0l through $\tilde{f} = 2k/l$,
related to the time of loop production t_i through $l = \alpha t_i - \Gamma G\mu(\tilde{t} - t_i)$),

t_i the time of loop production,
(related to observed frequency and emission time $\tilde{t}$ through
$t_i(f, \tilde{t}) = \frac{1}{\alpha+\Gamma G\mu}\left[\frac{2k}{f}\frac{a(\tilde{t})}{a(t_0)} + \Gamma G\mu\, \tilde{t}\right]$),

t_0 the time today,

t_{osc} the time at which the long strings start oscillating, $t_{osc} = \mathrm{Max}[t_{fric}, t_F]$,
t_F is the time of CS network formation, defined as $\sqrt{\rho_{tot}(t_F)} \equiv \mu$ where ρ_{tot} is
the universe total energy density. In the presence of friction, at high temperature,
the string motion is damped until the time t_{fric}, computed in Appendix 8.D.4,

l_* l_c for cusps and l_k for kinks in Eqs. (8.28) and (8.29)
(critical length below which the emission of massive radiation
is more efficient than the gravitational emission, cf. Sect. 8.3.1).

A first look at the GW spectrum:
Figure 8.4 shows the GW spectrum computed with Eq. (8.38). The multiple frequency cut-offs visible on the figure, follow from the Heaviside functions in Eq. (8.38), which subtract loops formed before network formation, cf. Eq. (8.12), or when thermal friction freezes the network, cf. Appendix 8.D.4, or which subtract loops decaying via massive particle emission from cusps and kinks instead of GW, cf. Sect. 8.3.1. We indicate separately the contributions from the emission occurring before and after the matter-radiation equality. One can see that loops emitting during the radiation era contribute to a flat spectrum whereas loops emitting during the matter era lead to a slope decreasing as $f^{-1/3}$. Similarly, the high-frequency cut-offs due to particle production, thermal friction, network formation, but also due to a second period of inflation (discussed in Sect. 8.7), give a slope $f^{-1/3}$. In Appendix 8.B.6, we show that the presence of high-frequency modes are responsible for changing the slope f^{-1}, expected from the ($k = 1$)-spectrum, to $f^{-1/3}$.

Impact of the cosmology on the GW spectrum:
In a nutshell, the frequency dependence of the GW spectrum receives two contributions, a red-tilt coming from the redshift of the GW energy density and a blue-tilt coming from the loop-production rate $\propto t_i^{-4}$. On the one hand, the higher the frequency the earlier the GW emission, so the larger the redshift of the GW energy density and the more suppressed the spectrum. On the other hand, high frequencies correspond to loops formed earlier, those being more numerous, this increases the GW amplitude.

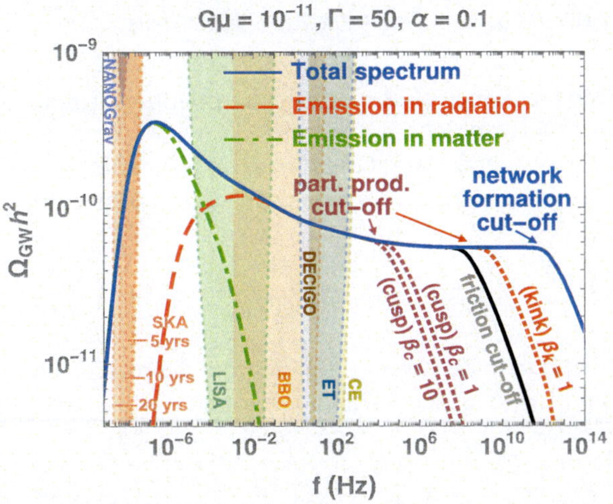

Fig. 8.4 GW spectrum from the scaling cosmic-string network evolving in a standard cosmology. Contributions from GW emitted during radiation and matter eras are shown with red and green dashed lines respectively. The high-frequency cut-offs correspond to either the time of formation of the network, cf. Eq. (8.12), the time when friction-dominated dynamics become irrelevant, cf. Appendix 8.D.4, or the time when gravitational emission dominates over massive particle production, for either kink or cusp-dominated small-scale structures, cf. Sect. 8.3.1. The cut-offs are described by Heaviside functions in the master formula in Eq. (8.38). In Appendix 8.B.6, we show that the slopes beyond the high-frequency cut-offs are given by $f^{-1/3}$. Colored regions indicate the integrated power-law sensitivity of future experiments, as described in Appendix 6.A

Interestingly, during radiation-domination the two contributions exactly cancel such that the spectrum is flat. As explained in more details in Appendix 8.B.5, the flatness of the GW spectrum during radiation is intimately related to the independence of the GW emission power on the loop length. In the same appendix, we show that a change in the equation of state of the universe impacts the GW spectrum if it modifies at least one of the two following redshift factors: the redshift of the number of emitting loops and the redshift of the emitted GW.

For instance, when GW emission occurs during radiation but loop formation occurs during matter, the loop density redshifts faster. Then, the larger the frequency, the earlier the loop formation, and the more suppressed the GW spectrum (as f^{-1} for $k = 1$ and as $f^{-1/3}$ when taking into account high-k modes). Conversely, if loop formation occurs during kination, the loop density redshift slower and the GW gets enhanced at large frequency (as f^1).

8.3.4 The Frequency—Temperature Relation

Relation between frequency of observation and temperature of loop formation:

In Appendix 8.C, we derive the relation between a detected frequency f and the temperature of the universe when the loops, mostly responsible for f, are formed

$$f = (6.7 \times 10^{-2}\,\text{Hz}) \left(\frac{T}{\text{GeV}}\right) \left(\frac{0.1 \times 50 \times 10^{-11}}{\alpha\,\Gamma G\mu}\right)^{1/2} \left(\frac{g_*(T)}{g_*(T_0)}\right)^{1/4}. \tag{8.40}$$

We emphasize that Eq. (8.40) is very different from the relation obtained in the case of GW generated by a first-order cosmological phase transition. In the latter case, the emitted frequency corresponds to the Hubble scale at T_p [2]

$$f = (19 \times 10^{-3}\,\text{mHz}) \left(\frac{T_p}{100\,\text{GeV}}\right) \left(\frac{g_*(T_p)}{100}\right)^{1/4}. \tag{8.41}$$

In the case of cosmic strings, instead of being set by the Hubble scale at the loop-formation time t_i, the emitted frequency is further suppressed by a factor $(\Gamma G\mu)^{-1/2}$, which we now explain. From the scaling law $\propto t_i^{-4}$ of the loop-production function in Eq. (8.35), one can understand that the most numerous population of emitting loops at a given time $\tilde{t}$ is the population of loops created at the earliest epoch. They are the oldest loops.[4] Hence, a loop created at time t_i contributes to the SGWB much later, at a time given by the loop half-lifetime $\tilde{t}_M = \alpha\, t_i/2\Gamma G\mu$, cf. Eq. (8.24). Therefore, the emitted frequency is dispensed from the redshift factor $a(\tilde{t}_M)/a(t_i) = (\tilde{t}_M/t_i)^{1/2} \sim (\Gamma G\mu)^{-1/2}$, and so, is higher. See Appendix 8.C.1 and its Fig. 8.21 for more details.

The detection of a non-standard cosmology:
During a change of cosmology, e.g. a change from a matter to a radiation-dominated era, the long-string network evolves from one scaling regime to the other. The response of the network to the change of cosmology is quantified by the VOS equations, which are presented in Sect. 8.4. As a result of the transient evolution towards the new scaling regime, the turning-point frequency Eq. (8.118) associated to the change of cosmology is lower in VOS than in the scaling network. The detection of a turning-point in a GW spectrum from CS by a future interferometer would be a smoking-gun signal for non-standard cosmology. Particularly, in Fig. 8.10, we show that LISA can probe a non-standard era ending around the QCD scale, ET/CE can probe a non-standard era ending around the TeV scale whereas DECIGO/BBO can probe the intermediate range. We focus on the particular case of a short matter era in Sect. 8.6 and a short inflation era in Sect. 8.7, respectively. In the latter case, the turning-point frequency is even further decreased due to the string stretching which we explain in the next paragraph.

[4] Note that they are also the smallest loops, with a length given by the gravitational radiation scale $\Gamma G\mu\, t$.

The detection of a non-standard cosmology (intermediate-inflation case):
If the universe undergoes a period of inflation lasting N_e e-folds, the correlation length of the network is stretched outside the horizon. After inflation, the network achieves a long transient regime lasting $\sim N_e$ other e-folds until the correlation length re-enters the horizon. Hence, the turning-point frequency in the GW spectrum, cf. Eq. (8.64), receives a $\exp N_e$ suppression compared to Eq. (8.40) due to the duration of the transient. We give more details in Sect. 8.7.

Cut-off frequency from particle production:
As discussed in the Sect. 8.3.1, particle production is the main decay channel of loops shorter than

$$l_* = \beta_m \frac{\mu^{-1/2}}{(\Gamma G\mu)^m}, \tag{8.42}$$

where $m = 1$ or 2 for loops kink-dominated or cusp-dominated, respectively, and $\beta_m \sim \mathcal{O}(1)$. The corresponding characteristic temperature above which loops, decaying preferentially into particles, are produced, is

$$T_* \simeq \beta_m^{-1/2}\,\Gamma^{m/2}\,\sqrt{\alpha}\,(G\mu)^{(2m+1)/4}\,M_{\rm pl} \simeq \begin{cases} (0.2\ \text{EeV})\sqrt{\dfrac{\alpha}{0.1}}\sqrt{\dfrac{1}{\beta_c}}\left(\dfrac{G\mu}{10^{-15}}\right)^{3/4} & \text{for kinks,} \\ (1\ \text{GeV})\sqrt{\dfrac{\alpha}{0.1}}\sqrt{\dfrac{1}{\beta_k}}\left(\dfrac{G\mu}{10^{-15}}\right)^{5/4} & \text{for cusps.} \end{cases} \tag{8.43}$$

We have used $l_* = \alpha\, t_i$, $H = 1/(2t_i)$ and $\rho_{\rm rad} = 3M_{\rm pl}^2 H^2$. Upon using the frequency-temperature correspondence in Eq. (8.40), we get the cut-off frequencies due to particle production

$$f_* \simeq \begin{cases} (1\,\text{GHz})\sqrt{\dfrac{1}{\beta_c}}\left(\dfrac{G\mu}{10^{-15}}\right)^{1/4} & \text{for kinks,} \\ (31\ \text{Hz})\sqrt{\dfrac{1}{\beta_k}}\left(\dfrac{G\mu}{10^{-15}}\right)^{3/4} & \text{for cusps.} \end{cases} \tag{8.44}$$

and which we show in most of our plots with dotted red and purple lines. Particularly, in Fig. 8.6, we see that particle production in the cusp-dominated case would start suppressing the GW signal in the ET/CE windows for string tension lower than $G\mu \lesssim 10^{-15}$. However, in the kink-dominated case, the spectrum is only impacted at frequencies much higher than the interferometer windows. In Appendix 8.B.6, we show that the slope of the GW spectrum beyond the high-frequency cut-off f_* is given by $f^{-1/3}$.

8.3.5 The Astrophysical Foreground

Crucial for our analysis is the assumption that the stochastic GW foreground of astrophysical origin can be substracted.

LIGO/VIRGO has already observed three binary black hole (BH-BH) merging events [177–179] during the first 4-month observing run O1 in 2015, and seven additional BH-BH [180–183] as well as one binary neutron star (NS-NS) [184] merging events during the second 9-month observing run O2 in 2017. And more events might still be discovered in the O2 data [185]. According to the estimation of the NS merging rate following the detection of the first (and unique up to now) NS-NS merger event GW170817, NS-NS stochatisc background may be detectable after a 20-month observing run with the expected LIGO/VIRGO design sensistivity in 2022+ and in the most optimistic scenario, it might be detectable after 18-month of the third observing run O3 who began in April, 1st, 2019 [186]. Hence, one might worry about the possibility to distinguish the GW SGWB sourced by CS from the one generated by the astrophysical foreground. However, in the BBO and ET/CE windows, the NS and BH foreground might be substracted with respective reached sensibilities $\Omega_{\rm GW} \simeq 10^{-15}$ [187] and $\Omega_{\rm GW} \simeq 10^{-13}$ [188]. In the LISA window, the binary white dwarf (WD-WD) foreground dominates over the NS-NS and BH-BH foregrounds [189–191]. The WD-WD galactic foreground, one order of magnitude higher than the WD-WD extragalactic [192], might be substracted with reached sensibility $\Omega_{\rm GW} \simeq 10^{-13}$ at LISA [193, 194]. Hence, in the optimistic case where the foreground can be removed and the latter sensibility are reached one might be able to distinguished the signal sourced by CS from the one generated by the astrophysical foreground. Furthermore, the GW spectrum generated by the astrophysical foreground increased with frequency as $f^{2/3}$ [195], which is different from the GW spectrum generated by CS during radiation (flat), matter $f^{-1/3}$, inflation $f^{-1/3}$ or kination f^{1}. We refer to [65, 196] for the study of the sensitivity of future GW observatories when assuming that astrophysical foregrounds can not be subtracted.

8.4 The Velocity-Dependent One-Scale Model

The master formula (8.38) crucially depends on the loop-production efficiency encoded in $C_{\rm eff}$. In this section, we discuss its derivation within the framework of the Velocity-dependent One-Scale (VOS) model.

8.4.1 The Loop-Production Efficiency

In a correlation volume L^3, a segment of length L must travel a distance L before encountering another segment. L is the correlation length of the long-string network.

The collision rate, per unit of volume, is $\frac{\bar{v}}{L} \cdot \frac{1}{L^3} \sim \frac{\bar{v}}{L^4}$ where $\bar{v}$ is the long-string mean velocity. At each collision forming a loop, the network looses a loop energy $\mu\, L = \rho_\infty\, L^3$. Hence, the loop-production energy rate can be written as [97]

$$\left.\frac{d\rho_\infty}{dt}\right|_{\text{loop}} = \tilde{c}\,\bar{v}\frac{\rho_\infty}{L}, \tag{8.45}$$

where one can compute $\tilde{c} = 0.23 \pm 0.04$ from Nambu-Goto simulations in expanding universe [197]. $\tilde{c}$ is the only free parameter of the VOS model. Hence, the loop-formation efficiency, defined in Eq. (8.31), can be expressed as a function of the long-string parameters, $\bar{v}$ and $\xi \equiv L/t$,

$$\tilde{C}_{\text{eff}} \equiv \sqrt{2}\, C_{\text{eff}}(t) = \frac{\tilde{c}\,\bar{v}(t)}{\xi^3(t)}. \tag{8.46}$$

In Appendix 8.E, we discuss how our results are changed when considering a recent extension of the VOS model with more free parameters, fitted on Abelian-Higgs field theory numerical simulations [198], and taking into account the emission of massive radiation. Basically, the loop-formation efficiency C_{eff} is only decreased by a factor $\sim$2. In the following, we derive $\bar{v}$ and ξ as solutions of the VOS equations.

8.4.2 The VOS Equations

The VOS equations describe the evolution of a network of long strings in term of the mean velocity $\bar{v}$ and the correlation length $\xi = L/t$ [197, 199–201]. The latter is defined through the long string energy density $\rho_\infty \equiv \mu/L^2$. Starting from the equations of motion of the Nambu-Goto string in a FRW universe, we can derive the so-called VOS equations (see Appendix 8.D for a derivation)

$$\frac{dL}{dt} = HL\,(1+\bar{v}^2) + \frac{1}{2}\tilde{c}\,\bar{v}, \tag{8.47}$$

$$\frac{d\bar{v}}{dt} = (1-\bar{v}^2)\left[\frac{k(\bar{v})}{L} - 2H\bar{v}\right], \tag{8.48}$$

where

$$k(\bar{v}) = \frac{2\sqrt{2}}{\pi}(1-\bar{v}^2)(1+2\sqrt{2}\bar{v}^3)\frac{1-8\bar{v}^6}{1+8\bar{v}^6}, \tag{8.49}$$

is the so-called momentum parameter and is a measure of the deviation from the straight string, for which $k(\bar{v}) = 1$ [201]. The first VOS equation describes the evolution of the long string correlation length under the effect of Hubble expansion and loop chopping. The second VOS equation is nothing more than a relativistic generalization of Newton's law where the string is accelerated by its curvature $1/L$ but is damped by the Hubble expansion after a typical length H^{-1}.

Numerical simulations [111–115] have shown that a network of long strings is first subject to a transient regime before reaching a scaling regime, in which the long string mean velocity $\bar{v}$ is constant and the correlation length grows linearly with the Hubble horizon $L = \xi\, t$. The values of the quantities $\bar{v}$ and ξ depend on the cosmological background, namely the equation of state of the universe. Hence, when passing from a cosmological era 1 to era 2, the network accomplishes a transient evolution from the scaling regime 1 to the scaling regime 2. We use the VOS equations to compute the time evolution of $\bar{v}$ and ξ during the change of cosmology and then compute their impact on the CS SGWB.

8.4.3 *Scaling Regime Solution and Beyond*

Scaling solution versus VOS solution:
Figure 8.5 shows the evolutions of ξ, $\bar{v}$, and $C_{\rm eff}$, from solving the VOS equations in Eq. (8.47) with three equations of state, matter, radiation and kination. Regardless of the initial-condition choice, the network approaches a scaling solution where all parameters become constant. The energy scale of the universe has to decrease by some 4 orders of magnitude before reaching the scaling regime after the network formation. For a cosmological background evolving as $a \propto t^{2/n}$ with $n \geq 2$, the scaling regime solution is

$$\xi = \text{constant} \quad \text{and} \quad \bar{v} = \text{constant}, \tag{8.50}$$

with

$$\text{with} \quad \xi = \frac{n}{2}\sqrt{\frac{k(\bar{v})[k(\bar{v}) + \tilde{c}]}{2(n-2)}} \quad \text{and} \quad \bar{v} = \sqrt{\frac{n}{2}\frac{k(\bar{v})}{[k(\bar{v}) + \tilde{c}]}\left(1 - \frac{2}{n}\right)}. \tag{8.51}$$

In order to fix the notation used in our plots, we define

- **(Instantaneous) scaling network**: The loop-formation efficiency $C_{\rm eff}$, defined in Eq. (8.46), is taken at its steady state value, given by Eq. (8.51). In particular for matter, radiation and kination domination, one has

$$C_{\rm eff} \simeq 0.39,\ 5.4,\ 29.6 \qquad \text{for} \quad n = 3,\ 4,\ 6. \tag{8.52}$$

  During a change of era $1 \to 2$, $C_{\rm eff}$ is assumed to change instantaneously from the scaling regime of era 1 to the scaling regime of era 2. This is the assumption adopted in [22, 23].
- **VOS network**: The loop-formation efficiency $C_{\rm eff}$, defined in Eq. (8.46), is computed upon integrating the VOS equations in Eq. (8.47). During a change of cosmology, the long-string network experiences a transient regime.

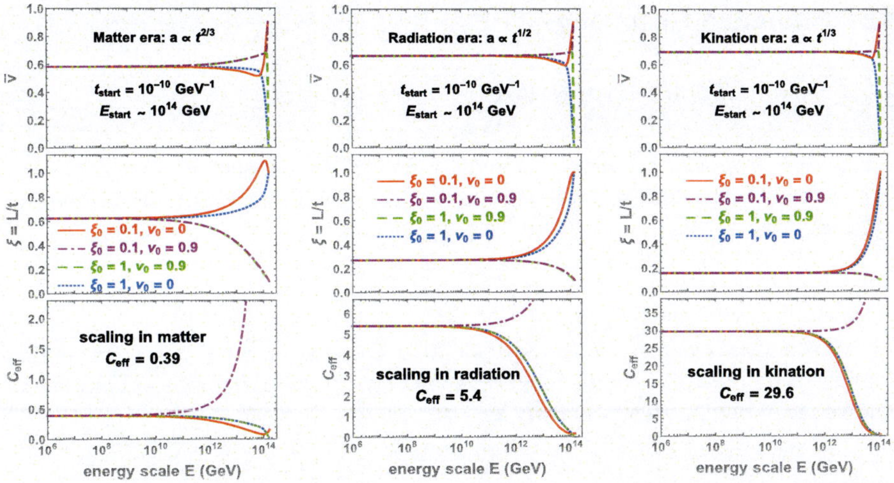

Fig. 8.5 Cosmic-string network evolving in the one-component universe with energy density $\rho \sim a^{-n}$ where $n = 3$, 4 and 6 correspond to matter, radiation and kination, respectively. The long-string-network mean velocity $\bar{v}$, the correlation length ξ and the corresponding loop-production efficiency $C_{\rm eff}$ reach the scale-invariant solutions after the Hubble expansion rate has dropped by 2 orders of magnitude, independently of the initial conditions

Beyond the scaling regime in standard cosmology:
In Figs. 8.6 and 8.7, we compare the GW spectra and the $C_{\rm eff}$ evolution, obtained with a scaling and VOS network. They are quite similar. The main difference arises from the change in relativistic degrees of freedom near the QCD confining temperature and from the matter-radiation transition. In contrast, predictions differ significantly when considering non-standard cosmology.

Beyond the scaling regime in non-standard cosmology:
In Fig. 8.9, in dashed vs solid, we compare the loop-production efficiency factor $C_{\rm eff}$ and the corresponding GW spectra for a scaling network and for a VOS network. The VOS frequency of the turning point due to the change of cosmology is shifted to a lower frequency by a factor ~ 22.5 with respect to the corresponding scaling frequency.[5] The shift results from the extra-time needed by the network to achieve its transient evolution to the new scaling regime. In the rest of this work, we go beyond the instantaneous scaling approximation used in [22, 23].

[5] The turning-point frequency can even be smaller by $\mathcal{O}(400)$ if in a far-future, a precision of the order of 1% can be reached in the measurement of the SGWB, cf. Eq. (8.118).

8.5 Standard Cosmology

8.5.1 The Cosmic Expansion

The SGWB from CS, cf. master formula in Eq. (8.38), depends on the cosmology through the scale factor a. We compute the later upon integrating the Friedmann equation

$$H^2 = \frac{\rho}{3M_{\rm pl}^2}, \tag{8.53}$$

for a given energy density ρ. In the standard ΛCDM scenario, the universe is first dominated by radiation, then a matter era, and finally the cosmological constant so that we can write the energy density as

$$\rho_{\rm ST,0}(a) = \rho_{r,0}\, \Delta_R(T(a), T_0) \left(\frac{a}{a_0}\right)^4 + \rho_{m,0} \left(\frac{a}{a_0}\right)^3 + \rho_{k,0} \left(\frac{a}{a_0}\right)^2 + \rho_{\Lambda,0}, \tag{8.54}$$

where r, m, k and Λ denote radiation, matter, curvature, and the cosmological constant, respectively. We take $\rho_i = \Omega_i h^2\, 3M_{\rm pl}^2 H_0^2$, where $H_0 = 100$ km/s/Mpc, $\Omega_r h^2 \simeq 4.2 \times 10^{-5}$, $\Omega_{\rm m} h^2 \simeq 0.14$, $\Omega_k \simeq 0$, $\Omega_\Lambda h^2 \simeq 0.31$ [202]. The presence of the function

$$\Delta_R = \left(\frac{g_*(T)}{g_*(T_0)}\right) \left(\frac{g_{*s}(T_0)}{g_{*s}(T)}\right)^{4/3}, \tag{8.55}$$

comes from imposing the conservation of the comoving entropy $g_{*s}\, T^3\, a^3$, where the evolutions of g_* and $g_{*,s}$ are taken from Appendix C of [17]. We discuss the possibility of adding an extra source of energy density in the next sections, intermediate matter in Sect. 8.6 and intermediate inflation in Sect. 8.7.

8.5.2 Gravitational Wave Spectrum

Figure 8.6 shows the dependence of the spectrum on the string tension. The amplitude decreases with $G\mu$ due to the lower energy stored in the strings. Moreover, at lower $G\mu$, the loops decaying slower, the GW are emitted later, implying a lower redshift factor and a global shift of the spectrum to higher frequencies. The figure also shows how the change in SM relativistic degrees of freedom introduces a small red-tilt which suppresses the spectrum by a factor $\Delta_R^{-1} \sim 2.5$ at high frequencies. We find that the amplitude of the GW spectrum at large frequency, assuming a standard cosmology, is given by

$$\Omega_{\rm GW} h^2 \simeq 15\pi\, \Delta_R\, \Omega_r h^2\, C_{\rm eff}(n=4)\, \mathscr{F}_\alpha\, (\alpha\, G\mu/\Gamma)^{1/2}, \tag{8.56}$$

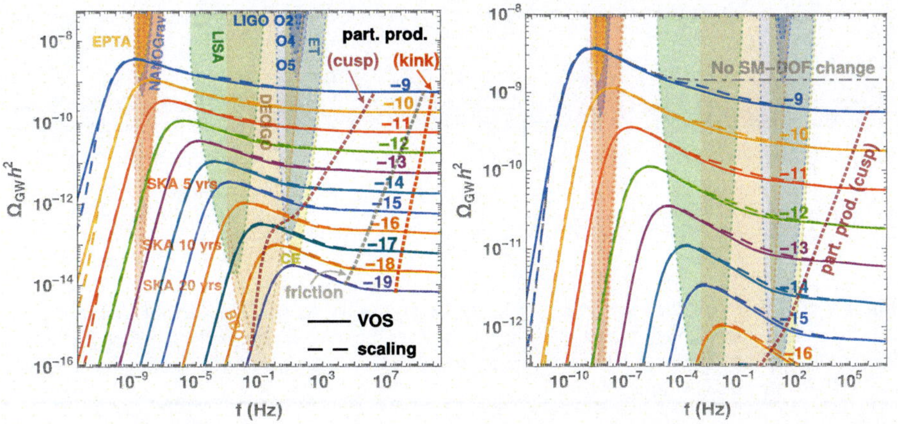

Fig. 8.6 Left: GW spectra from cosmic strings assuming either the scaling or VOS network, cf. Sect. 8.4.3, evolving in the standard cosmological background. Each line corresponds to string tension $G\mu = 10^x$, where x is specified by a number on each line. Dotted lines show the spectral cut-offs expected due to particle production, cf. Sect. 8.3.4 and thermal friction, cf. Sect. 8.B.4, which depend on the nature of the loop small-scale structures: cusp or kink-dominated. **Right:** The zoom-in plot of the left panel shows the effects from the change of SM degrees of freedom on the scaling and VOS networks

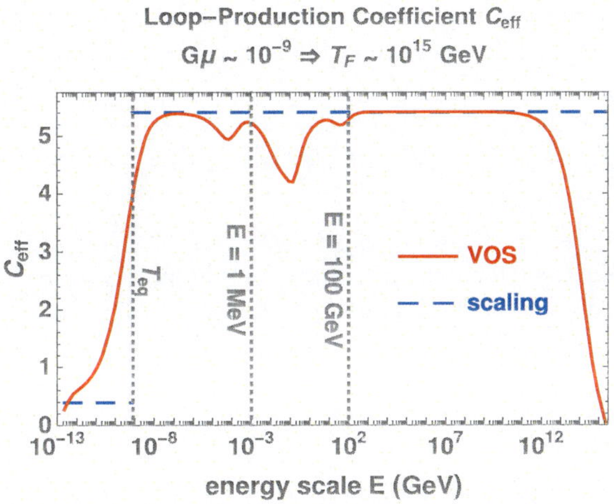

Fig. 8.7 Comparison of the loop-production efficiency under the scaling assumption, where the attractor solution of the VOS equations is assumed to be reached instantaneously, and under the VOS assumptions, where one integrates the VOS equations. A standard cosmology is assumed

where $\Omega_r h^2 \simeq 4.2 \times 10^{-5}$ is the present radiation energy density of the universe [202]. We provide an intuitive derivation based on the quadrupole formula in Appendix 8.B.5.

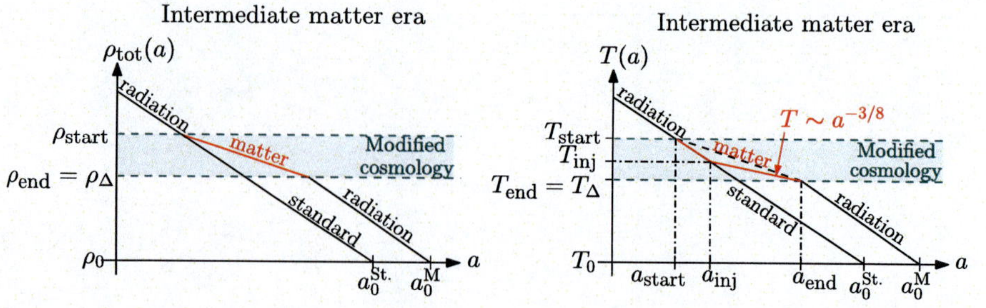

Fig. 8.8 Evolution of the total energy density **(left)** and the temperature **(right)** assuming the presence of an intermediate matter era. $T_{\rm inj}$ and $a_{\rm inj}$ are the temperature and scale factor at which the entropy injected by the decay of the matter content into radiation, starts to be effective, cf. Fig. 2 in [203]. St: standard; M: matter

8.5.3 Deviation from the Scaling Regime

Figure 8.7 shows how the loop-formation efficiency $C_{\rm eff}$ varies during the change of SM relativistic degrees of freedom and the matter-radiation equality, upon solving the VOS equations, cf. Sect. 8.4. We see the associated corrections to the spectrum in Fig. 8.6, and which were already pointed out in [24]. The spectrum is enhanced at low frequencies because more loops are produced than when assuming that the matter era is reached instantaneously, cf. Fig. 8.7.

8.5.4 Beyond the Nambu-Goto Approximation

Figure 8.6 shows the possibility of a cut-off at high frequencies due to particle production, for two different assumptions regarding the loop small-scale structures: cusps or kinks domination, cf. Sect. 8.3.1. Above these frequencies, loops decays into massive radiation before they have time to emit GW. For kinky loops, the cut-off is outside any future-planned observational bands, while for cuspy loops, the cut-off might be in the observed windows for $G\mu \lesssim 10^{-15}$.

8.6 Intermediate Matter Era

8.6.1 The Non-standard Scenario

In this section, we consider the existence of an early-intermediate-matter-dominated era, following an earlier radiation era and preceeding the standard radiation era. The intermediate matter-dominated era starts when the matter energy density $\rho_{\rm matter} \propto a^{-3}$ takes over the radiation energy density $\rho_{\rm radiation} \propto a^{-4}$ and ends when the matter content decays into radiation, cf. Fig. 8.8. The energy density profile is illustrated in Fig. 8.8 and can be written as

$$\rho_{\text{tot}}(a) = \begin{cases} \rho_{\text{rad}}^{\text{st}}(a) + \rho_{\text{late}}(a) & \text{for } \rho > \rho_{\text{start}}, \\ \frac{\rho_{\text{start}}}{a_{\text{start}}a^n + \rho_{\text{late}}(a)} & \text{for } \rho_{\text{start}} > \rho > \rho_{\text{end}}, \\ \frac{\rho_{\text{end}}\Delta_R(T_{\text{end}},T)}{a_{\text{end}}a^4 + \rho_{\text{late}}(a)} & \text{for } \rho < \rho_{\text{end}}. \end{cases} \tag{8.57}$$

where

$\rho_{\text{start}}, \rho_{\text{end}}$ the starting and ending energy density of the non-standard cosmology,
ρ_{late} the standard-cosmology energy density dominating at late times, e.g. the standard matter density, and cosmological constant.
Δ_R is given in Eq.(8.55).

8.6.2 *Impact on the Spectrum*

A low-pass filter
In the left panel of Fig. 8.9, we show that an intermediate matter era blue-tilts the spectral index of the spectrum. Furthermore, at higher frequencies, corresponding to loops produced during the radiation era preceding the matter era, the spectrum recovers a flat scaling but is suppressed by the duration r of the matter era

$$r = \frac{T_{\text{start}}}{T_\Delta}, \tag{8.58}$$

where $T_\Delta = T_{\text{end}}$. By suppressing the high-frequency part of the spectrum, an early matter era acts on the CS spectrum as a low-pass filter. The negative spectral index and the suppression can be understood from Fig. 8.8. Indeed, the universe, in the presence of an intermediate matter era, has expanded more than the standard universe. Hence at a fixed emitted frequency, loops are produced later and so are less numerous, implying less GW emission. In the right panel of Fig. 8.9, we show that for short intermediate matter era, $r = 2$ or $r = 10$, the scaling regime in the matter era, which is characterized by $C_{\text{eff}} = 0.39$, cf. Eq. (8.52), is not reached.

A turning-point:
A key observable is the frequency above which the GW spectrum differs from the one obtained in standard cosmology. This is the so-called *turning-point* frequency f_Δ. It corresponds to the redshifted-frequency emitted by the loops created during the change of cosmology at the temperature T_Δ. In the instantaneous scaling approximation, cf. dashed line in Fig. 8.9, the turning-point frequency f_Δ is given by the $(T,\ f)$-correspondence relation

$$f_\Delta^{\text{scaling}} = (4.5 \times 10^{-2}\ \text{Hz}) \left(\frac{T_\Delta}{\text{GeV}}\right) \left(\frac{0.1 \times 50 \times 10^{-11}}{\alpha\, \Gamma G\mu}\right)^{1/2} \left(\frac{g_*(T_\Delta)}{g_*(T_0)}\right)^{1/4}. \tag{8.59}$$

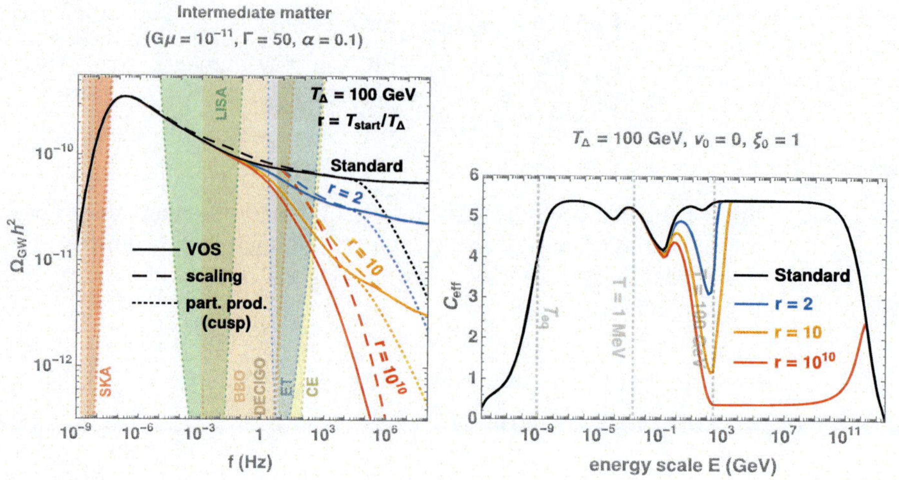

Fig. 8.9 GW spectrum from an intermediate matter era starting at the temperature $T_{\rm start}$ and ending at T_Δ. **Left:** The dashed-lines assume that the scaling regime in matter era switches instantaneously to the scaling regime in radiation era, meaning that $C_{\rm eff}$ varies discontinuously, whereas the plain lines incorporate the transient behavior solution of the VOS equations and shown on the right panel. The cut-offs due to particle production, cf. Sect. 8.3.4, are shown with dotted lines. **Right:** Time evolution of the loop-production efficiency $C_{\rm eff}$ after solving the VOS equations, cf. Sect. 8.4.3

However, the deviation from the scaling regime during the change of cosmology, cf. Sect. 8.4.3, implies a shift to lower frequencies of the $(T,\ f)$-correspondence, by a factor $\sim$ 22.5, cf. solid vs dashed lines in Fig. 8.9. The correct $(T,\ f)$-correspondence when applied to a change of cosmology is

$$f_\Delta^{\rm VOS} = (2\times 10^{-3}\ {\rm Hz})\left(\frac{T_\Delta}{\rm GeV}\right)\left(\frac{0.1\times 50\times 10^{-11}}{\alpha\,\Gamma G\mu}\right)^{1/2}\left(\frac{g_*(T_\Delta)}{g_*(T_0)}\right)^{1/4}. \qquad (8.60)$$

We fit the numerical factor in Eq. (8.60) (but also in Eq. (8.64)) by imposing[6] the non-standard-cosmology spectrum $\Omega_{\rm NS}$ to deviate from the standard-cosmology one $\Omega_{\rm ST}$ by 10% at the turning-point frequency,

$$\left|\frac{\Omega_{\rm NS}(f_\Delta)-\Omega_{\rm ST}(f_\Delta)}{\Omega_{\rm ST}(f_\Delta)}\right| \simeq 10\%. \qquad (8.61)$$

We are conservative here. Choosing 1% instead of 10% would lead to a frequency shift of the order of $\mathcal{O}(400)$, cf. Eq. (8.118). Note that our Eq. (8.60) is numerically very similar to the one in [22–24] although an instantaneous change of the loop-

[6] The coefficient in Eq. (8.60) has been fitted upon considering the matter case $\Omega_{\rm NS}=\Omega_{\rm matter}$. Note that the turning-point in the kination case is slightly higher frequency by a factor of order 1, cf. Fig. 8.9.

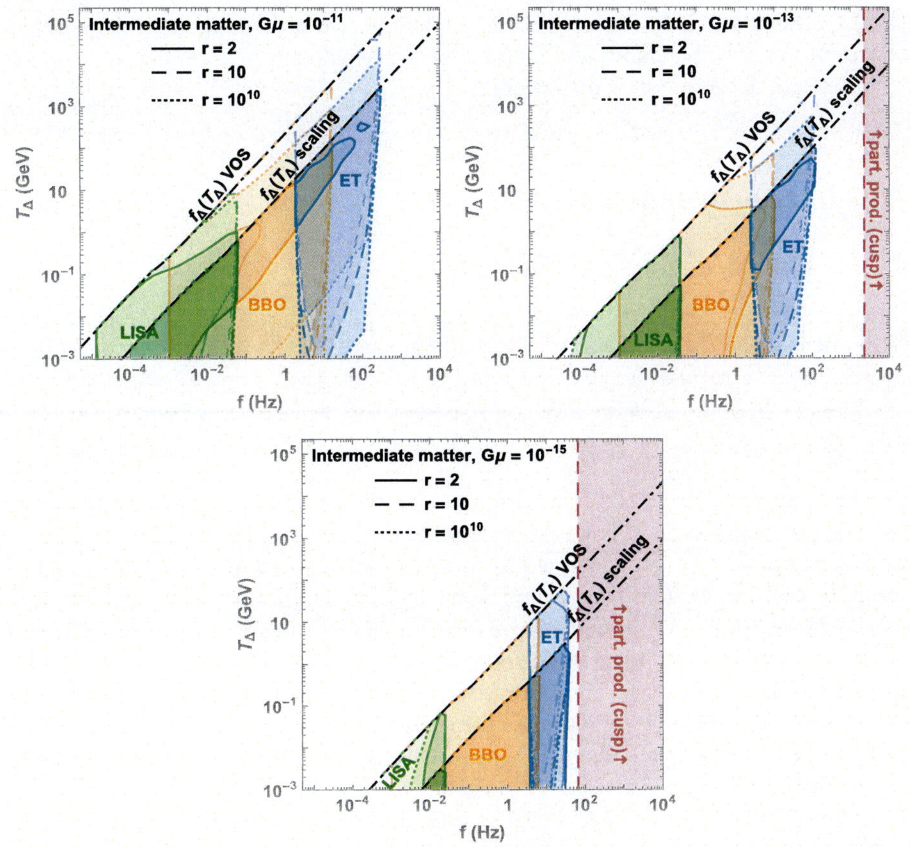

Fig. 8.10 The colored regions show the detectability of the spectral suppression, cf. *spectral-index prescription (Rx 2)* in Sect. 9.1.2., due to a NS intermediate matter era with duration $r = T_{\rm start}/T_\Delta$, assuming scaling and VOS networks, cf. Sect. 8.4.3. Limitation from particle production, cf. Sect. 8.3.4, is shown in purple

production efficiency $C_{\rm eff}$ at T_Δ is assumed in [22–24]. This can be explained if in Ref. [22–24], the criterion in Eq. (8.61) is smaller than the percent level.

8.6.3 *Constraints*

In Fig. 8.10, we show the constraints on the presence of an early-intermediate-non-standard-matter-dominated era starting at the temperature $r\, T_\Delta$ and ending at the temperature T_Δ. Matter eras as short as $r = 2$ and ending at temperature as large as 100 TeV could be probed by GW interferometers. We assume that an early-matter era is detectable if the spectral index is smaller than -0.15, cf. *spectral-index prescription*

(Rx 2) in Sect. 9.1.2. In the next chapter, Chap. 9, we provide model-independent constraints on the abundance and lifetime of an unstable particles giving rise to such a non-standard intermediate matter era.

8.7 Intermediate Inflation

8.7.1 The Non-standard Scenario

Next, we consider the existence of a short inflationary period with a number of e-folds

$$N_e \equiv \log\left(\frac{a_{\rm start}}{a_{\rm end}}\right), \tag{8.62}$$

smaller than $N_e \lesssim 20 \ll 60$, in order not to alter the predictions from the first inflation era regarding the CMB power spectrum. On the particle physics side, such a short inflationary period can be generated by a highly supercooled first-order phase transition. It was stressed that nearly-conformal scalar potentials naturally lead to such short, with $N_e \sim 1 - 15$, periods of inflation [56, 57, 59]. Those are well-motivated in new strongly interacting composite sectors arising at the TeV scale, as invoked to address the Higgs hierarchy problem and were first studied in a holographic approach [53, 54] (see also the review [204]). As the results on the scaling of the bounce action for tunnelling and on the dynamics of the phase transitions do essentially not depend on the absolute energy scale, but only on the shallow shape of the scalar potential describing the phase transition, those studies can thus be extended to a large class of confining phase transitions arising at any scale. In this section, we will take this inflationary scale as a free parameter.

We define the energy density profile as, cf. Fig. 8.11

$$\rho_{\rm tot}(a) = \begin{cases} \rho_{\rm rad}^{\rm st}(a) + \rho_{\rm late}(a) & \text{for } \rho > \rho_{\rm inf}, \\ \rho_{\rm inf} = E_{\rm inf}^4 & \text{for } \rho = \rho_{\rm inf}, \\ \frac{\rho_{\rm inf}\Delta_R(T_{\rm end},T)}{a_{\rm end}a^4 + \rho_{\rm late}(a)} & \text{for } \rho < \rho_{\rm inf}, \end{cases} \tag{8.63}$$

where $\rho_{\rm inf}$ is the total energy density of the universe during inflation and $E_{\rm inf} \equiv \rho_{\rm inf}^{1/4}$ is the corresponding energy scale. The function Δ_R is defined in (8.55) (Fig. 8.12).

8.7.2 The Stretching Regime and Its Impact on the Spectrum

Figure 8.13 shows how the fast expansion during inflation suppresses the GW spectrum for frequencies above a turning-point frequency f_Δ which depends on the number of e-folds. The larger the number of e-folds, the lower f_Δ. Indeed, dur-

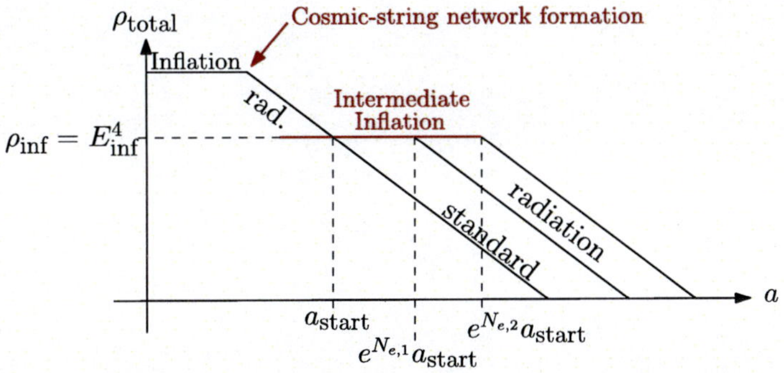

Fig. 8.11 Evolution of the total energy density assuming the presence of an intermediate inflationary era characterised by the energy density $\rho_{\rm inf}$, for two different durations (number of efolds), $N_{e,1}$ and $N_{e,2}$

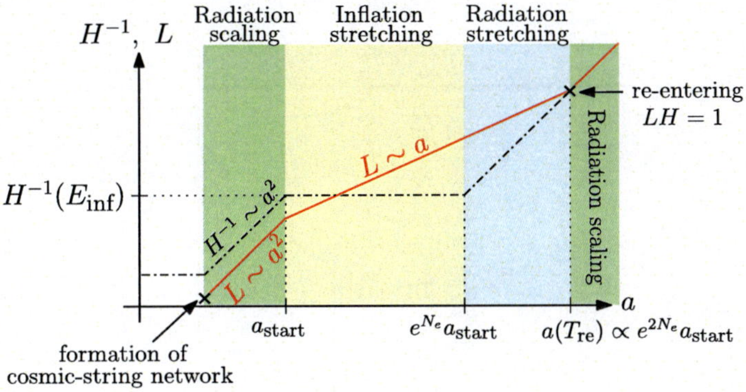

Fig. 8.12 After its formation, before inflation, the network enters the scaling regime with $L \sim a^2$ due to loop formation. During the N_e e-folds of inflation, the network correlation length gets stretched out of the horizon by the rapid expansion and loop formation stops, thus $L \sim a$. After inflation, during radiation, the correlation length starts to re-enter the horizon and scales again as $L \sim a^2$

ing inflation, the loop-production efficiency $C_{\rm eff} \propto \xi^{-3}$ is severely suppressed, cf. Fig. 8.13, by the stretching of the correlation length ξ beyond the Hubble horizon, and loop production freezes [25]. After the end of inflation, one must wait for the correlation length to re-enter the horizon in order to reach the scaling regime again. The duration of the transient regime receives an enhancement factor $\exp N_e$. As a result, the turning-point frequency f_Δ receives a suppression factor $\exp N_e$ as derived below:

$$f_\Delta = (1.5 \times 10^{-4}\ \text{Hz}) \left(\frac{T_{\rm re}}{\text{GeV}}\right) \left(\frac{0.1 \times 50 \times 10^{-11}}{\alpha\, \Gamma G\mu}\right)^{1/2} \left(\frac{g_*(T_{\rm re})}{g_*(T_0)}\right)^{1/4}, \quad (8.64)$$

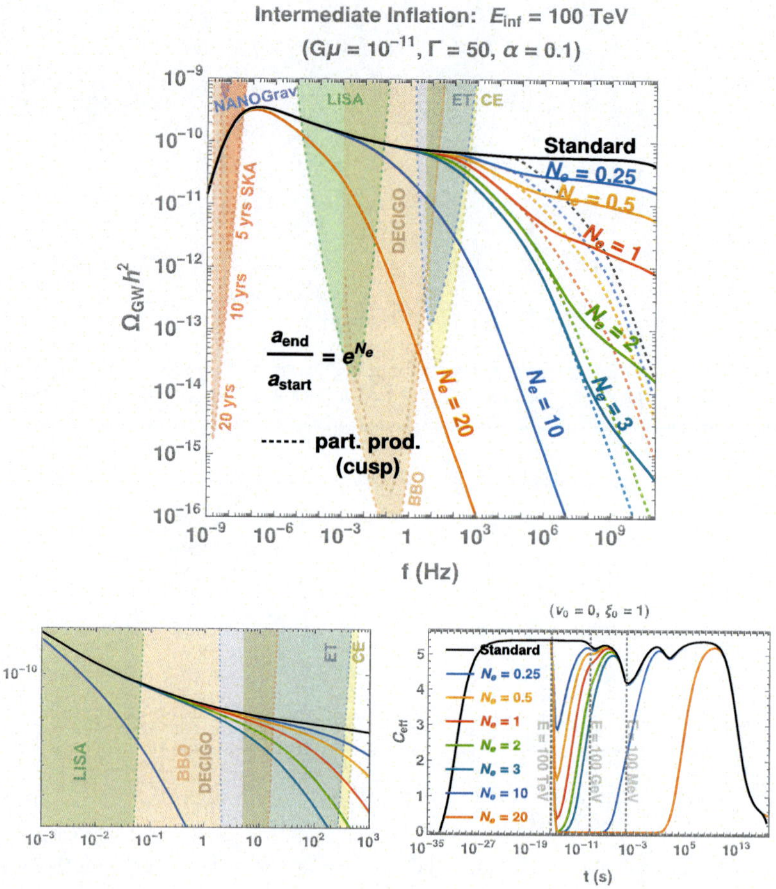

Fig. 8.13 **Top:** GW spectra from cosmic strings assuming either the scaling or the VOS network, evolved in the presence of a non-standard intermediate inflation era. Inflation directly affects the VOS parameters by stretching the strings beyond the horizon. The transition between the $f^{-1/3}$ scaling after the turning point, to the f^{-1} scaling at even larger frequencies, is an artefact due to total number of modes k being fixed to 2×10^4, see Fig. 8.14 for an extrapolation of the $f^{-1/3}$ behavior to arbitrary large frequencies and Appendix 8.B.6 for more details. **Bottom:** The loop-production is suppressed and only becomes significant again when the correlation length re-enters the horizon. Limitations due to particle production, cf. Sect. 8.3.4, are shown with dotted lines

with $T_{\rm re}$ the temperature at which the long-string network re-enters the Hubble horizon

$$T_{\rm re} \simeq \frac{E_{\rm inf}}{(0.1)\, g_*^{1/4}(T_{\rm re})\, \exp(N_e)}, \tag{8.65}$$

where (0.1) is the typical correlation length before the stretching starts. Note that the numerical factor in Eq. (8.64) comes from the demanded precision of 10% deviation,

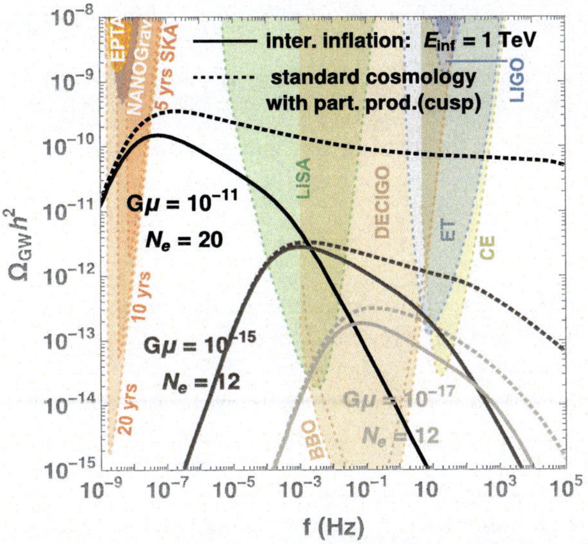

Fig. 8.14 In the case an intermediate inflationary era lasting for $\mathcal{O}(10)$ efolds, the SGWB from cosmic strings completely looses its scale invariant shape and has instead a peak structure. A TeV scale inflation era can lead to broad peaks either in the LISA or BBO band or even close to the SKA band, depending on the value of the string tensions $G\mu$, and the number of efolds N_e. At low $G\mu \lesssim 10^{-17}$, the spectrum manifests a peak structure even in standard cosmology because of the emission of massive particles at large frequencies, cf. Sect. 8.3.1. Here we extrapolate the $f^{-1/3}$ behavior to arbitrary large frequencies, which is equivalent to sum over an infinite number of proper modes k, see Appendix 8.B.6

cf. Eq. (8.61). It can be lower by a factor ~ 300 if the 1% precision is applied, as shown in Eq. (8.119).

Figure 8.14 shows how a sufficiently long period of intermediate inflation can lead to SGWB with peak shapes in the future GW interferometer bands. We emphasize that the change of the GW spectrum from CS in the presence of a non-standard matter-dominated era, a short inflation, and particle production look similar. Therefore, the question of how disentangling each effect from one another deserves further studies.

Interestingly, in contrast with the SGWB which is dramatically impacted by an intermediate period of inflation, the short-lasting GW burst signals [37, 38, 132, 133, 135] remain preserved if the correlation length re-enters the horizon at a red-shift higher than $\sim 5 \times 10^4$ [205]. Indeed, the bursts being generated by the small scale structures, they have higher frequencies and then are emitted later than the SGWB, cf. Fig. 2 in [133].

Derivation of the turning-point formula (inflation case):
Let us review the chronology of the network in the presence of an intermediate-inflation period (see Fig. 8.12) in order to derive Eq. (8.64). In the early radiation era, the network has already been produced and reached the scaling regime before

inflation starts. The correlation length scale is of order $(0.1)t$ or equivalently

$$L_{\text{start}} H_{\text{start}} \sim \mathcal{O}(0.1), \tag{8.66}$$

where L is the correlation length of strings, and H is the Hubble rate. When inflation begins, it stretches cosmic strings beyond the horizon with

$$L \propto a \quad \text{leading to} \quad LH \gg 1, \tag{8.67}$$

within a few e-folds. Later, the late-time energy density takes over inflation, but the network is still in the stretching regime $L \propto a$, i.e.

$$LH \propto t^{(2-n)/n} \quad \text{during the era with } \rho \propto a^n. \tag{8.68}$$

For $n > 2$, the Hubble horizon will eventually catch up with the string length, allowing them to re-enter, and initiate the loop production. We consider the case where the universe is radiation-dominated after the inflation period and define the temperature T_{re} of the universe when the long-string correlation length L re-enters the horizon

$$L_{\text{re}} H_{\text{re}} = 1, \tag{8.69}$$

where L_{re} and H_{re} are the correlation length and Hubble rate at the re-entering time. We can use Eq. (8.68) to evolve the correlation length, starting from the start of inflation up to the re-entering time

$$1 = L_{\text{re}} H_{\text{re}} = \left(\frac{t_{\text{re}}}{t_{\text{end}}}\right)^{-1/2} L_{\text{end}} H_{\text{end}}, \tag{8.70}$$

$$= \left(\frac{t_{\text{re}}}{t_{\text{end}}}\right)^{-1/2} \left(\frac{a_{\text{end}}}{a_{\text{start}}}\right) L_{\text{start}} H_{\text{start}}, \tag{8.71}$$

$$\simeq \left(\frac{T_{\text{re}}}{T_{\text{end}}}\right) e^{N_e} (0.1) \tag{8.72}$$

We have used $t \propto T^{-2}$ during the radiation era and introduced the number N_e of inflation e-folds. Finally, we obtain the re-entering temperature in terms of the number of e-folds N_e and the inflationary energy scale E_{inf} as

$$T_{\text{re}} \simeq \frac{E_{\text{inf}}}{(0.1)\, g_*^{1/4}(T_{\text{re}}) \exp(N_e)}. \tag{8.73}$$

After plugging Eq. (8.73) into the VOS turning-point formula Eq. (8.60), with $T_\Delta = T_{\text{re}}$, and adjusting the numerical factor with the GW spectrum computed numerically, we obtain the relation in Eq. (8.64) between the turning-point frequency and the inflation parameters N_e and E_{inf}.

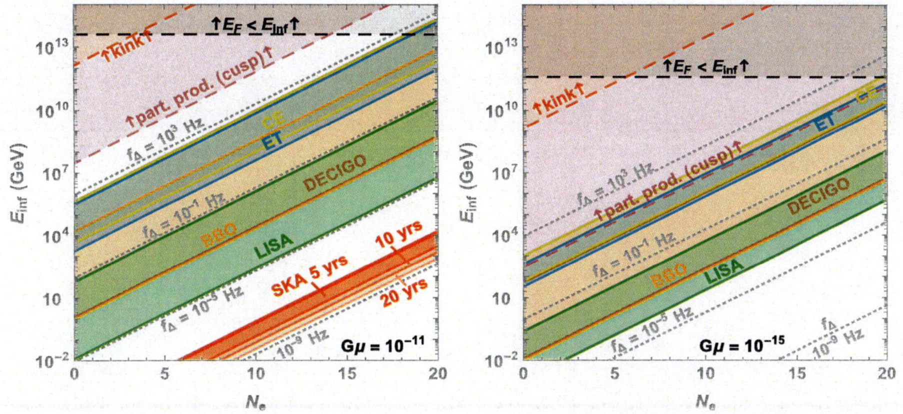

Fig. 8.15 Reach of future GW interferometers for probing an intermediate-inflation period with an energy scale E_{inf}, lasting N_e efolds. Colored regions correspond to the turning-points with amplitude higher than each power-law-sensitivity curve, cf. *turning-point prescription (Rx 1)* in Sect. 9.1.2. Gray dotted lines are turning-points, cf. Eq. (8.64), for given frequencies. Red and purple dashed lines are limitations from particle production, cf. Sect. 8.3.4

8.7.3 Model-Independent Constraints

In Sect. 8.7, we derived the imprint of a short period of inflation on the GW spectrum from CS. In Figs. 8.15, 8.16 and 8.17, we show the corresponding constraints—assuming one of the mentioned experiments detect a GW spectrum from CS in the future—on an intermediate short inflation period in the planes $E_{\text{inf}} - N_e$, $G\mu - E_{\text{inf}}$, and $G\mu - N_e$, respectively. We follow the *turning-point prescription (Rx 1)* defined in Sect. 9.1.2, which constrains a non-standard cosmology by using the detectability of the turning-point frequency defined by Eq. (8.64). The longer the intermediate inflation, the later the correlation length re-enters the horizon, the latter the long-string network goes back to the scaling regime, the lower the frequency of the turning-point and the larger the inflationary scale which can be probed. The detection of a GW spectrum generated by CS by future GW observatories would allow to probe an inflationary energy scale E_{inf} between 10^{-2} GeV and 10^{13} GeV assuming a number of e-folds $N_e \lesssim 20$ (Fig. 8.18).

8.8 Summary and Conclusion

In standard cosmology, the GW spectrum generated by a network of Nambu-Goto cosmic strings (and mainly due to emission by loops) is nearly scale-invariant. Its potential observation by third-generation interferometers would be a unique probe

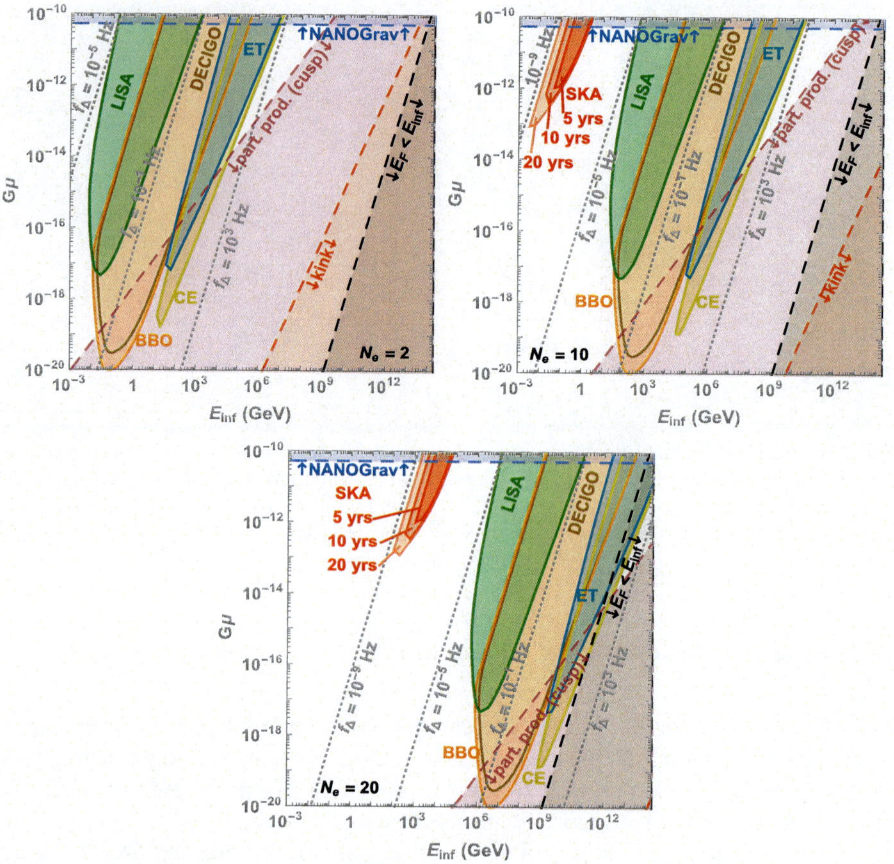

Fig. 8.16 Constraints on intermediate inflation from CS detection by future GW observatories. The longer the intermediate inflation, the later the correlation length re-enters the horizon, the more shifted to lower frequencies the turning-point and the larger the inflation scale which we can probe. Colored regions correspond to the turning-points with amplitude higher than each power-law-sensitivity curve, cf. *turning-point prescription (Rx 1)* in Sect. 9.1.2. The bound $E_F < E_{\text{inf}}$, where $E_F \sim m_{pl}\sqrt{G\mu}$ is the network-formation energy scale, guarantees that the CS network forms before the intermediate-inflation starts. Red and purple dashed lines are limitations from particle production, cf. Sect. 8.3.4

of new effects beyond the standard models of particle physics and cosmology. Such opportunity was pointed out in [22–25, 27].

Deriving firm conclusions is still premature as theoretical predictions of the GW spectrum from CS are subject to a number of large uncertainties. Still, we feel that the extraordinary potential offered by future GW observatories to probe high energy physics has not yet been explored, and in a series of papers, we are starting to scrutinise how much can be learnt, even if only in the far-future, after those planned

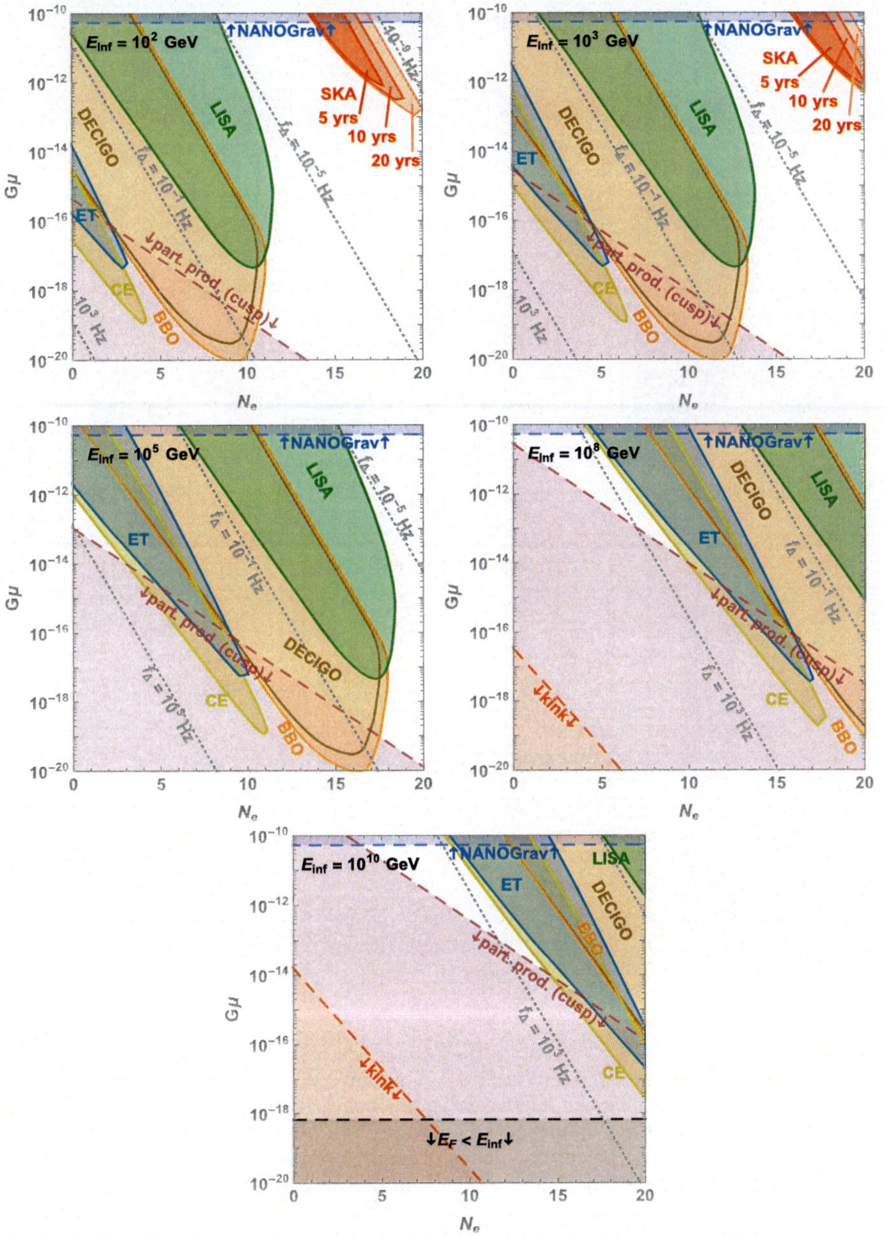

Fig. 8.17 Prospect constraints on intermediate inflation if a GW interferometer detects a SGWB from CS with tension $G\mu$. The freezing of the long-string network due to the stretching of the correlation length outside the horizon allows to probe large inflationary scale E_{inf} for large number of efolds N_e. Colored regions correspond to the turning-points with amplitude higher than each power-law-sensitivity curve, cf. *turning-point prescription (Rx 1)* in Sect. 9.1.2. Red and purple dashed lines are limitations from particle production, cf. Sect. 8.3.4

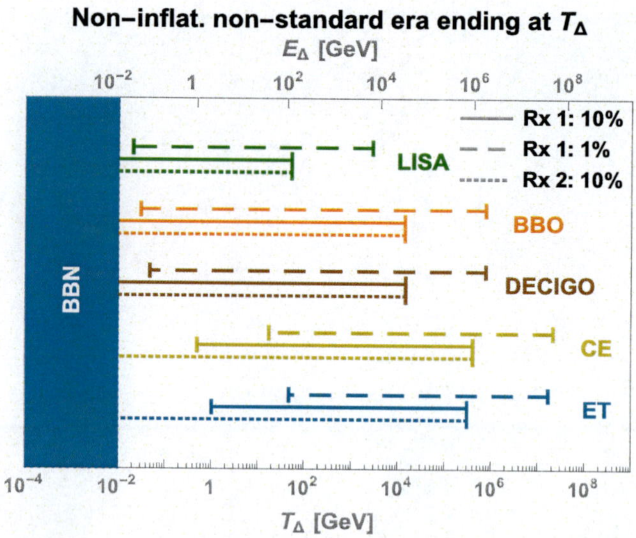

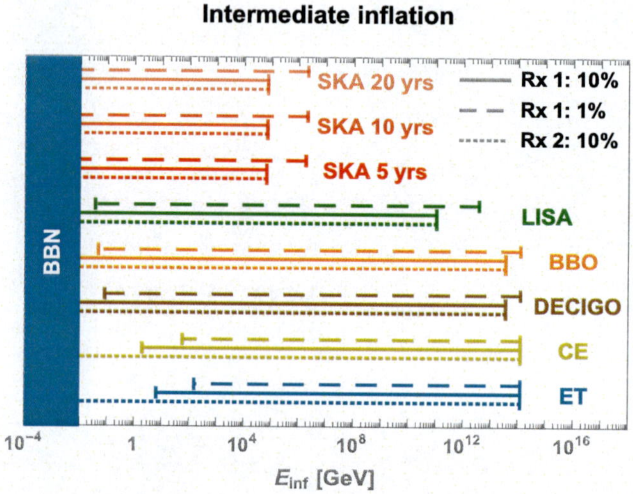

Fig. 8.18 **Top:** Sensitivity to the energy scale E_Δ of the universe at the end of any non-inflationary non-standard era for each future GW interferometer. The connection to E_Δ is given by the observation of the turning-point frequency defined in Eq. (8.118). The width of the bands includes varying the string tension for $G\mu < 10^{-10}$. The dotted, dashed and solid lines correspond to different observational prescriptions defined in Sect. 9.1.2. **Bottom:** Sensitivity to the energy scale $E_{\rm inf}$ of an intermediate inflationary era for each future GW interferometer as well as for future radio telescope SKA. The connection to $E_{\rm inf}$ is given by the observation of the turning-point frequency defined in Eq. (8.119). The width of the band also includes varying the number of efolds of inflation N_e up to 20

GW observatories will have reached their expected long-term sensitivity and the astrophysical foreground will have been fully understood.

Deviations in the cosmological history with respect to standard cosmology not only change the redshifting factor of GW but also modify the time of loop formation and the loop-production efficiency. We presented predictions for the resulting GW spectra under a number of assumptions which we have comprehensively reviewed.

We extend previous works in several directions, as listed in the introduction.

A particular feature of gravitational waves from cosmic strings is the relation between the observed frequency and the GW production mechanism. In contrast with short-lasting cosmological sources of gravitational waves, such as phase transitions, where the frequency is simply related to the Hubble radius at the time of GW emission, for cosmic strings the time of the dominant GW emission is much later than the time of loop production, by a factor $\sim 1/(G\mu)$, such that the observed frequency is higher due to a smaller redshift. We stressed that a given interferometer may be sensitive to very different energy scales, depending on the nature and duration of the non-standard era, and the value of the string tension. This goes against usual paradigms. For instance, it is customary to talk about LISA as a window on the EW scale [204, 206]. This does not apply for GW from cosmic strings, as LISA could either be a window on a non-standard matter era at the QCD scale or on a 10 TeV inflationary era, meaning that the GW observed in the LISA band have been emitted by loops that were created at the QCD epoch, or at a 10 TeV epoch depending on the nature of the new physics responsible for the non-standard era. Interestingly, the Einstein Telescope and Cosmic Explorer offer a window of observation on the highest scales, up to 10^{14} GeV inflationary eras. They can also be windows on the EW and TeV scales, as will be discussed in more details in [69]. BBO/DECIGO could probe new physics in an intermediate range. Finally, radio telescope SKA may be sensitive to a TeV scale inflationary era. The goal of this study was to stress the very rich variety of spectral shapes that can be obtained by combining different physical effects. In particular, we showed how peaked shapes can arise naturally in a large variety of models.

We apply these findings to probe well-motivated particle physics scenarios in [69]. Particularly generic are intermediate matter eras triggered by cold heavy particles arising in UV completions of the Standard Model. In [69], we show on specific models how a new uncharted particle theory space can be probed from analysis of SGWB from CS.

Finally, one important question will be to work out how to distinguish a stage of matter or inflationary expansion, which both lead to a suppression of the GW spectrum, from the cutoff induced by particle production from small loops. Both predict a cutoff at high frequencies and lead to similar spectra. Interestingly, particle production by cosmic string networks can be probed through cosmic rays and bring complementary non-gravitational information on the SGWB. Besides, the complementarity between different GW instruments will be crucial here as the detection of the low-frequency peak of the spectrum (due to the transition from the standard radiation to the standard matter era) can enable to probe the string tension and to break the degeneracy between different spectral predictions. The possibility to reconstruct the spectral shape of a SGWB was analysed in [207] using LISA data only. In the case of a SGWB generated by CS, which can span more than twenty decades in frequency, it will be crucial to use data from different interferometers (and even from radio telescopes) to probe the full spectrum.

Appendices

8.A Constraints on Cosmic Strings from BBN, Gravitational Lensing, CMB and Cosmic Rays

By confronting our theoretical predictions for the GW spectrum from CS with the sensitivity curves of EPTA [142] and NANOGrav [143] (which we take from [146]), we derived the respective bounds $G\mu \lesssim 2 \times 10^{-10}$ (EPTA) and $G\mu \lesssim 5 \times 10^{-11}$ (NANOGrav), as discussed in Sect. 8.2.4.[7] For this reason, we only considered in our analysis $G\mu$ values smaller than 5×10^{-11}. In Sect. 8.A.1, we give the constraints on the string tension from not changing the expansion rate of the universe at BBN. They are much weaker than the ones from Pulsar Timing Arrays but they can become relevant in the presence of kination.

In Sects. 8.A.2 and 8.A.3, we give bounds from gravitational lensing and CMB observables. They are also much weaker than the ones from Pulsar Timing Arrays but they have the strong advantage to be independent of our assumptions for the theoretical prediction of the GW background. Finally, in Sect. 8.A.4, we discuss the possibility of probing CS from the massive particle production in the presence of kinks and cusps.

8.A.1 GW Constraints from BBN

As a sub-component of the total energy density of the universe, the amount of GW can impact the expansion rate of the universe which is strongly constrained by BBN and CMB. More precisely, any non-standard energy density can act as an effective number of neutrino relics

$$N_{\rm eff} = \frac{8}{7}\left(\frac{\rho_{\rm tot} - \rho_\gamma}{\rho_\gamma}\right)\left(\frac{11}{4}\right)^{4/3}, \tag{8.74}$$

which is constrained by CMB measurements [209] to $N_{\rm eff} = 2.99^{+0.34}_{-0.33}$ and by BBN predictions [210, 211] to $N_{\rm eff} = 2.90^{+0.22}_{-0.22}$ whereas the SM prediction [212–214] is $N_{\rm eff} \simeq 3.045$. Using $\Omega_\gamma h^2 \simeq 2.47 \times 10^{-5}$ [202], we obtain the following bound on the GW spectrum

[7] A galaxy lensed by a cosmic string. Image credits: [208]

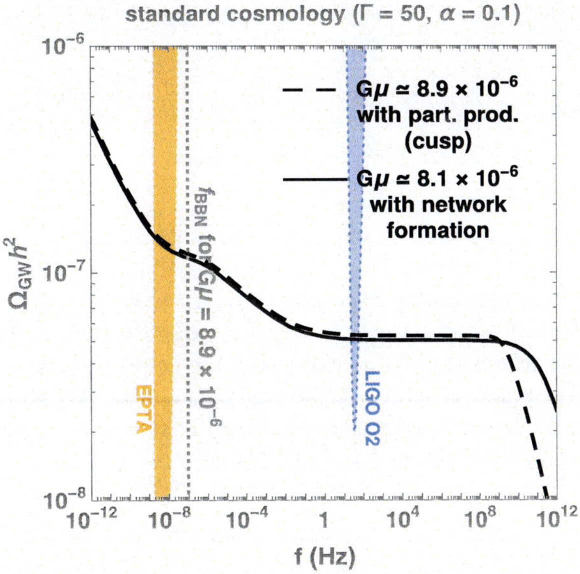

Fig. 8.19 Two GW spectra which saturate the BBN bounds, assuming a VOS string network, cf. Sect. 8.4.3, evolving in standard cosmology. The solid line assumes a cut-off due to network formation whereas the dashed line assumes a cut-off due to particle production from cusps. The dotted vertical line is the frequency emitted when BBN starts. We compare the BBN bounds to the bounds from EPTA and LIGO O2

$$\int_{f_{\rm BBN}}^{f_{\rm high}} \frac{df}{f} h^2 \Omega_{\rm GW}(f) \leq 5.6 \times 10^{-6}\, \Delta N_\nu, \tag{8.75}$$

where $f_{\rm high}$ is the frequency today of the first GW produced, $f_{\rm BBN}$ is the frequency today of the GW produced at BBN, and we set $\Delta N_\nu \leq 0.2$. The value of $f_{\rm BBN}$ depends on the source of GW. For CS, the temperature at BBN, $T_{\rm CMB} \simeq 1$ MeV, translates via Eq. (8.117) to the frequency

$$f_{\rm BBN} \simeq 8.9 \times 10^{-5}\,{\rm Hz} \left(\frac{0.1 \times 50 \times 10^{-11}}{\alpha \Gamma G\mu} \right)^{1/2}. \tag{8.76}$$

In Fig. 8.19, we show the GW spectra which saturate the BBN bound for two different high-frequency cut-offs. We can see that the lower the cut-off, the higher the upper bound on $G\mu$ due to less GW present at the time of BBN. Assuming the presence of the cut-off due to particle production from cusps, we obtain

$$\text{BBN:} \qquad h^2 \Omega_{\rm GW}(f) \lesssim 8.9 \times 10^{-6}. \tag{8.77}$$

We expect the BBN bounds to become softer in the presence of non-standard matter or inflation era but tighter in the presence of an early kination era. For instance, scenarios of inflation followed by a stiff equation of state (e.g. quintessential inflation [215]) are dramatically jeopardized by the BBN bounds [20]. Similarly, in the case of CS, we find that the maximally allowed string tension is $G\mu \simeq 3.9 \times 10^{-15}$, $3.8 \times$

10^{-17}, and 2.9×10^{-20} for long-lasting kination era ending at temperature $T_\Delta = 100$ TeV, 1 TeV, and 1 GeV, respectively.

8.A.2 *Gravitational Lensing*

The presence of energy confined within the core of CS affects the spacetime around them. The metric near a CS is locally flat but globally conical [216]. Photons from a distant celestial object travelling in the vicinity of a CS are subject to gravitational lensing effects. The corresponding constraint $G\mu \lesssim 3 \times 10^{-7}$ has been derived from the search of gravitational lensing signatures of CS in the high-resolution wide-field astronomical surveys GOODS [217] and COSMOS [218]. It has been claimed that constraints from gravitational lensing surveys at radio frequencies like LOFAR and SKA could reach $G\mu \lesssim 10^{-9}$ [219].

8.A.3 *Temperature Anisotropies in the CMB*

There are two possible effects from CS on temperature fluctuations in the CMB:

1. CS moving through the line-of-sight can induce Doppler shifts on the photons coming from the last scattering surface, known as the Kaiser-Stebbins-Gott effect [220–222], potentially leaving line-like discontinuities in the CMB.
2. A CS moving in the primordial plasma leaves surdensity perturbations, the so-called wakes [223], possibly imprinted in the CMB temperature anisotropy. Due to the stochastic behavior of the Kibble mechanism, these perturbations are decoherent and give rise to a CMB spectrum without acoustic peaks [224].

Lattice numerical computation of the temperature anisotropy in Abelian-Higgs [225, 226], Nambu-Goto [224, 227, 228] or global strings [229] have constrained the string tension to $G\mu \lesssim$ few $\times 10^{-7}$ [230]. Constraints of the same magnitude can be found from non-gaussianities [230–232]. Also, the same signatures as in the CMB can be imprinted in the 21 cm power spectrum, and an experiment with a collecting area of $10^4 - 10^6$ km^2 might constrain $G\mu \lesssim 10^{-10} - 10^{-12}$ [233].

8.A.4 *Non-gravitational Radiation*

As discussed in Sect. 8.3.1, the presence of small-scale structures on local strings, cusps and kinks, invalidates the Nambu-Goto approximation and implies the radiation of massive particles. Therefore, CS have been proposed as a possible mechanism for generating non-thermal Dark Matter [234–237].

At a cusp, the string can reach ultrarelativistic velocities. Therefore, CS have been pointed [238–241] as a possible candidate for the detection of ultra-high energy cosmic rays [242] above the Greisen-Zatsepin-Kuzmin (GZK) cut-off, around 10^{20} eV [243–245], even though the expected flux at earth is found to be too small for being detected [246–248].

More precisely, upon introducing a coupling between the SM and the dark $U(1)'$ from which the CS result, e.g. a Higgs portal or a kinetic mixing, an effective interaction between SM particles and the CS arises [249]. In that case, the formation of cusps and kinks on the string radiate SM particles [250]. The expected gamma-ray flux at the earth is too low to be observed by Fermi-Lat [251], even when assuming that all the massive particles radiated by CS subsequently decay into gamma-ray [67]. Note however [252] which finds particle production to be within the reach of Fermi-Lat for CS modelled according to Ringeval et al. [133, 149, 171]. For cusp domination and for large coupling between the SM and $U(1)'$, the flux of high-energy neutrino might be measured by the future experiments SKA and LOFAR for $G\mu \sim [10^{-14},\ 10^{-16}]$ [251]. Also, the distortions in the CMB may be detected by the future telescope PIXIE for $G\mu \sim [10^{-12},\ 10^{-14}]$ [251]. Finally, depending on the magnitude of the SM-$U(1)'$ coupling, the BBN constraints can already exclude values of string tensions between $10^{-8} \gtrsim G\mu \gtrsim 10^{-14}$ [251].

Constraints from particle emission apply on an interval of values for $G\mu$, and not as upper bound like for gravitational emission [246]. For longer lifetimes $\propto (\Gamma G\mu)^{-1}$, there are more loops and we expect a larger flux of emitted particles while gravitational emission grows with $G\mu$. At small $G\mu$, loops decay preferentially into particles, cf. Sect. 8.3.1. In that case, the expected flux of emitted particles increases with the string tension which controls the power of the particle emission. Therefore, there exists a value of $G\mu$ for which the expected flux of emitted particles is maximal. This is the value of $G\mu$ when particle production is as efficient as gravitational production. For example for loops created at the recombination time, the value of $G\mu$ maximizing the cosmic ray production is 10^{-18} [253].

Superconducting Cosmic Strings:
Another possibility for generating large particle production is to couple the CS with electromagnetic charge carriers and to spontaneously break electromagnetic gauge invariance inside the vortex [254]. Upon moving through cosmic magnetic fields, Superconducting Cosmic Strings (SCS) are able to develop a large electric current $\mathscr{I}$. The formation of cusps on SCS is expected to emit bursts of electromagnetic radiation [255–258], up to very high energies, set by the string tension $\sqrt{\mu} \sim 10^{13}\,\text{GeV}\,\sqrt{G\mu/10^{-15}}$, hence leading to high-energy gamma-rays [259–261]. Hence, SCS could be an explanation for the observed gamma-ray bursts at high redshifts, which depart from the predictions from star-formation-history [261]. However, the expected photon flux at earth is larger in the radio band than in the gamma-ray band [262–264] (but also mostly generated by kinks instead of cusps [265]). Thus, it has been proposed that SCS could be an explanation for the Fast-Radio-Burst events [266–268] for string tensions in the range $G\mu \sim [10^{-12},\ 10^{-14}]$ and string currents $\mathscr{I} \sim [10^{-1},\ 10^{2}]$ GeV [267]. Electromagnetic emission from SCS lead to CMB

distortions [269–271]. A next-generation telescope like PIXIE [272] would exclude string tensions $G\mu \sim 10^{-18}$, for string currents as low as $\mathcal{I} \sim 10^{-8}$ GeV [271]. Also, electromagnetic radiation, by increasing the ionization fraction of neutral hydrogen, can affect the CMB temperature and polarization correlation functions at large angular scales, leading to the contraint $\mathcal{I} \lesssim 10^7$ GeV [273]. Note that ionization of neutral hydrogen has been studied in [253] in the case of non-superconducting strings. Additionally, the radio emission from SCS can increase the depth of the 21 cm absorption signal, and EDGES data excludes the SCS tension $G\mu \sim 10^{-13}$ for string currents as low at $\mathcal{I} \sim 10$ GeV. Finally, emission of boosted charge carriers from SCS cusps moving in a cosmic magnetic field B, has been studied in [274], and provide a possible explanation for high-energy neutrino above 10^{20} eV, for $G\mu \sim [10^{-14},\ 10^{-20}]$.

8.B Derivation of the GW Spectrum from CS

In this appendix we provide the steps leading to Eq. (8.153).

8.B.1 From GW Emission to Detection

The GW energy density spectrum today is defined as

$$\Omega_{\rm GW}(f) = \frac{f}{\rho_c}\left|\frac{d\rho_{\rm GW}(f,\ t_0)}{df}\right|. \tag{8.78}$$

After emission, the GW energy density redshifts as radiation, $\rho_{\rm GW} \propto a^{-4}$, so the GW energy density per unit of frequency redshifts as

$$\frac{d\rho_{\rm GW}(f,\ t_0)}{df} = \frac{d\rho_{\rm GW}(\tilde{f},\ \tilde{t})}{d\tilde{f}}\left(\frac{a(\tilde{t})}{a(t_0)}\right)^3 \tag{8.79}$$

where the frequency at emission $\tilde{f}$ is related to the frequency today f through

$$\tilde{f} = \frac{a(t_0)}{a(\tilde{t})} f.$$

8.B.2 From Loop Production to GW Emission

After its formation at t_i, a loop shrinks through emission of GW with a rate $\Gamma G\mu$ so that its length evolves as, cf. Sect. 8.2.3

$$l(t) = \alpha t_i - \Gamma G\mu(t - t_i), \tag{8.80}$$

where α is the length at formation in units of the horizon size. The resulting GW are emitted at a frequency $\tilde{f}$ corresponding to one of the proper modes of the loop, i.e.

$$\tilde{f} = \frac{2k}{l}, \qquad k \in \mathbb{Z}^+. \tag{8.81}$$

The GW energy rate emitted by one loop through the mode k is, cf. Sect. 8.2.3

$$\frac{dE_{\rm GW}^{(k)}}{dt} = \Gamma^{(k)}\, G\mu^2, \qquad \text{with } \sum_k \Gamma^{(k)} = \Gamma, \tag{8.82}$$

where

$$\Gamma^{(k)} = \frac{\Gamma\, k^{-4/3}}{\sum_{p=1}^{\infty} p^{-4/3}} \simeq \frac{\Gamma\, k^{-4/3}}{3.60}, \tag{8.83}$$

which assumes that the GW emission is dominated by cusps. The GW energy density spectrum resulting from the emission of all the decaying loops until today is

$$\frac{d\rho_{\rm GW}(\tilde{f}, \tilde{t})}{d\tilde{f}} = \int_{t_F}^{t_0} d\tilde{t}\, \frac{dE_{\rm GW}}{d\tilde{t}}\, \frac{dn(\tilde{f}, \tilde{t})}{d\tilde{f}}, \tag{8.84}$$

where $dn(\tilde{f}, \tilde{t})/d\tilde{f}$ is the number density of loops emitting GW at frequency $\tilde{f}$ at time $\tilde{t}$ and t_0 is the age of the universe today. Loops start being created at time of CS network formation t_F, after the damped evolution has stopped, cf. Sect. 8.B.

8.B.3 The Loop Production

In Sect. 8.3.2, we assume the loop-formation rate to be

$$\frac{dn}{dt_i} = (0.1)\frac{C_{\rm eff}(t_i)}{\alpha\, t_i^4}, \tag{8.85}$$

where $C_{\rm eff}(t_i)$ is the loop-formation efficiency. We deduce the loop number density per unit of frequency

$$\frac{dn(\tilde{f},\tilde{t})}{d\tilde{f}} = \left[\frac{a(t_i)}{a(\tilde{t})}\right]^3 \frac{dn}{dt_i}\cdot\frac{dt_i}{dl}\cdot\frac{dl}{d\tilde{f}} \tag{8.86}$$

$$= \left[\frac{a(t_i)}{a(\tilde{t})}\right]^3 \sum_k (0.1)\frac{C_{\rm eff}(t_i)}{t_i^4}\cdot\frac{1}{\alpha\,(\alpha+\Gamma G\mu)}\cdot\frac{2k}{f^2}\left[\frac{a(\tilde{t})}{a(t_o)}\right]^2. \tag{8.87}$$

8.B.4 The Master Equation

Finally, we get the GW energy density spectrum

$$\Omega_{\rm GW}(f) = \sum_k \Omega_{\rm GW}^{(k)}(f)$$

$$= \sum_k \frac{1}{\rho_c}\frac{2k}{f}\frac{\mathcal{F}_\alpha\,\Gamma^{(k)}\,G\mu^2}{\alpha\,(\alpha+\Gamma\,G\mu)}\int_{t_{\rm osc}}^{t_0} d\tilde{t}\,\frac{C_{\rm eff}(t_i)}{t_i^4}\left[\frac{a(\tilde{t})}{a(t_0)}\right]^5\left[\frac{a(t_i)}{a(\tilde{t})}\right]^3 \theta(t_i-t_{\rm osc})\,\theta(t_i-\frac{l_*}{\alpha}). \tag{8.88}$$

The first Heaviside function stands for the time $t_{\rm osc}$ at which long-strings start oscillating, either just after formation of the long-string network or after that friction becomes negligible, cf. Sect. 8.B.4. The second Heaviside function stands for the energy loss into particle production which is more efficient than GW emission for loops of length smaller than a characteristic length l_*, which depends on the string small-scale structure, cf. Sect. 8.3.1. The time t_i of formation of the loops, which emit at time $\tilde{t}$ and which give the detected frequency f, can be determined from Eqs. (8.80) and (8.81)

$$t_i(f,\tilde{t}) = \frac{1}{\alpha+\Gamma G\mu}\left[\frac{2k}{f}\frac{a(\tilde{t})}{a(t_0)} + \Gamma G\mu\,\tilde{t}\right]. \tag{8.89}$$

Note that the contribution coming from the higher modes are related to the contribution of the first mode by

$$\Omega_{\rm GW}^{(k)}(f) = k^{-4/3}\,\Omega_{\rm GW}^{(1)}(f/k). \tag{8.90}$$

8.B.5 The GW Spectrum from the Quadrupole Formula

In standard cosmology: The scaling behavior $\Omega_{\rm GW} \propto \sqrt{G\mu}\times f^0$, e.g. discussed along Eq. (8.56), can be understood qualitatively from the quadrupole formula for the power emission of GW [30, 130]

$$P_{\rm GW} \sim N_{\rm loop}\,\frac{G}{5}\left(Q_{\rm loop}^{'''}\right)^2, \tag{8.91}$$

where the triple derivative of the quadrupole of a loop is simply the string tension

$$Q'''_{\text{loop}} \sim \text{mass} \times \text{length}^2/\text{time}^3 \sim \mu. \tag{8.92}$$

During the scaling regime, the number of loops formed at time t_i scales as t_i^{-3}. Hence, the number of loops formed at time t_i, evaluated at a later time $\tilde{t}$ is

$$N_{\text{loop}} \sim \left(\frac{\tilde{t}}{t_i}\right)^3 \left(\frac{t_i}{\tilde{t}}\right)^{3/2}, \tag{8.93}$$

where the second factor accounts for the redshift as a^{-3} of the loops between t_i and $\tilde{t}$ during radiation. Since GW redshift as radiation, their energy density today is

$$\Omega_{\text{GW}} \sim \Omega_{\text{rad}} \frac{\rho_{\text{GW}}(\tilde{t})}{\rho_{\text{rad}}(\tilde{t})} \sim \Omega_{\text{rad}} (G\mu)^2 \left(\frac{\tilde{t}}{t_i}\right)^{3/2}, \tag{8.94}$$

where we assumed radiation-domination at $\tilde{t}$

$$\rho_{\text{rad}}(\tilde{t}) \sim G^{-1} \tilde{H}^2 \frac{\rho_{\text{rad}}(\tilde{t})}{\rho_{\text{tot}}(\tilde{t})} \sim G^{-1} \tilde{t}^{-2}, \tag{8.95}$$

and where we used that the energy density of GW at $\tilde{t}$ is

$$\rho_{\text{GW}}(\tilde{t}) \sim \left(P_{\text{GW}}\, \tilde{t}\right) / \tilde{t}^3. \tag{8.96}$$

with the GW power P_{GW} defined in Eq. (8.91). From Eq. (8.93), one can see that, at a fixed formation time t_i, the later the GW emission, the more numerous the loops. Hence, the dominant contribution to the SGWB from a given population of loops formed at t_i occurs after one loop-lifetime, cf. Eq. (8.24), at

$$\tilde{t}_{\text{M}} \sim \frac{\alpha\, t_i}{\Gamma G\mu}. \tag{8.97}$$

Upon plugging Eq. (8.97) into Eq. (8.94), one gets

$$\Omega_{\text{GW}} \propto \sqrt{G\mu} \times f^0. \tag{8.98}$$

From Eq. (8.94), we can see that the GW spectrum during radiation is set by a combination of the strength of the GW emission from loops, $(G\mu)^2$, and the loop-lifetime $\tilde{t}_{\text{M}}/t_i$, cf. Eq. (8.97). Both are set by the triple derivative of the loop-quadrupole $Q'''_{\text{loop}} \sim \mu$. Hence we understand that the flatness in frequency during radiation is closely linked to the independence of the triple derivative of the loop-quadrupole,[8] cf. Eq. (8.92), on the loop length, and therefore on the frequency.

In non-standard cosmology:
In presence of non-standard cosmology, possibly being different during loop formation $a(t_i) \propto t_i^{2/m}$ and GW emission $a(\tilde{t}) \propto \tilde{t}^{2/n}$, one must modify how the number of emitting loops, c.f. Eq. (8.93), and the GW energy density, c.f. Eq. (8.95), get redshifted. Namely, Eqs. (8.93) and (8.95), here denoted by Std., become

$$N_{\text{loop}} \sim \left(\frac{\tilde{t}}{t_i}\right)^3 \left(\frac{a(t_i)}{a(\tilde{t})}\right)^3 \propto N_{\text{loop}}^{\text{Std.}} \times \frac{t_i^{\left(\frac{6}{m}-\frac{3}{2}\right)}}{\tilde{t}^{\left(\frac{6}{n}-\frac{3}{2}\right)}}, \tag{8.99}$$

and

$$\frac{\rho_{\text{rad}}(\tilde{t})}{\rho_{\text{tot}}(\tilde{t})} \sim \left(\frac{a(\tilde{t})}{a(t_{\text{end}})}\right)^{n-4} \propto \tilde{t}^{\frac{2(n-4)}{n}}, \tag{8.100}$$

where t_{end} is the ending time of the non-standard cosmology. The GW spectrum, c.f. Eq. (8.94), depends on the combination of the loop- and GW-redshift factors in Eqs. (8.99) and (8.100). Upon plugging the scaling $t_i \propto f^{\frac{n}{2-n}}$ and $\tilde{t}_M \propto f^{\frac{n}{2-n}}$ (which are themselves deduced from $t_i \propto a(\tilde{t})/f$, c.f. Eqs. (8.81), and (8.97)), we obtain

$$\Omega_{\text{GW}} \propto f^{-2\left(\frac{3n+m(n-7)}{m(n-2)}\right)}. \tag{8.101}$$

When loop formation at t_i occurs during matter/kination but GW emission at $\tilde{t}_{\text{M}}$ occurs during radiation, $(m,\ n) = (3,\ 4)/(6,\ 4)$, we find that the GW spectrum scales like f^{-1}/f^1. In Sect. 8.B.6, we show that the presence of high-frequency modes $k \gg 1$ turns the f^{-1} behavior to $f^{-1/3}$.

8.B.6 *Impact of the High-Frequency Proper Modes of the Loop*

The motivation:
When computing the GW spectrum from cosmic strings, given by Eq. (8.88), we are confronted with an infinite sum over the proper modes k of the loop. The number of modes before we violate the Nambu-Goto approximation is very large, as shown in the left panel of Fig. 8.20. In what follows, we study the impact of the high-frequency modes on the GW spectrum. From Eqs. (8.81), (8.83) and (8.88), we can see that the spectrum for the kth mode is related to the fundamental spectrum $k = 1$ through Eq. (8.90), which we rewrite here

---

[8] In contrast, the GW spectrum generated by domain walls during radiation is not flat since in that case the triple derivative of the wall-quadrupole depends on the emission time: $Q_{\text{DW}}''' \sim \sigma\,\tilde{t}$, where σ is the wall energy per unit of area. Hence, the energy density fraction in GW before wall annihilation [275] increases with time $\Omega_{\text{GW}}^{\text{DW}} \sim (G\,\sigma\,\tilde{t})^2$.

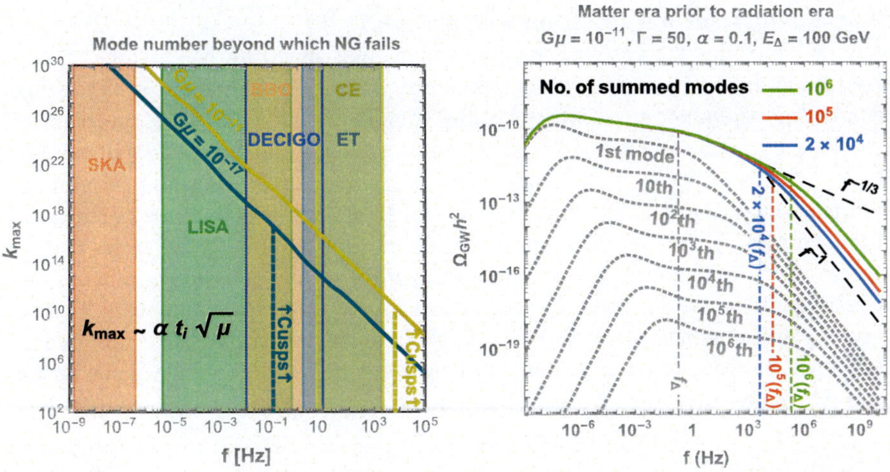

Fig. 8.20 **Left:** Maximal mode number $k_{\max}$ beyond which we can not trust the Nambu-Goto approximation anymore. It occurs when the wavelength of the oscillation, given by $\alpha\, t_i/k$ where t_i is the Hubble horizon when the loop forms, becomes of the order of the loop thickness $\mu^{-1/2}$. We can see that in the different interferometer windows, $k_{\max}$ is extremely large, often much larger than the maximal mode number tractable numerically $\sim 10^6$. **Right:** Decomposition of a GW spectrum under the contributions coming from the different proper modes of the loop. We can see that high-k modes are responsible for the change of slope $f^{-1/3} \rightarrow f^{-1}$ between the physical turning point frequency f_Δ and a second, artificial, turning-point $f_{\max}$, given by $f_{\max} = k_{\max}\, f_\Delta$, cf. Eq. (8.110), where $k_{\max}$ is the total number of modes chosen for doing the computation, here 2×10^4, 10^5 and 10^6. Except when explicitly specified, for technical reasons we fix $k_{\max} = 2 \times 10^4$ modes in all the plots of our study

$$\Omega_{\rm GW}^{(k)}(f) = k^{-\delta}\, \Omega_{\rm GW}^{(1)}(f/k), \tag{8.102}$$

In this study, we fix $\delta = 4/3$ since we assume that the small-scale structure is dominated by cusps. However, the results of the present section apply to any small-scale structure described by δ.

Case of a fundamental spectrum with a flat slope:
At first, if we assume that the one-mode spectrum is flat, $\Omega_{\rm GW}^{(1)}(f) \propto f^0$, then the total spectrum is a simple rescaling of the fundamental spectrum by the Riemann zeta function

$$\Omega_{\rm GW}(f) = \zeta\,(\delta)\; \Omega_{\rm GW}^{(1)}(f), \tag{8.103}$$

where in particular, $\zeta(4/3) = \sum_k k^{-4/3} \simeq 3.60$.

Case of a fundamental spectrum with a slope f^{-1}:
Now, we consider the case where the fundamental spectrum has a slope f^{-1}, as expected in thepresence of an early matter era, cf. Eq. (8.101), but also, in the presence of high-frequency cut-offs. The high-frequency cut-offs in the spectrum are described by Heaviside functions in the master formula in Eq. (8.88), of the type $\Theta(t_i - t_\Delta)$, where t_Δ is the cosmic time when loop formation starts, assuming it is suppressed before on. The time t_Δ can correspond to either the time of formation of the network, cf. Eq. (8.12), the time when friction-dominated dynamics become irrelevant, cf. Appendix 8.D.4, the time when gravitational emission dominates over massive particle production, cf. Sect. 8.3.1, or the time when the string correlation length re-enters the Hubble horizon after a short period of second inflation, cf. Sect. 8.7. The slope of the $k = 1$ spectrum beyond the cut-off frequency can be read from Eq. (8.88) after injecting Eq. (8.89) and $t_i = t_\Delta$, where we find

$$\Omega^{(1)}_{\rm GW}(f) = \Omega_\Delta \Theta\left(-f + f_\Delta\right) + \Omega_\Delta \frac{f_\Delta}{f} \Theta\left(f - f_\Delta\right). \tag{8.104}$$

The fundamental spectrum is flat until f_Δ and then shows a slope f^{-1} beyond. The total spectrum, summed over all the proper modes, can be obtained from Eqs. (8.102) and (8.104)

$$\Omega_{\rm GW}(f) = \sum_{k=1}^{k_\Delta} \frac{\Omega_\Delta}{k^\delta} k \frac{f_\Delta}{f} + \sum_{k=k_\Delta}^{k_{\rm max}} \frac{\Omega_\Delta}{k^\delta}, \tag{8.105}$$

where $k_{\rm max}$ is the maximal mode, chosen arbitrarily, and k_Δ is the critical mode defined such that modes with $k < k_\Delta$ have a slope f^{-1} whereas modes with $k > k_\Delta$ have a flat slope. For a given frequency f, the critical mode number k_Δ obeys

$$k_\Delta \simeq \frac{f}{f_\Delta}. \tag{8.106}$$

We now evaluate Eq. (8.105) in the large k_Δ limit, while still keeping $k_\Delta < k_{\rm max}$

$$\Omega_{\rm GW}^{1 \ll k_\Delta < k_{\rm max}}(f) \simeq \Omega_\Delta \frac{f_\Delta}{f} k_\Delta^{2-\delta} + \frac{1}{\delta - 1} \frac{\Omega_\Delta}{k_\Delta^{\delta-1}}, \tag{8.107}$$

where we have used the asymptotic expansion of the Euler-Maclaurin formula for the first term and the asymptotic expansion of the Hurwitz zeta function for the second term. Finally, after injecting Eq. (8.106), we get

$$\Omega_{\rm GW}^{1 \ll k_\Delta < k_{\rm max}}(f) \simeq \frac{\delta}{\delta - 1} \Omega_\Delta \left(\frac{f_\Delta}{f}\right)^{\delta - 1} \propto \begin{cases} f^{-1/3} & \text{for cusps}(\delta = 4/3) \\ f^{-2/3} & \text{kinks}(\delta = 5/3) \\ f^{-1} & \text{kink-kink collisions}(\delta = 2) \end{cases} \tag{8.108}$$

We conclude that the spectral index beyond a high-frequency turning point f_Δ due to an early matter era, a second inflation era, particle production, thermal friction domination, or the formation of the network, is modified by the presence of the high-k modes in a way that depends on the small-scale structure. Particularly, if the small-scale structure is dominated by cusps, we find a slope $-1/3$. We comment on the possibility to get information about the nature of the small-structure from detecting a GW spectrum from CS with a decreasing slope. The study [116] was the first one to point out the impact of the high-frequency modes on the value of a decreasing slope.

Impact of fixing the total number of proper modes:
For technical reasons we are unavoidably forced to choose a maximal number of modes $k_{\rm max}$. We now study the dependence of the GW spectrum on the choice of $k_{\rm max}$. The evaluation of Eq. (8.105) for $k_\Delta > k_{\rm max}$ leads to

$$\Omega_{\rm GW}^{1 \ll k_{\rm max} < k_\Delta}(f) = \zeta\,(\delta - 1)\,\Omega_\Delta\,\frac{f_\Delta}{f}. \tag{8.109}$$

Hence, in addition to the initial physical turning point f_Δ, where the slope changes from flat to $f^{-1/3}$, there is a second artificial turning point $f_{\rm max}$ given by

$$f_{\rm max} = k_{\rm max}\, f_\Delta, \tag{8.110}$$

where the slope changes from $f^{-1/3}$ to f^{-1}. We show the different behaviors in the right panel of Fig. 8.20.

Case of a fundamental spectrum with a slope f^{+1}:
As last, we comment on the case where the fundamental spectrum has a slope f^1, as in the case of an early kination era, cf. Eq. (8.101). Repeating the same steps as in Eq. (8.105), we obtain

$$\Omega_{\rm GW}(f) = \zeta\,(\delta + 1)\,\Omega_\Delta\,\frac{f}{f_\Delta}, \tag{8.111}$$

hence the slope of the full spectrum is the same as the slope of the fundamental spectrum.

8.C Derivation of the Frequency—Temperature Relation

In this appendix, we compute the correspondence between an observed frequency f and the temperature T of the universe when the loops responsible for that frequency have been formed.

8.C.1 In Standard Cosmology

According to the scaling of the loop-formation rate $dn/dt_i \propto t_i^{-4}$, the main contribution to the GW emission at time $\tilde{t}$ comes from the loops created at the earliest epoch. Correspondingly, loops created at t_i contribute to the spectrum as late as possible, at the *main emission time* $\tilde{t}_{\rm M}$. The latest emission time is set by the loop lifetime $\alpha\, t_i/\Gamma G\mu$, where α is the loop-length at formation in horizon unit, cf. Eq. (8.24). Hence, a loop produced at time t_i mainly contributes to the spectrum, much later cf. Fig. 8.21, at a time

$$\tilde{t}_{\rm M} \simeq \frac{\alpha\, t_i}{2\Gamma G\mu}, \tag{8.112}$$

where the factor $1/2$ is found upon maximizing the loop-formation rate $dn/dt_i \propto t_i^{-4}$ and upon assuming $\alpha \gg \Gamma G\mu$. The loop length after half the loop lifetime, in Eq. (8.112), is equal to half the length at formation $\alpha\, t_i/2$, cf. Eq. (8.80). Hence the emitted frequency is set by

$$\alpha\, t_i \simeq \frac{4}{f}\frac{a(\tilde{t}_M)}{a(t_0)}, \tag{8.113}$$

$$\simeq \frac{4}{f}\frac{a(\tilde{t}_M)}{a(t_{\rm eq})}\frac{a(t_{\rm eq})}{a(t_0)} \tag{8.114}$$

$$\simeq \frac{4}{f}\left(\frac{\tilde{t}_M}{t_{\rm eq}}\right)^{1/2}\left(\frac{t_{\rm eq}}{t_0}\right)^{2/3} \tag{8.115}$$

where we used $f\, a(t_0)/a(\tilde{t}) = 2k/l$ and only considered the first Fourier mode $k = 1$, cf. Eq. (8.20). By merging Eqs. (8.112) and (8.115), we obtain the relation between an observed frequency f and the time t_i of loop formation

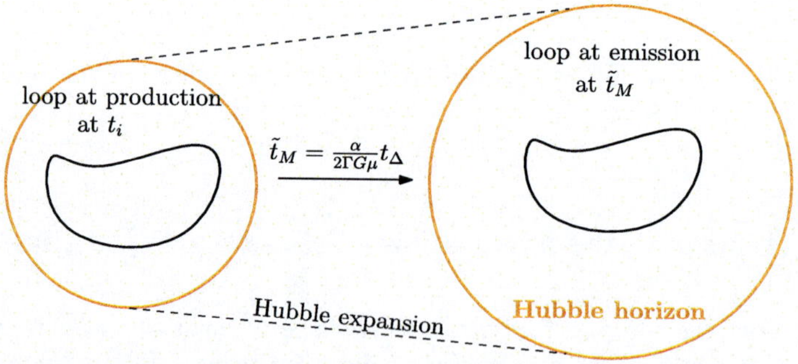

Fig. 8.21 Loops produced at time t_i contribute to the GW spectrum much later, when they have accomplished half of their lifetime, at $\tilde{t}_{\rm M} \simeq \alpha\, t_i/(2\Gamma G\mu)$. Hence GW emitted from cosmic-string loops are exempt from a redshift factor $a(\tilde{t}_{\rm M})/a(t_i)$ so have much higher frequency than GW produced from other sources at the same energy scale

$$f \simeq \sqrt{\frac{8 z_{\rm eq}}{\alpha \Gamma G\mu}} \left(\frac{t_{\rm eq}}{t_i}\right)^{1/2} t_0^{-1}, \tag{8.116}$$

where the redshift at matter-radiation equality is $z_{\rm eq} = \Omega_C/\Omega_\gamma \simeq 3360$, and $t_{\rm eq} \simeq$ 51.8 kyrs (from integrating Eq. (8.53)) and $t_0 \simeq 13.8$ Gyrs [202]. Finally, using entropy conservation, we obtain the relation between the frequency f at observation and the temperature T of the universe when the corresponding loops are formed

$$\begin{aligned} f &\simeq \sqrt{\frac{8}{z_{\rm eq}\alpha \Gamma G\mu}} \left(\frac{g_*(T)}{g_*(T_0)}\right)^{1/4} \left(\frac{T}{T_0}\right) t_0^{-1} \\ &\simeq (6.7 \times 10^{-2}\ {\rm Hz}) \left(\frac{T}{\rm GeV}\right) \left(\frac{0.1 \times 50 \times 10^{-11}}{\alpha \Gamma G\mu}\right)^{1/2} \left(\frac{g_*(T)}{g_*(T_0)}\right)^{1/4}. \end{aligned} \tag{8.117}$$

8.C.2 *During a Change of Cosmology*

The derivation of (8.117) does not take into account the time-variation of $C_{\rm eff}$. It assumes that loops are produced and decayed during the scaling regime in the radiation era. An observable to test the non-standard cosmology is the frequency f_Δ of the *turning-point* defined as the frequency at which the GW spectrum starts to deviate from the standard-cosmology behavior and the spectral index changes. We obtain different fitted values for this turning point frequency depending on the prescription. We quote below different expressions, depending whether we assume that the spectrum can be measured with a 10% precision, and 1% respectively. We compare the predictions obtained using a scaling and VOS network:

$$f_\Delta \simeq {\rm Hz} \left(\frac{T_\Delta}{\rm GeV}\right) \left(\frac{0.1 \times 50 \times 10^{-11}}{\alpha \Gamma G\mu}\right)^{1/2} \left(\frac{g_*(T_\Delta)}{g_*(T_0)}\right)^{1/4} \times \begin{cases} 2 \times 10^{-3} & \text{for VOS, 10\%} \\ 45 \times 10^{-3} & \text{for scaling, 10\%} \\ 0.04 \times 10^{-3} & \text{for VOS, 1\%} \\ 15 \times 10^{-3} & \text{for scaling, 1\%} \end{cases} \tag{8.118}$$

Therefore, the turning point frequency is lower in VOS than in scaling by a factor $\sim$22.5 if we define the turning-point frequency by an amplitude deviation of 10% with respect to standard cosmology, and by a factor $\sim$375 for a deviation of 1%. The loops contributing to this part of the spectrum have been formed at the time of the change of cosmology. When the cosmology changes, the network achieves a transient evolution in order to reach the new scaling regime. The long-string network needs extra time to transit from one scaling regime to the other, hence the shift in the relation between observed frequency and temperature of loop formation at the turning-point, cf. Sect. 8.6.2.

8.C.3 *In the Presence of an Intermediate Inflation Period*

The above derivation of the relation between the observed frequency and the time of loop production assumes that cosmic-string loops are constantly being produced throughout the cosmic history. It does not apply if the network experiences an intermediate era of inflation. This case is discussed in Sect. 8.7.2 and the turning-point formulae are, for a given precision

$$f_\Delta \simeq \text{Hz}\left(\frac{T_\Delta}{\text{GeV}}\right)\left(\frac{0.1\times 50\times 10^{-11}}{\alpha\Gamma G\mu}\right)^{1/2}\left(\frac{g_*(T_\Delta)}{g_*(T_0)}\right)^{1/4}\times\begin{cases}1.5\times 10^{-4} & \text{for } 10\% \\ 5\times 10^{-6} & \text{for } 1\%\end{cases} \tag{8.119}$$

8.C.4 *Cut-Off from Particle Production*

The cutoff frequency due to particle production is given in Sect. 8.3.4.

8.D Derivation of the VOS Equations

8.D.1 *The Nambu-Goto String in an Expanding Universe*

The Velocity-dependent One-Scale equations (VOS) in Eq. (8.47), describe the evolution of a network of long strings in term of the mean velocity $\bar{v}$ and the correlation length $\xi = L/t$, see the original papers [197, 199, 200] or the recent review [201]. The set of points visited by the Nambu-Goto string during its time evolution form a 2D manifold, called the world-sheet, described by time-like and space-like coordinates $\mathtt{t}$ and σ. The embedding of the 2D world-sheet in the 4D space-time is described by $x^\mu(\mathtt{t},\sigma)$ where $\mu = 1, 2, 3, 4$. The choice of the word-sheet coordinates being arbitrary, we have two gauge degrees of freedom which we can fix by imposing $\dot{\mathbf{x}}\cdot\mathbf{x}' = 0$ and $\mathtt{t} = \tau$ where τ is the conformal time of the expanding universe. The dot and prime denote the derivatives with respect to the time-like and space-like world-sheet coordinates, $\dot{\mathbf{x}} \equiv d\mathbf{x}/d\mathtt{t}$ and $\mathbf{x}' \equiv d\mathbf{x}/d\sigma$. Then, the equations of motion of the Nambu-Goto string in a FRW universe are [276]

$$\ddot{\mathbf{x}} + 2\mathcal{H}(1-\dot{\mathbf{x}}^2)\dot{\mathbf{x}} = \frac{1}{\varepsilon}\left(\frac{\mathbf{x}'}{\varepsilon}\right)', \tag{8.120}$$

$$\dot{\varepsilon} + 2\mathcal{H}\,\dot{\mathbf{x}}^2\,\varepsilon = 0, \tag{8.121}$$

where $\mathcal{H} \equiv \dot{a}/a = Ha$ and $\varepsilon \equiv \sqrt{\mathbf{x}'^2/(1-\dot{\mathbf{x}}^2)}$ is the coordinate energy per unit of length.

8.D.2 The Long-String Network

The macroscopic evolution of the long string network can be described by the energy density

$$\rho_\infty = \frac{E}{a^3} = \frac{\mu}{a^2(\tau)} \int \varepsilon \, d\sigma \equiv \frac{\mu}{L^2}, \tag{8.122}$$

and the root-mean-square averaged velocity

$$\bar{v}^2 \equiv \langle \dot{\mathbf{x}}^2 \rangle = \frac{\int \dot{\mathbf{x}}^2 \varepsilon \, d\sigma}{\int \varepsilon \, d\sigma}, \tag{8.123}$$

where we recall that μ is the CS linear mass density.

8.D.3 VOS 1: The Correlation Length

Differentiating Eq. (8.122) gives the evolution of the energy density in an expanding universe

$$\frac{d\rho_\infty}{dt} = \frac{d\rho_\infty}{d\tau} \cdot \frac{d\tau}{dt} = \frac{1}{a} \cdot \frac{d\rho_\infty}{d\tau}, \tag{8.124}$$

$$= \frac{\mu}{a}\left[\frac{d}{d\tau}\left(\frac{1}{a^2}\right)\int \varepsilon \, d\sigma + \frac{1}{a^2}\int \frac{d\varepsilon}{d\tau} d\sigma\right], \tag{8.125}$$

$$= -2\frac{\mu}{a^3}\mathcal{H}\left[\int \varepsilon \, d\sigma + \int \dot{\mathbf{x}}^2 \varepsilon \, d\sigma\right], \tag{8.126}$$

$$= -2H\rho_\infty \, (1 + \bar{v}^2). \tag{8.127}$$

Moreover, after each string crossing, the network transfers energy into loops with a rate given by Eq. (8.45) and we get

$$\frac{d\rho_\infty}{dt} = -2H\rho_\infty \, (1 + \bar{v}^2) - \tilde{c}\,\bar{v}\frac{\rho_\infty}{L}, \tag{8.128}$$

which after using Eq. (8.122), leads to the first VOS equation

$$\text{VOS 1:} \quad \frac{dL}{dt} = HL\,(1 + \bar{v}^2) + \frac{1}{2}\tilde{c}\,\bar{v}. \tag{8.129}$$

We neglect the back-reaction on long strings from gravitational emission which is suppressed with respect to the loop-chopping loss term by $O(G\mu)$. The case of global strings, for which however, the back-reaction due to particle production may play a role, is considered in Appendix 8.F.2.

8.D.4 Thermal Friction

In addition to the Hubble friction, there can be friction due to the interactions of the strings with particles of the plasma, leading to the retarding force [277]

$$F = \rho \sigma \bar{v} = \beta T^3 \bar{v}, \tag{8.130}$$

where $\rho \sim T^4$ is the plasma energy density and $\sigma \sim T^{-1}$ is the cross-section per unit of length. The effect of friction is to damp the string motion and to suppress the GW spectrum [278]. For gauge strings, a well-known realisation of friction is the interaction of charged particles with the pure gauge fields existing outside the string, the so-called Aharonov-Bohm effect [279]. In such a case, the friction coefficient β is given by [277]

$$\beta = 2\pi^{-2}\zeta(3) \sum_i g_i \sin^2(\pi \nu_i), \tag{8.131}$$

with

i	relativistic particle species ($m_i \ll T$),
g_i	number of relativistic degrees of freedom of $i \times \begin{cases} 3/4 \text{ (fermion)}, \\ 1 \text{ (boson)}, \end{cases}$
$2\pi\, \nu_i \equiv e_i\, \Phi$	phase-shift of the wave-function of particle i when transported on a close path around the string. e_i being its charge under the associated gauge group and Φ the magnetic field flux along the string.

The friction term in the first VOS equation, Eq. (8.129), becomes

$$2H\bar{v}^2 \longrightarrow \frac{\bar{v}^2}{l_d} \equiv 2H\bar{v}^2 + \frac{\bar{v}^2}{l_f}, \tag{8.132}$$

where we introduced a friction length due to particle scattering $l_f \equiv \mu/(\sigma\,\rho) = \mu/(\beta T^3)$ [199, 200], and the associated effective friction length l_d. At large temperature, the large damping due to the frictional force prevents the CS network to reach the scaling regime until it becomes sub-dominant when $2H \lesssim 1/l_f$, so after the time

$$t_{\rm fric} \simeq (2.5 \times 10^{-5}) \left(\frac{106.75}{g_*(t_{\rm fric})}\right)^{3/2} \beta^2\, (G\mu)^{-2}\, t_{\rm pl}, \tag{8.133}$$

$$\simeq (1.4 \times 10^{-4}) \left(\frac{g_*(t_{\rm F})}{106.65}\right)^{1/2} \left(\frac{106.75}{g_*(t_{\rm fric})}\right)^{3/2} \beta^2\, (G\mu)^{-1}\, t_F, \tag{8.134}$$

where $t_{\rm pl} \equiv \sqrt{G}$ and where the network formation time t_F is the cosmic time when the energy scale of the universe is equal to the string tension $\rho_{\rm tot}^{1/2}(t_F) \equiv \mu$. For friction coefficient $\beta = 1$, the friction becomes negligible at the temperatures $T_* \simeq 4\,$TeV for

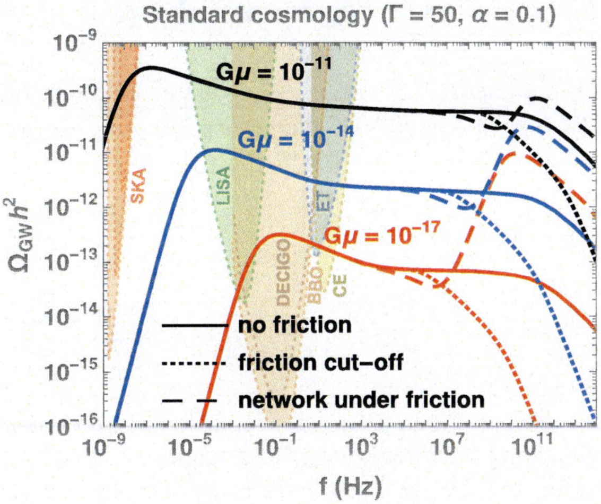

Fig. 8.22 GW spectrum from CS assuming no thermal friction (solid lines), thermal friction only at the level of the long-string network, i.e. upon including Eq. (8.132) in the VOS equations (dashed lines) or thermal friction taken at the loop level, i.e. by removing GW emissions anterior to t_{fric} defined in Eq. (8.133) (dotted lines). See text for more details. A standard cosmology is assumed

$G\mu = 10^{-17}$, $T_* \simeq 400$ TeV for $G\mu = 10^{-15}$, $T_* \simeq 40$ PeV for $G\mu = 10^{-13}$, hence respectively impacting the SGWB only above the frequencies 20 kHz, 200 kHz, 2 MHz, cf. Eq. (8.60), which are outside the GW interferometer windows, cf. Fig. 8.6.

In Fig. 8.22, we show the impact of thermal friction on the GW spectrum from CS in two different ways.

- **Network under friction** *(dashed lines in Fig. 8.22)*: the thermal friction is only taken into account at the level of the long-string network. Concretely, by simply including the friction term in Eq. (8.132) in the VOS equations. The GW peak at high frequency is due to the loop over-production by the frozen network, followed by a fast relaxation (with a little oscillatory behavior) to the scaling regime once friction becomes negligible with respect to Hubble expansion. This approach is insufficient since it assumes that the GW power emitted by loops is still given by $\Gamma G\mu^2$ with $\Gamma \simeq 50$ and therefore it does not take into account the damping of the oscillations of the loops under which we expect $\Gamma \to 0$.
- **GW emission cut-off** *(dotted lines in Fig. 8.22)*: The damping of the loop oscillations is now taken into account by discarding all GW emissions happening earlier than t_{fric} in Eq. (8.133), when thermal friction is larger than Hubble friction. Technically, the time t_{osc} of first loop oscillations in Eq. (8.153) is set equal to t_{fric} in Eq. (8.133).

In many of our plots, e.g. Figs. 8.4 or 8.6, we show the GW spectrum in the presence of thermal friction with a gray line, computed according to the second prescription above, entitled 'GW emission cut-off'. Note that in most cases, the effect of friction manifests itself at very high frequencies, outside the observability band of planned interferometers. It could however become relevant if those high frequencies could be probed in future experiments.

8.D.5 VOS 2: The Mean Velocity

Differentiating Eq. (8.123) gives the evolution of the averaged velocity, which constitutes the second VOS equation

$$\text{VOS 2:} \qquad \frac{d\bar{v}}{dt} = (1-\bar{v}^2)\left[\frac{k(\bar{v})}{L} - \frac{\bar{v}}{l_d}\right], \tag{8.135}$$

with

$$k(\bar{v}) \equiv \frac{\langle (1-\dot{\mathbf{X}}^2)\,(\dot{\mathbf{X}}\cdot\mathbf{u})\rangle}{\bar{v}\,(1-\bar{v}^2)}, \tag{8.136}$$

where $\mathbf{u}$ is the unit vector aligned with the radius of curvature $\propto d^2\mathbf{x}/d\sigma^2$. $k(\bar{v})$ indicates the degree of wiggliness of the string. More precisely, $k(\bar{v}) = 1$ for a straight string and $k(\bar{v}) \lesssim 1$ once we add small-scale structures. We use the results from numerical simulations [197]

$$k(\bar{v}) = \frac{2\sqrt{2}}{\pi}(1-\bar{v}^2)(1+2\sqrt{2}\bar{v}^3)\frac{1-8\bar{v}^6}{1+8\bar{v}^6}. \tag{8.137}$$

Equation (8.135) is a relativistic generalization of Newton's law where the string is accelerated by its curvature $1/L$ but is damped by the Hubble expansion and plasma friction after a typical length $1/l_d$. The Eq. (8.135) neglects the change in long string velocity $\bar{v}$ due to loop formation as proposed in [280].

8.E Extension of the Original VOS Model

8.E.1 VOS Model from Nambu-Goto Simulations

In our study, we describe the evolution of the long-string network through the VOS model, defined by the equations in Eq. (8.47). The only free parameter of the model is the loop-chopping efficiency $\tilde{c}$, which is computed to be

$$\textbf{NG:} \qquad \tilde{c} = 0.23 \pm 0.04 \tag{8.138}$$

from Nambu-Goto network simulations in an expanding universe [197].

8.E.2 VOS Model from Abelian-Higgs Simulations

Abelian-Higgs (AH) field theory simulations in both expanding and flat spacetime suggest a larger value [155, 281]

$$\textbf{AH:} \qquad \tilde{c} = 0.57 \pm 0.04. \tag{8.139}$$

Indeed, in Abelian-Higgs simulations, no loops are produced below the string core size so the energy loss into loop formation is lower. Consequently, the loop-chopping efficiency must be increased to maintain scaling.

8.E.3 VOS Model from Abelian-Higgs Simulations with Particle Production

In Abelian-Higgs simulations, the loops produced at the string core scale are non-linear lumps of field, called "proto-loops", which decay fast into massive radiation. Therefore, a recent work [198] extends the VOS model by including a term in Eq. (8.129) to account for the emission of massive radiation at the string core scale. The energy-loss function $F(v)$ is modified as

$$F(\bar{v})|_{\text{original}} = \frac{\tilde{c}\bar{v}}{2} \quad \Rightarrow \quad F(\bar{v})|_{\text{extended}} = \frac{\tilde{c}\bar{v} + d[k_0 - k(\bar{v})]^r}{2}, \tag{8.140}$$

and the momentum operator $k(v)$, cf. Eq. (8.137), accounting for the amount of small-scale structures in the string, is modified to

$$k(\bar{v}) = \frac{2\sqrt{2}}{\pi}\frac{1-8\bar{v}^6}{1+8\bar{v}^6} \quad \Rightarrow \quad k(\bar{v})|_{\text{extended}} = k_0\frac{1-(q\bar{v}^2)^\beta}{1+(q\bar{v}^2)^\beta}, \tag{8.141}$$

where more free parameters have been introduced. With Abelian-Higgs simulations, one finds [198]

$$\textbf{AH extended:} \qquad \tilde{c} = 0.31, \tag{8.142}$$

as well as $d = 0.26, k_0 = 1.27, r = 1.66, q = 2.27$, and $\beta = 1.54$. In Abelian-Higgs extended, the loop-chopping efficiency, cf. Eq. (8.142), is smaller than the one in the original Abelian-Higgs model, cf. Eq. (8.139). Indeed, because of the additional energy loss through massive-radiation, less energy loss via loop-chopping is needed to maintain scaling.

In Fig. 8.23, we compare the GW spectra in the different VOS models. The difference in amplitude comes from the difference in the number of loops, set by C_{eff}. The larger the loop-chopping efficiency $\tilde{c}$, the smaller the loop-formation efficiency C_{eff}. This counter-intuitive result can be better understood by looking at Table 8.1. A

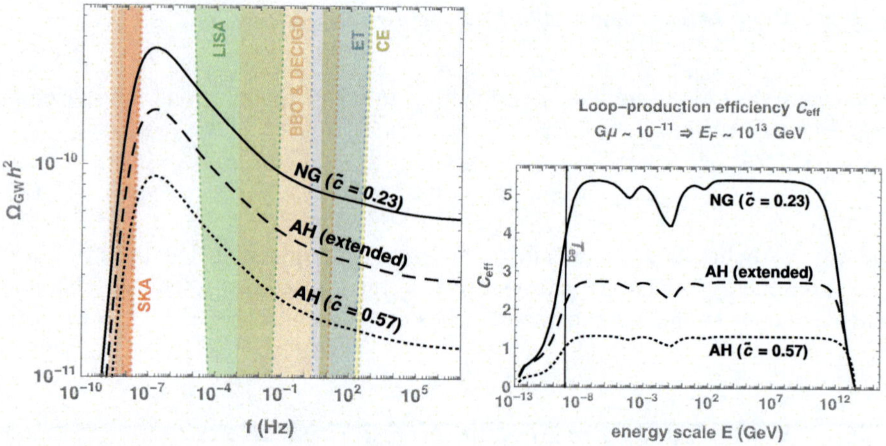

Fig. 8.23 **Left:** GW spectra with different VOS modellings of the long-string network evolution. The VOS models are either based on Nambu-Goto simulations (solid line—$\tilde{c} = 0.23$) [197] or abelian-Higgs (AH) field theory simulations (dashed line—$\tilde{c} = 0.57$) [155, 281], possibly extended to include particle production [198] (dotted line). **Right:** The corresponding loop-production efficiency for each VOS model

Table 8.1 Values of mean velocity $\bar{v}$, correlation length ξ, and loop-production efficiency $C_{\rm eff}$ in radiation scaling regime with different VOS calibrations

Scaling in radiation dominated universe	NG $\tilde{c} = 0.23$	AH $\tilde{c} = 0.57$	AH extended $\tilde{c} = 0.31$ $(d,\ k_0,\ r,\ q,\ \beta)$
$\bar{v}$	0.66	0.62	0.59
ξ	0.27	0.57	0.36
$C_{\rm eff}$	5.4	1.3	2.8

larger loop-chopping efficiency $\tilde{c}$ implies a larger loop formation rate only during the transient regime. In the scaling regime, a larger loop-chopping efficiency $\tilde{c}$ implies a more depleted long-string network and then a larger correlation length ξ. Hence, the long-string network is more sparse and so the rate of loop formation via string crossing is lower.

8.F GW Spectrum from Global Strings

The main distinction with loops from global strings is that they are short-lived whereas loops from local strings are long-lived. This results in different GW spectra in both frequency and amplitude as we discuss in detail below.

8.F.1 The Presence of a Massless Mode

For global string, the absence of gauge field implies the existence of a massless Goldstone mode, with logarithmically-divergent gradient energy, hence leading to the tension μ_g, cf. Eq. (8.6)

$$\mu_g \equiv \mu_l \ln\left(\frac{H^{-1}}{\delta}\right) \simeq \mu_l \ln(\eta t), \quad \text{with} \quad \mu_l \equiv 2\pi\eta^2, \tag{8.143}$$

where η is the scalar field VEV, μ_l is the tension of the would-be local string (when the gauge coupling is switched on) and $\delta \sim \eta^{-1}$ is the string thickness. Goldstones are efficiently produced by loop dynamics with the power

$$P_{\text{Gold}} = \Gamma_{\text{Gold}}\, \eta^2, \tag{8.144}$$

where $\Gamma_{\text{Gold}} \approx 65$ [27, 90], causing loops to decay with a rate

$$\frac{dl_g}{dt} = \frac{dE}{dt}\frac{dl}{dE} \equiv \kappa \equiv \frac{\Gamma_{\text{Gold}}}{2\pi\, \ln(\eta t)}. \tag{8.145}$$

Therefore, the string length evolving upon both GW and Goldstone bosons emission reads

$$l_g(t) = \alpha t_i - \Gamma G\mu_g(t - t_i) - \kappa(t - t_i). \tag{8.146}$$

8.F.2 Evolution of the Global Network

The Velocity-One-Scale equations, presented in Appendix 8.D,

$$\frac{dL}{dt} = HL\,(1 + \bar{v}^2) + F(\bar{v})|_{\text{global}}, \tag{8.147}$$

$$\frac{d\bar{v}}{dt} = (1 - \bar{v}^2)\left[\frac{k(\bar{v})}{L} - \frac{\bar{v}}{l_d}\right], \tag{8.148}$$

are modified to include the additional energy-loss due to Goldstone production. Namely, the energy-loss coefficient $F(\bar{v})$, cf. Eq. (8.140), becomes

$$F(\bar{v})|_{\text{local}} = \frac{\tilde{c}\bar{v} + d[k_0 - k(\bar{v})]^r}{2} \quad \Rightarrow \quad F(\bar{v})|_{\text{global}} = F(\bar{v})|_{\text{local}} + \frac{s\, v^6}{2\ln(\eta t)}, \tag{8.149}$$

where the constant s controlling the efficiency of the Goldstone production, is inferred from lattice simulations [282], to be $s \simeq 70$ [109]. However, the momentum operator $k(v)$ in Eq. (8.148), is unchanged with respect to the local case

$$k(\bar{v}) = k_0 \frac{1-(q\bar{v}^2)^\beta}{1+(q\bar{v}^2)^\beta}. \tag{8.150}$$

with $k_0 = 1.37$, $q = 2.3$, $\beta = 1.5$, $\tilde{c} = 0.34$, $d = 0.22$, $r = 1.8$ [198]. Here through Eqs. (8.149) and (8.150), we follow [283] and consider the extended VOS model based on Abelian-Higgs simulations, proposed in [198] and already discussed in Appendix 8.E.3, in which we have simply added the backreaction of Goldstone production on long strings in Eq. (8.149). We have checked that we can neglect the thermal friction due to the contact interaction of the particles in the plasma with the string, cf. Eq. (8.132) for which the interaction cross-section is given by the Everett formula in [72].

In order to later compute the GW spectrum, we defined the loop-formation efficiency, analog of the local case in Eq. (8.46)

$$C_{\rm eff}^{\rm g} = \tilde{c}\,\bar{v}/\xi^3, \tag{8.151}$$

with $\bar{v}$ and $\xi \equiv L/t$ obeying the VOS equations in Eqs. (8.147) and (8.148). Due to the logarithmic dependence of the string tension on the cosmic time, the scaling regime is slightly violated. Consequently, the loop-formation efficiency plotted in right panel of Fig. 8.25, never reaches a constant value. Hence, in this study we model the network based on VOS evolution, rather than using the scaling solutions. Only for enormous value of $\ln(\eta\, t)$ corresponding to cosmic times much larger than the age of the Universe today, we find that the solutions to the modified VOS equations in Eqs. (8.147) and (8.148) reach a scaling regime $C_{\rm eff} = 0.46, 2.24, 6.70$ for matter-, radiation-, and kination-dominated universe, respectively. By comparing to the values found in [27], our results agree only for the radiation case. Note that the dependence of the string network parameters $\bar{v}$ and $\xi \equiv L/t$ on the logarithmically-time-dependent string tension arises only through the term of Goldstone production energy loss in Eq. (8.149).

8.F.3 The GW Spectrum

Finally, the GW spectrum generated by loops of global strings is given by a similar expression as for the local case, cf. the mattress equation in Eq. (8.38),

$$\Omega_{\rm GW}^{\rm g}(f) \equiv \frac{f}{\rho_c}\left|\frac{d\rho_{\rm GW}^{\rm g}}{df}\right| = \sum_k \Omega_{\rm GW}^{(k),\,{\rm g}}(f), \tag{8.152}$$

where

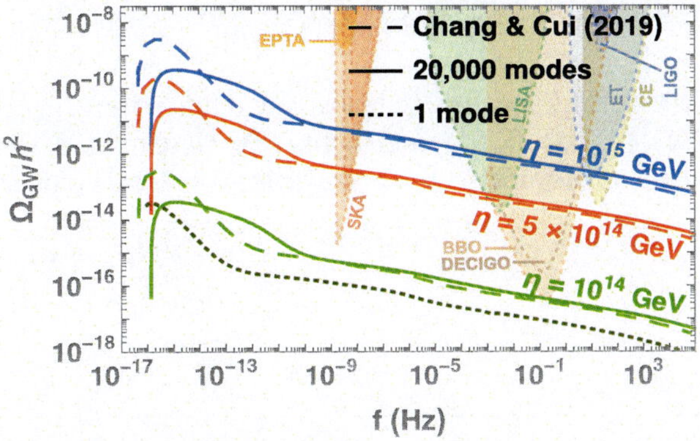

Fig. 8.24 GW spectrum from global cosmic strings assuming VOS network formed at energy scale η, evolved in the standard cosmology. The shape of the GW spectrum computed in [27] resembles the first mode $k = 1$ of our spectrum

$$\Omega_{\rm GW}^{(k),\,\rm g}(f) = \frac{1}{\rho_c} \cdot \frac{2k}{f} \cdot \frac{\mathcal{F}_\alpha\, \Gamma^{(k)} G\mu_{\rm g}^2}{\alpha(\alpha + \Gamma G\mu_{\rm g} + \kappa)} \int_{t_F}^{t_0} d\tilde{t}\; \frac{C_{\rm eff}^{\rm g}(t_i^{\rm g})}{(t_i^{\rm g})^4} \left[\frac{a(\tilde{t})}{a(t_0)}\right]^5 \left[\frac{a(t_i^{\rm g})}{a(\tilde{t})}\right]^3 \Theta(t_i^{\rm g} - t_F). \tag{8.153}$$

We have checked that we can safely neglect massive radiation. The loop formation time $t_i^{\rm g}$ is related to the emission time $\tilde{t}$ after using $l_{\rm g}(\tilde{t}) = 0$ in Eq. (8.146)

$$t_i^{\rm g} = \frac{\Gamma G\mu_{\rm g} + \kappa}{\alpha + \Gamma G\mu_{\rm g} + \kappa}\tilde{t}. \tag{8.154}$$

We plot the GW spectrum from local strings in Fig. 8.24 and compare to the spectrum computed in [27]. With respect to [27], we find a lower value for $C_{\rm eff}$ during the late matter-dominated universe (0.46 instead of 1.32) which implies a smaller spectral bump, while the radiation contributions are considerably the same. Moreover, the shapes of the spectra are different. An explanation could be the summation over the high-frequency modes (up to $k = 2 \times 10^4$ in our case) which smoothens the spectrum. The spectrum from [27] resembles to the first-mode of our spectrum.

The constraints on the inflation scale, 6×10^{13} GeV, from the non-detection of the fundamental B-mode polarization patterns in the CMB [284, 285], implies the upper bound $T_{\rm reh} \lesssim 5 \times 10^{16}$ GeV on the reheating temperature, assuming instantaneous reheating. Hence, assuming that the network is generated from a thermal phase transition, we are invited to impose $\eta \lesssim 5 \times 10^{16}$. However, a stronger restriction arises because of the CMB constraint on strings tensions

$$G\mu_{\rm g}\big|_{\rm CMB} = 2\pi \left(\frac{\eta}{m_{\rm pl}}\right)^2 \log(\eta\, t_{\rm CMB}) \lesssim 10^{-7} \quad \rightarrow \quad \eta \lesssim 1.4 \times 10^{15}, \tag{8.155}$$

where we use $t_{\rm CMB} \simeq 374$ kyr. Hence, we restrict to $\eta \lesssim 10^{15}$ GeV as in [27].

8.F.4 Global Versus Local Strings

The local spectrum is recovered upon setting $\ln(\eta\, t) = 1$, $\kappa = 0$ and $s = 0$ in Eqs. (8.143), (8.146) and (8.149). Because of the faster decay, in the global case, the main emission time $\tilde{t}_M^g$ is shorter than its local counterpart $\tilde{t}_M^l$, cf. Eq. (8.112),

$$\tilde{t}_M^g \simeq \frac{\alpha + \Gamma G\mu_g + \kappa}{\Gamma G\mu_g + \kappa}\frac{t_i}{2}, \tag{8.156}$$

$$\sim \frac{\Gamma G\mu_l}{\alpha}\,\tilde{t}_M^l, \tag{8.157}$$

where in the last line, we have assumed $\kappa \gg \alpha, \tilde{t}_M^l$. Therefore, the frequency f of a GW emitted by a loop produced at the temperature T, is lowered compared to the local case computed in Eq. (8.117), by a factor

$$f\Big|_{\text{global}} \simeq (4.7 \times 10^{-6}\,\text{Hz})\left(\frac{T}{\text{GeV}}\right)\left(\frac{0.1}{\alpha}\right)\left(\frac{g_*(T_i)}{g_*(T_0)}\right)^{1/4}, \tag{8.158}$$

$$\sim \left(\frac{\Gamma G\mu_l}{\alpha}\right)^{1/2} f\Big|_{\text{local}}. \tag{8.159}$$

In contrast to the local case, the frequency is independent of the string scale and we explain its origin at the end of the section. We can rewrite the GW spectrum in Eq. (8.152) as

$$\Omega_{GW}^g(f) \simeq \frac{1}{\rho_c}\cdot\frac{2}{f}\cdot\frac{\mathcal{F}_\alpha\,\Gamma G\mu_g^2}{\alpha(\alpha + \Gamma G\mu_g + \kappa)}\cdot C_{\text{eff}}^g \cdot t_0^{-5/2}\cdot(\tilde{t}_M^g)^{1+\frac{5}{2}-\frac{3}{2}}\cdot t_i^{-4+\frac{3}{2}}. \tag{8.160}$$

Then, from Eq. (8.156) and $\alpha\, t_i^g \simeq 4\, a(\tilde{t}_M^g)/a(t_0)\, f^{-1}$ in Eq. (8.113), one obtains

$$\tilde{t}_M^g \simeq \frac{1}{t_0}\frac{4}{\alpha^2}\left(\frac{1}{f}\right)^2\left(\frac{\alpha + \Gamma G\mu_g + \kappa}{\Gamma G\mu_g + \kappa}\right)^2,$$

$$t_i^g \simeq \frac{1}{t_0}\frac{8}{\alpha^2}\left(\frac{1}{f}\right)^2\left(\frac{\alpha + \Gamma G\mu_g + \kappa}{\Gamma G\mu_g + \kappa}\right).$$

From comparing the global GW spectrum Ω_{GW}^g to the local GW spectrum $\Omega_{GW}^l = \Omega_{GW}^g(\ln(\eta\, t) = 1,\ \kappa = 0,\ s = 0)$, one obtains

$$\frac{\Omega_{GW}^g}{\Omega_{GW}^l} \sim \frac{\alpha}{\kappa}\left(\frac{\Gamma G\mu_l}{\alpha}\right)^{3/2}\left(\frac{\mu_g}{\mu_l}\right)^2\frac{C_{\text{eff}}^g}{C_{\text{eff}}^l}, \tag{8.161}$$

where we have assumed $\kappa \gg \alpha,\ \Gamma G\mu_{\rm g}$. Upon using Eqs. (8.56) and (8.161), we get[9]

$$\Omega^{\rm l}_{\rm GW} \simeq \Omega_r h^2 \frac{\eta}{M_{\rm pl}}, \qquad \text{and} \qquad \Omega^{\rm g}_{\rm GW} \simeq \Omega_r h^2 \log^3\left(\eta\, \tilde{t}^{\rm g}_M\right) \frac{\eta^4}{M^4_{\rm pl}}, \tag{8.162}$$

where $\Omega_r h^2 \simeq 4.2 \times 10^{-5}$ is the present radiation energy density of the universe [202].

As shown in Fig. 8.24, in consequence of the strong dependence of the GW amplitude on the string scale η, only global networks above $\eta \gtrsim 5 \times 10^{14}$ GeV can be detected by LISA or CE whereas EPTA or BBO/DECIGO can probe $\eta \gtrsim 10^{14}$ GeV. Also note the logarithmic spectral tilt of the GW spectrum due to the logarithmic dependence of the global string tension on the cosmic time.

In summary, GW spectra from local and global strings manifest differences in frequency, Eq. (8.159), and amplitude, Eq. (8.162), because local loops are long-lived whereas global loops are short-lived. In the local case, the dominant GW emission at $\tilde{t}^{\rm l}_{\rm M} \sim t^{\rm l}_i / G\mu_{\rm l}$ occurs much after the loop formation time $t^{\rm l}_i$, after one loop lifetime. However, the emitted frequency is fixed by $(t^{\rm l}_i)^{-1}$. Hence, as discussed in Sect. 8.3.4, the observed frequency is exempted from a redshift factor given by $\sqrt{G\mu_{\rm l}}$. In the global case, the loops are short-lived and the time of dominant GW emission coincides with the time of loop formation. Hence, the emitted frequency redshifts more and the spectrum is shifted to the left. The GW spectrum is also reduced by the redshift factor $(G\mu)^{3/2}$.

8.F.5 As a Probe of Non-standard Cosmology

The impact of non-standard cosmology on the GW spectra of global strings is shown in Fig. 8.25. The frequency of the turning point corresponding to a change of cosmology at a temperature T_Δ is given by Eq. (8.158). We report here a numerically-fitted version

$$f^{\rm glob}_\Delta \simeq {\rm Hz} \left(\frac{T}{\rm GeV}\right) \left(\frac{0.1}{\alpha}\right) \left(\frac{g_*(T)}{g_*(T_0)}\right)^{1/4} \times \begin{cases} 8.9 \times 10^{-7} & \text{for } 10\% \\ 7.0 \times 10^{-8} & \text{for } 1\% \end{cases}, \tag{8.163}$$

where the detection criterion is defined as in Eq. (8.61). In contrast to local strings, the turning-point is independent of the string tension.

We now consider the reach of global strings for probing a non-standard cosmology. Fig. 8.26 shows the detectability of the turning-points by future GW experiments.

[9] Upon restoring the dependence on the different parameters appearing in Eq. (8.160), we obtain

$$\Omega^{\rm g}_{\rm GW} \simeq 25 \Delta_R\, \Omega_r h^2\, C^{\rm g}_{\rm eff}(n=4)\, \mathscr{F}_\alpha \frac{\Gamma}{\Gamma_{\rm Gold}} \log^3\left(\eta\, \tilde{t}^{\rm g}_M\right) \frac{\eta^4}{M^4_{\rm pl}}, \tag{8.162}$$

which can be compared with its local equivalent in Eq. (8.56).

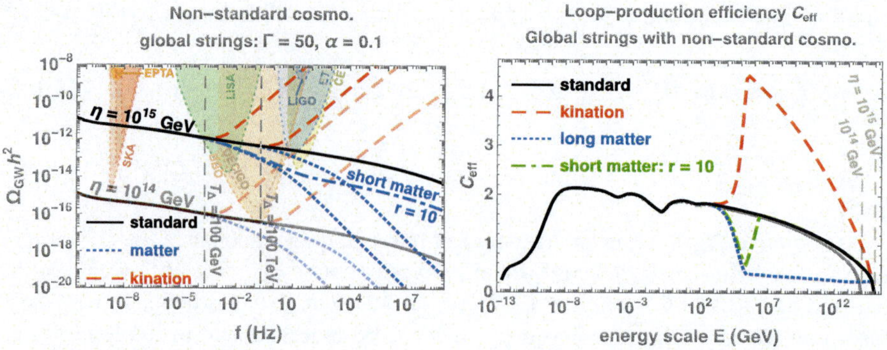

Fig. 8.25 **Left:** GW spectrum from the global cosmic strings assuming VOS network, evolving in the presence of a non-standard era, either long-lasting matter (dotted), intermediate matter (dot-dashed), or kination (dashed), ending at the temperature $T_\Delta = 100$ GeV or 100 TeV. The turning-point frequency is independent of the string tension. **Right:** The evolution of the loop-production efficiency for each cosmological background never reaches a plateau, in contrast to local strings in which case the scaling regime is an attactor solution, cf. right panel of Fig. 8.9. Indeed, for global strings the scaling behavior is logarithmically violated due to the inclusion of energy loss through Goldstone production in the VOS equations, cf. Sect. 8.B.2

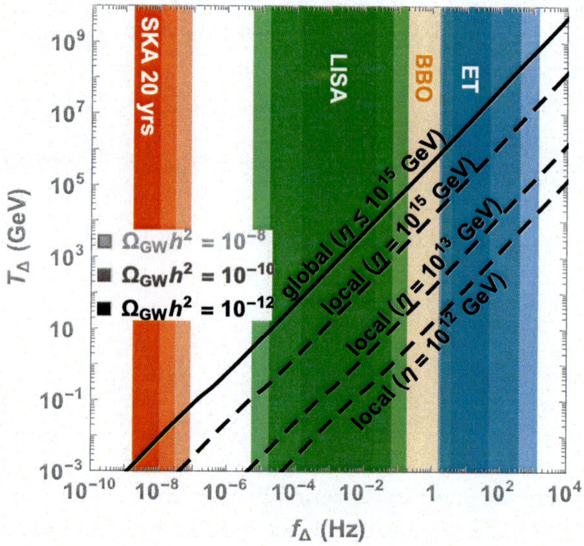

Fig. 8.26 Detectability of the turning-points in the $T_\Delta - f_\Delta$ plane. The solid line represents turning-points for global strings formed at any energy scale η, while dashed lines are turning points from local strings. Shaded areas correspond to the frequencies probed by each observatory assuming SGWB of amplitudes $\Omega_{\rm GW}h^2 = 10^{-8}$, 10^{-10} and 10^{-12}. The plot is totally inspired from [27]

Due to the string-scale independence, the global-string detectability collapses onto a line. Because of the shift of the spectrum to lower frequencies by a factor $\sim \sqrt{G\mu} \sim \eta/M_{\rm pl}$, cf. Eq. (8.159), GW from global string networks can probe earlier non-standard-cosmology and larger energy scales with respect to GW from local strings.

References

1. Gouttenoire Y, Servant G, Simakachorn P (2020) Beyond the standard models with cosmic strings. JCAP 07: 032. arxiv:1912.02569
2. Caprini C, Figueroa DG (2018) Cosmological backgrounds of gravitational waves. Class Quant Grav 35:163001. arxiv:1801.04268
3. Giovannini M (1998) Gravitational waves constraints on postinflationary phases stiffer than radiation. Phys Rev D 58:083504. arxiv:hep-ph/9806329
4. Giovannini M (2010) Stochastic backgrounds of relic gravitons: a theoretical appraisal. PMC Phys A 4:1. arxiv:0901.3026
5. Riazuelo A, Uzan J-P (2000) Quintessence and gravitational waves. Phys Rev D 62:083506. arxiv:astro-ph/0004156
6. Sahni V, Sami M, Souradeep T (2002) Relic gravity waves from brane world inflation. Phys Rev D 65:023518. arxiv:gr-qc/0105121
7. Seto N, Yokoyama J (2003) Probing the equation of state of the early universe with a space laser interferometer. J Phys Soc Jpn 72:3082–3086. arxiv:gr-qc/0305096
8. Tashiro H, Chiba T, Sasaki M (2004) Reheating after quintessential inflation and gravitational waves. Class Quant Grav 21:1761–1772. arxiv:gr-qc/0307068
9. Nakayama K, Saito S, Suwa Y, Yokoyama J (2008) Space laser interferometers can determine the thermal history of the early Universe. Phys Rev D 77:124001. arxiv:0802.2452
10. Nakayama K, Saito S, Suwa Y, Yokoyama J (2008) Probing reheating temperature of the universe with gravitational wave background. JCAP 0806:020. arxiv:0804.1827
11. Durrer R, Hasenkamp J (2011) Testing superstring theories with gravitational waves. Phys Rev D 84:064027. arxiv:1105.5283
12. Kuroyanagi S, Nakayama K, Saito S (2011) Prospects for determination of thermal history after inflation with future gravitational wave detectors. Phys Rev D 84:123513. arxiv:1110.4169
13. Kuroyanagi S, Chiba T, Takahashi T (2018) Probing the universe through the stochastic gravitational wave background. JCAP 11:038. arxiv:1807.00786
14. Jinno R, Moroi T, Nakayama K (2012) Probing dark radiation with inflationary gravitational waves. Phys Rev D 86:123502. arxiv:1208.0184
15. Lasky PD et al (2016) Gravitational-wave cosmology across 29 decades in frequency. Phys Rev X6:011035. arxiv:1511.05994
16. Li B, Shapiro PR, Rindler-Daller T (2017) Bose-Einstein-condensed scalar field dark matter and the gravitational wave background from inflation: new cosmological constraints and its detectability by LIGO, Phys Rev D96:063505. arxiv:1611.07961
17. Saikawa K, Shirai S (2018) Primordial gravitational waves, precisely: the role of thermodynamics in the standard model, JCAP 1805:035. arxiv:1803.01038
18. Caldwell RR, Smith TL, Walker DGE, Using a primordial gravitational wave background to illuminate new physics. arxiv:1812.07577
19. Bernal N, Hajkarim F, Primordial gravitational waves in non-standard cosmologies. arxiv:1905.10410
20. Figueroa DG, Tanin EH, Ability of LIGO and LISA to probe the equation of state of the early Universe. arxiv:1905.11960

21. D'Eramo F, Schmitz K, Imprint of a scalar era on the primordial spectrum of gravitational waves. arxiv:1904.07870
22. Cui Y, Lewicki M, Morrissey DE, Wells JD (2018) Cosmic archaeology with gravitational waves from cosmic strings, Phys Rev D97:123505. arxiv:1711.03104
23. Cui Y, Lewicki M, Morrissey DE, Wells JD (2019) Probing the pre-BBN universe with gravitational waves from cosmic strings. JHEP 01:081. arxiv:1808.08968
24. Auclair P et al, Probing the gravitational wave background from cosmic strings with LISA. arxiv:1909.00819
25. Guedes GSF, Avelino PP, Sousa L (2018) Signature of inflation in the stochastic gravitational wave background generated by cosmic string networks. Phys Rev D98:123505. arxiv:1809.10802
26. Ramberg N, Visinelli L (2019) Probing the early universe with axion physics and gravitational waves. Phys Rev D99:123513. arxiv:1904.05707
27. Chang C-F, Cui Y, Stochastic gravitational wave background from global cosmic strings. arxiv:1910.04781
28. Vilenkin A (1981) Gravitational radiation from cosmic strings. Phys Lett 107B:47–50
29. Hogan CJ, Rees MJ (1984) Gravitational interactions of cosmic strings. Nature 311:109–113
30. Vachaspati T, Vilenkin A (1985) Gravitational radiation from cosmic strings. Phys Rev D 31:3052
31. Accetta FS, Krauss LM (1989) The stochastic gravitational wave spectrum resulting from cosmic string evolution. Nucl Phys B 319:747–764
32. Bennett DP, Bouchet FR (1991) Constraints on the gravity wave background generated by cosmic strings. Phys Rev D 43:2733–2735
33. Caldwell RR, Allen B (1992) Cosmological constraints on cosmic string gravitational radiation. Phys Rev D 45:3447–3468
34. Allen B, Shellard EPS (1992) Gravitational radiation from cosmic strings. Phys Rev D 45:1898–1912
35. Battye RA, Caldwell RR, Shellard EPS (1997) Gravitational waves from cosmic strings. In: Topological defects in cosmology, pp 11–31. arxiv:astro-ph/9706013
36. DePies MR, Hogan CJ (2007) Stochastic gravitational wave background from light cosmic strings. Phys Rev D 75:125006. arxiv:astro-ph/0702335
37. Siemens X, Mandic V, Creighton J (2007) Gravitational wave stochastic background from cosmic (super)strings. Phys Rev Lett 98:111101. arxiv:astro-ph/0610920
38. Olmez S, Mandic V, Siemens X (2010) Gravitational-wave stochastic background from kinks and cusps on cosmic strings. Phys Rev D 81:104028. arxiv:1004.0890
39. Regimbau T, Giampanis S, Siemens X, Mandic V (2012) The stochastic background from cosmic (super)strings: popcorn and (Gaussian) continuous regimes. Phys Rev D 85:066001. arxiv:1111.6638
40. Sanidas SA, Battye RA, Stappers BW (2012) Constraints on cosmic string tension imposed by the limit on the stochastic gravitational wave background from the European Pulsar Timing Array. Phys Rev D 85:122003. arxiv:1201.2419
41. Sanidas SA, Battye RA, Stappers BW (2013) Projected constraints on the cosmic (super)string tension with future gravitational wave detection experiments. Astrophys J 764:108. arxiv:1211.5042
42. Binetruy P, Bohe A, Caprini C, Dufaux J-F (2012) Cosmological backgrounds of gravitational waves and eLISA/NGO: phase transitions. cosmic strings and other sources. JCAP 1206:027. arxiv:1201.0983
43. Kuroyanagi S, Miyamoto K, Sekiguchi T, Takahashi K, Silk J (2012) Forecast constraints on cosmic string parameters from gravitational wave direct detection experiments. Phys Rev D 86:023503. arxiv:1202.3032
44. Kuroyanagi S, Miyamoto K, Sekiguchi T, Takahashi K, Silk J (2013) Forecast constraints on cosmic strings from future CMB, pulsar timing and gravitational wave direct detection experiments. Phys Rev D 87:023522. arxiv:1210.2829

45. Sousa L, Avelino PP (2016) Probing cosmic superstrings with gravitational waves. Phys Rev D94:063529. arxiv:1606.05585
46. Sousa L, Avelino PP, Guedes GS (2020) Full analytical approximation to the stochastic gravitational wave background generated by cosmic string networks. Phys Rev D 101:103508. arxiv:2002.01079
47. Figueroa DG, Hindmarsh M, Lizarraga J, Urrestilla J (2020) Irreducible background of gravitational waves from a cosmic defect network: update and comparison of numerical techniques. Phys Rev D 102:103516. arxiv:2007.03337
48. Hindmarsh M, Lizarraga J, Lopez-Eiguren A, Urrestilla J, Approach to scaling in axion string networks. arxiv:2102.07723
49. Spokoiny B (1993) Deflationary universe scenario. Phys Lett B 315:40–45. arxiv:gr-qc/9306008
50. Joyce M (1997) Electroweak baryogenesis and the expansion rate of the universe. Phys Rev D 55:1875–1878. arxiv:hep-ph/9606223
51. Gouttenoire Y, Servant G, Simakachorn P, Revealing the primordial irreducible inflationary gravitational-wave background with a spinning peccei-quinn axion. arxiv:2108.10328
52. Gouttenoire Y, Servant G, Simakachorn P, Kination cosmology from scalar fields and gravitational-wave signatures. arxiv:2111.01150
53. Creminelli P, Nicolis A, Rattazzi R (2002) Holography and the electroweak phase transition. JHEP 03:051. arxiv:hep-th/0107141
54. Randall L, Servant G (2007) Gravitational waves from warped spacetime. JHEP 05:054. arxiv:hep-ph/0607158
55. Nardini G, Quiros M, Wulzer A (2007) A confining strong first-order electroweak phase transition. JHEP 09:077. arxiv:0706.3388
56. Konstandin T, Servant G (2011) Cosmological consequences of nearly conformal dynamics at the TeV scale. JCAP 12:009. arxiv:1104.4791
57. von Harling B, Servant G (2018) QCD-induced electroweak phase transition. JHEP 01:159. arxiv:1711.11554
58. Iso S, Serpico PD, Shimada K (2017) QCD-electroweak first-order phase transition in a supercooled universe, Phys Rev Lett 119:141301. arxiv:1704.04955
59. Bruggisser S, Von Harling B, Matsedonskyi O, Servant G (2018) Electroweak phase transition and baryogenesis in composite higgs models, JHEP 12:099. arxiv:1804.07314
60. Baratella P, Pomarol A, Rompineve F (2019) The supercooled universe. JHEP 03:100. arxiv:1812.06996
61. Agashe K, Du P, Ekhterachian M, Kumar S, Sundrum R, Cosmological phase transition of spontaneous confinement. arxiv:1910.06238
62. Delle Rose L, Panico G, Redi M, Tesi A (2020) Gravitational waves from supercool axions. JHEP 04:025. arxiv:1912.06139
63. Von Harling B, Pomarol A, Pujolas O, Rompineve F, Peccei-quinn phase transition at LIGO. arxiv:1912.07587
64. Baldes I, Gouttenoire Y, Sala F (2021) String fragmentation in supercooled confinement and implications for dark matter. JHEP 04:278. arxiv:2007.08440
65. Baldes I, Gouttenoire Y, Sala F, Servant G, Supercool composite dark matter beyond 100 TeV. arxiv:2110.13926
66. Matsunami D, Pogosian L, Saurabh A, Vachaspati T (2019) Decay of cosmic string loops due to particle radiation. Phys Rev Lett 122:201301. arxiv:1903.05102
67. Auclair P, Steer DA, Vachaspati T, Particle emission and gravitational radiation from cosmic strings: observational constraints. arxiv:1911.12066
68. Kawasaki M, Miyamoto K, Nakayama K, Cosmological effects of decaying cosmic string loops with TeV scale width. arxiv:1105.4383
69. Gouttenoire Y, Servant G, Simakachorn P (2020) BSM with cosmic strings: heavy, up to EeV mass, unstable particles. JCAP 07:016. arxiv:1912.03245
70. Kibble TWB (1976) Topology of cosmic domains and strings. J Phys A 9:1387–1398

71. Hindmarsh MB, Kibble TWB (1995) Cosmic strings. Rep Prog Phys 58:477–562. arxiv:hep-ph/9411342
72. Vilenkin A, Shellard EPS (2000) Cosmic strings and other topological defects. Cambridge University Press
73. Vachaspati T, Pogosian L, Steer D (2015) Cosmic strings. Scholarpedia 10:31682. arxiv:1506.04039
74. Witten E (1985) Cosmic superstrings. Phys Lett 153B:243–246
75. Dvali G, Vilenkin A (2004) Formation and evolution of cosmic D strings. JCAP 0403:010. arxiv:hep-th/0312007
76. Copeland EJ, Myers RC, Polchinski J (2004) Cosmic F and D strings. JHEP 06:013. arxiv:hep-th/0312067
77. Polchinski J (2004) Introduction to cosmic F- and D-strings. In: String theory: From gauge interactions to cosmology. Proceedings, NATO Advanced Study Institute. Cargese, France, pp 229–253. arxiv:hep-th/0412244
78. Sakellariadou M (2008) Cosmic superstrings. Philso Trans R Soc Lond A 366:2881–2894. arxiv:0802.3379
79. Davis A-C, Brax P, van de Bruck C (2008) Brane inflation and defect formation. Philso Trans R Soc Lond A 366:2833–2842. arxiv:0803.0424
80. Sakellariadou M (2009) Cosmic strings and cosmic superstrings. Nucl Phys Proc Suppl 192–193:68–90. arxiv:0902.0569
81. Copeland EJ, Kibble TWB (2010) Cosmic strings and superstrings. Proc R Soc Lond A 466:623–657. arxiv:0911.1345
82. BuchmÃŒller W, Domcke V, Kamada K, Schmitz K (2013) The gravitational wave spectrum from cosmological $B - L$ breaking. JCAP 1310:003. arxiv:1305.3392
83. Dror JA, Hiramatsu T, Kohri K, Murayama H, White G, Testing seesaw and leptogenesis with gravitational waves. arxiv:1908.03227
84. Jeannerot R, Rocher J, Sakellariadou M (2003) How generic is cosmic string formation in SUSY GUTs. Phys Rev D 68:103514. arxiv:hep-ph/0308134
85. Sakellariadou M (2008) Production of topological defects at the end of inflation. Lect Notes Phys 738:359–392. arxiv:hep-th/0702003
86. Nielsen HB, Olesen P (1973) Vortex line models for dual strings. Nucl Phys B 61:45–61
87. Laguna P, Matzner RA (1989) Peeling U(1) gauge cosmic strings. Phys Rev Lett 62:1948–1951
88. Maeda K-I, Turok N (1988) Finite width corrections to the nambu action for the Nielsen-Olesen string. Phys Lett B 202:376–380
89. Forster D (1974) Dynamics of relativistic vortex lines and their relation to dual theory. Nucl. Phys B 81:84–92
90. Vilenkin A, Vachaspati T (1987) Radiation of goldstone Bosons from cosmic strings. Phys Rev D 35:1138
91. Scherrer RJ, Frieman JA (1986) Cosmic strings as random walks. Phys Rev D 33:3556
92. Shellard EPS (1987) Cosmic string interactions. Nucl Phys B 283:624–656
93. Eto M, Hashimoto K, Marmorini G, Nitta M, Ohashi K, Vinci W (2007) Universal reconnection of non-abelian cosmic strings. Phys Rev Lett 98:091602. arxiv:hep-th/0609214
94. Jackson MG, Jones NT, Polchinski J (2005) Collisions of cosmic F and D-strings. JHEP 10:013. arxiv:hep-th/0405229
95. Copeland EJ, Kibble TWB, Steer DA (2007) Constraints on string networks with junctions. Phys Rev D 75:065024. arxiv:hep-th/0611243
96. Achucarro A, de Putter R (2006) Effective non-intercommutation of local cosmic strings at high collision speeds. Phys Rev D 74:121701. arxiv:hep-th/0605084
97. Kibble TWB (1985) Evolution of a system of cosmic strings. Nucl Phys B 252:227
98. Bennett DP, Bouchet FR (1988) Evidence for a scaling solution in cosmic string evolution. Phys Rev Lett 60:257
99. Bennett DP, Bouchet FR (1989) Cosmic string evolution. Phys Rev Lett 63:2776
100. Albrecht A, Turok N (1989) Evolution of cosmic string networks. Phys Rev D 40:973–1001

101. Allen B, Shellard EPS (1990) Cosmic string evolution: a numerical simulation. Phys Rev Lett 64:119–122
102. Klaer VB, Moore GD (2017) The dark-matter axion mass. JCAP 1711:049. arxiv:1708.07521
103. Gorghetto M, Hardy E, Villadoro G (2018) Axions from strings: the attractive solution. JHEP 07:151. arxiv:1806.04677
104. Gorghetto M, Hardy E, Villadoro G, More axions from strings. arxiv:2007.04990
105. Gorghetto M, Hardy E, Nicolaescu H (2021) Observing invisible axions with gravitational waves. JCAP 06:034. arxiv:2101.11007
106. Kawasaki M, Sekiguchi T, Yamaguchi M, Yokoyama J (2018) Long-term dynamics of cosmological axion strings. PTEP 2018:091E01. arxiv:1806.05566
107. Vaquero A, Redondo J, Stadler J, Early seeds of axion miniclusters. arxiv:1809.09241
108. Buschmann M, Foster JW, Safdi BR (2020) Early-universe simulations of the cosmological axion. Phys Rev Lett 124:161103. arxiv:1906.00967
109. Martins CJAP (2019) Scaling properties of cosmological axion strings. Phys Lett B788:147–151. arxiv:1811.12678
110. Hindmarsh M, Lizarraga J, Lopez-Eiguren A, Urrestilla J, The scaling density of axion strings. arxiv:1908.03522
111. Ringeval C, Sakellariadou M, Bouchet F (2007) Cosmological evolution of cosmic string loops. JCAP 0702:023. arxiv:astro-ph/0511646
112. Vanchurin V, Olum KD, Vilenkin A (2006) Scaling of cosmic string loops. Phys Rev D 74:063527. arxiv:gr-qc/0511159
113. Martins CJAP, Shellard EPS (2006) Fractal properties and small-scale structure of cosmic string networks. Phys Rev D 73:043515. arxiv:astro-ph/0511792
114. Olum KD, Vanchurin V (2007) Cosmic string loops in the expanding Universe. Phys Rev D 75:063521. arxiv:astro-ph/0610419
115. Blanco-Pillado JJ, Olum KD, Shlaer B (2011) Large parallel cosmic string simulations: new results on loop production. Phys Rev D 83:083514. arxiv:1101.5173
116. Blasi S, Brdar V, Schmitz K, Fingerprint of low-scale leptogenesis in the primordial gravitational-wave spectrum. arxiv:2004.02889
117. LIGO Scientific, VIRGO collaboration, Aasi J et al (2015) Characterization of the LIGO detectors during their sixth science run. Class Quant Grav 32:115012. arxiv:1410.7764
118. Yagi K, Seto N (2011) Detector configuration of DECIGO/BBO and identification of cosmological neutron-star binaries. Phys Rev D 83:044011. arxiv:1101.3940
119. LISA collaboration, Audley H et al, Laser interferometer space antenna. arxiv:1702.00786
120. LISA Cosmology Working Group collaboration, Auclair P et al, Cosmology with the laser interferometer space antenna. arxiv:2204.05434
121. Hild S et al (2011) Sensitivity studies for third-generation gravitational wave observatories. Class Quant Grav 28:094013. arxiv:1012.0908
122. Punturo M et al (2010) The Einstein telescope: a third-generation gravitational wave observatory. Class Quant Grav 27:194002
123. LIGO Scientific collaboration, Abbott BP et al (2017) Exploring the sensitivity of next generation gravitational wave detectors. Class Quant Grav 34:044001. arxiv:1607.08697
124. Blanco-Pillado JJ, Olum KD, Shlaer B (2014) The number of cosmic string loops. Phys Rev D 89:023512. arxiv:1309.6637
125. Krauss LM (1992) Gravitational waves from global phase transitions. Phys Lett B 284:229–233
126. Jones-Smith K, Krauss LM, Mathur H (2008) A nearly scale invariant spectrum of gravitational radiation from global phase transitions. Phys Rev Lett 100:131302. arxiv:0712.0778
127. Fenu E, Figueroa DG, Durrer R, Garcia-Bellido J (2009) Gravitational waves from self-ordering scalar fields. JCAP 0910:005. arxiv:0908.0425
128. Figueroa DG, Hindmarsh M, Urrestilla J (2013) Exact scale-invariant background of gravitational waves from cosmic defects. Phys Rev Lett 110:101302. arxiv:1212.5458
129. Blanco-Pillado JJ, Olum KD, Siemens X (2018) New limits on cosmic strings from gravitational wave observation. Phys Lett B778:92–396. arxiv:1709.02434

130. Maggiore M (2008) Gravitational waves: volume 1: theory and experiments, vol 1. Oxford University Press
131. Kibble TWB, Turok N (1982) Selfintersection of cosmic strings. Phys Lett 116B:141–143
132. Damour T, Vilenkin A (2001) Gravitational wave bursts from cusps and kinks on cosmic strings. Phys Rev D 64:064008. arxiv:gr-qc/0104026
133. Ringeval C, Suyama T (2017) Stochastic gravitational waves from cosmic string loops in scaling. JCAP 1712:027. arxiv:1709.03845
134. Burden CJ (1985) Gravitational radiation from a particular class of cosmic strings. Phys Lett 164B:277–281
135. Damour T, Vilenkin A (2000) Gravitational wave bursts from cosmic strings. Phys Rev Lett 85:3761–3764. arxiv:gr-qc/0004075
136. Saurabh A, Vachaspati T, Pogosian L, Decay of cosmic global string loops. arxiv:2001.01030
137. Bettoni D, DomÃšnech G, Rubio J (2019) Gravitational waves from global cosmic strings in quintessential inflation. JCAP 1902:034. arxiv:1810.11117
138. Vilenkin A, Everett AE (1982) Cosmic strings and domain walls in models with goldstone and pseudogoldstone Bosons. Phys Rev Lett 48:1867–1870
139. Vilenkin A (1985) Cosmic strings and domain walls. Phys Rep 121:263–315
140. Sikivie P (1982) Of axions, domain walls and the early universe. Phys Rev Lett 48:1156–1159
141. Blanco-Pillado JJ, Olum KD (1999) The form of cosmic string cusps. Phys Rev D 59:063508. arxiv:gr-qc/9810005
142. Lentati L et al (2015) European pulsar timing array limits on an isotropic stochastic gravitational-wave background. Mon Not R Astron Soc 453:2576–2598. arxiv:1504.03692
143. NANOGRAV collaboration, Arzoumanian Z et al (2018) The NANOGrav 11-year data set: pulsar-timing constraints on the stochastic gravitational-wave background. Astrophys J 859:47. arxiv:1801.02617
144. Sakellariadou M (2005) A Note on the evolution of cosmic string/superstring networks. JCAP 0504:003. arxiv:hep-th/0410234
145. Damour T, Vilenkin A (2005) Gravitational radiation from cosmic (super)strings: Bursts, stochastic background, and observational windows. Phys Rev D 71:063510. arxiv:hep-th/0410222
146. Breitbach M, Kopp J, Madge E, Opferkuch T, Schwaller P (1907) Dark, cold, and noisy: constraining secluded hidden sectors with gravitational waves. JCAP 1907:007. arxiv:1811.11175
147. LIGO Scientific, Virgo collaboration, Abbott BP et al (2018) Constraints on cosmic strings using data from the first Advanced LIGO observing run. Phys Rev D97:102002. arxiv:1712.01168
148. LIGO Scientific, Virgo collaboration, Abbott BP et al, All-sky search for short gravitational-wave bursts in the second advanced LIGO and advanced Virgo run. arxiv:1905.03457
149. Lorenz L, Ringeval C, Sakellariadou M (2010) Cosmic string loop distribution on all length scales and at any redshift. JCAP 1010:003. arxiv:1006.0931
150. Vincent G, Antunes ND, Hindmarsh M (1998) Numerical simulations of string networks in the Abelian Higgs model. Phys Rev Lett 80:2277–2280. arxiv:hep-ph/9708427
151. Hindmarsh M, Stuckey S, Bevis N (2009) Abelian Higgs cosmic strings: small scale structure and loops. Phys Rev D 79:123504. arxiv:0812.1929
152. Hindmarsh M, Lizarraga J, Urrestilla J, Daverio D, Kunz M (2017) Scaling from gauge and scalar radiation in Abelian Higgs string networks. Phys Rev D96:023525. arxiv:1703.06696
153. Moore JN, Shellard EPS, On the evolution of Abelian Higgs string networks. arxiv:hep-ph/9808336
154. Olum KD, Blanco-Pillado JJ (2000) Radiation from cosmic string standing waves. Phys Rev Lett 84:4288–4291. arxiv:astro-ph/9910354
155. Moore JN, Shellard EPS, Martins CJAP (2002) On the evolution of Abelian-Higgs string networks. Phys Rev D 65:023503. arxiv:hep-ph/0107171
156. Blanco-Pillado JJ, Olum KD, Shlaer B (2015) Cosmic string loop shapes. Phys Rev D92:063528. arxiv:1508.02693

157. Matsunami D, Pogosian L, Saurabh A, Vachaspati T, Decay of cosmic string loops due to particle radiation. https://www.ayushsaurabh.home.blog
158. Rocha JV, Analytic approaches to the study of small scale structure on cosmic string networks. arxiv:0812.4020
159. Quashnock JM, Spergel DN (1990) Gravitational selfinteractions of cosmic strings. Phys Rev D 42:2505–2520
160. Blanco-Pillado JJ, Olum KD, Wachter JM (2018) Gravitational backreaction near cosmic string kinks and cusps. Phys Rev D98:123507. arxiv:1808.08254
161. Blanco-Pillado JJ, Olum KD, Wachter JM (2019) Gravitational backreaction simulations of simple cosmic string loops. Phys Rev D100:023535. arxiv:1903.06079
162. Wachter JM, Olum KD (2017) Gravitational backreaction on piecewise linear cosmic string loops. Phys Rev D95:023519. arxiv:1609.01685
163. Wachter JM, Olum KD (2017) Gravitational smoothing of kinks on cosmic string loops. Phys Rev Lett 118:051301. arxiv:1609.01153
164. Olum KD, Blanco-Pillado JJ (1999) Field theory simulation of Abelian Higgs cosmic string cusps. Phys Rev D 60:023503. arxiv:gr-qc/9812040
165. Elghozi T, Nelson W, Sakellariadou M (2014) Cusps and pseudocusps in strings with Y-junctions. Phys Rev D 90:123517. arxiv:1403.3225
166. Stott MJ, Elghozi T, Sakellariadou M (2017) Gravitational wave bursts from cosmic string cusps and pseudocusps, Phys Rev D96:023533. arxiv:1612.07599
167. Binetruy P, Bohe A, Hertog T, Steer DA (2010) Gravitational wave signatures from kink proliferation on cosmic (super-) strings. Phys Rev D 82:126007. arxiv:1009.2484
168. Binetruy P, Bohe A, Hertog T, Steer DA (2010) Proliferation of sharp kinks on cosmic (super-)string loops with junctions. Phys Rev D 82:083524. arxiv:1005.2426
169. Hogan CJ (2006) gravitational waves from light cosmic strings: backgrounds and bursts with large loops. Phys Rev D 74:043526. arxiv:astro-ph/0605567
170. Blanco-Pillado JJ, Olum KD (2017) Stochastic gravitational wave background from smoothed cosmic string loops. Phys Rev D96:104046. arxiv:1709.02693
171. Auclair P, Ringeval C, Sakellariadou M, Steer D (2019) Cosmic string loop production functions. JCAP 06:015. arxiv:1903.06685
172. Polchinski J, Rocha JV (2006) Analytic study of small scale structure on cosmic strings. Phys Rev D 74:083504. arxiv:hep-ph/0606205
173. Polchinski J, Rocha JV (2007) Cosmic string structure at the gravitational radiation scale. Phys Rev D 75:123503. arxiv:gr-qc/0702055
174. Dubath F, Polchinski J, Rocha JV (2008) Cosmic string loops, large and small. Phys Rev D 77:123528. arxiv:0711.0994
175. Rocha JV (2008) Scaling solution for small cosmic string loops. Phys Rev Lett 100:071601. arxiv:0709.3284
176. Blanco-Pillado JJ, Olum KD, Wachter JM, Energy-conservation constraints on cosmic string loop production and distribution functions. arxiv:1907.09373
177. LIGO Scientific, Virgo collaboration, Abbott BP et al (2016) GW150914: First results from the search for binary black hole coalescence with Advanced LIGO. Phys Rev D93:122003. arxiv:1602.03839
178. LIGO Scientific, Virgo collaboration, Abbott BP et al (2016) GW151226: observation of gravitational waves from a 22-solar-mass binary black hole coalescence. Phys Rev Lett 116:241103. arxiv:1606.04855
179. LIGO Scientific, Virgo collaboration, Abbott BP et al (2016) Binary black hole mergers in the first advanced LIGO observing run. Phys Rev X6:041015. arxiv:1606.04856
180. LIGO Scientific, VIRGO collaboration, Abbott BP et al (2017) GW170104: observation of a 50-solar-mass binary black hole coalescence at redshift 0.2. Phys Rev Lett 118:221101. arxiv:1706.01812
181. LIGO Scientific, Virgo collaboration, Abbott BP et al (2017) GW170608: observation of a 19-solar-mass binary black hole coalescence. Astrophys J 851:L35. arxiv:1711.05578

182. LIGO Scientific, Virgo collaboration, Abbott BP et al (2017) GW170814: a three-detector observation of gravitational waves from a binary black hole coalescence. Phys Rev Lett 119:141101. arxiv:1709.09660
183. LIGO Scientific, Virgo collaboration, Abbott BP et al, GWTC-1: A gravitational-wave transient catalog of compact binary mergers observed by LIGO and virgo during the first and second observing runs. arxiv:1811.12907
184. LIGO Scientific, Virgo collaboration, Abbott BP et al (2017) GW170817: observation of gravitational waves from a binary neutron star inspiral. Phys Rev Lett 119:161101. arxiv:1710.05832
185. Venumadhav T, Zackay B, Roulet J, Dai L, Zaldarriaga M, New binary black hole mergers in the second observing run of advanced LIGO and advanced virgo. arxiv:1904.07214
186. Virgo, LIGO Scientific collaboration, Abbott BP et al (2018) GW170817: implications for the stochastic gravitational-wave background from compact binary coalescences. Phys Rev Lett 120:091101. arxiv:1710.05837
187. Cutler C, Harms J (2006) BBO and the neutron-star-binary subtraction problem. Phys Rev D 73:042001. arxiv:gr-qc/0511092
188. Regimbau T, Evans M, Christensen N, Katsavounidis E, Sathyaprakash B, Vitale S (2017) Digging deeper: observing primordial gravitational waves below the binary black hole produced stochastic background. Phys Rev Lett 118:151105. arxiv:1611.08943
189. Farmer AJ, Phinney ES (2003) The gravitational wave background from cosmological compact binaries. Mon Not R Astron Soc 346:1197. arxiv:astro-ph/0304393
190. Rosado PA (2011) Gravitational wave background from binary systems. Phys Rev D 84:084004. arxiv:1106.5795
191. Moore CJ, Cole RH, Berry CPL (2015) Gravitational-wave sensitivity curves. Class Quant Grav 32:015014. arxiv:1408.0740
192. Kosenko DI, Postnov KA (1998) On the gravitational wave noise from unresolved extragalactic binaries. Astron Astrophys 336:786. arxiv:astro-ph/9801032
193. Adams MR, Cornish NJ (2010) Discriminating between a stochastic gravitational wave background and instrument noise. Phys Rev D 82:022002. arxiv:1002.1291
194. Adams MR, Cornish NJ (2014) Detecting a stochastic gravitational wave background in the presence of a galactic foreground and instrument noise. Phys Rev D 89:022001. arxiv:1307.4116
195. Zhu X-J, Howell EJ, Blair DG, Zhu Z-H (2013) On the gravitational wave background from compact binary coalescences in the band of ground-based interferometers. Mon Not R Astron Soc 431:882–899. arxiv:1209.0595
196. Lewicki M, Vaskonen V, Impact of LIGO-Virgo binaries on gravitational wave background searches. arxiv:2111.05847
197. Martins CJAP, Shellard EPS (2002) Extending the velocity dependent one scale string evolution model. Phys Rev D 65:043514. arxiv:hep-ph/0003298
198. Correia JRCCC, Martins JAP (2019) Extending and calibrating the velocity dependent one-scale model for cosmic strings with one thousand field theory simulations. Phys Rev D100:103517. arxiv:1911.03163
199. Martins CJAP, Shellard EPS (1996) String evolution with friction. Phys Rev D 53:575–579. arxiv:hep-ph/9507335
200. Martins CJAP, Shellard EPS (1996) Quantitative string evolution. Phys Rev D 54:2535–2556. arxiv:hep-ph/9602271
201. Martins CJ (2016) Defect evolution in cosmology and condensed matter: quantitative analysis with the velocity-dependent one-scale model. Springer, Berlin
202. Particle Data Group collaboration, Tanabashi M et al (2018) Review of particle physics. Phys Rev D98:030001
203. Cirelli M, Gouttenoire Y, Petraki K, Sala F (2019) Homeopathic dark matter, or how diluted heavy substances produce high energy cosmic rays. JCAP 02:014. arxiv:1811.03608
204. Caprini C et al (2020) Detecting gravitational waves from cosmological phase transitions with LISA: an update. JCAP 03:024. arxiv:1910.13125

205. Cui Y, Lewicki M, Morrissey DE, Gravitational wave bursts as harbingers of cosmic strings diluted by inflation. arxiv:1912.08832
206. Grojean C, Servant G (2007) Gravitational waves from phase transitions at the electroweak scale and beyond. Phys Rev D 75:043507. arxiv:hep-ph/0607107
207. Caprini C, Figueroa DG, Flauger R, Nardini G, Peloso M, Pieroni M et al, Reconstructing the spectral shape of a stochastic gravitational wave background with LISA. arxiv:1906.09244
208. Sazhin M, Sazhina O, Capaccioli M, Longo G, Paolillo M, Riccio G (2010) Gravitational lens images generated by cosmic strings. Open Astron J 3:200–206
209. Planck collaboration, Aghanim N et al, Planck 2018 results. VI. Cosmological parameters. arxiv:1807.06209
210. Mangano G, Serpico PD (2011) A robust upper limit on $N_{\rm eff}$ from BBN, circa. Phys Lett B 701:296–299. arxiv:1103.1261
211. Peimbert A, Peimbert M, Luridiana V (2016) The primordial helium abundance and the number of neutrino families. Rev Mex Astron Astrofis 52:419. arxiv:1608.02062
212. Mangano G, Miele G, Pastor S, Pinto T, Pisanti O, Serpico PD (2005) Relic neutrino decoupling including flavor oscillations. Nucl Phys B 729:221–234. arxiv:hep-ph/0506164
213. de Salas PF, Pastor S (2016) Relic neutrino decoupling with flavour oscillations revisited. JCAP 1607:051. arxiv:1606.06986
214. Escudero Abenza M, Precision early universe thermodynamics made simple: $N_{\rm eff}$ and neutrino decoupling in the standard model and beyond. arxiv:2001.04466
215. Peebles PJE, Vilenkin A (1999) Quintessential inflation. Phys Rev D 59:063505. arxiv:astro-ph/9810509
216. Vilenkin A (1981) Gravitational field of vacuum domain walls and strings. Phys Rev D 23:852–857
217. Christiansen JL, Albin E, James KA, Goldman J, Maruyama D, Smoot GF (2008) Search for cosmic strings in the GOODS survey. Phys Rev D 77:123509. arxiv:0803.0027
218. Christiansen JL, Albin E, Fletcher T, Goldman J, Teng IPW, Foley M et al (2011) Search for cosmic strings in the COSMOS survey. Phys Rev D 83:122004. arxiv:1008.0426
219. Mack KJ, Wesley DH, King LJ (2007) Observing cosmic string loops with gravitational lensing surveys. Phys Rev D 76:123515. arxiv:astro-ph/0702648
220. Gott JR III (1985) Gravitational lensing effects of vacuum strings: exact solutions. Astrophys J 288:422–427
221. Kaiser N, Stebbins A (1984) Microwave anisotropy due to cosmic strings. Nature 310:391–393
222. Bouchet FR, Bennett DP, Stebbins A (1988) Microwave anisotropy patterns from evolving string networks. Nature 335:410–414
223. Silk J, Vilenkin A (1984) Cosmic strings and galaxy formation. Phys Rev Lett 53:1700–1703
224. Pogosian L, Vachaspati T (1999) Cosmic microwave background anisotropy from wiggly strings. Phys Rev D 60:083504. arxiv:astro-ph/9903361
225. Lizarraga J, Urrestilla J, Daverio D, Hindmarsh M, Kunz M, Liddle AR (2014) Constraining topological defects with temperature and polarization anisotropies. Phys Rev D 90:103504. arxiv:1408.4126
226. Lizarraga J, Urrestilla J, Daverio D, Hindmarsh M, Kunz M (2016) New CMB constraints for Abelian Higgs cosmic strings. JCAP 10:042. arxiv:1609.03386
227. Battye R, Moss A (2010) Updated constraints on the cosmic string tension. Phys Rev D 82:023521. arxiv:1005.0479
228. Charnock T, Avgoustidis A, Copeland EJ, Moss A (2016) CMB constraints on cosmic strings and superstrings. Phys Rev D93:123503. arxiv:1603.01275
229. Lopez-Eiguren A, Lizarraga J, Hindmarsh M, Urrestilla J (2017) Cosmic microwave background constraints for global strings and global monopoles. JCAP 1707:026. arxiv:1705.04154
230. Planck collaboration, Ade PAR et al (2014) Planck 2013 results. XXV. Searches for cosmic strings and other topological defects. Astron Astrophys 571:A25. arxiv:1303.5085

231. Ringeval C, Bouchet FR (2012) All sky CMB map from cosmic strings integrated Sachs-Wolfe effect. Phys Rev D 86:023513. arxiv:1204.5041
232. Ciuca R, Hernández OF, Information theoretic bounds on cosmic string detection in CMB maps with noise. arxiv:1911.06378
233. Khatri R, Wandelt BD (2008) Cosmic (super)string constraints from 21 cm radiation. Phys Rev Lett 100:091302. arxiv:0801.4406
234. Jeannerot R, Zhang X, Brandenberger RH (1999) Non-thermal production of neutralino cold dark matter from cosmic string decays. JHEP 12:003. arxiv:hep-ph/9901357
235. Matsuda T (2006) Dark matter production from cosmic necklaces. JCAP 0604:005. arxiv:hep-ph/0509064
236. Cui Y, Morrissey DE (2009) Non-thermal dark matter from cosmic strings. Phys Rev D 79:083532. arxiv:0805.1060
237. Long AJ, Wang L-T (2019) Dark photon dark matter from a network of cosmic strings. Phys Rev D99:063529. arxiv:1901.03312
238. MacGibbon JH, Brandenberger RH (1990) High-energy neutrino flux from ordinary cosmic strings. Nucl Phys B 331:153–172
239. Bhattacharjee P, Sigl G (2000) Origin and propagation of extremely high-energy cosmic rays. Phys Rep 327:109–247. arxiv:astro-ph/9811011
240. Berezinsky V, Blasi P, Vilenkin A (1998) Ultrahigh-energy gamma-rays as signature of topological defects. Phys Rev D 58:103515. arxiv:astro-ph/9803271
241. Berezinsky V, Sabancilar E, Vilenkin A (2011) Extremely high energy neutrinos from cosmic strings. Phys Rev D 84:085006. arxiv:1108.2509
242. Bird D, Corbato S, Dai H, Dawson B, Elbert J, Emerson B et al (1994) The cosmic-ray energy spectrum observed by the fly's eye. Astrophys J 424:491–502
243. Greisen K (1966) End to the cosmic-ray spectrum? Phys Rev Lett 16:748–750
244. Zatsepin GT, Kuzmin VA (1966) Upper limit of the spectrum of cosmic rays. J Exp Theor Phys Lett (JETP Letters) 4:78–80
245. HiRes collaboration, Abbasi RU et al (2008) First observation of the Greisen-Zatsepin-Kuzmin suppression. Phys Rev Lett 100:101101. arxiv:astro-ph/0703099
246. Bhattacharjee P (1989) Cosmic strings and ultrahigh-energy cosmic rays. Phys Rev D 40:3968
247. Srednicki M, Theisen S (1987) Nongravitational decay of cosmic strings. Phys Lett B 189:397
248. Gill AJ, Kibble TWB (1994) Cosmic rays from cosmic strings. Phys Rev D 50:3660–3665. arxiv:hep-ph/9403395
249. Hyde JM, Long AJ, Vachaspati T (2014) Dark strings and their couplings to the standard model. Phys Rev D 89:065031. arxiv:1312.4573
250. Long AJ, Hyde JM, Vachaspati T (2014) Cosmic strings in hidden sectors: 1. radiation of standard model particles. JCAP 1409:030. arxiv:1405.7679
251. Long AJ, Vachaspati T (2014) Cosmic strings in hidden sectors: 2. cosmological and astrophysical signatures. JCAP 1412:040. arxiv:1409.6979
252. Auclair P, Leyde K, Steer DA, A window for cosmic strings. arxiv:2112.11093
253. Laliberte S, Brandenberger R, Ionization from cosmic strings at cosmic dawn. arxiv:1907.08022
254. Witten E (1985) Superconducting strings. Nucl Phys B 249:557–592
255. Vilenkin A, Vachaspati T (1987) Electromagnetic radiation from superconducting cosmic strings. Phys Rev Lett 58:1041–1044
256. Spergel DN, Piran T, Goodman J (1987) Dynamics of superconducting cosmic strings. Nucl Phys B 291:847–875
257. Copeland EJ, Haws D, Hindmarsh M, Turok N (1988) Dynamics of and radiation from superconducting strings and springs. Nucl Phys B 306:908–930
258. Blanco-Pillado JJ, Olum KD (2001) Electromagnetic radiation from superconducting string cusps. Nucl Phys B 599:435–445. arxiv:astro-ph/0008297
259. Babul A, Paczynski B, Spergel D (1987) Gamma-ray bursts from superconducting cosmic strings at large redshifts. Astrophys J 316:L49–L54

260. Berezinsky V, Hnatyk B, Vilenkin A (2001) Gamma-ray bursts from superconducting cosmic strings. Phys Rev D 64:043004. arxiv:astro-ph/0102366
261. Cheng KS, Yu Y-W, Harko T (2010) High redshift gamma-ray bursts: observational signatures of superconducting cosmic strings? Phys Rev Lett 104:241102. arxiv:1005.3427
262. Vachaspati T (2008) Cosmic sparks from superconducting strings. Phys Rev Lett 101:141301. arxiv:0802.0711
263. Zadorozhna LV, Hnatyk BI (2009) Electromagnetic emission bursts from the near-cusp regions of superconducting cosmic strings. Ukr J Phys 54:1149–1156
264. Cai Y-F, Sabancilar E, Vachaspati T (2012) Radio bursts from superconducting strings. Phys Rev D 85:023530. arxiv:1110.1631
265. Cai Y-F, Sabancilar E, Steer DA, Vachaspati T (2012) Radio broadcasts from superconducting strings. Phys Rev D 86:043521. arxiv:1205.3170
266. Yu Y-W, Cheng K-S, Shiu G, Tye H (2014) Implications of fast radio bursts for superconducting cosmic strings. JCAP 1411:040. arxiv:1409.5516
267. Ye J, Wang K, Cai Y-F (2017) Superconducting cosmic strings as sources of cosmological fast radio bursts. Eur Phys J C77:720. arxiv:1705.10956
268. Brandenberger R, Cyr B, Iyer AV, Fast radio bursts from the decay of cosmic string cusps. arxiv:1707.02397
269. Sanchez NG, Signore M (1989) The cosmological microwave background radiation, cosmic and superconducting strings. Phys Lett B 219:413–418
270. Sanchez NG, Signore M (1990) The absence of distortion in the cosmic microwave background spectrum and superconducting cosmic strings. Phys Lett B 241:332–335
271. Tashiro H, Sabancilar E, Vachaspati T (2012) CMB distortions from superconducting cosmic strings. Phys Rev D 85:103522. arxiv:1202.2474
272. Kogut A et al (2011) The Primordial Inflation Explorer (PIXIE): A nulling polarimeter for cosmic microwave background observations. JCAP 1107:025. arxiv:1105.2044
273. Tashiro H, Sabancilar E, Vachaspati T (2012) Constraints on superconducting cosmic strings from early reionization. Phys Rev D 85:123535. arxiv:1204.3643
274. Berezinsky V, Olum KD, Sabancilar E, Vilenkin A (2009) UHE neutrinos from superconducting cosmic strings. Phys Rev D 80:023014. arxiv:0901.0527
275. Saikawa K (2017) A review of gravitational waves from cosmic domain walls. Universe 3:40. arxiv:1703.02576
276. Turok N, Bhattacharjee P (1984) Stretching cosmic strings. Phys Rev D 29:1557
277. Vilenkin A (1991) Cosmic string dynamics with friction. Phys Rev D 43:1060–1062
278. Garriga J, Sakellariadou M (1993) Effects of friction on cosmic strings. Phys Rev D 48:2502–2515. arxiv:hep-th/9303024
279. Aharonov Y, Bohm D (1959) Significance of electromagnetic potentials in the quantum theory. Phys Rev 115:485–491
280. Avelino P (2020) Assessing and improving on current semi-analytical models for cosmic defect network evolution. JCAP 04:012. arxiv:2001.06318
281. Martins CJAP, Moore JN, Shellard EPS (2004) A unified model for vortex string network evolution. Phys Rev Lett 92:251601. arxiv:hep-ph/0310255
282. Klaer VB, Moore GD (2017) How to simulate global cosmic strings with large string tension. JCAP 1710:043. arxiv:1707.05566
283. Klaer VB, Moore GD, Global cosmic string networks as a function of tension. arxiv:1912.08058
284. BICEP2, Keck Array collaboration, Ade PAR et al (2018) BICEP2/Keck Array x: constraints on primordial gravitational waves using planck, WMAP, and New BICEP2/Keck observations through the 2015 season. Phys Rev Lett 121:221301. arxiv:1810.05216
285. Planck collaboration, Akrami Y et al (2020) Planck 2018 results. X. Constraints on inflation. Astron Astrophys 641:A10. arxiv:1807.06211

Chapter 9
Probe Heavy DM with GW from CS

In the previous chapter, we have computed the GW spectrum from CS in non-standard cosmology and we have shown the possibility to use next generation of GW experiments to probe the existence of a early matter era or an early second period of inflation, In this chapter, we use those findings to constrain particle physics models leading to such a change of cosmology with a particular focus on models of heavy Dark Matter.

In Sect. 9.1, we assume the existence of heavy unstable particles that can temporarily dominate the energy density of the universe, and therefore induce a period of matter domination within the radiation era after post-inflationary reheating. This leads to a modified expansion of the universe compared to the usually assumed single radiation era. Such modified cosmological history can be probed if during this period, there is an active source of gravitational waves, in which case the resulting GW spectrum would imprint any modification of the equation of state of the universe. Particularly well-motivated are the long-lasting GW production from cosmic string networks. We derive the improvement by many orders of magnitude of the current model-independent BBN constraints on the abundance and lifetime of a particle, cf. Fig. 9.3, that can be inferred from the detection of GW produced by CS. In Sect. 9.1.4, we provide unprecedented exclusion bounds on the model of $U(1)_D$ DM, introduced in Chap. 5. We also study the scenario where the dark photon mass and the cosmic string network are generated by the spontaneous breaking of the same $U(1)$ symmetry.

In Sect. 8.7.3 of the previous chapter, we have derived model-independent constraints on the energy scale and number of e-folds of a second period of inflation due to its imprint on the GW spectrum from CS. In Sect. 9.2 of the present chapter, we are able to exclude regions of parameter space in the model of Supercooled Composite DM, introduced in Chap. 7, which predicts a short period of inflation.

Y. Gouttenoire, *Beyond the Standard Model Cocktail*, Springer Theses,
https://doi.org/10.1007/978-3-031-11862-3_9

9.1 The Imprints of an Early Era of Matter Domination

9.1.1 Modified Spectral Index

The part of the spectrum coming from loops produced and emitting during radiation is flat since there is an exact cancellation between the red-tilted red-shift factor and the blue-tilted loop number density. However, in the case of a matter era, a mismatch induces a slope f^{-1}. The impact of a non-standard matter era is shown in Fig. 9.1. The frequency detected today f_Δ of the turning point between the end of the matter domination and the beginning of the radiation-domination can be related to the temperature of the universe T_Δ when the change of cosmology occurs

$$f_\Delta = (2 \times 10^{-3}\,\mathrm{Hz}) \left(\frac{T_\Delta}{\mathrm{GeV}}\right) \left(\frac{0.1 \times 50 \times 10^{-11}}{\alpha \Gamma G_\mu}\right)^{1/2} \left(\frac{g_*(T_\Delta)}{g_*(T_0)}\right)^{1/4}. \tag{9.1}$$

The GW measured with frequency f_Δ have been emitted by loops produced during the change of cosmology at T_Δ. An extensive discussion of this frequency-temperature

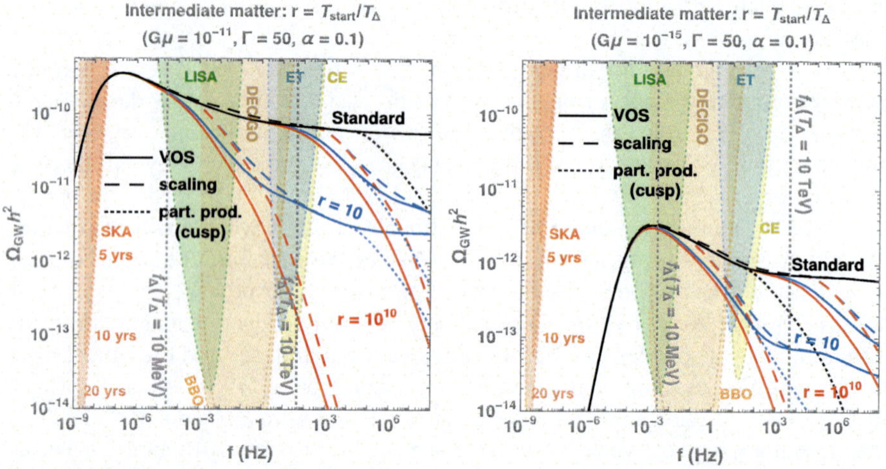

Fig. 9.1 SGWB generated by the gravitational decay of cosmic strings compared to the reach of different GW interferometers. We show the impact of a long (red) or a short (blue) intermediate matter era, starting at the temperature $r\,T_\Delta$ and ending at $T_\Delta = 10$ MeV or $T_\Delta = 10$ TeV. Black lines show the results obtained assuming standard cosmological evolution. The dashed-lines assume that the scaling regime switches on instantaneously during the change of cosmology whereas the solid lines incorporate the transient behavior, solution of the VOS equations, as discussed in [2]. Limitations due to particle production assuming that the small-scale structures are dominated by cusps are shown with dotted lines [2]. The dotted vertical lines indicate the relation in Eq. (9.1) between the temperature T_Δ and the frequency f_Δ of the turning point, where the matter-era-tilted spectrum meets the radiation-era-flat spectrum

relation as provided in [2]. The above formula entirely relies on the assumptions that the back-reaction scale is $\Gamma G\mu$ as claimed by Blanco-Pillado et al. [3–5] and not much lower as claimed by Ringeval et al. [6, 7] .

9.1.2 How to Detect a Matter Era with a GW Interferometer

For a first qualitative analysis, we start with two simple prescriptions for detecting a matter era from the measurement of a SGWB from CS by a GW interferometer.

- **Rx 1** *(turning-point prescription)*: The turning point, namely the frequency at which the spectral index of the GW spectrum changes, corresponding to the transition from the matter to the radiation era, defined in Eq. (9.1), must be inside the interferometer window, as shown for instance in Fig. 9.1.
- **Rx 2** *(spectra-index prescription)*: The measured spectral index must be smaller than −0.2, namely $\beta < -0.2$ where $\Omega_{\rm GW}h^2 \propto f^{\beta}$.

The same prescriptions can be applied to detect the presence of an inflationary era and more generally to detect any effect which induces a slope $\beta < -0.2$ in the GW spectrum from CS. In Fig. 9.2, we compare the above two prescriptions. The prescription Rx 1 is more conservative but enough to measure the lifetime of the particle. In our study, we use the prescription Rx 1 and, in right panel of Sect. 9.1.2, we show how to extend the constraints with Rx 2.

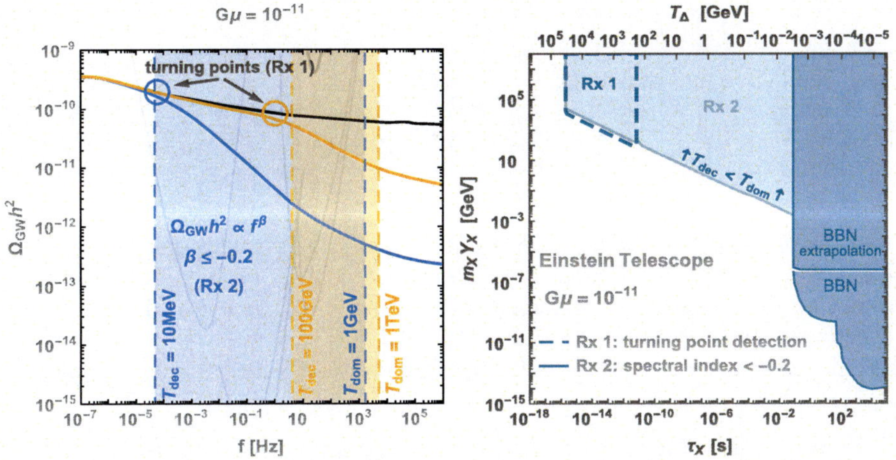

Fig. 9.2 **Left**: SGWB for $G\mu = 10^{-11}$ assuming that a heavy cold particle dominates the energy density of the universe at the temperature $T_{\rm dom}$ and decays at the temperature $T_\Delta = T_{\rm dec}$. **Right**: Considering the particular case of the Einstein Telescope, we illustrate how the constraints on the abundance and lifetime of a heavy relic depend on the choice of the prescription, Rx 1 or Rx 2 defined in Sect. 9.1.2

We note here that the presence of the turning point and the changed spectral index at high frequencies would be similar in the case of a long intermediate inflation era instead of an intermediate matter era. Disentangling the two effects deserves further studies. Interestingly, high-frequency burst signals due to cusp formation could be a way-out [2]. In the analysis of this work, we interpret the suppression of the GW spectrum as due to an intermediate matter era.

9.1.3 Model-Independent Constraints on Particle Physics Parameters

A matter-dominated era may result from an oscillating scalar field [11], such as a moduli field, or a relativistic plasma with a non-vanishing tensor bulk viscosity [12], or simply a massive particle dominating the energy density of the universe. A matter-dominated era may be motivated by the possibillity to enhance structure growth at small scales, since density perturbations start to grow linearly earlier [13, 14], hence boosting the dark matter indirect detection signals [15, 16], or the possibility to enhance the primordial black holes production [17–19].

We suppose an early-matter era is caused by the energy density of a cold particle X, meaning that X is non-relativistic and is decoupled chemically and kinetically from the visible sector. The energy density $m_X n_X$ of X dominates over the energy density of the SM radiation, with entropy $s_{\rm SM}$, at the temperature $T_{\rm dom}$

$$T_{\rm dom} = \frac{4}{3} m_X Y_X, \qquad Y_X \equiv n_X / s_{\rm SM}. \tag{9.2}$$

Then, the cold relic decays when its lifetime τ_X is equal to the age of the universe, corresponding to the temperature

$$T_{\rm dec} = 1\ {\rm GeV} \left(\frac{80}{g_{\rm SM}}\right)^{1/4} \left(\frac{2.7 \times 10^{-7}\ {\rm s}}{\tau_X}\right)^{1/2}. \tag{9.3}$$

Note that the above relation between $T_{\rm dec}$ and τ_X only assumes that the decay is followed by a radiation dominated era and is independent of the previous thermal history of the universe. $T_{\rm dec}$ is sometimes referred, mistakenly though [20], as the reheating temperature following the decay. We propose to use the third generation of GW interferometers to constrain cold relics responsible for early-matter domination. The constraints we will derive rely on the following assumptions:

(1) A SGWB from CS with tension $G\mu$ is measured by a GW interferometer i.
(2) The cold particle is abundant enough to lead to a matter-dominated era before it decays

$$T_{\rm dom} > T_{\rm dec}\,, \tag{9.4}$$

where $T_{\rm dom}$ and $T_{\rm dec}$ satisfy Eq. (9.2) and Eq. (9.3).

(3) The prescription Rx 1 of Sect. 9.1.2 is used, i.e. the turning point in the GW spectrum is in the observation window of the detector and

$$\Omega_{\rm GW}(f_\Delta(T_{\rm dec},\, G\mu),\, G\mu)h^2 \,>\, \Omega^{(i)}_{\rm sens}h^2, \tag{9.5}$$

where $\Omega_{\rm GW}(f,\, G\mu)h^2$ is the predicted scale-invariant GW spectrum from Eq. (8.38), and $\Omega^{(i)}_{\rm sens}h^2$ is the power-law sensitivity curve of the detector i.

Figure 9.3 shows these new constraints in comparison with the current complementary constraints from BBN, usually represented in the plane $(\tau_X,\, m_X Y_X)$ [8–10]. We can translate the sensitivity of each interferometer to probe the particle lifetime into typical mass windows, assuming some decay width.

9.1.4 Heavy Dark Photons

The $U(1)_D$ dark photon:

We consider a $U(1)_{\rm D}$ gauge boson, V_μ, the dark photon, of mass m_V, kinematically coupled to the $U(1)_{\rm Y}$ gauge boson of the SM [21, 22]

$$\mathscr{L} \supset -\frac{\epsilon}{2c_w} F_{{\rm Y}\mu\nu} F_{\rm D}^{\mu\nu}, \tag{9.6}$$

where c_w is the cosine of the weak angle and ϵ is the dark-SM coupling constant. The decay width into SM, Γ_V, is computed in [23]. We here report the expression for $m_V \gtrsim 2m_Z$

$$\Gamma_V \simeq \left(3\times 10^{-8}\ {\rm s}\right)^{-1} \left(\frac{\epsilon}{10^{-9}}\right)^2 \left(\frac{m_V}{1\ {\rm TeV}}\right). \tag{9.7}$$

The dark photon leads to an early-matter-dominated era if it has a large energy density $m_V Y_V \gtrsim 10$ GeV and a long lifetime $\tau_V \sim 10^{-8}$ s, cf. Fig. 9.3 at $G\mu = 10^{-11}$. Supposing that the dark photon abundance is close to thermal, $Y_{\rm V} \sim 0.02$, cf. Eq. (9.9), this implies $\epsilon \lesssim 10^{-9}$. At such a low ϵ, the dark sector and the SM sector may have never been at thermal equilibrium (cf. [24] or footnote 8 in [25]) and may have their own distinct temperature. We assume that the dark sector and the SM have a different temperature by introducing the dark-to-SM temperature ratio [25]

$$\tilde{r} \equiv \frac{\tilde{T}_D}{\tilde{T}_{SM}}, \tag{9.8}$$

where quantities with a $\sim$ on top are evaluated at some high temperature $\tilde{T}$. Thus, the dark photon abundance before its decay is given by

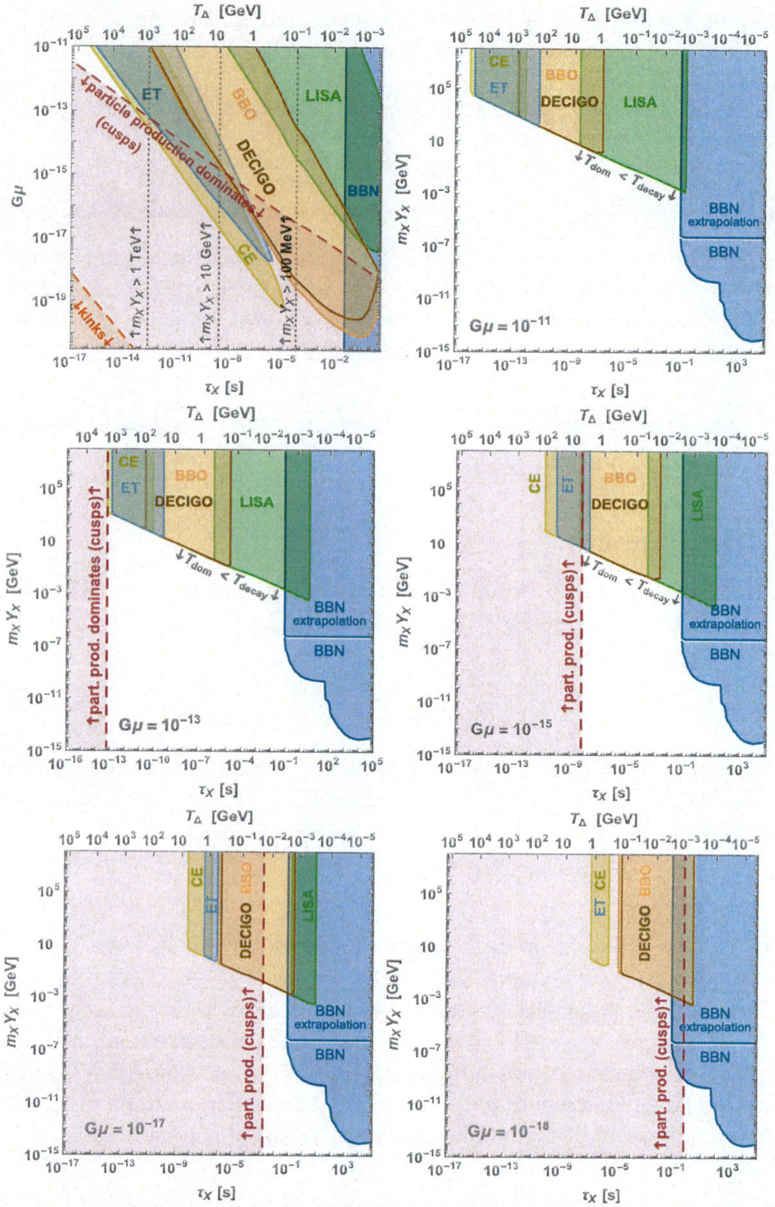

Fig. 9.3 Constraints on the lifetime τ_X and would-be abundance $m_X Y_X$ of a heavy unstable particle inducing an early-matter era, assuming the observation of a SGWB from CS by a GW interferometer, cf. Sect. 9.1.3. We compare the new prospects with the current limits inferred from BBN [8–10]. We assume the detectability of the turning point in the GW spectrum at the frequency f_Δ, induced by the decay of the particle at $T_{\rm dec} = T_\Delta$, cf. *turning-point description Rx 1* in Sect. 9.1.2. Limitation due to particle production in the cusp-domination case [2] are shown in purple

$$Y_V = \frac{n_V}{s_{SM}} = \frac{45\zeta(3)}{2\pi^4}\frac{\tilde{g}_D}{\tilde{g}_{SM}}\tilde{r}^3 \simeq 0.0169\left(\frac{\tilde{g}_D}{6.5}\right)\tilde{r}^3, \tag{9.9}$$

where $\tilde{g}_D$ and $\tilde{g}_{SM}$ are the relativistic number of degrees of freedom in the dark sector and the SM at temperature $\tilde{T}$. Plugging Eq. (9.9) into Eq. (9.2) implies a simple relation between the temperature at which the dark photon dominates the universe T_{dom} and its mass m_V. We choose to be agnostic about the mechanism setting the abundances in the dark sector and we enclose all possibilities by introducing a dark-to-SM temperature ratio $\tilde{r}$.[1]

As shown in the left panel of Fig. 9.4, low kinetic mixing ϵ, large mass m_V or large dark-to-SM temperature ratio $\tilde{r}$ lead to an early-matter-dominated era, triggered when $T_{dom} \gtrsim T_{dec}$. The non-detection with a future GW interferometer, of the imprint left by such a matter era, in the GW spectrum from CS, would exclude the existence of the dark photon for given values of the kinetic mixing, the dark photon mass and the dark-to-SM temperature ratio $(\epsilon,\ m_V,\ \tilde{r})$. We show the GW-from-CS constraints on the dark photon in the right panel of Fig. 9.4, together with existing constraints coming from supernova SN1987 [33, 34] and beam-dump experiments [25]. Other constraints on lighter dark photons do not appear on the plot and are summarized in the reviews [35–37]. We also include the BBN constraint which imposes the dark photon to decay before $\tau_V \lesssim 0.1$ s [8–10] or later if the energy density fraction carried by the dark photon is smaller than $\sim 10\%$ [25]. Note, that only the BBN and the GW-from-CS constraints depend on the dark-to-SM temperature ratio $\tilde{r}$ which fixes the abundance of the dark photon in the early universe.

We can appreciate the complementarity between the well-established supernova, beam dump, BBN constraints, and the expected constraints assuming the detection of a SGWB from CS by the GW interferometers. Indeed, whereas supernova and beam dump do not really constrain above $m_V \gtrsim 0.1$ GeV, the detection of a SGWB from CS with a string tension $G\mu \simeq 10^{-11}$ would exclude dark photon masses up to the maximal reheating temperature $m_V \sim 10^{16}$ GeV allowed by the maximal inflation scale $H_{inf} \lesssim 6 \times 10^{13}$ GeV [39, 40], and kinetic mixing as low as $\epsilon \sim 10^{-18}$.

The dark photon as a dark matter mediator:
An interesting motivation for the dark photon is that it can play the role of a dark matter mediator. We can suppose that the dark sector also contains a Dirac fermion χ_D charged under $U(1)_D$, playing the role of DM [23, 25, 41–49]

$$L \supset \quad \bar{\chi}_D i \not{D} \chi_D - M_{DM}\bar{\chi}_D\chi_D, \tag{9.10}$$

---

[1] Production of the dark photon in the early universe has been studied in the literature. For a small kinetic mixing ϵ, the abundance of the dark sector can be set non-thermally either by freeze-in [24, 26–28], or by a separate reheating mechanism. In the latter case, the temperature asymmetry in Eq. (9.8) results from an asymmetric reheating [29–32]. For moderate kinetic mixing $\epsilon \gtrsim 10^{-6}\sqrt{M_{DM}/\text{TeV}}$ [24], the dark sector may have been at thermal equilibrium with the SM, but asymmetric temperatures can result from asymmetric changes in relativisitic degrees of freedom [25]. On the other hand, a possibility for thermally equilibrating the $U(1)_D$ sector and the SM in the case of a small kinetic mixing ϵ would be to introduce a dark Higgs ϕ, mixing with the SM Higgs, which once at thermal equilibrium with SM, decays into dark photons.

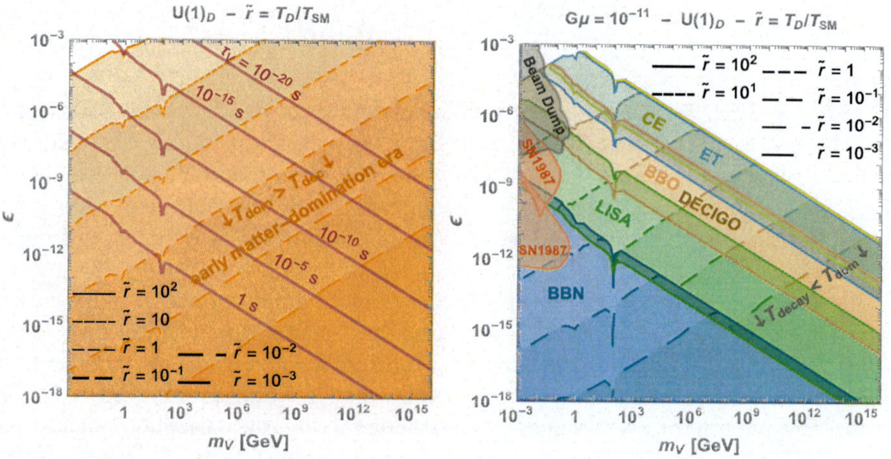

Fig. 9.4 **Left**: Constant dark photon lifetime τ_V contours. For a given dark-to-SM temperature ratio $\tilde{r} \equiv T_D/T_{\mathrm{SM}}$, a non-standard early matter domination is induced below the corresponding orange line where the dark photon dominates the universe before it decays. **Right**: Expected constraints on the dark photon mass m_V and kinetic mixing ϵ, assuming the measurement of a GW spectrum from CS with tension $G\mu = 10^{-11}$ by future GW interferometers. We use the *turning-point prescription (Rx 1)* discussed in Sect. 9.1.2

where $D_\mu = \partial_\mu + i g_{\mathrm{D}} V_{\mathrm{D}\mu}$ is the covariant derivative with g_{D} the $U(1)_{\mathrm{D}}$ gauge coupling constant. We suppose that the DM freezes-out by annihilating into pairs of dark photons, we impose $m_V < M_{\mathrm{DM}}$. We assume the dark photon to be non-relativistic when it decays but relativistic when it is produced, therefore, we set $\tilde{g}_D = 3 + \frac{7}{8} \cdot 4 = 6.5$ in Eq. (9.9).

The unitarity bound on the DM mass M_{DM} can be applied to the dark photon mass m_V upon assuming $m_V < M_{\mathrm{DM}}$. In the standard paradigm, the unitarity bound on s-wave annihilating dirac fermion DM is $M_{\mathrm{DM}} \lesssim 140$ TeV [38, 50]. However, if long-lived and heavy, the decay of the mediator can, by injecting entropy, dilute the DM abundance and relax the unitarity bound to [23]

$$M_{\mathrm{DM}} \lesssim 140\ \mathrm{TeV}\ \sqrt{D}, \tag{9.11}$$

where D is the dilution factor $D \simeq T_{\mathrm{dom}}/T_{\mathrm{dec}}$, T_{dom} and T_{dec} are as defined in Eq. (9.2) and Eq. (9.3).

In Fig. 9.5, we add the contraints on the dark photon when the later plays the role of the mediator of DM. They come from direct detection [25], CMB [23], indirect detection, using neutrino, gamma-rays, positrons-electrons and anti-protons [23], as well as from unitarity [23]. They are complemented by the GW-from-CS constraints. For $\epsilon \lesssim 10^{-10}$, all the traditional indirect detection constraints evaporate and the unitarity bound is pushed to larger masses due to the entropy dilution following the dark photon decay such that the model is then currently only constrained by BBN. It is remarkable that GW interferometers could probe this unconstrained region where

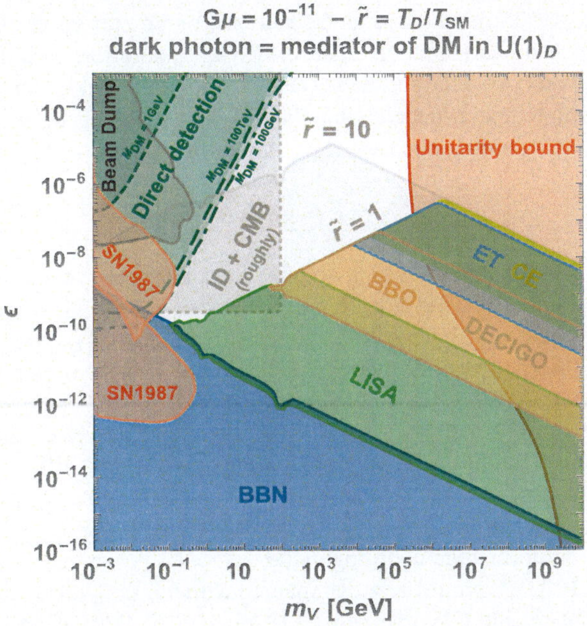

Fig. 9.5 Additional constraints when the dark photon is embedded in a DM model as the mediator of $U(1)_D$-charged DM (see text). We compare the expected GW constraints from cosmic strings with the existing constraints on the $U(1)_{\rm D}$ DM model: Supernovae bounds from [33, 34], direct detection bounds from [25] and the indirect detection + CMB constraints are a rough estimate from [23]. Beam dump constraints are also taken from [25]. The unitarity bound on the DM mass $M_{\rm DM}$ [38] can also be applied on the mediator mass because of the kinematic condition $m_V < M_{\rm DM}$. The unitarity bound gets relaxed at small ϵ because of the larger entropy injection following the dark photon decay [23]

$\epsilon < 10^{-10}$ and $m_V > 1$ GeV. In Figs. 9.4 and 9.5, we use the *turning-point prescription (Rx 1)* discussed in Sect. 9.1.2. Constraints are stronger when considering the *spectral-index prescription (Rx 2)*, as shown in Fig. 9.2.

Scenario where the cosmic string network and the dark photon mass have the same origin:
As a last remark, we comment on the case where the spontaneous breaking of the $U(1)_D$ symmetry would be responsible for the formation of the cosmic string network, so that the dark photon mass is no longer a free parameter but is related to the string tension μ, through the Abelian-Higgs relations [52]

$$\mu = 2\pi < \phi_{\rm D} >, \tag{9.12}$$

$$m_V^2 = 2g_{\rm D}^2 < \phi_{\rm D} >^2, \tag{9.13}$$

where $\phi_{\rm D}$ is the scalar field whose vacuum expectation value $< \phi_{\rm D} >$ breaks the $U(1)_D$ symmetry spontaneously.

In this case, we find that most of the relevant parameter space is ruled out due to overabundance of dark matter.[2] The only viable solution would be to assume that the states which are charged under $U(1)_D$ and stable under decay, are heavier than the reheating temperature such that they are never produced. The Weak Gravity Conjecture (WGC) requires the existence of a charged state with mass smaller than [53]

$$m_{\rm X} \lesssim g_{\rm D} M_{\rm pl}. \tag{9.14}$$

Hence, $g_{\rm D} M_{\rm pl}$ sets the maximal reheating temperature, above which charged states responsible for universe overclosure might be produced. Therefore, we should exclude the parameter space where the temperature of the $U(1)_D$ spontaneous breaking, taken as $\sim \langle \phi_{\rm D} \rangle$, is heavier than $g_{\rm D} M_{\rm pl}$, cf. pale sky blue region in right plot of Fig. 9.6. Note that the WGC does not specify if the suggested charged state is stable under decay or not. For instance, it would be stable and overclose the universe if it is a $U(1)_D$ fermion but not if it is a $U(1)_D$ Higgs, which can still decay into a dark photon pair when $m_{\phi_{\rm D}} \gtrsim 2m_V$. Hence, the WGC constraint in our parameter space has to be taken with a grain of salt.

Assuming a natural gauge coupling value, $g_{\rm D} = 10^{-1}$, we find that dark photons heavier than $\gtrsim$ 100 PeV would be accompanied with a $U(1)_D$ cosmic string network producing an observable GW spectrum, see left plot of Fig. 9.6. In the case where $g_{\rm D} = 10^{-4}$, we could probe dark photon masses down to $\gtrsim$ 100 TeV, see right plot of Fig. 9.6.

On the same plot, we superpose the constraints, shown with pale colors, coming from the simple observation of the GW spectrum with future experiments (except NANOGrav which is already operating), and the constraints, shown with opaque colors, coming from the detection of the turning point where the spectral index of the GW spectrum changes due to the decay of the dark photon which was dominating the energy density of the universe. The former detection would allow to measure the dark photon mass whereas the latter detection would allow to access its lifetime.

9.2 Supercooled Composite Dark Matter

In Figs. 8.15, 8.16 and 8.17 of the previous chapter, we were able to provide the energy scale and number of e-fold of a short period of inflation which can be probed by detecting the presence of a turning point in the GW spectrum from CS. In Fig. 9.7, we give the corresponding parameter space of the model of Supercooled Composite DM, studied in Sect. 6.4 and published in [54]. Indeed, in this model, DM of mass $M_{\rm DM} \simeq 4\pi f$ is formed during a supercooled confining first-order phase transition with confining scale f and nucleation temperature $T_{\rm nuc}$. The corresponding

[2] The cross-section of a pair of $U(1)_D$ fermions annihilating into dark photons is given by $\sigma v \simeq \pi \alpha_{\rm D}^2 / m_\psi^2$ with $\alpha_{\rm D} = g_{\rm D}^2/4\pi$. It is way too weak to prevent universe overclosure, except if we tune the Yukawa coupling of the fermion, λ, defined by $m_\psi = \lambda \langle \phi \rangle / \sqrt{2}$, to very small values.

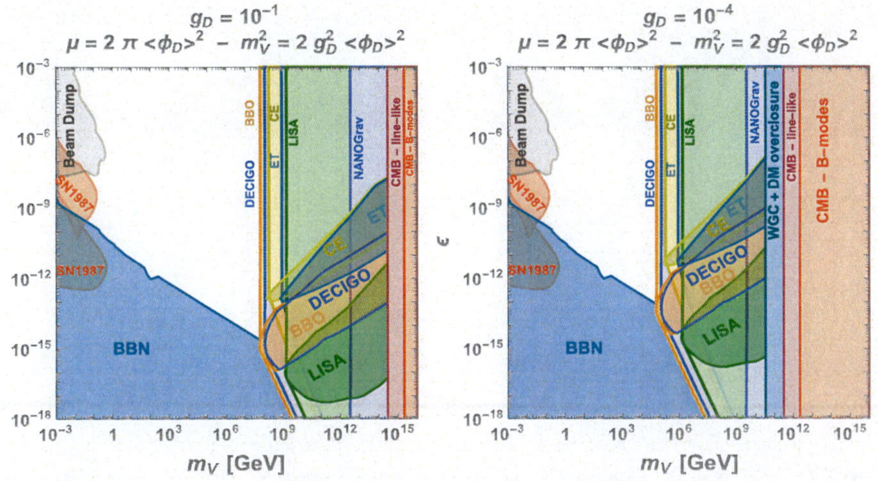

Fig. 9.6 Scenario where the dark photon mass m_V and the cosmic string network are generated by the spontaneous breaking of the same $U(1)$ symmetry, such that m_V is related to the string tension μ. **Pale colors**: Constraints on the dark photon parameter space assuming the mere detection of the GW spectrum from CS by NANOGrav, LISA, ET, CE, DECIGO and BBO. **Opaque colors**: Constraints assuming the detection of the turning point in the GW spectrum induced by the transition from matter to radiation when the heavy dark photon decays. When combining Eqs. (9.1) and (9.3), this last detection allows to measure the dark photon lifetime. The constraints described in the following part of this caption are independent of the GW emission. **Pale red**: The non-observation of the fundamental tensor B-modes in the CMB imposes the stringest upper bound on the energy density scale of inflation [40], $V_{\rm inf} \lesssim 1.6 \times 10^{16}$ GeV. This provides an upper-bound on the reheating temperature, which also must satisfy $\langle \phi_{\rm D} \rangle \lesssim T_{\rm reh}$ in order for the string network to be formed. Thus, we impose the CS formation to occur after the end of inflation with the following criteria: $\langle \phi_{\rm D} \rangle \lesssim V_{\rm inf}$. **Pale purple**: Constraints from the non-observation of line-like temperature anisotropies in the CMB, e.g. [51], $G\mu \lesssim 2 \times 10^{-7}$. **Pale sky blue**: In order to prevent DM overclosure, we must assume the $U(1)_D$ charged states to be heavier than the reheating temperature such that they are never produced. A possibility which is constrained by the Weak Gravity Conjecture, cf. main text, thus we impose $\langle \phi_{\rm D} \rangle \lesssim g_{\rm D} M_{\rm pl}$. The last inequality implicitly assumes $\langle \phi_{\rm D} \rangle \lesssim T_{\rm reh}$. Note however that such a charged state could be unstable, e.g. if it is a dark Higgs, in which case the WGC constraint is relaxed

inflationary scale and number of e-folds are

$$E_{\rm inf} \simeq f, \tag{9.15}$$

and

$$N_e = \log \frac{f}{T_{\rm nuc}}. \tag{9.16}$$

We can see that a part of the parameter space leading to the correct DM abundance could be probed with GW interferometers if a GW spectrum from CS is detected. We stress that those constraints assume the turning-point prescription (Rx 1) defined in

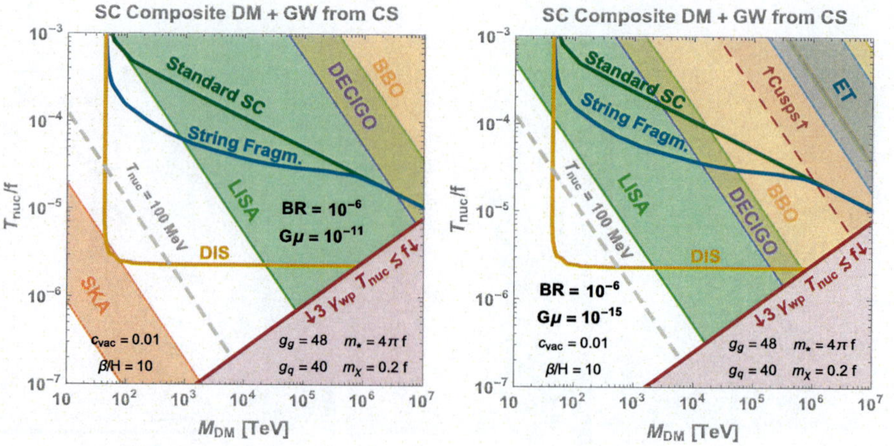

Fig. 9.7 Regions of the parameter space of Supercooled Composite DM, cf. Fig. 9.7, which can be probed with the detection of a GW spectrum from CS with tension $G\mu = 10^{-11}$ (left) and $G\mu = 10^{-15}$ (right) by future GW observatories. We have considered the detectability of the turning point due to the change of cosmology, cf. turning-point prescription (Rx 1) in Sect. 9.1.2. If instead, we use the detectability of the slope of the SGWB, cf. spectral-index description (R x 2), the reach would be considerably enlarged to lower DM masses $M_{\rm DM}$. This figure is not contained in [1] but instead it has been realized during the preparation of the thesis manuscript

Sect. 9.1.2 and would be considerably enlarged to smaller DM masses if instead one assumes the spectral-index prescription (Rx 2). The advantage of the first prescription Rx 1 is that it allows to measure the temperature when the strings re-enter the Hubble horizon, cf. Eq. (8.73)

$$T_{\rm re} \simeq \frac{E_{\rm inf}}{(0.1)\, g_*^{1/4}(T_{\rm re})\, \exp(N_e)} \simeq \frac{T_{\rm nuc}}{(0.1)\, g_*^{1/4}(T_{\rm re})}, \tag{9.17}$$

and then the nucleation temperature $T_{\rm nuc}$.[3] Some of the reachable parameter space falls in the region where the techni-quark can not trivially enter inside the bubbles of confined phase and where our modeling does not apply (purple). We leave the study of this regime for further studies. The detectability of SGWB generated by bubble collision are planned for a future study [55]. The present section Sect. 9.2 was not included in any of the publications but instead it has been added during the preparation of the manuscript.

[3] In the presence of multi-detection by different GW observatories of a SGWB from CS, the spectral-index prescription (Rx 2), defined in Sect. 9.1.2, could be enough to measure $T_{\rm re}$.

9.3 Summary

If future GW observatories have the sensitivity to detect stochastic GW backgrounds of primordial origin and to measure precise features in this spectrum, they can reveal very unique information about very high scale physics. Particularly relevant sources of GW are cosmic strings. Cosmic strings are almost ubiquitous in many Grand-Unified Theories. As they keep emitting throughout the whole cosmological history of the universe, the resulting GW spectrum covers a wide range of frequencies and can be detected either by space-based or ground-based observatories. An early era of matter domination due to new heavy particles or an intermediate period of inflation due to a supercool first-order phase transition generate clear signatures in the GW spectrum of cosmic strings.

First, in this study we assume the existence of an early matter era due to the presence of a cold particle X temporarily dominating the energy density of the universe and decaying before the onset of BBN. We compute its impact on the GW spectrum of CS beyond the scaling regime. We show that detecting such a feature and interpreting it in terms of a new heavy relic can lead to unparalleled constraints in the $(\tau_X, m_X Y_X)$ (lifetime, yield) plane. In Fig. 9.3, we provide model-independent constraints which extend the usual BBN constraints on the lifetime τ_X by 15 orders of magnitude for $G\mu = 10^{-11}$, as we are able to constrain early matter dominated era ending when the temperature of the universe is between 50 TeV and 1 MeV.

We show that this new search strategy is likely to provide unprecedented constraints on a model of dark photon kinematically coupled to the Standard Model hypercharge, possibly embedded in the $U(1)_D$ dark-photon-mediated dark matter model. The constraints we obtain from GW on $U(1)_D$ dark matter falls in the large mass/small kinetic mixing ballpark which is otherwise unreachable by any current and probably future direct/indirect detection and CMB constraints, cf. Fig. 9.5. At last, we consider the possibility that the dark photon mass and the cosmic string network are generated by the spontaneous breaking of the same $U(1)$ symmetry and show that we can use future GW interferometers to probe dark photon masses above 100 PeV, or even down to the TeV scale if we tune the gauge coupling to small values, see Fig. 9.6. These are only a few minimal examples of particle physics models generating early matter eras. There are many other well-motivated models which would deserve consideration in this respect. We will present the corresponding constraints on axion-like-particles and primordial black holes in a separate study.

Finally, we show that the period of vacuum domination preceding the formation of supercooled composite DM in Chap. 7, can leave a detectable imprint on the GW spectrum of the would-be CS network present at the same time.

References

1. Gouttenoire Y, Servant G, Simakachorn P (2020) BSM with cosmic strings: heavy, up to EeV mass, unstable particles. JCAP 07:016. arXiv:1912.03245, https://doi.org/10.1088/1475-7516/2020/07/016
2. Gouttenoire Y, Servant G, Simakachorn P, Beyond the standard models with cosmic strings. arXiv:1912.02569
3. Lorenz L, Ringeval C, Sakellariadou M (2010) Cosmic string loop distribution on all length scales and at any redshift. JCAP 1010:003. arXiv:1006.0931, https://doi.org/10.1088/1475-7516/2010/10/003
4. Ringeval C, Suyama T (2017) Stochastic gravitational waves from cosmic string loops in scaling. JCAP 1712:027. arXiv:1709.03845, https://doi.org/10.1088/1475-7516/2017/12/027
5. Auclair P, Ringeval C, Sakellariadou M, Steer D (2019) Cosmic string loop production functions. JCAP 06:015. arXiv:1903.06685, https://doi.org/10.1088/1475-7516/2019/06/015
6. Blanco-Pillado JJ, Olum KD, Shlaer B (2014) The number of cosmic string loops. Phys Rev D89:023512. arXiv:1309.6637, https://doi.org/10.1103/PhysRevD.89.023512
7. Blanco-Pillado JJ, Olum KD (2017) Stochastic gravitational wave background from smoothed cosmic string loops. Phys Rev D96:104046. arXiv:1709.02693, https://doi.org/10.1103/PhysRevD.96.104046
8. Jedamzik K (2006) Big bang nucleosynthesis constraints on hadronically and electromagnetically decaying relic neutral particles. Phys Rev D74:103509. arXiv:hep-ph/0604251, https://doi.org/10.1103/PhysRevD.74.103509
9. Jedamzik K, Pospelov M (2009) Big bang nucleosynthesis and particle dark matter. New J Phys 11:105028. arXiv:0906.2087, https://doi.org/10.1088/1367-2630/11/10/105028
10. Kawasaki M, Kohri K, Moroi T, Takaesu Y (2018) Revisiting big-bang nucleosynthesis constraints on long-lived decaying particles. Phys Rev D97:023502. arXiv:1709.01211, https://doi.org/10.1103/PhysRevD.97.023502
11. Turner MS (1983) Coherent scalar field oscillations in an expanding universe. Phys Rev D28:1243. https://doi.org/10.1103/PhysRevD.28.1243
12. Boyle LA, Buonanno A (2008) Relating gravitational wave constraints from primordial nucleosynthesis, pulsar timing, laser interferometers, and the CMB: implications for the early Universe. Phys Rev D78:043531. arXiv:0708.2279, https://doi.org/10.1103/PhysRevD.78.043531
13. Erickcek AL, Sigurdson K (2011) Reheating effects in the matter power spectrum and implications for substructure. Phys Rev D84:083503. arXiv:1106.0536, https://doi.org/10.1103/PhysRevD.84.083503
14. Fan J, Özsoy O, Watson S (2014) Nonthermal histories and implications for structure formation. Phys Rev D 90:043536. arXiv:1405.7373, https://doi.org/10.1103/PhysRevD.90.043536
15. Sten Delos M, Linden T, Erickcek AL, Breaking a dark degeneracy: the gamma-ray signature of early matter domination. arXiv:1910.08553
16. Blanco C, Delos MS, Erickcek AL, Hooper D, Annihilation signatures of hidden sector dark matter within early-forming microhalos. arXiv:1906.00010
17. Polnarev AG, Yu M (1985) Khlopov, cosmology, primordial black holes, and supermassive particles. Sov Phys Usp 28:213–232. https://doi.org/10.1070/PU1985v028n03ABEH003858
18. Green AM, Liddle AR, Riotto A (1997) Primordial black hole constraints in cosmologies with early matter domination. Phys Rev D56:7559–7565. arXiv:astro-ph/9705166, https://doi.org/10.1103/PhysRevD.56.7559
19. Georg J, Şengör G, Watson S (2016) Nonthermal WIMPs and primordial black holes. Phys Rev D93:123523. arXiv:1603.00023, https://doi.org/10.1103/PhysRevD.93.123523
20. Scherrer RJ, Turner MS (1985) Decaying particles do not heat up the universe. Phys Rev D31:681. https://doi.org/10.1103/PhysRevD.31.681
21. Holdom B (1986) Two U(1)'s and Epsilon Charge Shifts. Phys Lett 166B:196–198.https://doi.org/10.1016/0370-2693(86)91377-8
22. Foot R, He X-G (1991) Comment on Z Z-prime mixing in extended gauge theories. Phys Lett B267:509–512. https://doi.org/10.1016/0370-2693(91)90901-2

$T_{\mathrm{dec},\,\nu} \sim 1$ MeV. In that case the abundance grows linearly with the DM mass, cf. Eq. (3.43) in Chap. 3

$$\Omega_{\mathrm{DM}} h^2 \simeq \frac{M_{\mathrm{DM}}}{94\ \mathrm{eV}} \frac{11}{4} \left(\frac{T_{\mathrm{DM}}}{T_\gamma} \right)^3 . \tag{4.31}$$

The ratio of temperatures accounts for any possible asymmetric entropy injection in one sector and not the other. We suppose that kinetic decoupling takes place during the relativistic phase

$$T_{\mathrm{KD}} > T_{\mathrm{NR}}, \tag{4.32}$$

such that there is no extra heating due to interaction with the thermal bath in the non-relativistic phase, cf. second line in Eq. (4.26). This implies that the DM temperature can evolve differently from the SM temperature due to non-adiabatic processes in one of the two sectors, see e.g. neutrinos heating in Sect. 3.2.2 of Chap. 3. Hence we denote the DM temperature with a superscript 'DM'. Fixing $T_{\mathrm{KD}}^{\mathrm{DM}} = T_{\mathrm{NR}}^{\mathrm{DM}}$ in Eq. (4.30) leads to

$$\lambda_{\mathrm{FS}}(t_{\mathrm{eq}}) \simeq 2 \frac{t_{\mathrm{eq}}}{a_{\mathrm{eq}}} \frac{T_{\mathrm{eq}}^{\mathrm{DM}}}{T_{\mathrm{NR}}^{\mathrm{DM}}} \left[1 + \ln \frac{T_{\mathrm{NR}}^{\mathrm{DM}}}{T_{\mathrm{eq}}^{\mathrm{DM}}} \right] . \tag{4.33}$$

If the DM is thermally produced then the average DM momentum follows

$$\langle p \rangle = 3.15\, T_{\mathrm{DM}} \tag{4.34}$$

and DM becomes non-relativistic at the temperature[6]

$$T_{\mathrm{NR}}^{\mathrm{DM}} \simeq M_{\mathrm{DM}}/3. \tag{4.35}$$

Upon plugging $a_{\mathrm{eq}}^{-1} = 3402(26)$, $t_{\mathrm{eq}} = 51.1(8)$ kyrs and $T_{\mathrm{eq}} \simeq 0.8$ eV [34] into Eq. (4.33), we obtain

$$\lambda_{\mathrm{FS}}(t_{\mathrm{eq}}) \simeq 0.69\ \mathrm{Mpc} \left(\frac{3\ \mathrm{keV}}{M_{\mathrm{DM}}} \right) \left(\frac{T_{\mathrm{eq}}^{\mathrm{DM}}}{T_{\mathrm{eq}}} \right) \left[1 + 0.12 \ln \left(\frac{M_{\mathrm{DM}}}{3\ \mathrm{keV}} \right) \left(\frac{T_{\mathrm{eq}}}{T_{\mathrm{eq}}^{\mathrm{DM}}} \right) \right] . \tag{4.36}$$

We can use Eq. (4.31) to relate the DM temperature $T_{\mathrm{eq}}^{\mathrm{DM}}$ to the photon temperature T_{eq}

$$T_{\mathrm{eq}}^{\mathrm{DM}} = \left(\frac{4}{11} \frac{94\ \mathrm{eV}}{M_{\mathrm{DM}}} \Omega_{\mathrm{DM}} h^2 \right)^{1/3} T_{\mathrm{eq}}, \tag{4.37}$$

[6] WDM bounds from structure-formation can also be applied to non-thermal DM candidate. For those cases, the thermal relation $\langle p \rangle \big|_{\mathrm{thermal}} = 3.15\, T_{\mathrm{DM}}$ and the temperature Eq. (4.35) when DM becomes non-relativistic must be replaced accordingly.

23. Cirelli M, Gouttenoire Y, Petraki K, Sala F (2019) Homeopathic dark matter, or how diluted heavy substances produce high energy cosmic rays. JCAP 02:014. arXiv:1811.03608, https://doi.org/10.1088/1475-7516/2019/02/014
24. Hambye T, Tytgat MHG, Vandecasteele J, Vanderheyden L (2019) Dark matter from dark photons: a taxonomy of dark matter production. Phys Rev D 100:095018. arXiv:1908.09864, https://doi.org/10.1103/PhysRevD.100.095018
25. Cirelli M, Panci P, Petraki K, Sala F, Taoso M (2017) Dark matter's secret liaisons: phenomenology of a dark U(1) sector with bound states. JCAP 05:036. arXiv:1612.07295, https://doi.org/10.1088/1475-7516/2017/05/036
26. Hall LJ, Jedamzik K, March-Russell J, West SM (2010) Freeze-in production of FIMP dark matter. JHEP 03:080. arXiv:0911.1120, https://doi.org/10.1007/JHEP03(2010)080
27. Chu X, Hambye T, Tytgat MHG (2012) The four basic ways of creating dark matter through a portal. JCAP 05:034. arXiv:1112.0493, https://doi.org/10.1088/1475-7516/2012/05/034
28. Berger J, Jedamzik K, Walker DGE (2016) Cosmological constraints on decoupled dark photons and dark higgs. JCAP 1611:032. arXiv:1605.07195, https://doi.org/10.1088/1475-7516/2016/11/032
29. Hodges HM (1993) Mirror baryons as the dark matter. Phys Rev D47:456–459. https://doi.org/10.1103/PhysRevD.47.456
30. Berezhiani ZG, Dolgov AD, Mohapatra RN (1996) Asymmetric inflationary reheating and the nature of mirror universe. Phys Lett B375:26–36. arXiv:hep-ph/9511221, https://doi.org/10.1016/0370-2693(96)00219-5
31. Feng JL, Tu H, Yu H-B (2008) Thermal relics in hidden sectors. JCAP 0810:043. arXiv:0808.2318, https://doi.org/10.1088/1475-7516/2008/10/043
32. Adshead P, Cui Y, Shelton J (2016) Chilly dark sectors and asymmetric reheating. JHEP 06:016. arXiv:1604.02458, https://doi.org/10.1007/JHEP06(2016)016
33. Kazanas D, Mohapatra RN, Nussinov S, Teplitz VL, Zhang Y (2014) Supernova bounds on the dark photon using its electromagnetic decay. Nucl Phys B890:17–29. arXiv:1410.0221, https://doi.org/10.1016/j.nuclphysb.2014.11.009
34. Chang JH, Essig R, McDermott SD (2017) Revisiting supernova 1987A constraints on dark photons. JHEP 01:107. arXiv:1611.03864, https://doi.org/10.1007/JHEP01(2017)107
35. Jaeckel J, Ringwald A (2010) The low-energy frontier of particle physics. Ann Rev Nucl Part Sci 60:405–437. arXiv:1002.0329, https://doi.org/10.1146/annurev.nucl.012809.104433
36. Essig R et al (2013) Working group report: new light weakly coupled particles. In: Community summer study 2013: snowmass on the Mississippi, p 10. arXiv:1311.0029
37. Alexander J et al (2016) Dark sectors 2016 workshop: community report, p 8. arXiv:1608.08632
38. Griest K, Kamionkowski M (1990) Unitarity limits on the mass and radius of dark matter particles. Phys Rev Lett 64:615. https://doi.org/10.1103/PhysRevLett.64.615
39. BICEP2, Keck Array collaboration, Ade PAR et al (2018) BICEP2/Keck Array x: constraints on primordial gravitational waves using planck, WMAP, and New BICEP2/Keck observations through the 2015 season. Phys Rev Lett 121:221301. arXiv:1810.05216, https://doi.org/10.1103/PhysRevLett.121.221301
40. Shape Planck collaboration, Planck YA et al (2018) results. X constraints on inflation. Astron Astrophys 641:A10. arXiv:1807.06211, https://doi.org/10.1051/0004-6361/201833887
41. Kors B, Nath P (2004) A stueckelberg extension of the standard model. Phys Lett B586:366–372. arXiv:hep-ph/0402047, https://doi.org/10.1016/j.physletb.2004.02.051
42. Feldman D, Kors B, Nath P (2007) Extra-weakly interacting dark matter. Phys Rev D75:023503. arXiv:hep-ph/0610133, https://doi.org/10.1103/PhysRevD.75.023503
43. Fayet P (2007) U-boson production in e+ e- annihilations, psi and Upsilon decays, and Light Dark Matter. Phys Rev D75:115017. arXiv:hep-ph/0702176, https://doi.org/10.1103/PhysRevD.75.115017
44. Ackerman L, Buckley MR, Carroll SM, Kamionkowski M (2009) Dark matter and dark radiation. Phys Rev D79:023519. arXiv:0810.5126, https://doi.org/10.1103/PhysRevD.79.023519, https://doi.org/10.1142/9789814293792_0021

45. Goodsell M, Jaeckel J, Redondo J, Ringwald A (2009) Naturally light hidden photons in large volume string compactifications. JHEP 11:027. arXiv:0909.0515, https://doi.org/10.1088/1126-6708/2009/11/027
46. Morrissey DE, Poland D, Zurek KM (2009) Abelian hidden sectors at a GeV. JHEP 07:050. arXiv:0904.2567, https://doi.org/10.1088/1126-6708/2009/07/050
47. Andreas S, Goodsell MD, Ringwald A (2013) Dark matter and dark forces from a supersymmetric hidden sector. Phys Rev D87:025007. arXiv:1109.2869, https://doi.org/10.1103/PhysRevD.87.025007
48. Goodsell M, Ramos-Sanchez S, Ringwald A (2012) Kinetic mixing of U(1)s in heterotic orbifolds. JHEP 01:021. arXiv:1110.6901, https://doi.org/10.1007/JHEP01(2012)021
49. Fayet P (2017) The light U boson as the mediator of a new force, coupled to a combination of Q, B, L and dark matter. Eur Phys J C77:53. arXiv:1611.05357, https://doi.org/10.1140/epjc/s10052-016-4568-9
50. von Harling B, Petraki K (2014) Bound-state formation for thermal relic dark matter and unitarity. JCAP 12:033. arXiv:1407.7874, https://doi.org/10.1088/1475-7516/2014/12/033
51. Lizarraga J, Urrestilla J, Daverio D, Hindmarsh M, Kunz M (2016) New CMB constraints for Abelian Higgs cosmic strings. JCAP 10:042. arXiv:1609.03386, https://doi.org/10.1088/1475-7516/2016/10/042
52. Vilenkin A, Shellard EPS (2000) Cosmic strings and other topological defects. Cambridge University Press
53. Arkani-Hamed N, Motl L, Nicolis A, Vafa C (2007) The string landscape, black holes and gravity as the weakest force. JHEP 06:060. arXiv:hep-th/0601001, https://doi.org/10.1088/1126-6708/2007/06/060
54. Baldes I, Gouttenoire Y, Sala F (2021) String fragmentation in supercooled confinement and implications for dark matter. JHEP 04:278. arXiv:2007.08440, https://doi.org/10.1007/JHEP04(2021)278
55. Baldes I, Gouttenoire Y, Sala F, Servant G, Supercool composite dark matter beyond 100 TeV. arXiv:2110.13926

Chapter 10
Conclusion

In the last two decades, the LEP, Tevatron and LHC on one side, and WMAP and the Planck satellites on the other side, have confirmed the validity of the Standard Models of Elementary Particle and Cosmology with an unprecedented level of precision. However, the big questions, among which the nature of Dark Matter (DM) and the solution to the electroweak hierarchy problem, are left unanswered.

The absence of experimental evidences in colliders, telescopes and laboratories for Weakly-Interacting Massive Particles (WIMPs) in the naturalness window [100 GeV, few TeV] encourages searches beyond the TeV scale. A quest for such heavy particles is happening with currect telescopes Hess II, Hawc, Veritas, Magic, Taiga, Antares, IceCube, Baikal, Ams, Calet, Dampe and will intensify with the future development of numerous multi-TeV telescopes, e.g. Cta, Lhaaso, Km3net, Herd, Iss-Cream, and the high-sensitivity direct detection experiment Lux-Zeplin.

The detection of Gravitational Waves (GW) by LIGO in 2015 has opened new directions for investigating compact massive objects in our universe such as black holes and neutron stars and for testing the laws of gravity in a new regime. At the same time, this new GW era is enabling new ways to explore high-energy particle physics beyond the Standard Model. LISA with peak sensitivity around 3 mHz will offer a window on first-order phase transitions occurring around the electroweak scale while Einstein Telescope and Cosmic Explorer with peak sensitivity around 10 Hz will offer views on the PeV scale. Additionally, the future detection of GW generated by a network of Cosmic Strings (CS) would allow to infer the equation of state of the universe up to 10^{-17} seconds after the Big-Bang.

On the theoretical side, new models motivating indirect-detection experiments beyond the WIMP windows but also GW experiments beyond the millihertz window are welcome.

Homeopathic Dark Matter.

We explored the possibility to have thermal Dark Matter with mass beyond the unitarity bound $\gtrsim 100$ TeV. We introduced a heavy unstable DM mediator which induces

Y. Gouttenoire, *Beyond the Standard Model Cocktail*, Springer Theses,
https://doi.org/10.1007/978-3-031-11862-3_10

an early matter era and injects entropy when it decays into Standard Model. The dilution of the DM abundance allows to evade the unitarity bound. As a case of study, we considered the $U(1)_D$ model and studied the indirect-detection constraints, see Fig. 5.7. The use of a mediator avoids the necessity to resum electroweak Sudakov logarithms for computing cosmic-ray predictions from DM annhihilation. We showed that the indirect-detection constraints are limited by the data range used by collaborations, and we hope that our study will motivate data analysis beyond the 100 TeV threshold. We found that DM masses up to EeV are realizable, larger mass scales being constrained by big-bang nucleosynthesis. For such large mass scales, DM is far from being reached by standard methods: colliders, telescopes and laboratories. This motivates the use of GW from CS to detect the presence of an early matter era in the early universe.

Supercooled Composite Dark Matter.

We have explored an other possibility for having thermal (or would-be thermal) DM beyond the unitarity bound. Instead of using the entropy injection due to reheating after a matter-dominated era, we have used the entropy injection due to reheating after a vacuum-dominated era. A well-known possibility to generate a short period of vacuum-domination relies on supercooled first-order phase transitions. Initially motivated by the electroweak hierarchy problem, we have considered a confining phase transition arising from a strongly-interacting sector. We pointed out for the first time that when considering a confining phase transition with $T_{\rm nuc}/f \lesssim 0.1$ where f and $T_{\rm nuc}$ are the confining scale and nucleation temperature respectively, cf. Fig. 7.6, the final abundance of composite states is enhanced by three new effects. First, the fragmentation of the string binding the incoming quark and the wall. Second, the ejection of a quark from the bubble in order to satisfy charge conservation. And third, deep-inelastic-scattering between the string fragments and the preheated soup arising from the decay of the scalar field after bubble wall collision. Supposing that DM is one of the composite states of the strong sector, we found that the needed amount of supercooling in order to fix the correct abundance is larger than in the non-confining scenario, see Fig. 7.9. We obtained that DM masses up to EeV are realizable. The scenario with supra-EeV mass scales, where quarks are not energetic enough to enter the bubbles is left for further studies. In the follow-up [1], we have considered the case where the strong sector is successfully described by a light-dilaton effective-field-theory and we have studied the phenomenology of dilaton-mediated DM, including indirect detection, direct detection and GW. Far from the TeV scale, the indirect-detection and direct-detection constraints evaporate and GW is the only probe. It can either come from bubble collision or from the imprint of the vacuum-dominated era in the GW spectrum from CS.

Beyond the Standard Model with Cosmic Strings.

The presence of cosmic strings in our universe is motivated by many Grand Unification Theories, but also superstring theories. They are well-known to generate a GW spectrum over many decades in frequency which is flat during radiation-domination. We hope the paper [2] to become of standard reference for computing the GW spectrum from CS in standard cosmology or in the presence of an early matter-, kination- and inflation-era. As an input for the spectrum we computed the loop formation rate by solving numerically the evolution equation of the string network. We accounted for the deviation from the Nambu-Goto approximation due to the presence of small-scale structures which allow for the decay of string loops via particle production. We showed that the observation of a GW spectrum from CS could give information on the equation of state of the universe up to the PeV scale and even up to 10^{14} GeV in the special case of an inflationary era, see Fig. 8.18.

Probe Heavy Dark Matter with Gravitational Waves from Cosmic Strings.

The first two directions of research, homeopathic DM and supercooled composite DM, used the entropy injection following a matter- and vacuum-dominated era in order to have DM heavier than the unitarity bound at 100 TeV. We showed that we can use the imprint of such non-standard cosmologies on the would-be GW spectrum from CS detected by future GW interferometers in order to probe models of heavy DM, see Figs. 9.3 and 9.5 and 9.7. This new method of detection can probe DM mass regions far beyond the reach of standard techniques based on colliders, indirect detection and direct detection.

References

1. Baldes I, Gouttenoire Y, Sala F, Servant G, Supercool composite dark matter beyond 100 TeV. arxiv:abs/2110.13926
2. Gouttenoire Y, Servant G, Simakachorn P (2020) Beyond the standard models with cosmic strings. JCAP 07:032. arxiv:abs/1912.02569, https://doi.org/10.1088/1475-7516/2020/07/032